Molecular Machines Involved in Peroxisome Biogenesis and Maintenance

Cécile Brocard • Andreas Hartig
Editors

Molecular Machines Involved in Peroxisome Biogenesis and Maintenance

Springer

Editors
Cécile Brocard
Mauerbach
Austria

Andreas Hartig
Max F. Perutz Laboratories
University of Vienna
Vienna
Austria

ISBN 978-3-7091-1787-3 ISBN 978-3-7091-1788-0 (eBook)
DOI 10.1007/978-3-7091-1788-0
Springer Wien Heidelberg New York Dordrecht London

Library of Congress Control Number: 2014944309

Printed on acid-free paper

Springer is part of Springer Science+Business Media (www.springer.com)

Preface

Peroxisomes are essential for life and mutations in genes coding for biogenesis factors are often associated with lethal disorders. Proteins required for the biogenesis of this organelle typically assemble in large molecular complexes, which participate in membrane formation, protein transport, peroxisome duplication, and inheritance during cell division. With their species-specific pathways such as the synthesis of ether lipid in mammals or the glyoxylate cycle in plants and yeasts, peroxisomes are multitasking organelles. The establishment of experimental strategies using several yeast species as model organisms allowed most breakthroughs in the field of peroxisome research.

Ever since their discovery by Rhodin in 1954 peroxisomes have been perceived and presented as little sisters of mitochondria. Certainly their characteristics are reminiscent of the rebellious nature of a younger sibling often despised. On the one hand, peroxisomes employ of a great variety of natural protein complexes also present in mitochondria. On the other hand, peroxisomes have evolved their own processes to cope with specific challenges. Protein import across the peroxisomal membrane now represents the epitome for trafficking of fully folded proteins through membranes. This groundbreaking discovery profoundly changed the mindset on protein translocation. Protein monoubiquitylation as initial step in the recycle mechanism for membrane proteins was a hallmark discovery for studies focused protein translocation across membranes. Peroxisomes like mitochondria are prone to self-renewal. Simultaneously, however, peroxisomes necessitate the membrane generating system of the endoplasmic reticulum. Both processes were long thought to be mutually exclusive and discrimination was made between autonomous organelles and those belonging to the endomembrane system. That organelle abundance results from a competition between proliferation, regeneration and inheritance vs. degradation was discovered for the first time in peroxisomes. The novelty of several recent findings stimulated us to edit a book on the molecular machines functioning in peroxisome biogenesis and maintenance highlighting the relevance of organelle number and morphology for health and disease.

The reports presented herein highlight state-of-the-art technologies to study organelle maintenance and function including quantitative proteomics and live-cell imaging to analyze the molecular networks involved in the regulation of peroxisome formation. The findings of many studies point to the mode of action of several important factors and illustrate crosstalk with other organelles including protein interface between peroxisomes and the endoplasmic reticulum. In addition, this book presents future possibilities for peroxisomal research and highlights the relevance of this field for the development of biotechnology as well as therapeutic strategies.

This book would not have been possible without the generous help of many people. Authors devoted time and efforts to present their newest ideas and data, and comments of many reviewers have been invaluable to improve of the whole manuscript.

Mauerbach, Austria Cécile Brocard
Vienna, Austria Andreas Hartig

Contents

List of Contributors

Y. Abe Department of Biology, Faculty of Sciences, Kyushu University Graduate School, Fukuoka, Japan

Oksana Apanasets Laboratory of Lipid Biochemistry and Protein Interactions, Department of Cellular and Molecular Medicine, KU Leuven, Leuven, Belgium

C. Argyriou Department of Human Genetics, McGill University, Montreal, QC, Canada

Kyaw Aung Department of Energy Plant Research Laboratory, Michigan State University, East Lansing, MI, USA

Jorge E. Azevedo Organelle Biogenesis and Function Group, Instituto de Biologia Celular e Molecular (IBMC), Universidade do Porto, Porto, Portugal

Instituto de Ciências Biomédicas Abel Salazar (ICBAS), Universidade do Porto, Porto, Portugal

Myriam Baes Laboratory of Cell Metabolism, Department of Pharmaceutical and Pharmacological Sciences, KU Leuven, Leuven, Belgium

Bonnie Bartel Department of Biochemistry and Cell Biology, Rice University, Houston, TX, USA

Kiah A. Barton Laboratory of Plant Development and Interactions, Department of Molecular and Cellular Biology, University of Guelph, Guelph, ON, Canada

Johannes Berger Department of Pathobiology of the Nervous System, Center for Brain Research, Medical University of Vienna, Vienna, Austria

Nancy Braverman Department of Human Genetics, McGill University, Montreal, QC, Canada

Department of Pediatrics, McGill University, Montreal, QC, Canada

Cécile Brocard Tissue Med Biosciences, Krems, Austria

MFPL, University of Vienna, Vienna, Austria

Sarah E. Burkhart Department of Biochemistry and Cell Biology, Rice University, Houston, TX, USA

Andreia F. Carvalho Organelle Biogenesis and Function Group, Instituto de Biologia Celular e Molecular (IBMC), Universidade do Porto, Porto, Portugal

Inês Castro College of Life and Environmental Sciences, Biosciences, University of Exeter, Exeter, UK

Daniel Effelsberg Institute for Physiological Chemistry, Ruhr-University Bochum, Bochum, Germany

Birgit Eisenhaber Bioinformatics Institute (BII), Agency for Science, Technology and Research (A*STAR), Matrix, Singapore

Frank Eisenhaber Bioinformatics Institute (BII), Agency for Science, Technology and Research (A*STAR), Matrix, Singapore

Department of Biological Sciences (DBS), National University of Singapore (NUS), Singapore, Singapore

School of Computer Engineering (SCE), Nanyang Technological University (NTU), Singapore, Singapore

Ralf Erdmann Institute for Physiological Chemistry, Ruhr-University Bochum, Bochum, Germany

Systembiochemie, Ruhr-Universität Bochum, Bochum, Germany

H. Dariush Fahimi Division of Medical Cell Biology, Department of Anatomy and Cell Biology, University of Heidelberg, Heidelberg, Germany

Jean-Claude Farré Section of Molecular Biology, Division of Biological Sciences, UC San Diego, La Jolla, CA, USA

Sacha Ferdinandusse Laboratory Genetic Metabolic Diseases, Departments of Paediatrics and Clinical Chemistry, Emma Children's Hospital, Academic Medical Center, Amsterdam, The Netherlands

Wendell A. Fleming Department of Biochemistry and Cell Biology, Rice University, Houston, TX, USA

Tânia Francisco Organelle Biogenesis and Function Group, Instituto de Biologia Celular e Molecular (IBMC), Universidade do Porto, Porto, Portugal

Instituto de Ciências Biomédicas Abel Salazar (ICBAS), Universidade do Porto, Porto, Portugal

Marc Fransen Laboratory of Lipid Biochemistry and Protein Interactions, Department of Cellular and Molecular Medicine, KU Leuven, Leuven, Belgium

Y. Fujiki Department of Biology, Faculty of Sciences, Kyushu University Graduate School, Fukuoka, Japan

Hongbo Gao Biosciences, College of Life and Environmental Sciences, University of Exeter, Exeter, Devon, UK

Nigel Griffiths Laboratory of Plant Development and Interactions, Department of Molecular and Cellular Biology, University of Guelph, Guelph, ON, Canada

Cláudia P. Grou Organelle Biogenesis and Function Group, Instituto de Biologia Celular e Molecular (IBMC), Universidade do Porto, Porto, Portugal

Stefanie Hagen Systembiochemie, Ruhr-Universität Bochum, Bochum, Germany

M. Honsho Department of Biology, Faculty of Sciences, Kyushu University Graduate School, Fukuoka, Japan

Jianping Hu Department of Energy Plant Research Laboratory, Michigan State University, East Lansing, MI, USA

Plant Biology Department, Michigan State University, East Lansing, MI, USA

Markus Islinger Department of Neuroanatomy, Center for Biomedicine and Medical Technology Mannheim, University of Heidelberg, Mannheim, Germany

A. Itoyama Department of Biology, Faculty of Sciences, Kyushu University Graduate School, Fukuoka, Japan

Erica-Ashley Jaipargas Laboratory of Plant Development and Interactions, Department of Molecular and Cellular Biology, University of Guelph, Guelph, ON, Canada

Wilhelm Just Biochemisches Zentrum der Universität Heidelberg (BZH), Heidelberg, Germany

Navneet Kaur Department of Energy Plant Research Laboratory, Michigan State University, East Lansing, MI, USA

Tom Kettelhut Division of Neuropediatrics, Department of Child and Adolescent Health, University Medical Center, University of Göttingen, Göttingen, Germany

Wolf-H. Kunau Institut für Physiologische Chemie, Ruhr-Universität Bochum, Bochum, Germany

Markus Kunze Department of Pathobiology of the Nervous System, Center for Brain Research, Medical University of Vienna, Vienna, Austria

Igor Kurochkin Bioinformatics Institute (BII), Agency for Science, Technology and Research (A*STAR), Matrix, Singapore

Jaideep Mathur Laboratory of Plant Development and Interactions, Department of Molecular and Cellular Biology, University of Guelph, Guelph, ON, Canada

Sebastian Maurer-Stroh Bioinformatics Institute (BII), Agency for Science, Technology and Research (A*STAR), Matrix, Singapore

School of Biological Sciences (SBS), Nanyang Technological University (NTU), Singapore, Singapore

Ann Moser Department of Neurogenetics, Kennedy Krieger Institute, Baltimore, MD, USA

S. Mukai Faculty of Sciences, Department of Biology, Kyushu University Graduate School, Fukuoka, Japan

Taras Y. Nazarko Section of Molecular Biology, Division of Biological Sciences, UC San Diego, La Jolla, CA, USA

Marcus Nordgren Laboratory of Lipid Biochemistry and Protein Interactions, Department of Cellular and Molecular Medicine, KU Leuven, Leuven, Belgium

Silke Oeljeklaus Institute of Biology II and BIOSS—Centre for Biological Signalling Studies, University of Freiburg, Freiburg, Germany

K. Okumoto Faculty of Sciences, Department of Biology, Kyushu University Graduate School, Fukuoka, Japan

Manuel P. Pinto Organelle Biogenesis and Function Group, Instituto de Biologia Celular e Molecular (IBMC), Universidade do Porto, Porto, Portugal

Harald W. Platta Biochemie Intrazellulärer Transportprozesse, Ruhr-Universität Bochum, Bochum, Germany

Tony A. Rodrigues Organelle Biogenesis and Function Group, Instituto de Biologia Celular e Molecular (IBMC), Universidade do Porto, Porto, Portugal

Instituto de Ciências Biomédicas Abel Salazar (ICBAS), Universidade do Porto, Porto, Portugal

Wolfgang Schliebs Institute for Physiological Chemistry, Ruhr-University Bochum, Bochum, Germany

Michael Schrader College of Life and Environmental Sciences, Biosciences, University of Exeter, Exeter, UK

Department of Biology, Centre for Cell Biology, University of Aveiro, Aveiro, Portugal

Andreas Schummer Institute of Biology II and BIOSS—Centre for Biological Signalling Studies, University of Freiburg, Freiburg, Germany

Andriy A. Sibirny Institute of Cell Biology, NAS of Ukraine, Lviv, Ukraine

University of Rzeszow, Rzeszow, Poland

Poonam Singh Bioinformatics Institute (BII), Agency for Science, Technology and Research (A*STAR), Matrix, Singapore

Imogen Sparkes Biosciences, College of Life and Environmental Sciences, University of Exeter, Exeter, Devon, UK

S. Tamura Faculty of Sciences, Department of Biology, Kyushu University Graduate School, Fukuoka, Japan

Sven Thoms Division of Neuropediatrics, Department of Child and Adolescent Health, University Medical Center, University of Göttingen, Göttingen, Germany

Paul P. Van Veldhoven Laboratory of Lipid Biochemistry and Protein Interactions, Department of Cellular and Molecular Medicine, KU Leuven, Leuven, Belgium

Ida J. van der Klei Molecular Cell Biology, Groningen Biomolecular Sciences and Biotechnology Institute (GBB), University of Groningen, Groningen, The Netherlands

Simon Verheijden Laboratory of Cell Metabolism, Department of Pharmaceutical and Pharmacological Sciences, KU Leuven, Leuven, Belgium

Ronald J. A. Wanders Laboratory Genetic Metabolic Diseases, Departments of Paediatrics and Clinical Chemistry, Emma Children's Hospital, Academic Medical Center, Amsterdam, The Netherlands

Bo Wang Laboratory of Lipid Biochemistry and Protein Interactions, Department of Cellular and Molecular Medicine, KU Leuven, Leuven, Belgium

Bettina Warscheid Institute of Biology II and BIOSS—Centre for Biological Signalling Studies, University of Freiburg, Freiburg, Germany

Hans R. Waterham Laboratory Genetic Metabolic Diseases, Departments of Paediatrics and Clinical Chemistry, Emma Children's Hospital, Academic Medical Center, Amsterdam, The Netherlands

Chris Williams Molecular Cell Biology, Groningen Biomolecular Sciences and Biotechnology Institute (GBB), University of Groningen, Groningen, The Netherlands

Part I

Peroxisome Biogenesis and Function

History and Discovery of Peroxins

1

Wilhelm Just and Wolf-H. Kunau

Abstract

Peroxisomes started their life relatively late in comparison to many of their earlier discovered siblings among the set of cell organelles such as the nuclei, the mitochondria, the chloroplasts, or the ER. In addition, the metabolic pathways first described to this organelle did not immediately suggest the biological significance of peroxisomes for cells, distinct tissues, or even a total organism as we know it today. The major unexpected breakthrough bringing them to the limelight of cell biology came with the discovery of the peroxisomal diseases. For 25 years now, they triggered research on peroxisome biogenesis, a scientific journey full of excitement and unexpected results, as the authors experienced themselves.

Keywords

Microbodies • Glyoxysomes • Glycosomes • Peroxisome biogenesis • Peroxins • Hydrogen peroxide-based respiration

1.1 Introduction

A scientific discovery is frequently the combined result of both the different lines of observations and the development of new techniques. The identification of peroxisomes is a striking example of this. The organelle was first described as a

W. Just (✉)
Biochemisches Zentrum der Universität Heidelberg (BZH), Im Neuenheimer Feld 328, 69120 Heidelberg, Germany
e-mail: Wilhelm.just@bzh.uni-heidelberg.de

W.-H. Kunau (✉)
Institut für Physiologische Chemie, Ruhr-Universität Bochum, Universitätstr. 150, 44780 Bochum, Germany
e-mail: Wolf-h.kunau@rub.de

C. Brocard and A. Hartig (eds.), *Molecular Machines Involved in Peroxisome Biogenesis and Maintenance*, DOI 10.1007/978-3-7091-1788-0_1, © Springer-Verlag Wien 2014

distinct structure by Rhodin (1954) in mouse kidney cells. These organelles were surrounded by a single membrane and contained a finely granular matrix. Due to their small size and nondescript appearance he called them 'microbodies'.

This observation was facilitated by electron microscopy, which at that time was starting to become available as a tool to study the morphological organization of tissues and cells. It was not until 1960 and later that de Duve and coworkers using newly developed cell fractionation techniques demonstrated that certain enzymes, namely urate oxidase, catalase, and D-amino acid oxidase, were localized in this organelle and that these enzymes performed a hydrogen peroxide-based respiration (DeDuve and Baudhuin 1966). Oxygen is reduced by the oxidases to hydrogen peroxide, which is decomposed by catalase to water and oxygen. This discovery led de Duve to propose that the new particle should be named the 'peroxisome' (DeDuve and Baudhuin 1966). Another independent line of observation started with the investigation of the glyoxylate cycle in germinating castor beans by Beevers and coworkers. They located the enzymes catalyzing this metabolic pathway to an apparently novel kind of cytoplasmic particle, which they called the 'glyoxysome' (Breidenbach and Beevers 1967). This particle was found to have properties very similar to peroxisomes. From these earlier studies they demonstrated that the glyoxysome of castor beans also contains the complete β-oxidation pathway oxidizing fatty acids to acetyl-CoA (Cooper and Beevers 1969). Eight years later, Lazarow in de Duve's laboratory discovered the same pathway in rat liver peroxisomes (Lazarow and De Duve 1976). These initial and seminal findings have clearly demonstrated that these two closely related organelles with overlapping properties have more than a 'fossil role' as originally suspected by de Duve for the mammalian peroxisomes.

1.2 Metabolic Pathways Found in Peroxisomes

Subsequent studies in numerous laboratories during the last three decades have confirmed and greatly enlarged these original discoveries and conclusions. Today, we know that peroxisomes are probably present in almost all lower and higher eukaryotes. Central aspects of them are their diversity and plasticity. Their size, number, and protein content vary according to tissue and cell type, and in addition to environmental stimuli. Characteristic to almost all peroxisomes is the presence of hydrogen peroxide producing oxidases and catalase as well as a β-oxidation system. Many other functions assigned to the organelle vary in different cell types and organisms. This is especially apparent in lower eukaryotes, such as yeasts and fungi, where in accordance with the metabolic need of a cell peroxisomes can be markedly induced in number and distinct activities are greatly stimulated (van der Klei and Veenhuis 2006). By the year 1981, 40 enzymes had been localized to the peroxisome (Tolbert 1981) and the number has increased considerably since then. These findings led Opperdoes to label peroxisomes as a 'multi-purpose organelle' (Opperdoes 1988). The structural similarities and broad biochemical specification led to the overlapping terminology of microbodies, peroxisomes, and glyoxysomes.

While the first term is morphologically defined and comprises the other two, the last term is more restrictive and refers to a special class of peroxisomes that stresses the metabolism of glyoxylate leaving peroxisomes as the most general name. More recently, glycosomes became known as another member of the microbody family. These unique organelles of trypanosomes compartmentalize most of the glycolytic pathway (Opperdoes 1988).

1.3 Peroxisomal Diseases: A Turning Point for Peroxisome Research

A turning point in the history of peroxisomes was the discovery of the peroxisomal diseases. It began in 1973 with the observation that peroxisomes were absent in cells from Zellweger patients (Goldfischer et al. 1973). These patients suffer from the Cerebro-Hepato-Renal syndrome, which is a rare familial malady and manifested by severe hypotonia, central nervous system abnormalities, hepatomegaly, renal cortical cysts, and skeletal malformations. Afflicted infants do not thrive and usually die within the first 6 months of life (Zellweger 1965). This observation led to the notion that, although dispensable in single cells, peroxisomes are essential for the smooth and continued function of complex, multicellular, mammalian systems.

Since this first description of a peroxisomal disorder, many more peroxisomal functions and states of diseases have been recognized (Moser 1993; Schutgens et al. 1986; Waterham and Ebberink 2012). All of these disorders are genetically determined, cause serious disabilities, and can be identified by noninvasive diagnostic assays. These are classified into two major categories: disorders that are based on single-enzyme defects and disorders underlying a defect in the biogenetic assembly of the organelle (Steinberg et al. 2006; Wanders and Waterham 2006). In case of single-enzyme defects, single or multiple peroxisomal activities are deficient, while the peroxisomal structures as visualized by different microscopic techniques appear to be intact. Prominent examples of single-enzyme defects are the X-linked adrenoleukodystrophy (ALD) (Berger and Gärtner 2006) and the rhizomelic chondrodysplasia punctata (RCDP) type 2 and 3 (Samsom et al. 1992; Tager et al. 1994). From the biochemical point of view the single enzyme defects are well understood. Patients with X-linked ALD accumulate very long-chain fatty acids due to the defect of the peroxisomal ABC transporter D1, a membrane protein that is required to transport these fatty acids into human peroxisomes. RCDP type 2 and 3 are caused by defects in the two key enzymes of plasmalogen biosynthesis, dihydroxyacetonephosphate acyltransferase and alkyl-dihydroxyacetonephosphate synthase, respectively, both leading to a deficiency of plasmalogens.

Disorders of peroxisome biogenesis are based on pathways defective in peroxisome assembly and hence are more complex in their etiology than single-enzyme deficiencies. According to their clinical phenotype they belong to the Zellweger spectrum including the Cerebro-Hepato-Renal (Zellweger) syndrome, the neonatal ALD, and the infantile Refsum's disease (Steinberg et al. 2006; Waterham and

Ebberink 2012). The disorders of peroxisome biogenesis have been instrumental in discovery of the genes involved in this process. Skin fibroblasts of patients suffering from peroxisome biogenesis disorders were collected all over the world (Europe, USA, and Japan) and analysed for complementation of their defects (Fujiki et al. 2012). Consequently, 13 complementation groups (CG) were identified clearly indicating that defects in different genes may cause the same clinical phenotype.

The unifying concept that peroxisomal diseases could be traced to dysfunction or absence of the peroxisomes had arisen in concurrence with the localization of specific metabolic pathways to this organelle. The pathology of the peroxisomal diseases underlined the importance of at least some of these pathways for the overall cellular metabolism of higher eukaryotic organisms. In the second half of the 1980s, this concept greatly stimulated investigations on the biogenesis of peroxisomes especially as many of these disorders appeared to represent failures of the mechanisms normally involved in the import of proteins into the peroxisome. Investigations of this specific protein sorting pathway were facilitated by the molecular characterization of several of the peroxisomal enzymes, e.g. catalase, luciferase, and the ß-oxidation enzymes, at that time. Thus these proteins provided the proper tools to study peroxisomal protein import.

1.4 Peroxisome Biogenesis

Concepts about the biogenesis of peroxisomes had undergone a controversial evolution when in 1985 Lazarow and Fujiki published the growth and division model, which then for more than 20 years became the prevalent concept (Lazarow and Fujiki 1985). It postulated that peroxisomes grow by post-translational import of matrix and membrane proteins and eventually divide into two or more organelles. By the end of the 1980s two distinct targeting signals were identified within the amino acid sequence of peroxisomal matrix proteins, denoted as PTS1 and PTS2, comprising the last three amino acids at the C terminus and a degenerated nonapeptide near the amino terminus, respectively (de Hoop and Ab 1992; Gould et al. 1987; Swinkels et al. 1991). Either of them is necessary and sufficient for the import of the respective proteins. In this respect the peroxisomal matrix protein import followed the rules already established before in other biological compartments, such as the ER, mitochondria, and chloroplasts (Baker et al. 1992). However, in one key aspect it did not. The observation that folded, co-factor bound and/or even oligomeric proteins rather than unfolded polypeptide chains transverse the peroxisomal membrane was quite an unexpected result and at that time violated a dogma of cell biology (McNew and Goodman 1996; Purdue and Lazarow 2001).

1.4.1 Discovery of Peroxins in Lower Eukaryotes

Studies investigating the molecular mechanisms of peroxisomal biogenesis were especially successful in those model systems, which allowed the application of techniques of molecular biology and genetics. These were first of all four different fungi: *Saccharomyces cerevisiae* (Erdmann et al. 1989; van der Leij et al. 1992; Zhang et al. 1991), *Pichia pastoris* (Gould et al. 1992; Liu et al. 1992), *Hansenula polymorpha* (Cregg et al. 1990), and later also *Yarrowia lipolytica* (Szilard et al. 1995). In all these fungi the number of peroxisomes increases dramatically in response to distinct carbon or nitrogen sources. This property turned out to be of key importance for the identification of genes required for the formation and maintenance of this organelle. Forward genetic approaches enabled different laboratories to isolate mutants defective in this process and to identify the affected genes. Originally these mutants and genes were given different names in each organism, e.g. pas/*PAS* (Erdmann et al. 1989; Gould et al. 1992; van der Leij et al. 1992), per/*PER* (Cregg et al. 1990; Liu et al. 1992), pay/*PAY* (Nuttley et al. 1995; Szilard et al. 1995), and peb/*PEB* (Zhang et al. 1991), before. Fortunately, a common nomenclature pex/*PEX* was agreed upon by the community (Distel et al. 1996). Within a few years these common efforts led to the identification of about 20 different *PEX* genes and their corresponding proteins, the peroxins (Kunau 1998). Today, 33 *PEX* genes are known and the corresponding peroxins are involved in widely different aspects of organelle formation such as matrix protein import, membrane biogenesis, proliferation, and inheritance (see Table 1.1; Rucktäschel et al. 2011).

1.4.2 Conservation of Peroxins in Higher Eukaryotic Species

Studies applying the fungal systems were complemented by a genetic approach using Chinese hamster ovary cells that directly led to the mammalian *PEX* genes and to the orthologues of the fungal peroxins (Tsukamoto et al. 1990). More recently, the elucidation of the genome of *Arabidopsis thaliana*, the analyses of *A. thaliana* mutants obtained by forward genetic screening and the subsequent identification of plant *PEX* genes demonstrated that also plants possess a very similar set of peroxins (Hayashi and Nishimura 2006). Comparison of the results from the lower and higher eukaryotic model systems allowed the important conclusion to be drawn that, with the exceptions of minor variations, the principle mechanisms are conserved across all eukaryotes. Thus, the choice of fungi as model systems to determine which proteins and molecular mechanisms are required to form a peroxisome, based on the hope that this will also apply to higher eukaryotes was greatly rewarded. This approach together with the analyses of CHO mutants allowed first insights into the molecular causes of the defects in peroxisomal diseases. For most of the fungal *PEX* genes counterparts were identified in humans (Fujiki et al. 2012; Waterham and Ebberink 2012, see also Chaps. 4–8). Furthermore, for each of the known 13 complementation groups of the

Table 1.1 List of peroxin genes identified and function of the encoded proteins

Process	Gene	Human gene locus	Genes coding for functional orthologs	Feature	Molecular complex involved
PMP Import	***PEX3***	6q24.2	–		PMP-import complex; Pex19p anchoring
	PEX16	11p11.2	–		PMP-import complex
	PEX19	1q23.1	–	CAAX-box; farnesylated	PMP-import complex; PMP-class I receptor and chaperone
Matrix protein import	***PEX5***	12q13.31	–	WxxxF-motifs; TPR; ubiquit	Import receptor for PTS1-containing proteins
	PEX7	6q23.3	–	WD40-domain	Import receptor for PTS2-containing proteins
	PEX18[a]	–	*PEX20*	WxxxF-motifs; ubiquitinated	PTS2-co-receptor in *S.c.*
	PEX20[a]	–	*PEX18/ PEX21*	WxxxF-motifs; ubiquitinated	PTS2-co-receptor in most fungi
	PEX21[a]	–	*PEX20*	WxxxF-motifs; ubiquitinated (?)	PTS2-co-receptor in *S.c.*
Receptor docking	***PEX13***	2p16.1	–	SH3-domain	Matrix protein docking complex
	PEX14	1p36.22	–	PxxP-motif, hosphorylated	Matrix protein docking complex
	PEX17[a]	–	–		Matrix protein docking complex
	PEX33[b]	–	*PEX17*		Matrix protein docking complex
Ubiquitination	***PEX8***[a]	–	–	Coiled-coil domain, leu-zipper	Connecting component for docking- and ubiquitin ligase complexes
	PEX2	8q21.1	–	RING-finger	Ubiquitin ligase complex
	PEX4[a]	–	–	Ubc	Receptor ubiquitination complex
	PEX10	1q36.32	–	RING-finger	Ubiquitin ligase complex
	PEX12	17q12	–	RING-finger	Ubiquitin ligase complex
	PEX22[a]	–	–		Receptor ubiquitination complex; Pex4p anchoring in *S.c.*
Receptor regeneration	***PEX1***	7q21.2	–	AAA-type ATPase	ATP-dependent export complex
	PEX6	6q21.1	–	AAA-type ATPase	ATP-dependent export complex
	PEX15[a]	–	*PEX26*	Phosphorylated	ATP-dependent export complex; Pex6p anchoring in *S.c.*
	PEX26	22q11.21	*PEX15*		ATP-dependent export complex; Pex6p anchoring in *H.s.*

(continued)

Table 1.1 (continued)

Process	Gene	Human gene locus	Genes coding for functional orthologs	Feature	Molecular complex involved
Proliferation	***PEX11***	(α) 15q26.1 (β) 1q21.1 (γ) 19q13.2	*PEX25*/ *PEX27*	Amphipathic α-helix	Membrane elongation complex
	PEX23[a]	–	*PEX30*/*31*/*32*	DysF	Proliferation complex in *Y.l.*
	PEX24[a]	–	*PEX28*/*29*		Proliferation complex in *Y.l.*
	PEX25[a]	–	*PEX11*		Membrane elongation complex in *S.c.* and *H.p.*
	PEX27[a]	–	*PEX11*		Membrane elongation complex in *S.c.* and *H.p.*
	PEX28[a]	–	*PEX24*		Proliferation complex in *S.c.*
	PEX29[a]	–	*PEX24*		Proliferation complex in *S.c.*
	PEX30[a]	–	*PEX23*	DysF	Proliferation complex; ER tethering in *S.c.*
	PEX31[a]	–	*PEX23*	DysF	Proliferation complex; ER tethering in *S.c.*
	PEX32[a]	–	*PEX23*	DysF	Proliferation complex; ER tethering in *S.c.*
	PEX34[a]	–	–		Membrane fission complex in and *S.c.*
–	***PEX9***	*Eliminated*			*Wrongly annotated ORF*

Note: *PEX16* was first identified in *Y.l.* and does not exist in *S.c.*
ORF open reading frame, *PTS* peroxisomal targeting signal, *H.p. Hansenula polymorpha*, *S.c. Saccharomyces cerevisiae*, *N.c. Neurospora crassa*, *Y.l. Yarrowia lipolytica*
[a]Only identified in yeast species whereby *PEX23 and PEX24 were identified in Y.l. and PEX28, PEX29, PEX30, PEX31, PEX32 and PEX34 in S.c.*
[b]*Only identified in N.c.*

peroxisomal diseases it could be demonstrated that mutations in a distinct human *PEX* gene cause the observed defects of the respective complementation group (Table 1.2; Fujiki et al. 2012; see also Chap. 4–8).

Table 1.2 Complementation groups (CG) and afflicted PEX genes of peroxisomal biogenesis disorders

CG		
US/EU	Japan	Gene mutated
1	E	*PEX1*
2		*PEX5*
3		*PEX12*
4 (6)	C	*PEX6*
7 (5)	B	*PEX10*
8	A	*PEX26/PEX15*
9	D	*PEX16*
10	F	*PEX2*
11	R	*PEX7*
12	G	*PEX3*
13	H	PEX13
14	J	*PEX19*
15	K	*PEX14*

1.5 Protein Import into the Peroxisome Matrix

Since the newly identified peroxins turned out to be unknown proteins and their amino acid sequences did not, in most cases, allow their function to be deduced, it took several years to get first insights into their mechanistic roles and to understand their essential character in the biogenesis of peroxisomes. Even today this process is far from being completed.

Many efforts were made to understand how folded proteins were imported from the cytosol into the peroxisomal matrix. The receptor cycle was discovered as a central mechanistic feature of this process. Two soluble import receptors namely, Pex5p and Pex7p recognize and bind PTS1- and PTS2-containing proteins, respectively, in the cytosol. A first simple model proposed that the receptors only deliver their cargo to the outer face of the membrane before returning to the cytosol (Dodt and Gould 1996; Marzioch et al. 1994). Further observations from studies with Pex5p subsequently demonstrated that the receptor cycle involves additional membrane-bound steps (Gouveia et al. 2000) by which the receptor becomes an integral membrane protein and is accessible from the inside of the peroxisome (Dammai and Subramani 2001). Even receptor release to the matrix prior to its return to the cytosol was considered (Lazarow 2006). However, as evidence for such an export pathway has not been presented so far the current prevailing extended model predicts that the receptors insert deeply into the membrane prior to their return back to the cytosol (Kunau 2001; see also Chaps. 13–16).

Depending on the organism the membrane-bound steps of the receptor cycle require besides the two receptors at least 10–12 additional peroxins. Predominantly, these are integral membrane proteins, can be isolated as sub-complexes, and appear to function in a consecutive order (Agne 2002; Collins et al. 2000; Hazra et al. 2002). Together they are referred to as importomer (Agne 2002; see also Chap. 13). In lower and higher eukaryotes, the lack of any of these components is characterized by the cytosolic mislocalization of peroxisomal matrix proteins.

In *S. cerevisiae*, Pex13p, Pex14p, and Pex17p form the docking complex; Pex2p, Pex10p, and Pex12p are organized in a RING finger complex, while Pex8p bridges these two sub-complexes (Table 1.1). Release of the receptor is initiated by Pex4p, an ubiquitin-conjugating enzyme that is anchored via Pex22p to the outer face of the membrane. Central players in this export part of the cycle are the two AAA proteins Pex1p and Pex6p that bound to the membrane protein Pex15p release the receptor from the membrane (Platta et al. 2005; see also Chaps. 13–16). The concept of the export-driven import proposes that the energy for this receptor cycle is provided by these last steps of the cycle (Schliebs et al. 2010; see also Chaps. 15 and 16).

Although the receptor cycle and the participating peroxins form a framework, central questions as to the mechanistic understanding of the receptor cycle remain unanswered. Such a key question relates to the composition, the architecture, and the mechanism of the unique translocation pore, which has to facilitate the transport of folded proteins of very different sizes across the membrane without compromising its permeability properties. Evidence has been presented that the PTS1 receptors Pex5p is itself a constitutive part of the pore (Meinecke et al. 2010) and it is tempting to speculate that this also applies for the PTS2 receptor Pex7p and/or its co-receptors. It has been suggested that Pex5p and Pex14p represent the minimal translocation pore (Meinecke et al. 2010; Schliebs and Kunau 2006). In contrast to the PTS1 receptor Pex5p, the PTS2 receptor Pex7p requires a co-receptor (Schäfer et al. 2004). This difference may be explained by the fact that Pex5p comprises two different domains, the C-terminal half possessing the binding site for PTS1 proteins and the N-terminal half, which is supposed to facilitate the membrane-bound steps of the receptor cycle. The observation that the PTS2 co-receptor, Pex18p, can substitute for the N-terminal half of Pex5p in a chimeric protein strongly suggests that in the PTS2 pathway the co-receptor mediates these steps (Ma et al. 2009). Another unresolved fundamental question concerns the release of the cargo proteins from the receptors. It is still unclear where and how the matrix proteins segregate from their receptors which transport them to and into the peroxisomal membrane.

1.6 On the Origin of Peroxisomes

Recent studies provided evidence that the import of matrix proteins into the organelle and the biogenesis of the peroxisomal membranes mechanistically follow different pathways. Deletion mutants of both PTS receptors and co-receptors as well as mutants with defective components of the translocation machinery are still able to insert proteins into residual peroxisomal membranes (Kunau and Erdmann 1998). Three peroxins, Pex3p, Pex16, and Pex19p, have been demonstrated to play a pivotal role in the latter process (Hettema et al. 2000). In the absence of either one of these peroxins, no peroxisomal structures are present and peroxisomal membrane proteins (PMP) are degraded. Pex19p is proposed to function as a soluble and recycling targeting receptor for newly synthesized PMPs, delivering its cargo to peroxisomes after docking on Pex3p (Fujiki et al. 2006; Hoepfner et al. 2005). Pex16p is absent in *S. cerevisiae* but required in mammalian cells for the recruitment of Pex3p to the peroxisomal membrane. While for a long time peroxisomes were considered to be autonomous organelles and their PMPs to be post-translationally imported as suggested by the growth and division model (Lazarow and Fujiki 1985), it is now widely accepted that peroxisomes originate in a maturation process from the ER (endoplasmic reticulum) in cells temporarily devoid of peroxisomes (Hoepfner et al. 2005; Nuttall et al. 2011). However, controversy exists as to how peroxisomes multiply in wild-type cells with preexisting organelles. There are conflicting views as to what extent the ER-to-peroxisome route exists in these cells. Is it the sole mechanism, does it constitute an alternative pathway besides the formation of peroxisomes by growth and division, or is it not present in these cells (Nuttall et al. 2011; Tabak et al. 2013)? These different notions have large implications as to the role of Pex3p, Pex16p, and Pex19p as well as to the question of how PMPs reach peroxisomes. In their extreme form the ER-to-peroxisome route and the growth and division model appear hard to reconcile but perhaps the difference is semantic rather than mechanistic (Theodoulou et al. 2013).

1.7 Future Aspects

Although during the last two decades unexpected and intriguing insights into the biogenesis of peroxisomes have been gained, such as the identification of essential proteins involved and the discovery of a receptor cycle leading to the import of folded proteins, questions concerning the mechanistic aspects of this process remain largely unsolved. For example, what is the composition and structure of the transient import channel in the peroxisomal membrane and how does it facilitate the transport of folded proteins across this membrane without compromising its permeability properties? When and how is the ER involved in providing essential membrane components to the peroxisomes? These and other central questions now move into the focus of research and will be addressed in the following chapters.

References

Agne B (2002) Untersuchungen zur Funktion von RING-Finger-Proteinen beim peroxisomalen Protein Import der Hefe Saccharomyces cerevisiae. Dissertation, Ruhr-Universität Bochum, Bochum

Baker A, Kaplan CP, Pool MR (1992) Protein targeting and translocation; a comparative survey. Biol Rev Camb Philos Soc 71:637–702

Berger J, Gärtner J (2006) X-linked adrenoleukodystrophy: clinical, biochemical and pathogenetic aspects. Biochim Biophys Acta 1763:1721–1732

Breidenbach RW, Beevers H (1967) Association of the glyoxylate cycle enzymes in a novel subcellular particle from castor bean endosperm. Biochem Biophys Res Commun 27:462–469

Collins CS, Kalish JE, Morrell JC, McCaffery JM, Gould SJ (2000) The peroxisome biogenesis factors Pex4p, Pex22p, Pex1p, and Pex6p act in the terminal steps of peroxisomal matrix protein import. Mol Cell Biol 20:7516–7526

Cooper TG, Beevers H (1969) Mitochondria and glyoxysomes from castor bean endosperm. Enzyme constitutents and catalytic capacity. J Biol Chem 244:3507–3513

Cregg JM, van der Klei IJ, Sulter GJ, Veenhuis M, Harder W (1990) Peroxisome-deficient mutants of Hansenula polymorpha. Yeast 6:87–97

Dammai V, Subramani S (2001) The human peroxisomal targeting signal receptor, Pex5p, is translocated into the peroxisomal matrix and recycled to the cytosol. Cell 105:187–196

de Hoop MJ, Ab G (1992) Import of proteins into peroxisomes and other microbodies. Biochem J 286:657–669

DeDuve C, Baudhuin P (1966) Peroxisomes (microbodies and related particles). Physiol Rev 46:323–357

Distel B, Erdmann R, Gould SJ, Blobel G, Crane DI, Cregg JM, Dodt G, Fujiki Y, Goodman JM, Just WW, Kiel JAKW, Kunau W-H, Lazarow PB, Mannaerts GP, Moser HW, Osumi T, Rachubinski RA, Roscher A, Subramani S, Tabak HF, Tsukamoto T, Valle D, van der Klei I, van Veldhoven PP, Veenhuis M (1996) A unified nomenclature for peroxisome biogenesis factors. J Cell Biol 135:1–3

Dodt G, Gould SJ (1996) Multiple PEX genes are required for proper subcellular distribution and stability of Pex5p, the PTS1 receptor: evidence that PTS1 protein import is mediated by a cycling receptor. J Cell Biol 135:1763–1774

Erdmann R, Veenhuis M, Mertens D, Kunau W-H (1989) Isolation of peroxisome-deficient mutants of Saccharomyces cerevisiae. Proc Natl Acad Sci USA 86:5419–5423

Fujiki Y, Matsuzono Y, Matsuzaki T, Fransen M (2006) Import of peroxisomal membrane proteins: the interplay of Pex3p- and Pex19p-mediated interactions. Biochim Biophys Acta 1763:1639–1646

Fujiki Y, Nashiro C, Miyata N, Tamura S, Okumoto K (2012) New insights into dynamic and functional assembly of the AAA peroxins, Pex1p and Pex6p, and their membrane receptor Pex26p in shuttling of PTS1-receptor Pex5p during peroxisome biogenesis. Biochim Biophys Acta 1823:145–149

Goldfischer S, Moor CL, Johnson AB, Spiro AJ, Valsamis MP, Wisniewski HK, Ritch RH, Norton WT, Rapin I, Gerner LM (1973) Peroxisomal and mitochondrial defects in cerebrohepatorenal syndrome. Science 182:62–64

Gould SJ, Keller GA, Subramani S (1987) Identification of a peroxisomal targeting signal at the carboxy terminus of firefly luciferase. J Cell Biol 105:2923–2931

Gould SJ, McCollum D, Spong AP, Heyman JA, Subramani S (1992) Development of the yeast Pichia pastoris as a model organism for a genetic and molecular analysis of peroxisome assembly. Yeast 8:613–628

Gouveia AM, Reguenga C, Oliveira ME, Sa-Miranda C, Azevedo JE (2000) Characterization of peroxisomal Pex5p from rat liver: Pex5p in the Pex5p-Pex14p membrane complex is a transmembrane protein. J Biol Chem 275:32444–32451

Hayashi M, Nishimura M (2006) Arabidopsis thaliana–a model organism to study plant peroxisomes. Biochim Biophys Acta 1763:1382–1391

Hazra PP, Suriapranata I, Snyder WB, Subramani S (2002) Peroxisome remnants in pex3Delta cells and the requirement of Pex3p for interactions between the peroxisomal docking and translocation subcomplexes. Traffic 3:560–574

Hettema EH, Girzalsky W, van Den Berg M, Erdmann R, Distel B (2000) Saccharomyces cerevisiae Pex3p and Pex19p are required for proper localization and stability of peroxisomal membrane proteins. EMBO J 19:223–233

Hoepfner D, Schildknegt D, Braakman I, Philippsen P, Tabak HF (2005) Contribution of the endoplasmic reticulum to peroxisome formation. Cell 122:89–95

Kunau W-H (1998) Peroxisome biogenesis: from yeast to man. Curr Opin Microbiol 1:232–237
Kunau W-H, Erdmann R (1998) Peroxisome biogenesis: back to the endoplasmic reticulum? Curr Biol 8:299–302
Kunau W (2001) Peroxisomes: the extended shuttle to the peroxisome matrix. Curr Biol 11:R659–R662
Lazarow PB (2006) The import receptor Pex7p and the PTS2 targeting sequence. Biochim Biophys Acta 1763:1599–1604
Lazarow PB, De Duve C (1976) A fatty acyl-CoA oxidizing system in rat liver peroxisomes; enhancement by clofibrate, a hypolipidemic drug. Proc Natl Acad Sci USA 73(6):2043–2046
Lazarow PB, Fujiki Y (1985) Biogenesis of peroxisomes. Annu Rev Cell Biol 1:489–530
Liu H, Tan X, Veenhuis M, McCollum D, Cregg JM (1992) An efficient screen for peroxisome-deficient mutants of Pichia pastoris. J Bacteriol 174:4943–4951
Ma C, Schumann U, Rayapuram N, Subramani S (2009) The peroxisomal matrix import of Pex8p requires only PTS receptors and Pex14p. Mol Biol Cell 20:3680–3689
Marzioch M, Erdmann R, Veenhuis M, Kunau W-H (1994) PAS7 encodes a novel yeast member of the WD-40 protein family essential for import of 3-oxoacyl-CoA thiolase, a PTS2-containing protein, into peroxisomes. EMBO J 13:4908–4918
McNew JA, Goodman JM (1996) The targeting and assembly of peroxisomal proteins: some old rules do not apply. Trends Biochem Sci 21:54–58
Meinecke M, Cizmowski C, Schliebs W, Kruger V, Beck S, Wagner R, Erdmann R (2010) The peroxisomal importomer constitutes a large and highly dynamic pore. Nat Cell Biol 12:273–277
Moser HW (1993) Peroxisomal diseases. Adv Hum Genet 21:443–451
Nuttall JM, Motley A, Hettema EH (2011) Peroxisome biogenesis: recent advances. Curr Opin Cell Biol 23:421–426
Nuttley WM, Szilard RK, Smith JJ, Veenhuis M, Rachubinski RA (1995) The PAH2 gene is required for peroxisome assembly in the methylotrophic yeast Hansenula polymorpha and encodes a member of the tetratricopeptide repeat family of proteins. Gene 160:33–39
Opperdoes FR (1988) Glycosomes may provide clues to the import of peroxisomal proteins. Trends Biochem Sci 13:255–260
Platta HW, Grunau S, Rosenkranz K, Girzalsky W, Erdmann R (2005) Functional role of the AAA peroxins in dislocation of the cycling PTS1 receptor back to the cytosol. Nat Cell Biol 7:817–822
Purdue PE, Lazarow PB (2001) Peroxisome biogenesis. Annu Rev Cell Dev Biol 17:701–752
Rhodin J (1954) Correlation of ultrastructural organization and function in normal and experimentally changed peroxisomal convoluted tubule cells of the mouse kidney. Dissertation, Aktiebolaget Godvil, Stockholm
Rucktäschel R, Girzalsky W, Erdmann R (2011) Protein import machineries of peroxisomes. Biochim Biophys Acta 1808:892–900
Samsom JF, Jakobs C, van de Klei-van Moorsel J, Smit LM, Schutgens RB, Wanders RJ (1992) Zellweger syndrome in a preterm, small for gestational age infant. J Inherit Metab Dis 15:75–83
Schäfer A, Kerssen D, Veenhuis M, Kunau WH, Schliebs W (2004) Functional similarity between the peroxisomal PTS2 receptor binding protein Pex18p and the N-terminal half of the PTS1 receptor Pex5p. Mol Cell Biol 24:8895–8906
Schliebs W, Girzalsky W, Erdmann R (2010) Peroxisomal protein import and ERAD: variations on a common theme. Nat Rev Mol Cell Biol 11:885–890
Schliebs W, Kunau WH (2006) PTS2 co-receptors: diverse proteins with common features. Biochim Biophys Acta 1763:1605–1612
Schutgens RB, Heymans HS, Wanders RJ, van den Bosch H, Tager JM (1986) Peroxisomal disorders: a newly recognised group of genetic diseases. Eur J Pediatr 144:430–440
Steinberg SJ, Dodt G, Raymond GV, Braverman NE, Moser AB, Moser HW (2006) Peroxisome biogenesis disorders. Biochim Biophys Acta 1763:1733–1748

Swinkels BW, Gould SJ, Bodnar AG, Rachubinski RA, Subramani S (1991) A novel, cleavable peroxisomal targeting signal at the amino-terminus of the rat 3-ketoacyl-CoA thiolase. EMBO J 10:3255–3262

Szilard RK, Titorenko VI, Veenhuis M, Rachubinski RA (1995) Pay32p of the Yeast Yarrowia lipolytica is an intraperoxisomal component of the matrix protein translocation machinery. J Cell Biol 131:1453–1469

Tabak HF, Braakman I, van der Zand A (2013) Peroxisome formation and maintenance are dependent on the endoplasmic reticulum. Annu Rev Biochem 82:723–744

Tager JM, Aerts JM, van den Bogert C, Wanders RJ (1994) Signals on proteins, intracellular targeting and inborn errors of organellar metabolism. J Inherit Metab Dis 17:459–468

Theodoulou FL, Bernhardt K, Linka N, Baker A (2013) Peroxisome membrane proteins: multiple trafficking routes and multiple functions? Biochem J 451:345–352

Tolbert NE (1981) Metabolic pathways in peroxisomes and glyoxysomes. Annu Rev Biochem 50:133–157

Tsukamoto T, Yokota S, Fujiki Y (1990) Isolation and characterization of Chinese hamster ovary cell mutants defective in assembly of peroxisomes. J Cell Biol 110:651–660

van der Klei IJ, Veenhuis M (2006) Yeast and filamentous fungi as model organisms in microbody research. Biochim Biophys Acta 1763:1364–1373

van der Leij I, van den Berg M, Boot R, Franse MM, Distel B, Tabak HF (1992) Isolation of peroxisome assembly mutants from Saccharomyces cerevisiae with different morphologies using a novel positive selection procedure. J Cell Biol 119:153–162

Wanders RJ, Waterham HR (2006) Peroxisomal disorders: the single peroxisomal enzyme deficiencies. Biochim Biophys Acta 1763:1707–1720

Waterham HR, Ebberink MS (2012) Genetics and molecular basis of human peroxisome biogenesis disorders. Biochim Biophys Acta 1822:1430–1441

Zellweger H (1965) Genetic Aspects of Neurological Disease. Arch Intern Med 115:387–397

Zhang JW, Han Y, Lazarow PB (1991) Novel peroxisome clustering mutants and peroxisome biogenesis mutants of Saccharomyces cerevisiae. J Cell Biol 123:1133–1147

Mouse Models with Peroxisome Biogenesis Defects

2

Myriam Baes, Simon Verheijden, and Paul P. Van Veldhoven

Abstract

Peroxisomes are ubiquitous organelles in mammalian cells but it is still unclear how they contribute to normal development and tissue homeostasis. To address this question, gene targeting techniques have been applied on several peroxins to interfere with peroxisome biogenesis in mice. Both peroxins involved in peroxisomal matrix import and peroxins necessary for peroxisome division were inactivated. Besides generalized knockouts, mice were created with conditional inactivation of *PEX* genes either in certain cell types or induced in adulthood. Defective matrix import generates empty peroxisomal ghosts and metabolic derangements that are a direct consequence of peroxisome inactivity. In addition, ablation of functional peroxisomes from hepatocytes affects other cellular compartments such as mitochondria and the endoplasmic reticulum. Peroxisome inactivity in the central nervous system causes both developmental and degenerative pathologies. The impairment of peroxisome division in mice also results in cerebral and hepatic pathologies although peroxisomal metabolites are unaffected.

Keywords

Peroxisomes • Mouse models • Peroxisome biogenesis • Peroxisome proliferation

---

M. Baes (✉) • S. Verheijden
Laboratory of Cell Metabolism, Department of Pharmaceutical and Pharmacological Sciences, KU Leuven, Leuven, Belgium
e-mail: Myriam.Baes@pharm.kuleuven.be

P.P. Van Veldhoven
Laboratory of Lipid Biochemistry and Protein Interactions, Department of Cellular and Molecular Medicine, KU Leuven, Leuven, Belgium

C. Brocard and A. Hartig (eds.), *Molecular Machines Involved in Peroxisome Biogenesis and Maintenance*, DOI 10.1007/978-3-7091-1788-0_2, © Springer-Verlag Wien 2014

2.1 Introduction

The insights in the processes of peroxisomal matrix import, membrane formation, and proliferation in vertebrates have been greatly advanced by using diverse cell lines in which a peroxin was eliminated. Hereto, fibroblast cells from peroxisome biogenesis disorder (PBD) patients (Steinberg et al. 2006) and CHO cells in which *PEX* genes were inactivated have been used (Fujiki et al. 2006). From the pathological point of view, it is more informative to hinder the peroxisome biogenesis process in other cell types, e.g., hepatocytes or neural cells. To this end, several mouse models were generated in which *PEX* genes were targeted. In this chapter the cellular and tissue phenotypes will be discussed that result from the inactivation of peroxins either involved in peroxisomal matrix import (Pex2p, Pex5p, Pex13p, and Pex7p) or in peroxisomal proliferation (Pex11αp and Pex11βp). In view of the perinatal lethality of most mouse models with generalized gene inactivation, and with the purpose to explore peroxisome function in specific cell types and tissues, several cell-type selective (also denoted as conditional) *Pex5*- and *Pex13*-knockouts were created using Cre-loxP technology (Baes et al. 2002; Bjorkman et al. 2002; summarized in Table 2.1).

2.2 Mouse Models with Defects in Peroxisomal Matrix Import

2.2.1 Role of Pex2p, Pex5p, Pex7p, and Pex13p in Matrix Import

The four peroxins that were inactivated in mice take part in the early steps of peroxisomal matrix import (Girzalsky et al. 2010). Pex5p is the receptor recognizing newly synthesized proteins carrying a C-terminal peroxisome-targeting signal 1 (PTS1) consisting of serine–lysine–leucine or a conserved variant. The amino acids preceding this tripeptide may modulate the efficiency of peroxisomal import (Brocard and Hartig 2006). Pex5p occurs in two isoforms in vertebrates, a short (Pex5Sp) and a long form (Pex5Lp), the latter containing an additional stretch of 37 amino acids encoded by an extra exon, that allows it to bind Pex7p. The latter is the receptor for an N-terminal PTS2 sequence, consisting of a nonapeptide that occurs in a minority of peroxisomal matrix proteins. In mammals, these include enzymes of the α-oxidation (phytanoyl-CoA α-hydroxylase), β-oxidation (thiolase) and ether lipid synthesis pathways (alkyl-dihydroxyacetone-phosphate synthase). The cargo–Pex5p or cargo–Pex7p–Pex5Lp complexes dock at the peroxisomal membrane and bind Pex14p complexed with Pex13p. In contrast to yeast, Pex5p is thus not only essential for the import of PTS1 but also of PTS2 proteins. The precise role of Pex13p in the docking and translocation process is not elucidated, but it is an indispensable link in the import chain. According to recent evidence Pex5p and Pex14p form a flexible pore in the membrane through which the cargo is translocated (Meinecke et al. 2010). Subsequently, Pex5p is either monoubiquitinated and recovered to the cytosol for a subsequent import cycle or polyubiquitinated and targeted for degradation (Rucktaschel et al. 2011). It was

Table 2.1 Phenotype of mice with peroxisome biogenesis defects

Targeted peroxin gene	Model	Life span	Growth	Nervous tissue	Other pathologies	Biochemical parameters	References
Pex2	General	Embryonic lethality[a] <24 h–5 weeks[b]	Retarded	Cortex malformation Cerebellar malformation	Liver (cholestasis, steatosis, steatohepatitis) Hypotonia	Plasmalogens ↓↓ VLCFA ↑ Bile acids ↓↓	Faust and Hatten (1997), Faust (2003), Keane et al. (2007), Kovacs et al. (2009, 2012)
Pex13	General	<12 h	Retarded	Cortex malformation	Hypotonia	Plasmalogens ↓↓ VLCFA ↑	Maxwell et al. (2003), Nguyen et al. (2006)
Pex13-loxP Nestin-cre	Conditional neural cells	10–150 days	Retarded	Cerebellar malformation		Plasmalogens ↓↓ VLCFA normal	Muller et al. (2011)
Pex5	General	<24 h	Retarded	Cortex malformation	Hypotonia	Plasmalogens ↓↓ VLCFA ↑	Baes et al. (1997), Janssen et al. (2000), Baumgart et al. (2001), Janssen et al. (2003)
Pex5-loxP Alfp-cre	Conditional fetal liver	8–16 days	Postnatally retarded	Cortex malformation	Liver (steatosis)	Plasmalogens normal	Krysko et al. (2007)
Pex5-loxP Alb-cre	Conditional adult liver	>12 months	Postnatally retarded	Normal	Liver (hepatomegaly, steatosis, fibrosis, tumors from 12 months)	Plasmalogens normal VLCFA normal Phytanic/pristanic acid ↑ C27/C24 bile acids↑ Mitochondrial ATP production↓	Dirkx et al. (2005), Peeters et al. (2011a, b)

(continued)

Table 2.1 (continued)

Targeted peroxin gene	Model	Life span	Growth	Nervous tissue	Other pathologies	Biochemical parameters	References
Pex5-loxP aP2-cre	Conditional adipose[c]	Normal	Lower body weight after weaning	Normal	Increased fat mass Shivering thermogenesis↓ Motor performance↓ Insulin resistance↑	Plasmalogens ↓ Lipolysis ↓ Plasma adrenaline ↓↓	Martens et al. (2010, 2012)
Pex5-loxP Nestin-cre	Conditional neural cells	3 weeks–6 months	Postnatally retarded	Cerebellar malformation Demyelination, inflammation, axon loss Decline in motor and cognitive abilities	Male and female infertility	VLCFA ↑ PUFA ↓ (at young age) Plasmalogens ↓↓ Neutral lipid accumulations in astroglia and in ependymal cells	Krysko et al. (2007), Hulshagen et al. (2008), Bottelbergs et al. (2012)
Pex5-loxP Nex-cre	Conditional forebrain neurons	Normal	Normal	Normal	None	No abnormalities	Bottelbergs et al. (2010)
Pex5-loxP Gfap-cre	Conditional astrocytes	Normal	Normal	Normal	None	Plasmalogens ↓ VLCFA ↑ Neutral lipid accumulations in astroglia and in ependymal cells	Bottelbergs et al. (2010)
Pex5-loxP Cnp-cre	Conditional oligodendro cytes	6–12 months	Normal	Axonal degeneration; subcortical demyelination; neuroinflammation; impaired motor nerve conductance	None	Plasmalogens (in myelin) ↓ VLCFA ↑	Kassmann et al. (2007)

Pex5-loxP Wnt-cre	Conditional peripheral nervous system	Normal	Normal	Normal	None	ND	Martens et al. (2012)
Pex5-loxP Amh-cre	Conditional Sertoli cells	ND	Normal	Normal	Male infertility	Lipid droplets in Sertoli cells	Huyghe et al. (2006)
Pex5-loxP Cmv-cre-ER	Inducible[d] all cells	8 months	ND	Demyelination, neuroinflammation, axon loss	ND	ND	Bottelbergs et al. (2012)
Pex7	General	<24 h–>1 year	Retarded	Cortex malformation mild astrocytosis	Testis (infertile) Eyes (cataracts) Adipose (reduced) Harderian gland (atrophy)	Plasmalogens ↓↓ VLCFA↑ (certain organs)	Brites et al. (2003, 2011)
	Hypomorphic allel	Normal	Retarded	ND	Eyes (cataracts)	Plasmalogens ↓ DHA in RBC↓	Braverman et al. (2010)
Pex11α	General	Normal	Normal (Li) increased BW (Weng)	Normal	None Hepatic steatosis	Plasmalogens and VLCFA normal (Li) Liver triglycerides↑ (Weng)	Li et al. (2002a), Weng et al. (2013)
Pex11β	General	<24 h	Retarded	Cortex malformation	Hypotonia	Plasmalogens and VLCFA normal	Li et al. (2002b), Ahlemeyer et al. (2012)

(continued)

Table 2.1 (continued)

Targeted peroxin gene	Model	Life span	Growth	Nervous tissue	Other pathologies	Biochemical parameters	References
Pex11α + *Pex11β*	General	<24 h	Retarded	Cortex malformation	Hypotonia	Normal	Li et al. (2002a)

General: refers to whole animal knockout, conditional: refers to cell type selective knockouts. In the latter, the biochemical parameters were determined in the targeted tissue; *alfp* alphafetoprotein, *alb* albumin, *BW* body weight, *ND* not determined, *RBC* red blood cells

[a]In 129SV background

[b]In 129SV/Swiss Webster background

[c]Also in ganglia of the peripheral nervous system (PNS), in adrenal medulla, and in neurons throughout the central nervous system (CNS)

[d]*PEX5* was deleted after tamoxifen administration at 1–2 months

shown in yeast that Pex2p is a ubiquitin ligase that is crucial for polyubiquitination of Pex5p (Rucktaschel et al. 2011). In all organisms ranging from yeast to plants and mammals, Pex2p, Pex5p, Pex7p, and Pex13p are indispensable for peroxisomal matrix protein import.

2.2.2 Macroscopic Phenotypes of Mice with Peroxisome Biogenesis Defects

Inactivation of both PTS1 and PTS2 import throughout the body (further called generalized) in *Pex2*- (Faust and Hatten 1997), *Pex5*- (Baes et al. 1997), and *Pex13*- (Maxwell et al. 2003) knockout mice causes severe developmental abnormalities with intrauterine growth retardation and brain malformations. Depending on the genetic background of the mice, this results in fetal lethality, neonatal death due to severe hypotonia, or survival for a few weeks. Mice with loss of functional peroxisomes from all neural cells (neurons, astrocytes, oligodendrocytes) (Nestin-*Pex5−/−* and Nestin-*Pex13−/−*) are indistinguishable from wild-type mice at birth but they become growth retarded within the first postnatal week. Several succumb after weaning, whereas the survivors develop motor and cognitive deficits and die before the age of 6 months, proving that peroxisomes are indispensable for the integrity of the central nervous system (CNS) in adulthood (Hulshagen et al. 2008; Bottelbergs et al. 2012). Mice with cell type selective inactivation of Pex5p in hepatocytes, testis, or adipose tissue become adult but develop diverse pathologies as summarized in Table 2.1. *Pex7* knockouts are compromised at birth, and the majority dies before weaning although others survive into adulthood and beyond (Brites et al. 2003). A hypomorphic *PEX7*-deficient mouse model expressing less than 5 % of wild-type transcript levels has a normal life span (Braverman et al. 2010). All *PEX7*-deficient mice show pre- and postnatal growth impairment, but only the full male knockouts are infertile.

2.2.3 Peroxisomal Matrix Import Defects Give Rise to Peroxisomal Ghosts

Peroxisomes can arise by division of preexisting peroxisomes, presumably the major pathway (Huybrechts et al. 2009), or by de novo formation from the ER. In both cases, the process of peroxisomal membrane formation and membrane protein import precedes the import of matrix proteins. It can therefore be expected that inactivation of key components of the matrix import machinery yields empty peroxisomes also called "ghosts." In *Pex2−/−*, *Pex5−/−*, and *Pex13−/−* cells (Baes et al. 1997; Faust and Hatten 1997; Maxwell et al. 2003), vesicles were indeed identified that contain integral membrane proteins such as PMP70 but are devoid of catalase and other matrix proteins. These are however less numerous and larger in size than regular peroxisomes. In *PEX13*-deficient fibroblasts, neurons, and astrocytes, it was shown that trafficking of peroxisomal ghosts is perturbed resulting in an altered cytoplasmic distribution as compared to mature peroxisomes in wild-type cells. The ghosts cluster and are not aligned along peripheral microtubules (Nguyen et al. 2006). The impairment of peroxisomal matrix import was also proven by performing Western blots on the enzymes acyl-CoA oxidase (PTS1 containing enzyme) and thiolase (PTS2 containing enzyme). These proteins are normally split by the protease TYSND1 after their import into peroxisomes, but they occur in their unprocessed forms in knockout cells. As can be anticipated, in *PEX7*-deficient cells, only the PTS2 containing proteins are mislocalized to the cytosol (Brites et al 2003; Braverman et al. 2010).

2.2.4 Metabolic Consequences of Peroxisomal Matrix Import Defects

The reported biochemical consequences of import incompetent peroxisomes in mice are related to deficient peroxisomal β-oxidation and ether lipid synthesis. α-oxidation is another pathway that exclusively takes place in peroxisomes but when fed a normal diet, mice are not exposed to the substrate, phytanic acid. Therefore, in order to prove defects in this pathway, mice need to be supplemented with phytol, a precursor of phytanic acid. In the mouse models with generalized inactivation of both PTS1 and PTS2 import (*Pex2*, *Pex5*, and *Pex13* knockouts) plasmalogens are reduced to less than 10 % of normal values in brain and liver (Baes et al. 1997; Faust and Hatten 1997; Maxwell et al. 2003). In mice with conditional *PEX5* gene inactivation in peripheral tissues (liver, adipose) plasmalogens are not or only mildly reduced in the targeted tissues (Dirkx et al. 2005; Martens et al. 2012), whereas they are profoundly depleted in brain of Nestin-*Pex5* and Nestin-*Pex13* knockout mice. These data can be explained by transfer of ether lipids from peroxisome-bearing cells to peroxisome-deficient cells in the periphery but not through the blood brain barrier. This is in line with reports that supplementation of rodents with ether lipids can elevate plasmalogens in the periphery but not or hardly in the CNS (Das and Hajra 1988; Brites et al. 2011; Wood et al. 2011). Levels of C26:0 are three- to tenfold increased in tissues of generalized *Pex* knockouts. Remarkably, in brain of the conditional *Pex5* knockouts, C26:0 accumulate to similar extents whether all neural cells (Nestin-*Pex5*) (Hulshagen et al. 2008) or only oligodendrocytes (Cnp-*Pex5*) (Kassmann et al. 2007) or astrocytes (Gfap-*Pex5*) (Bottelbergs et al. 2010) are targeted. In contrast, loss of peroxisomes from neurons does not affect VLCFA levels (Bottelbergs et al. 2010). Surprisingly, in brain of juvenile Nestin-*Pex13* mice C26:0 levels are normal (Muller et al. 2011), which is in discordance with the findings in Nestin-*Pex5* mice. Docosahexaenoic acid (DHA, C22:6n-3), a polyunsaturated fatty acid that requires peroxisomal β-oxidation for its synthesis, is reduced in brain of newborn *PEX5* knockouts (Janssen et al. 2000) and in cerebellum of Nestin-*Pex5−/−* mice (Krysko et al. 2007).

In mouse models with *PEX* gene inactivation in liver, bile acid metabolism is severely affected (Dirkx et al. 2005; Keane et al. 2007). The ratio of immature C27/

mature C24 bile acids is increased in liver, bile and plasma, and bile acids appear predominantly in their unconjugated form.

The metabolic derangements in *Pex7* knockout mice pertain to the mislocalization of three enzymes that are part of three different pathways. The defect in ether lipid synthesis results in a profound deficiency of plasmalogens in total *Pex7* knockouts (Brites et al. 2003), whereas they are less drastically reduced in the hypomorphic counterparts (Braverman et al. 2010). As a result of deficient α-oxidation, phytanic acid accumulates in plasma after feeding a phytol-enriched diet (Brites et al. 2003). The consequences of the import defect of the β-oxidation enzyme thiolase, which is involved in breakdown of straight chain substrates, are more variable. C26:0 accumulates in some tissues (brain of newborn mice, spleen, kidney) but not in adult brain, testis, and liver (Brites et al. 2003, 2009). Possibly, the other peroxisomal enzyme with thiolytic activity, SCPx, takes over the function in the tissues with normal VLCFA levels.

2.2.5 Lipid Homeostasis Is Disturbed in Hepatocytes, Sertoli Cells and Astrocytes

Besides the direct consequences of peroxisomal dysfunction, it is remarkable that in certain cell types with peroxisome biogenesis defects lipid droplets accumulate. In *Pex2−/−* and *Pex5−/−* hepatocytes triglyceride concentrations are increased and microvesicular steatosis develops (Dirkx et al. 2005; Keane et al. 2007, Fig. 2.1). This is unexpected since *Pex5−/−* hepatocytes exhibit enhanced mitochondrial β-oxidation (Dirkx et al. 2007). In these peroxisome-deficient livers, PPARα is significantly activated which is likely due to the accumulation of ligands of this nuclear receptor that cannot be degraded (Peeters et al. 2011b). Possibly, the marked upregulation of the PPARα target CD36 causes an enhanced influx of fatty acids that may contribute to increased lipid stores. Interestingly, lipid droplets were also seen in the brain of Nestin-*Pex5−/−* mice, in which peroxisome inactivation is restricted to neural cells (Hulshagen et al. 2008). Lipids were particularly found in Bergmann glia astrocytes in the cerebellum and in ependymal cells lining the ventricles, also a glial cell type. Lipid storage was also found in an astrocyte selective (Gfap-*Pex5−/−*; Bottelbergs et al. 2010) but not in neuron (Nex-*Pex5−/−*; Bottelbergs et al. 2010) nor in oligodendrocyte selective (Cnp-*Pex5−/−*; Kassmann et al. 2007) *Pex5* knockout mice, confirming the cell type specificity. In view of the normal life span and absence of major neurological problems of Gfap-*Pex5−/−* mice, it seems that the accumulating lipids are rather harmless. The situation is different in the testis, where similar increased stores of neutral lipids were noticed when functional peroxisomes were deleted from Sertoli cells (Huyghe et al. 2006). This was accompanied with degeneration of the tubuli seminipheri and complete loss of spermatogenesis. It was later shown that Cre expression in Sertoli cells has a widespread impact on immunological signaling, oxidative stress, peroxisomal protein expression and on other features (Xiao et al. 2012). It can however be excluded that Sertoli cell lipid storage and testicular degeneration are artefacts of

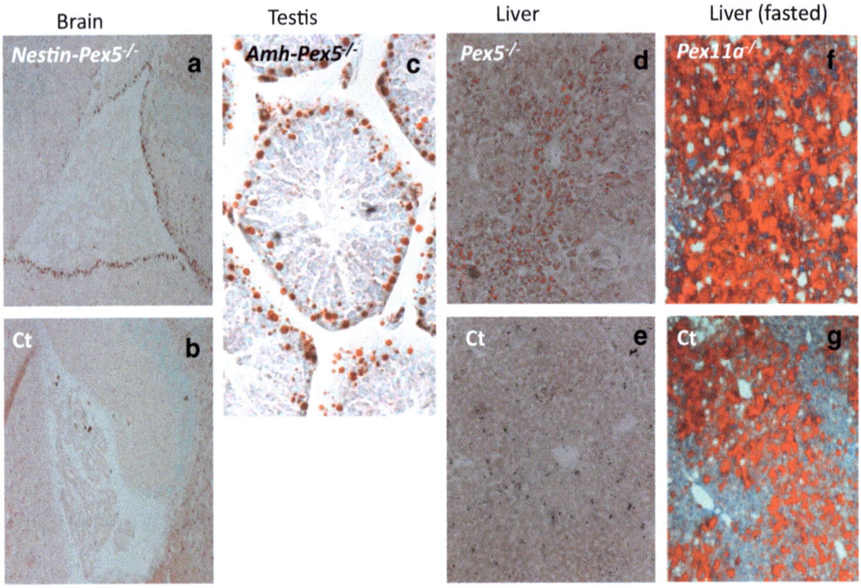

Fig. 2.1 Lipid accumulations in peroxin-deficient cells. Neutral lipids were visualized by Oil Red O staining on frozen sections. Lipid droplets accumulate in ependymal cells lining the fourth ventricle in Nestin-*Pex5−/−* mice (**a**, **b**), in Sertoli cells of Amh-*Pex5−/−* mice (selective Sertoli cell knockout) (**c**), and in hepatocytes depleted of *PEX5* (**d**, **e**). Furthermore, *Pex11α−/−* hepatocytes store more lipids than control mice (shown in **f**, **g** after a 48-h fasting period; reproduced from Weng et al. (2013) with permission from the American Physiological Society). Ct = sections from control mice

Cre expression because a similar pathology is observed in mice with a generalized peroxisomal β-oxidation defect (*Mfp2−/−* mice; Huyghe et al. 2006). This also indicates that the neutral lipid accumulation in peroxisome-deficient cells is a consequence of impaired β-oxidation although the precise origin and nature of the lipids needs to be determined.

2.2.6 Other Compartments Are Affected in Peroxisome-Deficient Hepatocytes

The absence of mature peroxisomes has also repercussions on other subcellular compartments. Most obvious are alterations in *PEX5*- and *PEX2*-deficient hepatocytes in which mitochondrial and ER structure and function are severely compromised (Baumgart et al. 2001; Dirkx et al. 2005; Kovacs et al. 2009). By ultrastructural analysis mitochondria appear swollen and the cristae of the inner mitochondrial membrane are rare and malformed (Baumgart et al. 2001; Dirkx et al. 2005). The function of the electron transport chain is severely affected, whereby in particular the activities of complex I, III and V are reduced. This results in impaired synthesis of mitochondrial ATP and, despite an upregulation of glycolysis, to a reduced AMP/ATP ratio and the activation of AMPK (Peeters et al. 2011a). Through the latter kinase, the activity of catabolic processes (such as glycolysis) is increased and of anabolic processes (gluconeogenesis, glycogen synthesis and fatty acid synthesis) is decreased in *Pex5* knockout hepatocytes (Peeters et al. 2011a, b). An additional compensatory mechanism is the proliferation of mitochondria as shown by increased numbers of mitochondria and elevated matrix enzyme activities. Of interest, not all mitochondria within a cell are affected and the mechanism is cell autonomous. Indeed, the few hepatocytes that escape the recombination process in hepatocyte selective *Pex5* knockouts, show normal mitochondria.

Both in *PEX2*- and *PEX5*-deficient liver, mediators of ER stress signaling pathways (PERK, ATF4) are upregulated (Kovacs et al. 2009; Peeters et al. 2011a). Similar to the mitochondrial alterations, the ER stress is a very early event as it already occurs in newborn mice (Baumgart et al. 2001; Kovacs et al. 2012). Microscopic investigations showed ER dilation and proliferation. It was proposed that these ER perturbations cause oxidative stress (see below) and deregulation of SREBP controlled cholesterol homeostasis.

At present, the mechanisms through which peroxisome inactivity affect mitochondria and ER are still obscure. It is also unclear whether these cellular compartments are altered in other cell types in which peroxisomes are inactive.

2.2.7 Does Peroxisome Deficiency Affect the Cellular and Organellar Redox State?

Peroxisomes play a pivotal role in cellular redox metabolism as they harbor both several oxygen radical generating and degrading enzymes (Fransen et al. 2012). When both categories are unable to enter the organelle, it is not unequivocal to predict how this affects the cellular redox homeostasis. Importantly, it has become clear that under normal conditions, peroxisomes are not an isolated compartment with regard to ROS generation and degradation. Peroxide generated in peroxisomes leaks into other cellular compartments and vice versa peroxisomal catalase degrades peroxide that is generated elsewhere in the cell (Fransen et al. 2012). In fact, diverse results were obtained with regard to the oxidative stress state depending on the cell types and *Pex* knockouts investigated. Different approaches were used including evaluation of oxidative damage to proteins, lipids or DNA, transcript levels of anti-oxidant enzymes or direct measurement of reactive oxygen species (ROS)(H_2O_2, $O_2^{\bullet -}$) in cultured cells.

In hepatocytes from newborn *Pex5* and *Pex2* knockout mice, increased expression of anti-oxidant proteins and of genes related to oxidative stress generation such as NADP oxidase were reported (Baumgart et al. 2001; Kovacs et al. 2012). As a potential mechanism, it was proposed that induction of the ER stress-related protein PERK phosphorylates the transcription factor NRF2 that further activates oxidative stress genes (Kovacs et al. 2012). However, studies on liver and on cultured hepatocytes from adult conditional hepatocyte selective *Pex5* knockouts did neither reveal upregulation of antioxidative enzymes nor oxidative damage to proteins, nor increased cytosolic peroxide (Dirkx et al. 2005). A complicating matter is that mitochondria are also severely affected at the level of the electron transport chain in peroxisome-deficient hepatocytes, as mentioned above, which could also affect the redox state of the cell.

In cerebella and in cultured cerebellar neurons of Nestin-*Pex13* mice, MnSOD2 levels are elevated (Muller et al. 2011). Furthermore, in the latter neurons, increased levels of superoxide but not of H_2O_2 were detected, rather compatible with a mitochondrial origin of oxidative stress. This was accompanied by increased apoptotic death of these neurons. In the equivalent Nestin-*Pex5* knockout mice, signs of oxidative stress were exclusively found in cerebellar Purkinje cells based on immunohistochemical detection of nitrotyrosine and 4-hydroxynonenal (Bottelbergs et al. 2012). Catalase was also markedly increased but this was confined to astrocytes including Bergmann glia in the cerebellum. When using a biochemical approach on several brain regions, there was no evidence for lipid peroxidation (Bottelbergs et al. 2012). Because anti-oxidative treatment could not prevent the severe neurodegeneration of Nestin-*Pex5* knockout mice (see below), it is improbable that excessive ROS underlies the brain phenotype (Bottelbergs et al. 2012).

Taken together, it is likely that the diverse cellular circumstances (in vivo versus in vitro, culture media (M. Fransen, personal communication), antioxidative capacity of different cell types) have a major impact on the redox state of cells lacking functional peroxisomes. When also mitochondria or the ER are affected in peroxisome-deficient cells, it is difficult to distinguish between primary and secondary consequences of peroxisome deficiency on the redox state.

2.2.8 Peroxisome Ablation Causes Developmental and Degenerative Neuropathologies

The CNS seems to be particularly vulnerable to peroxisome dysfunction, whereby both developmental and degenerative pathologies arise. Full deletion of peroxisomal function in *Pex2*, *Pex5*, and *Pex13* knockouts invariably causes cortical neuronal migration defects during fetal development resulting in abnormal lamination of the cortex at birth (Faust and Hatten 1997; Baes et al. 1997; Maxwell et al. 2003). It is not clear yet whether the severe neonatal hypotonia in these mice is related to these CNS or to peripheral anomalies. By using liver and brain selective *Pex5* rescue and knockout mice, it was shown that both peroxisome dysfunction in the liver and in the brain contribute to the hampered cortical development in *Pex5−/−* mice (Janssen et al. 2003; Krysko et al. 2007). The cerebellum that matures between birth and weaning, is also affected as shown in *Pex2* knockouts (Faust 2003) and in mice with neural inactivation of *PEX5* (Nestin-*Pex5*; Krysko et al. 2007) or *PEX13* (Nestin-*Pex13*; Muller et al. 2011). Multiple processes are disorganized including migration of granular cells, Purkinje cell arborization, and cerebellar foliation.

Additional pathologies in mice with selective deletion of functional peroxisomes from neural cells (Nestin-*Pex5*) are dysmyelination in cerebellum in juvenile mice and demyelination at later ages throughout the brain (Hulshagen et al. 2008; Bottelbergs et al. 2012). This is accompanied with severe inflammation, whereby microgliosis precedes astroglia activation. Axons degenerate and are lost but there is no evidence for neural cell death.

The question whether this detrimental neuropathology originates from a particular cell type was addressed by creating cell type selective (neurons, astrocytes, oligodendrocytes) *Pex5* knockouts. The oligodendrocyte (Cnp-*Pex5*) knockout clearly showed the most severe phenotype with similar pathologies as in Nestin-*Pex5−/−* mice but with later onset and slower progression (Kassmann et al. 2007). This suggests that besides cell autonomous functions of peroxisomes in oligodendrocytes, other mechanisms cause the early lethality of Nestin-*Pex5−/−* knockouts. In contrast, ablation of peroxisomal function in forebrain neurons (Nex-*Pex5−/−*) or in astrocytes does not cause obvious neurological symptoms nor impaired life span. Nevertheless, anomalies occurred in brains of Gfap-*Pex5−/−* mice (Bottelbergs et al. 2010) such as increased levels of C26:0 and overexpression of catalase in astrocytes, as previously mentioned. The same neurodegenerative sequence could also be provoked by deletion of *PEX5* in postnatal life excluding that the neurodegenerative phenotype is merely a result of developmental problems (Bottelbergs et al. 2012).

Pex7−/− mice exhibit more moderate developmental and degenerative brain pathology including anomalies in neuronal migration and mild astrocytosis in adulthood (Brites et al. 2003, 2009). Myelination in different brain regions seemed normal and no microgliosis was observed throughout the brain of *Pex7* knockouts.

2.3 Mouse Models with Impairment of Peroxisome Proliferation

2.3.1 Role of Pex11 Proteins in Peroxisome Proliferation

Peroxins of the Pex11p family are involved in the early steps of peroxisome proliferation (Schrader et al. 2012). As their overexpression elevates and their inactivation lowers peroxisome abundance in cells of species ranging from plants to mammals, it is clear that they regulate peroxisome number. There are three isoforms in mammals that share 40–60 % sequence similarity. Whereas *PEX11α* and *PEX11γ* show tissue specific expression and are primarily found in liver, *PEX11β* is ubiquitously expressed. Another distinguishing feature is that only the *PEX11α* gene is regulated by PPARα (Schrader et al. 2012). It is however less clear whether the isoforms exhibit also functional differences, analogous to the three yeast Pex11p homologues that have partially divergent functions (Huber et al. 2012). All the Pex11 proteins harbor an amphipathic helix that is essential for integration in and deformation of the peroxisomal membrane (Koch and Brocard 2011). In subsequent steps the Fis1 and Mff1 proteins are attracted that in turn recruit DLP1, the executor of peroxisome fission (Koch and Brocard 2012). The inactivation of *PEX11α* and *PEX11β* genes in mice results in strongly differing phenotypes, reflecting the difference in expression pattern and possibly the distinct function.

2.3.2 Pex11α Knockout Mice Display a Mild Hepatic Phenotype

Two different mouse models lacking *PEX11α* were generated (Li et al. 2002a; Weng et al. 2013). Although in both models a complete loss of *PEX11α* mRNA was confirmed, researchers came to different conclusions with regard to peroxisome numbers in cells. As Pex11αp is primarily expressed in hepatocytes, liver sections were investigated. Gould and coworkers did not detect a reduction in peroxisome abundance but there was a tendency for peroxisome clustering in *Pex11α−/−* hepatocytes (Li et al. 2002a). In contrast, according to Weng et al. (2013) immunofluorescently labeled peroxisomes were 20–30 % lower in number in liver sections. By EM analysis they further showed that small and rounded peroxisomes were more prevalent in *Pex11α−/−* hepatocytes both under normal diet and after treatment with fibrates, whereas in controls a higher frequency of elongated and irregular peroxisomes were present. As there is no clear explanation for the discrepant findings between the research groups, it remains inconclusive whether for basal maintenance of peroxisome numbers, the function of Pex11αp in hepatocytes can be taken over by the family members Pex11βp and/or Pex11γp. It was surprising that, although Pex11αp is the sole isoform inducible by PPARα, *Pex11α* knockout mice display normal peroxisome proliferation when treated with classical PPARα ligands (Li et al. 2002a). It was suggested that the elevation of peroxisome numbers by PPARα might be mediated by altered metabolite levels. On the other hand, Pex11αp was required for peroxisome proliferation induced by 4-phenylbutyrate, a compound that acts independently of PPARα (Li et al. 2002a).

Interestingly, Weng et al. 2013 showed that *Pex11α* knockout hepatocytes store more triglyceride containing lipid droplets in basal conditions than wild types, which is exacerbated after challenging the mice with high fat diets or by fasting (Fig. 2.1). Body weights of *Pex11α* knockout mice were significantly increased as compared to controls. This shows that Pex11αp is not fully redundant with Pex11βp and Pex11γp isoforms. The lipid accumulations are reminiscent of those in *PEX5*-deficient hepatocytes and indicate that lipid homeostasis in hepatocytes can be perturbed even when peroxisomes are only marginally affected. It was proposed that the reduced peroxisomal surface area in *Pex11α−/−* hepatocytes impairs the uptake of fatty acids from the cytosol leading to reduced peroxisomal β-oxidation and lipid accumulation. This explanation however implies that hepatic peroxisomes significantly contribute to long chain fatty acid degradation, whereas it is currently believed that this only accounts for less than 10 % (Mannaerts et al. 1979) or 25 % (Grum et al. 1994). It is unfortunate that peroxisomal metabolic activities were not determined in *Pex11α−/−* liver by any of the research groups. Gould and coworkers did analyze C26:0 in plasma of *Pex11α−/−* mice and the activity of α- and β-oxidation and plasmalogen synthesis in *Pex11α−/−* fibroblasts, none of which were altered compared to wild-type controls (Li et al. 2002a). Weng et al. reported deregulation of transcript levels of enzymes related to peroxisomal β-oxidation in liver (Weng et al. 2013) but as there is no concerted change, this does not clarify the observed lipid accumulations.

2.3.3 Deletion of a Single or Both PEX11β Alleles in Mice Impairs Neural Function

The ubiquitous expression of *PEX11β* is indicative of its housekeeping role to maintain peroxisome numbers in all cells in basal conditions. In agreement, in *Pex11β−/−* mouse fibroblasts, cultured neurons and astrocytes abundance of peroxisomes was halved as compared to wild-type cells (Li et al. 2002b; Li and Gould 2002). Also in brain sections peroxisomes were reduced by 30–50 %. Unexpectedly, newborn *Pex11β−/−* mice show a phenotype that closely resembles the mouse models with peroxisome matrix import defects including a cortical migration defect, hypotonia, a prominent growth retardation and neonatal lethality (Li et al. 2002b).

It is most intriguing that this severe pathology is not accompanied with defects in peroxisomal protein import and consequently not with changes in peroxisomal metabolites. Catabolism of C24:0, phytanic and pristanic acid and synthesis of plasmalogens were unaltered in fibroblasts and no significant changes in plasmalogen and C26:0 concentrations were observed in liver and brain. *PEX11β* deficiency in men likewise has no effect on biochemical parameters, but in contrast to mice, this was associated with a mild neurological phenotype (Ebberink et al. 2012). It was therefore proposed that the pathologies in PBD could be caused by mechanisms independent of metabolic perturbations (Li et al. 2002b). Because the pathogenic metabolites fail to be identified in PBD in men and mice, this is an attractive alternative consideration. On the other hand, we should bear in mind that very similar pathologies as those in PBD occur in patients suffering from peroxisome single enzyme disorders (such as MFP2/D-BP deficiency, ACOX1 deficiency) in which metabolic deficits are expected to be the prime cause (Van Veldhoven 2010).

More in-depth analysis revealed increased oxidative stress in neuronal cultures and in brain sections of *Pex11β−/−* mice, which was accompanied with increased neuronal cell death by apoptosis (Ahlemeyer et al. 2012). Neuronal differentiation was impaired as shown by reduced synaptophysin expression. Interestingly, haploinsufficiency of *PEX11β* caused similar but less extensive neural abnormalities that were easier to detect in vitro than in vivo. Strikingly, these anomalies were not related to effects on proliferation as peroxisome numbers were neither reduced in cultured neurons nor in brain tissue. In view of unchanged metabolism in homozygous knockouts, no biochemical alterations are expected to occur in the heterozygous mice. The latter however display deregulated expression of several genes at the mRNA and protein level. The cellular changes in *Pex11β*+/− mice did not markedly affect CNS functioning because the mice are macroscopically indistinguishable from wild types and they are fertile. This is in sharp contrast with the phenotype of mice with peroxisome import defects in neural cells that show reduced growth, fertility, coordination and motor defects as described above (Hulshagen et al. 2008; Muller et al. 2011).

It remains enigmatic how a 50 % loss of Pex11β protein levels can affect oxidative stress and survival of neural cells in the absence of functional and structural peroxisome defects. As Pex11βp is the isoform with the highest expression level in the majority of tissues, it would be interesting to investigate whether its loss from other cell types has similar consequences.

Conclusions and Perspectives

Although peroxisomes are ubiquitous cell organelles, many of their secrets still need to be uncovered. To investigate the necessity of peroxisomes in diverse cell types and tissues, analysis of mouse models with peroxisome deficiencies is a powerful approach. In the first generation, generalized knockouts with a complete *PEX* gene inactivation were created and phenotyped. It is striking that impeding peroxisomal matrix import versus peroxisome proliferation has a different impact on peroxisomal metabolite levels but causes similar developmental problems. These include not only brain malformations but also intrauterine growth retardation. The latter may complicate the investigation of pathologies in newborn mice as they can either be a direct consequence of peroxisome dysfunction or rather a nonspecific result of the developmental delay.

In the second generation of mice, conditional inactivation of *PEX* genes was achieved, allowing the study of the function of peroxisomes in adult tissues. Some pitfalls should be taken in consideration when analyzing cell type selective knockouts, e.g., not all pathologies might be mediated via cell autonomous defects and could therefore be missed in these conditional knockouts. Furthermore, when using Cre loxP technology, one should be aware of potential effects of Cre activity on the functioning of the targeted cells. It was indeed shown that Cre expression in Sertoli cells causes oxidative stress and deregulation of gene expression (Xiao et al. 2012). In addition, the success of this approach relies on the cell type specificity of gene inactivation that fully depends on the promoter that is used to drive Cre expression. There are several examples of promoters that were presumed to be cell type selective but induce Cre expression during fetal development in several other cell types. For example, aP2-Cre mice, which were thought to only drive gene recombination in adipocytes, also show Cre activity in cells sharing a common lineage with adipocytes such as chondrocytes, myocytes, neurons and osteocytes (Martens et al. 2010, 2012). Without any doubt, taking these limitations into account, additional cell type selective peroxin knockouts will reveal the importance of peroxisomes in tissues thus far not investigated.

In the future, besides the consequences of complete peroxisome dysfunction, also the impact of milder peroxisome biogenesis impairments will need to be explored. Indeed, increasing numbers of patients with mild dysfunction are diagnosed with a PBD in adulthood (Steinberg et al. 2009; Regal et al. 2010; Ebberink et al. 2010; Sevin et al. 2011; Mignarri et al. 2012; Matsui et al. 2012). Therefore, mouse models should be generated with point mutations resulting in partial peroxisome dysfunction. Attention should also be paid to the possibility that heterozygous null mutations of certain *PEX* genes could trigger pathologies, as shown for *PEX11β* (Ahlemeyer et al. 2012).

References

Ahlemeyer B, Gottwald M, Baumgart-Vogt E (2012) Deletion of a single allele of the Pex11β gene is sufficient to cause oxidative stress, delayed differentiation and neuronal death in mouse brain. Dis Model Mech 5:125–140

Baes M, Gressens P, Baumgart E et al (1997) A mouse model for Zellweger syndrome. Nat Genet 17:49–57

Baes M, Dewerchin M, Janssen A, Collen D, Carmeliet P (2002) Generation of Pex5-IoxP mice allowing the conditional elimination of peroxisomes. Genesis 32:177–178

Baumgart E, Vanhorebeek I, Grabenbauer M et al (2001) Mitochondrial alterations caused by defective peroxisomal biogenesis in a mouse model for Zellweger syndrome (PEX5 knockout mouse). Am J Pathol 159:1477–1494

Bjorkman J, Tonks I, Maxwell MA, Paterson C, Kay GF, Crane DI (2002) Conditional inactivation of the peroxisome biogenesis Pex13 gene by Cre-IoxP excision. Genesis 32:179–180

Bottelbergs A, Verheijden S, Hulshagen L et al (2010) Axonal integrity in the absence of functional peroxisomes from projection neurons and astrocytes. GLIA 58:1532–1543

Bottelbergs A, Verheijden S, Van Veldhoven PP, Just W, Devos R, Baes M (2012) Peroxisome deficiency but not the defect in ether lipid synthesis causes activation of the innate immune system and axonal loss in the central nervous system. J Neuroinflammation 9:61

Braverman N, Zhang R, Chen L et al (2010) A Pex7 hypomorphic mouse model for plasmalogen deficiency affecting the lens and skeleton. Mol Genet Metab 99:408–416

Brites P, Motley AM, Gressens P et al (2003) Impaired neuronal migration and endochondral ossification in PEX7 knockout mice: a model for rhizomelic chondrodysplasia punctata. Hum Mol Genet 12:2255–2267

Brites P, Mooyer PAW, el Mrabet L, Duran M, Waterham HR, Wanders RJA (2009) Plasmalogens participate in very-long-chain fatty acid-induced pathology. Brain 132:482–492

Brites P, Ferreira AS, da Silva TF et al (2011) Alkyl-glycerol rescues plasmalogen levels and pathology of ether-phospholipid deficient mice. PLoS One 6:e28539

Brocard C, Hartig A (2006) Peroxisome targeting signal 1: is it really a simple tripeptide? Biochim Biophys Acta 1763:1565–1573

Das AK, Hajra AK (1988) High incorporation of dietary 1-O-heptadecyl glycerol into tissue plasmalogens of young rats. FEBS Lett 227:187–190

Dirkx R, Vanhorebeek I, Martens K et al (2005) Absence of peroxisomes in hepatocytes causes mitochondrial and ER abnormalities. Hepatology 41:868–878

Dirkx R, Meyhi E, Asselberghs S, Reddy J, Baes M, Van Veldhoven PP (2007) β-oxidation in hepatocyte cultures from mice with peroxisomal gene knockouts. Biochem Biophys Res Commun 357:718–723

Ebberink MS, Csanyi B, Chong WK et al (2010) Identification of an unusual variant peroxisome biogenesis disorder caused by mutations in the PEX16 gene. J Med Genet 47:608–615

Ebberink MS, Koster J, Visser G et al (2012) A novel defect of peroxisome division due to a homozygous non-sense mutation in the PEX11β gene. J Med Genet 49:307–313
Faust PL (2003) Abnormal cerebellar histogenesis in Pex2 Zellweger mice reflects multiple neuronal defects induced by peroxisome deficiency. J Comp Neurol 461:394–413
Faust PL, Hatten ME (1997) Targeted deletion of the PEX2 peroxisome assembly gene in mice provides a model for Zellweger syndrome, a human neuronal migration disorder. J Cell Biol 139:1293–1305
Fransen M, Nordgren M, Wang B, Apanasets O (2012) Role of peroxisomes in ROS/RNS-metabolism: implications for human disease. Biochim Biophys Acta 1822:1363–1373
Fujiki Y, Okumoto K, Kinoshita N, Ghaedi K (2006) Lessons from peroxisome-deficient Chinese hamster ovary (CHO) cell mutants. Biochim Biophys Acta 1763:1374–1381
Girzalsky W, Saffian D, Erdmann R (2010) Peroxisomal protein translocation. Biochim Biophys Acta 1803:724–731
Grum DE, Hansen LR, Drackley JK (1994) Peroxisomal β-oxidation of fatty acids in bovine and rat liver. Comp Biochem Physiol B Biochem Mol Biol 109:281–292
Huber A, Koch J, Kragler F, Brocard C, Hartig A (2012) A subtle interplay between three Pex11 proteins shapes de novo formation and fission of peroxisomes. Traffic 13:157–167
Hulshagen L, Krysko O, Bottelbergs A et al (2008) Absence of functional peroxisomes from mouse CNS causes dysmyelination and axon degeneration. J Neurosci 28:4015–4027
Huybrechts SJ, Van Veldhoven PP, Brees C, Mannaerts GP, Los GV, Fransen M (2009) Peroxisome dynamics in cultured mammalian cells. Traffic 10:1722–1733
Huyghe S, Schmalbruch H, De Gendt K et al (2006) Peroxisomal multifunctional protein 2 is essential for lipid homeostasis in Sertoli cells and for male fertility in mice. Endocrinology 147:2228–2236
Janssen A, Baes M, Gressens P, Mannaerts GP, Declercq P, Van Veldhoven PP (2000) Docosahexaenoic acid deficit is not a major pathogenic factor in peroxisome-deficient mice. Lab Invest 80:31–35
Janssen A, Gressens P, Grabenbauer M et al (2003) Neuronal migration depends on intact peroxisomal function in brain and in extraneuronal tissues. J Neurosci 23:9732–9741
Kassmann CM, Lappe-Siefke C, Baes M et al (2007) Axonal loss and neuroinflammation caused by peroxisome-deficient oligodendrocytes. Nat Genet 39:969–976
Keane MH, Overmars H, Wikander TM et al (2007) Bile acid treatment alters hepatic disease and bile acid transport in peroxisome-deficient PEX2 Zellweger mice. Hepatology 45:982–997
Koch J, Brocard C (2011) Membrane elongation factors in organelle maintenance: the case of peroxisome proliferation. Biomol Concepts 2:353–364
Koch J, Brocard C (2012) PEX11 proteins attract Mff and human Fis1 to coordinate peroxisomal fission. J Cell Sci 125:3813–3826
Kovacs WJ, Tape KN, Shackelford JE et al (2009) Peroxisome deficiency causes a complex phenotype because of hepatic SREBP/Insig dysregulation associated with endoplasmic reticulum stress. J Biol Chem 284:7232–7245
Kovacs WJ, Charles KN, Walter KM et al (2012) Peroxisome deficiency-induced ER stress and SREBP-2 pathway activation in the liver of newborn PEX2 knock-out mice. Biochim Biophys Acta 1821:895–907
Krysko O, Hulshagen L, Janssen A et al (2007) Neocortical and cerebellar developmental abnormalities in conditions of selective elimination of peroxisomes from brain or from liver. J Neurosci Res 85:58–72
Li X, Gould SJ (2002) PEX11 promotes peroxisome division independently of peroxisome metabolism. J Cell Biol 156:643–651
Li X, Baumgart E, Dong G-X et al (2002a) PEX11a is required for peroxisome proliferation in response to 4-phenylbutyrate but is dispensable for peroxisome proliferator-activated receptor alpha-mediated peroxisome proliferation. Mol Cell Biol 22:8226–8240

Li X, Baumgart E, Morrell JC, Jimenez-Sanchez G, Valle D, Gould SJ (2002b) PEX11b deficiency is lethal and impairs neuronal migration but does not abrogate peroxisome function. Mol Cell Biol 22:4358–4365

Mannaerts GP, Debeer LJ, Thomas J, De Schepper PJ (1979) Mitochondrial and peroxisomal fatty acid oxidation in liver homogenates and isolated hepatocytes from control and clofibrate-treated rats. J Biol Chem 254:4585–4595

Martens K, Bottelbergs A, Baes M (2010) Ectopic recombination in the central and peripheral nervous system by aP2/FABP4-Cre mice: implications for metabolism research. FEBS Lett 584:1054–1058

Martens K, Bottelbergs A, Peeters A et al (2012) Peroxisome deficient aP2-Pex5 knockout mice display impaired white adipocyte and muscle function concomitant with reduced adrenergic tone. Mol Genet Metab 107:735–747

Matsui S, Funahashi M, Honda A, Shimozawa N (2012) Newly identified milder phenotype of peroxisome biogenesis disorder caused by mutated PEX3 gene. Brain Dev 35:842–848. doi:10.1016/j.braindev.2012.10.017

Maxwell M, Bjorkman J, Nguyen T et al (2003) Pex13 inactivation in the mouse disrupts peroxisome biogenesis and leads to a Zellweger syndrome phenotype. Mol Cell Biol 23:5947–5957

Meinecke M, Cizmowski C, Schliebs W et al (2010) The peroxisomal importomer constitutes a large and highly dynamic pore. Nat Cell Biol 12:273–277

Mignarri A, Vinciguerra C, Giorgio A et al (2012) Zellweger spectrum disorder with mild phenotype caused by PEX2 gene mutations. JIMD Rep 6:43–46

Muller CC, Nguyen TH, Ahlemeyer B et al (2011) PEX13 deficiency in mouse brain as a model of Zellweger syndrome: abnormal cerebellum formation, reactive gliosis and oxidative stress. Dis Model Mech 4:104–119

Nguyen T, Bjorkman J, Paton BC, Crane DI (2006) Failure of microtubule-mediated peroxisome division and trafficking in disorders with reduced peroxisome abundance. J Cell Sci 119:636–645

Peeters A, Fraisl P, van den Berg S et al (2011a) Carbohydrate metabolism is perturbed in peroxisome deficient hepatocytes due to mitochondrial dysfunction, AMP activated protein kinase (AMPK) activation and peroxisome proliferator-activated receptor {gamma} coactivator 1{alpha} (PGC-1{alpha}) suppressi. J Biol Chem 286:42162–42179

Peeters A, Swinnen JV, Van Veldhoven PP, Baes M (2011b) Hepatosteatosis in peroxisome deficient liver despite increased β-oxidation capacity and impaired lipogenesis. Biochimie 93:1828–1838

Regal L, Ebberink MS, Goemans N et al (2010) Mutations in PEX10 are a cause of autosomal recessive ataxia. Ann Neurol 68:259–263

Rucktaschel R, Girzalsky W, Erdmann R (2011) Protein import machineries of peroxisomes. Biochim Biophys Acta 1808:892–900

Schrader M, Bonekamp NA, Islinger M (2012) Fission and proliferation of peroxisomes. Biochim Biophys Acta 1822:1343–1357

Sevin C, Ferdinandusse S, Waterham HR, Wanders RJ, Aubourg P (2011) Autosomal recessive cerebellar ataxia caused by mutations in the PEX2 gene. Orphanet J Rare Dis 10:6–8

Steinberg SJ, Dodt G, Raymond GV, Braverman NE, Moser AB, Moser HW (2006) Peroxisome biogenesis disorders. Biochim Biophys Acta 1763:1733–1748

Steinberg SJ, Snowden A, Braverman NE et al (2009) A PEX10 defect in a patient with no detectable defect in peroxisome assembly or metabolism in cultured fibroblasts. J Inherit Metab Dis 32:109–119

Van Veldhoven PP (2010) Biochemistry and genetics of inherited disorders of peroxisomal fatty acid metabolism. J Lipid Res 51:2863–2895

Weng H, Ji X, Naito Y et al (2013) Pex11alpha deficiency impairs peroxisome elongation and division and contributes to nonalcoholic fatty liver in mice. Am J Physiol Endocrinol Metab 304:E187–E196

Wood PL, Khan MA, Smith T et al (2011) In vitro and in vivo plasmalogen replacement evaluations in rhizomelic chrondrodysplasia punctata and Pelizaeus-Merzbacher disease using PPI-1011, an ether lipid plasmalogen precursor. Lipids Health Dis 10:182

Xiao Y, Karnati S, Qian G et al (2012) Cre-mediated stress affects sirtuin expression levels, peroxisome biogenesis and metabolism, antioxidant and proinflammatory signaling pathways. PLoS One 7:e41097

Peroxisomes in Humans: Metabolic Functions, Cross Talk with Other Organelles, and Pathophysiology of Peroxisomal Disorders

3

Ronald J.A. Wanders, Sacha Ferdinandusse, and Hans R. Waterham

Abstract

Peroxisomes play a crucial role in cellular metabolism as exemplified by the devastating consequences caused by deficiencies of one or more peroxisomal enzymes in humans. The major metabolic functions of peroxisomes in humans include fatty acid beta-oxidation, etherphospholipid biosynthesis, fatty acid alpha-oxidation; glyoxylate detoxification, bile acid synthesis, L-pipecolic acid oxidation, and docosahexaenoic acid (DHA) formation. Except from the bile acids which are true metabolic end products of bile acid formation in the liver as generated in peroxisomes, all the other products of peroxisome metabolism are not true end products but require continued metabolism in other organelles to reach their final fate. This explains the crosstalk between peroxisomes and other subcellular organelles notably mitochondria and the endoplasmic reticulum. In this review we will discuss the metabolic functions of peroxisomes in humans and the crosstalk with other subcellular organelles. In addition we will discuss the pathophysiological consequences of genetic defects in peroxisome metabolism.

Keywords

Peroxisomes • Peroxisomal disorders • Lipidomics • Lipids • Fatty acids • Mitochondria

R.J.A. Wanders (✉) • S. Ferdinandusse • H.R. Waterham
Laboratory Genetic Metabolic Diseases, Departments of Paediatrics and Clinical Chemistry, Emma Children's Hospital, Academic Medical Center, 1105 Amsterdam, The Netherlands
e-mail: r.j.wanders@amc.uva.nl

C. Brocard and A. Hartig (eds.), *Molecular Machines Involved in Peroxisome Biogenesis and Maintenance*, DOI 10.1007/978-3-7091-1788-0_3, © Springer-Verlag Wien 2014

3.1 Introduction

Metabolism requires the catalysis of a large number of chemical reactions which allow cells to grow and reproduce, feed and excrete wastes, and move and communicate with their environment for the greater benefit of the cell itself whether part of a multicellular organism or not. This highly integrated network of chemical reactions is catalysed by enzymes that themselves are often organized in distinct multi-enzyme networks. An additional level of complexity is that, at least in higher eukaryotes, metabolism requires the participation of various membrane-bound compartments called organelles which each catalyse their own specific set of chemical reactions. The physical compartmentalization of the cytoplasm allows the coexistence of a diverse range of chemical micro-environments which are each precisely tailored to allow proper execution of a defined set of chemical reactions. There is constant communication and crosstalk between organelles which ensures an effective and cooperative division of metabolic labor. The fact that metabolism is divided over multiple subcellular organelles also requires the involvement of multiple transport proteins to allow exchange of metabolites between the different subcellular compartments.

In principle, metabolism can be subdivided into three distinct stages. In the first stage, larger molecules are broken down into smaller ones such as amino acids, single sugar molecules, and fatty acids. In the second stage, these small molecules are converted into a few single units that have a unique and pervasive role in metabolism. This includes acetyl-CoA as an obligatory intermediate linking the second and third stages of catabolism. The third stage includes the citric acid cycle and the oxidative phosphorylation system, which allow complete oxidation to CO_2 and H_2O with molecular oxygen as ultimate electron acceptor and ATP as universal form of energy.

Peroxisomes also play a major role in cellular metabolism by catalyzing a number of essential metabolic functions. Since peroxisomes lack a citric acid (Krebs) cycle and an oxidative phosphorylation system, they only contribute to the second, intermediate stage of metabolism.

In this review we will discuss the metabolic functions of peroxisomes in humans and the crosstalk with other subcellular organelles. In addition we will discuss the pathophysiological consequences of genetic defects causing deficiencies of different enzymes involved in peroxisome metabolism.

3.2 Metabolic Functions and Enzymology of Peroxisomes in Humans

Peroxisomes play an indispensable role in human metabolism as exemplified by the often devastating consequences of genetic defects causing deficiencies of one of the peroxisomal enzymes. Elucidation of the metabolic functions of peroxisomes was greatly helped by studies on a rare genetic disease, named the cerebro-hepato-renal syndrome, better known as Zellweger syndrome (ZS) in which peroxisomes and most of the enzymes normally contained within this organelle are missing as a consequence of mutations in genes (PEX) coding for proteins (peroxins) required for the proper synthesis of peroxisomes (Waterham and Ebberink 2012).

It all started with the discovery of Moser and coworkers (Brown et al. 1982) that the levels of certain fatty acids (FAs), notably the very long-chain fatty acids (VLCFAs) tetracosanoic acid (C24:0) and hexacosanoic acid (C26:0), were greatly elevated in plasma from ZS patients, whereas the levels of long-chain FAs, including oleate and palmitate, were normal. This suggested that oxidation of VLCFAs required the active participation of peroxisomes. At that time, peroxisomes were already known to contain a fatty acid beta-oxidation machinery, but until then the general notion was that peroxisomal beta-oxidation was just an auxiliary system assisting mitochondrial beta-oxidation in times of fatty acid overload (Lazarow and De Duve 1976). In subsequent years, additional FAs were discovered to accumulate in plasma from Zellweger patients including pristanic acid (2,6,10,14-tetramethylpentadecanoic acid) which turned out to be an exclusive substrate for peroxisomal beta-oxidation (Poulos et al. 1988). Furthermore, the findings by Hanson and co-workers who discovered the accumulation of the bile acid intermediates di- and trihydroxycholestanoic acid in plasma from Zellweger patients, which were originally thought to be due to the mitochondrial abnormalities in ZS (Hanson et al. 1979), could now be reinterpreted. Indeed, work by Pedersen and co-workers (see Ferdinandusse and Houten (2006) for review) has shown that the beta-oxidative cleavage of di- and trihydroxycholestanoic acid to generate chenodeoxycholic acid and cholic acid respectively, occurs solely in peroxisomes.

Parallel to the work done by Moser and co-workers, Heymans and co-workers reported the deficiency of plasmalogens in 1983, which indicated that peroxisomes also play a key role in the formation of etherphospholipids (Heymans et al. 1983). Similarly, the discovery that peroxisomes also catalyze the alpha-oxidation of fatty acids was based on the finding of greatly increased levels of phytanic acid (3,7,11,15-tetramethylhexadecanoic acid) in plasma from Zellweger patients (Poulos et al. 1985). Below, we will describe the major metabolic pathways in peroxisomes in humans and the interaction with other subcellular organelles. Figure 3.1 depicts the four major metabolic functions of peroxisomes in humans which include.

3.2.1 Peroxisomal Fatty Acid Beta-Oxidation

Isolated peroxisomes can handle a large variety of fatty acids for beta-oxidation including saturated FAs, mono- and poly-unsaturated FAs, 2- and 3-methyl branched-chain FAs, hydroxylated FAs notably 2-hydroxy-FAs, dicarboxylic FAs, and others. There are only a few fatty acid substrates which can be beta-oxidized by peroxisomes only. These include (1) the very long-chain fatty acids, notably C22:0, C24:0 and C26:0; (2) the 2-methyl branched-chain FA pristanic acid (2,6,10,14-tetramethylpentadecanoic acid); (3) di- and trihydroxycholestanoic acid;

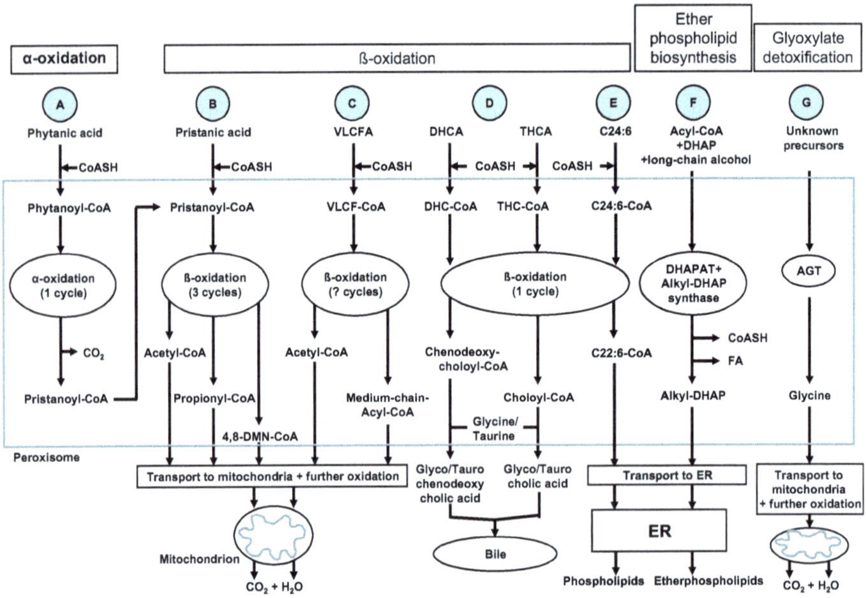

Fig. 3.1 Metabolic functions of peroxisomes in humans and the interaction between peroxisomes and other subcellular organelles in order to ensure the continued metabolism of the end products of peroxisome metabolism notably in mitochondria and the endoplasmic reticulum (see text for background information. 4,8-DMN-CoA, 4,8-dimethylnonanoyl-CoA; *ER* endoplasmic reticulum; *AGT* alanine glyoxylate aminotransferase

(4) the poly-unsaturated FA tetracosahexaenoic acid (C24:6$n-3$); (5) long-chain dicarboxylic acids, (6) 2-hydroxy-FAs, and (7) a number of prostanoids. The prostanoids that are currently known to require peroxisomal beta-oxidation for side-chain cleavage include PGF2-alpha (Diczfalusy et al. 1991), 8-iso-PGF2-alpha (Tsikas et al. 1998), thromboxane-B2 (de Waart et al. 1994; Diczfalusy et al. 1993), monohydroxy-eicosatrienoic acid (12-HETE, 15-HETE) (Spector et al. 2004), leukotriene-B4 (Jedlitschky et al. 1993; Mayatepek and Flock 1999), and the cysteinyl leukotriene LTE4 (Mayatepek et al. 2004) (see Van Veldhoven (2010) for review). In this way prostanoids are inactivated by cleavage of their carboxy-side chains.

The general theme of peroxisomal beta-oxidation is that peroxisomes perform one or more cycles of beta-oxidation and release the products to the cytosol for subsequent metabolism. In some cases peroxisomes catalyse a single cycle of beta-oxidation. This is true for the bile acid intermediates di- and trihydroxycholestanoic acid which undergo a single cycle of beta-oxidation generating cholic acid and chenodeoxycholic acid in their CoA-ester form. For most other FAs, including the VLCFAs C22:0, C24:0, and C26:0, the number of cycles of beta-oxidation in peroxisomes remains to be established definitively, although the general notion is that beta-oxidation in peroxisomes continues until a C8- or C6-medium-chain acyl-CoA has been generated. Exceptions to this rule are (1) pristanic acid which undergoes three cycles of beta-oxidation in peroxisomes to produce one unit of acetyl-CoA, two propionyl-CoA units plus 4,8-dimethylnonanoyl-CoA (Verhoeven et al. 1998) and (2) some of the prostanoids like PGF2-alpha and 8-iso-PGF2-alpha which undergo one or more cycles of beta-oxidation to produce the corresponding dinor and/or tetranor compounds.

In humans, the four steps of peroxisomal beta-oxidation are catalysed by two different acyl-CoA oxidases, two distinct bifunctional proteins, and two different thiolases. The physiological roles of the two acyl-CoA oxidases and to a lesser extent for the two bifunctional proteins and two thiolases has been resolved in recent years. The identification of patients with acyl-CoA oxidase 1 (ACOX1 deficiency) already in 1988 (Poll-The et al. 1988), in whom there was only accumulation of very long-chain fatty acids but not of pristanic acid and di- and trihydroxycholestanoic acid, indicated that there had to be a different oxidase with specificity for 2-methyl branched-chain acyl-CoAs. This enzyme was identified, purified, and cloned a few years later by Mannaerts and co-workers (Vanhove et al. 1993). Although so far no patients with a deficiency of branched-chain acyl-CoA oxidase (ACOX2) have been described and no mouse model with branched-chain acyl-CoA oxidase deficiency has been generated, the generally accepted view is that ACOX1 is the principal enzyme handling saturated acyl-CoAs including C22:0-CoA, C24:0-CoA, and C26:0-CoA, whereas ACOX2 is the prime oxidase dehydrogenating pristanoyl-CoA and di- and trihydroxycholestanoyl-CoA. It should be mentioned that there is also a third peroxisomal oxidase with high specificity for pristanoyl-CoA which is highly expressed in rat liver but hardly expressed in human liver. Interestingly, it has been shown that ACOX3 mRNA expression is upregulated in prostate and breast cancer (Zha et al. 2005). With respect to the two bifunctional proteins which have different names ranging from multifunctional proteins-1 and 2 (MFP1 and MFP2), multifunctional enzymes-1 and -2 (MFE-1 and MFE-2) and the L- and D-bifunctional proteins (LBP and DBP), it is clear that the D-specific enzyme plays an obligatory role in the beta-oxidation of very long-chain fatty acids, pristanic acid, and the bile acid intermediates. This is concluded from studies in patients suffering from D-bifunctional protein deficiency (Ferdinandusse et al. 2006a; Huyghe et al. 2006) and DBP-deficient mice (Baes et al. 2000). Recent work from our laboratory has shown that the L-specific enzyme is the major enzyme involved in the beta-oxidation of long-chain dicarboxylic acids (Houten et al. 2012). Earlier work in peroxisome-deficient fibroblasts had already shown that oxidation of long-chain dicarboxylic acids is primarily peroxisomal (Ferdinandusse et al. 2004). Finally, with respect to the two thiolases in human peroxisomes, the physiological role of the branched-chain specific thiolase has been elucidated again as a corollary of the identification of a human deficiency of this enzyme (Ferdinandusse et al. 2006b) and a mutant mouse model generated by Seedorf and coworkers (1998). The thiolase involved is called sterol carrier protein X (SCPx) because it also contains a sterol-carrier-binding unit and is able to cleave the 3-keto-acyl-CoA esters of pristanic acid and di- and trihydroxycholestanoic acid. SCPx thus plays a central role in the beta-oxidation of these FAs. Straight-chain FAs are handled both by SCPx and the other peroxisomal thiolase, called 3-ketoacyl-CoA thiolase (Wanders and Waterham 2006a).

For the beta-oxidation of mono- and poly-unsaturated FAs auxiliary enzymes, which include different isomerases and di-enoyl-CoA reductases are required (Van Veldhoven 2010; Wanders and Waterham 2006a). In contrast to the detailed knowledge about the enzymes of the peroxisomal beta-oxidation machinery itself, less information is available on the identity and catalytic properties of the latter enzymes.

3.2.2 Etherphospholipid Biosynthesis

Peroxisomes play a crucial role in the synthesis of etherphospholipids since the first part of the biosynthetic pathway is solely peroxisomal. This includes the enzyme alkyldihydroxyacetone phosphate synthase (ADHAPS) encoded by AGPS which is responsible for the generation of the characteristic ether bond. The two substrates required in the ADHAPS enzyme reaction, i.e., a long-chain alcohol and acyldihydroxyacetone phosphate (acyl-DHAP) are also synthesized by peroxisomes via the enzymes acyl-CoA: NADPH oxidoreductase and dihydroxyacetone phosphate acyltransferase (DHAPAT), respectively. DHAPAT and ADHAPS form a complex bound to the inner face of the peroxisomal membrane. The end product of the ADHAPS reaction, i.e. alkyl-DHAP is converted into alkylglycerol-3-phosphate (alkylG3P) either within peroxisomes or at the ER membrane (Brites et al. 2004). All the subsequent steps required for the synthesis of etherphospholipids are catalysed by ER enzymes (da Silva et al. 2012; Fig. 3.1).

3.2.3 Fatty Acid Alpha-Oxidation

In contrast to FA beta-oxidation with two distinct systems in mitochondria and peroxisomes, there is only one single alpha-oxidation machinery in human cells which is localized in peroxisomes. For some FAs, notably those with a methyl-group at the 3-position, alpha-oxidation is the main mechanism by which these FAs can be oxidized with oxidation from the omega-end as only alternative (Wanders et al. 2011b).

Phytanic acid (3,7,11,15-tetramethylhexadecanoic acid) is by far the best known FA undergoing alpha-oxidation resulting in the production of pristanic acid (2,6,10,14-tetramethylpentadecanoic acid), which is one carbon atom shorter. Indeed, studies in plasma and tissues from Refsum disease patients in whom alpha-oxidation is fully deficient (see below) have shown that phytanic acid is the major metabolite which accumulates. Fatty acid alpha-oxidation involves a 5-step pathway which includes (1) activation to an acyl-CoA; (2) hydroxylation of the acyl-CoA to a 2-hydroxyacyl-CoA; (3) cleavage of the 2-hydroxyacyl-CoA into an $(n-1)$ aldehyde plus formyl-CoA; (4) oxidation of the aldehyde to the corresponding acid; and (5) formation of an acyl-CoA ester. Although the basic enzymatic machinery involved in alpha-oxidation has been resolved in recent years, the identity of some of the enzymes from the pathway has remained ill defined. This does not apply to the enzymes catalyzing steps 2 and 3 of the pathway which includes phytanoyl-CoA hydroxylase and hydroxyacyl-CoA lyase which have been well characterized (see Van Veldhoven (2010) and Wanders et al. (2011a) for review), with phytanoyl-CoA hydroxylase being the enzyme deficient in Refsum disease.

3.2.4 Glyoxylate Detoxification

Peroxisomes also play a key role in the breakdown of glyoxylate, because the main enzyme involved in the detoxification of glyoxylate, i.e. alanine glyoxylate aminotransferase (AGT) is solely peroxisomal in humans (Salido et al. 2012).

3.3 Cross Talk with Other Subcellular Organelles

3.3.1 Cross Talk Between Peroxisomes and Mitochondria

As already eluded to above, peroxisomes rely on the interaction with other organelles to sustain their own intrinsic role in metabolism. Mitochondria are the ultimate site of oxidation of many intermediates in metabolism that are generated at other subcellular locations. This is not only true for the pyruvate and NADH produced during glycolysis in the cytosol but also for the NADH and acyl-CoA esters produced in peroxisomes.

Upon FA beta oxidation, peroxisomes produce NADH which needs to be reoxidized back to NAD+. Since peroxisomes lack a respiratory chain as present in mitochondria, and NAD+ and NADH are not able to traverse the peroxisomal membrane, the involvement of NAD(H)-redox shuttles has been proposed (see discussion below). Furthermore, since NADPH is required for the removal of the double bonds present in polyunsaturated FAs in the di-enoyl-CoA reductase reaction, the product NADP+ needs to be reduced back to NADPH. Also in this case, the involvement of an NADP(H)-linked redox shuttle has been proposed. Reoxidation of cytosolic NADH back to NAD+ is mediated by so-called redox shuttles which require the participation of two isoenzymes in the cytosol and mitochondrion respectively, plus metabolite transporters present in the mitochondrial membrane although there are exceptions to this rule. The malate/aspartate shuttle is the best known among the mitochondrial NAD(H)-linked redox shuttles. Our own work in the yeast *Saccharomyces cerevisiae* (van Roermund et al. 1995) has clearly established the presence of such a NAD(H)-linked redox-shuttle in peroxisomes which involves the cytosolic and peroxisomal isoforms of malate dehydrogenase. Human peroxisomes contain both malate dehydrogenase (Gronemeyer et al. 2013) and lactate dehydrogenase activity (Baumgart et al. 1996) which points to the existence of both a malate- as well as lactate-based redox shuttle.

Another aspect, in which peroxisomes heavily rely on the interaction with mitochondria in terms of their metabolism, is the degradation of the end products of peroxisomal beta-oxidation. Indeed, different CoA-esters produced upon fatty acid alpha- and beta-oxidation in peroxisomes cannot be degraded in peroxisomes any further and need to be exported from the peroxisomes to avoid built-up of intermediates. The generally accepted idea is that the acyl-CoA esters produced in peroxisomes may follow two different routes in order to achieve export from the peroxisome. The first one is the carnitine-dependent route which requires conversion of the different acyl-CoA esters into the corresponding acylcarnitines. Peroxisomes harbour at least two different carnitine acyltransferases with specificity for short-chain and medium-/long-chain acyl-CoA esters named carnitine acetyltransferase (CRAT) and carnitine octanoyltransferase (CROT), respectively. The mechanism by which acylcarnitines are transported across the peroxisomal membrane, however, has not been established yet. The acylcarnitines produced in peroxisomes can be taken up into mitochondria by the carnitine/acylcarnitine transporter (CACT) encoded by SLC25A2 and located in the mitochondrial inner-membrane. CACT catalyses the one-to-one exchange between free carnitine and a whole range of different acylcarnitines at the opposite side of the inner mitochondrial membrane (Indiveri et al. 2011). The importance of CACT for peroxisomal fatty acid metabolism has been established for the beta-oxidation of very long-chain fatty acids (Jakobs and Wanders 1991) and pristanic acid (Verhoeven et al. 1998) using fibroblasts from patients with CACT deficiency due to mutations in the SLC25A2 gene. In contrast to control fibroblasts in which C26:0 can be oxidized down to CO_2 and H_2O, the formation of CO_2 was fully deficient in CACT-deficient cells indicating that the acetyl-CoA units produced in peroxisomes require a functional CACT for their complete oxidation (Jakobs and Wanders 1991) (Fig. 3.2a). Verhoeven and co-workers showed the accumulation of several carnitine esters if CACT-deficient fibroblasts were incubated with pristanic acid (Verhoeven et al. 1998). One of these carnitine-esters was 4,8-dimethylnonanoylcarnitine indicating that this is the end product of pristanic acid beta-oxidation in peroxisomes which normally enters the mitochondria via CACT for subsequent oxidation (Fig. 3.2b). Taken together, these results clearly point to the existence of a carnitine-mediated export pathway of acyl-CoA esters from peroxisomes to mitochondria.

The other route by which acyl-CoAs may get out of the peroxisome is the thioesterase-dependent pathway which involves hydrolytic cleavage of the CoA-ester bond followed by export of the free FAs across the peroxisomal membrane. The acyl-CoA thioesterases ACOT4 and ACOT8 have been identified in peroxisomes in humans and have been proposed to play a role in this pathway (Hunt et al. 2012). Although so far no human patients with an inherited defect in CRAT, CROT, or any of the thioesterases have been identified, there is good evidence in literature, pointing to the true existence of these pathways both in human cells (Jakobs and Wanders 1991; Verhoeven et al. 1998) as well as in yeasts (van Roermund et al. 1995) (Fig. 3.2a, b).

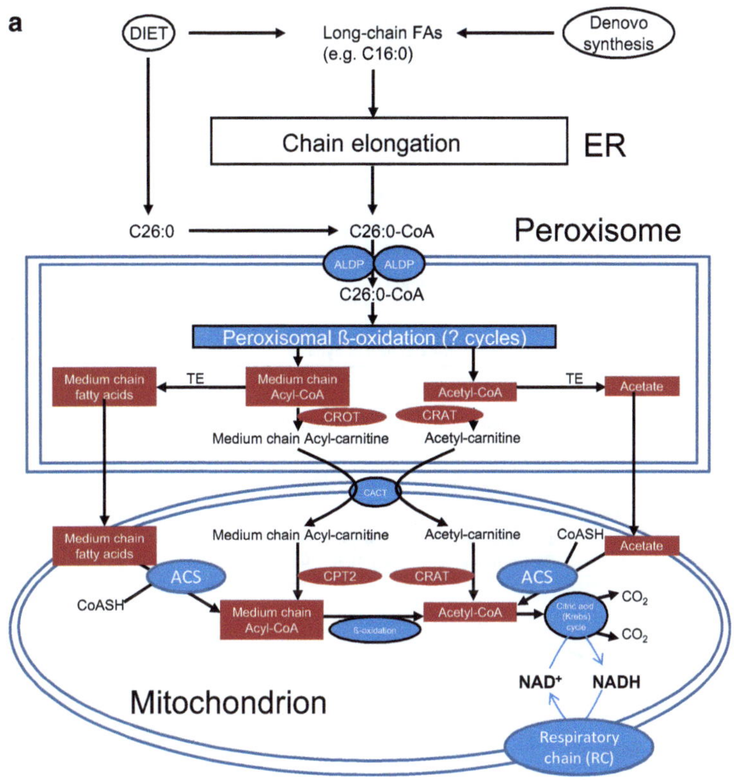

Fig. 3.2 (continued)

Crosstalk between peroxisomes and mitochondria is not restricted to the beta-oxidation of FAs. Indeed, other metabolic pathways in peroxisomes also require the mitochondrion for subsequent metabolism. This is true for glyoxylate detoxification, which in humans involves the peroxisomal enzyme AGT that converts glyoxylate and alanine into glycine and pyruvate. Reconversion of pyruvate back into alanine is catalysed by the cytosolic enzyme glutamate pyruvate aminotransferase (GPT), whereas glycine is transported out of the peroxisome to mitochondria to be degraded to CO_2 and H_2O by the glycine cleavage system (Fig. 3.1).

Phytanic acid alpha-oxidation also relies on mitochondria for its proper functioning. Indeed, the enzyme phytanoyl-CoA hydroxylase catalyses the hydroxylation of phytanoyl-CoA into 2-hydroxyphytanoyl-CoA using 2-oxoglutarate as co-substrate with succinate plus CO_2 as products. Indeed, 2-oxoglutarate can be regenerated from succinate—a 4-carbon metabolite—via part of the citric acid (Krebs) cycle which again shows the trafficking of metabolites between peroxisomes and mitochondria (see Wanders et al. (2011a) for review). 2-Oxoglutarate may also be produced from L-glutamate by transamination.

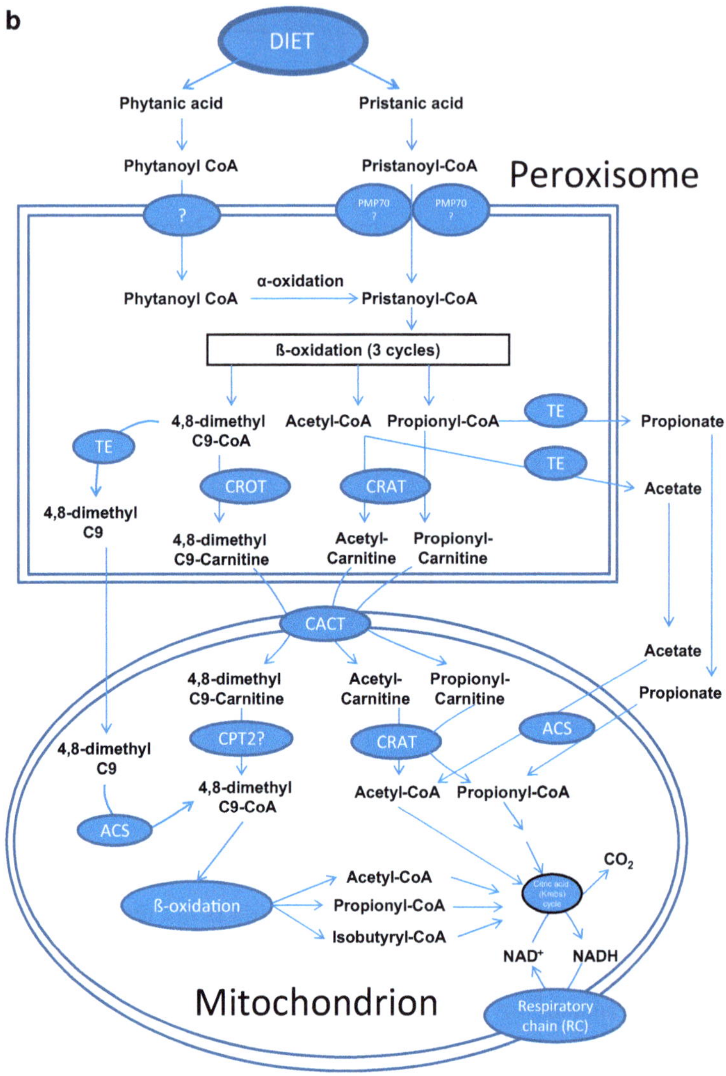

Fig. 3.2 (**a**) Interplay between the endoplasmic reticulum (ER), peroxisomes and mitochondria with respect to the biosynthesis and oxidation of very long-chain fatty acids, notably C26:0. *ALDP* adrenoleukodystrophy protein, *TE* acyl-CoA thioesterase, CROT carnitine octanoyltransferase, CRAT carnitine acetyltransferase. (**b**) Degradation of pristanic acid (2,6,10,14-tetramethylpentanoic acid) and the interplay between peroxisomes and mitochondria. *TE* acyl-CoA thioesterase, *ACS* acyl-CoA synthetase, *CRAT* carnitine acetyltransferase, *CROT* carnitine octanoyltransferase, *CPT2* carnitine palmitoyltransferase 2, *RC* respiratory chain

Continued alpha-oxidation of FAs in peroxisomes also requires ATP within peroxisomes which is most likely synthesized by the mitochondrial oxidative phosphorylation system and thus also comes from mitochondria (see Wanders et al. (2011a) for review).

3.3.2 Cross Talk Between Peroxisomes and the Endoplasmic Reticulum

Crosstalk also occurs between peroxisomes and the ER. The synthesis of etherphospholipids is a typical example of the crosstalk between peroxisomes and the ER, since the first steps of etherphospholipid synthesis take place in peroxisomes up to the level of alkyl-DHAP. All subsequent steps of etherphospholipid synthesis take place at the ER.

Another example is bile acid biosynthesis. Bile acids are synthesized from cholesterol in the liver via a series of reactions involving many different enzymes located throughout the cell (Russell 2003, 2009) with the bile acid intermediates di- and trihydroxycholestanoic acid (DHCA/THCA) as obligatory intermediates. DHCA and THCA are activated to CoA-esters at the cytosolic side of the ER by the enzyme very long-chain acyl-CoA synthetase (ACSVL1) (Steinberg et al. 1999) and then transported into the peroxisome for formation of the primary bile acids via one cycle of peroxisomal beta-oxidation. The site of activation of very long-chain fatty acids is currently less clear, but most likely ACSVL1 also plays a key-role in case of these peroxisomal substrates. Recently it has been established definitively that the CoA-esters of very long-chain fatty acids are transported by ABCD1 into the peroxisome where they are beta-oxidized (van Roermund et al. 2008; Wiesinger et al. 2013). The first steps in the biosynthesis of DHA (C22:6$n-3$) from dietary linolenic acid (C18:3$n-3$) also take place at the ER. After a series of reactions C24:6$n-3$ is formed which is then beta-oxidized in the peroxisome with DHA as product (Ferdinandusse et al. 2001). After its formation, DHA is transported back to the ER where it is esterified into membrane lipids (Fig. 3.1).

3.3.3 Pathophysiological Consequences of Defects in Peroxisome Metabolism

The importance of peroxisomes for whole organ and whole body metabolism in humans is stressed by the existence of a group of genetic diseases in which a particular peroxisomal enzyme or metabolite transporter is deficient. Indeed, single gene defects causing isolated peroxisomal enzyme deficiencies have been identified affecting peroxisomal beta-oxidation, etherphospholipid biosynthesis, peroxisomal fatty acid alpha oxidation, and glyoxylate metabolism (Braverman et al. 2013; Wanders and Waterham 2006b). Clinical signs and symptoms and the pathology observed in the patients involved allows conclusions to be drawn about the pathophysiological consequences caused by the accumulation or deficiency of certain metabolites as described below (Table 3.1).

Fatty acid beta-oxidation: At present five different disorders have been identified in which peroxisomal beta-oxidation is impaired (Table 3.1). These include (1) - X-linked adrenoleukodystrophy (X-ALD); (2) acyl-CoA oxidase (ACOX) deficiency; (3) D-bifunctional protein (DBP) deficiency; (4) sterol carrier protein X (SCPx) deficiency; and (5) 2-methylacyl-CoA racemase (AMACR) deficiency.

Table 3.1 Lipid metabolite abnormalities in the different human deficiencies of peroxisomal lipid metabolism

					Lipid abnormalities				
Species	Pathway affected	Disorders	Enzyme defect	Gene	VLCFA*	Pris*	Phyt*	D/THCA*	PL*
Human	Peroxisomal-beta-oxidation	X-ALD	ALDP	*ABCD1*	↑	N	N	N	N
		ACOX1-deficiency	Acyl-CoA oxidase	*ACOX1*	↑	N	N	N	N
		DBP-deficiency	D-bifunctional protein	*HSD17B4*	↑	↑	N-↑	↑	N
		SCPx-deficiency	Sterol carrier protein X	*SCP2*	N	↑	N-↑	↑	N
		AMACR-deficiency	2-Methylacyl-CoA racemase	*AMACR*	N	↑	N-↑	↑	N
	Peroxisomal-alpha-oxidation	Refsum disease	Phytanoyl-CoA hydroxylase	*PHYH/PAHX*	N	N	↑	N	N
	Etherphospholipid biosynthesis	RCDP Type2	Dihydroxyacetone-phosphate acyltransferase	*GNPAT*	N	N	N	N	↓
		RCDP Type3	AlkylDHAP synthase	*AGPS*	N	N	N	N	↓

VLCFA very-long-chain fatty acids, *Pris* pristanic acid, *Phyt* phytanic acid, *D/THCA* Di/trihydroxycholestanoic acid, *PL* plasmalogens

The clinical signs and symptoms associated with each of these disorders vary widely not only between the different disorders but also within each specific deficiency. Among the different beta-oxidation disorders, DBP deficiency is most severe with clinical signs and symptoms resembling those observed in Zellweger patients. In patients with DBP deficiency, the peroxisomal beta-oxidation of all major substrates is impaired which results in the accumulation of very long-chain fatty acids, pristanic acid, as well as di- and trihydroxycholestanoic acid in tissues and plasma. At the other end of the disease spectrum is AMACR deficiency which is usually associated with a relatively mild clinical phenotype resembling Refsum disease. AMACR-deficient patients only accumulate pristanic acid and the bile acid intermediates DHCA and THCA (Table 3.1).

Most frequent among the disorders of peroxisomal beta-oxidation is X-linked adrenoleukodystrophy which comes in two major phenotypes. The childhood cerebral form (CCALD) affects boys who develop normally for the first few years of life and then rapidly deteriorate, followed by early death. In adrenomyeloneuropathy the disease progressives slowly with main involvement of the spinal cord. In all forms of X-linked adrenoleukodystrophy there is only accumulation of very long-chain fatty acids. An important difference between the different disorders of peroxisomal beta-oxidation is that different metabolites accumulate as described above for X-linked adrenoleukodystrophy on the one hand and D-bifunctional protein deficiency on the other hand (see Table 3.1). Furthermore, the extent of the accumulation of certain metabolites may vary widely as a consequence of the nature of the block in the peroxisomal beta-oxidation pathway. A good example in this respect is the accumulation of VLCFAs in patients suffering from acyl-CoA oxidase deficiency and D-bifunctional protein deficiency. In the latter two disorders VLCFAs accumulate too much higher levels as compared to X-linked adrenoleukodystrophy (X-ALD). The underlying basis for this phenomenon resides in the fact that oxidation of VLCFAs is fully deficient if acyl-CoA oxidase and D-bifunctional protein are lacking since branched-chain acyl-CoA oxidase (ACOX2) and L-bifunctional protein (LBP) contribute very little to the beta-oxidation of VLCFAs. In contrast, in case of X-linked adrenoleukodystrophy the deficiency of ALDP (encoded by ABCD1) only leads to a partial deficiency of C26:0 beta-oxidation amounting 20–30 % of control. This is due to the fact that transport of C26:0 in its CoA ester form across the peroxisomal membrane is not only catalysed by ALDP but also by the other two peroxisomal half ABC transporters including ALDR and PMP70 as encoded by ABCD2 and ABCD3 respectively (van Roermund et al. 2011; Wiesinger et al. 2013). Recent work by Berger and co-workers has shown that ALDP and PMP70 contribute for 70 % and 30 % respectively to overall C26:0 beta-oxidation in human skin fibroblasts (Wiesinger et al. 2013).

3.3.4 Pathophysiology Associated with the Accumulation of the Individual Peroxisomal Substrates

3.3.4.1 Very Long-Chain Fatty Acids

The ultra-structural demonstration of specific cytoplasmic inclusions in the central nervous system of X-ALD patients by Schaumburg in 1974 (Schaumburg et al. 1974) and the subsequent demonstration of increased very long-chain fatty acids (VLCFAs) in these patients first by Garashi et al. (1976), prompted the hypothesis that the toxicity of VLCFAs would be due to their incorporation into membrane constituents and perturbation of membrane structure and function. Experimental support for this hypothesis came from work of Knazek et al. (1983) who reported on the increased membrane microviscosity in erythrocytes of adrenoleukodystrophy and adrenomyeloneuropathy patients. Direct evidence that abnormally high VLCFA levels can indeed disturb membrane function was provided by Whitcomb et al. (1988) who assessed ACTH-stimulated cortisol release in cultured human adrenocortical cells. The addition of C26:0 or C24:0 but not C16:0 to the culture medium in concentrations equivalent to those in X-ALD plasmas increased the microviscosity of adrenocortical cell membranes and decreased ACTH-stimulated cortisol secretion. It was speculated that analogous effects could occur in neural membranes such as the perikarya or axons of the ascending and descending spinal tracts in AMN. These data prompted Powers and Moser (Powers and Moser 1998) to propose a unifying hypothesis for most of the lesions observed in peroxisomal patients which reads: 'abnormal fatty acids, particularly VLCFA and phytanic acid accumulate in the peroxisomal disorders and are incorporated into cell membranes resulting in a perturbation of these membranes' microenvironments and the dysfunction, atrophy, and death of vulnerable cells'.

Following up on the earlier work by Garashi et al. (1976), Theda and co-workers (1992) took this type of lipid analysis one step further by reasoning that abnormalities in trigger molecules should precede histopathological abnormalities. To this end they performed detailed lipid analyses in white matter samples from a patient who died from AMN. These samples were either normal, showed earlier demyelination or were gliotic. In active demyelinating lesions the cholesterol ester fraction was found to contain the greatest excess of VLCFA. This appears to be a consequence rather than a cause of demyelination, because the fatty acid composition of this fraction was normal in regions of X-ALD brain in which myelin was still intact. In addition, it is unlikely that cholesterol esterified with these VLCFAs, acts as a trigger for the inflammatory response. The gangliosides in X-ALD brain contain 28–50 % of fatty acids with chain lengths exceeding 21 (Garashi et al. 1976). Such fatty acids are virtually absent in normal brain gangliosides. It should be noted that the immunological properties of gangliosides vary with their fatty acid composition and have been implicated in a variety of immunological brain diseases (see Singh and Pujol 2010). Importantly, Theda and colleagues (1992) found the greatest VLCFA excess (16-fold and higher) in the phosphatidylcholine fraction in intact regions of brain white matter from patients. These observations led Theda et al. (1992) to conclude that the phosphatidylcholine abnormalities in X-ALD may well be a critical player in the pathogenesis of the disease. More recently, highest concentrations of VLCFA were also observed in the lysophosphatidylcholine fraction of brain from childhood cerebral ALD patients (Hubbard et al. 2006). Interestingly, Eichler et al. (2008) has shown that injection of lysophosphatidylcholine esterified with C24:0 in mice led to wide-spread microglial activation and apoptosis. The structural abnormalities as described above in the myelin lipid components are supposed to cause destabilization of the myelin sheath and onset of demyelinating pathology. The same principle may underlie the axonal degeneration as observed in AMN because proper integrity of the cellular membrane is a prerequisite for myelin-axonal interactions.

In more recent years the problem of the VLCFA-induced pathology has received renewed attention and the concept of oxidative stress as the key-driver of the VLCFA-induced pathophysiology of X-ALD has received widespread experimental support. Indeed, Gilg and co-workers (2000) reported increased expression of inducible NO-synthase and increased nitrotyrosine levels in the central nervous system (CNS) of X-ALD patients. Furthermore, Vargas et al. (2004) reported increased chemiluminescence and thiobarbituric acid reactive substances (TBARS) levels and decreased total antioxidant reactivity (TAR) in plasma from X-ALD patients. Oxidative damage has also been observed in fibroblasts (Fourcade et al. 2008; Vargas et al. 2004), plasma (Deon et al. 2007; Rockenbach et al. 2012), and lymphoblasts (Uto et al. 2008) of X-ALD patients suggesting that the enhanced oxidative stress is systemic rather than localized. Moreover, Powers and co-workers reported oxidative modifications in post-mortem X-ALD brains with lipid peroxidation products predominantly present in the inflammatory demyelinative lesions and the adrenal cortex. Recent work by Bertini and co-workers (Petrillo et al. 2013) has provided additional evidence in favour of oxidative stress in X-ALD by demonstrating major abnormalities at the level of the glutathione system characterized by significant decreases in total and reduced glutathione in lymphocytes and especially erythrocytes from patients.

Work by Pujol and co-workers in a mouse model for X-ALD which develops a late onset phenotype resembling adrenomyeloneuropathy (AMN) rather than CCALD, has also brought oxidative stress to the forefront of the pathophysiological process in X-ALD. Indeed, Fourcade and co-workers (2008) reported signs of oxidative damage in spinal cords from Abcd1-null mice with oxidative, glycoxidative, and lipoxidative damage to proteins and altered expression levels of antioxidant enzymes including glutathione peroxidase 1 (GPX1) and superoxide dismutases 1 and 2 (SOD1 and SOD2). Second, the same group reported that oxidative damage affects key enzymes of glycolysis and the tricarboxylic acid cycle in Abcd1 (−/−) mice as well as in human X-ALD fibroblasts. Third, Galino and co-workers (2011) reported reduced NADH and ATP levels with decreased pyruvate kinase activities and GSH levels. Fourth, and most importantly, antioxidant treatment was found to reverse most signs of oxidative damage and to halt the axonal degeneration in Abcd1-null mice (Lopez-Erauskin et al. 2011). Finally,

Schluter and co-workers (2012) followed a transcriptomic approach and obtained evidence for a common signature comprising dysregulation of (1) oxidative phosphorylation, (2) adipocytokine and insulin signalling pathways, and (3) protein synthesis and suggested that X-ALD is a metabolic/inflammatory syndrome. Based on these data Pujol and co-workers have proposed that VLCFAs in the mitochondrial membrane might induce a stress response leading to the production of ROS and oxidation of major protein components including Krebs cycle enzymes, which together contribute to mitochondrial dysfunction, decreased ATP levels and a collapse of energy homeostasis. The exact underlying mechanism for the VLCFA induced stress response has remained unresolved so far.

3.3.4.2 Di- and Trihydroxycholestanoic Acid

The bile acid intermediates DHCA and THCA accumulate in several of the peroxisomal disorders including the peroxisome biogenesis defects and the single enzyme deficiencies D-bifunctional protein deficiency, SCPx deficiency, and AMACR deficiency (Wanders and Waterham 2006a).

The pathophysiological consequences associated with the accumulation of DHCA and THCA can probably best be deduced from the observations in patients affected by either AMACR deficiency or SCPx deficiency because in both these deficiencies there is accumulation of DHCA and THCA as well as pristanic acid (Table 3.1). So far, SCPx deficiency has been described in a single case only (Ferdinandusse et al. 2006b), whereas AMACR deficiency has been identified in more than ten patients. Interestingly, two different phenotypic presentations have been described for AMACR deficiency. Indeed, most patients with AMACR deficiency present in adolescence with a Refsum-like phenotype, whereas other patients have presented with severe, early-onset cholestatic liver disease as described by Setchell and co-workers (2003). As described in detail elsewhere (Wanders and Ferdinandusse 2012), DHCA and THCA are cytotoxic to cells especially in their unconjugated form. Furthermore, the deficiency of the primary bile acids cholic acid and chenodeoxycholic acid as a consequence of a block in peroxisomal beta-oxidation directly reduces bile flow causing cholestasis. Indeed, bile flow is determined by an ABC transporter known as the bile salt export pump (BSEP, ATP8B1) which is localized in the hepatocyte canalicular membrane and is critical to the formation of bile salt-dependent bile flow and a normal enterohepatic circulation of bile acids from the distal intestine back to the liver. Since the glycine and taurine conjugates of cholic acid and chenodeoxycholic acid are the primary substrates of BSEP, flux through BSEP and subsequently bile flow will be low if the intrahepatic levels of these glycine and taurine conjugates are reduced. Most likely, the glycine and taurine conjugates of DHCA and THCA are also substrates for BSEP as concluded from the fact that these bile acids have been identified in bile from Zellweger patients and other patients in which DHCA and THCA accumulate. However, due to the fact that the CoA esters of DHCA and THCA are poor substrates for the glycine/taurine-conjugating enzyme, i.e. bile acid: amino acid *N*-acyltransferase (BAAT), they will undergo thioesterase-mediated hydrolytic cleavage which explains why DHCA and THCA are mostly present in the unconjugated form and hence will be poorly excreted into the canalicular space which contributes to the hepatotoxicity.

3.3.4.3 Pristanic Acid

So far only few studies have focused on the cytotoxic effects of pristanic acid. First, Idel and co-workers (2002) showed that pristanic acid produced the activation of inducible nitric oxide synthase (iNOS) leading to reactive nitric oxide (NO) formation and apoptotic cell death via activation and secretion of tumour necrosis factor-alpha (TNFalpha). This work was followed up in recent years by the group of Wajner (Busanello et al. 2012; Leipnitz et al. 2010) showing that pristanic acid promotes oxidative stress in brain cortex of young rats, inhibits various complexes of the respiratory chain, and reduces GSH-levels. As discussed in more detail below, it remains to be established what the relevance of these in vitro results is for the in vivo situation. The availability of different mouse models as generated in recent years (Baes and Van Veldhoven 2012) may help to resolve this issue.

3.3.4.4 Phytanic Acid

In order to explain the toxic properties of phytanic acid two different hypotheses were originally conceived including the molecular distortion hypothesis and the anti-metabolite hypothesis (see Wanders et al. (2010) for review). In more recent years interest has shifted to the effects of phytanic acid on (1) signal transduction since phytanic acid is a ligand for different transcription factors and (2) the effects of phytanic acid on mitochondria.

In 1976 Kitareewan et al. (1996) and Lemotte et al. (1996) published their remarkable finding showing that phytanic acid is able to activate the nuclear receptor RXR. This was soon followed by reports from Ellinghaus et al. (1999) and Zomer et al. (2000) which revealed that phytanic acid is also a powerful ligand for another nuclear hormone receptor, i.e. PPARalpha. These findings prompted new pathophysiological concepts centred around the changes in gene expression caused by phytanic acid. Along the same lines Idel et al. (2002) reported that phytanic acid is able to induce iNOS expression in smooth muscle cells which is associated with NO-dependent apoptotic cell death via the activation and secretion of TNFalpha. The signalling pathway involved turned out to be independent of PPARalpha and RXR.

Inspired by the fact that mitochondrial dysfunction has been implicated in several neurodegenerative diseases, more recent studies notably by the group of Schonfeld and Reiser, have focused on the effect of phytanic acid on mitochondria. Firstly, Schonfeld and Struy (1999) reported that phytanic acid affects the dynamics of phospholipids in membranes as well as the physical state of membrane proteins, which prompted the authors to propose a novel mechanism for the uncoupler-like effect of phytanic acid (Schonfeld and Struy 1999). In agreement with its uncoupler-like action, phytanic acid was subsequently found to de-energize rat brain mitochondria respiring under resting (state 4) conditions, i.e. in the absence of ADP. This occurred already at low concentrations (Schonfeld et al. 2004).

Interestingly, similar low concentrations of phytanic acid turned out to inhibit, rather than to stimulate, mitochondrial respiration especially when ADP was used to induce respiration maximally (state 3). The fact that the inhibition of the respiratory chain by phytanic acid was less under uncoupler-stimulated versus ADP-stimulated conditions was explained from the observation that phytanic acid also turned out to inhibit the adenine nucleotide carrier, which exchanges extramitochondrial ADP for intramitochondrial ATP. Finally, at higher concentrations phytanic acid was also found to inhibit complex I of the respiratory chain which explains the inhibition of uncoupler-stimulated respiration by phytanic acid at concentrations >10 μmol/L. In our hands (Komen et al. 2007) the uncoupler-like action of phytanic acid predominates over its inhibitory effect on complex 1, at least in mitochondria from human fibroblasts (Komen et al. 2007). It should be noted that Bunik and co-workers (Bunik et al. 2006) have shown that phytanoyl-CoA is also a powerful inhibitor of both the pyruvate and 2-oxoglutarate dehydrogenase complexes which may also contribute to the generation of ROS (Starkov et al. 2004). However, it is important to emphasize that it is not very likely that phytanoyl-CoA actually enters the mitochondrial matrix simply because phytanoyl-CoA is not a substrate of carnitine palmitoyltransferase 1 (CPT1).

Studies in rat hippocampal astrocytes (Schonfeld and Reiser 2006) confirmed the marked toxicity of phytanic acid, as concluded from the following findings: (1) disturbed cytosolic calcium homeostasis, (2) de-energization of the mitochondria, and (3) increased ROS-production. Induction of oxidative stress by phytanic acid was also observed in isolated rat brain mitochondria, both under state 4 as well as state 3 conditions (Schonfeld and Reiser 2006) and was largely attributed to the interference of phytanic acid with the respiratory chain at the level of complex I. Interestingly the stimulation of superoxide production by phytanic acid was associated with the inactivation of mitochondrial aconitase and oxidation of the mitochondrial glutathion pool. Similar work has been done by Wajner and co-workers. First, Busanello et al. (2010) provided in vitro evidence indicating that phytanic acid inhibits the Na/K-ATPase as well as several respiratory chain complexes in brain cortex from young rats. Furthermore, the same group (Leipnitz et al. 2010) reported increased oxidative damage and reduced anti-oxidant defences in the cerebellum and cerebral cortex of rats. Recently, two additional reports from the same group (Busanello et al. 2013a, b) have appeared showing marked inhibition of the Na/K-ATPase and the respiratory chain in the cerebellum from young rats. Furthermore, in agreement with earlier work, phytanic acid was found to inhibit mitochondrial oxidative phosphorylation and uncouple the respiratory chain (Busanello et al. 2013a).

The problem with most of the studies cited above, which includes our own (Komen et al. 2007), is that in virtually all cases phytanic acid was added to the medium of the cells or to the homogenates in its free fatty acid form. This is in sharp contrast to the in vivo situation in which phytanic acid as "seen" by cells is not in its free acid form but incorporated into different lipid structures including triglycerides, phospholipids, and other lipid species. It remains to be established whether the in vitro results with phytanic acid also hold up under in vivo conditions.

In order to bridge the gap between the in vitro and in vivo studies, we have generated a mouse model for Refsum disease which has been generated by disrupting the gene coding for phytanoyl-CoA hydroxylase (Ferdinandusse et al. 2008). The mice were completely free of symptoms when fed a standard laboratory diet which appeared to contain very little phytanic acid. However, when the knock-out mice were fed a phytol-rich diet there was accumulation of phytanic acid in plasma and tissues and in addition, the mice developed distinct abnormalities, including reduction in body weight, hepatic steatosis, and ataxia.

In summary, much has been learned in recent years about the metabolic functions of peroxisomes and the interplay with other subcellular organelles. Furthermore, substantial progress has been made with respect to the question of the pathophysiological consequences of defects in peroxisome metabolism. The availability of an increasing number of mutant mouse models often with organ-specific disruption of specific peroxisomal genes holds great promise for the future in terms of the pathophysiological consequences per organ.

Acknowledgment The authors gratefully acknowledge the expert help of Mrs. Maddy Festen and Mr. Jos Ruiter in preparation of the manuscript. Work by the authors is supported by different grants including LeukoTreat (HEALTH-F2-2010-241622), Hersenstichting Nederland: F2012(1)-102, and MetaKids.

References

Baes M, Van Veldhoven PP (2012) Mouse models for peroxisome biogenesis defects and beta-oxidation enzyme deficiencies. Biochim Biophys Acta 1822:1489–1500

Baes M, Huyghe S, Carmeliet P, Declercq PE, Collen D, Mannaerts GP, Van Veldhoven PP (2000) Inactivation of the peroxisomal multifunctional protein-2 in mice impedes the degradation of not only 2-methyl-branched fatty acids and bile acid intermediates but also of very long chain fatty acids. J Biol Chem 275:16329–16336

Baumgart E, Fahimi HD, Stich A, Volkl A (1996) L-lactate dehydrogenase A4- and A3B isoforms are bona fide peroxisomal enzymes in rat liver. Evidence for involvement in intraperoxisomal NADH reoxidation. J Biol Chem 271:3846–3855

Braverman NE, D'Agostino MD, Maclean GE (2013) Peroxisome biogenesis disorders: Biological, clinical and pathophysiological perspectives. Dev Disabil Res Rev 17:187–196

Brites P, Waterham HR, Wanders RJ (2004) Functions and biosynthesis of plasmalogens in health and disease. Biochim Biophys Acta 1636:219–231

Brown FR III, McAdams AJ, Cummins JW, Konkol R, Singh I, Moser AB, Moser HW (1982) Cerebro-hepato-renal (Zellweger) syndrome and neonatal adrenoleukodystrophy: similarities in phenotype and accumulation of very long chain fatty acids. Johns Hopkins Med J 151:344–351

Bunik VI, Raddatz G, Wanders RJ, Reiser G (2006) Brain pyruvate and 2-oxoglutarate dehydrogenase complexes are mitochondrial targets of the CoA ester of the Refsum disease marker phytanic acid. FEBS Lett 580:3551–3557

Busanello EN, Viegas CM, Moura AP, Tonin AM, Grings M, Vargas CR, Wajner M (2010) In vitro evidence that phytanic acid compromises Na(+), K(+)-ATPase activity and the electron flow through the respiratory chain in brain cortex from young rats. Brain Res 1352:231–238

Busanello EN, Amaral AU, Tonin AM, Grings M, Moura AP, Eichler P, Vargas CR, Wajner M (2012) Experimental evidence that pristanic acid disrupts mitochondrial homeostasis in brain of young rats. J Neurosci Res 90:597–605

Busanello EN, Amaral AU, Tonin AM, Zanatta A, Viegas CM, Vargas CR, Wajner M (2013a) Disruption of mitochondrial homeostasis by phytanic acid in cerebellum of young rats. Cerebellum 12:362–369

Busanello EN, Zanatta A, Tonin AM, Viegas CM, Vargas CR, Leipnitz G, Ribeiro CA, Wajner M (2013b) Marked inhibition of Na+, K(+)- ATPase activity and the respiratory chain by phytanic acid in cerebellum from young rats: possible underlying mechanisms of cerebellar ataxia in Refsum disease. J Bioenerg Biomembr 45:137–144

da Silva TF, Sousa VF, Malheiro AR, Brites P (2012) The importance of ether-phospholipids: a view from the perspective of mouse models. Biochim Biophys Acta 1822:1501–1508

de Waart DR, Koomen GC, Wanders RJ (1994) Studies on the urinary excretion of thromboxane B2 in Zellweger patients and control subjects: evidence for a major role for peroxisomes in the beta-oxidative chain-shortening of thromboxane B2. Biochim Biophys Acta 1226:44–48

Deon M, Sitta A, Barschak AG, Coelho DM, Pigatto M, Schmitt GO, Jardim LB, Giugliani R, Wajner M, Vargas CR (2007) Induction of lipid peroxidation and decrease of antioxidant defenses in symptomatic and asymptomatic patients with X-linked adrenoleukodystrophy. Int J Dev Neurosci 25:441–444

Diczfalusy U, Kase BF, Alexson SE, Bjorkhem I (1991) Metabolism of prostaglandin F2 alpha in Zellweger syndrome. Peroxisomal beta-oxidation is a major importance for in vivo degradation of prostaglandins in humans. J Clin Invest 88:978–984

Diczfalusy U, Vesterqvist O, Kase BF, Lund E, Alexson SE (1993) Peroxisomal chain-shortening of thromboxane B2: evidence for impaired degradation of thromboxane B2 in Zellweger syndrome. J Lipid Res 34:1107–1113

Eichler FS, Ren JQ, Cossoy M, Rietsch AM, Nagpal S, Moser AB, Frosch MP, Ransohoff RM (2008) Is microglial apoptosis an early pathogenic change in cerebral X-linked adrenoleukodystrophy? Ann Neurol 63:729–742

Ellinghaus P, Wolfrum C, Assmann G, Spener F, Seedorf U (1999) Phytanic acid activates the peroxisome proliferator-activated receptor alpha (PPARalpha) in sterol carrier protein 2-/ sterol carrier protein x-deficient mice. J Biol Chem 274:2766–2772

Ferdinandusse S, Houten SM (2006) Peroxisomes and bile acid biosynthesis. Biochim Biophys Acta 1763:1427–1440

Ferdinandusse S, Denis S, Mooijer PA, Zhang Z, Reddy JK, Spector AA, Wanders RJ (2001) Identification of the peroxisomal beta-oxidation enzymes involved in the biosynthesis of docosahexaenoic acid. J Lipid Res 42:1987–1995

Ferdinandusse S, Denis S, Van Roermund CW, Wanders RJ, Dacremont G (2004) Identification of the peroxisomal beta-oxidation enzymes involved in the degradation of long-chain dicarboxylic acids. J Lipid Res 45:1104–1111

Ferdinandusse S, Denis S, Mooyer PA, Dekker C, Duran M, Soorani-Lunsing RJ, Boltshauser E, Macaya A, Gartner J, Majoie CB, Barth PG, Wanders RJ, Poll-The BT (2006a) Clinical and biochemical spectrum of D-bifunctional protein deficiency. Ann Neurol 59:92–104

Ferdinandusse S, Kostopoulos P, Denis S, Rusch H, Overmars H, Dillmann U, Reith W, Haas D, Wanders RJ, Duran M, Marziniak M (2006b) Mutations in the gene encoding peroxisomal sterol carrier protein X (SCPx) cause leukencephalopathy with dystonia and motor neuropathy. Am J Hum Genet 78:1046–1052

Ferdinandusse S, Zomer AW, Komen JC, van den Brink CE, Thanos M, Hamers FP, Wanders RJ, van der Saag PT, Poll-The BT, Brites P (2008) Ataxia with loss of Purkinje cells in a mouse model for Refsum disease. Proc Natl Acad Sci USA 105:17712–17717

Fourcade S, Lopez-Erauskin J, Galino J, Duval C, Naudi A, Jove M, Kemp S, Villarroya F, Ferrer I, Pamplona R, Portero-Otin M, Pujol A (2008) Early oxidative damage underlying neurodegeneration in X-adrenoleukodystrophy. Hum Mol Genet 17:1762–1773

Galino J, Ruiz M, Fourcade S, Schluter A, Lopez-Erauskin J, Guilera C, Jove M, Naudi A, Garcia-Arumi E, Andreu AL, Starkov AA, Pamplona R, Ferrer I, Portero-Otin M, Pujol A (2011) Oxidative damage compromises energy metabolism in the axonal degeneration mouse model of X-adrenoleukodystrophy. Antioxid Redox Signal 15:2095–2107

Garashi M, Belchis D, Suzuki K (1976) Brain gangliosides in adrenoleukodystrophy. J Neurochem 27:327–328

Gilg AG, Singh AK, Singh I (2000) Inducible nitric oxide synthase in the central nervous system of patients with X-adrenoleukodystrophy. J Neuropathol Exp Neurol 59:1063–1069

Gronemeyer T, Wiese S, Ofman R, Bunse C, Pawlas M, Hayen H, Eisenacher M, Stephan C, Meyer HE, Waterham HR, Erdmann R, Wanders RJ, Warscheid B (2013) The proteome of human liver peroxisomes: identification of five new peroxisomal constituents by a label-free quantitative proteomics survey. PLoS One 8:e57395

Hanson RF, Szczepanik-VanLeeuwen P, Williams GC, Grabowski G, Sharp HL (1979) Defects of bile acid synthesis in Zellweger's syndrome. Science 203:1107–1108

Heymans HS, Schutgens RB, Tan R, van den Bosch H, Borst P (1983) Severe plasmalogen deficiency in tissues of infants without peroxisomes (Zellweger syndrome). Nature 306:69–70

Houten SM, Denis S, Argmann CA, Jia Y, Ferdinandusse S, Reddy JK, Wanders RJ (2012) Peroxisomal L-bifunctional enzyme (Ehhadh) is essential for the production of medium-chain dicarboxylic acids. J Lipid Res 53:1296–1303

Hubbard WC, Moser AB, Tortorelli S, Liu A, Jones D, Moser H (2006) Combined liquid chromatography-tandem mass spectrometry as an analytical method for high throughput screening for X-linked adrenoleukodystrophy and other peroxisomal disorders: preliminary findings. Mol Genet Metab 89:185–187

Hunt MC, Siponen MI, Alexson SE (2012) The emerging role of acyl-CoA thioesterases and acyltransferases in regulating peroxisomal lipid metabolism. Biochim Biophys Acta 1822:1397–1410

Huyghe S, Mannaerts GP, Baes M, Van Veldhoven PP (2006) Peroxisomal multifunctional protein-2: the enzyme, the patients and the knockout mouse model. Biochim Biophys Acta 1761:973–994

Idel S, Ellinghaus P, Wolfrum C, Nofer JR, Gloerich J, Assmann G, Spener F, Seedorf U (2002) Branched chain fatty acids induce nitric oxide-dependent apoptosis in vascular smooth muscle cells. J Biol Chem 277:49319–49325

Indiveri C, Iacobazzi V, Tonazzi A, Giangregorio N, Infantino V, Convertini P, Console L, Palmieri F (2011) The mitochondrial carnitine/acylcarnitine carrier: function, structure and physiopathology. Mol Asp Med 32:223–233

Jakobs BS, Wanders RJ (1991) Conclusive evidence that very-long-chain fatty acids are oxidized exclusively in peroxisomes in human skin fibroblasts. Biochem Biophys Res Commun 178:842–847

Jedlitschky G, Mayatepek E, Keppler D (1993) Peroxisomal leukotriene degradation: biochemical and clinical implications. Adv Enzym Regul 33:181–194

Kitareewan S, Burka LT, Tomer KB, Parker CE, Deterding LJ, Stevens RD, Forman BM, Mais DE, Heyman RA, McMorris T, Weinberger C (1996) Phytol metabolites are circulating dietary factors that activate the nuclear receptor RXR. Mol Biol Cell 7:1153–1166

Knazek RA, Rizzo WB, Schulman JD, Dave JR (1983) Membrane microviscosity is increased in the erythrocytes of patients with adrenoleukodystrophy and adrenomyeloneuropathy. J Clin Invest 72:245–248

Komen JC, Distelmaier F, Koopman WJ, Wanders RJ, Smeitink J, Willems PH (2007) Phytanic acid impairs mitochondrial respiration through protonophoric action. Cell Mol Life Sci 64:3271–3281

Lazarow PB, De Duve C (1976) A fatty acyl-CoA oxidizing system in rat liver peroxisomes; enhancement by clofibrate, a hypolipidemic drug. Proc Natl Acad Sci USA 73:2043–2046

Leipnitz G, Amaral AU, Zanatta A, Seminotti B, Fernandes CG, Knebel LA, Vargas CR, Wajner M (2010) Neurochemical evidence that phytanic acid induces oxidative damage and reduces the antioxidant defenses in cerebellum and cerebral cortex of rats. Life Sci 87:275–280
Lemotte PK, Keidel S, Apfel CM (1996) Phytanic acid is a retinoid X receptor ligand. Eur J Biochem 236:328–333
Lopez-Erauskin J, Fourcade S, Galino J, Ruiz M, Schluter A, Naudi A, Jove M, Portero-Otin M, Pamplona R, Ferrer I, Pujol A (2011) Antioxidants halt axonal degeneration in a mouse model of X-adrenoleukodystrophy. Ann Neurol 70:84–92
Mayatepek E, Flock B (1999) Increased urinary excretion of LTB4 and omega-carboxy-LTB4 in patients with Zellweger syndrome. Clin Chim Acta 282:151–155
Mayatepek E, Ferdinandusse S, Meissner T, Wanders RJ (2004) Analysis of cysteinyl leukotrienes and their metabolites in bile of patients with peroxisomal or mitochondrial beta-oxidation defects. Clin Chim Acta 345:89–92
Petrillo S, Piemonte F, Pastore A, Tozzi G, Aiello C, Pujol A, Cappa M, Bertini E (2013) Glutathione imbalance in patients with X-linked adrenoleukodystrophy. Mol Genet Metab 109:366–370
Poll-The BT, Roels F, Ogier H, Scotto J, Vamecq J, Schutgens RB, Wanders RJ, van Roermund CW, van Wijland MJ, Schram AW et al (1988) A new peroxisomal disorder with enlarged peroxisomes and a specific deficiency of acyl-CoA oxidase (pseudo-neonatal adrenoleukodystrophy). Am J Hum Genet 42:422–434
Poulos A, Sharp P, Fellenberg AJ, Danks DM (1985) Cerebro-hepato-renal (Zellweger) syndrome, adrenoleukodystrophy, and Refsum's disease: plasma changes and skin fibroblast phytanic acid oxidase. Hum Genet 70:172–177
Poulos A, Sharp P, Fellenberg AJ, Johnson DW (1988) Accumulation of pristanic acid (2, 6, 10, 14 tetramethylpentadecanoic acid) in the plasma of patients with generalised peroxisomal dysfunction. Eur J Pediatr 147:143–147
Powers JM, Moser HW (1998) Peroxisomal disorders: genotype, phenotype, major neuropathologic lesions, and pathogenesis. Brain Pathol 8:101–120
Rockenbach FJ, Deon M, Marchese DP, Manfredini V, Mescka C, Ribas GS, Habekost CT, Castro CG Jr, Jardim LB, Vargas CR (2012) The effect of bone marrow transplantation on oxidative stress in X-linked adrenoleukodystrophy. Mol Genet Metab 106:231–236
Russell DW (2003) The enzymes, regulation, and genetics of bile acid synthesis. Annu Rev Biochem 72:137–174
Russell DW (2009) Fifty years of advances in bile acid synthesis and metabolism. J Lipid Res 50 (Suppl):S120–S125
Salido E, Pey AL, Rodriguez R, Lorenzo V (2012) Primary hyperoxalurias: disorders of glyoxylate detoxification. Biochim Biophys Acta 1822:1453–1464
Schaumburg HH, Powers JM, Suzuki K, Raine CS (1974) Adreno-leukodystrophy (sex-linked Schilder disease). Ultrastructural demonstration of specific cytoplasmic inclusions in the central nervous system. Arch Neurol 31:210–213
Schluter A, Espinosa L, Fourcade S, Galino J, Lopez E, Ilieva E, Morato L, Asheuer M, Cook T, McLaren A, Reid J, Kelly F, Bates S, Aubourg P, Galea E, Pujol A (2012) Functional genomic analysis unravels a metabolic-inflammatory interplay in adrenoleukodystrophy. Hum Mol Genet 21:1062–1077
Schonfeld P, Reiser G (2006) Rotenone-like action of the branched-chain phytanic acid induces oxidative stress in mitochondria. J Biol Chem 281:7136–7142
Schonfeld P, Struy H (1999) Refsum disease diagnostic marker phytanic acid alters the physical state of membrane proteins of liver mitochondria. FEBS Lett 457:179–183
Schonfeld P, Kahlert S, Reiser G (2004) In brain mitochondria the branched-chain fatty acid phytanic acid impairs energy transduction and sensitizes for permeability transition. Biochem J 383:121–128
Seedorf U, Raabe M, Ellinghaus P, Kannenberg F, Fobker M, Engel T, Denis S, Wouters F, Wirtz KW, Wanders RJ, Maeda N, Assmann G (1998) Defective peroxisomal catabolism of branched

fatty acyl coenzyme A in mice lacking the sterol carrier protein-2/sterol carrier protein-x gene function. Genes Dev 12:1189–1201

Setchell KD, Heubi JE, Bove KE, O'Connell NC, Brewsaugh T, Steinberg SJ, Moser A, Squires RH Jr (2003) Liver disease caused by failure to racemize trihydroxycholestanoic acid: gene mutation and effect of bile acid therapy. Gastroenterology 124:217–232

Singh I, Pujol A (2010) Pathomechanisms underlying X-adrenoleukodystrophy: a three-hit hypothesis. Brain Pathol 20:838–844

Spector AA, Fang X, Snyder GD, Weintraub NL (2004) Epoxyeicosatrienoic acids (EETs): metabolism and biochemical function. Prog Lipid Res 43:55–90

Starkov AA, Fiskum G, Chinopoulos C, Lorenzo BJ, Browne SE, Patel MS, Beal MF (2004) Mitochondrial alpha-ketoglutarate dehydrogenase complex generates reactive oxygen species. J Neurosci 24:7779–7788

Steinberg SJ, Wang SJ, Kim DG, Mihalik SJ, Watkins PA (1999) Human very-long-chain acyl-CoA synthetase: cloning, topography, and relevance to branched-chain fatty acid metabolism. Biochem Biophys Res Commun 257:615–621

Theda C, Moser AB, Powers JM, Moser HW (1992) Phospholipids in X-linked adrenoleukodystrophy white matter: fatty acid abnormalities before the onset of demyelination. J Neurol Sci 110:195–204

Tsikas D, Schwedhelm E, Fauler J, Gutzki FM, Mayatepek E, Frolich JC (1998) Specific and rapid quantification of 8-iso-prostaglandin F2alpha in urine of healthy humans and patients with Zellweger syndrome by gas chromatography-tandem mass spectrometry. J Chromatogr B Biomed Sci Appl 716:7–17

Uto T, Contreras MA, Gilg AG, Singh I (2008) Oxidative imbalance in nonstimulated X-adrenoleukodystrophy-derived lymphoblasts. Dev Neurosci 30:410–418

van Roermund CW, Elgersma Y, Singh N, Wanders RJ, Tabak HF (1995) The membrane of peroxisomes in Saccharomyces cerevisiae is impermeable to NAD(H) and acetyl-CoA under in vivo conditions. EMBO J 14:3480–3486

van Roermund CW, Visser WF, Ijlst L, van Cruchten A, Boek M, Kulik W, Waterham HR, Wanders RJ (2008) The human peroxisomal ABC half transporter ALDP functions as a homodimer and accepts acyl-CoA esters. FASEB J 22:4201–4208

van Roermund CW, Visser WF, Ijlst L, Waterham HR, Wanders RJ (2011) Differential substrate specificities of human ABCD1 and ABCD2 in peroxisomal fatty acid beta-oxidation. Biochim Biophys Acta 1811:148–152

Van Veldhoven PP (2010) Biochemistry and genetics of inherited disorders of peroxisomal fatty acid metabolism. J Lipid Res 51:2863–2895

Vanhove GF, Van Veldhoven PP, Fransen M, Denis S, Eyssen HJ, Wanders RJ, Mannaerts GP (1993) The CoA esters of 2-methyl-branched chain fatty acids and of the bile acid intermediates di- and trihydroxycoprostanic acids are oxidized by one single peroxisomal branched chain acyl-CoA oxidase in human liver and kidney. J Biol Chem 268:10335–10344

Vargas CR, Wajner M, Sirtori LR, Goulart L, Chiochetta M, Coelho D, Latini A, Llesuy S, Bello-Klein A, Giugliani R, Deon M, Mello CF (2004) Evidence that oxidative stress is increased in patients with X-linked adrenoleukodystrophy. Biochim Biophys Acta 1688:26–32

Verhoeven NM, Roe DS, Kok RM, Wanders RJ, Jakobs C, Roe CR (1998) Phytanic acid and pristanic acid are oxidized by sequential peroxisomal and mitochondrial reactions in cultured fibroblasts. J Lipid Res 39:66–74

Wanders RJ, Ferdinandusse S (2012) Peroxisomes, peroxisomal diseases, and the hepatotoxicity induced by peroxisomal metabolites. Curr Drug Metab 13:1401–1411

Wanders RJ, Waterham HR (2006a) Biochemistry of mammalian peroxisomes revisited. Annu Rev Biochem 75:295–332

Wanders RJ, Waterham HR (2006b) Peroxisomal disorders: the single peroxisomal enzyme deficiencies. Biochim Biophys Acta 1763:1707–1720

Wanders RJ, Ferdinandusse S, Brites P, Kemp S (2010) Peroxisomes, lipid metabolism and lipotoxicity. Biochim Biophys Acta 1801:272–280

Wanders RJ, Komen J, Ferdinandusse S (2011a) Phytanic acid metabolism in health and disease. Biochim Biophys Acta 1811:498–507
Wanders RJ, Komen J, Kemp S (2011b) Fatty acid omega-oxidation as a rescue pathway for fatty acid oxidation disorders in humans. FEBS J 278:182–194
Waterham HR, Ebberink MS (2012) Genetics and molecular basis of human peroxisome biogenesis disorders. Biochim Biophys Acta 1822:1430–1441
Whitcomb RW, Linehan WM, Knazek RA (1988) Effects of long-chain, saturated fatty acids on membrane microviscosity and adrenocorticotropin responsiveness of human adrenocortical cells in vitro. J Clin Invest 81:185–188
Wiesinger C, Kunze M, Regelsberger G, Forss-Petter S, Berger J (2013) Impaired very long-chain acyl-CoA beta-oxidation in human X-linked adrenoleukodystrophy fibroblasts is a direct consequence of ABCD1 transporter dysfunction. J Biol Chem 288:19269–19279
Zha S, Ferdinandusse S, Hicks JL, Denis S, Dunn TA, Wanders RJ, Luo J, De Marzo AM, Isaacs WB (2005) Peroxisomal branched chain fatty acid beta-oxidation pathway is upregulated in prostate cancer. Prostate 63:316–323
Zomer AW, van Der Burg B, Jansen GA, Wanders RJ, Poll-The BT, van Der Saag PT (2000) Pristanic acid and phytanic acid: naturally occurring ligands for the nuclear receptor peroxisome proliferator-activated receptor alpha. J Lipid Res 41:1801–1807

Part II

Peroxisomal Diseases

Human Disorders of Peroxisome Biogenesis: Zellweger Spectrum and Rhizomelic Chondrodysplasia Punctata

4

Nancy Braverman, Catherine Argyriou, and Ann Moser

Abstract

The human peroxisome biogenesis disorders (PBD) are caused by autosomal recessive mutations in any of the 14 *PEX* genes, which encode peroxins, or PEX proteins, that act cooperatively to assemble functional peroxisomes. Mutations in *PEX* genes affect the import of peroxisome matrix enzymes and formation of new peroxisomes. The consequences are multiple enzyme deficiencies causing developmental malformations and progressive postnatal tissue injury. The PBD are classified into two distinct groups: Zellweger spectrum disorders (ZSD) and Rhizomelic Chondrodysplasia Punctata spectrum type 1 (RCDP1). The term "spectrum" is preferred because there is a gradation of phenotypes from severe to mild in these disorders, as well as newly recognized atypical phenotypes. In general, the milder the phenotype, the greater are the residual functions of the defective peroxin.

Tissues most affected in the PBD include the brain, peripheral nerves, eye, liver, kidney, heart, adrenal glands, bone, and lung. The severe form is distinguished by developmental malformations and early demise. In the intermediate and milder phenotypes, malformations may not be present and prominent disease features reflect postnatal peroxisome dysfunction over time. It is this latter group of patients that would benefit most from targeted therapies. In this chapter, we will review clinical phenotypes, diagnoses, supportive management, and research approaches to developing targeted therapies. In the overall theme of

N. Braverman (✉)
Department of Human Genetics, McGill University, Montreal, QC, Canada

Department of Pediatrics, McGill University, Montreal, QC, Canada
e-mail: nancy.braverman@mcgill.ca

C. Argyriou
Department of Human Genetics, McGill University, Montreal, QC, Canada

A. Moser
Department of Neurogenetics, Kennedy Krieger Institute, Baltimore, MD, USA

C. Brocard and A. Hartig (eds.), *Molecular Machines Involved in Peroxisome Biogenesis and Maintenance*, DOI 10.1007/978-3-7091-1788-0_4, © Springer-Verlag Wien 2014

this chapter, we will highlight how patient studies have contributed to our knowledge of peroxisome disorders and normal peroxisome biology.

Keywords
Peroxisome Biogenesis Disorders • Zellweger Spectrum Disorder • Rhizomelic Chondrodysplasia Punctata

Abbreviations

PBD	Peroxisome biogenesis disorders
ZSD	Zellweger spectrum disorder
ZS	Zellweger syndrome
NALD	Neonatal adrenoleu kodystrophy
IRD	Infantile Refsum disease
RCDP	Rhizomelic chondro dysplasia punctata

4.1 Historical Background

More than half a century has passed since the 1960s, when biochemist Christian de Duve first isolated the peroxisome: a newly discovered organelle containing hydrogen peroxide-generating oxidases (De Duve and Baudhuin 1966). A few years later, Hans Zellweger and other clinicians described Zellweger syndrome (ZS), a malformation syndrome involving the brain, liver, and kidneys and with characteristic facial features (reviewed in Zellweger et al. 1988). However, an association between ZS and peroxisome dysfunction was not made until 1973 when pathologist Sidney Goldfischer reported the absence of morphologically identifiable peroxisomes in hepatocytes and renal tubular cells of ZS patients after applying catalase markers (Goldfischer et al. 1973). The etiological significance of this observation became clear a decade later, when peroxisomal enzymes were found to be deficient in ZS and related disorders of neonatal adrenoleukodystrophy (NALD), infantile Refsum disease (IRD), and rhizomelic chondrodysplasia punctata (RCDP) (Datta et al. 1984; Moser et al. 1984; Heymans et al. 1985). ZS became the paradigm for metabolic malformation syndromes, in which metabolic disturbances cause organ malformations. This highlighted a role for lipid metabolic pathways in fetal development and elevated the status of peroxisomes to organelles with vital functions in human development.

With foresight, investigators collected fibroblast cell lines from these patients and performed somatic cell complementation studies, ultimately showing that peroxisome biogenesis disorders (PBD) are heterogeneous and that disease severity can vary within each complementation group (Moser et al. 1995). Using the sequence of peroxisomal biogenesis proteins identified in yeast and the newly created cDNA databases at the US National Center for Biotechnology Information, investigators were able to select candidate human *PEX* genes that could then be tested for rescue in each established complementation group (Dodt et al. 1996). In a parallel fashion, cDNA library rescue of peroxisome-deficient Chinese hamster ovary (CHO) cell lines was performed (Fujiki et al. 2006). In fact, the first human *PEX* gene identified, *PEX2*, was by CHO cell rescue (Shimozawa et al. 1992). These two strategies soon led to the identification of the mammalian genes responsible for each complementation group. All of this work was pioneering, as new technologies introduced in lipid biochemistry, genetics, and genomics, were applied to peroxisome disorders. In the past decade, peroxin functions were determined, different model organisms were engineered, and pathophysiological and therapeutic studies initiated.

From these groundwork investigations, we know that the PBD are autosomal recessive disorders caused by defects in any one of multiple *PEX* genes, whose protein products are required for peroxisome assembly. Peroxisome assembly includes the import of matrix enzymes, peroxisome division by fission, and addition of new membrane material from the ER. Currently, there are 16 known mammalian *PEX* genes and defects in 14 of these cause PBD (see Table 4.1). No disease has yet been associated with defects in *PEX11α* and *PEX11γ*. The proteins encoded by *PEX* genes are alternatively called PEX proteins or peroxins (Distel et al. 1996). In addition to these peroxisome assembly factors, peroxisome dynamics involve proteins that are shared in peroxisome and mitochondrial division (Schrader et al. 2012), proteins involved in autophagy of senescent peroxisomes (Nordgren et al. 2013) and proteins that regulate, through signaling pathways, peroxisome number and enzyme content according to cell and tissue requirements (Fransen 2012; Chang et al. 1999). The recent report of an infant with a dominant negative mutation in *DLP1*, encoding a dynamin-like protein required for peroxisome and mitochondrial fission (Waterham et al. 2007), suggests that novel disorders of peroxisome dynamics will continue to be defined and contribute to our knowledge of peroxisome biology.

The PBD include Zellweger spectrum disorder (ZSD) and RCDP type 1. The overall incidence in North America for ZSD is around 1/50,000 and for RCDP, 1/100,000 (Steinberg et al. 2006). RCDP1 is caused by defects in PEX7, whereas ZSD can result from defects in any of the remaining 13 *PEX* genes. In addition, around 10 % of clinically suspected cases of PBD are caused by peroxisome single enzyme defects. In the case of ZSD, defects in very long-chain fatty acid ß-oxidation cause pseudo-NALD (acyl CoA oxidase, ACOX1) or pseudo-ZSD (D-bifunctional protein, DBP). For RCDP, defects in plasmalogen synthesis cause RCDP2 (glyceronephosphate *O*-acyltransferase, GNPAT) and RCDP3 (alkylglycerol phosphate synthase, AGPS). These single enzyme defects are distinguished by biochemical (Table 4.1) and/or gene sequencing analysis. The clinical similarity between single enzyme defects and PBD suggests that defects in very long-chain fatty acid ß-oxidation cause the major pathology in ZSD and that defects in plasmalogen synthesis cause the RCDP phenotype. For RCDP we know this to be true (Braverman and Moser 2012; Braverman et al. 2010). However, this is not as straightforward in ZSD due to the large clinical variation and multiple enzyme

Table 4.1 Biochemical features of peroxisome disorders

Group	Peroxisome biogenesis disorders (PBD)				Single enzyme/protein/other defect									
	Zellweger spectrum disorder (ZSD)			RCDP										
Disorder	ZS, NALD, IRD	Atypical ZSD[a]		RCDP1	RCDP2	RCDP3	Adult refsum disease	Pseudo-ZS, DBP, Perrault syndrome	AMACR	LDMN[b]	Pseudo-NALD	X-ALD	CADDS
Gene	*PEX 1, 2, 3, 5, 6, 10, 12, 13, 14, 16, 19, 26*	*PEX 2, 10, 12, 16*	*PEX11B*[c]	*PEX7*	*GNPAT*	*AGPS*	*PhyH, PEX7*	*HSD17B4*	*AMCR*	*SCP2*	*ACOX1*	*ABCD1*	*ABCD1-BAP31 deletion*[d]
Metabolites (blood, urine)													
VLCFA	↑	↑, N	N	N	N	N	N	↑	N	N	↑	↑	↑
Methyl-branched FA:													
Phytanic	↑	↑	N	↑	N	N	↑	↑	N, ↑	↑	N	N	N
Pristanic	↑	↑	N	↑	N	N	↑	↑	↑	↑	N	N	N
C27 bile acids	↑	↑	N	N	N	N	N	↑	↑	↑	N	N	N
Pipecolic acid	↑	↑	N	N	N	N	N	N	N	N	N	N	N
Plasmalogens	↓	N	N	↓	↓	↓	N	N	N	N	N	N	N
Fibroblast analysis													
Plasmalogen synthesis	↓	N	N	↓	↓	↓	N	N	N	N	N	N	N
C26:0 ß-oxidation	↓	N, ↓	N	N	N	N	N	↓	N	N	↓	↓	↓
Pristanic acid ß-oxidation	↓	N, ↓	N	N	N	N	N	↓	↓	↓	N	N	N

Phytanic acid α-oxidation	↓	N, ↓	N	↓	N	N	↓	↓	↓	↓	N	N	N
Catalase solubility	↑	N, ↑	↑	N	N	N	N	N	N	N	N	N	N
Peroxisome number	Reduced	Reduced (PEX16)	Reduced	N	N	N	N	N			N	N	N
Peroxisome morphology	Enlarged	Enlarged (PEX16), mosaic for catalase (PEX2, 12)	Enlarged and elongated	N	N	N	N	N			Enlarged	N	N

[a]Refers to mutations in the ring finger domains of PEX2 (Sevin 2011), PEX 10 (Steinberg 2009), PEX12 (Zeharia 2007), and the C terminus of PEX16 (Ebberink 2010

[b]LDMN, Leukencephalopathy with Dystonia and Motor Neuropathy

[c]Peroxisome profile at 37 °C, vs. increased catalase solubility and reduced peroxisome ß-oxidationt at 40 °C (Ebberink 2012)

[d]Minmal critical deletion (Corzo 2002)

deficiencies. Other single enzyme defects in branched chain fatty acid metabolism, such as phytanyl-CoA hydroxylase (PhyH; Wierzbicki et al. 2002), alpha-methylacyl-CoA racemase (AMACR; Ferdinandusse et al. 2000), and sterol carrier protein 2 (SCP2; Ferdinandusse et al. 2006) deficiency, present mainly in adults but have clinical and biochemical features overlapping that of ZSD (see Table 4.1 for biochemical features). Finally, secondary oxidant injury and mitochondrial dysfunction, reported in original descriptions of Zellweger syndrome (Goldfischer et al. 1973; Kelley 1983), are now being re-explored and may play a role in pathophysiology in both ZSD and RCDP, as well as more common neurodegenerative disorders of aging (Braverman and Moser 2012; Fransen et al. 2011). Taken together, peroxisomes must now be considered vital organelles in development, as well as ongoing tissue and organ homeostasis over our full lifespan.

4.2 Clinical Features-ZSD

The clinical phenotypes of ZSD form an overlapping spectrum of severity. For ZSD, the phenotypes of ZS, NALD, and IRD were described before a peroxisome etiology was known. These phenotypes currently fit a description of severe, intermediate, and milder forms, respectively. Since considerable overlap occurs, the current preference is to use the broader term, ZSD in order to appreciate the wide variations in presentation and natural history. The majority of ZSD patients fit the less severe forms.

As in other inborn errors of metabolism, the severe form of the disease is the one first described, the most homogenous and the most predictable. These infants are born with a characteristic malformation syndrome, initially described in case reports (Bowen et al. 1964; Smith et al. 1965) and then followed by a series of similar cases and more detailed descriptions (Passarge and McAdams 1967). Cerebro-hepato-renal syndrome was coined by Passarge and McAdams (1967) and Zellweger syndrome by Opitz (Opitz et al. 1969). These original clinical descriptions remain true to the mark today. In fact, the major organ systems involved, combined by the term "cerebro-hepato-renal" also remain a starting point to describe all ZSD forms. Currently, we suggest the extended mnemonic "cerebro-hepato-renal-adrenal" to any clinician considering this diagnosis, as adrenal dysfunction can occur in all severity groups.

4.2.1 Severe ZSD

Severe ZSD, or Zellweger syndrome, is distinguished by a neonatal presentation that typically includes seizures, profound hypotonia, and areflexia consequent to cerebral dysgenesis. As clinical geneticists are taught, the face predicts the brain (Demyer et al. 1964), and infants with severe ZSD have a prominent forehead, large fontanelles and separated sutures, flattened facial profile with hypoplastic supraorbital ridges, low nasal bridge, hypertelorism, epicanthic folds, high arched palate,

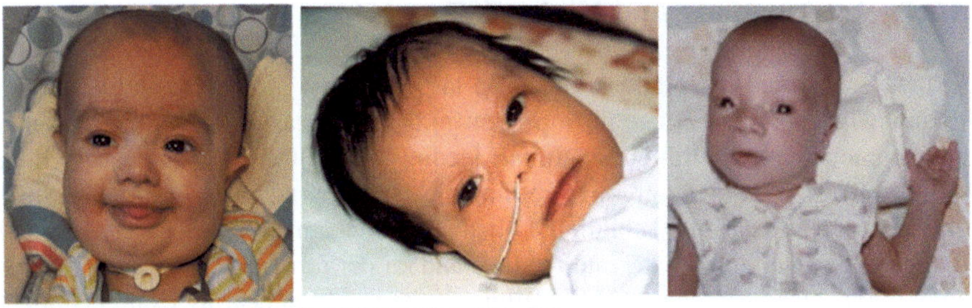

Fig. 4.1 Three infants with Zellweger syndrome show classic facial features, including a prominent forehead, low nasal bridge, hypertelorism and small chin. *Left*, this infant (PEX26p.Leu12fs/?) required a tracheostomy tube because of severe laryngomalacia. *Center*, this infant (PEX1p.Ile700fs/Ile700fs) is jaundiced due to liver disease and has a nasogastric feeding tube. *Right*, this infant is showing her single transverse palmer crease, commonly seen in ZSD

micrognathia, and redundant nuchal folds. This pattern of facial features is observed in the infants shown in Fig. 4.1.

Brain magnetic resonance imaging (MRI) reveals characteristic defects in neuronal migration. Polymicrogyria in the frontal and perisylvian regions and pachygyria in the perirolandic and occipital regions is a distinguishing feature of ZS, as well as heterotopic, subcortical collections of neurons that have arrested along their migratory path (Barkovich and Peck 1997). Other findings include hypomyelination of white matter, cerebral atrophy with ventricular dilation, and subependymal germinolytic cysts.

Eye malformations include congenital glaucoma, corneal opacities, cataract, and optic nerve hypoplasia (Cohen et al. 1983). Sensorineural deafness is present. Renal malformations consist of micronodular cortical cysts apparent on ultrasound examination. Chondrodysplasia punctata (discussed in RCDP section) is observed, predominantly in the knees (patella) and hips (acetabulum and greater trochanter of femurs). Liver disease manifests with hepatomegaly, elevated transaminases, coagulopathy, cholestasis, and bile acid abnormalities. During infancy, elevated serum iron with hepatic siderosis is present. Adrenal dysfunction can occur. The liver and adrenal disease are not secondary to malformations, but to progressive tissue injury from peroxisome enzyme deficiencies. The liver disease is thought to be secondary to defects in mature bile acid synthesis leading to the elevation of intermediary bile acids, which are toxic (Bove et al. 2004). Adrenal dysfunction occurs secondary to accumulation of VLCFA in the adrenal gland. Pathological studies of brain, liver, kidney, and adrenal glands on Zellweger syndrome patients have been reviewed (Gilchrist et al. 1976; Danks et al. 1975; Powers et al. 1985). Overall, developmental milestones are not obtained, lifespan is shortened due to cerebral dysgenesis and most infants do not survive beyond 1 year of age.

4.2.2 Intermediate and Milder ZSD

In the less severe presentations there is a large element of overlap and variation. Infants with the intermediate form present in the first year of life with hypotonia, developmental delays, failure to thrive, feeding difficulties, and cholestatic liver disease. They have a progressive retinopathy and hearing loss, leading to their classification as deaf–blind children. Leber congenital amaurosis has been an initial diagnosis reported in both the intermediate and milder forms of ZSD (Michelakakis et al. 2004; Majewski et al. 2011). Failure of the newborn hearing exam occurs in many patients and thus may also be a first indication of ZSD. The facial dysmorphic features of Zellweger syndrome may or may not be present. They are at high risk for developing seizures, white matter changes, and adrenal dysfunction, which is what led to their past classification as NALD. Liver disease is more apparent early on, but subsides later. Nevertheless, it may contribute to failure to thrive and feeding issues. Adrenal insufficiency can develop. Renal cysts and chondrodysplasia punctata are absent.

Brain MRI in the intermediate form can show limited neuronal migration defects and delayed myelination. Early onset of a progressive diffuse demyelinating leukodystrophy involving the cerebrum, mid-brain, and cerebellar white matter predicts an early demise.

A patient was recently reported with demyelination occurring predominantly in the brainstem and showing features of an active inflammatory process, similar to X-linked adrenoleukodystrophy (X-ALD) (Kulkarni et al. 2011). In contrast, most pathological descriptions of NALD depict inflammation, but to a degree less severe than that seen in X-ALD (Kulkarni et al. 2011; Aubourg et al. 1986). Brain MRI imaging can also be normal, but there is an unknown risk for the occurrence of demyelination over time, which can stabilize or progress (Barth et al. 2004). It will be important to determine if the leukodystrophy in ZSD is similar on a mechanistic level to that present in X-ALD, as bone marrow transplant can halt demyelination in X-ALD.

In later childhood, the clinical presentation is dominated by psychomotor retardation, hypotonia, visual loss from retinal degeneration, and sensorineural hearing loss. Again, liver disease may be prominent early on, adrenal insufficiency can develop, and leukodystrophy can occur. Early on, it can be difficult to prognosticate which patient will fit an intermediate or milder phenotype, and only time and developmental progress can distinguish these patients. Most children obtain developmental milestones at later ages. The degree of peroxisome deficiency measured by biochemical testing and mutation analysis can assist in prognosis. Figure 4.2 shows a group of intermediate ZSD children and their personal vignettes.

In both the intermediate and milder forms of ZSD, dental enamel defects mainly affecting the secondary teeth have been reported (Tran et al. 2011; Acharya et al. 2012). Physician experience indicates that these patients develop osteopenia that can lead to fractures. In addition, renal calcium oxalate stones can occur (van Woerden et al. 2006). It is now apparent that the preservation of intellect does not

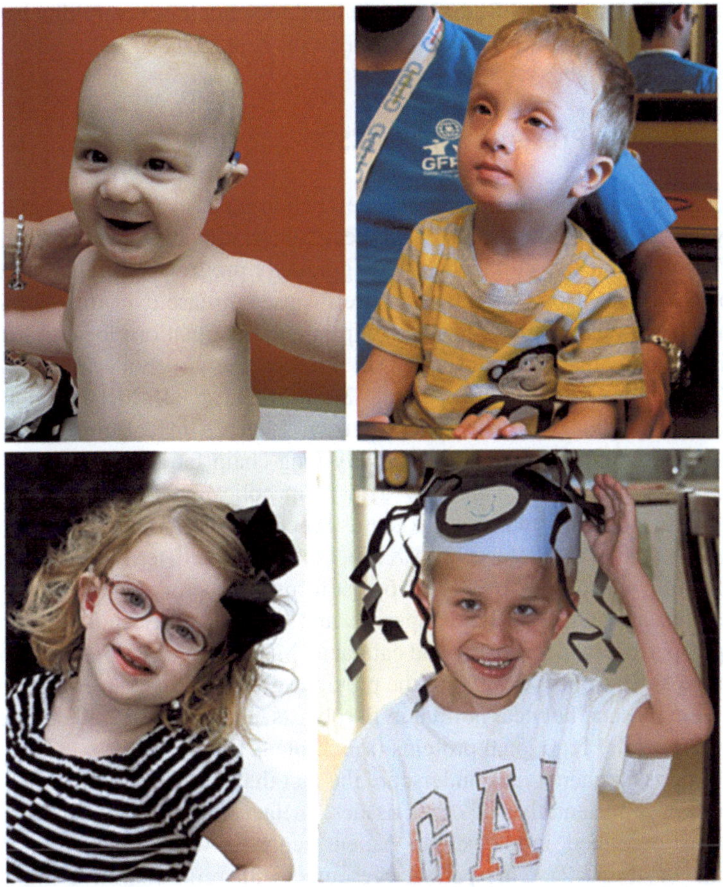

Photos courtesy of the Global Foundation for Peroxisomal Disorders

Fig. 4.2 Children with intermediate ZSD. *Upper left*, this is a 12-month-old (PEX1p.Gln128*/Gly843Asp), diagnosed at 1 month of age. He has hearing and visual loss, a normal brain MRI and is starting to sit unsupported. Currently 3 years old, he has a cochlear implant and has developed a seizure disorder and adrenal insufficiency. He is starting to walk and communicates with signs. *Upper right*, this is a 4-year-old (PEX1p.Pro162Thr/Gly843Asp) diagnosed at age 2 years. He is deaf and has visual loss from retinal degeneration. His brain MRI showed subtle dysmyelination. He is nonverbal, pulls to stand and walks with a walker. He is now 5 years of age, has a cochlear implant and adrenal insufficiency. *Lower left*, this is a 4-year-old (PEX1p.Ile700fs/Gly843Asp), diagnosed at age 2 years. She is deaf, has optic nerve hypoplasia, adrenal insufficiency, and patchy hypomyelination on brain MRI. She is now 5 years old, has a cochlear implant, walks, and communicates with speech and signs. *Lower right*, this is a 5-year-old (PEX1p.Ile700fs/Gly843Asp) diagnosed at 4 months of age. He is deaf, has a cochlear implant and visual loss from retinal degeneration. Brain MRI was normal at age 2 years. He walks and communicates with signs. He continued to make developmental progress until 7 years old, when he was treated for leukemia. Brain MRI at that time showed white matter changes. He is now 13 years old and has recovered some developmental milestones.

exclude a PBD. Patients with hearing and visual loss, but intact cognition have been reported (Raas-Rothschild et al. 2002; Majewski et al. 2011).

4.2.3 Atypical Presentations of ZSD

A novel group of patients present with cerebellar ataxia, variable peripheral neuropathy, and relative preservation of intellect. Sensory deficits are not predominant. Age of onset is in early childhood and is gradually progressive highlighting again that postnatal peroxisome dysfunction itself causes disease over time. Brain MRI may show cerebellar atrophy. In this group, missense mutations located within the zing finger domains were described in PEX2p (Sevin et al. 2011; Mignarri et al. 2012), PEX10p (Steinberg et al. 2009; Regal et al. 2010), and PEX12p (Gootjes et al. 2004) and the N-terminus of PEX12p (MacCollin et al. 1990; Zeharia et al. 2007). Although plasma very long-chain fatty acids can be normal, in most cases branched chain fatty acids, pipecolic and bile acid levels were abnormal. In addition, many peroxisome functions in fibroblasts were normal. Another new group had defects in PEX16p, located at the C-terminus of the protein and affecting one of the two PEX16p isoforms (Ebberink et al. 2010). These patients presented with spastic paraparesis and ataxia in early childhood and later developed peripheral neuropathy and cataracts. MRI showed cerebellar atrophy and cerebral leukodystrophy. Similar to the previous group, peroxisome functions were abnormal in plasma, but near normal in fibroblasts, and the mutant alleles identified expressed partially functional proteins (see Table 4.1).

These unusual phenotypes underscore the fact that peroxisome functions in skin cells are not representative of other tissues. In fact, they may also be normal in plasma, as described in the single patient reported with a defect in *PEX11ß* (Ebberink et al. 2012). This patient presented with mild cognitive impairment, congenital cataracts and poor vision, childhood progressive hearing loss, and peripheral neuropathy. The neurological symptoms worsened with intercurrent illness. In this case, all biochemical profiles were normal in both plasma and fibroblast cultures; in fibroblasts however, peroxisome morphology was abnormal and there was increased cytosolic catalase (see Table 4.1).

CADDS, or contiguous deletion syndrome consisting at minimum, of the X-linked adrenoleukodystrophy gene (*ABCD1*) and D17XS1357E (*BAP31*), was described thus far in four patients in 2002 (Corzo et al. 2002; Iwasa et al. 2012). This is a lethal disorder in infancy with features that overlap those of ZS. Patients present as neonates with hypotonia, seizures, severe developmental delays, deafness, and cholestatic liver disease. Autopsy shows small adrenal glands, hypomyelination, and in one case, neuronal migration defects (Iwasa et al. 2012). Peroxisome biochemistry shows defects of only VLCFA ß-oxidation, which are more severe than found in X-ALD. This condition suggests that the deletion of ABCD1 causes a more severe impact on VLCFA metabolism when combined with a deletion of BAP31. BAP31 is an abundant ER chaperone protein required for proper ER-to-golgi protein trafficking (Cacciagli et al. 2013). The role it may play in VLCFA ß-oxidation will be interesting to determine, as this may be one example of a modifier effect in peroxisome disorders.

4.2.4 Clinical Features-RCDP1

RCDP1 also has a clinical spectrum, although in contrast to ZSD, the majority of cases have been severe. This may be secondary to lack of recognition of milder phenotypes, or the fact that the common mutation in the *PEX7* gene, encoding PEX7p.Leu292*, encodes a nonfunctional protein (Braverman et al. 2002). The phenotypes associated with *PEX7* deficiency include classic RCDP and milder variants, as well as adult Refsum disease.

Classically these patients are recognized in the newborn period with typical facial dysmorphic features that include frontal bossing, depressed nasal bridge, small upturned nose, and hypoplastic midface (Fig. 4.3, right panel). There is a severe skeletal dysplasia that includes rhizomelia in the humeri and femur, although more so in the humeri. On X-ray, there is generalized chondrodysplasia punctata of all of the long bones epiphyses, often involving the shoulders, elbows hips, knees, costochondral junctions, and lateral vertebral bodies (Fig. 4.4). Punctate calcifications can also be present in the hyoid bone and tracheal cartilage; the intervertebral discs can be calcified. Metaphyseal splaying and other metaphyseal irregularities of the long bones are present. In addition, there is coronal clefting of the vertebrae-translucent bands on X-ray, due to delayed mineralization of the vertebral bodies (Wells et al. 1992). Cleft palate occurs in around 25 % of children (White et al. 2003). These findings indicate premature, abnormal mineralization of the growth plates, and cartilaginous structures, as well as delayed mineralization in the vertebral bodies. Although the epiphyseal punctata are less apparent by X-ray as the growth plates mineralize over time, the structure of the epiphyses and metaphyses remain remarkably abnormal. The skeletal involvement results in profound growth failure (White et al. 2003). It also incurs a risk for cervical spine stenosis that may be apparent on imaging studies in the newborn period or later (Khanna et al. 2001).

Bilateral cataracts are typically noted at birth or within 6 months of life. There is profound psychomotor delay and seizures develop in the majority of patients. There is an increased risk for congenital heart disease, and in one study, cardiac malformations were present in 52 % of RCDP1 patients (Huffnagel et al. 2013). Brain MRI shows a general decrease in neuron number and white matter, and progressive cerebellar degeneration (Bams-Mengerink et al. 2006). The striking neuronal migration abnormalities seen in ZS are not typically observed in RCDP. Ichthyosiform skin lesions are found in less than 30 % of patients. The etiology maybe secondary to phytanic acid accumulation, similar to the ichthyosis observed in adult Refsum disease and/or accumulation of fatty alcohols, similar to Sjögren–Larsson syndrome (Rizzo 1998).

Although patients with RCDP have severe handicaps, lifespan is broader than previously noted. An unknown proportion of these infants die in the neonatal

Photos courtesy of RhizoKids International

Fig. 4.3 Children with RCDP1. *Left*, 18-month-old with a mild RCDP phenotype (PEX7p. Tyr40*/Ile245Phe) who was diagnosed at 3 months of age with cataracts and chondrodysplasia punctata. She does not have rhizomelia. Cataracts were extracted. She is now 4 years old and had surgery for cervical spine stenosis and hip dysplasia, and recently developed hypothyroidism. She walks, speaks in sentences, and has had only mild developmental delays. *Right*, four children with classic RCDP diagnosed as neonates and showing frontal bossing, depressed nasal bridge, small upturned nose, hypoplastic midface, and shortened humeri. All have had cataract extractions, are unable to sit unsupported, and are nonverbal. From the *left*, a 20-month-old with seizures and aortic coarctation; 2½ year old (PEX7p.Tyr40*/L292X) with seizures, atrial septal defects, scoliosis, brain MRI showed agenesis of the corpus collosum and colpocephaly; 3 year old with kyphosis and prior cecal volvulus, no seizures and normal cardiac ultrasound; 3½ year old with seizures and tetrology of Fallot

period, secondary to cardiac defects or lung disease (Oswald et al. 2011). Of those who survive, 50 % are alive at 6 years of age, and most succumb by 10 years of age due to respiratory complications (White et al. 2003). The mechanism for the recurrent pulmonary dysfunction observed in these children may be secondary to the skeletal dysplasia, which causes a small chest, restricted ventilation due to costochondral punctata, and lack of mobility. However, it is also likely that plasmalogens play a role in pulmonary development and postnatal pulmonary functions and therefore plasmalogen deficiency itself may contribute to the lung disease observed in RCDP pateints (Oswald et al. 2011; Braverman and Moser 2012).

Milder forms of RCDP are now being recognized and correlate with less severe plasmalogen deficiency than classic RCDP. This category includes patients without rhizomelia and with less severe developmental delays (Fig. 4.3, left panel). In these cases, the presence of congenital cataracts and X-ray evidence of chondrodysplasia punctata has led to the diagnosis (Braverman et al. 2002). However, a patient with normal cognition and congenital cataracts, but without chondrodysplasia punctata was reported (Braverman et al. 2002) and, as well, two families with cataracts and autism spectrum disorder (Yu et al. 2013). Mild PEX7p deficiency can also cause a phenotype identical to adult Refsum disease (van den Brink et al. 2003; Horn et al. 2007) and as well a crossover between adult Refsum disease and RCDP

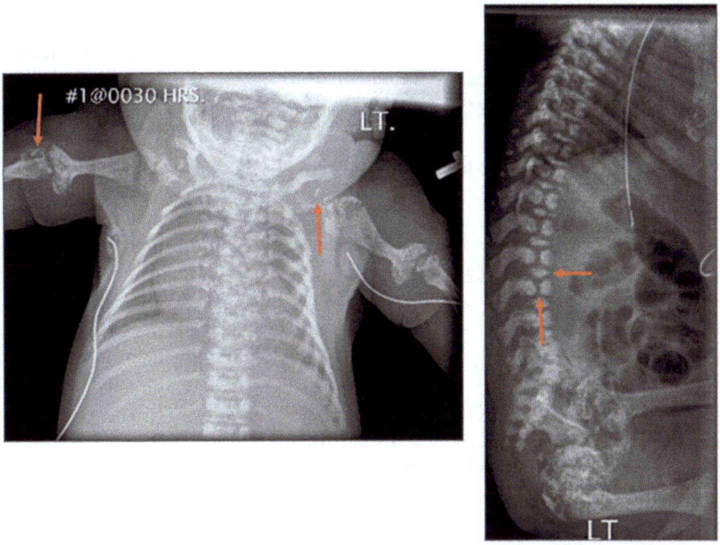

Fig. 4.4 X-ray images of classic RCDP1. *Left*, AP chest film shows rhizomelia, punctate calcifications in the epiphyses of the elbow and shoulder (*arrows*), metaphyseal widening of the long bones and ribs, and a small thorax. Lateral thoracolumbar spine shows radiolucent longitudinal vertebral clefts that represent unossified cartilage (*vertical arrow*) and mineralization of intervertebral discs (*longitudinal arrow*) and punctata in the proximal femur (*arrowhead*)

(Braverman et al. 2002). In these cases, erythrocyte plasmalogen levels and plasmalogen synthesis in skin fibroblasts were near normal, and disease was mainly due to phytanic acid accumulation over time.

There are a limited number of pathological reports in RCDP (Poulos et al. 1988; Viseskul et al. 1974; Sugarman 1974; Gilbert et al. 1976). CNS findings include decreased brain volume, with deficiency of both neurons and white matter, and a progressive cerebellar degeneration that includes the loss of Purkinje and granular cells. Neuronal migration defects and demyelination are not typically observed.

4.3 Diagnosis

4.3.1 Biochemical

Peroxisomes contain numerous enzymes involved in multiple metabolic pathways. Defects in peroxisome assembly result in peroxisomal enzyme deficiencies and tissue-specific pathology due to the accumulation of toxic metabolites that are not broken down, or deficiencies of products not synthesized (see Table 4.2). In order to better understand and treat these disorders, it will be important to uncover the cellular responses to the abnormal peroxisome metabolites, which result in the observed pathology.

Table 4.2 Peroxisome functions relevant to PBD

Biochemical pathway	Enzymes	Features	Potential disease relevance
β-oxidation of straight, very long chain (≥C22) fatty acids	ACOX1, DBP, ACAA1/ SCPX	VLCFA oxidation, synthesis of docosohexanoic acid (DHA)	Tissue accumulation of VLCFA causes brain, peripheral nerve and adrenal damage; deficiency of DHA affects brain and vision (X-ALD)
β-oxidation of methyl-branched fatty acids and C27 bile acids (di- and tri-hydroxycholestanoic acid)	ACOX2, DBP, SCPX	Oxidation of pristanic acid, synthesis of mature C24 bile acids; utilizes a different oxidase than VLCFA oxidation	Accumulation of pristanic acid affects brain. Increased bile acid intermediates cause liver toxicity
β-oxidation of dicarboxylic fatty acids	ACOX1, DBP/LBP, ACAA1/ SCPX	Oxidation of dicarboxylic acids utilizes a different dehydrogenase/hydratase than VLCFA oxidation	Unknown
Auxiliary pathways prior to β-oxidation: Fatty acid α-oxidation	PhyH	Methyl-branched phytanic acid requires an α-oxidation step before entering β-oxidation pathway as pristanoyl-CoA	Tissue accumulation of phytanic acid affects retina, cerebellum and peripheral nerves (adult Refsum disease)
Auxiliary pathways prior to β-oxidation: Fatty acid racemization	AMACR	Pristanoyl-CoA and C27-bile acyl-CoAs are converted to the (S)-stereoisomer before β-oxidation	Tissue accumulation of phytanic and pristanic acids, and bile acid intermediates (AMACR deficiency)
Ether phospholipid (plasmalogen) biosynthesis	GNPAT, AGPS	Commiting steps for the formation of plasmalogens, a specialized class of membrane phospholipids	Deficiency affects brain, lens, bone and heart development (RCDP)
Glyoxylate detoxification	AGXT	Prevents the conversion of glyoxylate into the toxic metabolite, oxalate	Accumulation of oxalate results in calcium oxalate renal stones (Hyperoxaluria Type I)
L-lysine oxidation	Pipecolic acid oxidase	Major pathway for lysine degradation in the brain	Pipecolic acid, a peroxisome biomarker, accumulates in ZSD and pyridoxine dependent epilepsy
Hydrogen peroxide detoxification	Catalase	Required for degradation of hydrogen peroxide, produced as a by-product of peroxisome β-oxidation reactions	Increased oxidant damage

Full enzyme names: ACOX1, acyl-CoA oxidase 1; ACOX2, acyl-CoA oxidase 2; DBP, D-bifunctional protein; LBP, L-bifunctional protein; ACAA1, acetyl-CoA acyltransferase 1 (peroxisomal thiolase); SCPX, peroxisomal thiolase; PhyH, phytanyl-CoA hydroxylase; AMACR, α-methylacyl-CoA racemase; GNPAT, glyceronephosphate O-acyltransferase; AGPS, alkylglycerone phosphate synthase; AGT, alanine-glyoxylate aminotransferase

After clinically suspecting ZSD in a patient, peroxisome functions are evaluated by metabolite profiles in blood and urine. Metabolite testing should represent at least two or more different peroxisomal pathways and usually include levels of plasma VLCFA (straight chain saturated and monounsaturated), phytanic, pristanic, bile acids (branched chain fatty acids), pipecolic acid, and erythrocyte plasmalogens (Steinberg et al. 2008). Note that peroxisomal metabolite profiles are not always conclusive. Phytanic acid and its oxidized product pristanic acid are exclusively dietary in origin and thus can be normal in patients that are breast or formula fed, or restricted in foods rich in phytanic acid. In patients with milder or atypical forms of ZSD, plasma VLCFA and erythrocyte plasmalogens can be normal, and peroxisome functions can also be normal in fibroblasts (Table 4.1).

In RCDP1, peroxisome dysfunction is limited to deficiency of plasmalogens and elevation of phytanic acid. Fibroblast analysis shows defects in plasmalogen synthesis and phytanic acid oxidation. In milder cases, plasmalogen synthesis is near normal. Single enzyme defects that mimic ZSD or RCDP can usually be distinguished in the biochemical profile, by enzymatic testing of fibroblast cultures or identifying mutations in the suspected genes (Table 4.1).

4.3.2 Cellular

Metabolite testing is confirmed by enzymatic testing in fibroblast cultures and includes assays for ß-oxidation of VLCFA and pristanic acid, α-oxidation of phytanic acid, plasmalogen synthesis, and activity of GNPAT. Cell cultures can also be evaluated for other measures of peroxisome function, thus providing additional clues for diagnosis. These include peroxisome size, shape and number, catalase solubility, and ability to import PTS1 and PTS2 enzymes.

4.3.3 Molecular

Identification of *PEX* gene mutations in patients provides prognostic information and allows for carrier detection and expanded prenatal diagnostic options. In difficult cases, it may be the only means by which a diagnosis can be made (Zeharia et al. 2007; Steinberg et al. 2009; Ebberink et al. 2010; McMillan et al. 2012; Pierce et al. 2010).

Studies on more than 800 patients show that defects in *PEX1* are responsible for around 60–70 % of ZSD, and the majority of *PEX1* mutations encode PEX1p, Gly843Asp, and PEX1p.Ile700fs (Steinberg et al. 2004; Yik et al. 2009; Ebberink et al. 2011). In studies on more than 100 RCDP1 patients, 50 % of all *PEX7* mutations encode PEX7p.Leu292* (Braverman et al. 2002; Motley et al. 2002). The high frequency of these alleles relate to founder effects in patients of Northern European ancestry (Braverman et al. 2000; Collins and Gould 1999). In addition, a founder mutation in *PEX6* is responsible for a very high incidence of ZS in Saguenay-Lac-Saint-Jean region of Quebec (Levesque et al. 2012). Overall, mutations in five genes, *PEX1*, *PEX6*, *PEX26*, *PEX10*, and *PEX12*, account for more than 90 % of all ZSD alleles (Steinberg et al. 1993). Currently, *PEX* gene sequencing is being performed on a clinical basis using a hierarchal approach, or after the specific gene is determined by cDNA complementation of fibroblast lines (Steinberg et al. 2004; Ebberink et al. 2011). It is anticipated that next generation sequencing approaches will soon be replacing these techniques, and have already been made available (Levesque et al. 2012; Sun et al. 2012)

4.3.4 Prenatal Diagnosis

PBD can be identified prenatally in the first or second trimester of pregnancy, using cultured amniocytes or chorionic villus cells and measuring VLCFA, plasmalogen synthesis, and phytanic acid oxidation. Alternatively, if the *PEX* gene mutations are known, a molecular diagnosis can be made on these cells or on the pre-implantation embryo after an in vitro fertilization (preimplantation genetic diagnosis). Chondrodysplasia punctata has been observed on prenatal ultrasounds as early as 19 weeks gestation (Zwijnenburg et al. 2010).

4.3.5 Genotype–Phenotype Correlations in the PBD

In general, milder phenotypes are associated with *PEX* gene mutations that encode peroxins with residual function. These mutations are associated with higher residual peroxisome biochemical functions in blood and skin cells and better import of peroxisomal matrix proteins at the cellular level than *PEX* gene mutations that encode null alleles.

The high frequency of the PEX1p.Gly843Asp and PEX1p.Ile700fs alleles has enabled phenotype correlations. The presence of at least one PEX1p.Gly843Asp allele, which encodes a protein with residual function, predicts a non-severe phenotype. In fact, patients homozygous for PEX1p.Gly843Asp have some of the mildest phenotypes reported, including preservation of intellect (Poll-The et al. 2004; Rosewich et al. 2005; Majewski et al. 2011). In contrast, homozygosity for PEX1p.Ile700fs, a null allele, predicts Zellweger syndrome (Maxwell et al. 1999). Compound heterozygosity for PEX1p.Gly843Asp/PEX1p.Ile700fs typically presents as an intermediate phenotype. Still, the variations seen among the PEX1p.Gly843Asp homozygotes and PEX1p.Gly843Asp/PEX1p.Ile700fs compound heterozygotes suggest the additional influence of modifier genes on the phenotype. Other genotype–phenotype correlations in ZSD include the atypical phenotypes associated with mutations in the zinc finger domains of PEX2p, PEX10p, and PEX12p and C terminus of PEX16p.

For RCDP, milder phenotypes are associated with higher residual plasmalogen levels. Skeletal involvement may be absent; cataracts and mild cognitive delay may be the only manifestation of such residual plasmalogen levels. In *PEX7* deficiency, near normal plasmalogen levels with elevated phytanic acid levels are associated with a phenotype similar to adult Refsum disease. Interestingly, mutations reported in classic RCDP1 patients, including missense mutations, are predicted to disrupt the core ß-propeller structure of PEX7p and lead to loss of function (Braverman et al. 2002; Motley et al. 2002). A few missense mutations that are located on the surfaces of the ß-propeller structure are associated with residual import function and milder phenotypes. Reduced amounts of wild-type PEX7 protein are also associated with milder phenotypes (Braverman et al. 2002).

In addition to these direct genotype–phenotype correlations, investigations of ZSD patient cell lines have provided important insights into mammalian peroxisome biology. These include the initial observation of temperature sensitivity in cell lines with *PEX* missense alleles in which peroxisome functions could be recovered at 30 °C (Osumi et al. 2000). This paved the way for discovering chaperone drugs (see below), and as well, examining the loss of peroxisome functions at 40 °C, which helped to establish mutation pathology in cell lines with normal functions at 37 °C (Gootjes et al. 2004; Ebberink et al. 2012). Furthermore, the study of cultured patient cell lines with milder ZSD phenotypes show "mosaicism" in peroxisome import that is recapitulated in liver biopsy tissue (Roels et al. 2003). Mosaicism is defined by the observation of cell groups that import peroxisome matrix proteins, adjacent to cell groups that do not import (Steinberg et al. 2006). The mechanism(s) underlying this observation is unknown, but may involve inherent instability of defective peroxins with residual function, and some unknown component of the cellular milieu or cell cycle that stabilizes the defective peroxin. A better understanding of mosaicism could potentially lead to therapeutic interventions.

Characterization of a fibroblast cell line from the only patient reported with a *PEX5* null mutation, encoding homozygous PEX5p.Arg390*, demonstrated that PEX5p was required for the import of both PTS1 and PTS2 proteins in mammalian cells, a finding that was not anticipated from yeast peroxisome assembly models (Braverman et al. 1998; Otera et al. 1998). The finding that *PEX3* deficient cell lines without peroxisome membranes could reconstitute peroxisomes upon introduction of wild-type *PEX3* challenged the existing model that peroxisomes could arise only from fission of existing peroxisomes (Muntau et al. 2000). Characterization of the cell line from a single adult patient homozygous for the *PEX11ß* null mutation, PEX11ßp.Gln22*, showed that *PEX11γ* was able to partially compensate for the cellular defect at 37 °C, which may underlie the mild phenotype (Ebberink et al. 2012). In addition, the study of a patient cell line homozygous for PEX13p. Trp313Gly revealed that this substitution affects PEX13p homo-oligomerization and the import of PTS1, but not PTS2 proteins, thus dissecting the two import processes at the level of the peroxisome membrane (Krause et al. 2013). These examples highlight the contributions of PBD patients to our ongoing knowledge of peroxisome biology.

4.4 Management

4.4.1 Zellweger Spectrum

After extent of disease is determined through clinical, radiology, and laboratory examinations, patient management is multidisciplinary and based on surveillance of systems including nutrition, hearing, vision, neurologic, liver, renal, and adrenal assessments. Intervention is supportive and based on symptoms. Families may elect palliative care for those with severe disease. Genetic counseling and parent advocacy groups provide crucial assistance to parents in their decision making process. Prompt diagnosis of ZS is important, not only for providing proper care to the patient but also for counseling regarding family planning.

Interventions targeted to abnormal peroxisome biochemistry include restriction of peroxisome metabolites that accumulate and replacement of those that are deficient. Reports of these approaches are mostly anecdotal. Oral bile acid therapy improved hepatobiliary function in several infants with ZS (Maeda et al. 2002; Setchell et al. 1992). Patients have received DHA supplementation, with controversial clinical benefit (Noguer and Martinez 2010; Paker et al. 2010). Nevertheless, considering that ZSD patients have reduced synthesis of DHA and mature bile acids, these metabolites could be supplemented. Due to the benefit of dietary phytanic acid restriction in adult Refsum disease (Baldwin et al. 2010), this is often considered for ZSD patients after weaning from formula. A single ZSD patient received an orthotopic liver transplant (Van Maldergem et al. 2005) and another, a hepatocyte transplant (Sokal et al. 2003), resulting in decreased VLCFA and pipecolic acid and improved bile acid profiles. Long term objective clinical improvement has not yet been reported.

4.4.2 RCDP Spectrum

After extent of disease is determined, patient management is focused on nutrition, vision, cardiac, neurologic, and orthopedic assessment. Attention is also focused on pulmonary status to prevent respiratory complications, which can be lethal. Dietary interventions have included supplementation with the oral plasmalogen precursor, batyl alcohol. In a small case series, erythrocyte plasmalogen levels improved but clinical outcome did not (Das et al. 1992; Wilson et al. 1986). Considering that plasmalogens may be depots for DHA storage (Braverman and Moser 2012), DHA levels in RCDP children could be supplemented if deficient. Phytanic acid can be restricted in RCDP1 patients advancing to a regular diet.

Thus far, there is no evidence that any of the dietary interventions reported for ZSD and RCDP improve the ultimate course of the disease, but these observations would benefit by systematic, properly designed studies in order to make valid conclusions. Understanding tissue requirements for peroxisome functions will help distinguish which organs can respond to dietary interventions and which organs cannot, because of cell autonomous mechanisms of disease.

4.5 Challenges and Future Directions

Expanded newborn screening (NBS) for a variety of genetic disorders has been in place for more than a decade and continues to expand as new disorders are being included. Recently, a NBS protocol to detect disorders of VLCFA metabolism by measuring C26:0 lysophosphatidylcholine using existing LC/MSMS technology was piloted (Hubbard et al. 2009). It is anticipated that this screen will be included in NBS programs in the USA in the near future. Although the primary disease identification is X-ALD, most cases of ZSD, ACOX1, DBP, and CADDS will also be identified. For the latter group, this will enable more accurate incidence figures and help to delineate the disease spectrum. In addition, it will further challenge investigators to identify best management protocols and treatment options, as relatively asymptomatic newborns will be identified and physicians and families will want to know how to best manage this condition.

To design effective therapies, it is useful to know the natural history of the disorder, the pathophysiology, and molecular mechanisms of disease. For PBD and related peroxisome defects, there is an urgent need to systematically document the natural history in order to provide better prognostic information and best management practices. Natural history studies are also needed to determine efficacy of a clinical intervention in PBD, which has high clinical variability, and to develop reliable endpoints for clinical trials. Considering that these are rare diseases, an international effort is required to expedite this, and these studies are now underway (http://clinicaltrials.gov/ct2/show/NCT01668186?term=peroxisome&rank=1).

Molecular diagnosis of PBD patients will be improved by next gen sequencing approaches. Pathophysiological studies will depend on further investigations in animal models, which are also underway.

4.5.1 Drug Therapies Under Investigation

Reports from different investigators have shown that defective PEX proteins are amenable to therapy at the cellular level. For ZSD patient cell lines with different missense alleles, culture at 30 °C improves peroxisome biogenesis and peroxisome functions consistent with reduced peroxin misfolding at lower temperature. Peroxisome functions also recover when PEX6 is overexpressed in PEX1p.Gly843Asp cells, consistent with conformational rescue by a binding partner (Geisbrecht et al. 1998). Furthermore, recovery has also been shown in *Pex6*-deficient plants by either overexpression of *Pex5* (Zolman and Bartel 2004; Ratzel et al. 2010) or reduction of *Pex13* (Ratzel et al. 2010), suggesting that restoring the balance between import and export can suppress the effect of the original mutation. Finally, when ZSD fibroblasts, including those containing PEX1p.Gly843Asp, are cultured in 4-phenylbutyrate, peroxisome numbers increase two- to threefold and measurements of peroxisome enzymatic functions improve (Wei et al. 2000). Overall, these observations implicate peroxin stabilization, suppressor or modifier effects, as well as peroxisome proliferation as mechanisms that drugs could recapitulate.

Chaperones PEX1p.Gly843Asp, the most common defective PEX protein, has been studied by several investigators. PEX1p.Gly843Asp is markedly reduced (85–95 %) compared to wild-type *PEX1* in fibroblast lysates from patients homozygous for this mutation and increases two- to threefold at 30 °C (Walter et al. 2001). Since *PEX1* transcript level in these patients is normal (Maxwell et al. 1999), it is likely that PEX1p.Gly843Asp is misfolded and degraded. It was recently shown that peroxisome functions could be improved in PEX1p.Gly843Asp cell lines by nonspecifc chemical chaperones, as well as drugs that might act as more specific pharmacological chaperones (Zhang et al. 2010; Berendse et al. 2013). Chemical chaperones are small molecules that nonselectively stabilize mutant proteins by affecting the protein environment, while pharmacologic chaperones selectively bind target proteins and therefore require lower concentrations for effect. The latter include enzyme substrates or inhibitors that bind to and facilitate folding of non-native protein intermediates to their native state. One of the chemical chaperones identified, betaine, is now being used in a clinical trial (http://clinicaltrials.gov/ct2/show/NCT01838941?term=peroxisome&rank=2). The platform for drug screening utilized a patient cell line containing PEX1p.Gly843Asp and expressing a GFP-PTS1 reporter. A phenotype-based assay evaluated the redistribution of the reporter from the cytosol to the peroxisome, shown to accurately reflect recovery of peroxisome functions (Zhang et al. 2010). The endpoint of import recovery, downstream from the defective protein, provides for an unbiased screen that can identify any drug that can recover import. It is anticipated that a collection of drugs will be identified that can improve peroxisome functions in patient cell lines by different mechanisms. These can then be tested for clinical efficacy in animal models.

Nonsense Suppressors These are drugs that promote translational read-through of premature stop codons. During suppression, the amino acids inserted in place of a premature stop codon are likely donated by aminoacyl-tRNAs that base pair with two nucleotides in any position of the stop codon. They are exemplified by aminoglycoside antibiotics, such as gentamicin and geneticin (G418) and work best on stable nonsense transcripts. The ability of G418 to restore peroxisome assembly and biochemical functions was tested in patient cell lines with stable nonsense transcripts that encode PEX2p.Arg1119*, PEX2p.Arg125*, PEX12p.Arg180*, and PEX7p.Leu292* (Dranchak et al. 2011). G418 was able to recover peroxisome functions in the *PEX2* and *PEX12*, but not *PEX7*, mutant cell lines tested. Although aminoglycosides are too toxic for continuous therapy, these studies provide proof-of principal that nonsense suppressor therapies could be an effective treatment for a subgroup of ZSD.

Antioxidants Peroxisomes generate reactive oxygen and nitrogen species as byproducts of metabolism and simultaneously have various mechanisms to neutralize oxidative stress and maintain cellular redox balance (Fransen et al. 2011).

Fibroblast cultures and blood samples from patients with PBD and related single enzyme defects, as well as mouse models for these diseases, have shown increased levels of cellular oxidative stress markers, which could also contribute to disease pathogenesis (El-Bassyouni et al. 2012; Ferdinandusse et al. 2003; Muller et al. 2010). Interestingly, the mitochondrial abnormalities historically noted in tissues from ZS patients, including pleomorphic mitochondria with distorted cristae and variable reduction in activity of respiratory chain complexes, may reflect increased oxidative stress from peroxisome dysfunction (Fourcade et al. 2014). Studies in the *Abcd1* null mouse model have shown that oxidative damage to tissues can be relieved by treatment with anti-oxidants (Lopez-Erauskin et al. 2011). A clinical trial using antioxidants in X-ALD adults with adrenomyeloneuropathy is now underway (http://clinicaltrials.gov/ct2/show/NCT01495260?term=Xlinked+adrenoleukodystrophy&rank=7).

Plasmalogen Precursors Although peroxisomes are required for the committing steps of plasmalogen synthesis, the formation of alkylglycerols, the subsequent maturation of these compounds takes place in the ER (Braverman and Moser 2012). Since plasmalogen deficiency directly causes RCDP, the provision of alkylglycerols, or any of the downstream plasmalogen precursors, could theoretically halt progression of the disease. In *Pex7* deficient mouse models (Brites et al. 2011; Braverman et al. 2010), oral supplementation with the alkylglycerol, batyl alcohol, was shown to increase plasmalogen levels in somatic tissues, but the compound poorly enters the central and peripheral nervous system. More mature plasmalogen precursors are now being developed in hope of obtaining better recovery of brain plasmalogens (Wood et al. 2011).

Conclusions and Perspectives

The majority of patients with ZSD have a progressive disorder of ongoing peroxisome dysfunction. New phenotypes are emerging. Advancement in NBS will identify these patients at early points in their disease. The natural history of ZSD and RCDP disorders needs to be documented in order to propose better management plans and set clinical endpoints for treatment trials. Therapeutic interventions are now being studied at the bench and in animal models. If treatments are proposed, these need to be studied systematically and not as anecdotal cases, so that the community can benefit. Since these are rare diseases, international collaborations will be required. The identification of patients by physicians, followed by scientific investigation of biochemical, cellular, and molecular alterations has led to enormous insight into peroxisome disorders. This cooperative engagement that requires physicians, scientists and families has empowered this field and should continue to do so.

Acknowledgments CA is a PhD candidate in Human Genetics at McGill University, NB is funded by the Canadian Institutes of Health Research and Alberta Collaborative Research and Innovation Opportunities. AM is supported by the Peroxisomal Diseases Laboratory at the

Kennedy Krieger Institute and the Neurosciences Core C of the Intellectual and Developmental Disabilities Research Center, NIH/NICHD 5P30HD024061-22. The authors are grateful to all the physicians, scientists, and families who work together to advance knowledge and treatment of peroxisomal diseases.

References

Acharya BS, Ritwik P, Velasquez GM, Fenton SJ (2012) Medi-cal-dental findings and management of a child with infantile Refsum disease: a case report. Spec Care Dentist 32:112–117

Aubourg P, Scotto J, Rocchiccioli F, Feldmann-Pautrat D, Robain O (1986) Neonatal adrenoleukodystrophy. J Neurol Neurosurg Psychiatry 49:77–86

Baldwin EJ, Gibberd FB, Harley C, Sidey MC, Feher MD, Wierzbicki AS (2010) The effectiveness of long-term dietary therapy in the treatment of adult Refsum disease. J Neurol Neurosurg Psychiatry 81:954–957

Bams-Mengerink AM, Majoie CB, Duran M, Wanders RJ, Van Hove J, Scheurer CD, Barth PG, Poll-The BT (2006) MRI of the brain and cervical spinal cord in rhizomelic chondrodysplasia punc-tata. Neurology 66:798–803, discussion 789

Barkovich AJ, Peck WW (1997) MR of Zellweger syndrome. AJNR Am J Neuroradiol 18:1163–1170

Barth PG, Majoie CB, Gootjes J, Wanders RJ, Waterham HR, van der Knaap MS, de Klerk JB, Smeitink J, Poll-The BT (2004) Neuroimaging of peroxisome biogenesis disorders (Zellweger spectrum) with prolonged survival. Neurology 62:439–444

Berendse K, Ebberink MS, Ijlst L, Poll-The BT, Wanders RJ, Waterham HR (2013) Arginine improves peroxisome functioning in cells from patients with a mild peroxisome biogenesis disorder. Orphanet J Rare Dis 8:138

Bove KE, Heubi JE, Balistreri WF, Setchell KD (2004) Bile acid synthetic defects and liver disease: a comprehensive review. Pediatr Dev Pathol 7:315–334

Bowen P, Lee CS, Zellweger H, Lindenberg R (1964) A familial syndrome of multiple congenital defects. Bull Johns Hopkins Hosp 114:402–414

Braverman N, Moser A (2012) Functions of plasmalogen lipids in health and disease. Biochim Biophys Acta 1822(9):1442–1452

Braverman N, Dodt G, Gould SJ, Valle D (1998) An isoform of pex5p, the human PTS1 receptor, is required for the import of PTS2 proteins into peroxisomes. Hum Mol Genet 7:1195–1205

Braverman N, Steel G, Lin P, Moser A, Moser H, Valle D (2000) PEX7 gene structure, alternative transcripts, and evidence for a founder haplotype for the frequent RCDP allele, L292ter. Genomics 63:181–192

Braverman N, Chen L, Lin P, Obie C, Steel G, Douglas P, Chakraborty PK, Clarke JT, Boneh A, Moser A, Moser H, Valle D (2002) Mutation analysis of PEX7 in 60 probands with rhizomelic chondrodysplasia punctata and functional correlations of genotype with phenotype. Hum Mutat 20:284–297

Braverman N, Zhang R, Chen L, Nimmo G, Scheper S, Tran T, Chaudhury R, Moser A, Steinberg S (2010) A Pex7 hypomorphic mouse model for plasmalogen deficiency affecting the lens and skeleton. Mol Genet Metab 99:408–416

Brites P, Ferreira AS, da Silva TF, Sousa VF, Malheiro AR, Duran M, Waterham HR, Baes M, Wanders RJ (2011) Alkyl-glycerol rescues plasmalogen levels and pathology of ether-phospholipid deficient mice. PLoS One 6:e28539

Cacciagli P, Sutera-Sardo J, Borges-Correia A, Roux JC, Dorboz I, Desvignes JP, Badens C, Delepine M, Lathrop M, Cau P, Levy N, Girard N, Sarda P, Boespflug-Tanguy O, Villard L (2013) Mutations in BCAP31 cause a severe X-linked phenotype with deafness, dystonia, and central hypomyelination and disorganize the Golgi apparatus. Am J Hum Genet 93:579–586

Chang CC, South S, Warren D, Jones J, Moser AB, Moser HW, Gould SJ (1999) Metabolic control of peroxisome abundance. J Cell Sci 112(Pt 10):1579–1590

Cohen SM, Brown FR 3rd, Martyn L, Moser HW, Chen W, Kistenmacher M, Punnett H, de la Cruz ZC, Chan NR, Green WR (1983) Ocular histopathologic and biochemical studies of the cerebrohepatorenal syndrome (Zellweger's syndrome) and its relationship to neonatal adrenoleukodystrophy. Am J Ophthalmol 96:488–501

Collins CS, Gould SJ (1999) Identification of a common PEX1 mutation in Zellweger syndrome. Hum Mutat 14:45–53

Corzo D, Gibson W, Johnson K, Mitchell G, LePage G, Cox GF, Casey R, Zeiss C, Tyson H, Cutting GR, Raymond GV, Smith KD, Watkins PA, Moser AB, Moser HW, Steinberg SJ (2002) Contig-uous deletion of the X-linked adrenoleukodystrophy gene (ABCD1) and DXS1357E: a novel neonatal phenotype similar to peroxisomal biogenesis disorders. Am J Hum Genet 70:1520–1531

Danks DM, Tippett P, Adams C, Campbell P (1975) Cerebro-hepato-renal syndrome of Zellweger. J Pediatr 86:382–387

Das AK, Holmes RD, Wilson GN, Hajra AK (1992) Dietary ether lipid incorporation into tissue plasmalogens of humans and rodents. Lipids 27:401–405

Datta NS, Wilson GN, Hajra AK (1984) Deficiency of enzymes catalyzing the biosynthesis of glycerol-ether lipids in Zellweger syn-drome. A new category of metabolic disease involving the absence of peroxisomes. N Engl J Med 311:1080–1083

De Duve C, Baudhuin P (1966) Peroxisomes (microbodies and related particles). Physiol Rev 46:323–357

Demyer W, Zeman W, Palmer CG (1964) The face predicts the brain: diagnostic significance of median facial anomalies for holo-prosencephaly (arhinencephaly). Pediatrics 34:256–263

Distel B, Erdmann R, Gould SJ, Blobel G, Crane DI, Cregg JM, Dodt G, Fujiki Y, Goodman JM, Just WW, Kiel JA, Kunau WH, Lazarow PB, Mannaerts GP, Moser HW, Osumi T, Rachubinski RA, Roscher A, Subramani S, Tabak HF, Tsukamoto T, Valle D, van der Klei I, van Veldhoven PP, Veenhuis M (1996) A unified nomenclature for peroxisome biogenesis factors. J Cell Biol 135:1–3

Dodt G, Braverman N, Valle D, Gould SJ (1996) From ex-pressed sequence tags to peroxisome biogenesis disorder genes. Ann N Y Acad Sci 804:516–523

Dranchak PK, Di Pietro E, Snowden A, Oesch N, Braverman NE, Steinberg SJ, Hacia JG (2011) Nonsense suppressor therapies rescue peroxisome lipid metabolism and assembly in cells from patients with specific PEX gene mutations. J Cell Biochem 112:1250–1258

Ebberink MS, Csanyi B, Chong WK, Denis S, Sharp P, Mooijer PA, Dekker CJ, Spooner C, Ngu LH, De Sousa C, Wanders RJ, Fietz MJ, Clayton PT, Waterham HR, Ferdinandusse S (2010) Identification of an unusual variant peroxisome biogenesis disorder caused by mutations in the PEX16 gene. J Med Genet 47:608–615

Ebberink MS, Mooijer PA, Gootjes J, Koster J, Wanders RJ, Waterham HR (2011) Genetic classification and mutational spectrum of more than 600 patients with a Zellweger syndrome spectrum disorder. Hum Mutat 32:59–69

Ebberink MS, Koster J, Visser G, Spronsen F, Stolte-Dijkstra I, Smit GP, Fock JM, Kemp S, Wanders RJ, Waterham HR (2012) A novel defect of peroxisome division due to a homozygous non-sense mutation in the PEX11beta gene. J Med Genet 49:307–313

El-Bassyouni HT, Abdel Maksoud SA, Salem FA, Badr El-Deen R, Abdel Aziz H, Thomas MM (2012) Evidence of oxidative stress in peroxisomal disorders. Singapore Med J 53:608–614

Ferdinandusse S, Denis S, Clayton PT, Graham A, Rees JE, Al-len JT, McLean BN, Brown AY, Vreken P, Waterham HR, Wan-ders RJ (2000) Mutations in the gene encoding peroxisomal alpha-methylacyl-CoA racemase cause adult-onset sensory motor neuropathy. Nat Genet 24:188–191

Ferdinandusse S, Finckh B, de Hingh YC, Stroomer LE, Denis S, Kohlschutter A, Wanders RJ (2003) Evidence for increased oxidative stress in peroxisomal D-bifunctional protein deficiency. Mol Genet Metab 79:281–287

Ferdinandusse S, Kostopoulos P, Denis S, Rusch H, Overmars H, Dillmann U, Reith W, Haas D, Wanders RJ, Duran M, Marziniak M (2006) Mutations in the gene encoding peroxisomal sterol carrier protein X (SCPx) cause leukencephalopathy with dystonia and motor neuropathy. Am J Hum Genet 78:1046–1052

Fourcade S, Lopez-Erauskin J, Ruiz M, Ferrer I, Pujol A (2014) Mitochondrial dysfunction and oxidative damage cooperatively fuel axonal degeneration in X-linked adrenoleukodystrophy. Biochimie 98:143–149

Fransen M (2012) Peroxisome dynamics: molecular players, mechanisms, and (Dys)functions. ISRN Cell Biol 2012:24 p

Fransen M, Nordgren M, Wang B, Apanasets O (2011) Role of peroxisomes in ROS/RNS-metabolism: implications for human disease. Biochim Biophys Acta 1822:1363–1373

Fujiki Y, Okumoto K, Kinoshita N, Ghaedi K (2006) Lessons from peroxisome-deficient Chinese hamster ovary (CHO) cell mutants. Biochim Biophys Acta 1763:1374–1381

Geisbrecht BV, Collins CS, Reuber BE, Gould SJ (1998) Disruption of a PEX1-PEX6 interaction is the most common cause of the neurologic disorders Zellweger syndrome, neonatal adrenoleu-kodystrophy, and infantile Refsum disease. Proc Natl Acad Sci USA 95:8630–8635

Gilbert EF, Opitz JM, Spranger JW, Langer LO Jr, Wolfson JJ, Viseskul C (1976) Chondrodysplasia punctata–rhizomelic form. Pathologic and radiologic studies of three infants. Eur J Pediatr 123:89–109

Gilchrist KW, Gilbert EF, Goldfarb S, Goll U, Spranger JW, Opitz JM (1976) Studies of malformation syndromes of man XIB: the cerebro-hepato-renal syndrome of Zellweger: comparative pathology. Eur J Pediatr 121:99–118

Goldfischer S, Moore CL, Johnson AB, Spiro AJ, Valsamis MP, Wisniewski HK, Ritch RH, Norton WT, Rapin I, Gartner LM (1973) Peroxisomal and mitochondrial defects in the cerebro-hepato-renal syndrome. Science 182:62–64

Gootjes J, Schmohl F, Mooijer PA, Dekker C, Mandel H, Topcu M, Huemer M, Von Schutz M, Marquardt T, Smeitink JA, Waterham HR, Wanders RJ (2004) Identification of the molecular defect in patients with peroxisomal mosaicism using a novel method involving culturing of cells at 40 degrees C: implications for other inborn errors of metabolism. Hum Mutat 24:130–139

Heymans HS, Oorthuys JW, Nelck G, Wanders RJ, Schutgens RB (1985) Rhizomelic chondrodysplasia punctata: another peroxisomal disorder. N Engl J Med 313:187–188

Horn MA, van den Brink DM, Wanders RJ, Duran M, Poll-The BT, Tallaksen CM, Stokke OH, Moser H, Skjeldal OH (2007) Phenotype of adult Refsum disease due to a defect in peroxin 7. Neurology 68:698–700

Hubbard WC, Moser AB, Liu AC, Jones RO, Steinberg SJ, Lorey F, Panny SR, Vogt RF Jr, Macaya D, Turgeon CT, Tortorelli S, Raymond GV (2009) Newborn screening for X-linked adrenoleu-kodystrophy (X-ALD): validation of a combined liquid chromatog-raphy-tandem mass spectrometric (LC-MS/MS) method. Mol Genet Metab 97:212–220

Huffnagel IC, Clur SA, Bams-Mengerink AM, Blom NA, Wanders RJ, Waterham HR, Poll-The BT (2013) Rhizomelic chondro-dysplasia punctata and cardiac pathology. J Med Genet 50:419–424

Iwasa M, Yamagata T, Mizuguchi M, Itoh M, Matsumoto A, Hironaka M, Honda A, Momoi MY, Shimozawa N (2012) Contiguous ABCD1 DXS1357E deletion syndrome: report of an autopsy case. Neuropathology 33:292–298

Kelley RI (1983) Review: the cerebrohepatorenal syndrome of Zellweger, morphologic and metabolic aspects. Am J Med Genet 16:503–517

Khanna AJ, Braverman NE, Valle D, Sponseller PD (2001) Cervical stenosis secondary to rhizomelic chondrodysplasia punctata. Am J Med Genet 99:63–66

Krause C, Rosewich H, Woehler A, Gartner J (2013) Functional analysis of PEX13 mutation in a Zellweger syndrome spectrum patient reveals novel homooligomerization of PEX13 and its role in human peroxisome biogenesis. Hum Mol Genet 22:3844–3857

Kulkarni KS, Baranano KW, Lin DD, Raymond GV (2011) Contrast enhancement of brainstem tracts in Zellweger spectrum disorder: evidence of inflammatory demyelination? Neuropediatrics 42:32–34

Levesque S, Morin C, Guay SP, Villeneuve J, Marquis P, Yik WY, Jiralerspong S, Bouchard L, Steinberg S, Hacia JG, Dewar K, Braverman NE (2012) A founder mutation in the PEX6 gene is re-sponsible for increased incidence of Zellweger syndrome in a French Canadian population. BMC Med Genet 13:72

Lopez-Erauskin J, Fourcade S, Galino J, Ruiz M, Schluter A, Naudi A, Jove M, Portero-Otin M, Pamplona R, Ferrer I, Pujol A (2011) Antioxidants halt axonal degeneration in a mouse model of X-adrenoleukodystrophy. Ann Neurol 70:84–92

MacCollin M, De Vivo DC, Moser AB, Beard M (1990) Ataxia and peripheral neuropathy: a benign variant of peroxisome dysgenesis. Ann Neurol 28:833–836

Maeda K, Kimura A, Yamato Y, Nittono H, Takei H, Sato T, Mitsubuchi H, Murai T, Kurosawa T (2002) Oral bile acid treatment in two Japanese patients with Zellweger syndrome. J Pediatr Gastroenterol Nutr 35:227–230

Majewski J, Wang Z, Lopez I, Al Humaid S, Ren H, Racine J, Bazinet A, Mitchel G, Braverman N, Koenekoop RK (2011) A new ocular phenotype associated with an unexpected but known systemic disorder and mutation: novel use of genomic diagnostics and exome sequencing. J Med Genet 48:593–596

Maxwell MA, Nelson PV, Chin SJ, Paton BC, Carey WF, Crane DI (1999) A common PEX1 frameshift mutation in patients with disorders of peroxisome biogenesis correlates with the severe Zellweger syndrome phenotype. Hum Genet 105:38–44

McMillan HJ, Worthylake T, Schwartzentruber J, Gottlieb CC, Lawrence SE, Mackenzie A, Beaulieu CL, Mooyer PA, Wanders RJ, Majewski J, Bulman DE, Geraghty MT, Ferdinandusse S, Boycott KM (2012) Specific combination of compound heterozygous mutations in 17beta-hydroxysteroid dehydrogenase type 4 (HSD17B4) defines a new subtype of D-bifunctional protein deficiency. Orphanet J Rare Dis 7:90

Michelakakis HM, Zafeiriou DI, Moraitou MS, Gootjes J, Wanders RJ (2004) PEX1 deficiency presenting as Leber congenital amaurosis. Pediatr Neurol 31:146–149

Mignarri A, Vinciguerra C, Giorgio A, Ferdinandusse S, Waterham H, Wanders R, Bertini E, Dotti MT, Federico A (2012) Zellweger spectrum disorder with mild phenotype caused by PEX2 gene mutations. JIMD Rep 6:43–46

Moser AE, Singh I, Brown FR III, Solish GI, Kelley RI, Benke PJ, Moser HW (1984) The cerebrohepatorenal (Zellweger) syndrome. Increased levels and impaired degradation of very-long-chain fatty acids and their use in prenatal diagnosis. N Engl J Med 310:1141–1146

Moser AB, Rasmussen M, Naidu S, Watkins PA, McGuinness M, Hajra AK, Chen G, Raymond G, Liu A, Gordon D et al (1995) Phenotype of patients with peroxisomal disorders subdivided into six-teen complementation groups. J Pediatr 127:13–22

Motley AM, Brites P, Gerez L, Hogenhout E, Haasjes J, Benne R, Tabak HF, Wanders RJ, Waterham HR (2002) Mutational spectrum in the PEX7 gene and functional analysis of mutant alleles in 78 patients with rhizomelic chondrodysplasia punctata type 1. Am J Hum Genet 70:612–624

Muller CC, Nguyen TH, Ahlemeyer B, Meshram M, Santram-purwala N, Cao S, Sharp P, Fietz PB, Baumgart-Vogt E, Crane DI (2010) PEX13 deficiency in mouse brain as a model of Zellweger syndrome: abnormal cerebellum formation, reactive gliosis and oxida-tive stress. Dis Model Mech 4:104–119

Muntau AC, Mayerhofer PU, Paton BC, Kammerer S, Roscher AA (2000) Defective peroxisome membrane synthesis due to mutations in human PEX3 causes Zellweger syndrome, complementation group G. Am J Hum Genet 67:967–975

Noguer MT, Martinez M (2010) Visual follow-up in peroxisomal-disorder patients treated with docosahexaenoic acid ethyl ester. Invest Ophthalmol Vis Sci 51:2277–2285

Nordgren M, Wang B, Apanasets O, Fransen M (2013) Peroxisome degradation in mammals: mechanisms of action, recent advances, and perspectives. Front Physiol 4:145

Opitz J, Rhein GZ, Vitale L, Shahidi N, Howe J, Chou S, Shanklin D, Sybers H, Dood A, Gerritsen T (1969) The Zellweger syndrome (cerebro-hepato-renal syndrome). Birth Defects 2:1–15844

Osumi T, Imamura A, Tsukamoto T, Fujiwara C, Hashiguchi N, Shimozawa N, Suzuki Y, Kondo N (2000) Temperature sensitivity in peroxisome assembly processes characterizes milder forms of pe-roxisome biogenesis disorders. Cell Biochem Biophys 32(Spring):165–170

Oswald G, Lawson C, Raymond G, Golden WC, Braverman N (2011) Rhizomelic chondrodysplasia punctata type 1 and fulminant neonatal respiratory failure, a case report and discussion of patho-physiology. Am J Med Genet A 155A:3160–3163

Otera H, Okumoto K, Tateishi K, Ikoma Y, Matsuda E, Nishi-mura M, Tsukamoto T, Osumi T, Ohashi K, Higuchi O, Fujiki Y (1998) Peroxisome targeting signal type 1 (PTS1) receptor is involved in import of both PTS1 and PTS2: studies with PEX5-defective CHO cell mutants. Mol Cell Biol 18:388–399

Paker AM, Sunness JS, Brereton NH, Speedie LJ, Albanna L, Dharmaraj S, Moser AB, Jones RO, Raymond GV (2010) Docosahexaenoic acid therapy in peroxisomal diseases: results of a double-blind, randomized trial. Neurology 75:826–830

Passarge E, McAdams AJ (1967) Cerebro-hepato-renal syndrome. A newly recognized hereditary disorder of multiple congenital defects, including sudanophilic leukodystrophy, cirrhosis of the liver, and polycystic kidneys. J Pediatr 71:691–702

Pierce SB, Walsh T, Chisholm KM, Lee MK, Thornton AM, Fiumara A, Opitz JM, Levy-Lahad E, Klevit RE, King MC (2010) Mutations in the DBP-deficiency protein HSD17B4 cause ovarian dysgenesis, hearing loss, and ataxia of Perrault Syndrome. Am J Hum Genet 87:282–288

Poll-The BT, Gootjes J, Duran M, De Klerk JB, Wenniger-Prick LJ, Admiraal RJ, Waterham HR, Wanders RJ, Barth PG (2004) Peroxisome biogenesis disorders with prolonged survival: phenotypic expression in a cohort of 31 patients. Am J Med Genet A 126A:333–338

Poulos A, Sheffield L, Sharp P, Sherwood G, Johnson D, Beckman K, Fellenberg AJ, Wraith JE, Chow CW, Usher S et al (1988) Rhizomelic chondrodysplasia punctata: clinical, pathologic, and biochemical findings in two patients. J Pediatr 113:685–690

Powers JM, Moser HW, Moser AB, Upshur JK, Bradford BF, Pai SG, Kohn PH, Frias J, Tiffany C (1985) Fetal cerebrohepato-renal (Zellweger) syndrome: dysmorphic, radiologic, biochemical, and pathologic findings in four affected fetuses. Hum Pathol 16:610–620

Raas-Rothschild A, Wanders RJ, Mooijer PA, Gootjes J, Waterham HR, Gutman A, Suzuki Y, Shimozawa N, Kondo N, Eshel G, Espeel M, Roels F, Korman SH (2002) A PEX6-defective peroxi-somal biogenesis disorder with severe phenotype in an infant, versus mild phenotype resembling Usher syndrome in the affected parents. Am J Hum Genet 70:1062–1068

Ratzel SE, Lingard MJ, Woodward AW, Bartel B (2010) Reducing PEX13 expression ameliorates physiological defects of late-acting peroxin mutants. Traffic 12:121–134

Regal L, Ebberink MS, Goemans N, Wanders RJ, De Meirleir L, Jaeken J, Schrooten M, Van Coster R, Waterham HR (2010) Mutations in PEX10 are a cause of autosomal recessive ataxia. Ann Neurol 68:259–263

Rizzo WB (1998) Inherited disorders of fatty alcohol metabolism. Mol Genet Metab 65:63–73

Roels F, Saudubray JM, Giros M, Mandel H, Eyskens F, Sara-cibar N, Atares Pueyo B, Prats JM, De Prest B, De Preter K, Pineda M, Krystkowiak P, Gootjes J, Wanders RJ, Espeel M, Poll-The BT (2003) Peroxisome mosaics. Adv Exp Med Biol 544:97–106

Rosewich H, Ohlenbusch A, Gartner J (2005) Genetic and clini-cal aspects of Zellweger spectrum patients with PEX1 mutations. J Med Genet 42:e58

Schrader M, Bonekamp NA, Islinger M (2012) Fission and proliferation of peroxisomes. Biochim Biophys Acta 1822:1343–1357

Setchell KD, Bragetti P, Zimmer-Nechemias L, Daugherty C, Pelli MA, Vaccaro R, Gentili G, Distrutti E, Dozzini G, Morelli A et al (1992) Oral bile acid treatment and the patient with Zellweger syn-drome. Hepatology 15:198–207

Sevin C, Ferdinandusse S, Waterham HR, Wanders RJ, Aubourg P (2011) Autosomal recessive cerebellar ataxia caused by muta-tions in the PEX2 gene. Orphanet J Rare Dis 6:8

Shimozawa N, Tsukamoto T, Suzuki Y, Orii T, Fujiki Y (1992) Animal cell mutants represent two complementation groups of peroxi-some-defective Zellweger syndrome. J Clin Invest 90:1864–1870

Smith DW, Opitz JM, Inhorn SL (1965) A syndrome of multiple developmental defects including polycystic kidneys and intrahepatic biliary dysgenesis in 2 siblings. J Pediatr 67:617–624

Sokal EM, Smets F, Bourgois A, Van Maldergem L, Buts JP, Reding R, Bernard Otte J, Evrard V, Latinne D, Vincent MF, Moser A, Soriano HE (2003) Hepatocyte transplantation in a 4-year-old girl with peroxisomal biogenesis disease: technique, safety, and metabolic follow-up. Transplantation 76:735–738

Steinberg SJ, Raymond GV, Braverman NE, Moser AB (1993) Peroxisome biogenesis disorders, Zellweger syndrome spectrum. In: Pagon RA, Adam MP, Bird TD, Dolan CR, Fong CT, Smith RJH, Stephens K, editors. GeneReviews® [Internet]. Seattle (WA): University of Washington, Seattle; [updated 2012 May 10]

Steinberg S, Chen L, Wei L, Moser A, Moser H, Cutting G, Braverman N (2004) The PEX Gene Screen: molecular diagnosis of peroxisome biogenesis disorders in the Zellweger syndrome spectrum. Mol Genet Metab 83:252–263

Steinberg SJ, Dodt G, Raymond GV, Braverman NE, Moser AB, Moser HW (2006) Peroxisome biogenesis disorders. Biochim Biophys Acta 1763:1733–1748

Steinberg S, Jones R, Tiffany C, Moser A (2008) Investigational methods for peroxisomal disorders. Curr Protoc Hum Genet, Chapter 17, Unit 17.6

Steinberg SJ, Snowden A, Braverman NE, Chen L, Watkins PA, Clayton PT, Setchell KD, Heubi JE, Raymond GV, Moser AB, Moser HW (2009) A PEX10 defect in a patient with no detectable de-fect in peroxisome assembly or metabolism in cultured fibroblasts. J Inherit Metab Dis 32:109–119

Sugarman GI (1974) Chondrodysplasia punctata (rhizomelic type): case report and pathologic findings. Birth Defects Orig Artic Ser 10:334–340

Sun Y, Wang L, Wei X, Zhu Q, Yang Y, Lan Z, Qu N, Chu Y, Wang Y, Yang S, Liang Y, Wang W, Yi X (2012) Analysis of a Chinese pedigree with Zellweger syndrome reveals a novel PEX1 mutation by next-generation sequencing. Clin Chim Acta 417:57–61

Tran D, Greenhill W, Wilson S (2011) Infantile refsum disease with enamel defects: a case report. Pediatr Dent 33:266–270

van den Brink DM, Brites P, Haasjes J, Wierzbicki AS, Mitchell J, Lambert-Hamill M, de Belleroche J, Jansen GA, Waterham HR, Wanders RJ (2003) Identification of PEX7 as the second gene in-volved in Refsum disease. Am J Hum Genet 72:471–477

Van Maldergem L, Moser AB, Vincent MF, Roland D, Reding R, Otte JB, Wanders RJ, Sokal E (2005) Orthotopic liver transplan-tation from a living-related donor in an infant with a peroxisome bio-genesis defect of the infantile Refsum disease type. J Inherit Metab Dis 28:593–600

van Woerden CS, Groothoff JW, Wijburg FA, Duran M, Wanders RJ, Barth PG, Poll-The BT (2006) High incidence of hyperox-aluria in generalized peroxisomal disorders. Mol Genet Metab 88:346–350

Viseskul C, Opitz JM, Spranger JW, Hartmann HA, Gilbert EF (1974) Pathology of chondrodysplasia punctata rhizomelic type. Birth Defects Orig Artic Ser 10:327–333

Walter C, Gootjes J, Mooijer PA, Portsteffen H, Klein C, Waterham HR, Barth PG, Epplen JT, Kunau WH, Wanders RJ, Dodt G (2001) Disorders of peroxisome biogenesis due to mutations in PEX1: phenotypes and PEX1 protein levels. Am J Hum Genet 69:35–48

Waterham HR, Koster J, van Roermund CW, Mooyer PA, Wanders RJ, Leonard JV (2007) A lethal defect of mitochondrial and peroxisomal fission. N Engl J Med 356:1736–1741

Wei H, Kemp S, McGuinness MC, Moser AB, Smith KD (2000) Pharmacological induction of peroxisomes in peroxisome biogenesis disorders. Ann Neurol 47:286–296

Wells TR, Landing BH, Bostwick FH (1992) Studies of vertebral coronal cleft in rhizomelic chondrodysplasia punctata. Pediatr Pathol 12:593–600

White AL, Modaff P, Holland-Morris F, Pauli RM (2003) Natural history of rhizomelic chondrodysplasia punctata. Am J Med Genet A 118A:332–342

Wierzbicki AS, Lloyd MD, Schofield CJ, Feher MD, Gibberd FB (2002) Refsum's disease: a peroxisomal disorder affecting phytanic acid alpha-oxidation. J Neurochem 80:727–735

Wilson GN, Holmes RG, Custer J, Lipkowitz JL, Stover J, Datta N, Hajra A (1986) Zellweger syndrome: diagnostic assays, syndrome delineation, and potential therapy. Am J Med Genet 24:69–82

Wood PL, Khan MA, Smith T, Ehrmantraut G, Jin W, Cui W, Braverman NE, Goodenowe DB (2011) In vitro and in vivo plas-malogen replacement evaluations in rhizomelic chrondrodysplasia punctata and Pelizaeus-Merzbacher disease using PPI-1011, an ether lipid plasmalogen precursor. Lipids Health Dis 10:182

Yik WY, Steinberg SJ, Moser AB, Moser HW, Hacia JG (2009) Identification of novel mutations and sequence variation in the Zellweger syndrome spectrum of peroxisome biogenesis disorders. Hum Mutat 30:E467–E480

Yu TW, Chahrour MH, Coulter ME, Jiralerspong S, Okamura-Ikeda K, Ataman B, Schmitz-Abe K, Harmin DA, Adli M, Malik AN, D'Gama AM, Lim ET, Sanders SJ, Mochida GH, Partlow JN, Sunu CM, Felie JM, Rodriguez J, Nasir RH, Ware J, Joseph RM, Hill RS, Kwan BY, Al-Saffar M, Mukaddes NM, Hashmi A, Balkhy S, Gascon GG, Hisama FM, LeClair E, Poduri A, Oner O, Al-Saad S, Al-Awadi SA, Bastaki L, Ben-Omran T, Teebi AS, Al-Gazali L, Eapen V, Stevens CR, Rappaport L, Gabriel SB, Markianos K, State MW, Greenberg ME, Taniguchi H, Braverman NE, Morrow EM, Walsh CA (2013) Using whole-exome sequencing to identify inherited causes of autism. Neuron 77:259–273

Zeharia A, Ebberink MS, Wanders RJ, Waterham HR, Gutman A, Nissenkorn A, Korman SH (2007) A novel PEX12 mutation identified as the cause of a peroxisomal biogenesis disorder with mild clinical phenotype, mild biochemical abnormalities in fibroblasts and a mosaic catalase immunofluorescence pattern, even at 40 degrees C. J Hum Genet 52:599–606

Zellweger H, Maertens P, Superneau D, Wertelecki W (1988) History of the cerebrohepatorenal syndrome of Zellweger and other peroxisomal disorders. South Med J 81:357–364

Zhang R, Chen L, Jiralerspong S, Snowden A, Steinberg S, Braverman N (2010) Recovery of PEX1-Gly843Asp peroxisome dys-function by small-molecule compounds. Proc Natl Acad Sci USA 107:5569–5574

Zolman BK, Bartel B (2004) An Arabidopsis indole-3-butyric ac-id-response mutant defective in PEROXIN6, an apparent ATPase im-plicated in peroxisomal function. Proc Natl Acad Sci USA 101:1786–1791

Zwijnenburg PJ, Deurloo KL, Waterham HR, Meijers-Heijboer EJ, van Vugt JM, Tan-Sindhunata MB (2010) Second trimester prenatal diagnosis of rhizomelic chondrodysplasia punctata type 1 on ultrasound findings. Prenat Diagn 30:162–164

Molecular Basis for Peroxisome Biogenesis Disorders

5

Y. Fujiki, K. Okumoto, S. Mukai, and S. Tamura

Abstract

The functional importance of peroxisomes in humans is highlighted by peroxisome-deficient peroxisome biogenesis disorders (PBDs) such as Zellweger syndrome (ZS), autosomal recessive, and progressive disorders characterized by loss of multiple peroxisomal metabolic functions and defects in peroxisome assembly, consisting of 13 complementation groups (CGs). Two mutually distinct but complementary approaches, forward genetic approach using more than a dozen CGs of peroxisome-deficient Chinese hamster ovary (CHO) cell mutants and the homology search by screening the human expressed sequence tag (EST) database using yeast peroxin (*PEX*) genes, have been taken in order to isolate mammalian *PEX* genes. Search for pathogenic genes responsible for PBDs of all 13 CGs is now accomplished. Gene defects of peroxins required for both membrane assembly and matrix protein import are identified: ten mammalian pathogenic peroxins, Pex1p, Pex2p, Pex5p, Pex6p, Pex7p, Pex10p, Pex12p, Pex13p, Pex14p, and Pex26p, for 10 CGs of PBDs, are required for matrix protein import; three, Pex3p, Pex16p, and Pex19p, are essential for peroxisome membrane assembly and responsible for the most severe ZS in PBDs of three CGs, 12, 9, and 14, respectively; *PEX11β* mutation causes dysmorphogenesis of peroxisomes in ZS-like phenotype of CG16. Patients with severe ZS with defects of *PEX3*, *PEX16*, and *PEX19* tend to carry severe mutation such as nonsense mutations, frameshifts, and deletions. Prenatal DNA diagnosis using *PEX* genes is now possible for PBDs of all 13 CGs.

Keywords

Peroxisome biogenesis disorders • Membrane biogenesis • Peroxins • CHO cell mutants • Membrane protein topogenesis • Classes I and II import pathways

Y. Fujiki (✉) • K. Okumoto • S. Mukai • S. Tamura
Faculty of Sciences, Department of Biology, Kyushu University Graduate School, Fukuoka 812-8581, Japan
e-mail: yfujiki@kyudai.jp

C. Brocard and A. Hartig (eds.), *Molecular Machines Involved in Peroxisome Biogenesis and Maintenance*, DOI 10.1007/978-3-7091-1788-0_5, © Springer-Verlag Wien 2014

Abbreviations

CG	Complementation group
CHO	Chinese hamster ovary
ER	Endoplasmic reticulum
PBD	Peroxisome biogenesis disorder
PMP	Peroxisomal membrane protein
PTS	Peroxisomal targeting signal

5.1 Introduction

5.1.1 Peroxisome Functions

Peroxisome is a single-membrane-bounded ubiquitous organelle containing a hundred different enzymes that catalyze various metabolic pathways, including β-oxidation of very long-chain fatty acids, the synthesis of ether lipids such as plasmalogens and bile-acid metabolism (Wanders and Waterham 2006). Peroxisome was discovered as a microbody in 1954 and functionally named peroxisomes in 1965. In all types of cells, peroxisome was defined to contain one or more enzymes that use molecular oxygen to remove hydrogen atoms and form hydrogen peroxide from organic substrates. Catalase, a typical marker enzyme of peroxisomal matrix, degrades hydrogen peroxide.

5.2 Peroxisome Biogenesis Disorders

5.2.1 Biogenesis

Over the past three decades, much has been learned about how peroxisomes arise (Kunau 2006). The biogenesis of this organelle can be viewed as the integration of multiple processes: peroxisomal membrane biosynthesis, import of matrix proteins, and peroxisome growth and division (Lazarow and Fujiki 1985; Sacksteder and Gould 2000). The phospholipids of the peroxisome membrane are synthesized in the endoplasmic reticulum (ER) and transported to the peroxisome membrane by an unknown mechanism (Lazarow 2003). Newly synthesized peroxisomal membrane proteins (PMPs) can be targeted directly from the cytoplasm to peroxisomes (Matsuzaki and Fujiki 2008; Sparkes et al. 2005) or travel to peroxisomes via the ER (Tabak et al. 2003). The mechanisms responsible for the targeting of these proteins to the peroxisomal membrane are still less understood (Matsuzaki and Fujiki 2008), as compared to those underlying the matrix protein import (Platta and Erdmann 2007). What is clear is that PMPs do not use the sorting machinery of peroxisomal matrix proteins but follow an alternative pathway (Subramani 1998).

Such distinct protein import defects are best observed in yeast strains deficient in peroxisome biogenesis (Baerends et al. 1996; Götte et al. 1998; Hettema et al. 2000; Otzen et al. 2004; Shimozawa et al. 1998), Chinese hamster ovary (CHO) cell mutants (Fujiki et al. 2006b), and fibroblasts from patients with peroxisome biogenesis disorders (PBDs) (Matsumoto et al. 2003b; Weller et al. 2003; Fujiki 2011).

5.2.2 Abnormalities in Patients with PBDs

Inborn errors of peroxisomes are classified into two categories: PBDs and single-enzyme deficiencies (Gould et al. 2001; Fujiki 2003). Generalized PBDs include cerebrohepatorenal Zellweger syndrome (ZS), the most severe phenotype; neonatal adrenoleukodystrophy (NALD), the intermediate; and infantile Refsum disease (IRD), the least severe phenotype. Rhizomelic chondrodysplasia punctata (RCDP) is a partial PBD.

Biochemical abnormalities of PBDs include accumulation of very long-chain fatty acids (Singh et al. 1984), intermediate substances of bile acid synthesis (trihydroxycholestanoic acid, dihydroxycholestanoic acid), and branched-chain fatty acids (pristanic and phytanic acid); deficiency of plasmalogens and docosahexaenoic acid; and absent or severely decreased peroxisomes (Wanders and Waterham 2006). All phenotypes of PBDs show these findings; however, patients with milder phenotypes tend to have less severe abnormalities. Deficiency of peroxisomes in ZS was first identified in 1973 (Goldfischer et al. 1973), and then the accumulation of very long-chain fatty acids was defined in 1982. Peroxisomal membrane remnant structures called "ghosts" can be detected in many PBD cell lines (Gould et al. 2001; Fujiki 2003); several PBD cell lines lack ghosts, indicative of impaired peroxisomal membrane assembly (see below). In contrast to fibroblasts from ZS patients, cells from patients with milder phenotypes such as NALD and IRD are restored in peroxisome biogenesis at a lower cell-culture temperature, 30 °C (Imamura et al. 1998b). Such temperature sensitivity (*ts*) is useful for predicting the clinical severity. The PBDs are caused by a defect in *PEX* genes encoding peroxins, protein factors required for peroxisome biogenesis, and import of peroxisomal proteins (see below).

5.3 Genetic Heterogeneity

5.3.1 Complementation Group

Fused cells of PBD patient-derived cells such as skin fibroblasts each with the normal control cells show morphologically and biochemically normal peroxisomes, thereby indicating that lesions of alleles in the PBD patients are recessive. Genetic heterogeneity is seen in subjects with PBDs. Thirteen complementation groups (CGs) have been identified, mainly by pair-wise combination of cell-fusion analysis of the patient-derived fibroblasts (Fujiki et al. 2006b; Weller et al. 2003; Fujiki

Table 5.1 Complementation groups (CGs) and *PEX* genes of peroxisome deficiencies

Gene	Chromosome location	CG US/EU	CG Japan	Phenotype	CHO mutants	Peroxisome ghosts	Peroxin (kDa)	Characteristics
PEX1	*7q21.2*	1	E	ZS, NALD[a], IRD[a]	Z24, ZP107	+	143	AAA family
PEX2	*8q21.1*	10	F	ZS, IRD[a]	Z65	+	35	PMP, RING
PEX3	*6q24.2*	12	G	ZS	ZPG208	–	42	PMP, PMP-DP
PEX5	*12p13.3*	2		ZS, NALD	ZP105[a], ZP139	+	68	PTS1 receptor, TPR family
PEX6	*6p21.1*	4(6)	C	ZS, NALD[a]	ZP92	+	104	AAA family
PEX7	*6q23.3*	11	R	RCDP	ZPG207	+	36	PTS2 receptor, WD motif
PEX10	*1p36.32*	7(5)	B	ZS, NALD		+	37	PMP, RING
PEX11β	*1q21.1*	16		ZS		+	28	PMP
PEX12	*17q21.1*	3		ZS, NALD, IRD	ZP109	+	40	PMP, RING
PEX13	*2p16.1*	13	H	ZS, NALD[a]	ZP128	+	44	PMP, PTS1-DP, SH3
PEX14	*1p36.22*	15	K	ZS	ZP110	+	41	PMP, PTS1-DP, PTS2-DP
PEX16	*11p11.11*	9	D	ZS		–	39	PMP, PMP-DP
PEX19	*1q23.1*	14	J	ZS	ZP119	–	33	CAAX motif, PMP receptor
PEX26	*22q11.21*	8	A	ZS, NALD[a], IRD[a]	ZP124, ZP167	+	34	PMP, Pex1p–Pex6p recruiter

[a]Temperature-sensitive phenotype

ZS Zellweger syndrome, *IRD* infantile Refsum disease, *NALD* neonatal adrenoleukodystrophy, *RCDP* rhizomelic chondrodysplasia punctata, *DP* docking protein, *PMP* peroxisome membrane protein, *TPR* tetratricopeptide repeat

2011) (Table 5.1). Upon cell fusion between cells of mutually distinct CGs, numerous peroxisomes will be formed and impaired peroxisomal functions are restored in the hybrids. Thereby, more than 13 genes are likely to be involved in human peroxisome biogenesis. More than half the numbers of PBD patients belong to the largest incidence CG1 (E).

Cells from patients classified in most CGs of the Zellweger spectrum (CGs 1, 2, 3, 4, 7, 8, 10, 11, 13, and 15) are characterized by the presence of peroxisomal remnant structures that contain a number of PMPs but no or only residual amounts of matrix proteins (Table 5.1) (Fujiki et al. 2006b; Sacksteder and Gould 2000; Santos et al. 1988). This suggests that the genes defective in these CGs are involved in peroxisomal matrix protein import. In contrast, cells from patients classified in CG9, CG12, or CG14 are devoid of peroxisomal remnant structures (Honsho et al. 1998; Matsuzono et al. 1999; Sacksteder et al. 2000; South and Gould 1999; Ghaedi et al. 2000a; South et al. 2000). This indicates that the genes defective in these CGs are required for peroxisome membrane biogenesis.

5.3.2 Pathogenic Genes

Expression Cloning of Peroxin cDNAs Genetic phenotype complementation of peroxisome assembly-defective mutants of mammalian somatic cells such as CHO cells and of several yeast species including *Saccharomyces cerevisiae*, *Pichia pastoris*, *Hansenula polymorpha*, and *Yarrowia lipolytica* have led to identification and characterization of numerous genes, termed *PEX*s, encoding peroxins that are essential for peroxisome biogenesis (Distel et al. 1996; Subramani et al. 2000; Fujiki et al. 2006b). Two mutually distinct but complementary approaches have been taken in order to isolate mammalian *PEX* genes.

A direct cloning approach has been taken by means of genetic complementation with peroxin cDNA essential for assembly of peroxisomes in CHO cells. Establishment of an effective method, termed P12 (12-(1′-pyrene)dodecanoic acid)/ultraviolet selection method, made it feasible to isolate revertant (transfectant) cells showing a morphologically and biochemically normal peroxisome-phenotype, whereby *PEX2* (formerly *PAF-1*) encoding the 35-kDa membrane peroxin Pex2p with RING zinc-finger motif was cloned for the first time (Tsukamoto et al. 1991). Expression of *PEX2* in fibroblasts from a ZS patient of CG10 (F) complemented the impaired peroxisome assembly (Shimozawa et al. 1992). Dysfunction of *PEX2* caused by a homozygous nonsense point mutation at R119ter was shown for the first time to be responsible for ZS, a prototype of the PBDs (Shimozawa et al. 1992). A transient expression assay has also been developed for further isolation of *PEX* cDNAs. A cDNA library divided into small pools was transfected into peroxisome assembly-defective CHO cell mutants, and positive clones were searched for by immunocytochemical staining of restored peroxisomes, or more recently by using a readily visible GFP-peroxisome targeting signal (PTS) fusion protein. Nine other peroxin cDNAs, *PEX1*, *PEX3*, *PEX5*, *PEX6*, *PEX12*, *PEX13*, *PEX14*, *PEX19*, and *PEX26* have been cloned by the transient functional phenotype-complementation

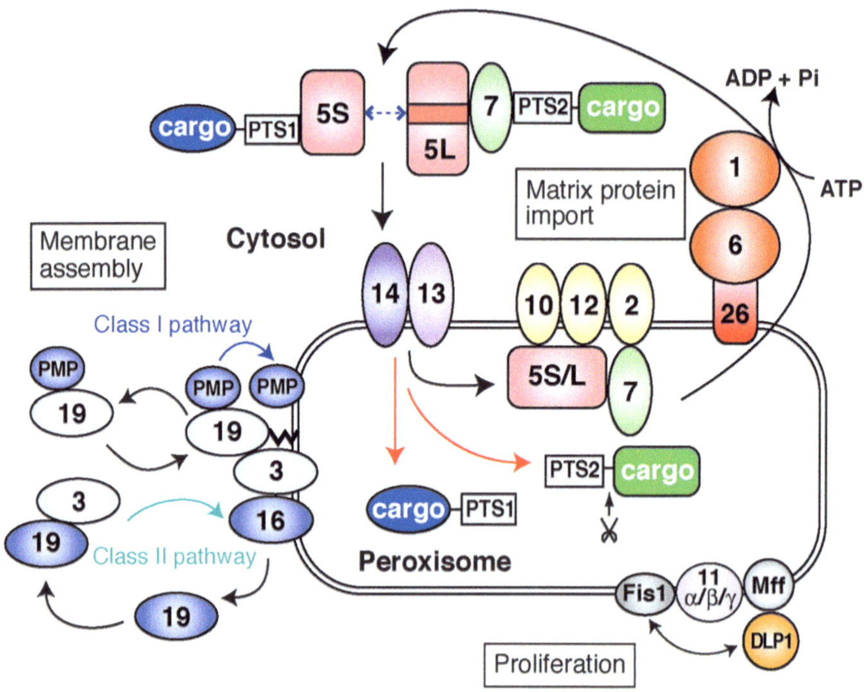

Fig. 5.1 A schematic view of peroxisome biogenesis in mammals. The subcellular localization and molecular characteristics of peroxins are shown. Peroxins are divided into three groups (1) peroxins that are required for matrix protein import; (2) those including Pex3p, Pex16p, and Pex19p, responsible for peroxisome membrane assembly via classes I and II pathways (see Fig. 5.2); (3) those such as three forms of Pex11p, Pex11pα, Pex11pβ, and Pex11pγ, apparently involved in peroxisome proliferation where DLP1, Mff, and Fis1 coordinately function. PTS1 and PTS2 proteins are recognized by Pex5p and Pex7p, respectively, in the cytoplasm. Two isoforms, Pex5pS and Pex5pL, of Pex5p are identified in mammals. PTS1 proteins are transported by homo- and hetero-oligomers of Pex5pS and Pex5pL to peroxisomes, where Pex14p functions as a convergent, initial docking site of the "protein import machinery" translocon. Pex5pL directly interacts with the PTS2 receptor, Pex7p, carrying its cargo PTS2 protein in the cytosol and translocates the Pex7p–PTS2 protein complex to Pex14p. PTS1 and PTS2 proteins are then released at the inner surface and/or inside of peroxisomes, downstream Pex14p and upstream Pex13p. Pex5p and Pex7p subsequently translocate to other translocon components, termed translocation complex comprising the RING peroxins, Pex2p, Pex10p, and Pex12p. Both Pex5p and Pex7p finally shuttle back to the cytosol. In regard to peroxisome-cytoplasmic shuttling of Pex5p, Pex5p initially targets to an 800-kDa docking complex containing Pex14p and then translocates to a 500-kDa translocation complex comprising RING peroxins. At the terminal step of the protein import reaction, Pex1p and Pex6p of the AAA family catalyze the export of Pex5p, where Cys-ubiquitination of Pex5p is prerequisite to the Pex5p exit

assay on CHO cell mutants (Fujiki 2003; Fujiki et al. 2006b) (Fig. 5.1). These *PEX*s were shown to be the pathogenic genes responsible for PBDs of nine CGs (Fujiki et al. 2006b; Weller et al. 2003; Fujiki 2011) (Table 5.1).

Expressed Sequence Tag Homology Search As an alternative strategy, the homology search by screening the human expressed sequence tag (EST) database using yeast *PEX* genes has successfully led to isolation of human ortholog genes responsible for PBDs: *PEX1*, *PEX3*, *PEX5*, *PEX6*, *PEX7*, *PEX10*, *PEX12*, *PEX13*, and *PEX16* (Weller et al. 2003; Fujiki et al. 2006b).

Within about 10 years after the first isolation of the ZS gene, *PEX2*, all of pathogenic genes responsible for PBDs of currently identified 13 CGs have been successfully cloned by such extensive search using the mutually complementary methods.

Gene Mutations of Peroxins Involved in Import of Matrix Proteins In regard to the transport of PTS1 and PTS2 proteins, Pex5p and Pex7p respectively function as a cytosolic signal receptor in mammals (Fransen et al. 1995; Otera et al. 2000, 2002; Dodt et al. 1995; Wiemer et al. 1995; Subramani et al. 2000). In mammals, two isoforms of Pex5p, termed Pex5pS and Pex5pL with an internal 37-amino-acid insertion, have been identified (Otera et al. 1998). Pex5pL transports PTS1 proteins and Pex7p–PTS2 cargo complexes to the initial Pex5p-docking site Pex14p on peroxisome membranes, whereas Pex5pS translocates only PTS1 cargoes (Fransen et al. 1995; Otera et al. 2000, 2002) (Fig. 5.1). Deficiency of Pex5p, a member of the tetratricopeptide repeat (TPR) family, causes PBDs of CG2 manifesting protein-import defects (Dodt et al. 1995; Wiemer et al. 1995; Subramani et al. 2000) (Table 5.1). A *PEX5*-defective ZS patient of CG2 carried a nonsense mutation R390ter, and an NALD patient had a missense mutation N489K (Dodt et al. 1995). Dysfunction of Pex7p containing six WD motifs is responsible for CG11 RCDP showing a cell phenotype of impaired PTS2-import (Braverman et al. 1997; Motley et al. 1997; Purdue et al. 1997; Weller et al. 2003; Fujiki 2003). Pex7p with one site-mutation at G217R, A218V, or L292Ter, frequently seen in RCDP1, is impaired in binding to both PTS2 cargo protein and Pex5pL (Mukai et al. 2002). Mutations of *PEX1* encoding Pex1p of the AAA ATPase family are the cause of PBDs of the largest group CG1 (E) (Reuber et al. 1997; Portsteffen et al. 1997; Tamura et al. 1998a). A frequent one-base insertion 2097insT was found in severe ZS patients; a common missense mutation G843D was identified among patients with milder phenotypes such as IRD, as a *ts* mutation (Imamura et al. 1998a; Geisbrecht et al. 1998; Walter et al. 2001). *PEX6*, CG4 (C) PBD gene, encodes AAA ATPase Pex6p (Tsukamoto et al. 1995; Fukuda et al. 1996; Yahraus et al. 1996). Pex1p interacts with Pex6p (Tamura et al. 1998b, 2001; Geisbrecht et al. 1998).

Import of PTS1 and PTS2 proteins most likely shares the common translocon, as inferred from the common phenotype, the impaired import of both PTS1 and PTS2, in mammalian cell mutants, *pex2*, *pex10*, *pex12*, *pex13*, *pex14*, and *pex26*. PBD patients defective in *PEX14* encoding the initial docking site for Pex5p and Pex7p was identified (Shimozawa et al. 2004; Huybrechts et al. 2008). Expression of *PEX13* encoding an SH3 membrane protein of peroxisomes restores the biogenesis of peroxisomes in CG13 (H) PBDs (Liu et al. 1999; Shimozawa et al. 1999). A ZS patient was a homozygote for nonsense mutation at W234ter that completely eliminated the SH3 domain (Liu et al. 1999; Shimozawa et al. 1999), and an NALD patient had a homozygous missense mutation at I326T (Liu et al. 1999; Shimozawa et al. 1999), with a *ts* phenotype (Liu et al. 1999; Shimozawa et al. 1999). *PEX10* is the gene responsible for CG7 (B) PBDs and encodes another RING-finger integral-membrane protein (Okumoto et al. 1998a; Warren et al. 1998). A ZS patient possessed a homozygous, inactivating mutation: a two-base deletion in a codon for Leu^{272} immediately upstream the RING resulted in a frameshift. *PEX12* also codes for a peroxisomal integral RING membrane protein (Okumoto and Fujiki 1997; Chang et al. 1997; Okumoto et al. 1998b). *PEX12* expression restored peroxisome assembly in fibroblasts from three ZS patients of CG3, where homozygote alleles for K231ter, R180ter and two-base deletion resulting in S292ter were identified, hence implying that the C-terminal part, including RING finger, is required for the biological function of Pex12p (Chang et al. 1997; Okumoto et al. 1998b). RING peroxins, Pex2p, Pex10p, and Pex12p, are considered to function as a translocation complex for Pex5p and Pex7p in protein import, downstream of their docking complexes consisting of Pex14p and Pex13p (Chang et al. 1999; Okumoto et al. 2000; Otera et al. 2002; Miyata and Fujiki 2005). *PEX26* encoding a type II membrane peroxin, the recruiter of AAA ATPase complexes comprising Pex1p and Pex6p, was cloned and shown to be responsible for CG8 PBD (Matsumoto et al. 2003a, b; Weller et al. 2005; Fig. 5.1).

5.3.3 Peroxisome-Cytoplasmic Shuttling of Import Receptors

Ten peroxins are involved in matrix protein import (Fig. 5.1). PTS1 and PTS2 proteins are recognized by Pex5p and Pex7p, respectively, in the cytoplasm. PTS1 proteins are transported by homo- and hetero-oligomers of Pex5pS and Pex5pL to peroxisomes, where Pex14p functions as the initial site of an 800-kDa "docking complex". Pex5pL translocates the Pex7p–PTS2 protein complex to Pex14p (Otera et al. 2002; Miyata and Fujiki 2005). After releasing the cargoes, Pex5p and Pex7p translocate to a 500-kDa "translocation complex" comprising the RING peroxins, Pex2p, Pex10p, and Pex12p (Miyata and Fujiki 2005). Both Pex5p and Pex7p finally translocate back to the cytosol (Dammai and Subramani 2001; Gouveia et al. 2003; Miyata and Fujiki 2005; Platta et al. 2005; Miyata et al. 2009; Nair et al. 2004). At the terminal step of the protein import reaction, AAA peroxins, Pex1p and Pex6p, anchored to Pex26p (Pex15p in yeast) on peroxisomes catalyze the ATP-dependent export of Pex5p (Miyata and Fujiki 2005; Platta et al. 2005). Ubiquitination of Pex5p is prerequisite for the Pex5p exit (Carvalho et al. 2007; Williams et al. 2007; Okumoto et al. 2011).

Monoubiquitination of the conserved cysteine residue at position 11 in the N-terminal region of mammalian Pex5p plays an essential role in the recycling, especially in the export step from peroxisomes to the cytosol (Grou et al. 2009; Okumoto et al. 2011; Miyata et al. 2012), as in yeast (Platta et al. 2009). A cytosolic factor, AWP1/ZFAND6, involved in the recycling of Pex5p is recently identified in mammals (Miyata et al. 2012); USP9X and Ubp15 are suggested as a potential deubiquitinase in mammals (Grou et al. 2012) and yeast (Debelyy et al. 2011), respectively. Moreover, the redox state of the cytosol of *pex* cell mutants is more reductive than that of the wild-type CHO cells (Yano et al. 2010). Such distinct redox state may affect the recycling of Pex5p requiring Cys-ubiquitination, thereby leading as a possible cause to the phenotype of deficiency in protein import in *PEX*-defective cells including cell lines from patients with PBDs.

5.3.4 Peroxins Essential for Membrane Assembly of Peroxisomes

Of 13 peroxins of which mutations are responsible for PBDs, three peroxins—Pex3p, Pex16p, and Pex19p—have been identified as essential factors for PMP assembly in several species including humans (Ghaedi et al. 2000a; Sacksteder et al. 2000; South et al. 2000; Honsho et al. 1998; South and Gould 1999; Matsuzono et al. 1999; Baerends et al. 1996; Götte et al. 1998; Hettema et al. 2000; Otzen et al. 2004) (Fig. 5.1). Pex19p is a predominantly cytoplasmic protein that shows a broad PMP-binding specificity; Pex3p serves as the membrane-anchoring site for Pex19p (Class I pathway) and Pex16p—a protein absent in most yeasts (Eitzen et al. 1997; South and Gould 1999) is shown to be the receptor for Pex19p complexes with newly synthesized Pex3p (Matsuzaki and Fujiki 2008) (Class II pathway) (Fig. 5.1). The function of Pex16p is not conserved between different species. In addition, disagreement exists about whether Pex19p has a chaperone-like role in the cytosol or at the peroxisome membrane and/or functions as a cycling import receptor for newly synthesized PMPs (Fujiki et al. 2006a).

Gene Defects of Peroxins Required for Both Membrane Assembly and Matrix Protein Import Three mammalian pathogenic peroxins, Pex3p, Pex16p, and Pex19p, have been isolated and their mutations are shown to be responsible for ZS of three CGs, CG12 (G), CG9 (D), and CG14 (J), respectively (Fujiki et al. 2006b, 2012; Weller et al. 2003; Fujiki 2011).

Pex3p *PEX3* encodes 42-kDa integral membrane protein of peroxisomes (Ghaedi et al. 2000a, b). Upon expression of *PEX3* in a CHO *pex3* mutant (Ghaedi et al. 2000b) and fibroblasts from three ZS patients of CG12 (G) (Ghaedi et al. 2000a), peroxisomal membrane vesicles were assembled before the import of soluble proteins such as PTS1 and PTS2 proteins (Ghaedi et al. 2000a; Muntau et al. 2000; Shimozawa et al. 2000; South et al. 2000; Fujiki et al. 2006a; Fujiki 2011), as in *pex19* and *pex16* patients-derived cells (see below), implying the temporally differentiated translocation of matrix proteins into peroxisomal membrane vesicles. Two types of inactivating mutations, exon 11 deletion and a single-nucleotide insertion in the codon for Val^{182} in exon 7, in *PEX3* were identified in the ZS patients (Ghaedi et al. 2000a; Muntau et al. 2000; Shimozawa et al. 2000).

Pex16p Fibroblasts from a ZS patient of CG9 (D) are defective in peroxisomal membrane biogenesis and morphologically devoid of peroxisomal remnants, as in *PEX3*- and *PEX19*-defective fibroblasts of CG12 (G) and CG14 (J), respectively. Expression of *PEX16* encoding 336-amino-acid peroxisomal membrane protein restored peroxisomal membrane assembly and matrix protein import in CG9 (D) fibroblasts (Honsho et al. 1998; South and Gould 1999), of which mutation was identified to be a homozygous nonsense mutation R176ter (Honsho et al. 1998). More recently, other mutations are identified: exon 10 skip (Shimozawa et al. 2002) and five novel parent homozygous mutations (Ebberink et al. 2010).

Pex19p *PEX19* encodes 33-kDa farnesylated protein partly anchoring to peroxisomal membranes (Matsuzono et al. 1999). *PEX19* expression complemented impaired peroxisome biogenesis in fibroblasts from a patient with CG14 (J) PBD. This patient possessed a homozygous, inactivating mutation: a one-base insertion, A764, in a codon for Met^{255}, resulted in a frameshift. Upon transfection of *PEX19* into a CHO *pex19* mutant ZP119 devoid of peroxisomal ghosts, most striking was formation of peroxisomal membranes, followed by import of matrix proteins (Matsuzono et al. 1999; Sacksteder et al. 2000). This was the first observation of the membrane assembly process during peroxisome biogenesis, particularly differentiated from the import of soluble proteins.

Collectively, Pex3p, Pex16p, and Pex19p are categorized as a peroxin essential for the assembly of peroxisome membranes. They function as essential factors required for the translocation process of membrane proteins and membrane vesicle assembly in a concerted manner. Two distinct pathways have recently been suggested for the import of PMPs: a Pex19p- and Pex3p-dependent class I pathway for PMP-import complex, except for Pex3p (Fang et al. 2004; Matsuzono et al. 2006) and a Pex19p- and Pex16p-dependent class II pathway for Pex3p (Matsuzaki and Fujiki 2008) (Fig. 5.2). It is noteworthy that C-tailed anchor-type peroxin Pex26p, the recruiter of Pex1p–Pex6p complex, is transported in a Pex19p-dependent (Halbach et al. 2006), class I pathway (Yagita et al. 2013), which is distinct from the GET3-dependent topogenesis of yeast Pex15p, a functional homologue of Pex26p (Schuldiner et al. 2008).

5.3.5 Involvement of ER in Peroxisome Assembly

In regard to involvement of the ER in peroxisome biogenesis, ER was postulated to provide the initial "seed" for recruiting other components required for peroxisome assembly (Kim et al. 2006; Ma et al. 2011; Tabak et al. 2013). Recently, several groups suggested a different view of peroxisomal membrane biogenesis that peroxisomes are formed from ER upon induction of Pex3p (Hoepfner et al. 2005; Kim et al. 2006; Kragt et al. 2005), although the significance of such observations remains under debate. More recently, a study (Motley and Hettema 2007) suggest

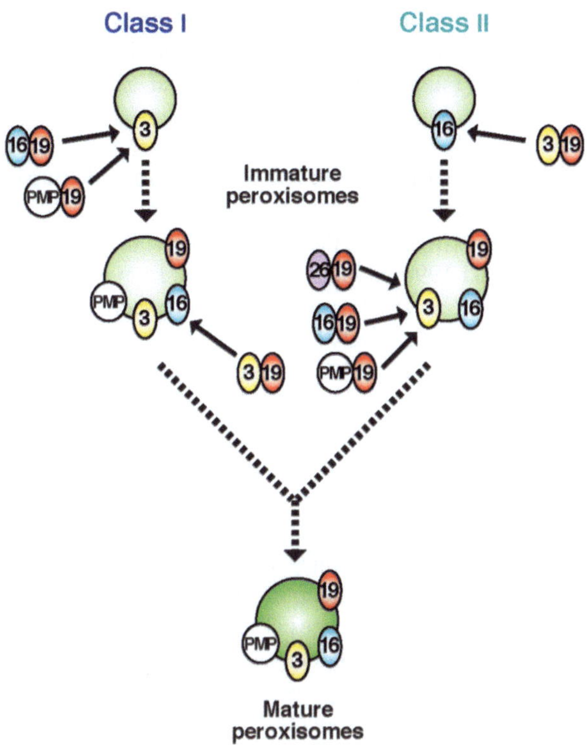

Fig. 5.2 A model for early stages of peroxisomal membrane biogenesis involving mutually dependent targeting of Pex3p and Pex16p, termed classes I and II pathways. The initial membranes harboring Pex3p or Pex16p culminate in indistinguishable, matured peroxisomes. Pex19p forms complexes in the cytosol with newly synthesized PMPs including Pex16p and C-tailed anchored membrane proteins such as Pex26p and transports them to the membrane protein receptor Pex3p, whereby peroxisome membrane is assembled (Class I pathway). With respect to biogenesis of Pex3p, Pex19p likewise forms a complex with newly synthesized Pex3p and translocates it to the Pex3p receptor, Pex16p (Class II pathway). Of note, peroxisomes are assembled no matter which pathway initially proceeds

that peroxisomes are generally formed by growth and division under normal conditions and that only under a condition where no peroxisome is present in a cell, they can be formed from the ER after expression of the complementing *PEX* gene, while another study (van der Zand et al. 2010) proposes that all peroxisomal membrane proteins are transported via ER. Meanwhile, we demonstrated that Pex3p, the membrane receptor for Pex19p-complexes with PMPs including Pex16p, is directly targeted to peroxisomes in a Pex19p–Pex16p dependent class II pathway in mammalian cells such as CHO and human cell lines (Matsuzaki and Fujiki 2008). Moreover, it is noteworthy that several peroxisomal membrane proteins might be translocated to peroxisomes via ER (Agrawal et al. 2011; Lam et al. 2010; Yonekawa et al. 2011), likely implying a sort of semi-autonomous property of peroxisomes. At any event, the issue with respect to how peroxisome membrane is assembled is one of the important and of high interest problems to be tackled (Ma et al. 2011; Fujiki et al. 2012; Tabak et al. 2013).

5.3.6 Gene Defects of Proteins for Peroxisomal Morphogenesis

Three isoforms of Pex11p family, Pex11pα (Abe et al. 1998; Li et al. 2002a), Pex11pβ (Abe and Fujiki 1998; Schrader et al. 1998; Li et al. 2002b), and Pex11pγ (Tanaka et al. 2003; Li et al. 2002a), are identified as factors involved in morphogenesis of peroxisomes in mammals (Kobayashi et al. 2007; Delille et al. 2010; Koch et al. 2010; Itoyama et al. 2013). In mammalian cells, dynamin-like protein 1 (DLP1) (Koch et al. 2003; Li and Gould 2003; Tanaka et al. 2006; Waterham et al. 2007), fission 1 (Fis1) (Koch et al. 2005; Kobayashi et al. 2007), and mitochondrial fission factor (Mff) (Gandre-Babbe and van der Bliek 2008; Otera et al. 2010; Koch and Brocard 2012; Itoyama et al. 2013) are shown to be involved in peroxisomal fission.

In regards to peroxisomal dysmorphogenesis in humans, only two patients have been identified with a different defect in any of the proteins involved in the proliferation and division of peroxisomes. The first reported patient was a severely affected female patient, who died 1 month after birth and postmortally was found to have a dominant-negative heterozygous mutation in the dynamin-like protein 1 (DLP1) gene, which resulted in a severe fission defect of both peroxisomes and mitochondria (Waterham et al. 2007). More recently, the first patient with a defect of peroxisomal division due to a homozygous non-sense mutation in the *PEX11β* gene was reported as the 14th CG (CG16) of PBDs (Ebberink et al. 2012).

5.3.7 Genotype–Phenotype Relationships

Patients with milder form of PBDs, NALD, and IRD tend to have less severe biochemical abnormalities, whose specimen such as skin fibroblasts likely contain residual peroxisomes, occasionally termed mosaicism. However, clinical severity or prognosis of a patient with PBDs cannot be easily predicted solely based on biochemical analyses. Various types of mutations such as nonsense point mutations, missense mutations, insertion and deletion of nucleotides mostly with concomitant frameshift, splicing defect, etc., in both homozygotic and heterozygotic alleles have been identified in PBD patients (Weller et al. 2003; Steinberg et al. 2006; Fujiki 2011). Patients with severe ZS tend to carry severe mutation such as nonsense mutations, frameshifts, and deletions, while many of NALD or IRD patients frequently possess missense mutations (Weller et al. 2003; Steinberg et al. 2006; Fujiki 2011). There is also a relationship between severe phenotype and absence of ghosts. Defects of *PEX3*, *PEX16*, and *PEX19* encoding membrane-assembly peroxins lead to absence of ghosts and cause ZS phenotypes (Fujiki et al. 2012). Many cell lines from milder PBD patients with missense *PEX* mutations showed a *ts* phenotype, restoration of peroxisome biogenesis at 30 °C (Imamura et al. 1998a, b; Geisbrecht et al. 1998; Walter et al. 2001).

Animal Models for PBDs Search for pathogenic genes responsible for all CGs of PBDs is accomplished. Prenatal DNA diagnosis using *PEX* genes is now possible for PBDs of all 14 CGs.

Currently ongoing and future investigations using the cloned peroxins and *pex* mutants including fibroblasts from patients with PBDs, CHO cell mutants, and *PEX* gene knockout mice such as *PEX2* (Faust and Hatten 1997), *PEX5* (Baes et al. 1997; Kassmann et al. 2007), *PEX11α* (Li et al. 2002a), *PEX11β* (Li et al. 2002b), and *PEX13* (Maxwell et al. 2003) shed light on the mechanisms underlying peroxisome biogenesis and pathogenesis of PBDs. However, gene-knockout mice for *PEX3*, *PEX16*, and *PEX19* have not been established yet.

It is noteworthy that invertebrate models including *Caenorhabditis elegans* (Petriv et al. 2002; Thieringer et al. 2003) and *Drosophila melanogaster* (Chen et al. 2010; Nakayama et al. 2011; Faust et al. 2012) for PBDs have recently been established.

Acknowledgments We thank K. Shimizu for figure illustrations. This work was supported in part by a CREST grant (to Y.F.) from the Science and Technology Agency of Japan, Grants-in-Aid for Scientific Research (19058011, 20370039, 24247038, 25112518, and 25116717 to Y.F.; 21770117 and 24770130 to K.O.; 21570116 and 24570134 to S.T.), Global Center of Excellence (GCOE) Program, Grants for Excellent Graduate Schools from the Ministry of Education, Culture, Sports, Science and Technology of Japan, and grants (to Y.F.) from Takeda Science Foundation and Japan Foundation for Applied Enzymology.

References

Abe I, Fujiki Y (1998) cDNA cloning and characterization of a constitutively expressed isoform of the human peroxin Pex11p. Biochem Biophys Res Commun 252:529–533

Abe I, Okumoto K, Tamura S, Fujiki Y (1998) Clofibrate-inducible, 28-kDa peroxisomal integral membrane protein is encoded by *PEX11*. FEBS Lett 431:468–472

Agrawal G, Joshi S, Subramani S (2011) Cell-free sorting of peroxisomal membrane proteins from the endoplasmic reticulum. Proc Natl Acad Sci USA 108:9113–9118

Baerends RJS, Rasmussen SW, Hilbrands RE, van der Heide M, Faber KN, Reuvekamp PTW, Klei JAKW, Cregg JM, van der Klei IJ, Veenhuis M (1996) The *Hansenula polymorpha PER9* gene encodes a peroxisomal membrane protein essential for peroxisome assembly and integrity. J Biol Chem 271:8887–8894

Baes M, Gressens P, Baumgart E, Carmeliet P, Casteels M, Fransen M, Evrard P, Fahimi D, Declercq PE, Collen D, van Veldhoven PP, Mannaerts GP (1997) A mouse model for Zellweger syndrome. Nat Genet 17:49–57

Braverman N, Steel G, Obie C, Moser A, Moser H, Gould SJ, Valle D (1997) Human *PEX7* encodes the peroxisomal PTS2 receptor and is responsible for rhizomelic chondrodysplasia punctata. Nat Genet 15:369–376

Carvalho AF, Pinto MP, Grou CP, Alencastre IS, Fransen M, Sá-Miranda C, Azevedo JE (2007) Ubiquitination of mammalian Pex5p, the peroxisomal import receptor. J Biol Chem 282:31267–31272

Chang C-C, Lee W-H, Moser H, Valle D, Gould SJ (1997) Isolation of the human *PEX12* gene, mutated in group 3 of the peroxisome biogenesis disorders. Nat Genet 15:385–388

Chang C-C, Warren DS, Sacksteder KA, Gould SJ (1999) PEX12 interacts with PEX5 and PEX10 and acts downstream of receptor docking in peroxisomal matrix protein import. J Cell Biol 147:761–774

Chen H, Liu Z, Huang X (2010) *Drosophila* models of peroxisomal biogenesis disorder: peroxins are required for spermatogenesis and very-long-chain fatty acid metabolism. Hum Mol Genet 19:494–505

Dammai V, Subramani S (2001) The human peroxisomal targeting signal receptor, Pex5p, is translocated into the peroxisomal matrix and recycled to the cytosol. Cell 105:187–196

Debelyy MO, Platta HW, Saffian D, Hensel A, Thoms S, Meyer HE, Warscheid B, Girzalsky W, Erdmann R (2011) Ubp15p, a ubiquitin hydrolase associated with the peroxisomal export machinery. J Biol Chem 286:28223–28234

Delille HK, Agricola B, Guimaraes SC, Borta H, Lüers GH, Fransen M, Schrader M (2010) Pex11pβ-mediated growth and division of mammalian peroxisomes follows a maturation pathway. J Cell Sci 123:2750–2762

Distel B, Erdmann R, Gould SJ, Blobel G, Crane DI, Cregg JM, Dodt G, Fujiki Y, Goodman JM, Just WW, Kiel JAKW, Kunau W-H, Lazarow PB, Mannaerts GP, Moser H, Osumi T, Rachubinski RA, Roscher A, Subramani S, Tabak HF, Tsukamoto T, Valle D, van der Klei I, van Veldhoven PP, Veenhuis M (1996) A unified nomenclature for peroxisome biogenesis factors. J Cell Biol 135:1–3

Dodt G, Braverman N, Wong CS, Moser A, Moser HW, Watkins P, Valle D, Gould SJ (1995) Mutations in the PTS1 receptor gene, *PXR1*, define complementation group 2 of the peroxisome biogenesis disorders. Nat Genet 9:115–125

Ebberink MS, Csanyi B, Chong WK, Denis S, Sharp P, Mooijer PAW, Dekker CJM, Spooner C, Ngu LH, De Sousa C, Wanders RJA, Fietz MJ, Clayton PT, Waterham HR, Ferdinandusse S (2010) Identification of an unusual variant peroxisome biogenesis disorder caused by mutations in the *PEX16* gene. J Med Genet 47:608–615

Ebberink MS, Koster J, Visser G, van Spronsen F, Stolte-Dijkstra I, Smit GPA, Fock JM, Kemp S, Wanders RJA, Waterham HR (2012) A novel defect of peroxisome division due to a homozygous non-sense mutation in the *PEX11β* gene. J Med Genet 49:307–313

Eitzen GA, Szilard RK, Rachubinski RA (1997) Enlarged peroxisomes are present in oleic acid-grown *Yarrowia lipolytica* overexpressing the *PEX16* gene encoding an intraperoxisomal peripheral membrane peroxin. J Cell Biol 137:1265–1278

Fang Y, Morrell JC, Jones JM, Gould SJ (2004) PEX3 functions as a PEX19 docking factor in the import of class I peroxisomal membrane proteins. J Cell Biol 164:863–875

Faust PL, Hatten ME (1997) Targeted deletion of the PEX2 peroxisome assembly gene in mice provides a model for Zellweger syndrome, a human neuronal migration disorder. J Cell Biol 139:1293–1305

Faust JE, Verma A, Peng C, McNew JA (2012) An inventory of peroxisomal proteins and pathways in *Drosophila melanogaster*. Traffic 13:1378–1392

Fransen M, Brees C, Baumgart E, Vanhooren JC, Baes M, Mannaerts GP, Veldhoven PPV (1995) Identification and characterization of the putative human peroxisomal C-terminal targeting signal import receptor. J Biol Chem 270:7731–7736

Fujiki Y (2003) Peroxisome biogenesis disorders. In: Cooper DN (ed) Nature encyclopedia of the human genome, vol 4. Nature Publishing Group, London, pp 541–547

Fujiki Y (2011) Peroxisome biogenesis disorders. Encyclopedia of Life Sciences. Wiley, Chichester. doi:10.1002/9780470015902.a0006109.pub2

Fujiki Y, Matsuzono Y, Matsuzaki T, Fransen M (2006a) Import of peroxisomal membrane proteins: the interplay of Pex3p- and Pex19p-mediated interactions. Biochim Biophys Acta Mol Cell Res 1763:1639–1646

Fujiki Y, Okumoto K, Kinoshita N, Ghaedi K (2006b) Lessons from peroxisome-deficient Chinese hamster ovary (CHO) cell mutants. Biochim Biophys Acta Mol Cell Res 1763:1374–1381

Fujiki Y, Yagita Y, Matsuzaki T (2012) Peroxisome biogenesis disorders: molecular basis for impaired peroxisomal membrane assembly—in metabolic functions and biogenesis of peroxisomes in health and disease. Biochim Biophys Acta 1822:1337–1342

Fukuda S, Shimozawa N, Suzuki Y, Zhang Z, Tomatsu S, Tsukamoto T, Hashiguchi N, Osumi T, Masuno M, Imaizumi K, Kuroki Y, Fujiki Y, Orii T, Kondo N (1996) Human peroxisome assembly factor-2 (PAF-2): a gene responsible for group C peroxisome biogenesis disorder in humans. Am J Hum Genet 59:1210–1220

Gandre-Babbe S, van der Bliek AM (2008) The novel tail-anchored membrane protein Mff controls mitochondrial and peroxisomal fission in mammalian cells. Mol Biol Cell 19:2402–2412

Geisbrecht BV, Collins CS, Reuber BE, Gould SJ (1998) Disruption of a PEX1-PEX6 interaction is the most common cause of the neurologic disorders Zellweger syndrome, neonatal adrenoleukodystrophy, and infantile Refsum disease. Proc Natl Acad Sci USA 95:8630–8635

Ghaedi K, Honsho M, Shimozawa N, Suzuki Y, Kondo N, Fujiki Y (2000a) *PEX3* is the causal gene responsible for peroxisome membrane assembly-defective Zellweger syndrome of complementation group G. Am J Hum Genet 67:976–981

Ghaedi K, Tamura S, Okumoto K, Matsuzono Y, Fujiki Y (2000b) The peroxin Pex3p initiates membrane assembly in peroxisome biogenesis. Mol Biol Cell 11:2085–2102

Goldfischer S, Moore CL, Johnson AB, Spiro AJ, Valsamis MP, Wisniewski HK, Ritch RH, Norton WT, Rapin I, Gartner LM (1973) Peroxisomal and mitochondrial defects in the cerebro-hepato-renal syndrome. Science 182:62–64

Götte K, Girzalsky W, Linkert M, Baumgart E, Kammerer S, Kunau W-H, Erdmann R (1998) Pex19p, a farnesylated protein essential for peroxisome biogenesis. Mol Cell Biol 18:616–628

Gould SJ, Raymond GV, Valle D (2001) The peroxisome biogenesis disorders. The metabolic and molecular bases of inherited disease, 8th edn. McGraw Hill, New York

Gouveia AM, Guimaraes CP, Oliveira ME, Reguenga C, Sa-Miranda C, Azevedo JE (2003) Characterization of the peroxisomal cycling receptor, Pex5p, using a cell-free in vitro import system. J Biol Chem 278:226–232

Grou CP, Carvalho AF, Pinto MP, Huybrechts SJ, Sá-Miranda C, Fransen M, Azevedo JE (2009) Properties of the ubiquitin-Pex5p thiol ester conjugate. J Biol Chem 284:10504–10513

Grou CP, Francisco T, Rodrigues TA, Freitas MO, Pinto MP, Carvalho AF, Domingues P, Wood SA, Rodríguez-Borges JE, Sá-Miranda C, Fransen M, Azevedo JE (2012) Identification of ubiquitin-specific protease 9X (USP9X) as a deubiquitinase acting on ubiquitin-peroxin 5 (PEX5) thioester conjugate. J Biol Chem 287:12815–12827

Halbach A, Landgraf C, Lorenzen S, Rosenkranz K, Volkmer-Engert R, Erdmann R, Rottensteiner H (2006) Targeting of the tail-anchored peroxisomal membrane proteins PEX26 and PEX15 occurs through C-terminal PEX19-binding sites. J Cell Sci 119:2508–2517

Hettema EH, Girzalsky W, van den Berg M, Erdmann R, Distel B (2000) *Saccharomyces cerevisiae* Pex3p and Pex19p are required for proper localization and stability of peroxisomal membrane proteins. EMBO J 19:223–233

Hoepfner D, Schildknegt D, Braakman I, Philippsen P, Tabak HF (2005) Contribution of the endoplasmic reticulum to peroxisome formation. Cell 122:85–95

Honsho M, Tamura S, Shimozawa N, Suzuki Y, Kondo N, Fujiki Y (1998) Mutation in *PEX16* is causal in the peroxisome-deficient Zellweger syndrome of complementation group D. Am J Hum Genet 63:1622–1630

Huybrechts SJ, Van Veldhoven PP, Hoffman I, Zeevaert R, de Vos R, Demaerel P, Brams M, Jaeken J, Fransen M, Cassiman D (2008) Identification of a novel *PEX14* mutation in Zellweger syndrome. J Med Genet 45:376–383

Imamura A, Tamura S, Shimozawa N, Suzuki Y, Zhang Z, Tsukamoto T, Orii T, Kondo N, Osumi T, Fujiki Y (1998a) Temperature-sensitive mutation in *PEX1* moderates the phenotypes of peroxisome deficiency disorders. Hum Mol Genet 7:2089–2094

Imamura A, Tsukamoto T, Shimozawa N, Suzuki Y, Zhang Z, Imanaka T, Fujiki Y, Orii T, Kondo N, Osumi T (1998b) Temperature-sensitive phenotypes of peroxisome assembly processes represent the milder forms of human peroxisome-biogenesis disorders. Am J Hum Genet 62:1539–1543

Itoyama A, Michiyuki S, Honsho M, Yamamoto T, Moser A, Yoshida Y, Fujiki Y (2013) Mff functions with Pex11pβ and DLP1 in peroxisomal fission. Biol Open 2(10):998–1006

Kassmann CM, Lappe-Siefke C, Baes M, Brügger B, Mildner A, Werner HB, Natt O, Michaelis T, Prinz M, Frahm J, Nave K-A (2007) Axonal loss and neuroinflammation caused by peroxisome-deficient oligodendrocytes. Nat Genet 39:969–976

Kim PK, Mullen RT, Schumann U, Lippincott-Schwartz J (2006) The origin and maintenance of mammalian peroxisomes involves a de novo PEX16-dependent pathway from the ER. J Cell Biol 173:521–532

Kobayashi S, Tanaka A, Fujiki Y (2007) Fis1, DLP1, and Pex11p coordinately regulate peroxisome morphogenesis. Exp Cell Res 313:1675–1686

Koch J, Brocard C (2012) PEX11 proteins attract Mff and hFis1 to coordinate peroxisomal fission. J Cell Sci 125:3813–3826

Koch A, Thiemann M, Grabenbauer M, Yoon Y, McNiven MA, Schrader M (2003) Dynamin-like protein 1 is involved in peroxisomal fission. J Biol Chem 278:8597–8605

Koch A, Yoon Y, Bonekamp NA, McNiven MA, Schrader M (2005) A role for Fis1 in both mitochondrial and peroxisomal fission in mammalian cells. Mol Biol Cell 16:5077–5086

Koch J, Pranjic K, Huber A, Ellinger A, Hartig A, Kragler F, Brocard C (2010) PEX11 family members are membrane elongation factors that coordinate peroxisome proliferation and maintenance. J Cell Sci 123:3389–3400

Kragt A, Voorn-Brouwer T, van den Berg M, Distel B (2005) Endoplasmic reticulum-directed Pex3p routes to peroxisomes and restores peroxisome formation in a *Saccharomyces cerevisiae pex3*Δ strain. J Biol Chem 280:34350–34357

Kunau W-H (2006) Peroxisomes: morphology, function, biogenesis and disorders. Biochim Biophys Acta Mol Cell Res 1763(Special issue):1363–1808

Lam SK, Yoda N, Schekman R (2010) A vesicle carrier that mediates peroxisome protein traffic from the endoplasmic reticulum. Proc Natl Acad Sci USA 107:21523–21528

Lazarow PB (2003) Peroxisome biogenesis: advances and conundrums. Curr Opin Cell Biol 15:489–497

Lazarow PB, Fujiki Y (1985) Biogenesis of peroxisomes. Annu Rev Cell Biol 1:489–530

Li X, Gould SJ (2003) The dynamin-like GTPase DLP1 is essential for peroxisome division and is recruited to peroxisomes in part by PEX11. J Biol Chem 278:17012–17020

Li X, Baumgart E, Dong G-X, Morrell JC, Jimenez-Sanchez G, Valle D, Smith KD, Gould SJ (2002a) PEX11α is required for peroxisome proliferation in response to 4-phenylbutyrate but is dispensable for peroxisome proliferator-activated receptor alpha-mediated peroxisome proliferation. Mol Cell Biol 22:8226–8240

Li X, Baumgart E, Morrell JC, Jimenez-Sanchez G, Valle D, Gould SJ (2002b) PEX11β deficiency is lethal and impairs neuronal migration but does not abrogate peroxisome function. Mol Cell Biol 22:4358–4365

Liu Y, Bjoerkman J, Urquhart A, Wanders RJA, Crane DI, Gould SJ (1999) *PEX13* is mutated in complementation group 13 of the peroxisome-biogenesis disorders. Am J Hum Genet 65:621–634

Ma C, Agrawal G, Subramani S (2011) Peroxisome assembly: matrix and membrane protein biogenesis. J Cell Biol 193:7–16

Matsumoto N, Tamura S, Fujiki Y (2003a) The pathogenic peroxin Pex26p recruits the Pex1p-Pex6p AAA ATPase complexes to peroxisomes. Nat Cell Biol 5:454–460

Matsumoto N, Tamura S, Furuki S, Miyata N, Moser A, Shimozawa N, Moser HW, Suzuki Y, Kondo N, Fujiki Y (2003b) Mutations in novel peroxin gene *PEX26* that cause peroxisome biogenesis disorders of complementation group 8 provide a genotype-phenotype correlation. Am J Hum Genet 73:233–246

Matsuzaki T, Fujiki Y (2008) The peroxisomal membrane-protein import receptor Pex3p is directly transported to peroxisomes by a novel Pex19p- and Pex16p-dependent pathway. J Cell Biol 183:1275–1286

Matsuzono Y, Kinoshita N, Tamura S, Shimozawa N, Hamasaki M, Ghaedi K, Wanders RJA, Suzuki Y, Kondo N, Fujiki Y (1999) Human *PEX19*: cDNA cloning by functional complementation, mutation analysis in a patient with Zellweger syndrome and potential role in peroxisomal membrane assembly. Proc Natl Acad Sci USA 96:2116–2121

Matsuzono Y, Matsuzaki T, Fujiki Y (2006) Functional domain mapping of peroxin Pex19p: interaction with Pex3p is essential for function and translocation. J Cell Sci 119:3539–3550

Maxwell M, Bjorkman J, Nguyen T, Sharp P, Finnie J, Paterson C, Tonks I, Paton BC, Kay GF, Crane DI (2003) *Pex13* inactivation in the mouse disrupts peroxisome biogenesis and leads to a Zellweger syndrome phenotype. Mol Cell Biol 23:5947–5957

Miyata N, Fujiki Y (2005) Shuttling mechanism of peroxisome targeting signal type 1 receptor, Pex5: ATP-independent import and ATP-dependent export. Mol Cell Biol 25:10822–10832

Miyata N, Hosoi K, Mukai S, Fujiki Y (2009) In vitro import of peroxisome-targeting signal 2 (PTS2) receptor Pex7p into peroxisomes. Biochim Biophys Acta Mol Cell Res 1793:860–870

Miyata N, Okumoto K, Mukai S, Noguchi M, Fujiki Y (2012) AWP1/ZFAND6 functions in Pex5 export by interacting with Cys-monoubiquitinated Pex5 and Pex6 AAA ATPase. Traffic 13:168–183

Motley AM, Hettema EH (2007) Yeast peroxisomes multiply by growth and division. J Cell Biol 178:399–410

Motley AM, Hettema EH, Hogenhout EM, Brites P, ten Asbroek ALMA, Wijburg FA, Baas F, Heijmans HS, Tabak HF, Wanders RJA, Distel B (1997) Rhizomelic chondrodysplasia punctata is a peroxisomal protein targeting disease caused by a non-functional PTS2 receptor. Nat Genet 15:377–380

Mukai S, Ghaedi K, Fujiki Y (2002) Intracellular localization, function, and dysfunction of the peroxisome-targeting signal type 2 receptor, Pex7p, in mammalian cells. J Biol Chem 277:9548–9561

Muntau AC, Mayerhofer PU, Paton BC, Kammerer S, Roscher AA (2000) Defective peroxisome membrane synthesis due to mutations in human *PEX3* causes Zellweger syndrome, complementation group G. Am J Hum Genet 67:967–975

Nair DM, Purdue PE, Lazarow PB (2004) Pex7p translocates in and out of peroxisomes in *Saccharomyces cerevisiae*. J Cell Biol 167:599–604

Nakayama M, Sato H, Okuda T, Fujisawa N, Kono N, Arai H, Suzuki E, Umeda M, Ishikawa HO, Matsuno K (2011) *Drosophila* carrying Pex3 or Pex16 mutations are models of Zellweger syndrome that reflect its symptoms associated with the absence of peroxisomes. PLoS One 6: e22984

Okumoto K, Fujiki Y (1997) *PEX12* encodes an integral membrane protein of peroxisomes. Nat Genet 17:265–266

Okumoto K, Itoh R, Shimozawa N, Suzuki Y, Tamura S, Kondo N, Fujiki Y (1998a) Mutation in *PEX10* is the cause of Zellweger peroxisome deficiency syndrome of complementation group B. Hum Mol Genet 7:1399–1405

Okumoto K, Shimozawa N, Kawai A, Tamura S, Tsukamoto T, Osumi T, Moser H, Wanders RJA, Suzuki Y, Kondo N, Fujiki Y (1998b) *PEX12*, the pathogenic gene of group III Zellweger syndrome: cDNA cloning by functional complementation on a CHO cell mutant, patient analysis, and characterization of Pex12p. Mol Cell Biol 18:4324–4336

Okumoto K, Abe I, Fujiki Y (2000) Molecular anatomy of the peroxin Pex12p: RING finger domain is essential for Pex12p function and interacts with the peroxisome targeting signal type 1-receptor Pex5p and a RING peroxin, Pex10p. J Biol Chem 275:25700–25710

Okumoto K, Misono S, Miyata N, Matsumoto Y, Mukai S, Fujiki Y (2011) Cysteine ubiquitination of PTS1-receptor Pex5p regulates Pex5p recycling. Traffic 12:1067–1083

Otera H, Okumoto K, Tateishi K, Ikoma Y, Matsuda E, Nishimura M, Tsukamoto T, Osumi T, Ohashi K, Higuchi O, Fujiki Y (1998) Peroxisome targeting signal type 1 (PTS1) receptor is involved in import of both PTS1 and PTS2: studies with *PEX5*-defective CHO cell mutants. Mol Cell Biol 18:388–399

Otera H, Harano T, Honsho M, Ghaedi K, Mukai S, Tanaka A, Kawai A, Shimizu N, Fujiki Y (2000) The mammalian peroxin Pex5pL, the longer isoform of the mobile PTS1-transporter, translocates Pex7p-PTS2 protein complex into peroxisomes via its initial docking site, Pex14p. J Biol Chem 275:21703–21714

Otera H, Setoguchi K, Hamasaki M, Kumashiro T, Shimizu N, Fujiki Y (2002) Peroxisomal targeting signal receptor Pex5p interacts with cargoes and import machinery components in a spatiotemporally differentiated manner: conserved Pex5p WXXXF/Y motifs are critical for matrix protein import. Mol Cell Biol 22:1639–1655

Otera H, Wang C, Cleland MM, Setoguchi K, Yokota S, Youle RJ, Mihara K (2010) Mff is an essential factor for mitochondrial recruitment of Drp1 during mitochondrial fission in mammalian cells. J Cell Biol 191:1141–1158

Otzen M, Perband U, Wang D, Baerends RJ, Kunau WH, Veenhuis M, Van der Klei IJ (2004) *Hansenula polymorpha* Pex19p is essential for the formation of functional peroxisomal membranes. J Biol Chem 279:19181–19190

Petriv OI, Pilgrim DB, Rachubinski RA, Titorenko VI (2002) RNA interference of peroxisome-related genes in *C. elegans*: a new model for human peroxisomal disorders. Physiol Genomics 10:79–91

Platta HW, Erdmann R (2007) The peroxisomal protein import machinery. FEBS Lett 581:2811–2819

Platta HW, Grunau S, Rosenkranz K, Girzalsky W, Erdmann R (2005) Functional role of the AAA peroxins in dislocation of the cycling PTS1 receptor back to the cytosol. Nat Cell Biol 7:817–822

Platta HW, Magraoui FE, Bäumer BE, Schlee D, Girzalsky W, Erdmann R (2009) Pex2 and Pex12 function as protein-ubiquitin ligases in peroxisomal protein import. Mol Cell Biol 29:5505–5516

Portsteffen H, Beyer A, Becker E, Epplen C, Pawlak A, Kunau W-H, Dodt G (1997) Human *PEX1* is mutated in complementation group 1 of the peroxisome biogenesis disorders. Nat Genet 17:449–452

Purdue PE, Zhang JW, Skoneczny M, Lazarow PB (1997) Rhizomelic chondrodysplasia punctata is caused by deficiency of human *PEX7*, a homologue of the yeast PTS2 receptor. Nat Genet 15:381–384

Reuber BE, Germain-Lee E, Collins CS, Morrell JC, Ameritunga R, Moser HW, Valle D, Gould SJ (1997) Mutations in *PEX1* are the most common cause of peroxisome biogenesis disorders. Nat Genet 17:445–448

Sacksteder KA, Gould SJ (2000) The genetics of peroxisome biogenesis. Annu Rev Genet 34:623–652

Sacksteder KA, Jones JM, South ST, Li X, Liu Y, Gould SJ (2000) PEX19 binds multiple peroxisomal membrane proteins, is predominantly cytoplasmic, and is required for peroxisome membrane synthesis. J Cell Biol 148:931–944

Santos MJ, Imanaka T, Shio H, Small GM, Lazarow PB (1988) Peroxisomal membrane ghosts in Zellweger syndrome—aberrant organelle assembly. Science 239:1536–1538

Schrader M, Reuber BE, Morrell JC, Jimenez-Sanchez G, Obie C, Stroh TA, Valle D, Schroer TA, Gould SJ (1998) Expression of *PEX11*β mediates peroxisome proliferation in the absence of extracellular stimuli. J Biol Chem 273:29607–29614

Schuldiner M, Metz J, Schmid V, Denic V, Rakwalska M, Schmitt HD, Schwappach B, Weissman JS (2008) The GET complex mediates insertion of tail-anchored proteins into the ER membrane. Cell 134:634–645

Shimozawa N, Tsukamoto T, Suzuki Y, Orii T, Shirayoshi Y, Mori T, Fujiki Y (1992) A human gene responsible for Zellweger syndrome that affects peroxisome assembly. Science 255:1132–1134

Shimozawa N, Suzuki Y, Zhang Z, Imamura A, Kondo N, Kinoshita N, Fujiki Y, Tsukamoto T, Osumi T, Imanaka T, Orii T, Beemer F, Mooijer P, Dekker C, Wanders RJA (1998) Genetic basis of peroxisome-assembly mutants of humans, Chinese hamster ovary cells and yeast: identification of a new complementation group of peroxisome-biogenesis disorders apparently lacking peroxisomal-membrane ghosts. Am J Hum Genet 63:1898–1903

Shimozawa N, Suzuki Y, Zhang Z, Imamura A, Toyama R, Mukai S, Fujiki Y, Tsukamoto T, Osumi T, Orii T, Wanders RJA, Kondo N (1999) Nonsense and temperature-sensitive mutations in *PEX13* are the cause of complementation group H of peroxisome biogenesis disorders. Hum Mol Genet 8:1077–1083

Shimozawa N, Suzuki Y, Zhang Z, Imamura A, Ghaedi K, Fujiki Y, Kondo N (2000) Identification of *PEX3* as the gene mutated in a Zellweger syndrome patient lacking peroxisomal remnant structures. Hum Mol Genet 9:1995–1999

Shimozawa N, Nagase T, Takemoto Y, Suzuki Y, Fujiki Y, Wanders RJA, Kondo N (2002) A novel aberrant splicing mutation of the PEX16 gene in two patients with Zellweger syndrome. Biochem Biophys Res Commun 292:109–112

Shimozawa N, Tsukamoto T, Nagase T, Takemoto Y, Koyama N, Suzuki Y, Komori M, Osumi T, Jeannette G, Wanders RJA, Kondo N (2004) Identification of a new complementation group of the peroxisome biogenesis disorders and *PEX14* as the mutated gene. Hum Mutat 23:552–558

Singh I, Moser AE, Goldfischer S, Moser HW (1984) Lignoceric acid is oxidized in the peroxisome: implications for the Zellweger cerebro-hepato-renal syndrome and adrenoleukodystrophy. Proc Natl Acad Sci USA 81:4203–4207

South ST, Gould SJ (1999) Peroxisome synthesis in the absence of preexisting peroxisomes. J Cell Biol 144:255–266

South ST, Sacksteder KA, Li X, Liu Y, Gould SJ (2000) Inhibitors of COPI and COPII do not block *PEX3*-mediated peroxisome synthesis. J Cell Biol 149:1345–1360

Sparkes IA, Hawes C, Baker A (2005) *At*PEX2 and *At*PEX10 are targeted to peroxisomes independently of Known endoplasmic reticulum trafficking routes. Plant Physiol 139:690–700

Steinberg SJ, Dodt G, Raymond GV, Braverman NE, Moser AB, Moser HW (2006) Peroxisome biogenesis disorders. Biochim Biophys Acta Mol Cell Res 1763:1733–1748

Subramani S (1998) Components involved in peroxisome import, biogenesis, proliferation, turnover, and movement. Physiol Rev 78:171–188

Subramani S, Koller A, Snyder WB (2000) Import of peroxisomal matrix and membrane proteins. Annu Rev Biochem 69:399–418

Tabak HF, Murk JL, Braakman I, Geuze HJ (2003) Peroxisomes start their life in the endoplasmic reticulum. Traffic 4:512–518

Tabak HF, Braakman I, van der Zand A (2013) Peroxisome formation and maintenance are dependent on the endoplasmic reticulum. Annu Rev Biochem 82:723–744

Tamura S, Okumoto K, Toyama R, Shimozawa N, Tsukamoto T, Suzuki Y, Osumi T, Kondo N, Fujiki Y (1998a) Human *PEX1* cloned by functional complementation on a CHO cell mutant is responsible for peroxisome-deficient Zellweger syndrome of complementation group I. Proc Natl Acad Sci USA 95:4350–4355

Tamura S, Shimozawa N, Suzuki Y, Tsukamoto T, Osumi T, Fujiki Y (1998b) A cytoplasmic AAA family peroxin, Pex1p, interacts with Pex6p. Biochem Biophys Res Commun 245:883–886

Tamura S, Matsumoto N, Imamura A, Shimozawa N, Suzuki Y, Kondo N, Fujiki Y (2001) Phenotype-genotype relationships in peroxisome biogenesis disorders of *PEX1*-defective complementation group 1 are defined by Pex1p-Pex6p interaction. Biochem J 357:417–426

Tanaka A, Okumoto K, Fujiki Y (2003) cDNA cloning and characterization of the third isoform of human peroxin Pex11p. Biochem Biophys Res Commun 300:819–823

Tanaka A, Kobayashi S, Fujiki Y (2006) Peroxisome division is impaired in a CHO cell mutant with an inactivating point-mutation in dynamin-like protein 1 gene. Exp Cell Res 312:1671–1684

Thieringer H, Moellers B, Dodt G, Kunau W-H, Driscoll M (2003) Modeling human peroxisome biogenesis disorders in the nematode *Caenorhabditis elegans*. J Cell Sci 116:1797–1804

Tsukamoto T, Miura S, Fujiki Y (1991) Restoration by a 35K membrane protein of peroxisome assembly in a peroxisome-deficient mammalian cell mutant. Nature 350:77–81

Tsukamoto T, Miura S, Nakai T, Yokota S, Shimozawa N, Suzuki Y, Orii T, Fujiki Y, Sakai F, Bogaki A, Yasumo H, Osumi T (1995) Peroxisome assembly factor-2, a putative ATPase cloned by functional complementation on a peroxisome-deficient mammalian cell mutant. Nat Genet 11:395–401

van der Zand A, Braakman I, Tabak HF (2010) Peroxisomal membrane proteins insert into the endoplasmic reticulum. Mol Biol Cell 21:2057–2065

Walter C, Gootjes J, Mooijer PA, Portsteffen H, Klein C, Waterham HR, Barth PG, Epplen JT, Kunau W-H, Wanders RJA, Dodt G (2001) Disorders of peroxisome biogenesis due to mutations in *PEX1*: phenotypes and PEX1 protein levels. Am J Hum Genet 69:35–48

Wanders RJ, Waterham HR (2006) Peroxisomal disorders: the single peroxisomal enzyme deficiencies. Biochim Biophys Acta 1763:1707–1720

Warren DS, Morrell JC, Moser HW, Valle D, Gould SJ (1998) Identification of *PEX10*, the gene defective in complementation group 7 of the peroxisome-biogenesis disorders. Am J Hum Genet 63:347–359

Waterham HR, Koster J, van Roermund CWT, Mooyer PAW, Wanders RJA, Leonard JV (2007) A lethal defect of mitochondrial and peroxisomal fission. N Engl J Med 356:1736–1741

Weller S, Gould SJ, Valle D (2003) Peroxisome biogenesis disorders. Annu Rev Genomics Hum Genet 4:165–211

Weller S, Cajigas I, Morrell J, Obie C, Steel G, Gould SJ, Valle D (2005) Alternative splicing suggests extended function of *PEX26* in peroxisome biogenesis. Am J Hum Genet 76:987–1007

Wiemer EA, Nuttley WM, Bertolaet BL, Li X, Francke U, Wheelock MJ, Anne UK, Johnson KR, Subramani S (1995) Human peroxisomal targeting signal-1 receptor restores peroxisomal protein import in cells from patients with fatal peroxisomal disorders. J Cell Biol 130:51–65

Williams C, van den Berg M, Sprenger RR, Distel B (2007) A conserved cysteine is essential for Pex4p-dependent ubiquitination of the peroxisomal import receptor Pex5p. J Biol Chem 282:22534–22543

Yagita Y, Hiromasa T, Fujiki Y (2013) Tail-anchored PEX26 targets peroxisomes via a PEX19-dependent and TRC40-independent class I pathway. J Cell Biol 200:651–666

Yahraus T, Braverman N, Dodt G, Kalish JE, Morrell JC, Moser HW, Valle D, Gould SJ (1996) The peroxisome biogenesis disorder group 4 gene, *PXAAA1*, encodes a cytoplasmic ATPase required for stability of the PTS1 receptor. EMBO J 15:2914–2923

Yano T, Oku M, Akeyama N, Itoyama A, Yurimoto H, Kuge S, Fujiki Y, Sakai Y (2010) A novel fluorescent sensor protein for visualization of redox states in the cytoplasm and in peroxisomes. Mol Cell Biol 30:3758–3766

Yonekawa S, Furuno A, Baba T, Fujiki Y, Ogasawara Y, Yamamoto A, Tagaya M, Tani K (2011) Sec16B is involved in the endoplasmic reticulum export of the peroxisomal membrane biogenesis factor peroxin 16 (Pex16) in mammalian cells. Proc Natl Acad Sci USA 108:12746–12751

Expanding the Clinical Phenotypes of Peroxisome Biogenesis Disorders: PEX11 Function in Health and Disease

6

Tom Kettelhut and Sven Thoms

Abstract

Pex11p and related proteins, collectively termed PPPs for Pex11p-type peroxisome proliferators, constitute an unusual group of proteins among the other Pex proteins. PPPs are genuine peroxisome proliferators, although they are not directly involved in matrix and membrane protein import. PPPs directly shape the peroxisomal membrane, but they also cooperate with fission proteins located at the peroxisome. Only recently, a relatively mild form of human peroxisome biogenesis disorder (PBD) associated with a defect in *PEX11β* could be identified. This finding provides new insight into the function of PPPs, expands the spectrum of clinical PBD phenotypes, and stresses the need for research into PBD pathophysiology and therapy.

Keywords

PEX11 • PEX11p • PEX11p-type of peroxisome proliferators (PPP) • Peroxisome biogenesis disorders (PBD) • Zellweger syndrome • Dynamin-like protein 1 (DLP1) • Peroxisome proliferation • Membrane curvature • PEX11β – PBD therapy

6.1 Introduction

Pex11 proteins are outliers among the Pex proteins: They are not known to be involved in the formation of the major protein complexes required for the formation of peroxisomes, and they show few interactions with other Pex proteins. In agreement with their special status, until 2012, no complementation group could be associated with human Pex11 proteins. *PEX11β* is now the latest addition to the

T. Kettelhut • S. Thoms (✉)
Division of Neuropediatrics, Department of Child and Adolescent Health, University Medical Center, University of Göttingen, Robert Koch Strasse 40, 37075 Göttingen, Germany
e-mail: sven.thoms@med.uni-goettingen.de

C. Brocard and A. Hartig (eds.), *Molecular Machines Involved in Peroxisome Biogenesis and Maintenance*, DOI 10.1007/978-3-7091-1788-0_6, © Springer-Verlag Wien 2014

human *PEX* genes known to be associated with peroxisomal disorders. With a surprisingly mild phenotype, the *PEX11β* patient expands the spectrum of peroxisome biogenesis disorder (PBD). Not the classical biochemical parameters but peroxisome morphology played a decisive role in identification and diagnosis of the *PEX11β* disease—peroxisomes in the patient's fibroblasts appeared enlarged as if they were not divided.

In this chapter, we will introduce the Pex11 protein family and highlight structural and functional aspects of this remarkable group of peroxisome proliferators. We then discuss models for their mechanism of action and the relationship of Pex11 proteins with factors required for organelle division. In the second part of the chapter, we discuss PBD with a focus on the recently discovered *PEX11β* disease.

6.2 Pex11p and Pex11p-Type Peroxisome Proliferators

PEX11 is highly conserved in evolution. In all species with peroxisomes, at least three paralogues are found. To mention only the best-characterized members of the Pex11 protein family: Pex11p, Pex25p, and Pex27p in yeasts, Pex11a to Pex11e in plants, Pex11αp, Pex11βp, and Pex11γp in mammalian, and Pex11p together with two Gim5p proteins in the trypanosome. Interestingly, the evolutionary history of PPPs indicates that the gene multiplication leading to the various isoforms occurred several times independently in evolution (Orth et al. 2007).

Pex11 proteins are involved in peroxisome proliferation (Thoms and Erdmann 2005). Cells lacking Pex11p show fewer but larger peroxisomes while an overexpression of *PEX11* generally leads to peroxisome elongation and proliferation with smaller, in some cases aggregated peroxisomes. In the yeast *Saccharomyces cerevisiae*, where oleate causes only a modest increase in peroxisome number, *PEX11* is among the genes whose expression is strongly increased by oleate. The Pex11 protein is possibly the most important mediator of oleate function on peroxisome proliferation. It is the remarkable ability of Pex11p family members to stimulate peroxisome proliferation, that the protein family has been termed Pex11p-type peroxisome proliferators (PPPs; Thoms and Erdmann 2005). PPP-mediated peroxisome proliferation progresses through several steps: First, formation of a peroxisomal subdomain as the center for the formation of new peroxisomes mediated mainly by Pex11βp; second, membrane expansion and integration of most peroxisomal membrane proteins; third, constriction of the peroxisome probably mediated by PPPs, Mff and hFis1; fourth, assembly of the peroxisome matrix protein import machinery, and last, fission of the membrane catalyzed by the dynamin-like protein DLP1 (Delille et al. 2010).

In mammalian cells, only *PEX11α* expression is increased by external stimuli-like fibrates or phthalate plasticizers (Abe et al. 1998; Schrader et al. 1998). But those proliferators do not influence the expression of *PEX11β* and *PEX11γ* (Abe and Fujiki 1998; Tanaka et al. 2003). The mRNA levels of *PEX11α* vary widely between different tissues: *PEX11α* mRNA is abundant in kidney, lung, brain, liver, and testis, tissues sensitive to peroxisome-proliferating agents. Low levels of

Table 6.1 TMD prediction and experimental verification for the three human PPP proteins

TMD predictor	Hs PEX11α		Hs PEX11β		Hs PEX11γ	
HMMTOP[a]		220–239		230–251	125–142	209–226
Top Pred II/0.01[b]		219–239	168–188	232–252	125–145	209–229
TMpred[c]	94–114	220–239		230–255	127–149	209–227
TMHMM[d]		220–239		233–255		
ΔG-scale[e]	84–106	220–239	80–98	186–204 235–257	127–149	204–226
SCAMPI-seq[f]			79–99	183–203 235–255	128–148	212–232
SCAMPI-msa[f]			122–142	232–252	129–149	212–232
Experimental[g]			90–110	230–255		

The table lists amino acid positions of predicted and experimentally verified TMD in PPPs
[a]HMMTOP: http://www.enzim.hu/hmmtop/ (Tusnády and Simon 2001)
[b]Top Pred II/0.01: http://mobyle.pasteur.fr (Claros and von Heijne 1994)
[c]TMpred: http://www.ch.embnet.org/software/TMPRED_form.html (Hofmann and Stoffel 1993)
[d]TMHMM Server v. 2.0: http://www.cbs.dtu.dk/services/TMHMM/ (Krogh et al. 2001)
[e]ΔG-scale: http://dgpred.cbr.su.se/index.php?p=fullscan (Hessa et al. 2007)
[f]SCAMPI: http://scampi.cbr.su.se/ (Bernsel et al. 2008)
[g]Experimental: PEX11β (Bonekamp et al. 2013)

PEX11α were measured in spleen, heart, and skeletal muscle. In contrast, mRNA levels of *PEX11β* vary to a lesser extent between different tissues (Schrader et al. 1998).

Predictions of topology and transmembrane domains (TMDs) indicate that Pex11αp and Pex11βp have at least one TMD (Table 6.1), in contrast to Pex11γp which is predicted to have two TMDs. Recent experimental analysis of the topology of Pex11βp shows that this protein is an integral membrane protein with two TMDs (Bonekamp et al. 2013). One TMD is, as predicted, at position 230–255. The first TMD is experimentally determined to span amino acids 90–110 (Table 6.1). The two termini of the protein face the cytosol (Abe and Fujiki 1998; Bonekamp et al. 2013). An analysis of the topology of Pex11γp suggests that the two membrane-bound regions do not span the membrane, but insert only into one leaflet of the bilayer (Koch and Brocard 2012). In consequence, the loop domain between the two hydrophobic regions would be exposed to the cytosol (Fig. 6.1). Interestingly, the two membrane-bound stretches in Pex11γp surround a predicted amphipathic helix (Koch and Brocard 2012). Similarly, the N-termini of Pex11αp and Pex11βp, and Pex11 proteins from various yeast species contain putative amphipathic helices. These have been shown to be involved in membrane elongation (Opaliński et al. 2011; Koch and Brocard 2012). Although it is too early to speculate about how PPPs generally shape the membrane, it is fascinating that different PPPs evolved non-conserved amphipathic helices controlling membrane curvature and elongation. A detailed understanding of these functions will have to await X-ray structures of PPPs, which are not available at the time, because PPPs are difficult to purify.

Grouping Pex11p and its homologues under the term PPP is not to suggest that all PPPs essentially exert the same function as striking differences within the PPP paralogues are now emerging. For example in yeast, Pex25p is required for the

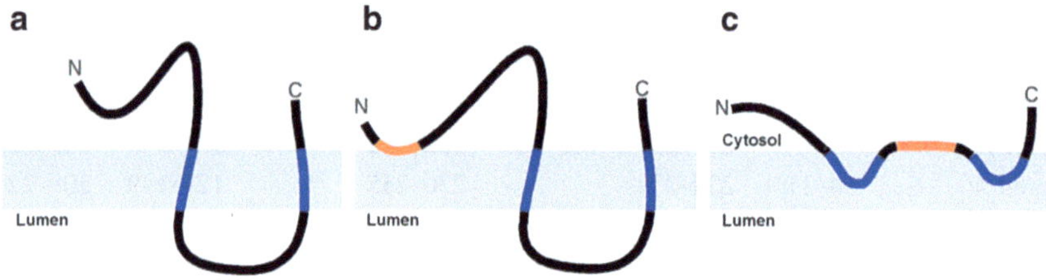

Fig. 6.1 Schematic representation of PPP topologies and amphipathic helices. (**a**) Pex11βp has two TMDs (transmembrane domains, *blue*). Both termini face the cytosol; the protein therefore has an intraluminal domain. The topology of Pex11αp is possibly similar. (**b**) The N-termini of Pex11αp and Pex11βp and Pex11p orthologues from various yeast species contain putative amphipathic helices (*orange*), which would allow them to bind to a lipid bilayer, partially integrate into the membrane, and promote membrane curvature and elongation. (**c**) Pex11γp has, like Pex11βp, two hydrophobic stretches (*blue*). An intervening loop contains a putative amphipathic helix. Like the N-terminal amphipathic helix in B, the partial introduction of the amphipathic helix into the membrane could selectively enlarge the outer membrane leaflet, generating membrane curvature needed for peroxisome expansion and/or peroxisome division

formation of new peroxisomes, at the ER, and Pex27p exerts a negative effect on peroxisome elongation, possibly by interfering with the binding of other PPPs (Saraya et al. 2011; Huber et al. 2012).

6.3 Mechanism of PPP Action

How, mechanistically, are PPPs involved in peroxisome proliferation? When the topic of the mechanistic action of PPPs was first reviewed a decade ago, four mutually non-exclusive modes of PPP functioning were suggested (Thoms and Erdmann 2005): (1) PPPs act as structural components of the peroxisomal membrane, directly affecting membrane structure and elongation. (2) PPPs are the recruitment proteins for other factors involved in organelle division. (3) PPPs are directly involved in fatty acid metabolism. (4) PPPs act like metabolite transporters. While the last two of these hypotheses have received little experimental support, there is now convincing evidence that PPPs can directly affect membrane structure and also recruit other proteins involved in peroxisome division. We support a model where PPPs exert direct effect on membrane elongation, and, in a subsequential step, recruit factors that further downstream catalyze membrane fission.

6.4 PPPs Shape the Peroxisomal Membrane

The first part of this model suggests that the membrane-to-be-expanded is covered with PPP species, and the extent of elongation is approximately proportional to the PPPs bound to the membrane. This model is in agreement with most of what is known about *PEX11*: First, overexpression of PPPs can expand and proliferate peroxisomes in all species; second, *PEX11* in yeast is the most strongly upregulated gene upon oleate treatment indicating that *PEX11* is functional on its own; and third, membrane-active, amphipathic helices have been identified that have the potential to insert into membranes influencing membrane curvature and elongate peroxisomes. The wedge-like partial integration of an amphipathic helix into the membrane expands one leaflet of the membrane, creating membrane curvature in a manner required for fission/fusion processes (McMahon et al. 2010). To regulate membrane curvature generation, elongation, constriction, and fission in a coordinated manner, PPPs have to form organized assemblies on the membrane. This is achieved by oligomerization. For example, yeast and mammalian PPPs have been shown to form homodimers (Marshall et al. 1996; Kobayashi et al. 2007; Koch et al. 2010; Bonekamp et al. 2013) and hetero-oligomerization has been described for the human PPPs (Koch et al. 2010).

6.5 PPPs Interact with Peroxisome Division Proteins

The PPPs are connected to the mitochondrial organelle division machinery. In mitochondria, the equilibrium between division and fusion is vital. If fission is inhibited, or fusion factors are overexpressed, mitochondria form a strongly interconnected network. If fission outweighs fusion, mitochondria vesiculate. Both can be detrimental to mitochondrial function and inheritance of the mitochondrial genome. Especially in nerve cells such abnormalities have constitute a special burden (Itoh et al. 2013). Interestingly, only the molecular machinery required for mitochondrial fission (but not fusion) is shared between peroxisomes and mitochondria, and there is evidence that peroxisomes cannot fuse with each other (Huybrechts et al. 2009; Bonekamp et al. 2012).

Dynamin-like protein 1 (DLP1), the mitochondrial fission protein 1 (hFIS1), and mitochondrial fission factor (Mff) are key components for the mitochondrial and peroxisomal fission (Schrader 2006; Otera et al. 2010). DLP1 is a member of the dynamin family of large GTPases. Like dynamins, it can oligomerize and form collars around membrane constrictions and further constrict them, ultimately leading to GTPase-driven membrane scission. Thus DLP1 carries out the actual fission of the membrane. A human disease caused by a mutation in the DLP1 gene has a combined mitochondrial and peroxisomal phenotype, illustrating the fact that DLP1 is involved in both, mitochondrial and peroxisomal fission (Waterham et al. 2007).

hFis1 is an integral transmembrane protein of mitochondria, which interacts with the DLP1-mediated fission pathway in mitochondria and peroxisomes. hFis1 is thought to play a similar role at peroxisomes like at mitochondria that is to act as a molecular adaptor for the fission machinery and recruit DLP1 (Yoon et al. 2003; Koch et al. 2005). The tail-anchored protein Mff is a third protein shared between peroxisomes and mitochondria, and is involved in fission of both organelles (Gandre-Babbe and van der Bliek 2008; Otera et al. 2010).

Other recruiting factors for DLP1 to mitochondria are MiD49 and MiD51/MIEF1 (mitochondrial elongation factor 1) (Palmer et al. 2011; Zhao et al. 2011;

Losón et al. 2013). These proteins recruit DLP1 to mitochondria independently of Mff and hFis1 and inhibit GTP-binding to DLP1. As a result, fission is inhibited, and the mitochondria show a strongly interconnected network due to an excess of fusion (Zhao et al. 2011). In contrast, Mff-dependent recruitment of DLP1 stimulates fission of mitochondria (Otera et al. 2010).

In yeast, Fis1 is the key component for recruitment of Dnm1, the DLP1 homolog, with help of the two proteins Mdv1 (Mitochondrial division protein 1) and Caf4 (CCR4 Associated Factor) (Tieu and Nunnari 2000; Griffin et al. 2005; Motley et al. 2008). In mammalian cells, the two proteins Mdv1 and Caf4 have not been identified but a direct interaction between hFis1 and DLP1 was shown (Kobayashi et al. 2007). hFis1 has also been shown to directly interact with the C-terminal domain of Pex11βp (Kobayashi et al. 2007; Koch and Brocard 2012). By chemical cross-linking, however, a complex with DLP1, hFis1 and Pex11βp was found (Kobayashi et al. 2007). Surprisingly—given a role of Fis1 proteins in peroxisome division—hFis1 knockout cells have normal peroxisome morphology (Otera et al. 2010). Therefore, more information regarding the role of hFis1 in peroxisomal division is needed. Also Mff interacts with Pex11βp, and with lower affinity to the other PPPs, suggesting that Pex11βp has a special role in recruiting the division factors (Koch and Brocard 2012). Mff is localized at peroxisomal constrictions where also Pex11βp is found. Because of the same cellular localization and the interaction of Mff with Pex11βp, and the information that Mff is an adaptor for DLP1, it is likely that these proteins form a ternary complex at peroxisomal division sites (Itoyama et al. 2013).

PPPs interact among each other through hydrophobic domains forming heteromeric pairs: Pex11αp–Pex11γp and Pex11βp–Pex11γp. These pairs could represent different spacial or temporal aspects of peroxisome proliferation, in a manner that Pex11γp inclusion in the heterodimer could inhibit the recruitment of the division machinery (Koch et al. 2010). In yeasts, Pex11p is regulated by phosphorylation. A Pex11p mutant mimicking the constitutively phosphorylated form resembles the phenotype of *PEX11* overexpression with hyperproliferated peroxisomes, suggesting phosphorylation is required for the proliferative function of Pex11p (Knoblach and Rachubinski 2010; Joshi et al. 2012). It is an open question if other PPPs are subject to regulation by phosphorylation. Several putative phosphorylation sites can be found in human Pex11βp, including conserved sites at positions Ser11 and Ser38, both in the N-terminal domain, facing the cytosol, and thus possibly accessible to cytosolic kinases. However, mutation of Ser11 and Ser38 have no effect on the ability of human Pex11βp to proliferate peroxisomes (Bonekamp et al. 2013), suggesting that mammalian and yeast PPPs might be differently regulated.

All these observations are in agreement with a peroxisome division cascade consisting of constriction, recruitment, and division, which is triggered by the upstream acting PPPs. Therefore, Pex11βp, by constriction, prepares the peroxisomal membrane for division. It then attracts Mff (or Fis1 in yeast) by direct interaction with the peroxisomal membrane. This is followed by recruitment of DLP1 through Mff, leading to the formation of a ternary complex consisting of

Pex11βp, Mff and DLP1. Then DLP1 triggers GTPase-dependent oligomerization and scission. This cascade could be under the control of Pex11γp, which interacts with and controls the other PPPs.

6.6 Zellweger Syndrome and Milder Forms of Peroxisome Biogenesis Disorders

Zellweger syndrome (ZS), the prototypical peroxisome biogenesis disorder (PBD), represents the severe end of the PBD spectrum or ZS spectrum (ZSS) with neurological dysfunctions, including cortical dysplasia, and leukoencephalopathy, together with hypotonia, craniofacial dysmorphism, hepatomegaly, renal microcysts, and calcification of the patella (Preuss et al. 2002; Weller et al. 2003).

PBD is caused by mutations in genes essential for cellular peroxisome formation. These genes are termed *PEX* genes (peroxin-encoding genes). In humans, 16 *PEX* genes are known, including the three PPPs. The only disease not included in the ZSS is rhizomelic chondrodysplasia punctata (RCDP), caused by mutations in *PEX7*. *PEX1* is the most commonly affected *PEX* gene in PBD, accounting for more than 50 % of PBD cases. The most common *PEX1* mutation is c.2528G > A leading to a single amino acid exchange at position 844 (G843D). This mutation is, in many patients, associated with a comparably mild phenotype with survival into the second decade of life (Gärtner et al. 1999; Thoms et al. 2011).

Diagnostically important, PBD patients are characterized by elevated serum levels of very long-chain fatty acids (VLCFA) and bile acid intermediates, due to a deficiency of peroxisomal β-oxidation, which is involved in the degradation of VLCFAs, and in the final steps of bile acid biosynthesis. Elevated serum levels of phytanic acid can be found in PBD, because α-oxidation is also peroxisomal. Plasmalogen levels are low in PBD, because essential steps of plasmalogen synthesis are in the peroxisome. Whether these accumulating and depleted metabolites are mainly of diagnostic importance, or if they can also account for the full spectrum of pathology is an open question (Thoms et al. 2009).

The last decade has seen a considerable expansion of PBD phenotypes beyond the classical ZS and the ZSS with the integration of milder, previously different clinical entities. Remarkably, these cases include the peroxisomal membrane proteins Pex3p and Pex16p, both indispensable for the formation of peroxisomal membrane structures, and the RING peroxins Pex2p and Pex10p. Mutations in *PEX2* and *PEX10* have been found in association with autosomal recessive cerebellar ataxia with gait problems, axonal motor neuropathy, and progressive cerebellar atrophy. Interestingly, peroxisomal biochemical parameters were close to normal (Régal et al. 2010; Sevin et al. 2011). Another comparably mild form of PBD is associated with a mutation in *PEX16*. These PBD patients suffer from progressive spastic paraparesis, ataxia, cataracts, progressive demyelinating neuropathy, and leukodystrophy with biochemical parameters that are not clearly pathological (Ebberink et al. 2010). Also a patient with mild PBD caused by a missense mutation (D347Y) in *PEX3* was reported (Matsui et al. 2012). The patient presented with peripheral neuropathy, psychomotor regression, renal hypertension, a renal cyst, hearing impairment, and late-onset leukodystrophy. Due to this comparably mild phenotype, and the rarity of PBD, the diagnosis was found when the patient was over 30 years old.

Mouse models have been generated to study the pathogenesis of PBD (Baes and Van Veldhoven 2006, 2012). Mice with complete deletions of *PEX2*, *PEX5*, *PEX11β*, or *PEX13* genes phenocopy many aspects of ZS, the severe end of the ZSS, including brain abnormalities, hypotonia, and early death. The *PEX11β−/−* mice are generally similar to other ZS mice, but the characteristic import defect for the peroxisomal matrix proteins is not present. Also the plasmalogen and VLCFA levels show little deviation from normal (Li et al. 2002). The *PEX11β−/−* mouse challenges the idea that biochemical alterations and defective protein import into peroxisomes are the sole factors contributing to ZS pathophysiology (Thoms et al. 2009).

6.7 The First Patient with a Defect in PEX11β

In 2012, a patient with a mutation in a PPP gene was presented (Ebberink et al. 2012). The 26-year old male patient presented with mild intellectual disability, migraine-like episodes, gastrointestinal problems, and skin abnormalities. His peroxisomal biochemical parameters were nearly normal. In cultured fibroblasts, peroxisomes appeared enlarged and elongated, especially when the cells were cultivated at 40 °C. The patient had a homozygous point mutation in *PEX11β* leading to a truncation of the Pex11β protein after 21 of 259 amino acids (Ebberink et al. 2012). In this patient, *DLP1*, *hFIS1*, *PEX11α*, *PEX11γ*, and the *PEX* genes known to be associated with PBD were unaffected. The temperature dependence of the peroxisomal phenotype could possibly be explained by a downregulation of *PEX11γ* expression at higher temperature, in both, control and in patient cells. This effect could correspond to the patient's generally slow recovery after episodes of fever and indirectly suggests that *PEX11γ* can partially compensate *PEX11β* function.

This relatively mild phenotype raises questions about the physiological function of Pex11βp and the pathomechanism of the Pex11βp disease (Thoms and Gärtner 2012). One could speculate that Pex11βp is only required for proliferation of peroxisomes upon special stimuli. These, however, would still have to be identified, because Pex11αp, not Pex11βp, is the PPP responsive to classical peroxisome-proliferating drugs. Secondly, the defect in *PEX11β* could possibly be partially compensated by *PEX11γ*, which is found elevated in Pex11βp-deficient fibroblasts (Ebberink et al. 2012). Interestingly, Pex11γp is also found elevated in embryonic fibroblasts from *PEX11β−/−* mice (Ahlemeyer et al. 2012), so that partial compensation of *PEX11β* deficiency by *PEX11γ* is likely occurring in humans and in mice. Could it be that the patient's fragment of Pex11βp comprising 21 amino acids retains residual function? This is unlikely in light of a recent study, introducing a alanine to proline mutation at position 21, which breaks helix 2 and blocks the ability of Pex11βp-myc to elongate peroxisomes (Bonekamp et al. 2013). Fourth and very likely, enlarged and elongated peroxisomes could also be caused by insufficient recruitment of division proteins, due to the absence of functional Pex11βp. In this scenario, not the membrane-active function, but the recruitment function of Pex11βp would be important, and, when absent, can be partially compensated by alternative pathways of division factor recruitment.

6.8 Perspectives for PBD Therapy

Mild forms of PBD have important implications for diagnosis and therapy. In all mild forms of PBDs, peroxisomal matrix protein import is compromised to a lesser extent than in ZS. Consequentially, these patients might escape diagnosis by the classical biochemical parameters, because VLCFAs and plasmalogens are in the normal range. Management of PBD involves many clinical disciplines and focuses on treatment of sensory, developmental, metabolic, and orthopedic symptoms (Braverman et al. 2013). Supplementation of major peroxisomal metabolites is one way to address PBD therapeutically. While the results of docosahexaenoic acid (DHA) supplementation are controversial (Noguer and Martinez 2010; Paker et al. 2010), oral bile acid treatment (Maeda et al. 2002), and, in *PEX7*-deficient RCDP, treatment with bile acid precursors (Wood et al. 2011) are more promising. Patients with mild PBDs might profit from therapeutic interventions aimed at stimulating peroxisome proliferation or specifically enhancing residual *PEX* functions. These and other therapeutic approaches are experimental and under development. 4-Phenylbutyrate has been reported to stimulate peroxisome proliferation in fibroblasts form ZS patients independently of PPARα (Wei et al. 2000). Of special interest are chemical chaperones that would stabilize the PEX1 allele c.2528G > A (G843D), the most common ZS mutation. A small-molecule screen for peroxisome restoration in a *PEX1*-G843D cell line identified several promising compounds that have to be further evaluated (Zhang et al. 2010). In cases where the *PEX* gene mutation is due to a premature stop codon, a general stimulation of translational read-through of stop codons by aminoglycoside antibiotics might be an option (Dranchak et al. 2011).

6.9 Outlook

The new *PEX11β* patient brings up a number of old and new questions revolving around PBD and PPPs. First, the discovery has reopened the quests for PPP function. While the ability to proliferate peroxisomes is fundamental to PPPs, the mechanism and the cellular integration of the individual members and their function still await more experimental input. If some PPPs are connected to the endoplasmic reticulum is a special intriguing question in this respect. Second, the diagnostic challenges associated with mild forms of PBD: how to diagnose when manifestation is through an alteration in peroxisome morphology and diagnosis ideally depends on microscopy on skin fibroblasts? Patients with normal or near-normal biochemical parameters are likely to be overlooked with the present diagnostic schemes. Third, when trying to understand the pathomechanisms involved in PBD, phenotypes with altered peroxisomes in the absence of gross biochemical changes shift the focus from the metabolic functions of peroxisomes to their cellular interactions with other organelles and cellular structures (Thoms et al. 2009). Importantly, the discovery of new mild forms of PBD should stimulate research into causal therapeutic interventions.

Acknowledgments We thank Rosemol George, Noa Lipstein, and Michael Schrader for comments on the manuscript.

References

Abe I, Fujiki Y (1998) cDNA cloning and characterization of a constitutively expressed isoform of the human peroxin Pex11p. Biochem Biophys Res Commun 252:529–533. doi:10.1006/bbrc.1998.9684

Abe I, Okumoto K, Tamura S, Fujiki Y (1998) Clofibrate-inducible, 28-kDa peroxisomal integral membrane protein is encoded by PEX11. FEBS Lett 431:468–472

Ahlemeyer B, Gottwald M, Baumgart-Vogt E (2012) Deletion of a single allele of the Pex11β gene is sufficient to cause oxidative stress, delayed differentiation and neuronal death in mouse brain. Dis Model Mech 5:125–140. doi:10.1242/dmm.007708

Baes M, Van Veldhoven PP (2006) Generalised and conditional inactivation of Pex genes in mice. Biochim Biophys Acta 1763:1785–1793. doi:10.1016/j.bbamcr.2006.08.018

Baes M, Van Veldhoven PP (2012) Mouse models for peroxisome biogenesis defects and β-oxidation enzyme deficiencies. Biochim Biophys Acta 1822:1489–1500. doi:10.1016/j.bbadis.2012.03.003

Bernsel A, Viklund H, Falk J et al (2008) Prediction of membrane-protein topology from first principles. Proc Natl Acad Sci USA 105:7177–7181. doi:10.1073/pnas.0711151105

Bonekamp NA, Sampaio P, de Abreu FV et al (2012) Transient complex interactions of mammalian peroxisomes without exchange of matrix or membrane marker proteins. Traffic 13:960–978. doi:10.1111/j.1600-0854.2012.01356.x

Bonekamp NA, Grille S, Cardoso MJ et al (2013) Self-interaction of human Pex11pβ during peroxisomal growth and division. PLoS One 8:e53424. doi:10.1371/journal.pone.0053424

Braverman NE, D'Agostino MD, Maclean GE (2013) Peroxisome biogenesis disorders: biological, clinical and pathophysiological perspectives. Dev Disabil Res Rev 17:187–196. doi:10.1002/ddrr.1113

Claros MG, von Heijne G (1994) TopPred II: an improved software for membrane protein structure predictions. Comput Appl Biosci 10:685–686

Delille HK, Agricola B, Guimaraes SC et al (2010) Pex11p-mediated growth and division of mammalian peroxisomes follows a maturation pathway. J Cell Sci 123:2750–2762. doi:10.1242/jcs.062109

Dranchak PK, Di Pietro E, Snowden A et al (2011) Nonsense suppressor therapies rescue peroxisome lipid metabolism and assembly in cells from patients with specific PEX gene mutations. J Cell Biochem 112:1250–1258. doi:10.1002/jcb.22979

Ebberink MS, Csanyi B, Chong WK et al (2010) Identification of an unusual variant peroxisome biogenesis disorder caused by mutations in the PEX16 gene. J Med Genet 47:608–615. doi:10.1136/jmg.2009.074302

Ebberink MS, Koster J, Visser G et al (2012) A novel defect of peroxisome division due to a homozygous nonsense mutation in the PEX11beta gene. J Med Genet 49(5):307–313

Gandre-Babbe S, van der Bliek AM (2008) The novel tail-anchored membrane protein Mff controls mitochondrial and peroxisomal fission in mammalian cells. Mol Biol Cell 19:2402–2412. doi:10.1091/mbc.E07-12-1287

Gärtner J, Preuss N, Brosius U, Biermanns M (1999) Mutations in PEX1 in peroxisome biogenesis disorders: G843D and a mild clinical phenotype. J Inherit Metab Dis 22:311–313

Griffin EE, Graumann J, Chan DC (2005) The WD40 protein Caf4p is a component of the mitochondrial fission machinery and recruits Dnm1p to mitochondria. J Cell Biol 170:237–248. doi:10.1083/jcb.200503148

Hessa T, Meindl-Beinker NM, Bernsel A et al (2007) Molecular code for transmembrane-helix recognition by the Sec61 translocon. Nature 450:1026–1030. doi:10.1038/nature06387

Hofmann K, Stoffel W (1993) TMbase—a database of membrane spanning proteins segments (Meeting Abstract). Biol Chem Hoppe Seyler 374:166

Huber A, Koch J, Kragler F et al (2012) A subtle interplay between three Pex11 proteins shapes de novo formation and fission of peroxisomes. Traffic 13:157–167. doi:10.1111/j.1600-0854.2011.01290.x

Huybrechts SJ, Van Veldhoven PP, Brees C et al (2009) Peroxisome dynamics in cultured mammalian cells. Traffic 10:1722–1733. doi:10.1111/j.1600-0854.2009.00970.x

Itoh K, Nakamura K, Iijima M, Sesaki H (2013) Mitochondrial dynamics in neurodegeneration. Trends Cell Biol 23:64–71. doi:10.1016/j.tcb.2012.10.006

Itoyama A, Michiyuki S, Honsho M et al (2013) Mff functions with Pex11pβ and DLP1 in peroxisomal fission. Biol Open 2:998–1006. doi:10.1242/bio.20135298

Joshi S, Agrawal G, Subramani S (2012) Phosphorylation-dependent Pex11p and Fis1p interaction regulates peroxisome division. Mol Biol Cell 23:1307–1315. doi:10.1091/mbc.E11-09-0782

Knoblach B, Rachubinski RA (2010) Phosphorylation-dependent activation of peroxisome proliferator protein PEX11 controls peroxisome abundance. J Biol Chem 285:6670–6680. doi:10.1074/jbc.M109.094805

Kobayashi S, Tanaka A, Fujiki Y (2007) Fis1, DLP1, and Pex11p coordinately regulate peroxisome morphogenesis. Exp Cell Res 313:1675–1686. doi:10.1016/j.yexcr.2007.02.028

Koch J, Brocard C (2012) PEX11 proteins attract Mff and human Fis1 to coordinate peroxisomal fission. J Cell Sci 125:3813–3826. doi:10.1242/jcs.102178

Koch A, Yoon Y, Bonekamp NA et al (2005) A role for Fis1 in both mitochondrial and peroxisomal fission in mammalian cells. Mol Biol Cell 16:5077–5086. doi:10.1091/mbc.E05-02-0159

Koch J, Pranjic K, Huber A et al (2010) PEX11 family members are membrane elongation factors that coordinate peroxisome proliferation and maintenance. J Cell Sci 123:3389–3400. doi:10.1242/jcs.064907

Krogh A, Larsson B, von Heijne G, Sonnhammer EL (2001) Predicting transmembrane protein topology with a hidden Markov model: application to complete genomes. J Mol Biol 305:567–580. doi:10.1006/jmbi.2000.4315

Li X, Baumgart E, Morrell JC et al (2002) PEX11 beta deficiency is lethal and impairs neuronal migration but does not abrogate peroxisome function. Mol Cell Biol 22:4358–4365

Losón OC, Song Z, Chen H, Chan DC (2013) Fis1, Mff, MiD49, and MiD51 mediate Drp1 recruitment in mitochondrial fission. Mol Biol Cell 24:659–667. doi:10.1091/mbc.E12-10-0721

Maeda K, Kimura A, Yamato Y et al (2002) Oral bile Acid treatment in two Japanese patients with Zellweger syndrome. J Pediatr Gastroenterol Nutr 35:227–230

Marshall PA, Dyer JM, Quick ME, Goodman JM (1996) Redox-sensitive homodimerization of Pex11p: a proposed mechanism to regulate peroxisomal division. J Cell Biol 135:123–137

Matsui S, Funahashi M, Honda A, Shimozawa N (2012) Newly identified milder phenotype of peroxisome biogenesis disorder caused by mutated PEX3 gene. Brain Dev 35(9):842–849. doi:10.1016/j.braindev.2012.10.017

McMahon HT, Kozlov MM, Martens S (2010) Membrane curvature in synaptic vesicle fusion and beyond. Cell 140:601–605. doi:10.1016/j.cell.2010.02.017

Motley AM, Ward GP, Hettema EH (2008) Dnm1p-dependent peroxisome fission requires Caf4p, Mdv1p and Fis1p. J Cell Sci 121:1633–1640. doi:10.1242/jcs.026344
Noguer MT, Martinez M (2010) Visual follow-up in peroxisomal-disorder patients treated with docosahexaenoic acid ethyl ester. Invest Ophthalmol Vis Sci 51:2277–2285. doi:10.1167/iovs.09-4020
Opaliński Ł, Kiel JAKW, Williams C et al (2011) Membrane curvature during peroxisome fission requires Pex11. EMBO J 30:5–16. doi:10.1038/emboj.2010.299
Orth T, Reumann S, Zhang X et al (2007) The PEROXIN11 protein family controls peroxisome proliferation in Arabidopsis. Plant Cell 19:333–350. doi:10.1105/tpc.106.045831
Otera H, Wang C, Cleland MM et al (2010) Mff is an essential factor for mitochondrial recruitment of Drp1 during mitochondrial fission in mammalian cells. J Cell Biol 191:1141–1158. doi:10.1083/jcb.201007152
Paker AM, Sunness JS, Brereton NH et al (2010) Docosahexaenoic acid therapy in peroxisomal diseases: results of a double-blind, randomized trial. Neurology 75:826–830. doi:10.1212/WNL.0b013e3181f07061
Palmer CS, Osellame LD, Laine D et al (2011) MiD49 and MiD51, new components of the mitochondrial fission machinery. EMBO Rep 12:565–573. doi:10.1038/embor.2011.54
Preuss N, Brosius U, Biermanns M et al (2002) PEX1 mutations in complementation group 1 of Zellweger spectrum patients correlate with severity of disease. Pediatr Res 51:706–714
Régal L, Ebberink MS, Goemans N et al (2010) Mutations in PEX10 are a cause of autosomal recessive ataxia. Ann Neurol 68:259–263. doi:10.1002/ana.22035
Saraya R, Krikken AM, Veenhuis M, van der Klei IJ (2011) Peroxisome reintroduction in Hansenula polymorpha requires Pex25 and Rho1. J Cell Biol 193:885–900. doi:10.1083/jcb.201012083
Schrader M (2006) Shared components of mitochondrial and peroxisomal division. Biochim Biophys Acta 1763:531–541. doi:10.1016/j.bbamcr.2006.01.004
Schrader M, Reuber BE, Morrell JC et al (1998) Expression of PEX11beta mediates peroxisome proliferation in the absence of extracellular stimuli. J Biol Chem 273:29607–29614
Sevin C, Ferdinandusse S, Waterham HR et al (2011) Autosomal recessive cerebellar ataxia caused by mutations in the PEX2 gene. Orphanet J Rare Dis 6:8. doi:10.1186/1750-1172-6-8
Tanaka A, Okumoto K, Fujiki Y (2003) cDNA cloning and characterization of the third isoform of human peroxin Pex11p. Biochem Biophys Res Commun 300:819–823
Thoms S, Erdmann R (2005) Dynamin-related proteins and Pex11 proteins in peroxisome division and proliferation. FEBS J 272:5169–5181. doi:10.1111/j.1742-4658.2005.04939.x
Thoms S, Gärtner J (2012) First PEX11β patient extends spectrum of peroxisomal biogenesis disorder phenotypes. J Med Genet 49:314–316. doi:10.1136/jmedgenet-2012-100899
Thoms S, Grønborg S, Gärtner J (2009) Organelle interplay in peroxisomal disorders. Trends Mol Med 15:293–302. doi:10.1016/j.molmed.2009.05.002
Thoms S, Grønborg S, Rabenau J et al (2011) Characterization of two common 5' polymorphisms in PEX1 and correlation to survival in PEX1 peroxisome biogenesis disorder patients. BMC Med Genet 12:109. doi:10.1186/1471-2350-12-109
Tieu Q, Nunnari J (2000) Mdv1p is a WD repeat protein that interacts with the dynamin-related GTPase, Dnm1p, to trigger mitochondrial division. J Cell Biol 151:353–366
Tusnády GE, Simon I (2001) The HMMTOP transmembrane topology prediction server. Bioinformatics 17:849–850
Waterham HR, Koster J, van Roermund CWT et al (2007) A lethal defect of mitochondrial and peroxisomal fission. N Engl J Med 356:1736–1741. doi:10.1056/NEJMoa064436
Wei H, Kemp S, McGuinness MC et al (2000) Pharmacological induction of peroxisomes in peroxisome biogenesis disorders. Ann Neurol 47:286–296
Weller S, Gould SJ, Valle D (2003) Peroxisome biogenesis disorders. Annu Rev Genomics Hum Genet 4:165–211. doi:10.1146/annurev.genom.4.070802.110424
Wood PL, Khan MA, Smith T et al (2011) In vitro and in vivo plasmalogen replacement evaluations in rhizomelic chrondrodysplasia punctata and Pelizaeus-Merzbacher disease

using PPI-1011, an ether lipid plasmalogen precursor. Lipids Health Dis 10:182. doi:10.1186/1476-511X-10-182

Yoon Y, Krueger EW, Oswald BJ, McNiven MA (2003) The mitochondrial protein hFis1 regulates mitochondrial fission in mammalian cells through an interaction with the dynamin-like protein DLP1. Mol Cell Biol 23:5409–5420

Zhang R, Chen L, Jiralerspong S et al (2010) Recovery of PEX1-Gly843Asp peroxisome dysfunction by small-molecule compounds. Proc Natl Acad Sci USA 107:5569–5574. doi:10.1073/pnas.0914960107

Zhao J, Liu T, Jin S et al (2011) Human MIEF1 recruits Drp1 to mitochondrial outer membranes and promotes mitochondrial fusion rather than fission. EMBO J 30:2762–2778. doi:10.1038/emboj.2011.198

Peroxisome Morphology in Pathologies 7

Michael Schrader, Inês Castro, H. Dariush Fahimi, and Markus Islinger

Abstract

Peroxisomes are ubiquitous and heterogeneous multi-purpose organelles, which are indispensable for human health and development. The invention of specific cytochemical staining methods for peroxisomes revealed their high plasticity and ability to alter their morphology in response to environmental cues. Peroxisome dynamics depend on peroxisomal morphology proteins such as Pex11p, DLP1/Drp1, Fis1, Mff, and GDAP1 which are partially shared with mitochondria. Here, we address variations of peroxisome morphology in the healthy organism and summarize findings on altered organelle morphology in peroxisomal disorders. We highlight recent insights in novel disorders with defects in peroxisome morphology proteins and alterations of peroxisomes during stress and signaling, as well as secondary alterations in liver disease and cancer.

M. Schrader (✉)
College of Life and Environmental Sciences, Biosciences, University of Exeter, Geoffrey Pope Building, Stocker Road, Exeter EX4 4QD, UK

Department of Biology, Centre for Cell Biology, University of Aveiro, Campus Universitário de Santiago, 3810-193 Aveiro, Portugal
e-mail: m.schrader@exeter.ac.uk

I. Castro
College of Life and Environmental Sciences, Biosciences, University of Exeter, Geoffrey Pope Building, Stocker Road, Exeter EX4 4QD, UK

H.D. Fahimi
Division of Medical Cell Biology, Department of Anatomy and Cell Biology, University of Heidelberg, 69120 Heidelberg, Germany

M. Islinger
Department of Neuroanatomy, Center for Biomedicine and Medical Technology Mannheim, University of Heidelberg, 68167 Mannheim, Germany

C. Brocard and A. Hartig (eds.), *Molecular Machines Involved in Peroxisome Biogenesis and Maintenance*, DOI 10.1007/978-3-7091-1788-0_7, © Springer-Verlag Wien 2014

Keywords
Peroxisomes • Organelle dynamics • Biogenesis disorders • Dynamin • Pex11 • Mff • GDAP1

Abbreviations

AOX	Acyl-CoA oxidase
CMT	Charcot-Marie-Tooth disease
D-BP	D-bifunctional protein
DAB	Diaminobenzidine
DHA	Docosahexaenoic acid
DLP1/Drp1	Dynamin-like/related protein 1
ER	Endoplasmic reticulum
GDAP1	Ganglioside-induced differentiation associated protein 1
Mff	Mitochondrial fission factor
PBD	Peroxisome biogenesis disorder
PEX	Peroxin
PMP	Peroxisomal membrane protein
PPAR	Peroxisome proliferator activated receptor
PTS	Peroxisomal targeting signal
ROS	Reactive oxygen species
SED	Single enzyme deficiency

7.1 Variation of Peroxisome Morphology in the Healthy Organism

Peroxisomes (microbodies) were first described morphologically in 1954 by the Swedish PhD student and later electron microscopy pioneer Johannes Rhodin who performed ultrastructural studies of the mouse kidney (Rhodin 1954). They belong to the basic equipment of the eukaryotic cell and are found in mammals, plants, and fungi. In ultrastructure studies, they present themselves as single-membrane bound subcellular compartments with a fine granular matrix harboring diverse enzymes and metabolic pathways for the detoxification of lipids and H_2O_2 (Fig. 7.1a) (Wanders and Waterham 2006). Peroxisomes represent a heterogeneous class of organelles which according to their evident differences in size in various tissues were already historically subdivided in core-containing peroxisomes from liver and kidney and smaller microperoxisomes, which were initially supposed to be specialized portions of the smooth ER and the precursors of the former (Novikoff and Novikoff 1973). In 1965, Christian de Duve (Nobel laureate in 1974) revealed the oxidative nature of the organelles and identified several H_2O_2-producing oxidases as well as catalase, a H_2O_2-degrading heme-containing enzyme, in the peroxisomal matrix. H. Dariush Fahimi and Alexander Novikoff independently

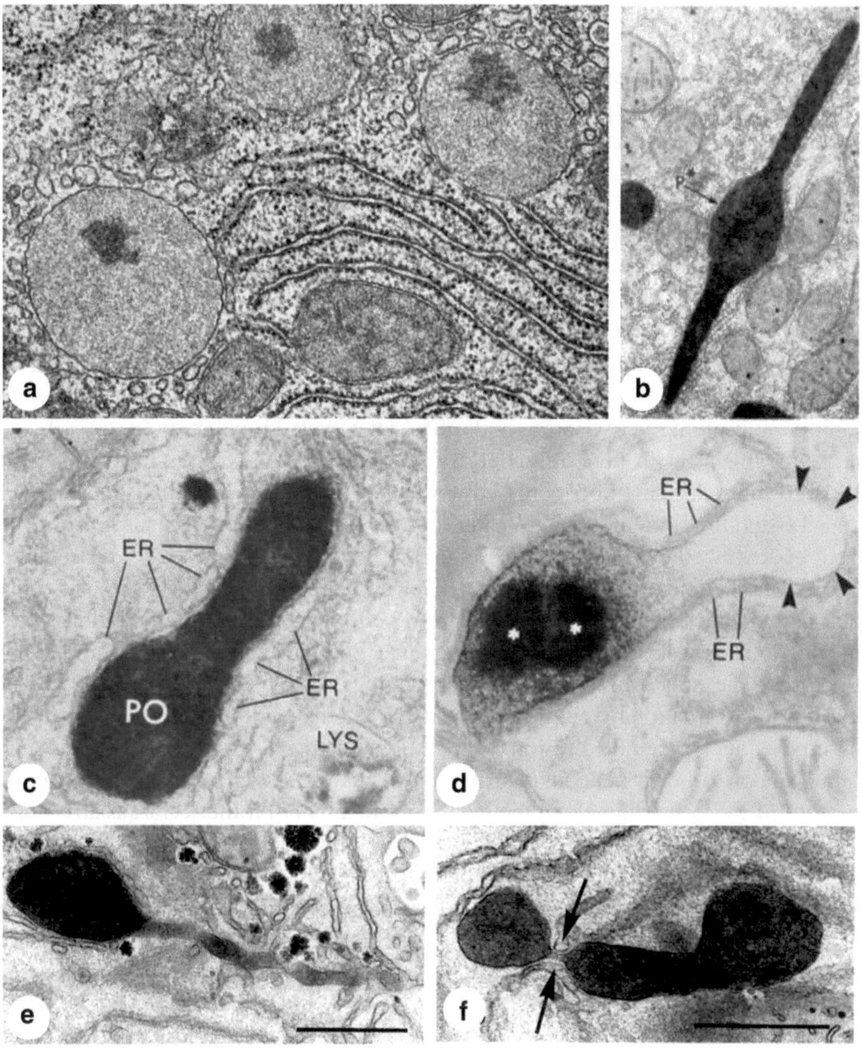

Fig. 7.1 Electron micrographs of peroxisomes from normal (**a**), bezafibrate-treated (**b**) and regenerating (**c–f**) rat livers. Some sections were stained cytochemically for catalase (Fahimi 1969) and urate oxidase (Angermuller and Fahimi 1986). (**a**) In this routine EM preparation of normal rat liver peroxisomes appear spherical with a distinct limiting membrane, a finely granular matrix and contain usually an electron dense crystalline inclusion (core). Magnification: ×30,000 [from Yokota and Fahimi (1978)]. (**b**) In rats treated for long periods with bezafibrate and stained for catalase peroxisomes proliferate and some of them exhibit an elongated biconcave shape. Such forms of peroxisomes have also been called "Phi bodies" and are observed in myelocytes in acute myeloid leukemias (Hanker and Romanovicz 1977) [from Fahimi et al. (1982)]. (**c**) A tubular peroxisome (PO) next to a lysosome (LYS) showing close association with endoplasmic reticulum (ER) from regenerating liver. Magnification: ×85,000. (**d**) In this elongated peroxisome, stained cytochemically for urate oxidase the staining is confined to the core region (*asterisks*) and does not extend to the expanding matrix (*arrowheads*). Magnification: ×88,000 [(**c**) and (**d**) from Fahimi et al. (1993)]. (**e**) A spherical peroxisome with an elongated tortuous tail-like extension.

exploited the peroxidatic reaction of catalase to develop a specific cytochemical staining method for peroxisomes (Fahimi 2009). This method is based on the reduction of diaminobenzidine (DAB) in an alkaline milieu resulting in the formation of an electron-dense, black precipitate, which is visible by light- and electron microscopy (Figs. 7.1 and 7.2). With the development of the alkaline DAB method the ultrastructure of peroxisomes was intensely studied during the 1970s and 1980s revealing an immense variation in form and structure. During these decades it became evident that peroxisomes are not merely spherical organelles but can also be found as tubules or elongated structures (Pipan and Psenicnik 1975; Pavelka et al. 1976; Hicks and Fahimi 1977; Roels et al. 1981; Gorgas and Volkl 1984; Gorgas and Zaar 1984; Gorgas 1984, 1985; Schrader et al. 1994) (Figs. 7.1 and 7.2). Often, elongated peroxisomes were observed during phases of enhanced cellular proliferation and differentiation suggesting that these structures mirror stages of rapidly multiplying peroxisomes and reflect an elevated production by growth and division. In line with this, Schrader et al. (1996, 1998) showed that mammalian cells go through phases of increased numbers of tubular peroxisomes at low cell density (and thus ongoing cell division) or upon induction of peroxisome proliferation by extracellular stimuli. These data were corroborated by ultrastructural 3D-reconstructions in regenerating rat liver and cultured hepatoma cells showing that peroxisomes under rapid cellular proliferation form elongated and more complex peroxisomal structures (Grabenbauer et al. 2000; Yamamoto and Fahimi 1987; Fig. 7.2). We now know that these morphologies reflect a multistep process of peroxisome formation by outgrowth of tubular extension, import of matrix proteins and subsequent organelle division (Fig. 7.3). Interestingly, in sebaceous glands such tubules show a heterogeneous distribution of matrix enzymes displaying discrete patches of DAB-staining along the length of the tubular profiles (Gorgas and Volkl 1984). Similarly, spherical peroxisomes were found to form small extensions devoid of catalase, which were interpreted as signs for budding of a new peroxisomal vesicle (Fahimi et al. 1993). These early findings already point to an asymmetric growth and division process for peroxisomes as later proposed (Delille et al. 2010; Schrader and Fahimi 2006a) when observing that tubular extensions induced by the expression of Pex11βp do not initially contain peroxisomal enzymes.

Elongated peroxisomes were also reported to associate into higher order structures, e.g., the peroxisomal fascicles and stacks described in different kinds of sebaceous glands (Gorgas and Zaar 1984; Gorgas and Volkl 1984), which most likely reflect functional specializations according to the tissues' physiological status. Accordingly, peroxisomes may change their morphology to meet the requirements for enhanced metabolite transport, membrane signaling or protection against reactive oxygen species (ROS).

Fig. 7.1 (continued) Magnification: ×32,000. (**f**) A pleomorphic peroxisome connected via a narrow bridge to a spherical one (*arrows*). Magnification: ×56,000 [(**e**) and (**f**) from Yamamoto and Fahimi (1987)]

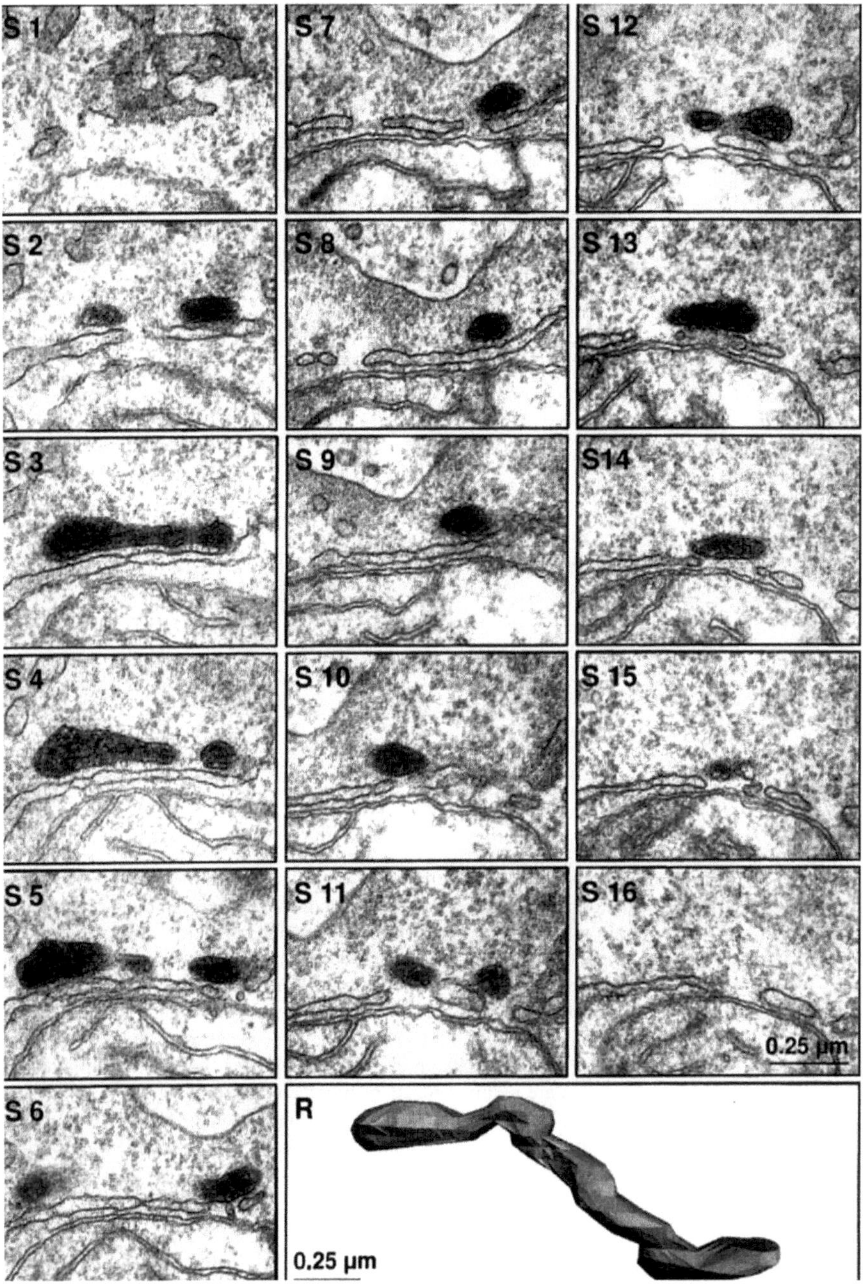

Fig. 7.2 Peroxisomes from a freshly seeded culture of HepG2 cells fixed and processed cytochemically for detection of catalase. Serial section reconstruction of a tubular peroxisome. The illustrations are from 16 consecutive serial sections (S1–S16) and show the final result of electronic image reconstruction (R). Note, that the individual small spherical and tubular peroxisomes in single sections are all interconnected, forming an elongated tubular peroxisome extending over 2 μm [from Grabenbauer et al. (2000)]

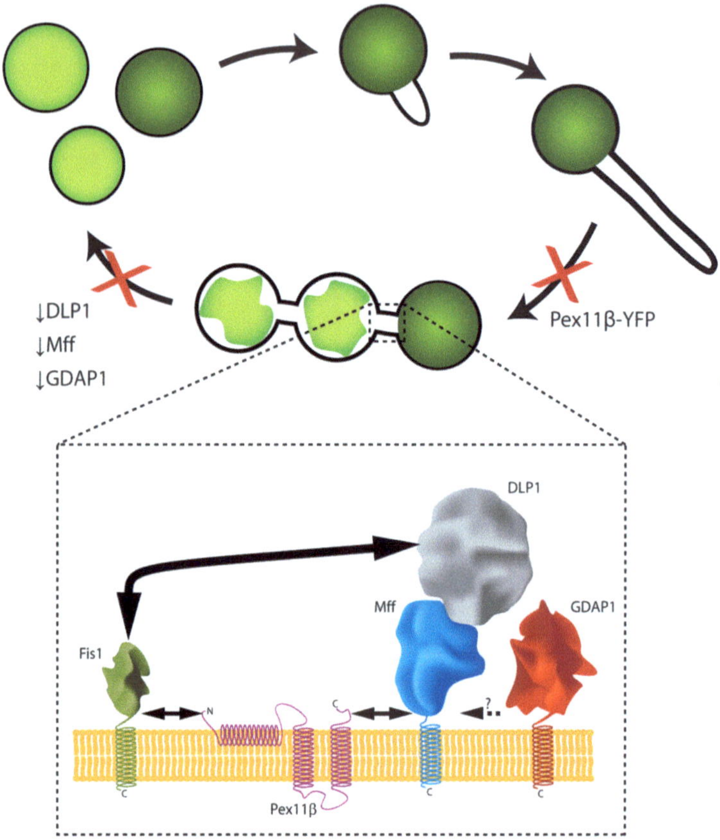

Fig. 7.3 Growth and division model of peroxisome formation. The formation of new peroxisomes is initiated by the deformation of the peroxisomal membrane and the generation of a tubular extension from a mature peroxisome. Upon growth, this extension segments and constricts and new proteins are imported to the forming peroxisomes (this step can be inhibited by the expression of a Pex11ß-YFPm fusion protein) (Delille et al. 2010). Final membrane fission is mediated by DLP1/Drp1, a dynamin-like protein which is recruited to the peroxisomal membrane by Mff and Fis1, two tail-anchored membrane proteins. Fis1 and Mff have been shown to interact with the membrane shaping protein Pex11ß. Di- and oligomerization of Pex11ßp and its amphipathic domains in the N-terminal region are required for its membrane elongating activity. GDAP1 can mediate peroxisome division in a DLP1 and Mff-dependent manner in neurons. Loss of DLP1 or Mff function results in the formation and accumulation of constricted, elongated peroxisomes ("beads on-a-string"-like morphology) (see Fig. 7.4)

The discovery of peroxisomal targeting signals (PTS) (Gould et al. 1987) and the availability of fluorescent proteins such as GFP in the 1990s enabled the expression of fluorescent fusion proteins with C-terminal signals for matrix protein import (e.g., GFP-PTS1) in eukaryotic cells and for the first time allowed life cell imaging of peroxisomes. Those studies confirmed the morphological heterogeneity and revealed that peroxisomes in living cells were much more interactive and dynamic

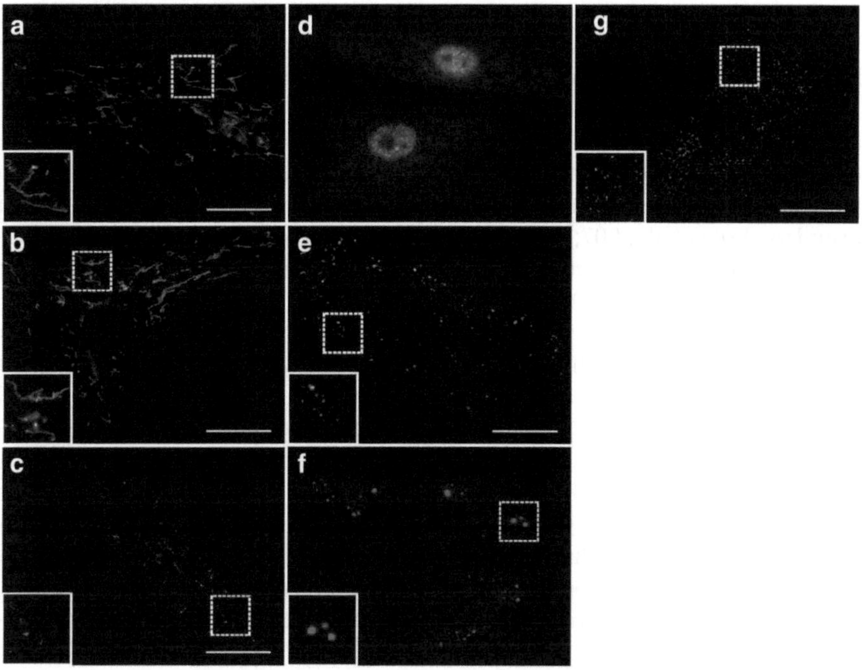

Fig. 7.4 Peroxisome morphologies in human skin fibroblasts under control and disease conditions. Cultured cells were processed for immunofluorescence microscopy and stained with anti-Pex14 antibodies (**a**, **c**, **e** and **g**), anti-PMP70 (**b**, **d**) and anti-catalase (**f**). (**a**) Loss of DLP1 function results in the formation of highly elongated, constricted peroxisomes, which are unable to divide. In this case, a mutation in the middle domain of DLP1 inhibits its oligomerization. (**b**) Similar to DLP1 deficiency, loss of Mff function results in the formation of highly elongated and often constricted peroxisomes, indicative of a block in peroxisomal division. (**c**) Loss of Pex11β function gives rise to reduced peroxisome numbers and enlarged and elongated peroxisomes, suggesting a defect in peroxisomal division or proliferation. In this case, a nonsense mutation leads to the absence of Pex11βp. (**d**) Defects in Pex19p result in a PBD with complete loss of peroxisomal structures. (**e**) Defects in Pex5p, the import receptor for peroxisomal matrix proteins, results in accumulation of the peroxisomal marker catalase in the cytosol. However, peroxisomal membranes (so-called ghosts) are formed. (**f**) Defects in AOX, which is involved in peroxisomal fatty acid β-oxidation, result in reduced peroxisome numbers and enlarged peroxisomes. (**g**) Control fibroblast from a healthy individual show a large number of small, spherical peroxisomes. Inserts show higher magnification views. Bars, 20 μm

than assumed (Schrader et al. 2000; Bonekamp et al. 2012). They allowed insights in the cytoskeleton-dependent movement, inheritance and membrane dynamics of the organelle. In mammalian cells, a microtubule-dependent transport of peroxisomes was demonstrated (Schrader et al. 2003), whereas in plants and yeast peroxisome motility is actomyosin-based. Peroxisomes were also observed to generate reticular structures and tubular protrusions, which may facilitate inter-organellar communication (Schrader et al. 2000; Sinclair et al. 2009), thus confirming the early ultrastructural observations.

Depending on species and tissue examined, peroxisomes can harbor crystalline inclusions of matrix enzymes (so called cores), which some-times result in bizarre hexagonal or triangular morphologies of the organelle (Zaar and Fahimi 1990; Fig. 7.1). Urate oxidase, which forms cores in peroxisomes of rat hepatocytes, is absent in primates, which renders us susceptible to human gout. Moreover, its crystalline nature may support channeling of peroxisome-derived ROS, which have been linked to signaling, oxidative stress, senescence, and age-related disorders (Fritz et al. 2007; see Chap. 11). Under certain disease conditions, cores have also been observed in human peroxisomes, which usually lack crystalline inclusions (Biempica 1966).

Finally, peroxisomes are well known to respond to changes in the extracellular environment with alterations in their number, size, and morphology and protein composition. Such a peroxisome proliferation in mammals is best studied in rodents, e.g., during cold adaptation, fatty acid diet, or treatment with hypolipidemic fibrates (Fig. 7.1) and is mediated by the peroxisome proliferator activated receptors (e.g., PPARα; for a recent review see Schrader et al. 2012). In this respect it should be highlighted that peroxisomes can appear as very heterologous organelles, depending on cell type and physiologic state of a cell. Thus, pathologic alterations of peroxisomes as described in the latter have to be cautiously evaluated and discriminated from naturally occurring variations in peroxisome morphology.

7.2 Peroxisome Abnormalities in Peroxisome Deficiencies

Common peroxisomal functions in many eukaryotes include the metabolism of H_2O_2 and the synthesis and breakdown of lipids, e.g., the β-oxidation of fatty acids (Wanders and Waterham 2006). Over the years, multiple other important peroxisomal functions have been discovered, which are highly diverse depending on organism, cell type, and developmental stage of the organism. Those include penicillin and biotin synthesis in fungi, glycolysis in parasitic trypanosomes (that cause sleeping sickness in man), photorespiration and glyoxylate cycle in plants (to convert fat into carbohydrates), and etherlipid biosynthesis in mammals (etherlipids, such as plasmalogens, are important components of the myelin sheaths of neurons, and may have antioxidative functions as well). Moreover, in mammals peroxisomes are involved in the synthesis of bile acids, inflammatory mediators (e.g., leukotrienes and docosahexaenoic acid, DHA), a modulator of neuronal function. In animals, peroxisomes further cooperate metabolically with mitochondria in the degradation of fatty acids by β-oxidation (see Chap. 3), and a closer, medically relevant interplay has been discovered (Schrader and Yoon 2007; Camões et al. 2009; Dixit et al. 2010). It also became evident that peroxisomes act as signaling platforms to modulate cellular functions and developmental decisions (Titorenko and Rachubinski 2004; del Rio 2013). It is thus not surprising that peroxisomal dysfunction has been linked to inherited disorders in man that are associated with multiple severe clinical symptoms (e.g., neonatal hypotonia, craniofacial dysmorphy, neuronal myelination and migration defects, seizures, hepatomegaly, liver cirrhosis, and renal cysts) and are often lethal (Waterham and Ebberink 2012; see Chaps. 1 and 2). Peroxisome biogenesis disorders (PBDs) are caused by defects in peroxins (Pex). *PEX* genes encode proteins which are essential for the biogenesis of peroxisomes, in particular the import of enzymes into the peroxisomal matrix (e.g., the import receptors Pex5p or Pex7p) or the biogenesis of the peroxisomal membrane which in animals depends on three peroxins (Pex3p, Pex16p, Pex19p; see Table 7.1, Chap. 1). Defects in the latter can result in the complete absence of peroxisomes with loss of all peroxisomal functions (Fig. 7.4). Peroxisomal membrane proteins in mammalian cells are often mistargeted to mitochondria under those conditions. Interestingly, re-introduction of the functional gene has been observed to restore peroxisome function in cells by the de novo formation of peroxisomes in association with the ER (Hoepfner et al. 2005; van der Zand et al. 2006). Defects in the matrix import machinery, on the other hand, leave the peroxisomal membranes intact, but impair import of matrix proteins. They remain in the cytosol, where they are non-functional and/or degraded leaving the cell with empty peroxisomal membranes, so-called ghosts, which have been described in Zellweger syndrome, a frequent PBD (see Chap. 2). It should be noted that in contrast to other subcellular organelles such as mitochondria or the ER, peroxisomes can import completely folded or oligomeric proteins via a unique import mechanism, which involves membrane insertion and pore formation of shuttling receptors (e.g., Pex5p), which are released by ubiquitination (see Chap. 4). Clinical features which are generally used as biomarkers for peroxisomal disorders include high plasma levels of very long-chain fatty acids and phytanic acid (a derivative of phytol from dairy products), which can solely be degraded by peroxisomal β- and α-oxidation, respectively. When accumulating, they are toxic for the cell and the organism. Furthermore, the synthesis of ether lipids/plasmalogens is decreased, which is thought to contribute to demyelination and neurodegeneration.

The other group of peroxisomal disorders, the single enzyme deficiencies (SEDs), is due to defects in a single protein or enzyme, which is not a peroxin, and thus not essential for peroxisome biogenesis. Defects are for example found in ABC transporter proteins required for the import of fatty acids into peroxisomes (e.g., in X-linked adrenoleukodystrophy; see http://www.myelin.org) or in one of the enzymes involved in peroxisomal fatty acid oxidation. Clinical features can resemble the PBDs, but may also result in milder symptoms which are more compatible with life (Waterham and Ebberink 2012).

Defects in acyl-CoA oxidase (ACOXI) or D-bifunctional protein (DBP), which catalyze the first two steps in peroxisomal fatty acid β-oxidation have been reported to result in fewer and enlarged peroxisomes in patient fibroblasts (Fig. 7.4; Chang et al. 1999; Ebberink et al. 2010). Interestingly, addition of docosahexaenoic acid (DHA), an essential polyunsaturated fatty acid and component of phospholipids and a major product of peroxisomal β-oxidation (Ferdinandusse et al. 2001), was recently reported to promote peroxisome elongation and proliferation in AOX or DBP-defective fibroblasts and to restore the normal peroxisome morphology

Table 7.1 Novel disorders with defects in peroxisome dynamics

Gene	Function	Mutation	Clinical features	Organelle morphology	References
DNM1L	Mitochondrial and peroxisomal membrane fission	c.1184C > A A395D Heterozygous Dominant-negative Oligomerization defect	Microcephaly, abnormal brain development, optic atrophy, hypoplasia; lactic acidosis, slightly elevated VLCFA levels	Elongated and constricted peroxisomes; elongated mitochondria with uneven diameter	Waterham et al. (2007)
MFF	Recruitment of DLP1 to mitochondria and peroxisomes	c.190C > T Q64X Homozygous Non-sense mutation Truncated protein lacking TMD	Developmental delay, abnormal intensity on brain MRI of the globus pallidus, motor and speech deficits, mild hypertonia, borderline microcephaly, pale optic discs	Elongated peroxisomes and mitochondria	Shamseldin et al. (2012)
GDAP1	GST-transferase; involved in mitochondrial and peroxisomal division	Heteroallelic Frameshift mutation c.863insA T288fs290X / Nonsense mutation c.487 T Q163X Truncated proteins	Charcot-Marie-Tooth neuropathy; Hoarse voice and vocal cord paresis, onset at childhood with weakness and hand wasting leading to disability at the end of the first decade, sensory nerve action potential decreased or absent	Inability to induce mitochondrial and peroxisomal fragmentation[a]	Cuesta et al. (2002) Niemann et al. (2005) Huber et al. (2013)
PEX11B	Regulation of peroxisome division and proliferation	c.64C > T Q22X Homozygous Non-sense mutation No functional protein	Congenital cataracts, mild intellectual disability, progressive hearing loss, gastrointestinal problems, recurrent migraine-like episodes; sensitivity to illness and long recovery times	Enlarged and elongated peroxisomes	Ebberink et al. (2012)

VLCFA very-long-chain fatty acids, *TMD* transmembrane domain
[a]Results from hippocampal cells expressing each mutation separately

(Itoyama et al. 2012). Intriguingly, these findings imply that peroxisome morphogenesis is modulated by phospholipids within the peroxisomal membrane, which in turn are influenced by peroxisomal lipid metabolism such as fatty acid β-oxidation. In line with this, cultivation of mammalian cells in lipid-free medium results in large spherical peroxisomes reminiscent of those observed in fibroblasts from patients with AOX deficiency (Bonekamp et al. 2013). Fewer and enlarged peroxisomes have also been observed in an unusual variant of PBD caused by mutations in *PEX16* (Ebberink et al. 2010). As mentioned above, Pex16p deficiency usually results in a complete loss of peroxisomes with a severe Zellweger syndrome phenotype, as it contributes to peroxisomal membrane assembly. Ebberink et al. (2010) describe six patients with mutations in the Pex16p C terminus, which displayed a relatively mild clinical phenotype with abnormal peroxisomal metabolites such as raised plasma levels of VLCFAs and phytanic acid. Patient fibroblasts unexpectedly contained peroxisomes, which were import competent for peroxisomal membrane and matrix proteins but enlarged in size and reduced in number. These findings are important for future diagnostics of peroxisomal disorders, and indicate that Pex16p, besides its function in membrane assembly, may have a role in peroxisome proliferation.

7.3 Defects in the Peroxisomal Growth and Division Machinery

In recent years much progress has been made in the understanding of the formation of peroxisomes. Peroxisomes can form by growth and division out of pre-existing organelles, which in mammalian cells comprises a multistep maturation process including elongation of the peroxisomal membrane, the formation of constriction sites, and final membrane fission into spherical organelles (Schrader et al. 2012; Schrader and Fahimi 2006a for recent reviews) (Fig. 7.3). Remarkably, peroxisomes are also supposed to form de novo in association with the ER via a maturation process, e.g., under conditions where pre-existing peroxisomes are initially absent (Hoepfner et al. 2005). Growth and division involves the formation of an early peroxisomal membrane compartment from a pre-existing peroxisome and its stepwise conversion into a mature, metabolically active peroxisome compartment by the recruitment of new membrane and matrix proteins (Delille et al. 2010) as well as membrane lipids which are supplied by the ER (Fig. 7.3). These findings support the view that peroxisomal division is an asymmetric rather than a symmetric process. It is far more complex than simple squeezing of a mother peroxisome into two equal daughter organelles (Huybrechts et al. 2009; Delille et al. 2010, 2011).

Formation of peroxisomes by growth and division is initiated by the deformation and tubular extension of the peroxisomal membrane which involves membrane proteins of the Pex11p family (see Chap. 19) (Schrader et al. 1998; Thoms and Erdmann 2005). The tubular membrane extensions, which are initially devoid of matrix proteins, enlarge and form constriction sites (Fig. 7.3). How these constrictions are formed is currently unknown. However, considerable progress has been made in the identification and characterization of the division machinery (Schrader et al. 2012). Final membrane fission depends on the large GTPase DLP1/Drp1, a dynamin-like mechanochemical enzyme, which forms higher ordered ring-like structures in a GTP-dependent manner that wrap around constricted membranes and sever them in a GTP-hydrolysis dependent process (Praefcke and McMahon 2004; see Chap. 20). Recruitment of DLP1 to the organelle membrane depends on the tail-anchored membrane receptors Fis1 (fission factor 1) and Mff (mitochondrial fission factor; Fig. 7.3). Remarkably, these key fission components are shared by peroxisomes and mitochondria, which appears to be an evolutionary conserved strategy among mammals, fungi, and plants (Delille et al. 2009; Schrader 2006). In this respect, Dnm1-dependent ordered mitochondrial and peroxisomal fission has also been reported in the unicellular red alga *Cyanidioschyzon merolae*, which only contains a single peroxisome and a single mitochondrion (Imoto et al. 2013). Interestingly, immuno-EM revealed dynamin-based and filamentous ring structures on dividing peroxisomes. In *C. merolae* peroxisomes do not form de novo from the ER in the peroxisomal division cycle.

Although peroxisomes and mitochondria share key division components, the key proteins for mitochondrial fusion (e.g., the dynamin-related GTPases Mfn1, Mfn2 or OPA1) are not present on peroxisomes. In contrast to mitochondria, whose morphologies are regulated by constant fusion and fission events, mature peroxisomes have not been observed to fuse (Huybrechts et al. 2009; Bonekamp et al. 2012). Notably, live cell imaging revealed peroxisome interactions with transient and long-term contacts, but without exchange of matrix or membrane markers (Bonekamp and Schrader 2012). In combination with microtubule-dependent transport these interactions may contribute to the equilibration of the peroxisomal compartment in mammalian cells.

7.3.1 DLP1 Deficiency

The first report on a patient with a novel disorder based on a defect in both peroxisomal and mitochondrial division was published in 2007 (Waterham et al. 2007). In previous studies, an involvement of DLP1 in the division of peroxisomes (and mitochondria) was reported (Koch et al. 2003, 2004; Li and Gould 2003). Silencing of DLP1 in mammalian cells gave rise to unusual peroxisome morphology with highly elongated, constricted peroxisomes (Fig. 7.4). This morphological hallmark was based on an inhibition of peroxisomal (and mitochondrial) division due a functional loss of DLP1. A similar peroxisome morphology was detected in the patients' skin fibroblasts and subsequent genetic analysis revealed a heterozygous, dominant-negative missense mutation (A395D) in the middle domain of DLP1 (Waterham et al. 2007; Table 7.1). It was later reported that this mutation inhibits oligomerization of DLP1 (Chang et al. 2010). The female patient died only few weeks after birth. Clinical features included microcephaly, abnormal brain development, optic atrophy, and hypoplasia. Magnetic resonance imaging (MRI) detected an abnormal gyral pattern in both frontal lobes and was associated with dysmyelination. Some of the abnormalities were similar to known disorders based on defects in mitochondrial dynamics (e.g., autosomal dominant optic atrophy, Charcot-Marie-Tooth neuropathy), but the clinical course was more severe implying a peroxisomal contribution. Similar abnormalities have now been observed in DLP1 knockout mice, which display developmental defects (e.g., in synapse formation and brain development) resulting in embryonic lethality (Ishihara et al. 2009; Wakabayashi et al. 2009). In contrast to PBDs, the metabolic functions of peroxisomes and biomarkers indicative for classical peroxisomal disorders were not or only slightly altered. Notably, elevated plasma levels of lactate and slightly elevated levels of VLCFA were reported, indicating defects in mitochondrial respiration and slightly altered peroxisomal ß-oxidation.

It is likely that other patients with mutations in DLP1 will be identified. It has already been experimentally demonstrated that other mutations in the middle domain or the GTPase domain of DLP1 (e.g., K38A, C452F, G363D) interfere with GTP hydrolysis and result in elongated peroxisomes (and mitochondria; Koch et al. 2003; Tanaka et al. 2006; Ashrafian et al. 2010). Furthermore, DLP1 activity is highly regulated through a number of posttranslational modifications including phosphorylation, ubiquitination, S-nitrosylation, and sumoylation (Santel and Frank 2008; Chang et al. 2010).

7.3.2 Mff Deficiency

With Mff deficiency, another member of the new group of combined peroxisomal-mitochondrial disorders was very recently identified through genomic analysis of a cohort of patients with suspected or confirmed mitochondrial encephalomyopathy (Shamseldin et al. 2012; Table 7.1). Like Fis1, Mff is a C-tail anchored membrane adaptor for DLP1 on both the peroxisomal and the mitochondrial membrane (Fig. 7.3). Its N-terminal portion is exposed to the cytosol and contains two short repeat motifs and a central coiled-coil domain (Gandre-Babbe and van der Bliek 2008). The human MFF gene contains nine coding exons, and at least eight MFF splice variants are predicted. Both Fis1 and Mff can bind DLP1, but Mff might represent the major receptor for DLP1 on peroxisomes and mitochondria in mammals (Otera et al. 2010; Itoyama et al. 2013). Furthermore, Mff is supposed to be part of a higher molecular mass complex which does not contain Fis1 (Gandre-Babbe and van der Bliek 2008). Similar to silencing of DLP1, loss of Mff function by siRNA mediated silencing in cultured mammalian cells results in elongated peroxisomes (and mitochondria; Gandre-Babbe and van der Bliek 2008; Otera et al. 2010; Fig. 7.4). The elongated peroxisomes observed after silencing of DLP1 or Mff have a constricted morphology, indicating that both proteins are likely not essential for membrane constriction prior to fission (Koch et al. 2004; Ribeiro et al. 2012; Itoyama et al. 2013).

Shamseldin et al. (2012) identified a homozygous truncating mutation (Q64X) in the MFF gene of two brothers, born of consanguineous Saudi Arabian parents, which is predicted to remove the C-terminal transmembrane domain. Similar to other tail-anchored membrane proteins, the transmembrane domain and C-terminal tail contain essential information for proper organelle targeting; removal results in a cytosolic localization of Mff, and loss of function at peroxisomes and mitochondria. Subsequently, an abnormal elongated morphology of mitochondria and peroxisomes was detected in patient skin fibroblasts (Fig. 7.4, Table 7.1), indicative for a block in organelle fission. The patients were initially suspected to have a mitochondrial encephalopathy. One proband, a 4.5-year-old boy, showed delayed psychomotor development, microcephaly, pale optic discs, and mild hypertonia. Brain MRI revealed abnormal intensity of the globus pallidus. The younger brother showed similar developmental delay. The severity of DLP1 deficiency (e.g., early death of the patient 37 days after birth) in contrast to Mff deficiency may be explained by findings indicating that DLP1 fulfils additional functions, e.g., in apical sorting at the trans-Golgi network (Bonekamp et al. 2010) or in the regulation of synaptic vesicle morphology and membrane dynamics during endocytosis in hippocampal neurons (Li et al. 2013).

Biochemical parameters such as lactate and VLCFA levels or mitochondrial respiratory chain complex profiles on skin fibroblasts from the Mff patient were normal. This is similar to the reported DLP1 patient indicating that metabolic parameters of the organelles are not or only slightly affected. Such a scenario complicates the detection of this group of disorders by the analysis of classical biomarkers such as VLCFA. The analysis of peroxisomal (and mitochondrial) morphology in patient cells or tissue biopsies is thus a valuable diagnostic tool for the determination of disorders based on defects in peroxisomal (mitochondrial) division.

7.3.3 GDAP1 Deficiency and Charcot-Marie-Tooth Disease

Very recently, it was discovered that the C-tail anchored protein GDAP1 (Ganglioside-induced differentiation-associated protein 1) localizes as well to both peroxisomes and mitochondria (Huber et al. 2013). Besides Fis1, DLP1, and Mff, this is now the fourth factor shared by the peroxisomal and the mitochondrial division machinery thus further supporting the concept of the "peroxisome-mitochondria connection" (Camões et al. 2009). Sharing of fission components has presumably developed by similar cellular demands, as peroxisomes and mitochondria are metabolically linked (Schrader and Yoon 2007; see Chap. 1).

GDAP1 is the founder of a new family of glutathione S-transferases, which in rat is expressed in the central and peripheral nervous system, particularly in Schwann cells (Niemann et al. 2005). Its expression level has been reported to influence the glutathione levels in cultured cells (Noack et al. 2012). Over 40 different mutations in the GDAP1 gene lead to Charcot-Marie-Tooth disease (CMT), the most common inherited peripheral neuropathy, and affect mitochondrial dynamics (Niemann et al. 2005, 2006, 2009). It became now evident that GDAP1 is also involved in peroxisomal division (Huber et al. 2013) (Fig. 7.3, Table 7.1). Its targeting to peroxisomes is mediated by the import receptor Pex19p. Silencing of GDAP1 leads to peroxisomal elongation (less prominent than after loss of DLP1 or Mff function) in hippocampal cells that can be rescued by re-expressing GDAP1, whereas overexpression results in peroxisomal (and mitochondrial) division. GDAP1-induced peroxisomal fission is dependent on the integrity of its hydrophobic domain 1 (HD1), and on DLP1 and Mff, as is mitochondrial fission (Fig. 7.3). Consistently, truncations that lack HD1 or the C-terminal tail, which is required for proper targeting (Wagner et al. 2009) no longer promote peroxisomal (and mitochondrial) fission. However, other autosomal recessively inherited disease mutations in the N-terminal cytosolic GDAP1 domain are still able to promote peroxisomal but not mitochondrial division (Huber et al. 2013; Niemann et al. 2009). These findings suggest overlapping regulatory mechanisms for GDAP1-induced fission at both organelles but reveal a difference in the cell biology of recessively inherited missense mutations and the more severe, recessively inherited C-terminal truncation mutations of GDAP1. Although peroxisomal functions have not yet been studied in patient cells, these findings suggest that peroxisomes may contribute to some degree to the clinical features of CMT (Table 7.1).

It is presently unclear to what extent defects in peroxisomal or in mitochondrial dynamics (and functions) contribute to the clinical phenotypes of DLP1- and Mff deficiency or to CMT. However, in light of the functional interplay between peroxisomes and mitochondria and the vital role of peroxisomes in brain development and neurodegenerative disorders, their contribution to the reported pathophysiologies should not be underestimated. Furthermore, peroxisome number, size, and shape are critical cellular parameters which may have an impact on metabolic function, motility, and distribution or autophagic degradation of the organelle. Very recently, it has been demonstrated that peroxisome numbers and subsequent alterations in ROS levels influence melanocortin tone in hypothalamic neurons and thus, feeding behavior in diet-induced obesity (Diano et al. 2011). Peroxisomal shape may also affect peroxisomal movement and distribution, e.g., in neurons or during inheritance, with larger peroxisomal structures being less motile and more stationary. In addition, an enlargement or elongation of peroxisomes might have an impact on proper organelle clearance by pexophagy (see Chap. 6).

7.3.4 Pex11β Deficiency

Pex11-proteins are unique membrane components of the peroxisomal growth and division machinery. They are thought to mediate the initial step(s) of peroxisomal division and proliferation including shaping and elongation of the peroxisomal membrane prior to fission as well as recruitment or assembly of components of the fission machinery (e.g., DLP1, Fis1, Mff; Fig. 7.3). A loss of Pex11p leads to a reduced peroxisome number and the formation of enlarged peroxisomes, whereas overexpression promotes peroxisome elongation and proliferation (Thoms and Erdmann 2005; Yan et al. 2005; Schrader and Fahimi 2006a; Delille et al. 2010; Koch et al. 2010; Opalinski et al. 2011; Fig. 7.4). Pex11βp, one of the three human isoforms, is an integral membrane protein with the N- and C-termini exposed to the cytosol (Schrader et al. 1998; Bonekamp et al. 2013). It is expressed in most tissues, forms homooligomers, and likely interacts with membrane lipids to deform and shape the peroxisomal membrane (Itoyama et al. 2012; Bonekamp et al. 2013). Furthermore, an interaction with Fis1 and Mff has been reported (Kobayashi et al. 2007; Koch and Brocard 2012; Itoyama et al. 2013). Its N-terminal domain contains amphi-pathic helices that are required for membrane elongation in vitro and in vivo as well as for dimerization (Opalinski et al. 2011; Bonekamp et al. 2013; see Chap. 5). Knockout of *PEX11β* in mice causes neonatal lethality and is accompanied by several defects reminiscent of PBDs (e.g., developmental delay, hypotonia, neuronal migration defects and neuronal apoptosis; Li et al. 2002). Peroxisome abundance is reduced in cultured mouse *PEX11β−/−* fibroblasts, but peroxisomal metabolism is only slightly affected.

The severe pathologies of *PEX11β* knockout mice are in contrast to a recent report on the identification of the first patient with a defect in *PEX11β* (Ebberink et al. 2012; see Chaps. 2 and 5 and Table 7.1). A homozygous non-sense mutation was identified in the *PEX11*β gene leading to a truncation of the protein after 21 amino acids (Q22X), while *DLP1*, *FIS1*, *PEX11α*, *PEX11γ* and the 13 *PEX* genes known to be associated with PBD were unaffected. The patient, a 26-year-old Dutch male displayed mild intellectual disability, congenital cataracts, progressive hearing loss, sensory nerve involvement, gastrointestinal problems, and recurrent migraine-like episodes. Both parents were heterozygous for this mutation. No peroxisomal biochemical abnormalities in plasma, erythrocytes and cultured skin fibroblasts were detected, but peroxisome morphology in skin fibroblasts was abnormal with enlarged and elongated peroxisomes, indicative of a defect in peroxisome division or proliferation (Fig. 7.4). Overexpression studies with Pex11γp suggest that the function of Pex11βp is partly redundant with that of Pex11γp.

Once more, these findings demonstrate that the analysis of peroxisomal morphology in patient cells is a valuable diagnostic tool for the determination of disorders based on defects in peroxisomal morphology proteins, especially when no (or only mild) biochemical alterations of peroxisome metabolism can be detected. Thorough analysis of peroxisome morphology, an as yet underappreciated cell biological parameter, may help to identify more patients with defects in Pex11p or other peroxisome morphology proteins. As those patients may not display the typical features of PBDs, a link to peroxisome dysfunction can be overlooked or misinterpreted, for example as a mitochondrial dysfunction.

7.4 Morphological Alterations of Peroxisomes During Stress and Signaling

Particularly in plants, peroxisomes are known to respond to certain stress conditions (e.g., wounding, pathogen attack, drought, osmotic stress, excess light) that generate H_2O_2 as a signaling molecule with morphological alterations and peroxisome proliferation (Sandalio et al. 2013). Peroxisomes were observed to rapidly form tubular extensions after exposure to H_2O_2 or excess light (Sinclair et al. 2009; see Chap. 5). Similar extending and retracting peroxisomal tubules were also monitored in mammalian cells (Schrader et al. 2000), and peroxisome elongation was inducible by exposure of cells to UV irradiation and ROS (Schrader et al. 1999). Based on the important role of peroxisomes in ROS metabolism and signaling (Schrader and Fahimi 2006b; see Chap. 3), peroxisome membrane elongation may have a protective function in the scavenging of ROS (e.g., by increasing the membrane surface); the membrane extensions might also contribute to inter-organellar communication. Furthermore, it has been demonstrated that peroxisomes contribute to antiviral signaling and defense and that viral stimulation causes elongation and tubule formation of peroxisomes (Dixit et al. 2010).

7.5 Defects in Autophagic Processes Are Associated with Changes in Peroxisome Morphology

In line with a continuously occurring formation, ageing peroxisomes exert increasing deficits in their metabolic capacity and have thus to be removed from the cell in order to guarantee a stable population of functional organelles (Nordgren et al. 2013; see Chap. 6). The controlled removal of organelles within a healthy cell is commonly accomplished by autophagy which can occur mainly in two separate forms named micro- and macroautophagy (Islinger et al. 2012). Microautophagy occurs via a direct engulfment and internalization of cytoplasm and organelles into lysosomes, where they are subsequently removed by digestive processes. In macroautophagy organelles to be removed are first enclosed in double membrane vesicles (autophagosomes), which transport the cargo to lysosomes. Both processes require highly specific recognition mechanisms which guarantee the selective removal of individual organelles and for peroxisomes are called micro- and macropexophagy (Nordgren et al. 2013). Currently, both macro- and micropexophagy are known to occur in parallel in yeast, whereas in mammals microautophagic degradation of organelles is supposed to play only a minor role and thus macroautophagy is regarded as the major pathway of organelle removal (Mizushima et al. 2011). At the molecular level, several genes have been associated with the recognition process between peroxisomes and the autophagic membrane (see Chap. 6). As peroxisomes are changing their shape and size according to the metabolic and proliferative state of the cell (Schrader et al. 1998; Funato et al. 2006), it is quite likely that impairment in their degradation may be reflected in their morphological appearance as well. Peroxisomes in endothelial cells of lysosome-defective Lyst-mice possess a significantly larger peroxisome volume than wild-type mice (Vasko et al. 2013). This volume change was accompanied by increases in peroxisomal protein amounts per cell. Catalase activities, however, were significantly lower in the lysosome-defective cells and resulted in increased ROS production, pointing to a dysfunctional organelle, reminiscent of peroxisomes in ageing cells (Legakis et al. 2002). Treatment with lipopolysaccharides (LPS), triggering endotoxin stress, led to higher rates of peroxisome recycling by induction of both peroxisome proliferation and pexophagy in wild-type endothelial cells; in Lyst-mice, however, peroxisomes were found to elongate in response to LPS treatment, reminiscent of a peroxisome morphology occurring in cells with a defect in peroxisome division. In this respect, a disturbed peroxisomal degradation pathway appears to have a direct impact on the regulative processes occurring during peroxisomal fission or may be a morphological sign for a de-stabilized autophagic pathway itself. Mitochondrial morphology was reported to have a direct impact on the cell's capacity for autophagy: mitochondrial tubulation protects the organelle from degradation (Rambold et al. 2011; Gomes et al. 2011). Currently, it is unclear whether peroxisomes are protected from degradation by similar processes; recently, however it was reported that larger peroxisomes require additional sets of autophagy-related (Atg) proteins for their degradation in *Pichia pastoris* (Nazarko et al. 2009) thus suggesting a relation between organelle size and the regulation of pexophagy. In summary, these data show that peroxisomes are likely affected in diseases with a lysosomal background (e.g., lysosomal storage disorders) thus leading to peroxisomal dysfunction (Platt et al. 2012) by obscuring autophagic processes.

7.6 Secondary Peroxisome Alterations in Disease

7.6.1 Peroxisomes in Hepatitis, Hepatic Steatosis and Liver Cirrhosis

In addition to diseases with a direct peroxisome relationship (e.g., PBDs, division defects), peroxisome morphology can be altered in disorders with a more complex etiology. Due to their abundance in liver, peroxisomes were primarily investigated in common liver diseases such as hepatitis, steatosis, and cirrhosis. Commonly, no major alterations in organelle morphology such as elongation or protrusions were observed, but frequently changes in number and size were reported (De Craemer 1995). In viral hepatitis peroxisomes appear to increase in number and to display irregular shapes (De Craemer 1995). Increased peroxisome numbers with a reduced size were observed in cholestasis (De Craemer et al. 1998) and the different stages of alcoholic liver disease (De Craemer et al. 1996). In non-alcoholic cirrhosis an even stronger proliferation of the peroxisomal compartment was reported (Litwin et al. 1999). These morphological changes very likely reflect an elevated peroxisomal turnover meeting the requirements for an induced lipid metabolism, which, however, ultimately fails to cope with the overload of substrates in alcoholic and non-alcoholic liver disease. Peroxisome proliferation and induction of fatty acid β-oxidation in peroxisomes and mitochondria is mediated by the nuclear receptor PPARα, which is activated by the binding of long-chain fatty acids (Issemann and Green 1990; Dreyer et al. 1993). Indeed, PPARα-null mice show an increased susceptibility for hepatic steatosis (Hashimoto et al. 2000) and when alcohol-fed, develop hepatomegaly, steatohepatosis, and fibrosis (Nakajima et al. 2004). Corroborating these findings, the administration of PPARα ligands was found to ameliorate alcohol-induced liver fibrosis in rodents (Nakajima et al. 2004; Toyama et al. 2004). Likewise, significantly increased levels of peroxisomal β-oxidation enzymes were found in mice resistant to diet-induced steatosis pointing to such a compensatory induction of peroxisomes in the disease state (Hall et al. 2010). It is discussed that alcohol inhibits PPARα expression as well as its binding to DNA, thus reducing the activation of PPARα-regulated genes (Fischer et al. 2003; Nanji et al. 2004). Thus, the lower peroxisome numbers observed in alcohol-induced versus non-alcohol-induced human cirrhosis may also reflect such a reduced capacity to respond to the elevated fatty acid levels in hepatic steatosis. It should be, however, taken into account that humans are much more refractory toward PPAR-activators than rodents and usually do not respond with a significant peroxisome proliferation in hepatocytes (reviewed in Islinger et al. 2010). Interestingly, ethanol treatment increased catalase expression and peroxisome numbers in the myocard of rats after 5–18 weeks of ethanol exposure but not in the hepatocytes of the animals (Fahimi et al. 1979), indicating that not only the PPARα signaling pathway influences peroxisome abundance. Thus, the elevated peroxisome numbers in cirrhotic human livers may be mediated by the interplay of numerous alternative factors and signaling pathways involved in the regulation of human lipid or ROS homeostasis (see Purohit et al. 2009).

7.6.2 Peroxisomes and Cancer

The mechanism causing the formation of liver tumors after feeding of PPAR-α-activating peroxisome proliferators to rodents has been studied to a high extend during the last decades (reviewed in Islinger et al. 2010; Misra et al. 2013). In humans, however, induction of peroxisome proliferation by PPARα-agonists and subsequent tumorigenesis has not been observed. In contrast, peroxisome numbers were found to be significantly decreased in human hepatocellular tumors (Litwin et al. 1999), and in colon tumors (Cablé et al. 1992; Lauer et al. 1999), or were even reported to be absent in malignant tissue (Frederiks et al. 2010). Early studies on peroxisome numbers in human glial tumors stated that a decrease in peroxisome numbers correlates with the grade of malignancy of the investigated tumor types (Sima 1980). A similar relation was also suggested for the growth rate and peroxisome number of different Morris hepatomas (De Duve and Baudhuin 1966; Kang et al. 1982). As a decline in peroxisomes appears to be a general feature in tumorigenesis of quite different origin, one may speculate that peroxisome abundance correlates with the overall degree of differentiation of a cell. Indeed, peroxisome numbers were found to increase along the crypt to villus axis of the guinea pig small intestine, reflecting the maturation from stem cells to mature epithelial gut cells (Phipps et al. 2000). Likewise, peroxisome numbers and size increase during differentiation of kidney and liver tissue (Espeel et al. 1997; Johkura et al. 1998), likely reflecting the increasing metabolic importance of this organelle in those organs. Contradicting these observations, Cimini and coworkers reported that an increase in peroxisome numbers positively correlates with the malignancy grade of glioblastomas and is paralleled by an activation of PPARα by the hypoxia-inducible factor HIF1α (Benedetti et al. 2010; Laurenti et al. 2011). As fatty acid β-oxidation is an O_2-consuming process, it is somehow surprising that PPARα was found to be induced under hypoxic conditions. Indeed, several studies reported that HIF1α and PPARα activation are negatively correlated processes (Zhou et al. 2012; Belanger et al. 2007; Narravula and Colgan 2001). As lipid droplets were also reported to be strongly enriched in the glioblastomas during tumorigenesis, peroxisome proliferation may be a secondary event caused by activation of PPARα through a cellular excess of fatty acids.

Taken together, the complex metabolic needs of eukaryotic cells during different stages of differentiation very likely also influence peroxisomal function and consequently also their morphology and abundance. In this respect, it should be noted that the manifestation of more rigid cellular contacts in differentiated tissue requires changes in the assembly of lipids in the cellular membranes, which may involve peroxisomes as key organelles of lipid homeostasis. Unfortunately, there is currently nearly a complete lack of knowledge about the number and morphology of peroxisomes in embryonic or adult stem cells and about alterations during development into differentiated tissue. Thus, it is still not fully understood if peroxisomes fulfill a significant role during differentiation processes and cancerogenesis.

Conclusions

In recent years, an increasing number of peroxisome morphology proteins have been identified in cell biological studies, which when nonfunctional, result in aberrant peroxisome morphology, particularly in elongated organelles due to a block in division. Such elongated organelles have now become a morphological hallmark in a novel group of disorders affecting peroxisomal (and mitochondrial) dynamics, division, and proliferation. As biochemical parameters and biomarkers for classical PBDs are not or only slightly altered under those conditions, the analysis of peroxisome morphology in patient cells turns out to be a valuable, yet underappreciated tool in supporting the diagnosis of disorders based on defects in peroxisomal morphology proteins. The recent observations also suggest that it is not only metabolic alterations of peroxisomes which can contribute to the pathology of peroxisome-based disorders. In addition, a closer functional interplay between peroxisomes and mitochondria becomes apparent. The restoration of peroxisome proliferation and normal peroxisome morphology by treatment of AOX- or D-BP-deficient patient cells with certain fatty acids such as docosahexaenoic acid (Itoyama et al. 2012) might point the way for the development of PBD therapy. It is very likely that due to an increased awareness of the morphological hallmarks more patients will be identified in the near future, which will further widen the spectrum of clinical and cellular phenotypes of diseases associated with defects in peroxisome dynamics and proliferation. The link between aberrant peroxisome morphology and disease also underlines the importance of cell biological studies to further elucidate the molecular mechanisms of organelle division.

Acknowledgments We would like to thank M. Ebberink and H. Waterham (Amsterdam Medical Center), H. Shamseddin and F.S. Alkuraya (King Faisal Specialist Hospital and Research Center, Riyadh) for kindly providing images of patient fibroblasts and members of the laboratory for stimulating discussions and comments on the manuscript. We apologize to those whose work has not been cited owing to space limitations. This work was generously supported by a Wellcome Trust Institutional Strategic Support Award (WT097835MF), BBSRC (BB/K006231/1), the Portuguese Foundation for Science and Technology (FCT) and FEDER/COMPETE (PTDC/SAU/OSM/103647/2008; PTDC/BIA-BCM/099613/2008; PTDC/BIA-BCM/118605/2010; SFRH/BPD/74428/2010 to M.I.). I.C. is supported by a fellowship from the University of Exeter, UK.

References

Angermuller S, Fahimi HD (1986) Ultrastructural cytochemical localization of uricase in peroxisomes of rat liver. J Histochem Cytochem 34:159–165

Ashrafian H, Docherty L, Leo V, Towlson C, Neilan M, Steeples V, Lygate CA, Hough T, Townsend S, Williams D, Wells S, Norris D, Glyn-Jones S, Land J, Barbaric I, Lalanne Z, Denny P, Szumska D, Bhattacharya S, Griffin JL, Hargreaves I, Fernandez-Fuentes N, Cheeseman M, Watkins H, Dear TN (2010) A mutation in the mitochondrial fission gene Dnm1l leads to cardiomyopathy. PLoS Genet 6:e1001000

Belanger AJ, Luo Z, Vincent KA, Akita GY, Cheng SH, Gregory RJ, Jiang C (2007) Hypoxia-inducible factor 1 mediates hypoxia-induced cardiomyocyte lipid accumulation by reducing the DNA binding activity of peroxisome proliferator-activated receptor alpha/retinoid X receptor. Biochem Biophys Res Commun 364:567–572

Benedetti E, Galzio R, D'Angelo B, Ceru MP, Cimini A (2010) PPARs in human neuroepithelial tumors: PPAR ligands as anticancer therapies for the most common human neuroepithelial tumors. PPAR Res 2010:427401

Biempica L (1966) Human hepatic microbodies with crystalloid cores. J Cell Biol 29:383–386

Bonekamp NA, Schrader M (2012) Transient complex peroxisomal interactions: A new facet of peroxisome dynamics in mammalian cells. Commun Integr Biol 5:534–537

Bonekamp NA, Vormund K, Jacob R, Schrader M (2010) Dynamin-like protein 1 at the Golgi complex: a novel component of the sorting/targeting machinery en route to the plasma membrane. Exp Cell Res 316:3454–3467

Bonekamp NA, Sampaio P, de Abreu FV, Luers GH, Schrader M (2012) Transient complex interactions of mammalian peroxisomes without exchange of matrix or membrane marker proteins. Traffic 13:960–978

Bonekamp NA, Grille S, Cardoso MJ, Almeida M, Aroso M, Gomes S, Magalhaes AC, Ribeiro D, Islinger M, Schrader M (2013) Self-interaction of human Pex11pbeta during peroxisomal growth and division. PLoS One 8:e53424

Cablé S, Keller JM, Colin S, Haffen K, Kedinger M, Parache RM, Dauca M (1992) Peroxisomes in human colon carcinomas. A cytochemical and biochemical study. Virchows Arch B 62: 221–226

Camões F, Bonekamp NA, Delille HK, Schrader M (2009) Organelle dynamics and dysfunction: a closer link between peroxisomes and mitochondria. J Inherit Metab Dis 32:163–180

Chang CC, South S, Warren D, Jones J, Moser AB, Moser HW, Gould SJ (1999) Metabolic control of peroxisome abundance. J Cell Sci 112:1579–1590

Chang CR, Manlandro CM, Arnoult D, Stadler J, Posey AE, Hill RB, Blackstone C (2010) A lethal de novo mutation in the middle domain of the dynamin-related GTPase Drp1 impairs higher order assembly and mitochondrial division. J Biol Chem 285:32494–32503

Cuesta A, Pedrola L, Sevilla T, Garcia-Planells J, Chumillas MJ, Mayordomo F, LeGuern E, Marin I, Vilchez JJ, Palau F (2002) The gene encoding ganglioside-induced differentiation-

associated protein 1 is mutated in axonal Charcot-Marie-Tooth type 4A disease. Nat Genet 30: 22–25
De Craemer D (1995) Secondary alterations of human hepatocellular peroxisomes. J Inherit Metab Dis 18:181–213
De Craemer D, Pauwels M, Van den Branden C (1996) Morphometric characteristics of human hepatocellular peroxisomes in alcoholic liver disease. Alcohol Clin Exp Res 20:908–913
De Craemer D, Van den Branden C, Pauwels M, Vamecq J (1998) Peroxisome-proliferating effects of fenoprofen in mice. Lipids 33:539–543
De Duve C, Baudhuin P (1966) Peroxisomes (microbodies and related particles). Physiol Rev 46: 323–357
del Rio LA (ed) (2013) Peroxisomes and their key role in cellular signaling and metabolism, vol 69, Subcellular biochemistry. Springer, Heidelberg
Delille HK, Alves R, Schrader M (2009) Biogenesis of peroxisomes and mitochondria: linked by division. Histochem Cell Biol 131:441–446
Delille HK, Agricola B, Guimaraes SC, Borta H, Luers GH, Fransen M, Schrader M (2010) Pex11pbeta-mediated growth and division of mammalian peroxisomes follows a maturation pathway. J Cell Sci 123:2750–2762
Delille HK, Dodt G, Schrader M (2011) Pex11pbeta-mediated maturation of peroxisomes. Commun Integr Biol 4:51–54
Diano S, Liu ZW, Jeong JK, Dietrich MO, Ruan HB, Kim E, Suyama S, Kelly K, Gyengesi E, Arbiser JL, Belsham DD, Sarruf DA, Schwartz MW, Bennett AM, Shanabrough M, Mobbs CV, Yang X, Gao XB, Horvath TL (2011) Peroxisome proliferation-associated control of reactive oxygen species sets melanocortin tone and feeding in diet-induced obesity. Nat Med 17:1121–1127
Dixit E, Boulant S, Zhang Y, Lee AS, Odendall C, Shum B, Hacohen N, Chen ZJ, Whelan SP, Fransen M, Nibert ML, Superti-Furga G, Kagan JC (2010) Peroxisomes are signaling platforms for antiviral innate immunity. Cell 141:668–681
Dreyer C, Keller H, Mahfoudi A, Laudet V, Krey G, Wahli W (1993) Positive regulation of the peroxisomal beta-oxidation pathway by fatty acids through activation of peroxisome proliferator-activated receptors (PPAR). Biol Cell 77:67–76
Ebberink MS, Csanyi B, Chong WK, Denis S, Sharp P, Mooijer PA, Dekker CJ, Spooner C, Ngu LH, De Sousa C, Wanders RJ, Fietz MJ, Clayton PT, Waterham HR, Ferdinandusse S (2010) Identification of an unusual variant peroxisome bio-genesis disorder caused by mutations in the PEX16 gene. J Med Genet 47:608–615
Ebberink MS, Koster J, Visser G, Spronsen F, Stolte-Dijkstra I, Smit GP, Fock JM, Kemp S, Wanders RJ, Waterham HR (2012) A novel defect of peroxisome division due to a homozygous non-sense mutation in the PEX11beta gene. J Med Genet 49:307–313
Espeel M, Depreter M, Nardacci R, D'Herde K, Kerckaert I, Stefanini S, Roels F (1997) Biogenesis of peroxisomes in fetal liver. Microsc Res Tech 39:453–466
Fahimi HD (1969) Cytochemical localization of peroxidatic activity of catalase in rat hepatic microbodies (peroxisomes). J Cell Biol 43:275–288
Fahimi HD (2009) Peroxisomes: 40 years of histochemical staining, personal reminiscences. Histochem Cell Biol 131:437–440
Fahimi HD, Kino M, Hicks L, Thorp KA, Abelman WH (1979) Increased myocardial catalase in rats fed ethanol. Am J Pathol 96:373–390
Fahimi HD, Reinicke A, Sujatta M, Yokota S, Ozel M, Hartig F, Stegmeier K (1982) The short- and long-term effects of bezafibrate in the rat. Ann N Y Acad Sci 386:111–135
Fahimi HD, Baumgart E, Volkl A (1993) Ultrastructural aspects of the biogenesis of peroxisomes in rat liver. Biochimie 75:201–208
Ferdinandusse S, Denis S, Mooijer PA, Zhang Z, Reddy JK, Spector AA, Wanders RJ (2001) Identification of the peroxisomal beta-oxidation enzymes involved in the biosynthesis of docosahexaenoic acid. J Lipid Res 42:1987–1995

Fischer M, You M, Matsumoto M, Crabb DW (2003) Peroxisome proliferator-activated receptor alpha (PPARalpha) agonist treatment reverses PPARalpha dysfunction and abnormalities in hepatic lipid metabolism in ethanol-fed mice. J Biol Chem 278:27997–28004

Frederiks WM, Bosch KS, Hoeben KA, van Marle J, Langbein S (2010) Renal cell carcinoma and oxidative stress: the lack of peroxisomes. Acta Histochem 112:364–371

Fritz R, Bol J, Hebling U, Angermuller S, Volkl A, Fahimi HD, Mueller S (2007) Compartment-dependent management of H(2)O(2) by peroxisomes. Free Radic Biol Med 42:1119–1129

Funato M, Shimozawa N, Nagase T, Takemoto Y, Suzuki Y, Imamura Y, Matsumoto T, Tsukamoto T, Kojidani T, Osumi T, Fukao T, Kondo N (2006) Aberrant peroxisome morphology in peroxisomal beta-oxidation enzyme deficiencies. Brain Dev 28:287–292

Gandre-Babbe S, van der Bliek AM (2008) The novel tail-anchored membrane protein Mff controls mitochondrial and peroxisomal fission in mammalian cells. Mol Biol Cell 19: 2402–2412

Gomes LC, Di Benedetto G, Scorrano L (2011) During autophagy mitochondria elongate, are spared from degradation and sustain cell viability. Nat Cell Biol 13:589–598

Gorgas K (1984) Peroxisomes in sebaceous glands. V. Complex peroxisomes in the mouse preputial gland: serial sectioning and three-dimensional reconstruction studies. Anat Embryol 169:261–270

Gorgas K (1985) Serial section analysis of mouse hepatic peroxisomes. Anat Embryol 172:21–32

Gorgas K, Volkl A (1984) Peroxisomes in sebaceous glands. IV. Aggregates of tubular peroxisomes in the mouse Meibomian gland. Histochem J 16:1079–1098

Gorgas K, Zaar K (1984) Peroxisomes in sebaceous glands. III. Morphological similarities of peroxisomes with smooth endoplasmic reticulum and Golgi stacks in the circumanal gland of the dog. Anat Embryol 169:9–20

Gould SG, Keller GA, Subramani S (1987) Identification of a peroxisomal targeting signal at the carboxy terminus of firefly luciferase. J Cell Biol 105:2923–2931

Grabenbauer M, Satzler K, Baumgart E, Fahimi HD (2000) Three-dimensional ultrastructural analysis of peroxisomes in HepG2 cells. Absence of peroxisomal reticulum but evidence of close spatial association with the endoplasmic reticulum. Cell Biochem Biophys 32:37–49

Hall D, Poussin C, Velagapudi VR, Empsen C, Joffraud M, Beckmann JS, Geerts AE, Ravussin Y, Ibberson M, Oresic M, Thorens B (2010) Peroxisomal and microsomal lipid pathways associated with resistance to hepatic steatosis and reduced pro-inflammatory state. J Biol Chem 285:31011–31023

Hanker JS, Romanovicz DK (1977) Phi bodies: peroxidatic particles that produce crystalloidal cellular inclusions. Science 197:895–898

Hashimoto T, Cook WS, Qi C, Yeldandi AV, Reddy JK, Rao MS (2000) Defect in peroxisome proliferator-activated receptor alpha-inducible fatty acid oxidation determines the severity of hepatic steatosis in response to fasting. J Biol Chem 275:28918–28928

Hicks L, Fahimi HD (1977) Peroxisomes (microbodies) in the myocardium of rodents and primates. A comparative Ultrastructural cytochemical study. Cell Tissue Res 175:467–481

Hoepfner D, Schildknegt D, Braakman I, Philippsen P, Tabak HF (2005) Contribution of the endoplasmic reticulum to peroxisome formation. Cell 122:85–95

Huber N, Guimaraes S, Schrader M, Suter U, Niemann A (2013) Charcot-Marie-Tooth disease-associated mutants of GDAP1 dissociate its roles in peroxisomal and mitochondrial fission. EMBO Rep 14:545–552

Huybrechts SJ, Van Veldhoven PP, Brees C, Mannaerts GP, Los GV, Fransen M (2009) Peroxisome dynamics in cultured mammalian cells. Traffic 10:1722–1733

Imoto Y, Kuroiwa H, Yoshida Y, Ohnuma M, Fujiwara T, Yoshida M, Nishida K, Yagisawa F, Hirooka S, Miyagishima SY, Misumi O, Kawano S, Kuroiwa T (2013) Single-membrane-bounded peroxisome division revealed by isolation of dynamin-based machinery. Proc Natl Acad Sci USA 110:9583–9588

Ishihara N, Nomura M, Jofuku A, Kato H, Suzuki SO, Masuda K, Otera H, Nakanishi Y, Nonaka I, Goto Y, Taguchi N, Morinaga H, Maeda M, Takayanagi R, Yokota S, Mihara K (2009)

Mitochondrial fission factor Drp1 is essential for embryonic development and synapse formation in mice. Nat Cell Biol 11:958–966

Islinger M, Cardoso MJ, Schrader M (2010) Be different–the diversity of peroxisomes in the animal kingdom. Biochim Biophys Acta 1803:881–897

Islinger M, Grille S, Fahimi HD, Schrader M (2012) The peroxisome: an update on mysteries. Histochem Cell Biol 137:547–574

Issemann I, Green S (1990) Activation of a member of the steroid hormone receptor superfamily by peroxisome proliferators. Nature 347:645–650

Itoyama A, Honsho M, Abe Y, Moser A, Yoshida Y, Fujiki Y (2012) Docosahexaenoic acid mediates peroxisomal elongation, a prerequisite for peroxisome division. J Cell Sci 125:589–602

Itoyama A, Michiyuki S, Honsho M, Yamamoto T, Moser A, Yoshida Y, Fujiki Y (2013) Mff functions with Pex11pβ and DLP1 in peroxisomal fission. Biol Open 2:998–1006

Johkura K, Usuda N, Liang Y, Nakazawa A (1998) Immunohistochemical localization of peroxisomal enzymes in developing rat kidney tissues. J Histochem Cytochem 46:1161–1173

Kang YH, Morris HP, Criss WE (1982) Correlation between growth rate and cytochemistry in Morris hepatomas. Anat Rec 202:209–219

Kobayashi S, Tanaka A, Fujiki Y (2007) Fis1, DLP1, and Pex11p coordinately regulate peroxisome morphogenesis. Exp Cell Res 313:1675–1686

Koch A, Thiemann M, Grabenbauer M, Yoon Y, McNiven MA, Schrader M (2003) Dynamin-like protein 1 is involved in peroxisomal fission. J Biol Chem 278:8597–8605

Koch A, Schneider G, Luers GH, Schrader M (2004) Peroxisome elongation and constriction but not fission can occur independently of dynamin-like protein 1. J Cell Sci 117:3995–4006

Koch J, Brocard C (2012) PEX11 proteins attract Mff and human Fis1 to coordinate peroxisomal fission. J Cell Sci 125:3813–3826

Koch J, Pranjic K, Huber A, Ellinger A, Hartig A, Kragler F, Brocard C (2010) PEX11 family members are membrane elongation factors that coordinate peroxisome proliferation and maintenance. J Cell Sci 123:3389–3400

Lauer C, Volkl A, Riedl S, Fahimi HD, Beier K (1999) Impairment of peroxisomal biogenesis in human colon carcinoma. Carcinogenesis 20:985–989

Laurenti G, Benedetti E, D'Angelo B, Cristiano L, Cinque B, Raysi S, Alecci M, Ceru MP, Cifone MG, Galzio R, Giordano A, Cimini A (2011) Hypoxia induces peroxisome proliferator-activated receptor alpha (PPARalpha) and lipid metabolism peroxisomal enzymes in human glioblastoma cells. J Cell Biochem 112:3891–3901

Legakis JE, Koepke JI, Jedeszko C, Barlaskar F, Terlecky LJ, Edwards HJ, Walton PA, Terlecky SR (2002) Peroxisome senescence in human fibroblasts. Mol Biol Cell 13:4243–4255

Li H, Alavian KN, Lazrove E, Mehta N, Jones A, Zhang P, Licznerski P, Graham M, Uo T, Guo J, Rahner C, Duman RS, Morrison RS, Jonas EA (2013) A Bcl-xL-Drp1 complex regulates synaptic vesicle membrane dynamics during endocytosis. Nat Cell Biol 15:773–785

Li X, Baumgart E, Morrell JC, Jimenez-Sanchez G, Valle D, Gould SJ (2002) PEX11 beta deficiency is lethal and impairs neuronal migration but does not abrogate peroxisome function. Mol Cell Biol 22:4358–4365

Li X, Gould SJ (2003) The dynamin-like GTPase DLP1 is essential for peroxisome division and is recruited to peroxisomes in part by PEX11. J Biol Chem 278:17012–17020

Litwin JA, Beier K, Volkl A, Hofmann WJ, Fahimi HD (1999) Immunocytochemical investigation of catalase and peroxisomal lipid beta-oxidation enzymes in human hepatocellular tumors and liver cirrhosis. Virchows Arch 435:486–495

Misra P, Viswakarma N, Reddy JK (2013) Peroxisome proliferator-activated receptor-alpha signaling in hepatocarcinogenesis. Subcell Biochem 69:77–99

Mizushima N, Yoshimori T, Ohsumi Y (2011) The role of Atg proteins in autophagosome formation. Annu Rev Cell Dev Biol 27:107–132

Nakajima T, Kamijo Y, Tanaka N, Sugiyama E, Tanaka E, Kiyosawa K, Fukushima Y, Peters JM, Gonzalez FJ, Aoyama T (2004) Peroxisome proliferator-activated receptor alpha protects against alcohol-induced liver damage. Hepatology 40:972–980

Nanji AA, Dannenberg AJ, Jokelainen K, Bass NM (2004) Alcoholic liver injury in the rat is associated with reduced expression of peroxisome proliferator-alpha (PPARalpha)-regulated genes and is ameliorated by PPARalpha activation. J Pharmacol Exp Ther 310:417–424

Narravula S, Colgan SP (2001) Hypoxia-inducible factor 1-mediated inhibition of peroxisome proliferator-activated receptor alpha expression during hypoxia. J Immunol 166:7543–7548

Nazarko TY, Farre JC, Subramani S (2009) Peroxisome size provides insights into the function of autophagy-related proteins. Mol Biol Cell 20:3828–3839

Niemann A, Ruegg M, La Padula V, Schenone A, Suter U (2005) Ganglioside-induced differentiation associated protein 1 is a regulator of the mitochondrial network: new implications for Charcot-Marie-Tooth disease. J Cell Biol 170:1067–1078

Niemann A, Berger P, Suter U (2006) Pathomechanisms of mutant proteins in Charcot-Marie-Tooth disease. Neuromolecular Med 8:217–242

Niemann A, Wagner KM, Ruegg M, Suter U (2009) GDAP1 mutations differ in their effects on mitochondrial dynamics and apoptosis depending on the mode of inheritance. Neurobiol Dis 36:509–520

Noack R, Frede S, Albrecht P, Henke N, Pfeiffer A, Knoll K, Dehmel T, Meyer Zu Horste G, Stettner M, Kieseier BC, Summer H, Golz S, Kochanski A, Wiedau-Pazos M, Arnold S, Lewerenz J, Methner A (2012) Charcot-Marie-Tooth disease CMT4A: GDAP1 increases cellular glutathione and the mitochondrial membrane potential. Hum Mol Genet 21:150–162

Nordgren M, Wang B, Apanasets O, Fransen M (2013) Peroxisome degradation in mammals: mechanisms of action, recent advances, and perspectives. Front Physiol 4:145

Novikoff AB, Novikoff PM (1973) Microperoxisomes. J Histochem Cytochem 11:963–966

Opalinski L, Kiel JA, Williams C, Veenhuis M, van der Klei IJ (2011) Membrane curvature during peroxisome fission requires Pex11. EMBO J 30:5–16

Otera H, Wang C, Cleland MM, Setoguchi K, Yokota S, Youle RJ, Mihara K (2010) Mff is an essential factor for mitochondrial recruitment of Drp1 during mitochondrial fission in mammalian cells. J Cell Biol 191:1141–1158

Pavelka M, Goldenberg H, Huttinger M, Kramar R (1976) Enzymic and morphological studies on catalase positive particles from brown fat of cold adapted rats. Histochemistry 50:47–55

Phipps AN, Connock MJ, Johnson P, Burdett K (2000) Peroxisome distribution along the crypt-villus axis of the guinea pig small intestine. Mol Cell Biochem 203:119–126

Pipan N, Psenicnik M (1975) The development of microperoxisomes in the cells of the proximal tubules of the kidney and epithelium of the small intestine during the embryonic development and postnatal period. Histochemistry 44:13–21

Platt FM, Boland B, van der Spoel AC (2012) The cell biology of disease: lysosomal storage disorders: the cellular impact of lysosomal dysfunction. J Cell Biol 199:723–734

Praefcke GJ, McMahon HT (2004) The dynamin superfamily: universal membrane tubulation and fission molecules? Nat Rev Mol Cell Biol 5:133–147

Purohit V, Gao B, Song BJ (2009) Molecular mechanisms of alcoholic fatty liver. Alcohol Clin Exp Res 33:191–205

Rambold AS, Kostelecky B, Elia N, Lippincott-Schwartz J (2011) Tubular network formation protects mitochondria from autophagosomal degradation during nutrient starvation. Proc Natl Acad Sci USA 108:10190–10195

Rhodin J (1954) Correlation of ultrastructural organization and function in normal experimentally changed convoluted tubule cells of the mouse kidney. Aktiebolaget Godvil, Stockholm

Ribeiro D, Castro I, Fahimi HD, Schrader M (2012) Peroxisome morphology in pathology. Histol Histopathol 27:661–676

Roels F, Geerts A, de Prest B, de Coster W, Plum J (1981) Absence of peroxisome turnover during inhibition of catalase synthesis. Verh Anat Ges 75:623–626

Sandalio LM, Rodriguez-Serrano M, Romero-Puertas MC, Del Rio LA (2013) Role of peroxisomes as a source of reactive oxygen species (ROS) signaling molecules. Subcell Biochem 69:231–255
Santel A, Frank S (2008) Shaping mitochondria: the complex posttranslational regulation of the mitochondrial fission protein DRP1. IUBMB Life 60:448–455
Schrader M (2006) Shared components of mitochondrial and peroxisomal division. Biochim Biophys Acta 1763:531–541
Schrader M, Fahimi HD (2006a) Growth and division of peroxisomes. Int Rev Cytol 255:237–290
Schrader M, Fahimi HD (2006b) Peroxisomes and oxidative stress. Biochim Biophys Acta 1763: 1755–1766
Schrader M, Baumgart E, Volkl A, Fahimi HD (1994) Heterogeneity of peroxisomes in human hepatoblastoma cell line HepG2. Evidence of distinct subpopulations. Eur J Cell Biol 64: 281–294
Schrader M, Burkhardt JK, Baumgart E, Luers G, Spring H, Volkl A, Fahimi HD (1996) Interaction of microtubules with peroxisomes. Tubular and spherical peroxisomes in HepG2 cells and their alterations induced by microtubule-active drugs. Eur J Cell Biol 69:24–35
Schrader M, Reuber BE, Morrell JC, Jimenez-Sanchez G, Obie C, Stroh TA, Valle D, Schroer TA, Gould SJ (1998) Expression of PEX11beta mediates peroxisome proliferation in the absence of extracellular stimuli. J Biol Chem 273:29607–29614
Schrader M, Wodopia R, Fahimi HD (1999) Induction of tubular peroxisomes by UV irradiation and reactive oxygen species in HepG2 cells. J Histochem Cytochem 47:1141–1148
Schrader M, King SJ, Stroh TA, Schroer TA (2000) Real time imaging reveals a peroxisomal reticulum in living cells. J Cell Sci 113:3663–3671
Schrader M, Thiemann M, Fahimi HD (2003) Peroxisomal motility and interaction with microtubules. Microsc Res Tech 61:171–178
Schrader M, Yoon Y (2007) Mitochondria and peroxisomes: are the 'big brother' and the 'little sister' closer than assumed? BioEssays 29:1105–1114
Schrader M, Bonekamp NA, Islinger M (2012) Fission and proliferation of peroxisomes. Biochim Biophys Acta 1822:1343–1357
Shamseldin HE, Alshammari M, Al-Sheddi T, Salih MA, Alkhalidi H, Kentab A, Repetto GM, Hashem M, Alkuraya FS (2012) Genomic analysis of mitochondrial diseases in a consanguineous population reveals novel candidate disease genes. J Med Genet 49:234–241
Sima AA (1980) Peroxisomes (microbodies) in human glial tumors. A cytochemical ultrastructural study. Acta Neuropathol 51:113–117
Sinclair AM, Trobacher CP, Mathur N, Greenwood JS, Mathur J (2009) Peroxule extension over ER-defined paths constitutes a rapid subcellular response to hydroxyl stress. Plant J 59: 231–242
Tanaka A, Kobayashi S, Fujiki Y (2006) Peroxisome division is impaired in a CHO cell mutant with an inactivating point-mutation in dynamin-like protein 1 gene. Exp Cell Res 312: 1671–1684
Thoms S, Erdmann R (2005) Dynamin-related proteins and Pex11 proteins in peroxisome division and proliferation. FEBS J 272:5169–5181
Titorenko VI, Rachubinski RA (2004) The peroxisome: orchestrating important developmental decisions from inside the cell. J Cell Biol 164:641–645
Toyama T, Nakamura H, Harano Y, Yamauchi N, Morita A, Kirishima T, Minami M, Itoh Y, Okanoue T (2004) PPARalpha ligands activate antioxidant enzymes and suppress hepatic fibrosis in rats. Biochem Biophys Res Commun 324:697–704
van der Zand A, Braakman I, Geuze HJ, Tabak HF (2006) The return of the peroxisome. J Cell Sci 119:989–994
Vasko R, Ratliff BB, Bohr S, Nadel E, Chen J, Xavier S, Chander P, Goligorsky MS (2013) Endothelial peroxisomal dysfunction and impaired pexophagy promotes oxidative damage in lipopolysaccharide-induced acute kidney injury. Antioxid Redox Signal 19:211–230

Wagner KM, Ruegg M, Niemann A, Suter U (2009) Targeting and function of the mitochondrial fission factor GDAP1 are dependent on its tail-anchor. PLoS One 4:e5160

Wakabayashi J, Zhang Z, Wakabayashi N, Tamura Y, Fukaya M, Kensler TW, Iijima M, Sesaki H (2009) The dynamin-related GTPase Drp1 is required for embryonic and brain development in mice. J Cell Biol 186:805–816

Wanders RJ, Waterham HR (2006) Biochemistry of mammalian peroxisomes revisited. Annu Rev Biochem 75:295–332

Waterham HR, Ebberink MS (2012) Genetics and molecular basis of human peroxisome biogenesis disorders. Biochim Biophys Acta 1822:1430–1441

Waterham HR, Koster J, van Roermund CW, Mooyer PA, Wanders RJ, Leonard JV (2007) A lethal defect of mitochondrial and peroxisomal fission. N Engl J Med 356:1736–1741

Yamamoto K, Fahimi HD (1987) Three-dimensional reconstruction of a peroxisomal reticulum in regenerating rat liver: evidence of interconnections between heterogeneous segments. J Cell Biol 105:713–722

Yan M, Rayapuram N, Subramani S (2005) The control of peroxisome number and size during division and proliferation. Curr Opin Cell Biol 17:376–383

Yokota S, Fahimi HD (1978) The peroxisome (microbody) membrane: effects of de-tergents and lipid solvents on its ultrastructure and permeability to catalase. Histochem J 10:469–487

Zaar K, Fahimi HD (1990) A freeze-etch study of angular marginal-plate-containing peroxisomes in the proximal tubes of bovine kidney. Cell Tissue Res 260:409–414

Zhou J, Zhang S, Xue J, Avery J, Wu J, Lind SE, Ding WQ (2012) Activation of peroxisome proliferator-activated receptor alpha (PPARalpha) suppresses hypoxia-inducible factor-1alpha (HIF-1alpha) signaling in cancer cells. J Biol Chem 287:35161–35169

Single Peroxisomal Enzyme and Transporter Deficiencies in Human Diseases and Mouse Models

8

Markus Kunze and Johannes Berger

Abstract

Peroxisomal single enzyme and transporter deficiencies are inherited human diseases caused by the absence of individual enzymatic activities exerted in peroxisomes. These deficiencies either cause an abnormal accumulation of substances normally degraded in peroxisomes or the lack of biomolecules that require peroxisomal function for their synthesis. Consequently, the symptoms observed in diverse tissues of affected patients can be interpreted as consequences of metabolic intoxications or deficiencies in essential biomolecules. Thus, these detrimental effects should add-up to a very severe pathology upon concomitant inactivation of all peroxisomal functions, which is actually observed in patients suffering from peroxisome biogenesis disorders. Interestingly, only a subset of peroxisomal enzymes has been associated with single enzyme and transporter deficiencies and the inactivation of enzymes participating in the same metabolic pathway sometimes presents with drastically different phenotypical outcomes. Moreover, a remarkable spectrum of clinical pictures has been observed in human patients suffering from the same single enzyme deficiency. The utilization of mice lacking specifically one gene encoding for a peroxisomal enzyme (knockout-mice) allows the reduction of the broad spectrum of pathologies observed in human patients, because these mice are genetically very similar and live under standardized housing conditions. Thereby, these mice can serve as valuable tools to confirm biochemical outcomes of enzyme inactivation, to attribute specific phenotypic aberrations to the absence of an individual enzyme, and to test the contribution of exogenously added compounds to the development of certain symptoms.

M. Kunze (✉) • J. Berger (✉)
Department of Pathobiology of the Nervous System, Center for Brain Research, Medical University of Vienna, Vienna, Austria
e-mail: markus.kunze@meduniwien.ac.at; johannes.berger@meduniwien.ac.at

C. Brocard and A. Hartig (eds.), *Molecular Machines Involved in Peroxisome Biogenesis and Maintenance*, DOI 10.1007/978-3-7091-1788-0_8, © Springer-Verlag Wien 2014

Keywords
Single enzyme deficiencies • Mouse models • Peroxisomes • Genotype • Phenotype • Environment • Peroxisome biogenesis disorder • Metabolism • Fatty acid degradation

Abbreviations

RCDP	Rhizomelic chondrodysplasia punctate
MRI	Magnetic resonance imaging
PAF	Platelet activating factor
PBD	Peroxisome biogenesis disorder
PUFA	Polyunsaturated fatty acid
VLCFA	Very long chain fatty acids

8.1 Introduction

Peroxisomes encase a variety of enzymatic reactions participating in biosynthetic and degrading metabolic pathways (Wanders and Waterham 2006a). The importance of peroxisomal functions for mammalian physiology is stressed by the severe phenotype of inherited human diseases due to dysfunctional peroxisomes that are summarized as peroxisome biogenesis disorders (PBD; Waterham and Ebberink 2012). However, even the loss of individual peroxisomal enzymatic activities or of transporter functions causes physiological dysfunctions of varying severity and diversity, such as progressive loss of liver- and kidney function, problems with bone formation, hearing loss, or inflammation of the central nervous system (Wanders and Waterham 2006b). Overall, 59 proteins have been attributed to human liver peroxisomes (Gronemeyer et al. 2013) and around 60 proteins to murine kidney peroxisomes (Wiese et al. 2007), but the enzymatic equipment of peroxisomes depends on the tissue, the developmental stage or external stimuli. Our knowledge about the functions of these peroxisomal proteins is based on a variety of different approaches such as the biochemical investigation of individual enzymes or of the metabolic conversion of radiolabelled precursors or from studies on cultured cells obtained from organisms lacking specific peroxisomal proteins. All these investigations provide overlapping, but also mutually confirming results. In this respect, the comparison of pathological signs observed in patients suffering from inherited diseases associated with peroxisomal dysfunction with the phenotype of mice, in which individual genes encoding peroxisomal enzymes have been inactivated (knock-out mice), appears promising, because the genetic homogeneity of the mice and the standardized living conditions allow the attribution of the observed pathological observations to the genetic difference of one inactivated gene, which is often not possible in human patients. Furthermore, such studies reveal the contribution of individual metabolic pathways to the symptoms observed in the different tissues of mammalian organisms lacking functional peroxisomes and suggest pathomechanisms for inherited human diseases.

In this review, we will summarize the current knowledge about the metabolic pathways exerted in peroxisomes and describe the symptoms of human patients with single enzyme deficiencies and the phenotype of the corresponding knock-out mice. Thereby, we will try to elucidate unifying patterns in the observations and reflect on advantages and disadvantages of these mouse models. Due to the limitations in space we will only briefly summarize the enzymatic pathways and the biochemical details, which have been extensively reviewed previously (Wanders and Waterham 2006a; Van Veldhoven 2010).

8.2 Single Enzyme and Transporter Deficiencies

Although united by the headline "single enzyme and transporter deficiencies," the lack of individual peroxisomal activities is accompanied by drastically diverse physiological consequences for the organism. The abrogation of a biosynthetic process prevents the provision of an important component to the cell, whereas the lack of a degrading activity results in the accumulation of potentially toxic compounds or a prolonged half-life of signaling intermediates. However, such discrimination is not distinct, because precursor or intermediates of biosynthetic pathways can exert negative functions (e.g., as competitors) and reactive intermediates can act as important second messengers (e.g., H_2O_2).

Mammalian peroxisomes are known to degrade structurally divergent fatty acid classes, namely very long-chain fatty acids (VLCFAs) ($C > 22{:}0$), polyunsaturated fatty acids (PUFAs), branched-chain fatty acids, and dicarboxylic acids, whereas mitochondria exert the degradation of the more abundant short, medium and long-chain fatty acids, although the degradation of the latter might also occur in peroxisomes (Van Veldhoven 2010; Wanders and Waterham 2006a). Furthermore, the degradation of H_2O_2, glyoxylate and D-amino acids occurs in peroxisomes. If these compounds are not degraded due to peroxisomal defects, they either accumulate locally in the tissue (e.g., liver) or are distributed by the body fluids across the organism and can damage other organs (e.g., kidney). Furthermore, peroxisomes exert the essential first steps in the biosynthesis of ether-phospholipids, the last steps in the production of the important PUFAs docosahexaenoic acid (DHA), and the side-chain shortening of bile acid precursors, such as trihydroxycholestanoic acid to cholic acid, and its conjugation with taurine or glycine. In the absence of these peroxisomal functions the organism lacks lipid structures that are normally enriched in the brain or in testis (plasmalogens and DHA) or play an important role in the emulsification of ingested lipids (bile acids).

To date, 11 peroxisomal enzymes and transporter proteins have been associated with inherited human diseases (see Table 8.1). The majority of these enzymes contribute to the degradation of fatty acids and side-chain shortening of bile acids (6), whereas others are involved in the maturation of bile acids (1), in the

Table 8.1 Metabolic pathways involving single enzyme deficiencies

Metabolic pathway	Gene	Protein	Human disease	OMIM	Affected pathways	LIT	Mouse model	LIT
Fatty acid degradation								
β-Oxidation	*Acox1*	**ACOX1**	Acyl-CoA deficiency	264470	SC-FA$\uparrow$, DHA$\uparrow$, DCA$\uparrow$	Poll-The (1988)	ACOX1	Fan (1996)
	Acox2	ACOX2						
	Acox3							
	Hsd17b4	**D-BP**	D-BP deficiency	261515	SC-FA$\uparrow$ BC-FA$\uparrow$, DHA$\uparrow$	Une (1997)	MFP2	Baes (2000)
	Echhadh	L-BP			DCA$\uparrow$	Nguyen (2008)	MFP1	Qi (1999)
	Acaa1	Thiolase					ACAA1b	Chevillard (2004)
	Scpx	**SCPx/ SCP2**	SCPx deficiency	613724	BC-FA$\uparrow$, BA$\uparrow$	Ferdinandusse (2006)	SCPx	Seedorf (1998)
Auxiliary enzymes	*Peci* *Ech1* *Decr2*	PECI ECH1 DECR2						
α-Oxidation	*Phyh*	**PHYH**	Refsum disease	266500	BC-FA$\uparrow$, BA$\uparrow$	Jansen (2004)	PHYH	Ferdinandusse (2008)
	Hacl	2-HACL						
Auxiliary enzymes	*Amacr* *Aldh3A2*	**AMACR** FALDH	Racemase deficiency Sjögren-Larsson syndrome[a]	604489 270200	BC-FA$\uparrow$ BA$\uparrow$	Ferdinandusse (2002), De Laurenzi (1996)	AMACR	Savolainen (2004)
	Pecr	PECR						
Termination	*Crat* *Crot* *Acot4* *Acot8*	CRAT CROT ACOT4 ACTO8						
Membrane transporter	*Abcd1*	**ABCD1**	X-linked adrenoleukodystrophy (X-ALD)	300100	SC-FA$\uparrow$	Mosser (1993)	ABCD1	Forss-Petter (1997) Liu (1997) Kobayashi (1997)
	Abcd2	ABCD2					ABCD2	Ferrer (2005)
	Abcd3	ABCD3						

Bile acid maturation								
Amidation	*Baat*	**BAAT**	Familiar hypercholanemia	607748	T-CH↑/G-CH↑	Setchell (2013)		
Ether-phospholipid biosynthesis								
	Gnpat	**GNPAT**	RCDP type II	222765	Plasmalogens ↓	Ofman (1998)	GNPAT	Rodemer (2003)
	Agps	**AGPS**	RCDP type III	600121	Plasmalogens ↓	DeVet (1998)	Blind sterile (hp)	Liegel (2011)
	Far1	FAR1						
	Far1	FAR2						
Degradation of small molecules								
Glyoxvlate degradation	*Agxt*	**AGT**	Primary hyperoxaluria (type I)	259900	Oxalata↑, glycolate↑	Danpure (1986)	AGT	Salido (2006)
Hydrogen peroxide degradation	*Cat*	**CAT**	Acatalasemia	115500	H_2O_2↑	Ogata (1991)	CAT	Ho (2004)
Lysine degradation	*Pipox*	PIPOX	Pipecolic acidemia[b]					

The enzymes participating in these pathways are listed and the protein affected in single enzyme deficiencies are bold; proteins: ACOX acyl-CoA oxidase, D-BP D-bifunctional protein, SCPx thioase with sterol carrier protein domain, PECI peroxisomal enoyl-CoA isomerase, ECH1 enoyl-CoA hydratase, DECR2 dienoyl-CoA reductase, PHYH phytanoyl-CoA hydroxylase, 2-HACL 2-hydroxyacly-CoA lyase, AMACR alpha-methylacyl-CoA reductase, FALDH fatty aldehyde dehydrogenase, PECR peroxisomal enoyl-CoA reductase, CRAT carnitine-acetyltransferase, CROT carnitine-octanoyltransferase, ACOT acyl-CoA thioesterase, ALDP adrenoleukodystrophy protein, ALDRP adrenoleukodystrophy related protein, PMP70 peroxisomal membrane protein of 70 kD, BAAT bile acid-CoA:amino acid *N*-acyltransferase, GNPAT glyceronephosphate *O*-acyltransferase, AGPS alkylglycerone phosphate synthase, FAR1/2 fatty acid reductase 1/2, AGT alanine-glyoxylate aminotransferase, CAT catalase, PIPOX pipecolic acid oxidase; Disease: RCDP rhizomelic chondrodysplasia punctate; Affected pathways: SC-FA short-chain fatty acids; BC-FA branched-chain fatty acids; BA bile acids; T-CH/G-CH taurocholate and glycocholate, DHA docosahexaenoic acid, DCA dicarboxylic acid; Mouse proteins: (hp) hypomorph

[a]SLS cannot be attributed to the peroxisomal fraction of FALDH

[b]Pipecolic academia is indicated as single enzyme deficiency due to an accumulation of pipecolic acid in the plasma of PDB-patients, but no patient with dysfunctional PIPOX has been identified until now

biosynthesis of plasmalogens (2), or in the detoxification of glyoxylate (1) and of hydrogen peroxide (1). Furthermore, an inherited human disease has been linked to dysfunctional fatty aldehyde dehydrogenase (FALDH), which partially occurs in peroxisomes. Finally, an individual defect in the peroxisomal enzyme pipecolic acid oxidase (IJlst et al. 2000; Dodt et al. 2000) has been suggested as single enzyme deficiency, because of the consistent observation that pipecolic acid is accumulated in the plasma of PBD-patients (Mihalik et al. 1989; Wanders et al. 1988), but no patient has been found. Thus, listing of peroxisomal fatty aldehyde dehydrogenase and pipecolic acid oxidase as single enzyme and transporter deficiencies is disputed.

In the following chapters the metabolic pathways are briefly described and the biochemical and physiological consequences of an interruption of these pathways for human patients and knock-out mice are described.

8.2.1 Degradation of Fatty Acids

All fatty acids that are degraded in peroxisomes repeatedly pass through a four-step process called β-oxidation with the sequential activity of an acyl-CoA oxidase (ACOX), a 2-enoyl-CoA hydratase, a 3-hydroxyacyl-CoA oxidase and a 3-oxo-acyl-CoA thiolase (THIO) that shortens the backbone of the fatty acid by two carbon atoms releasing acetyl-CoA (Fig. 8.1). However, the 2-enoyl-CoA hydratase and the 3-hydroxyacyl-CoA oxidase activities are exerted by one protein with two active centers, the bi- or multifunctional protein (BP, MFP). Although these three enzymes comprise the core-unit of β-oxidation for all the diverse fatty acids, mammalian genomes encode several isoenzymes with diverging, but sometimes overlapping substrate specificities (Van Veldhoven 2010). Extensive investigations suggest that in humans and in mice the degradation of straight VLCFAs is exerted by acyl-CoA oxidase 1 (ACOX1), by D-bifunctional protein (D-BP) and by either the straight chain thiolase (ACAA1) or the thiolase with a sterol carrier protein domain (SCPx). However, in the mouse two isoforms of ACAA1 exist that are encoded by Acaa1a and Acaa1b. Similarly, branched-chain fatty acids are degraded in both organisms by ACOX2, D-BP and SCPx, but only in rodents a third gene encoding acyl-CoA oxidase (ACOX3) is translated into functional protein. However, both degradation pathways involve D-BP, whereas L-BP appears not to be involved. Other enzymatic activities join this core-process to cope with the incompatibility of certain intermediates with structural restrictions of certain core-enzymes. This involves the position of additional double bond (s) relative to the carboxylate group within unsaturated fatty acids and the stereochemistry of the C_α position of branched chain fatty acids.

The major substrates for the β-oxidation of unsaturated fatty acids are long- and very long-chain PUFAs, which are characterized by the presence of several Z-double bonds that are separated by a methylene group ($-CH_2-$). For the degradation of these fatty acids the peroxisomal Δ3-Δ2-enoyl-CoA isomerase (PECI) (Geisbrecht et al. 1999) cooperates with Δ3,5- Δ2,4-dienoyl-CoA isomerase/

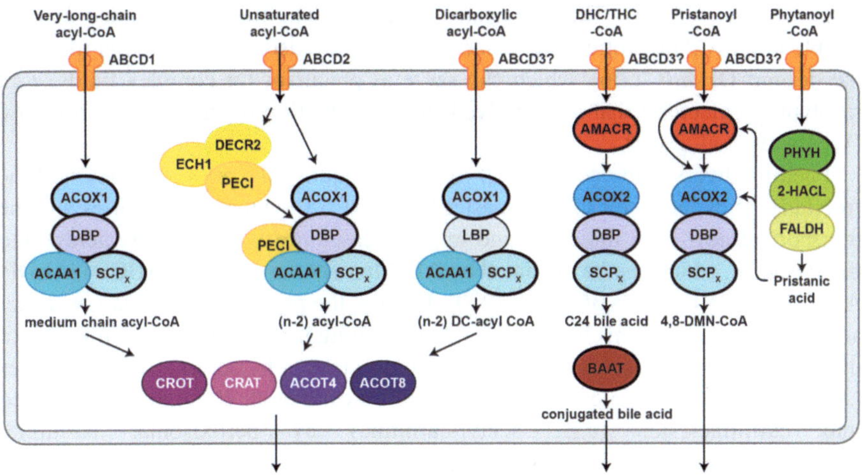

Fig. 8.1 Proteins involved in the degradation of different fatty acids and the shortening of the bile acid side chain: enzymes are indicated as *coloured ovals*, ACOX acyl-CoA oxidase, D-BP D-bifunctional protein, SCPx thiolase with sterol carrier protein domain, PECI peroxisomal enoyl-CoA isomerase, ECH1 enoyl-CoA hydratase, DECR2 dienoyl-CoA reductase, PHYH phytanoyl-CoA 2-hydroxylase, 2-HACL 2-hydroxyacly-CoA lyase, AMACR alpha-methylacyl-CoA reductase, FALDH fatty aldehyde dehydrogenase, PECR peroxisomal enoyl-CoA reductase, CRAT carnitine-acetyltransferase, CROT carnitine-octanoyltransferase, ACOT acyl-CoA thioesterase, ALDP adrenoleukodystrophy protein, ALDRP adrenoleukodystrophy-related protein, PMP70 peroxisomal membrane protein of 70 kD, BAAT bile acid-CoA:amino acid *N*-acyltransferase

enoyl-CoA hydratase (ECH1) (Filppula et al. 1998) and peroxisomal Δ2-Δ4-dienoyl-CoA reductase 2 (DECR2) to generate 2E-enoyl-CoA (Dommes et al. 1981), which is a standard intermediate of the core β-oxidation.

The change in the stereochemistry at the position C_α ($=C_2$) of branched-chain fatty acids is catalyzed by the enzyme alpha-methylacyl-CoA racemase (AMACR). However, the degradation of certain branched-chain fatty acids (e.g., phytanic acid) requires an additional process that changes the position of the extra methyl-side chain relative to the carboxylate by removal of the first carbon, a process named α-oxidation (Wanders and Waterham 2006a). This pathway consists of a hydroxylation of the C_α (next to the carboxylate carbon) exerted by phytanoyl-CoA-2-hydroxylase (PHYH) (Mihalik et al. 1995) and the breakage of the bond between these carbons by 2-hydroxyphytanoyl-CoA lyase (2-HPCL; Foulon et al. 1999). This generates formaldehyde and the alkyl-aldehyde shortened by one carbon (pristanal), which can be again oxidized by the peroxisomal fatty aldehyde dehydrogenase (FALD/ALDH3A2) to pristanic acid (Ashibe et al. 2007).

Furthermore, the synthesis of the very long-chain PUFAs docosahexaenoic acid (DHA, C22:6 $n-3$) involves peroxisomal β-oxidation to compensate for the lack of Δ4-desaturase in mammals. Therefore, the PUFA-precursor (C18:3) is first elongated until the chain length of the fatty acids exceeds the desired length by two carbon atoms (C24:5). Then this fatty acid is desaturated to introduce another double bond (C24:6) followed by one round of peroxisomal β-oxidation to obtain the desired DHA (C22:6), which can serve as precursor for complex lipids (e.g., resolvins, protectins) that are involved in the regulation of the immune response (Weylandt et al. 2012).

Fuelling diverse fatty acids into β-oxidation requires the prior conversion of the free fatty acids into their CoA-thioester by acyl-CoA synthetase activities and the transport of acyl-CoA moieties (and in some cases possibly free fatty acids) into peroxisomes, which can be mediated by ABC-transporter proteins (ABCD1/ALDP, ABCD2/ALDRP, and ABCD3/PMP70). Finally, the product of β-oxidation, either acetyl-CoA or a medium chain acyl-CoA originating from premature termination of the β-oxidation (e.g., octanoyl-CoA) are either cleaved by acyl-CoA thioesterases (ACOT4, ACOT8) to release free fatty acids (Westin et al. 2005) or are converted into carnitine-ester (e.g., acetyl-carnitine) by peroxisomal carnitine-acetyltransferase (CRAT; Westin et al. 2008) or carnitine-octanoyl-transferase (CROT; Ferdinandusse et al. 1999) to allow their export. The bile acid intermediate cholic acid (C24) is conjugated to glycine or taurine by the enzyme bile acid-CoA: amino acid *N*-acyltransferase (BAAT; Johnson et al. 1991).

8.2.2 Degradation of Straight-Chain Fatty Acids

8.2.2.1 Acyl-CoA Oxidase

Human patients with acyl-CoA deficiency show increased levels of VLCFAs in the plasma, but normal levels of branched-chain fatty acids, bile acids, and plasmalogens (Poll-The et al. 1988) (Poll-The and Gartner 2012). In the liver peroxisomes appear enlarged (Poll-The et al. 1988), and a similar observation was obtained in cultured fibroblasts although here the peroxisome number was decreased (El Hajj et al. 2012). Patients show severe symptoms comparable to the mild form of Zellweger syndrome with neonatal onset of muscular hypotonia, seizures, and a delay in psychomotor developmental (Poll-The et al. 1988). Furthermore, hepatomegaly, dysmorphism, white matter abnormalities, and impaired auditory and visionary capacities have been observed (Ferdinandusse et al. 2007). Recently, signs of neuroinflammation have been reported in brains from acyl-CoA-deficient patients (Ferdinandusse et al. 2010) and even in cultured fibroblasts an up-regulation in the synthesis and secretion of inflammatory cytokines was observed (El Hajj et al. 2012). Although acyl-CoA deficiency is a severe disease with short life expectation recently several patients with less severe symptoms have been described (Ferdinandusse et al. 2007, 2010).

Mice lacking ACOX1 also accumulate VLCFAs in the plasma comparable to human patients. These mice appear phenotypically relatively normal but are growth retarded and sterile (Fan et al. 1996). In the liver a steatosis and a drastically increased expression of PPARα regulated genes is observed in the absence of exogenous agonist (Fan et al. 1996). This PPARα-dependent upregulation was detrimental under standard conditions, because the steatosis and the sterility of

ACOX1 deficient mice were abrogated in ACOX1/PPARα double-deficient mice (Hashimoto et al. 1999). However, upon prolonged fasting the metabolic defect due to a low peroxisomal and mitochondrial β-oxidation activity overruled the protective effect of PPARα depletion and double deficient mice showed a more severe steatosis (Hashimoto et al. 2000). Moreover, in the liver of these mice a mosaicism is observed in which some cells have many peroxisomes and do not accumulate fatty acids, whereas other cells show no detectable immunoreactivity against peroxisomal proteins, but a massive accumulation of lipids (Fan et al. 1996, 1998). In an early phase (2–4 months), the proliferation rate of hepatic cells containing peroxisomes from ACOX1-deficient mice is much higher, later (6–8 months) peroxisome-containing cells overgrow and become the predominant cell type (Fan et al. 1998) and finally (10–15 months), all ACOX1-deficient mice develop adenomas that later transform into hepatocellular carcinomas (Fan et al. 1998). In these adenoma or carcinoma cells immunoreactivity against peroxisomal proteins is again hardly detectable, whereas the surrounding cells show strong immuno-staining (Fan et al. 1998).

Furthermore, in these mice a part of the ER-stress pathway is upregulated, with increasing severity over time (Huang et al. 2011). However, this can act also as protective preconditioning against obesity caused by leptin deficiency (*ob*/*ob*) and the pathological consequences thereof (Huang et al. 2012).

8.2.2.2 Multifunctional Proteins (MFP1 and MFP2)/Bifunctional Proteins (D-BP and L-BP)

Human patients suffering from D-BP-deficiency harbor mutations in the gene HSD17B4, which biochemically manifests by the accumulation of straight VLCFAs and of branched-chain fatty acids in the serum of the patients (Suzuki et al. 1997). C24-bile acids are reduced, whereas the enoyl-CoA-C27-bile acid intermediates appear increased (Une et al. 1997; Clayton et al. 1988) and in the urine increased levels of leukotriene E4 (LTE4) and ω-carboxy-LTE4 were observed (Ferdinandusse et al. 2002a). These patients are hypotonic, show seizures, and usually die within the first 2 years of life, although individual patients with a late onset have been described (Ferdinandusse et al. 2006a; Mizumoto et al. 2012). In the brain of these patients neuronal migration defects and sometimes neuronal heterotopia or cerebral demyelination can be observed. Furthermore, patients have dysmorphic features and suffer from hearing and vision problems (Ferdinandusse et al. 2007). The disease can be caused either by the complete absence of the protein or by mutations affecting only the hydratase domain (de Launoit and Adamski 1999) or only the dehydrogenase domain (Ferdinandusse et al. 2006c). Moreover, two patients with less severe symptoms have been observed that have compound heterozygous defects in their D-BP/HSD17B4 proteins (Pierce et al. 2010). These patients present with ovarian dysgenesis, hearing loss, and ataxia, which is summarized as Perrault syndrome (Perrault et al. 1951; Bösze et al. 1983). However, other patients with Perrault syndrome do not have mutations in the HSD17B4-gene, suggesting that this syndrome is genetically heterogeneous and that the clinical pictures overlap.

Mice lacking MFP2/D-BP/HSD17B4 are normal at birth but show a drastic delay of weight gain during lactation resulting in much smaller animals (Baes et al. 2000). About a third of the offspring are lost early in life (Baes et al. 2000), whereas the rest die around 6 months (Huyghe et al. 2006b). In these animals the β-oxidation activity toward straight-chain VLCFAs and branched-chain fatty acids is reduced and C26:0 levels are increased in tissues (especially in brain), whereas branched-chain fatty acids accumulate only upon phytol feeding (Baes et al. 2000). Furthermore, the amount of bile was markedly reduced, whereas the remaining bile consisted of residual C24-bile acids and various C27-bile acid derivatives (Baes et al. 2000). Interestingly, SREBP2-dependent expression and the rate of cholesterol synthesis was increased in the liver, which correlates with an increased cholesterol level in the stool (Martens et al. 2008), whereas the level of plasma triglycerides indicating the efficiency of lipid uptake was reduced (Martens et al. 2008). In the liver the expression of PPARα-dependent genes was upregulated similar to ACOX1-deficient mice (Baes et al. 2000), but peroxisomal proliferation was absent (Jia et al. 2003). Male mice were sterile with an accumulation of neutral lipids in Sertoli cells, although in contrast to ACOX1 deficient mice deletion of PPARα does not restore fertility (Huyghe et al. 2006a). Mice developed abnormal cramping of limbs (fore- and hindlimb-clasping) and motor deficits (Huyghe et al. 2006b), which was accompanied by an atrophy of the cerebellum, but peripheral nerves and muscle appeared normal. Interestingly, growth retardation, the development of motor disabilities, and signs of ataxia were observed even when MFP2 deficiency was restricted to the brain (Verheijden et al. 2013). These mice survive the first year which allowed the investigation of cerebellar degeneration progressing from primary axonal damage via a deterioration of Purkinje cells to a cerebellar atrophy. This was accompanied by progressive myelin loss (Verheijden et al. 2013), although the severe inflammation observed in complete MFP2-deficient mice was not observed (Verheijden et al. 2013).

In contrast, mice lacking MFP1/L-BP appear phenotypically normal and all symptoms observed in the liver of MFP2-deficient mice such as hepatic steatosis, spontaneous peroxisome proliferation, or an upregulation of PPARα-dependent gene expression were absent (Qi et al. 1999; Baes et al. 2000) and the bile acid profiles were normal (Ferdinandusse et al. 2005). Altogether these observations suggested that MFP1 does not contribute to the degradation of fatty acids or to bile acid side chain shortening. However, in these mice the formation of medium-chain dicarboxylic acid is ablated in liver, plasma, and urine upon fasting, although the protein level of the key enzyme of ω-oxidation, CYP4α10, is induced (Houten et al. 2012). Thus, apparently MFP1 is involved in the degradation of dicarboxylic acids, which is supported by the observation that the degradation of very long-chain dicarboxylic acids is severely compromised in hepatocytes from MFP1, but not in those from MFP2-deficient mice (Nguyen et al. 2008) and that human fibroblasts from D-BP-deficient patients still can degrade these fatty acids (Rizzo et al. 2003; Ferdinandusse et al. 2004).

Mice in which MFP1 and MFP2 have been concomitantly deleted show growth retardation already at birth and do not survive the lactating period (Jia et al. 2003).

Furthermore, these double-deficient mice recapitulate the hypotonia of patients lacking D-BP, but do not show signs of neuronal migration defects in the cortex (Baes et al. 2002) that have been observed in D-BP-deficiency patients and in mice with dysfunctional peroxisomes (Baes et al. 1997). In the liver of these mice steatosis and an increased level of PPARα-dependent gene expression were observed, but they lack peroxisomal structures which were accumulating in MFP2-deficient mice (Jia et al. 2003) and the mosaicism of ACOX1-deficient mice (Fan et al. 1996). Altogether, these results suggest that MFP2 is required for the degradation of straight- and branched-chain fatty acids including side chain shortening of bile acids, whereas MFP1 is required for the degradation of dicarboxylic acids.

8.2.2.3 Thiolase (THIO-B)

One human patient with a mutation in peroxisomal 3-oxo-acyl-CoA thiolase (ACAA1) has been described presenting with a number of severe symptoms (Schram et al. 1987), but this patient was reanalyzed and another defect in D-BP was identified (Ferdinandusse et al. 2002c). Consequently, no patient lacking thiolase-B has been identified until now.

Deletion of thiolase-B (thio-B/ACAA1b) in mice showed a very mild phenotype and hardly any accumulation of VLCFAs (Chevillard et al. 2004) indicating that other thiolases (ACAA1a or SCPx) can compensate for the lack of ACAA1b. The combination of a treatment with PPARα agonist (Wy14-653) and a fasting regiment increased the differences in the expression pattern between mice with a targeted deletion in Acaa1b and wild-type mice. Thereby, the expression of cholesterogenic enzymes was blunted and the expression of ACOX1 and D-BP was increased compared to wild-type animals (Fidaleo et al. 2011). Furthermore, the increased synthesis of stearoyl-CoA desaturase (SCD1) and fatty acid synthase (FAS) correlated with an increased level of monounsaturated fatty acids (Arnauld et al. 2009).

8.2.3 Degradation of Unsaturated Fatty Acids

Up to now, neither a patient with a mutation in any of the enzymes involved in the metabolization of unsaturated fatty acids (PECI, DECR2, or ECH1) nor mice with targeted inactivation of these genes have been described.

8.2.4 Degradation of Branched Chain Fatty Acids

The major source of branched chain fatty acids in the human body originates from a diet that contains phytol such as meat derived from ruminants or fish (Verhoeven et al. 1998b), but they can also be derived from oxidation of endogenous isoprenoid-derived compounds. Structurally, the intermediates of bile acid synthesis resemble branched-chain fatty acids and defects in the degradation of branched chain fatty acids also affect bile acid side chain shortening. Mechanistically, branched-chain fatty acid degradation simulates the degradation of straight chain fatty acids, but the enzymes exerting the oxidation are ACOX2 in humans and ACOX2 and ACOX3 in mice, whereas the thioclastic cleavage is exerted in both species by SCPx, a thiolase with a sterol carrier protein domain. D-BP contributes the hydratase and dehydrogenase activity for all fatty acids, whereas neither ACOX1 nor ACAA1 show activity toward the branched chain fatty acid derivatives. Furthermore, the degradation of branched chain fatty acids requires α-metylacyl-CoA racemase (AMACR), which inverts the stereochemistry at the chiral center involving the methyl-side chain, and the enzymes exerting α-oxidation, namely phytanoyl-CoA 2-hydroxylase (PHYH) and 2-hydroxyacyl-CoA lyase (2-HACL). Thus, the degradation of branched-chain fatty acids is a complex process involving a large number of enzymes (Wanders et al. 2011), which all can potentially be affected in single enzyme deficiencies.

8.2.4.1 Acyl-CoA Oxidase 2 and 3 (ACOX2/3)

Neither a patient with a mutation in the human enzyme ACOX2 nor mice with a targeted deletion of the homologous enzymes (Acox2 or Acox3) have been described.

8.2.4.2 Thiolase with Sterol Carrier Protein Domain (SCPx)

SCPx was described as thiolase for branched-chain fatty acids and bile acids (Wanders et al. 1998; Ferdinandusse et al. 2000c). This protein contains an additional sterol carrier protein (SCP)-domain at its C-terminus that can also be generated independently by alternative transcription initiation giving rise to SCP2.

One patient with a defect in SCPx has been investigated (Ferdinandusse et al. 2006b), which accumulates pristanic acid and partially phytanic acid in the plasma, whereas the level of VLCFAs was just above the normal range. Moreover, abnormal bile acid glucuronides were secreted in the urine. The patient appeared with dystonic head tremor (spasmodic torticollis), pathological saccadic eye movements (nystagmus) and hyposmia. Within the brain a leukoencephalopathy and slight cerebellar abnormalities were observed, whereas the lower extremities show a strong motor neuropathy and a weak sensory neuropathy. Furthermore, the patient showed hypergonadotrophic hypogonadism and azoospermia.

In mice lacking either both proteins SCPx/SCP2 or selectively only SCPx an increased level of phytanic acid was observed in the plasma, which was drastically increased upon phytol treatment (Seedorf et al. 1998; Atshaves et al. 2007). In the liver of these mice the expression of PPARα-dependent genes and the staining intensity for peroxisomes were drastically increased. Supplementing the chow of these mice with phytol revealed a severe phenotype including a loss of adipose tissue and body weight and changes in the liver appearance, which appears more severe in female mice (Atshaves et al. 2004).

8.2.4.3 α-Methylacyl-CoA Racemase (AMACR)

Racemase inverts the stereochemistry at the chiral C_2-position of branched-chain fatty acids from an R- to an S-conformation, which is required for the recognition by ACOX2 and thus for the degradation of branched-chain fatty acids and the side chain shortening of di- and trihydroxycholestanoic acid. The enzyme has been observed in peroxisomes and mitochondria (Amery et al. 2000; Ferdinandusse et al. 2000b; Kotti et al. 2000), which might render the phenotype a mixture of peroxisomal and mitochondrial defects.

Patients suffering from α-methylacyl-CoA racemase (AMACR) deficiency lack functional racemase due to a mutation in the AMACR gene. In the plasma of these patients a drastic accumulation of pristanic acid with a characteristic overrepresentation of its R-conformation was observed, whereas in patients with other peroxisomal deficiencies predominantly phytanic acid accumulates and comparable amounts of the R- and the S-stereoisomers of pristanic acid are observed (Ferdinandusse et al. 2002b). Furthermore, C24-bile acids were drastically reduced, whereas the C27-bile acid intermediates accumulated which appear exclusively in form of the R-isomer (Ferdinandusse et al. 2000a, b). General symptoms of patients with racemase deficiency are sensory motor neuropathy (Smith et al. 2010) or late onset cerebellar ataxia (Dick et al. 2011). Due to the mild general symptoms a variety of descriptions of individual patients can be found including tremor and white matter abnormalities (Clarke et al. 2004), relapsing rhabdomyolysis (Kapina et al. 2010) or relapsing encephalopathy (Thompson et al. 2009). This is corroborated by a recent description of a pair of siblings with a complex, adult-onset phenotype including peripheral neuropathy, epilepsy, relapsing encephalopathy, pigmentary retinopathy, tremor, and cerebellar abnormalities (Haugarvoll et al. 2013).

When the gene encoding racemase was inactivated by a targeted deletion in mice, low levels of C24-bile acids were found in plasma, bile, and liver, whereas various forms of C27-bile acids were increased (Savolainen et al. 2004). These mice appeared physiologically asymptomatic, which has been attributed to a compensatory mechanism that involves an increased rate of hepatic cholesterol biosynthesis and an increased cholesterol-secretion to the bile (Selkala et al. 2013). Furthermore, the amount of triacylglycerol in VLDL was drastically upregulated, whereas lipase activity was reduced (Selkala et al. 2013).

8.2.4.4 Phytanoyl-CoA 2-Hydroxylase (PHYH)

PHYH is the first enzyme in α-oxidation and induces the introduction of a hydroxyl-group at the C_α-position, which is prerequisite for the scission of chemical bond between the carboxylate carbon and this C_α.

Human patients described as Refsum disease patients (Jansen et al. 2004) were found to harbor defects in the gene encoding PHYH (Jansen et al. 1997; Mihalik et al. 1997). Biochemically, patients show an accumulation of phytanic acid, but not of pristanic acid in plasma and tissues (Ferdinandusse et al. 2002b). These patients show symptoms at variable times of onset, but their life expectation is drastically reduced, although late-onset patients have described as well (Ruether et al. 2010).

The spectrum of phenotypic aberrations involves cardiomyopathy, peripheral neuropathy, ataxia, ichtyosis, and an early-onset retinitis pigmentosa (RP), which can be extended by anosmia, deafness, and blindness (Wierzbicki et al. 2002), whereas skeletal malformation appears in adult onset patients (Ruether et al. 2010).

Mice lacking the enzyme PHYH present with an accumulation of phytanic acid in plasma and testis but also show marked accumulation in all tissues upon supplementation of the diet with phytol (Ferdinandusse et al. 2008). Only this feeding paradigm induces characteristic gait problems reflecting ataxia in the fore and hind-paws that is reflected by an ablation of Purkinje cells from the cerebellum and by an inflammation of the brain indicated by astrogliosis. Moreover, motor-nerve conductance velocity is reduced and also the myelination of the sciatic nerve appears reduced (Ferdinandusse et al. 2008).

8.2.4.5 Supplementary Enzymes

Two further peroxisomal enzymes have been linked to the degradation of phytol, namely peroxisomal 2E-enoyl-CoA reductase (PECR) (Gloerich et al. 2006), that reduces phytenic acid inside peroxisomes before α-oxidation occurs, although possibly an alternative mechanism for this reduction exists (van den Brink and Wanders 2006) and the fatty aldehyde dehydrogenase ALDH3A2/FALDH, which oxidizes medium and long-chain alkylaldehydes such as pristanal generated by 2-HACL. The pathology of patients summarized as Sjögren-Larsson syndrome (SLS) is associated with a mutation the gene ALDH3A2 (De Laurenzi et al. 1996) and fibroblasts from such patients show drastically reduced pristanal dehydrogenase activity in microsomes (Verhoeven et al. 1998a). However, this activity was also demonstrated in peroxisomes (Jansen et al. 2001) and this bilocalisation is due to alternative splicing (Ashibe et al. 2007). SLS patients show mental retardation, spasticity and ichtyosis and sometimes ocular abnormalities (Willemsen et al. 2001). Furthermore, SLS patients show pruritus and suffer from a neurocutanous disorder, in which tumors grow in various parts of the body.

8.2.5 Transport of Fatty Acids

The transport of different fatty acids across the peroxisomal membrane is essential to initiate their degradation. The peroxisomal membrane contains only a limited number of membrane proteins that are not involved in peroxisomal biogenesis. Among these, three proteins (ABCD1/ALDP, ABCD2/ALDRP, and ABCD3/PMP70) couple the consumption of ATP with the transport of substrates across membranes (Morita and Imanaka 2012). ABCD1/ALDP has been directly linked to the transport of a specific substrate, namely VLCFA-CoA (Wiesinger et al. 2013), whereas unsaturated fatty acids have been suggested as substrate for ABCD2 (van Roermund et al. 2011).

Human patients suffering from X-linked adrenoleukodystrophy (X-ALD) were found to harbor mutations in ABCD1/ALDP (Mosser et al. 1993). All these patients show an accumulation of VLCFAs in plasma and tissues, but have normal levels of branched-chain fatty acids in the plasma (ten Brink et al. 1991) and of dicarboxylic acids and LTE4 in the urine (Kodama et al. 1990). Fibroblasts from X-ALD patients show defects in β-oxidation of VLCFAs, but neither in the synthesis of DHA (Ferdinandusse et al. 2001) nor in the degradation of dicarboxylic acids (Ferdinandusse et al. 2004). However, the clinical picture of these patients appears heterogeneous, with two major representations. Adrenomyeloneuropathy (AMN) is characterized by axonopathy with microgliosis, but without cerebral demyelination (Powers et al. 2000), whereas in the cerebral form of the disease (CALD) a severe demyelination and inflammation of the central nervous system is observed, which appears commonly in young boys, but can also start later (Berger and Gartner 2006; van Geel et al. 2001). In the cerebral form, patients show early symptoms such as impaired auditory and visual processing, cerebellar ataxia, or seizures that precede intellectual deterioration (Berger and Gartner 2006). AMN patients develop sensory loss for vibrations and stiffening of the legs that can lead to severe motor disability. In contrast, in both forms of the disease a loss of endocrinological functions may occur, affecting the adrenal gland (Addison-disease; Dubey et al. 2005) or the testis (hypergonadotropic hypogonadism; Brennemann et al. 1997).

In contrast, mice with a disrupted Abcd1 gene (Forss-Petter et al. 1997; Kobayashi et al. 1997; Lu et al. 1997) show increased concentrations of VLCFAs in tissues, but also a higher level of plasma cholesterol was observed at young age (Weinhofer et al. 2005). Mice appear phenotypically normal and fertile as young adults, but at higher age start to show a mild AMN-like phenotype with minor abnormalities in myelin and axons of the spinal cord, with slower nerve conductance velocity and with motor deficits (Pujol et al. 2002). Furthermore, these mice show signs of oxidative stress in the spinal cord already at an early age (Fourcade et al. 2008). However, neither inflammation of the CNS nor severe functional disturbances of the adrenal gland or testis have been observed in these mice in spite of the accumulation of VLFCAs and morphological changes in testis and adrenal gland (Forss-Petter et al. 1997). Although ectopic expression of ABCD2 has been shown to functionally compensate for ABCD1 deficiency (Netik et al. 1999; Pujol et al. 2004), the symptoms observed in ABCD1-deficient mice are not aggravated by the concomitant deletion of the highly similar protein ABCD2, except for spinal cord pathology. In mice the deletion in the Abcd2 gene alone causes a slight accumulation of certain unsaturated fatty acids and late onset ataxia (Ferrer et al. 2005).

8.2.6 Bile Acid Conjugation

After side-chain shortening of trihydroxylcholestanoic acid the product cholic acid is conjugated to glycine or taurine by BAAT (Johnson et al. 1991), which renders the molecule more soluble. Patients with mutations in BAAT have been found in a cohort of patients with familiar hypercholanemia (FHC), but also another gene was found affected in this cohort excluding monogenice origin of the symptoms. Biochemically, the patients show increased serum bile concentrations (Carlton et al. 2003), whereas in the urine unconjugated bile acids dominate, and amidated forms are absent (Setchell et al. 2013). Furthermore, these patients have a reduced ability to take up hydrophobic compounds and thus show a deficiency in lipid soluble vitamins (e.g., Vitamin D, E, K). Young patients with defects in bile acid conjugation show jaundice due to a neonatal cholestatic hepatitis, but these episodes of cholestatic liver disease remain transient, which is in contrast to patients with impairment in cholic- and chenodeoxycholic acid biosynthesis that develop a well-defined progressive familial cholestatic liver disease (Setchell et al. 2013).

8.2.7 Biosynthesis of Ether-Phospholipids

Peroxisomes generate a biosynthetic intermediate in the production of ether-phospholipids, which represent a subtype of phospholipids that contain an alkyl-ether or a vinylic alkenyl-ether at the Sn1-position instead of the acyl-ester in other phospholipids. Ether-phospholipids serve as precursor molecules for the biosynthesis of seminolipids and platelet activating factor (PAF) and as part of the GPI-anchor of proteins (Kanzawa et al. 2009). However, the most abundant ether-phosopholipids are plasmalogens, which are characterized by a vinyl-ether bond at the Sn1 position. Plasmalogens are found in various cellular membranes, but are especially enriched in lipid rafts of the plasma membrane (Pike et al. 2002), where they could participate in membrane fusion (Glaser and Gross 1995) and endocytic processes (Rodemer et al. 2003). Interestingly, an intermediate of plasmalogen biosynthesis has been identified as stimulating lipid self-antigen required for the maturation of a subset of T-cells, the invariant natural killer T-cells (iNKT; Facciotti et al. 2012). Within the body, plasmalogens are ubiquitously distributed, but appear enriched in the central nervous system and there specifically in the white matter myelin sheets. During development the expression of the key enzymes of plasmalogen biosynthesis peak during myelination in the murine brain (Huyghe et al. 2001). The biosynthesis of plasmalogens initiates inside peroxisomes by the enzyme glyceronephosphate *O*-acyltransferase (GNPAT; Hajra 1997), which adds an acyl-group at the Sn1 position of dihydroxyacetone-phosphate. Next, the enzyme alkylglyceronephosphate-synthase (AGPS; van den Bosch and de Vet 1997) substitutes this acyl-group by an alkyl-group utilizing an alkyl-alcohol that has previously been generated by a reductase, FAR1 or FAR2 at the cytoplasmic side of peroxisomes (Cheng and Russell 2004). The product is then transferred from peroxisomes to the endoplasmic reticulum and all subsequent reactions of the biosynthetic pathway are exerted there (Wanders and Waterham 2006a).

All human patients suffering from rhizomelic chondrodysplasia punctate (RCDP) lack plasmalogens and genetic analysis divided these patients into three complementation groups. Patients with RCDP type II were found to harbor a mutation in GNPAT (Ofman et al. 1998) and patients with RCDP type III to harbor a mutation in AGPS (DeVet et al. 1998), whereas patients with RCDP type I harbor a mutation in PEX7, the receptor for type 2 peroxisomal targeting signals (PTS2; Braverman et al. 1997), which mediates the transport of AGPS to peroxisomes. All RCDP patients have profound growth and psychomotor retardation with signs of seizures and show the name-giving symptoms, a shortening of the proximal long bones (rhizomelia) and stippled foci of calcification on their bones (chondrodysplasia punctate; Braverman and Moser 2012). The life expectation of these patients is markedly reduced with an average of 5 years. Furthermore, facial dysmorphia, congenital cataracts, and abnormalities in kidney, lung, and heart are often observed (White et al. 2003; Huffnagel et al. 2013). A variety of abnormalities in the central nervous system have been observed as well, including microencephaly and cerebellar degeneration (Agamanolis and Novak 1995), cerebellar post-developmental degeneration with a loss of Purkinje cells and granular neurons (Powers et al. 1999) or abnormal myelin formation (Sztriha et al. 2000). Developmental defects of the brain and also signs of degeneration have also been observed in an MRI study (Bams-Mengerink et al. 2006).

In mice harboring a deletion in the *Gnpat* gene plasmalogens are absent in plasma and tissues, but most prominently in the myelin fraction of the brain, where this deficiency appears counterbalanced by other phospholipids. These mice are growth retarded and a fraction dies early in development (Rodemer et al. 2003), whereas those that reach adulthood live up to more than a year. Male mice are sterile due to an arrest in spermatogenesis (Rodemer et al. 2003) caused by impaired remodeling capacity of the blood/testis-barrier (Komljenovic et al. 2009). In the brain of these mice changes in cerebellar development were observed that affect the foliation pattern and the distribution of climbing fibers, which are narrowed around the somata of Purkinje cells. In addition, a marked demyelination of the cerebellum was observed which fits to the reduced conduction velocity of action potential in myelinated fibers of the corpus callosum (Teigler et al. 2009). The eyes are smaller (microphtalmia) and develop cataracts and also the optic nerve shows hypoplasia (Rodemer et al. 2003). Interestingly, a similar formation of cataracts and male sterility was discovered and investigated independently in a hypomorphic spontaneous *Agps* mutant mouse line, blind sterile (*bs2*; Liegel et al. 2011).

8.2.8 Degradation of Hydrogen Peroxide

A variety of enzymes compartmentalized within peroxisomes generate hydrogen peroxide (H_2O_2), which is efficiently degraded by the peroxisomal enzyme catalase, an enzyme that is known for its high turnover rate. Human patients lacking functional catalase have been observed and the pathological symptoms are summarized as acatalasemia (Ogata 1991). Patients are described with progressive oral gangrene and ulceration (Takahara's disease; Takahara 1952), but overall their symptoms are surprisingly mild. However, acatalasemia has been linked to an increased incidence of cancer and diabetes (Goth et al. 2004). Mice lacking catalase develop normally although the degradation rate of H_2O_2 in the liver is markedly decreased and the tissue concentration of H_2O_2 is concomitantly increased (Ho et al. 2004). However, these mice are more susceptible to photochemical damage and are more prone to nephropathy upon external stressors such as streptozin induced diabetes (Hwang et al. 2012) or operative removal of a large part of the kidney (Kobayashi et al. 2005).

8.2.9 Degradation of Glyoxylate

The peroxisomal enzyme alanine:glyoxylate aminotransferase (AGT) degrades glyoxylate (Danpure 2004) that can be generated within peroxisomes from glycine by D-amino acid oxidase (Ohide et al. 2011) or from glycolate by hydroxyacid oxidase 1 (glycolate oxidase/HAOX1; Jones et al. 2000). However, in contrast to plants, in which glyoxylate is generated and converted within peroxisomes by the photorespiratory process (Kisaki and Tolbert 1969) and the glyoxylate cycle (Breidenbach and Beevers 1967), the primary metabolic source of the pathologically relevant amount of glyoxylate in humans was enigmatic. However, it has been demonstrated recently that the degradation of hydroxyproline derived from processing of collagen/gelatin upon meat consumption is a major pathway for glyoxylate production (Knight et al. 2006; Salido et al. 2012).

In a subset of human patients that suffer from the severe form of primary hyperoxaluria mutations in the *AGTX* gene have been found, which led to the assignment of primary hyperoxaluria type 1 (PH1) to all patients with a defect in AGT (Danpure and Jennings 1986). These patients present with increased levels of oxalate and glycolate the urine, resulting in the progressive deposition of insoluble calcium oxalate precipitates (CaOx) in the kidney and the urinary tract (urolithiasis) (Danpure 2004), the latter frequently causes urinary tract obstructions and infections. The clinical picture is heterogenous, but successive development of urolithiasis and nephrocalcinosis that often ends in end stage renal disease (ESRD) is characteristic (Hoppe 2012; Leumann and Hoppe 2001). However, oxalosis can occur in many tissues of the body and causes there secondary defects and inflammation. Importantly, the metabolic cause of hyperoxaluria is the absence of the hepatic enzyme AGT, but the pathological consequences are primarily observed in the excretory organs.

Mice lacking the enzyme AGT appear normal, but show an increased level of oxalate in urine (Salido et al. 2006). About half of the male AGT-deficient mice develop urinary stones in the bladder after 6 months, but the more severe nephrocalcinosis and renal failure occurred only upon administration of the glyoxylate precursor polyethylene-glycol with the drinking water, which increased the amount of metabolically generated glyoxylate and thereby oxalate (Salido et al. 2006). The human disease was convincingly recapitulated by the addition of hydroxyproline to the chow, which resulted in the increase of renal oxalate concentration and of urinary glycolate levels only in AGT-deficient mice, although the hydroxyproline levels in the plasma were raised similarly in mice from both genotypes (Knight et al. 2012).

8.2.10 Knock-Out Mouse Models with Unclear Link to Diseases

In addition to the mouse models described above, several knock-out mouse lines have been reported that are not extensively described here, because the link to a corresponding human single enzyme deficiency is either unclear or lacking.

The enzyme D-amino acid oxidase (DAO) exerts the degradation of various D-amino acids that originate from the ingestion of exerternal compounds or has been synthesized endogenously such as the neuromolulatory amino acid D-serine that binds to the NMDA-type glutamate receptor (Oliet and Mothet 2009), but DAO can also convert glycine into glyoxylate (Ohide et al. 2011). In humans a mutation in D-amino acid oxidase that nearly abrogates the enzymatic activity has recently been associated with a familiar form of amyotrophic lateral sclerosis (FALS; Mitchell et al. 2010). In this neurodegenerative disorder, in which motor neurons deteriorate first the level of D-serine is a key determinant for the development of ALS via its modulatory activity on glutamate induced neurotoxicity (Sasabe et al. 2007). D-serine primarily affects motor neurons and thus the accumulation of D-serine upon inactivation of DAO appears consistent with the disease phenotype. This effect can be mimicked in a mouse model for ALS, in which the mouse develops an ALS-like phenotype upon ectopic overexpression of a variant of the human Cu/Zn-dependent SOD1 harboring a specific mutation (G93A; Sasabe et al. 2012). In this model the level of DAO appears reduced, whereas the level of D-serine is decreased in the spinal cord (Sasabe et al. 2012). Conversely, the lack of DAO-activity due to a deletion of Dao causes an increase in D-serine and this is detrimental for motor neurons in the spinal cord.

A mouse model for the peroxisomal membrane protein PMP22 has been described, which shows an accumulation of uric acid, but the interpretation is difficult due to the abnormalities in peroxisomal membrane permeability (Rokka et al. 2009). Furthermore, a mouse model for PMP34-deficiency has been indicated (van Ael, unpublished; Baes and Van Veldhoven 2012), which is the homologue of plant and yeast proteins that transport various cofactors (FAD, NAD) and nucleotides (AMP, ADP; Agrimi et al. 2012; Bernhardt et al. 2012).

8.3 Concluding Remarks

8.3.1 Comparison Between Human Disease and Mouse Models

The investigation of mouse models can also serve to attribute phenotypic aberrations that have been observed in human patients to the lack of a peroxisomal enzymatic activity. This is sometimes difficult in human patients due to the low number of patients, the variable consequences of individual point mutations, some of which preserve residual enzymatic activity and the distortion of the primary symptoms by external infections or therapeutic administrations. This has been well described for the phenotypic variability of patients all suffering from ACOX1- or D-BP-deficiency or from X-linked adrenoleukodystrophy. In these cases, the clinical picture varies from very severe forms with an early diagnosis and short life expectation to patients with a late onset of the disease and milder symptoms.

Overall, the mouse models for peroxisome single enzyme and transporter deficiencies appear quite successful in recapitulating the human diseases, because biochemical investigations in cultured cells obtained from patients were confirmed in these mice and many symptoms observed in human patients were found as well. Especially, studies on hepatic fatty acid metabolism proved invaluable for the elucidation of substrate specificity of fatty acid degrading enzymes, because the liver is especially rich in peroxisomes rendering this organ especially sensitive to the lack of individual enzymatic activities. Thereby, the induction of PPARα dependent gene expression upon interference with peroxisomal fatty acids degradation has been observed together with an increase in peroxisome number. Sometimes, further investigations of unexpected observations led to a deeper understanding of compensatory mechanisms such as the high cholesterol turnover in the liver of racemase deficient mice (Selkala et al. 2013). However, in several cases the human pathology is not well reflected in the mouse model. For instance, neither the adrenal dysfunction observed in human X-ALD patients nor the inflammatory aspect of the cerebral form of X-ALD has ever been observed in the mouse model. Certainly, the inability to recapitulate a clinical finding in the corresponding knock-out mouse line can originate from a plethora of different reasons such as the presence of an additional enzyme or differences in the expression pattern of human and murine genes encoding peroxisomal proteins.

However, the description of an observed phenotype as the product of the contributions of the genotype and of the environment (present and past; $P = G*E$, Beckers et al. 2009) can serve as a helpful framework to understand some of the differences between human patients and mouse models. In mouse models the variability in the phenotype (indicated by the variance) is usually reduced by an isogenic background and the low variability in the environment such as nutrition, living conditions, or exposure to infectious agents. Consequently, the statistical power of comparisons is higher in mice and, thus, small differences between wild-type and knock-out mice are easier to detect, such as in case of the increased plasma cholesterol level in young X-ALD mice (Weinhofer et al. 2005), which has been suggested for human patients as well, but could not be demonstrated due to the highly variable blood cholesterol levels within the healthy human population and the low number of X-ALD patients (our personal investigation). Contrarily, mice are maintained under standardized, sterile living conditions that minimize the variability in the environment. Thus, any pathology that requires an external trigger (e.g., infectious agents) or the (temporal) overrepresentation of specific food residues (e.g., contained within meat) will probably not precipitate. From this point of view, it is revealing that both, mice with defects in the degradation of branched chain fatty acids (SCPx-, AMACR- or PHYH-deficient mice) and mice with a defect in glyoxylate degradation (AGT-deficient mice) show mild symptoms compared to human patients, because human patients usually consume considerable amounts of meat and thereby ingest the precursors of branched chain fatty acids (phytol) and glyoxylate (collagen), whereas murine chow hardly contains these precursors. Accordingly, a diet strictly avoiding branched chain fatty acids and their precursors stopped the development of symptoms in patients with racemase deficiency (Smith et al. 2010) and a patient with SCPx-deficiency (Ferdinandusse et al. 2006b) and avoids some symptoms in patients with Refsum disease (PHYH-deficiency; Hungerbuhler et al. 1985; Ruether et al. 2010). Conversely, the symptoms of mice lacking these enzymes are aggravated by the supplementation of phytol to murine chow (Ferdinandusse et al. 2008; Savolainen et al. 2004; Seedorf et al. 1998). Similarly, the symptoms of AGT-deficient mice can be aggravated by feeding hydroxyproline and the avoidance of dietary hydroxyproline has been suggested as supportive regimen for PH1 (Knight et al. 2012).

8.3.2 Improving Mouse Models

The mouse models of single enzyme and transporter deficiencies also allow experimental approaches that take advantage of the possibility to further modulate either the genotype or the environment of these mice. Thereby, an aggravation of symptoms might be obtained without losing the advantages of a low variation in individual parameters.

The modulation of the genotype can be obtained by combining the knock-out mouse model for a single enzyme deficiency with other targeted deletions in genes of interest. Candidates for the latter are either homologous enzymes with suspected redundant enzymatic activity or enzymes required for a second peroxisomal metabolic pathway. Whereas the first approach tries to reveal a phenotype upon the complete loss of the enzymatic activity of interest (e.g., deletion of both multifunctional enzymes, MFP1 and MFP2 (Jia et al. 2003) or the deletion of the two homologous ABC-transporter proteins ABCD1 and ABCD2 (Pujol et al. 2004)), the second approach tries to obtain an intermediate phenotype between the single enzyme deficiency and a complete loss of all peroxisomal functions upon inactivation of the whole organelle (e.g., the concomitant ablation of MFP2 and GNPAT; Krysko et al. 2010). Furthermore, the deletion of the second gene can abrogate a signaling cascade associated with peroxisomal function, which is hyperactivated in the mouse model for a single enzyme deficiency (e.g., the PPARα-dependent signaling in the ACOX1-deficient mouse; Hashimoto et al. 1999, 2000).

The modulation of the context under well-defined experimental conditions can involve special feeding paradigms that may aggravate (e.g., phytol- or hydroxyproline feeding) or alleviate [e.g., modulation of individual symptoms in ABCD1-deficient mice by antioxidant supplementation or application of lipid lowering drugs (Lopez-Erauskin et al. 2011; Engelen et al. 2012)] the severity of the observed phenotype.

Altogether, the comparison of human symptoms and the murine phenotypes not only confirmed the biochemical contribution of individual peroxisomal enzymes, but revealed surprising similarities in the observable symptoms that can even converge upon specific treatments that compensate for differences in the living conditions. Moreover, putting together the knowledge from mouse models and human single enzyme and transporter deficiency pathologies we start to understand the metabolic basis of different symptoms observed in Zellweger spectrum, where all different peroxisomal pathways are missing.

Acknowledgment The authors thank Fabian Dorninger, Sonja Forss-Petter, and Christoph Wiesinger for helpful discussions and critically reading the manuscript and Christoph Wiesinger to provide Fig. 8.1.

References

Agamanolis DP, Novak RW (1995) Rhizomelic chondrodysplasia punctata: report of a case with review of the literature and correlation with other peroxisomal disorders. Pediatr Pathol Lab Med 15(3):503–513

Agrimi G, Russo A, Pierri CL, Palmieri F (2012) The peroxisomal NAD + carrier of Arabidopsis thaliana transports coenzyme A and its derivatives. J Bioenerg Biomembr 44(3):333–340. doi:10.1007/s10863-012-9445-0

Amery L, Fransen M, De Nys K, Mannaerts GP, Van Veldhoven PP (2000) Mitochondrial and peroxisomal targeting of 2-methylacyl-CoA racemase in humans. J Lipid Res 41(11):1752–1759

Arnauld S, Fidaleo M, Clemencet MC, Chevillard G, Athias A, Gresti J, Wanders RJ, Latruffe N, Nicolas-Frances V, Mandard S (2009) Modulation of the hepatic fatty acid pool in peroxisomal 3-ketoacyl-CoA thiolase B-null mice exposed to the selective PPARalpha agonist Wy14,643. Biochimie 91(11–12):1376–1386. doi:10.1016/j.biochi.2009.09.004

Ashibe B, Hirai T, Higashi K, Sekimizu K, Motojima K (2007) Dual subcellular localization in the endoplasmic reticulum and peroxisomes and a vital role in protecting against oxidative stress of fatty aldehyde dehydrogenase are achieved by alternative splicing. J Biol Chem 282 (28):20763–20773. doi:10.1074/jbc.M611853200

Atshaves BP, Payne HR, McIntosh AL, Tichy SE, Russell D, Kier AB, Schroeder F (2004) Sexually dimorphic metabolism of branched-chain lipids in C57BL/6J mice. J Lipid Res 45 (5):812–830. doi:10.1194/jlr.M300408-JLR200

Atshaves BP, McIntosh AL, Landrock D, Payne HR, Mackie JT, Maeda N, Ball J, Schroeder F, Kier AB (2007) Effect of SCP-x gene ablation on branched-chain fatty acid metabolism. Am J Physiol Gastrointest Liver Physiol 292(3):G939–G951. doi:10.1152/ajpgi.00308.2006

Baes M, Van Veldhoven PP (2012) Mouse models for peroxisome biogenesis defects and beta-oxidation enzyme deficiencies. Biochim Biophys Acta 1822(9):1489–1500. doi:10.1016/j.bbadis.2012.03.003

Baes M, Gressens P, Baumgart E, Carmeliet P, Casteels M, Fransen M, Evrard P, Fahimi D, Declercq PE, Collen D, van Veldhoven PP, Mannaerts GP (1997) A mouse model for Zellweger syndrome. Nat Genet 17(1):49–57. doi:10.1038/ng0997-49

Baes M, Huyghe S, Carmeliet P, Declercq PE, Collen D, Mannaerts GP, Van Veldhoven PP (2000) Inactivation of the peroxisomal multifunctional protein-2 in mice impedes the degradation of not only 2-methyl-branched fatty acids and bile acid intermediates but also of very long chain fatty acids. J Biol Chem 275(21):16329–16336. doi:10.1074/jbc.M001994200

Baes M, Gressens P, Huyghe S, De NK, Qi C, Jia Y, Mannaerts GP, Evrard P, Van VP, Declercq PE, Reddy JK (2002) The neuronal migration defect in mice with Zellweger syndrome (Pex5 knockout) is not caused by the inactivity of peroxisomal beta-oxidation. J Neuropathol Exp Neurol 61(4):368–374

Bams-Mengerink AM, Majoie CB, Duran M, Wanders RJ, Van Hove J, Scheurer CD, Barth PG, Poll-The BT (2006) MRI of the brain and cervical spinal cord in rhizomelic chondrodysplasia punctata. Neurology 66(6):798–803. doi:10.1212/01.wnl.0000205594.34647.d0, discussion 789

Beckers J, Wurst W, de Angelis MH (2009) Towards better mouse models: enhanced genotypes, systemic phenotyping and envirotype modelling. Nat Rev Genet 10(6):371–380. doi:10.1038/nrg2578

Berger J, Gartner J (2006) X-linked adrenoleukodystrophy: clinical, biochemical and pathogenetic aspects. Biochim Biophys Acta 1763(12):1721–1732. doi:10.1016/j.bbamcr.2006.07.010

Bernhardt K, Wilkinson S, Weber AP, Linka N (2012) A peroxisomal carrier delivers NAD(+) and contributes to optimal fatty acid degradation during storage oil mobilization. Plant J 69(1):1–13. doi:10.1111/j.1365-313X.2011.04775.x

Bösze P, Skripeczky K, Gaál M, Tóth A, László J (1983) Perrault's syndrome in two sisters. Am J Med Genet 16(2):237–241

Braverman NE, Moser AB (2012) Functions of plasmalogen lipids in health and disease. Biochim Biophys Acta 1822(9):1442–1452. doi:10.1016/j.bbadis.2012.05.008

Braverman N, Steel G, Obie C, Moser A, Moser H, Gould SJ, Valle D (1997) Human PEX7 encodes the peroxisomal PTS2 receptor and is responsible for rhizomelic chondrodysplasia punctata. Nat Genet 15(4):369–376

Breidenbach RW, Beevers H (1967) Association of the glyoxylate cycle enzymes in a novel subcellular particle from castor bean endosperm. Biochem Biophys Res Commun 27(4):462–469

Brennemann W, Kohler W, Zierz S, Klingmuller D (1997) Testicular dysfunction in adrenomyeloneuropathy. Eur J Endocrinol 137(1):34–39

Carlton VE, Harris BZ, Puffenberger EG, Batta AK, Knisely AS, Robinson DL, Strauss KA, Shneider BL, Lim WA, Salen G, Morton DH, Bull LN (2003) Complex inheritance of familial hypercholanemia with associated mutations in TJP2 and BAAT. Nat Genet 34(1):91–96. doi:10.1038/ng1147

Cheng JB, Russell DW (2004) Mammalian wax biosynthesis. I. Identification of two fatty acyl-Coenzyme A reductases with different substrate specificities and tissue distributions. J Biol Chem 279(36):37789–37797. doi:10.1074/jbc.M406225200

Chevillard G, Clemencet MC, Latruffe N, Nicolas-Frances V (2004) Targeted disruption of the peroxisomal thiolase B gene in mouse: a new model to study disorders related to peroxisomal lipid metabolism. Biochimie 86(11):849–856. doi:10.1016/j.biochi.2004.09.028

Clarke CE, Alger S, Preece MA, Burdon MA, Chavda S, Denis S, Ferdinandusse S, Wanders RJ (2004) Tremor and deep white matter changes in alpha-methylacyl-CoA racemase deficiency. Neurology 63(1):188–189

Clayton PT, Lake BD, Hjelm M, Stephenson JB, Besley GT, Wanders RJ, Schram AW, Tager JM, Schutgens RB, Lawson AM (1988) Bile acid analyses in "pseudo-Zellweger" syndrome; clues to the defect in peroxisomal beta-oxidation. J Inherit Metab Dis 11(Suppl 2):165–168

Danpure CJ (2004) Molecular aetiology of primary hyperoxaluria type 1. Nephron Exp Nephrol 98 (2):e39–e44. doi:10.1159/000080254

Danpure CJ, Jennings PR (1986) Peroxisomal alanine:glyoxylate aminotransferase deficiency in primary hyperoxaluria type I. FEBS Lett 201(1):20–24

de Launoit Y, Adamski J (1999) Unique multifunctional HSD17B4 gene product: 17beta-hydroxysteroid dehydrogenase 4 and D-3-hydroxyacyl-coenzyme A dehydrogenase/hydratase involved in Zellweger syndrome. J Mol Endocrinol 22(3):227–240

De Laurenzi V, Rogers GR, Hamrock DJ, Marekov LN, Steinert PM, Compton JG, Markova N, Rizzo WB (1996) Sjogren-Larsson syndrome is caused by mutations in the fatty aldehyde dehydrogenase gene. Nat Genet 12(1):52–57. doi:10.1038/ng0196-52

de Vet EC, Ijlst L, Oostheim W, Wanders RJ, van den Bosch H (1998) Alkyl-dihydroxyacetonephosphate synthase. Fate in peroxisome biogenesis disorders and identification of the point mutation underlying a single enzyme deficiency. J Biol Chem 273(17):10296–10301

Dick D, Horvath R, Chinnery PF (2011) AMACR mutations cause late-onset autosomal recessive cerebellar ataxia. Neurology 76(20):1768–1770. doi:10.1212/WNL.0b013e31821a4484

Dodt G, Kim DG, Reimann SA, Reuber BE, McCabe K, Gould SJ, Mihalik SJ (2000) L-Pipecolic acid oxidase, a human enzyme essential for the degradation of L-pipecolic acid, is most similar to the monomeric sarcosine oxidases. Biochem J 345(Pt 3):487–494

Dommes V, Baumgart C, Kunau WH (1981) Degradation of unsaturated fatty acids in peroxisomes. Existence of a 2,4-dienoyl-CoA reductase pathway. J Biol Chem 256 (16):8259–8262

Dubey P, Raymond GV, Moser AB, Kharkar S, Bezman L, Moser HW (2005) Adrenal insufficiency in asymptomatic adrenoleukodystrophy patients identified by very long-chain fatty acid screening. J Pediatr 146(4):528–532. doi:10.1016/j.jpeds.2004.10.067

El Hajj HI, Vluggens A, Andreoletti P, Ragot K, Mandard S, Kersten S, Waterham HR, Lizard G, Wanders RJ, Reddy JK, Cherkaoui-Malki M (2012) The inflammatory response in acyl-CoA oxidase 1 deficiency (pseudoneonatal adrenoleukodystrophy). Endocrinology 153(6):2568–2575. doi:10.1210/en.2012-1137

Engelen M, Schackmann MJ, Ofman R, Sanders RJ, Dijkstra IM, Houten SM, Fourcade S, Pujol A, Poll-The BT, Wanders RJ, Kemp S (2012) Bezafibrate lowers very long-chain fatty acids in X-linked adrenoleukodystrophy fibroblasts by inhibiting fatty acid elongation. J Inherit Metab Dis 35(6):1137–1145. doi:10.1007/s10545-012-9471-4

Facciotti F, Ramanjaneyulu GS, Lepore M, Sansano S, Cavallari M, Kistowska M, Forss-Petter S, Ni G, Colone A, Singhal A, Berger J, Xia C, Mori L, De Libero G (2012) Peroxisome-derived lipids are self antigens that stimulate invariant natural killer T cells in the thymus. Nat Immunol 13(5):474–480. doi:10.1038/ni.2245

Fan CY, Pan J, Chu R, Lee D, Kluckman KD, Usuda N, Singh I, Yeldandi AV, Rao MS, Maeda N, Reddy JK (1996) Targeted disruption of the peroxisomal fatty acyl-CoA oxidase gene: generation of a mouse model of pseudoneonatal adrenoleukodystrophy. Ann N Y Acad Sci 804:530–541

Fan CY, Pan J, Usuda N, Yeldandi AV, Rao MS, Reddy JK (1998) Steatohepatitis, spontaneous peroxisome proliferation and liver tumors in mice lacking peroxisomal fatty acyl-CoA oxidase. Implications for peroxisome proliferator-activated receptor alpha natural ligand metabolism. J Biol Chem 273(25):15639–15645

Ferdinandusse S, Mulders J, IJlst L, Denis S, Dacremont G, Waterham HR, Wanders RJ (1999) Molecular cloning and expression of human carnitine octanoyltransferase: evidence for its role in the peroxisomal beta-oxidation of branched-chain fatty acids. Biochem Biophys Res Commun 263(1):213–218. doi:10.1006/bbrc.1999.1340

Ferdinandusse S, Denis S, Clayton PT, Graham A, Rees JE, Allen JT, McLean BN, Brown AY, Vreken P, Waterham HR, Wanders RJ (2000a) Mutations in the gene encoding peroxisomal alpha-methylacyl-CoA racemase cause adult-onset sensory motor neuropathy. Nat Genet 24 (2):188–191. doi:10.1038/72861

Ferdinandusse S, Denis S, IJlst L, Dacremont G, Waterham HR, Wanders RJ (2000b) Subcellular localization and physiological role of alpha-methylacyl-CoA racemase. J Lipid Res 41 (11):1890–1896

Ferdinandusse S, Denis S, van Berkel E, Dacremont G, Wanders RJ (2000c) Peroxisomal fatty acid oxidation disorders and 58 kDa sterol carrier protein X (SCPx). Activity measurements in liver and fibroblasts using a newly developed method. J Lipid Res 41(3):336–342

Ferdinandusse S, Denis S, Mooijer PA, Zhang Z, Reddy JK, Spector AA, Wanders RJ (2001) Identification of the peroxisomal beta-oxidation enzymes involved in the biosynthesis of docosahexaenoic acid. J Lipid Res 42(12):1987–1995

Ferdinandusse S, Meissner T, Wanders RJ, Mayatepek E (2002a) Identification of the peroxisomal beta-oxidation enzymes involved in the degradation of leukotrienes. Biochem Biophys Res Commun 293(1):269–273. doi:10.1016/S0006-291X(02)00214-0

Ferdinandusse S, Rusch H, van Lint AE, Dacremont G, Wanders RJ, Vreken P (2002b) Stereochemistry of the peroxisomal branched-chain fatty acid alpha- and beta-oxidation systems in patients suffering from different peroxisomal disorders. J Lipid Res 43(3):438–444

Ferdinandusse S, van Grunsven EG, Oostheim W, Denis S, Hogenhout EM, IJlst L, van Roermund CW, Waterham HR, Goldfischer S, Wanders RJ (2002c) Reinvestigation of peroxisomal 3-ketoacyl-CoA thiolase deficiency: identification of the true defect at the level of d-bifunctional protein. Am J Hum Genet 70(6):1589–1593. doi:10.1086/340970

Ferdinandusse S, Denis S, Van Roermund CW, Wanders RJ, Dacremont G (2004) Identification of the peroxisomal beta-oxidation enzymes involved in the degradation of long-chain dicarboxylic acids. J Lipid Res 45(6):1104–1111. doi:10.1194/jlr.M300512-JLR200

Ferdinandusse S, Denis S, Overmars H, Van Eeckhoudt L, Van Veldhoven PP, Duran M, Wanders RJ, Baes M (2005) Developmental changes of bile acid composition and conjugation in L- and D-bifunctional protein single and double knockout mice. J Biol Chem 280(19):18658–18666. doi:10.1074/jbc.M414311200

Ferdinandusse S, Denis S, Mooyer PA, Dekker C, Duran M, Soorani-Lunsing RJ, Boltshauser E, Macaya A, Gartner J, Majoie CB, Barth PG, Wanders RJ, Poll-The BT (2006a) Clinical and biochemical spectrum of D-bifunctional protein deficiency. Ann Neurol 59(1):92–104. doi:10.1002/ana.20702

Ferdinandusse S, Kostopoulos P, Denis S, Rusch H, Overmars H, Dillmann U, Reith W, Haas D, Wanders RJ, Duran M, Marziniak M (2006b) Mutations in the gene encoding peroxisomal sterol carrier protein X (SCPx) cause leukencephalopathy with dystonia and motor neuropathy. Am J Hum Genet 78(6):1046–1052. doi:10.1086/503921

Ferdinandusse S, Ylianttila MS, Gloerich J, Koski MK, Oostheim W, Waterham HR, Hiltunen JK, Wanders RJ, Glumoff T (2006c) Mutational spectrum of D-bifunctional protein deficiency and structure-based genotype-phenotype analysis. Am J Hum Genet 78(1):112–124. doi:10.1086/498880

Ferdinandusse S, Denis S, Hogenhout EM, Koster J, van Roermund CW, IJlst L, Moser AB, Wanders RJ, Waterham HR (2007) Clinical, biochemical, and mutational spectrum of peroxisomal acyl-coenzyme A oxidase deficiency. Hum Mutat 28(9):904–912. doi:10.1002/humu.20535

Ferdinandusse S, Zomer AW, Komen JC, van den Brink CE, Thanos M, Hamers FP, Wanders RJ, van der Saag PT, Poll-The BT, Brites P (2008) Ataxia with loss of Purkinje cells in a mouse model for Refsum disease. Proc Natl Acad Sci USA 105(46):17712–17717. doi:10.1073/pnas.0806066105

Ferdinandusse S, Barker S, Lachlan K, Duran M, Waterham HR, Wanders RJ, Hammans S (2010) Adult peroxisomal acyl-coenzyme A oxidase deficiency with cerebellar and brainstem atrophy. J Neurol Neurosurg Psychiatry 81(3):310–312. doi:10.1136/jnnp.2009.176255

Ferrer I, Kapfhammer JP, Hindelang C, Kemp S, Troffer-Charlier N, Broccoli V, Callyzot N, Mooyer P, Selhorst J, Vreken P, Wanders RJ, Mandel JL, Pujol A (2005) Inactivation of the peroxisomal ABCD2 transporter in the mouse leads to late-onset ataxia involving mitochondria, Golgi and endoplasmic reticulum damage. Hum Mol Genet 14(23):3565–3577. doi:10.1093/hmg/ddi384

Fidaleo M, Arnauld S, Clemencet MC, Chevillard G, Royer MC, De Bruycker M, Wanders RJ, Athias A, Gresti J, Clouet P, Degrace P, Kersten S, Espeel M, Latruffe N, Nicolas-Frances V, Mandard S (2011) A role for the peroxisomal 3-ketoacyl-CoA thiolase B enzyme in the control of PPARalpha-mediated upregulation of SREBP-2 target genes in the liver. Biochimie 93 (5):876–891. doi:10.1016/j.biochi.2011.02.001

Filppula SA, Yagi AI, Kilpelainen SH, Novikov D, FitzPatrick DR, Vihinen M, Valle D, Hiltunen JK (1998) Delta3,5-delta2,4-dienoyl-CoA isomerase from rat liver. Molecular characterization. J Biol Chem 273(1):349–355

Forss-Petter S, Werner H, Berger J, Lassmann H, Molzer B, Schwab MH, Bernheimer H, Zimmermann F, Nave KA (1997) Targeted inactivation of the X-linked adrenoleukodystrophy gene in mice. J Neurosci Res 50(5):829–843

Foulon V, Antonenkov VD, Croes K, Waelkens E, Mannaerts GP, Van Veldhoven PP, Casteels M (1999) Purification, molecular cloning, and expression of 2-hydroxyphytanoyl-CoA lyase, a peroxisomal thiamine pyrophosphate-dependent enzyme that catalyzes the carbon-carbon bond cleavage during alpha-oxidation of 3-methyl-branched fatty acids. Proc Natl Acad Sci USA 96(18):10039–10044

Fourcade S, Lopez-Erauskin J, Galino J, Duval C, Naudi A, Jove M, Kemp S, Villarroya F, Ferrer I, Pamplona R, Portero-Otin M, Pujol A (2008) Early oxidative damage underlying neurodegeneration in X-adrenoleukodystrophy. Hum Mol Genet 17(12):1762–1773. doi:10.1093/hmg/ddn085

Geisbrecht BV, Zhang D, Schulz H, Gould SJ (1999) Characterization of PECI, a novel monofunctional Delta(3), Delta(2)-enoyl-CoA isomerase of mammalian peroxisomes. J Biol Chem 274(31):21797–21803

Glaser PE, Gross RW (1995) Rapid plasmenylethanolamine-selective fusion of membrane bilayers catalyzed by an isoform of glyceraldehyde-3-phosphate dehydrogenase: discrimination between glycolytic and fusogenic roles of individual isoforms. Biochemistry 34 (38):12193–12203

Gloerich J, Ruiter JP, van den Brink DM, Ofman R, Ferdinandusse S, Wanders RJ (2006) Peroxisomal trans-2-enoyl-CoA reductase is involved in phytol degradation. FEBS Lett 580 (8):2092–2096. doi:10.1016/j.febslet.2006.03.011

Goth L, Rass P, Pay A (2004) Catalase enzyme mutations and their association with diseases. Mol Diagn 8(3):141–149

Gronemeyer T, Wiese S, Ofman R, Bunse C, Pawlas M, Hayen H, Eisenacher M, Stephan C, Meyer HE, Waterham HR, Erdmann R, Wanders RJ, Warscheid B (2013) The proteome of human liver peroxisomes: identification of five new peroxisomal constituents by a label-free quantitative proteomics survey. PLoS One 8(2):e57395. doi:10.1371/journal.pone.0057395

Hajra AK (1997) Dihydroxyacetone phosphate acyltransferase. Biochim Biophys Acta 1348(1–2):27–34

Hashimoto T, Fujita T, Usuda N, Cook W, Qi C, Peters JM, Gonzalez FJ, Yeldandi AV, Rao MS, Reddy JK (1999) Peroxisomal and mitochondrial fatty acid beta-oxidation in mice nullizygous for both peroxisome proliferator-activated receptor alpha and peroxisomal fatty acyl-CoA oxidase. Genotype correlation with fatty liver phenotype. J Biol Chem 274(27):19228–19236

Hashimoto T, Cook WS, Qi C, Yeldandi AV, Reddy JK, Rao MS (2000) Defect in peroxisome proliferator-activated receptor alpha-inducible fatty acid oxidation determines the severity of hepatic steatosis in response to fasting. J Biol Chem 275(37):28918–28928. doi:10.1074/jbc.M910350199

Haugarvoll K, Johansson S, Tzoulis C, Haukanes BI, Bredrup C, Neckelmann G, Boman H, Knappskog PM, Bindoff LA (2013) MRI characterisation of adult onset alpha-methylacyl-coA racemase deficiency diagnosed by exome sequencing. Orphanet J Rare Dis 8:1. doi:10.1186/1750-1172-8-1

Ho YS, Xiong Y, Ma W, Spector A, Ho DS (2004) Mice lacking catalase develop normally but show differential sensitivity to oxidant tissue injury. J Biol Chem 279(31):32804–32812. doi:10.1074/jbc.M404800200

Hoppe B (2012) An update on primary hyperoxaluria. Nat Rev Nephrol 8(8):467–475. doi:10.1038/nrneph.2012.113

Houten SM, Denis S, Argmann CA, Jia Y, Ferdinandusse S, Reddy JK, Wanders RJ (2012) Peroxisomal L-bifunctional enzyme (Ehhadh) is essential for the production of medium-chain dicarboxylic acids. J Lipid Res 53(7):1296–1303. doi:10.1194/jlr.M024463

Huang J, Viswakarma N, Yu S, Jia Y, Bai L, Vluggens A, Cherkaoui-Malki M, Khan M, Singh I, Yang G, Rao MS, Borensztajn J, Reddy JK (2011) Progressive endoplasmic reticulum stress contributes to hepatocarcinogenesis in fatty acyl-CoA oxidase 1-deficient mice. Am J Pathol 179(2):703–713. doi:10.1016/j.ajpath.2011.04.030

Huang J, Jia Y, Fu T, Viswakarma N, Bai L, Rao MS, Zhu Y, Borensztajn J, Reddy JK (2012) Sustained activation of PPARalpha by endogenous ligands increases hepatic fatty acid oxidation and prevents obesity in ob/ob mice. FASEB J 26(2):628–638. doi:10.1096/fj.11-194019

Huffnagel IC, Clur SA, Bams-Mengerink AM, Blom NA, Wanders RJ, Waterham HR, Poll-The BT (2013) Rhizomelic chondrodysplasia punctata and cardiac pathology. J Med Genet 50 (7):419–424. doi:10.1136/jmedgenet-2013-101536

Hungerbuhler JP, Meier C, Rousselle L, Quadri P, Bogousslavsky J (1985) Refsum's disease: management by diet and plasmapheresis. Eur Neurol 24(3):153–159

Huyghe S, Casteels M, Janssen A, Meulders L, Mannaerts GP, Declercq PE, Van Veldhoven PP, Baes M (2001) Prenatal and postnatal development of peroxisomal lipid-metabolizing pathways in the mouse. Biochem J 353(Pt 3):673–680

Huyghe S, Schmalbruch H, De Gendt K, Verhoeven G, Guillou F, Van Veldhoven PP, Baes M (2006a) Peroxisomal multifunctional protein 2 is essential for lipid homeostasis in Sertoli cells and male fertility in mice. Endocrinology 147(5):2228–2236. doi:10.1210/en.2005-1571

Huyghe S, Schmalbruch H, Hulshagen L, Veldhoven PV, Baes M, Hartmann D (2006b) Peroxisomal multifunctional protein-2 deficiency causes motor deficits and glial lesions in the adult central nervous system. Am J Pathol 168(4):1321–1334. doi:10.2353/ajpath.2006.041220

Hwang I, Lee J, Huh JY, Park J, Lee HB, Ho YS, Ha H (2012) Catalase deficiency accelerates diabetic renal injury through peroxisomal dysfunction. Diabetes 61(3):728–738. doi:10.2337/db11-0584

Ijlst L, de Kromme I, Oostheim W, Wanders RJ (2000) Molecular cloning and expression of human L-pipecolate oxidase. Biochem Biophys Res Commun 270(3):1101–1105. doi:10.1006/bbrc.2000.2575

Jansen GA, Wanders RJ, Watkins PA, Mihalik SJ (1997) Phytanoyl-coenzyme A hydroxylase deficiency—the enzyme defect in Refsum's disease. N Engl J Med 337(2):133–134. doi:10.1056/NEJM199707103370215

Jansen GA, van den Brink DM, Ofman R, Draghici O, Dacremont G, Wanders RJ (2001) Identification of pristanal dehydrogenase activity in peroxisomes: conclusive evidence that the complete phytanic acid alpha-oxidation pathway is localized in peroxisomes. Biochem Biophys Res Commun 283(3):674–679. doi:10.1006/bbrc.2001.4835

Jansen GA, Waterham HR, Wanders RJ (2004) Molecular basis of Refsum disease: sequence variations in phytanoyl-CoA hydroxylase (PHYH) and the PTS2 receptor (PEX7). Hum Mutat 23(3):209–218. doi:10.1002/humu.10315

Jia Y, Qi C, Zhang Z, Hashimoto T, Rao MS, Huyghe S, Suzuki Y, Van Veldhoven PP, Baes M, Reddy JK (2003) Overexpression of peroxisome proliferator-activated receptor-alpha (PPARalpha)-regulated genes in liver in the absence of peroxisome proliferation in mice deficient in both L- and D-forms of enoyl-CoA hydratase/dehydrogenase enzymes of peroxisomal beta-oxidation system. J Biol Chem 278(47):47232–47239. doi:10.1074/jbc.M306363200

Johnson MR, Barnes S, Kwakye JB, Diasio RB (1991) Purification and characterization of bile acid-CoA:amino acid N-acyltransferase from human liver. J Biol Chem 266(16):10227–10233

Jones JM, Morrell JC, Gould SJ (2000) Identification and characterization of HAOX1, HAOX2, and HAOX3, three human peroxisomal 2-hydroxy acid oxidases. J Biol Chem 275(17):12590–12597

Kanzawa N, Maeda Y, Ogiso H, Murakami Y, Taguchi R, Kinoshita T (2009) Peroxisome dependency of alkyl-containing GPI-anchor biosynthesis in the endoplasmic reticulum. Proc Natl Acad Sci USA 106(42):17711–17716. doi:10.1073/pnas.0904762106

Kapina V, Sedel F, Truffert A, Horvath J, Wanders RJ, Waterham HR, Picard F (2010) Relapsing rhabdomyolysis due to peroxisomal alpha-methylacyl-coa racemase deficiency. Neurology 75 (14):1300–1302. doi:10.1212/WNL.0b013e3181f612a5

Kisaki T, Tolbert NE (1969) Glycolate and glyoxylate metabolism by isolated peroxisomes or chloroplasts. Plant Physiol 44(2):242–250

Knight J, Jiang J, Assimos DG, Holmes RP (2006) Hydroxyproline ingestion and urinary oxalate and glycolate excretion. Kidney Int 70(11):1929–1934. doi:10.1038/sj.ki.5001906

Knight J, Holmes RP, Cramer SD, Takayama T, Salido E (2012) Hydroxyproline metabolism in mouse models of primary hyperoxaluria. Am J Physiol Renal Physiol 302(6):F688–F693. doi:10.1152/ajprenal.00473.2011

Kobayashi T, Shinnoh N, Kondo A, Yamada T (1997) Adrenoleukodystrophy protein-deficient mice represent abnormality of very long chain fatty acid metabolism. Biochem Biophys Res Commun 232(3):631–636. doi:10.1006/bbrc.1997.6340

Kobayashi M, Sugiyama H, Wang DH, Toda N, Maeshima Y, Yamasaki Y, Masuoka N, Yamada M, Kira S, Makino H (2005) Catalase deficiency renders remnant kidneys more susceptible to oxidant tissue injury and renal fibrosis in mice. Kidney Int 68(3):1018–1031. doi:10.1111/j.1523-1755.2005.00494.x

Kodama S, Nishio H, Nakajima T, Nakamura H (1990) Urinary dicarboxylic acids in X-linked adrenoleukodystrophy. Acta Paediatr Jpn 32(3):257–261, Overseas edition

Komljenovic D, Sandhoff R, Teigler A, Heid H, Just WW, Gorgas K (2009) Disruption of blood-testis barrier dynamics in ether-lipid-deficient mice. Cell Tissue Res 337(2):281–299. doi:10.1007/s00441-009-0809-7

Kotti TJ, Savolainen K, Helander HM, Yagi A, Novikov DK, Kalkkinen N, Conzelmann E, Hiltunen JK, Schmitz W (2000) In mouse alpha -methylacyl-CoA racemase, the same gene product is simultaneously located in mitochondria and peroxisomes. J Biol Chem 275 (27):20887–20895. doi:10.1074/jbc.M002067200

Krysko O, Bottelbergs A, Van Veldhoven P, Baes M (2010) Combined deficiency of peroxisomal beta-oxidation and ether lipid synthesis in mice causes only minor cortical neuronal migration defects but severe hypotonia. Mol Genet Metab 100(1):71–76. doi:10.1016/j.ymgme.2010.02.003

Leumann E, Hoppe B (2001) The primary hyperoxalurias. J Am Soc Nephrol 12(9):1986–1993

Liegel R, Chang B, Dubielzig R, Sidjanin DJ (2011) Blind sterile 2 (bs2), a hypomorphic mutation in Agps, results in cataracts and male sterility in mice. Mol Genet Metab 103(1):51–59. doi:10.1016/j.ymgme.2011.02.002

Lopez-Erauskin J, Fourcade S, Galino J, Ruiz M, Schluter A, Naudi A, Jove M, Portero-Otin M, Pamplona R, Ferrer I, Pujol A (2011) Antioxidants halt axonal degeneration in a mouse model of X-adrenoleukodystrophy. Ann Neurol 70(1):84–92. doi:10.1002/ana.22363

Lu JF, Lawler AM, Watkins PA, Powers JM, Moser AB, Moser HW, Smith KD (1997) A mouse model for X-linked adrenoleukodystrophy. Proc Natl Acad Sci USA 94(17):9366–9371

Martens K, Loren V, van Themaat E, van Batenburg MF, Heinaniemi M, Huyghe S, Van Hummelen P, Carlberg C, Van Veldhoven PP, Van Kampen A, Baes M (2008) Coordinate induction of PPAR alpha and SREBP2 in multifunctional protein 2 deficient mice. Biochim Biophys Acta 1781(11–12):694–702. doi:10.1016/j.bbalip.2008.07.010

Mihalik SJ, Moser HW, Watkins PA, Danks DM, Poulos A, Rhead WJ (1989) Peroxisomal L-pipecolic acid oxidation is deficient in liver from Zellweger syndrome patients. Pediatr Res 25(5):548–552. doi:10.1203/00006450-198905000-00024

Mihalik SJ, Rainville AM, Watkins PA (1995) Phytanic acid alpha-oxidation in rat liver peroxisomes. Production of alpha-hydroxyphytanoyl-CoA and formate is enhanced by dioxygenase cofactors. Eur J Biochem 232(2):545–551

Mihalik SJ, Morrell JC, Kim D, Sacksteder KA, Watkins PA, Gould SJ (1997) Identification of PAHX, a Refsum disease gene. Nat Genet 17(2):185–189. doi:10.1038/ng1097-185

Mitchell J, Paul P, Chen HJ, Morris A, Payling M, Falchi M, Habgood J, Panoutsou S, Winkler S, Tisato V, Hajitou A, Smith B, Vance C, Shaw C, Mazarakis ND, de Belleroche J (2010) Familial amyotrophic lateral sclerosis is associated with a mutation in D-amino acid oxidase. Proc Natl Acad Sci USA 107(16):7556–7561. doi:10.1073/pnas.0914128107

Mizumoto H, Akashi R, Hikita N, Kumakura A, Yoshida Y, Honda A, Shimozawa N, Hata D (2012) Mild case of D-bifunctional protein deficiency associated with novel gene mutations. Pediatr Int 54(2):303–304. doi:10.1111/j.1442-200X.2012.03562.x

Morita M, Imanaka T (2012) Peroxisomal ABC transporters: structure, function and role in disease. Biochim Biophys Acta 1822(9):1387–1396. doi:10.1016/j.bbadis.2012.02.009

Mosser J, Douar AM, Sarde CO, Kioschis P, Feil R, Moser H, Poustka AM, Mandel JL, Aubourg P (1993) Putative X-linked adrenoleukodystrophy gene shares unexpected homology with ABC transporters. Nature 361(6414):726–730. doi:10.1038/361726a0

Netik A, Forss-Petter S, Holzinger A, Molzer B, Unterrainer G, Berger J (1999) Adrenoleukodystrophy-related protein can compensate functionally for adrenoleukodystrophy protein deficiency (X-ALD): implications for therapy. Hum Mol Genet 8(5):907–913

Nguyen SD, Baes M, Van Veldhoven PP (2008) Degradation of very long chain dicarboxylic polyunsaturated fatty acids in mouse hepatocytes, a peroxisomal process. Biochim Biophys Acta 1781(8):400–405. doi:10.1016/j.bbalip.2008.06.004

Ofman R, Hettema EH, Hogenhout EM, Caruso U, Muijsers AO, Wanders RJ (1998) Acyl-CoA: dihydroxyacetonephosphate acyltransferase: cloning of the human cDNA and resolution of the molecular basis in rhizomelic chondrodysplasia punctata type 2. Hum Mol Genet 7(5):847–853

Ogata M (1991) Acatalasemia. Hum Genet 86(4):331–340

Ohide H, Miyoshi Y, Maruyama R, Hamase K, Konno R (2011) D-Amino acid metabolism in mammals: biosynthesis, degradation and analytical aspects of the metabolic study. J Chromatogr B Anal Technol Biomed Life Sci 879(29):3162–3168. doi:10.1016/j.jchromb. 2011.06.028

Oliet SH, Mothet JP (2009) Regulation of N-methyl-D-aspartate receptors by astrocytic D-serine. Neuroscience 158(1):275–283. doi:10.1016/j.neuroscience.2008.01.071

Perrault M, Klotz B, Housset E (1951) Deux cas de syndrome de Turner avec surdi-mutite dans une meme fratrie. Bull Mem Soc Med Hop Paris 16:79–84, http://www.ncbi.nlm.nih.gov/pubmed/6650568

Pierce SB, Walsh T, Chisholm KM, Lee MK, Thornton AM, Fiumara A, Opitz JM, Levy-Lahad E, Klevit RE, King MC (2010) Mutations in the DBP-deficiency protein HSD17B4 cause ovarian dysgenesis, hearing loss, and ataxia of Perrault Syndrome. Am J Hum Genet 87(2):282–288. doi:10.1016/j.ajhg.2010.07.007

Pike LJ, Han X, Chung KN, Gross RW (2002) Lipid rafts are enriched in arachidonic acid and plasmenylethanolamine and their composition is independent of caveolin-1 expression: a quantitative electrospray ionization/mass spectrometric analysis. Biochemistry 41(6):2075–2088

Poll-The BT, Gartner J (2012) Clinical diagnosis, biochemical findings and MRI spectrum of peroxisomal disorders. Biochim Biophys Acta 1822(9):1421–1429. doi:10.1016/j.bbadis.2012.03.011

Poll-The BT, Roels F, Ogier H, Scotto J, Vamecq J, Schutgens RB, Wanders RJ, van Roermund CW, van Wijland MJ, Schram AW et al (1988) A new peroxisomal disorder with enlarged peroxisomes and a specific deficiency of acyl-CoA oxidase (pseudo-neonatal adrenoleukodystrophy). Am J Hum Genet 42(3):422–434

Powers JM, Kenjarski TP, Moser AB, Moser HW (1999) Cerebellar atrophy in chronic rhizomelic chondrodysplasia punctata: a potential role for phytanic acid and calcium in the death of its Purkinje cells. Acta Neuropathol 98(2):129–134

Powers JM, DeCiero DP, Ito M, Moser AB, Moser HW (2000) Adrenomyeloneuropathy: a neuropathologic review featuring its noninflammatory myelopathy. J Neuropathol Exp Neurol 59(2):89–102

Pujol A, Hindelang C, Callizot N, Bartsch U, Schachner M, Mandel JL (2002) Late onset neurological phenotype of the X-ALD gene inactivation in mice: a mouse model for adrenomyeloneuropathy. Hum Mol Genet 11(5):499–505

Pujol A, Ferrer I, Camps C, Metzger E, Hindelang C, Callizot N, Ruiz M, Pampols T, Giros M, Mandel JL (2004) Functional overlap between ABCD1 (ALD) and ABCD2 (ALDR) transporters: a therapeutic target for X-adrenoleukodystrophy. Hum Mol Genet 13(23):2997–3006. doi:10.1093/hmg/ddh323

Qi C, Zhu Y, Pan J, Usuda N, Maeda N, Yeldandi AV, Rao MS, Hashimoto T, Reddy JK (1999) Absence of spontaneous peroxisome proliferation in enoyl-CoA Hydratase/L-3-hydroxyacyl-

CoA dehydrogenase-deficient mouse liver. Further support for the role of fatty acyl CoA oxidase in PPARalpha ligand metabolism. J Biol Chem 274(22):15775–15780

Rizzo C, Boenzi S, Wanders RJ, Duran M, Caruso U, Dionisi-Vici C (2003) Characteristic acylcarnitine profiles in inherited defects of peroxisome biogenesis: a novel tool for screening diagnosis using tandem mass spectrometry. Pediatr Res 53(6):1013–1018

Rodemer C, Thai TP, Brugger B, Kaercher T, Werner H, Nave KA, Wieland F, Gorgas K, Just WW (2003) Inactivation of ether lipid biosynthesis causes male infertility, defects in eye development and optic nerve hypoplasia in mice. Hum Mol Genet 12(15):1881–1895

Rokka A, Antonenkov VD, Soininen R, Immonen HL, Pirila PL, Bergmann U, Sormunen RT, Weckstrom M, Benz R, Hiltunen JK (2009) Pxmp2 is a channel-forming protein in Mammalian peroxisomal membrane. PLoS One 4(4):e5090. doi:10.1371/journal.pone.0005090

Ruether K, Baldwin E, Casteels M, Feher MD, Horn M, Kuranoff S, Leroy BP, Wanders RJ, Wierzbicki AS (2010) Adult Refsum disease: a form of tapetoretinal dystrophy accessible to therapy. Surv Ophthalmol 55(6):531–538. doi:10.1016/j.survophthal.2010.03.007

Salido EC, Li XM, Lu Y, Wang X, Santana A, Roy-Chowdhury N, Torres A, Shapiro LJ, Roy-Chowdhury J (2006) Alanine-glyoxylate aminotransferase-deficient mice, a model for primary hyperoxaluria that responds to adenoviral gene transfer. Proc Natl Acad Sci USA 103 (48):18249–18254. doi:10.1073/pnas.0607218103

Salido E, Pey AL, Rodriguez R, Lorenzo V (2012) Primary hyperoxalurias: disorders of glyoxylate detoxification. Biochim Biophys Acta 1822(9):1453–1464. doi:10.1016/j.bbadis.2012.03.004

Sasabe J, Chiba T, Yamada M, Okamoto K, Nishimoto I, Matsuoka M, Aiso S (2007) D-serine is a key determinant of glutamate toxicity in amyotrophic lateral sclerosis. EMBO J 26(18):4149–4159. doi:10.1038/sj.emboj.7601840

Sasabe J, Miyoshi Y, Suzuki M, Mita M, Konno R, Matsuoka M, Hamase K, Aiso S (2012) D-amino acid oxidase controls motoneuron degeneration through D-serine. Proc Natl Acad Sci USA 109(2):627–632. doi:10.1073/pnas.1114639109

Savolainen K, Kotti TJ, Schmitz W, Savolainen TI, Sormunen RT, Ilves M, Vainio SJ, Conzelmann E, Hiltunen JK (2004) A mouse model for alpha-methylacyl-CoA racemase deficiency: adjustment of bile acid synthesis and intolerance to dietary methyl-branched lipids. Hum Mol Genet 13(9):955–965. doi:10.1093/hmg/ddh107

Schram AW, Goldfischer S, van Roermund CW, Brouwer-Kelder EM, Collins J, Hashimoto T, Heymans HS, van den Bosch H, Schutgens RB, Tager JM et al (1987) Human peroxisomal 3-oxoacyl-coenzyme A thiolase deficiency. Proc Natl Acad Sci USA 84(8):2494–2496

Seedorf U, Raabe M, Ellinghaus P, Kannenberg F, Fobker M, Engel T, Denis S, Wouters F, Wirtz KW, Wanders RJ, Maeda N, Assmann G (1998) Defective peroxisomal catabolism of branched fatty acyl coenzyme A in mice lacking the sterol carrier protein-2/sterol carrier protein-x gene function. Genes Dev 12(8):1189–1201

Selkala EM, Kuusisto SM, Salonurmi T, Savolainen MJ, Jauhiainen M, Pirila PL, Kvist AP, Conzelmann E, Schmitz W, Alexson SE, Kotti TJ, Hiltunen JK, Autio KJ (2013) Metabolic adaptation allows Amacr-deficient mice to remain symptom-free despite low levels of mature bile acids. Biochim Biophys Acta 1831(8):1335–1343. doi:10.1016/j.bbalip.2013.05.002

Setchell KD, Heubi JE, Shah S, Lavine JE, Suskind D, Al-Edreesi M, Potter C, Russell DW, O'Connell NC, Wolfe B, Jha P, Zhang W, Bove KE, Knisely AS, Hofmann AF, Rosenthal P, Bull LN (2013) Genetic defects in bile acid conjugation cause fat-soluble vitamin deficiency. Gastroenterology 144(5):945–955 e946; quiz e914–945. doi:10.1053/j.gastro.2013.02.004

Smith EH, Gavrilov DK, Oglesbee D, Freeman WD, Vavra MW, Matern D, Tortorelli S (2010) An adult onset case of alpha-methyl-acyl-CoA racemase deficiency. J Inherit Metab Dis 33(Suppl 3):S349–S353. doi:10.1007/s10545-010-9183-6

Suzuki Y, Jiang LL, Souri M, Miyazawa S, Fukuda S, Zhang Z, Une M, Shimozawa N, Kondo N, Orii T, Hashimoto T (1997) D-3-hydroxyacyl-CoA dehydratase/D-3-hydroxyacyl-CoA dehydrogenase bifunctional protein deficiency: a newly identified peroxisomal disorder. Am J Hum Genet 61(5):1153–1162. doi:10.1086/301599

Sztriha L, Al-Gazali LI, Wanders RJ, Ofman R, Nork M, Lestringant GG (2000) Abnormal myelin formation in rhizomelic chondrodysplasia punctata type 2 (DHAPAT-deficiency). Dev Med Child Neurol 42(7):492–495

Takahara S (1952) Progressive oral gangrene probably due to lack of catalase in the blood (acatalasaemia); report of nine cases. Lancet 2(6745):1101–1104

Teigler A, Komljenovic D, Draguhn A, Gorgas K, Just WW (2009) Defects in myelination, paranode organization and Purkinje cell innervation in the ether lipid-deficient mouse cerebellum. Hum Mol Genet 18(11):1897–1908. doi:10.1093/hmg/ddp110

ten Brink HJ, Wanders RJ, Stellaard F, Schutgens RB, Jakobs C (1991) Pristanic acid and phytanic acid in plasma from patients with a single peroxisomal enzyme deficiency. J Inherit Metab Dis 14(3):345–348

Thompson SA, Calvin J, Hogg S, Ferdinandusse S, Wanders RJ, Barker RA (2009) Relapsing encephalopathy in a patient with alpha-methylacyl-CoA racemase deficiency. BMJ Case Rep 2009. doi:10.1136/bcr.08.2008.0814

Une M, Konishi M, Suzuki Y, Akaboshi S, Yoshii M, Kuramoto T, Fujimura K (1997) Bile acid profiles in a peroxisomal D-3-hydroxyacyl-CoA dehydratase/D-3-hydroxyacyl-CoA dehydrogenase bifunctional protein deficiency. J Biochem 122(3):655–658

van den Bosch H, de Vet EC (1997) Alkyl-dihydroxyacetonephosphate synthase. Biochim Biophys Acta 1348(1–2):35–44

van den Brink DM, Wanders RJ (2006) Phytanic acid: production from phytol, its breakdown and role in human disease. Cell Mol Life Sci 63(15):1752–1765. doi:10.1007/s00018-005-5463-y

van Geel BM, Bezman L, Loes DJ, Moser HW, Raymond GV (2001) Evolution of phenotypes in adult male patients with X-linked adrenoleukodystrophy. Ann Neurol 49(2):186–194

van Roermund CW, Visser WF, Ijlst L, Waterham HR, Wanders RJ (2011) Differential substrate specificities of human ABCD1 and ABCD2 in peroxisomal fatty acid beta-oxidation. Biochim Biophys Acta 1811(3):148–152. doi:10.1016/j.bbalip.2010.11.010

Van Veldhoven PP (2010) Biochemistry and genetics of inherited disorders of peroxisomal fatty acid metabolism. J Lipid Res 51(10):2863–2895. doi:10.1194/jlr.R005959

Verheijden S, Bottelbergs A, Krysko O, Krysko DV, Beckers L, De Munter S, Van Veldhoven PP, Wyns S, Kulik W, Nave KA, Ramer MS, Carmeliet P, Kassmann CM, Baes M (2013) Peroxisomal multifunctional protein-2 deficiency causes neuroinflammation and degeneration of Purkinje cells independent of very long chain fatty acid accumulation. Neurobiol Dis 58C:258–269. doi:10.1016/j.nbd.2013.06.006

Verhoeven NM, Jakobs C, Carney G, Somers MP, Wanders RJ, Rizzo WB (1998a) Involvement of microsomal fatty aldehyde dehydrogenase in the alpha-oxidation of phytanic acid. FEBS Lett 429(3):225–228

Verhoeven NM, Wanders RJ, Poll-The BT, Saudubray JM, Jakobs C (1998b) The metabolism of phytanic acid and pristanic acid in man: a review. J Inherit Metab Dis 21(7):697–728

Wanders RJ, Waterham HR (2006a) Biochemistry of mammalian peroxisomes revisited. Annu Rev Biochem 75:295–332. doi:10.1146/annurev.biochem.74.082803.133329

Wanders RJ, Waterham HR (2006b) Peroxisomal disorders: the single peroxisomal enzyme deficiencies. Biochim Biophys Acta 1763(12):1707–1720. doi:10.1016/j.bbamcr.2006.08.010

Wanders RJ, Romeyn GJ, van Roermund CW, Schutgens RB, van den Bosch H, Tager JM (1988) Identification of L-pipecolate oxidase in human liver and its deficiency in the Zellweger syndrome. Biochem Biophys Res Commun 154(1):33–38

Wanders RJ, Denis S, van Berkel E, Wouters F, Wirtz KW, Seedorf U (1998) Identification of the newly discovered 58 kDa peroxisomal thiolase SCPx as the main thiolase involved in both pristanic acid and trihydroxycholestanoic acid oxidation: implications for peroxisomal beta-oxidation disorders. J Inherit Metab Dis 21(3):302–305

Wanders RJ, Komen J, Ferdinandusse S (2011) Phytanic acid metabolism in health and disease. Biochim Biophys Acta 1811(9):498–507. doi:10.1016/j.bbalip.2011.06.006

Waterham HR, Ebberink MS (2012) Genetics and molecular basis of human peroxisome biogenesis disorders. Biochim Biophys Acta 1822(9):1430–1441. doi:10.1016/j.bbadis.2012.04.006

Weinhofer I, Forss-Petter S, Kunze M, Zigman M, Berger J (2005) X-linked adrenoleukodystrophy mice demonstrate abnormalities in cholesterol metabolism. FEBS Lett 579 (25):5512–5516. doi:10.1016/j.febslet.2005.09.014

Westin MA, Hunt MC, Alexson SE (2005) The identification of a succinyl-CoA thioesterase suggests a novel pathway for succinate production in peroxisomes. J Biol Chem 280 (46):38125–38132. doi:10.1074/jbc.M508479200

Westin MA, Hunt MC, Alexson SE (2008) Short- and medium-chain carnitine acyltransferases and acyl-CoA thioesterases in mouse provide complementary systems for transport of beta-oxidation products out of peroxisomes. Cell Mol Life Sci 65(6):982–990. doi:10.1007/s00018-008-7576-6

Weylandt KH, Chiu CY, Gomolka B, Waechter SF, Wiedenmann B (2012) Omega-3 fatty acids and their lipid mediators: towards an understanding of resolvin and protectin formation. Prostaglandins Other Lipid Mediat 97(3–4):73–82. doi:10.1016/j.prostaglandins.2012.01.005

White AL, Modaff P, Holland-Morris F, Pauli RM (2003) Natural history of rhizomelic chondrodysplasia punctata. Am J Med Genet A 118A(4):332–342. doi:10.1002/ajmg.a.20009

Wierzbicki AS, Lloyd MD, Schofield CJ, Feher MD, Gibberd FB (2002) Refsum's disease: a peroxisomal disorder affecting phytanic acid alpha-oxidation. J Neurochem 80(5):727–735

Wiese S, Gronemeyer T, Ofman R, Kunze M, Grou CP, Almeida JA, Eisenacher M, Stephan C, Hayen H, Schollenberger L, Korosec T, Waterham HR, Schliebs W, Erdmann R, Berger J, Meyer HE, Just W, Azevedo JE, Wanders RJ, Warscheid B (2007) Proteomics characterization of mouse kidney peroxisomes by tandem mass spectrometry and protein correlation profiling. Mol Cell Proteomics 6(12):2045–2057. doi:10.1074/mcp.M700169-MCP200

Wiesinger C, Kunze M, Regelsberger G, Forss-Petter S, Berger J (2013) Impaired very long-chain acyl-CoA beta-oxidation in human X-linked adrenoleukodystrophy fibroblasts is a direct consequence of ABCD1 transporter dysfunction. J Biol Chem 288(26):19269–19279. doi:10.1074/jbc.M112.445445

Willemsen MA, IJlst L, Steijlen PM, Rotteveel JJ, de Jong JG, van Domburg PH, Mayatepek E, Gabreels FJ, Wanders RJ (2001) Clinical, biochemical and molecular genetic characteristics of 19 patients with the Sjogren-Larsson syndrome. Brain 124(Pt 7):1426–1437

Part III

New Technologies Applied to Study Peroxisomes

Understanding the Functions of Peroxisomal Proteins: The Peroxisomal Proteome, Peroxisomal Import, Proteases and Other Protein Families and Their Network Organization: What Has Computational Biology Contributed?

9

Poonam Singh, Sebastian Maurer-Stroh, Igor Kurochkin, Birgit Eisenhaber, and Frank Eisenhaber

Abstract

Computational studies based on high-throughput experimental datasets, some of which were even not generated in the context of peroxisome research, have considerably shaped the understanding of the peroxisomal proteome. Most importantly, this research revealed to a considerable extent how the total peroxisomal proteome is composed and what is its network and pathway structure. Computational prediction tools have been instrumental for finding proteins that are imported into peroxisomes via canonical import mechanisms. Based on sequence homology considerations, functions of many experimentally uncharacterized proteins have been suggested and subsequently verified experimentally. As an example, the case of peroxisomal proteases is analyzed in detail.

P. Singh • I. Kurochkin • B. Eisenhaber
Bioinformatics Institute (BII), Agency for Science, Technology and Research (A*STAR), 30 Biopolis Street, #07-01, Matrix, Singapore 138671

S. Maurer-Stroh
Bioinformatics Institute (BII), Agency for Science, Technology and Research (A*STAR), 30 Biopolis Street, #07-01, Matrix, Singapore 138671

School of Biological Sciences (SBS), Nanyang Technological University (NTU), 60 Nanyang Drive, Singapore, Singapore 637551

F. Eisenhaber (✉)
Bioinformatics Institute (BII), Agency for Science, Technology and Research (A*STAR), 30 Biopolis Street, #07-01, Matrix, Singapore 138671

Department of Biological Sciences (DBS), National University of Singapore (NUS), 8 Medical Drive, Singapore, Singapore 117597

School of Computer Engineering (SCE), Nanyang Technological University (NTU), 50 Nanyang Drive, Singapore, Singapore 637553
e-mail: franke@bii.a-star.edu.sg

C. Brocard and A. Hartig (eds.), *Molecular Machines Involved in Peroxisome Biogenesis and Maintenance*, DOI 10.1007/978-3-7091-1788-0_9,

Additionally, resources such databases or WWW servers dedicated to peroxisomal biology are reviewed.

Keywords
Peroxisomal proteome • peroxisomal import • peroxisomal proteases • PTS1 • PTS2 • Tysnd1 • Lonp2 • IDE

9.1 Introduction

Life sciences are not a truly theoretical discipline; the extrapolation depth is small as a consequence of the fragmentary knowledge of biomolecular mechanisms. All practically important things such as the dose of a drug are the result of measurement/experiment, not of computation with a theoretical argument. The theory of biomolecular sequence homology is historically the first and still one of the few achievements of theoretical biology that have significance for all fields of life science. The concept of sequence homology with its sequence–structure–function trinity states that the similarity of protein sequences implies common evolutionary origin and, subsequently, similarity of three-dimensional globular structure and biological function. Thus, sequence similarity in the regions of globular domains can be used in practical applications for predicting molecular and cellular functions (Bork et al. 1998; Bork and Koonin 1998; Wong et al. 2010) for uncharacterized genes by annotation transfer from well-studied homologues. In the era of cheap gene/protein sequencing, the homology concept is the key tool for understanding the biological meaning encoded in biomolecular sequences that have never been the target of any experimental study.

As many other areas of cell biology and biochemistry, peroxisomal biology considerably benefited and it will gain from the application of computational biology approaches in the analyses of high-throughput data obtained from sequencing, expression profiling, and biomacromolecular studies, even if the data were often not generated in context with peroxisome research (such as the full human genome sequence). This article will summarize how computational/theoretical approaches either alone or in context with dedicated experimental efforts following up computationally derived hypotheses have shaped our understanding of the peroxisome proteome (Sect. 9.3 and special emphasis on peroxisomal proteases in Sect. 9.4), the functions of peroxisomal proteins and the peroxisomal import mechanisms (Sect. 9.2). Additionally, database and WWW server resources will be reviewed (Sect. 9.5).

In the mid 1950s, peroxisomes were first discovered as microbodies from mouse kidney and rat liver cells as a particular structure visible in electron micrographs without clear understanding of their biological role (De 1969). In plants, peroxisomes were found due to their association with enzymes of photorespiration in the late 1960s (Tolbert et al. 1968; Tolbert and Yamazaki 1969) and, subsequently, they were detected in most eukaryotic organisms (De 1969).

Peroxisomes are single membrane-bound organelles in eukaryote cells harboring a broad variety of critical metabolic pathways (Wanders and Waterham 2006) like lipid metabolism, free radical detoxification, etc.; they compartmentalize the glyoxylate cycle in germinating seeds ("glyoxysomes"), the photorespiration in leaves, they harbor glycolysis in trypanosomes ("glycosomes"), and methanol and/or amine oxidation and assimilation pathways in some yeasts. Many peroxisomal functions are still expected to be discovered (see below). In the homeostatic context, peroxisomes appear to provide separated spaces for biochemical processes that might disturb the intracellular equilibrium otherwise. With such a wide-range participation in metabolism and signaling, it is not surprising that peroxisomes have an important role in development, differentiation and morphogenesis from yeasts to humans (Bonekamp et al. 2009; Schrader and Fahimi 2008; Wanders and Waterham 2006).

9.2 Peroxisomal Protein Import

9.2.1 Peroxisomal Targeting Signals in the Historical Perspective

In view of the facts that, on the one hand, peroxisomes lack their own genome and have no transcriptional machinery and, on the other hand, matrix- and membrane-destined peroxisomal proteins are encoded in the nuclear genome and synthesized on free cytosolic ribosomes (Baker and Sparkes 2005), these proteins have to be finally targeted posttranslationally to the organelle in a regulated manner. The import of proteins into the peroxisome is generally mediated by peroxisomal import signals encoded in non-globular segments of the respective target protein sequences. Specialized receptor and transporter proteins organized in complexes and pathways recognize these import signals and transport them to their destination (Fujiki et al. 2012; Girzalsky et al. 2009; Rayapuram and Subramani 2006). A characteristic difference to some other organelle import processes is that target proteins are being transported and imported into peroxisomes in their folded and sometimes even oligomeric state (Lanyon-Hogg et al. 2010; Mast et al. 2010; McNew and Goodman 1996; Meinecke et al. 2010). With the improving knowledge of both the targeting signals and their respective receptors, progress could be made in the endeavor to describe these sequence motifs as generalized patterns and to use this insight for predicting these signals in protein sequences and, thus, to generate increasingly exhaustive lists of target proteins for the respective import pathways.

Currently, three different canonical import mechanisms are coherently supported by extensive experimental evidence. They are called after their peroxisomal targeting signals PTS1, PTS2, and mPTS (see Fig. 9.1). PTS1, a C-terminal signal motif, and PTS2, an N-terminal signal motif, drive proteins into the peroxisomal matrix. mPTS, the signal for the least well studied import route among the three, directs target proteins into the peroxisomal membrane compartment.

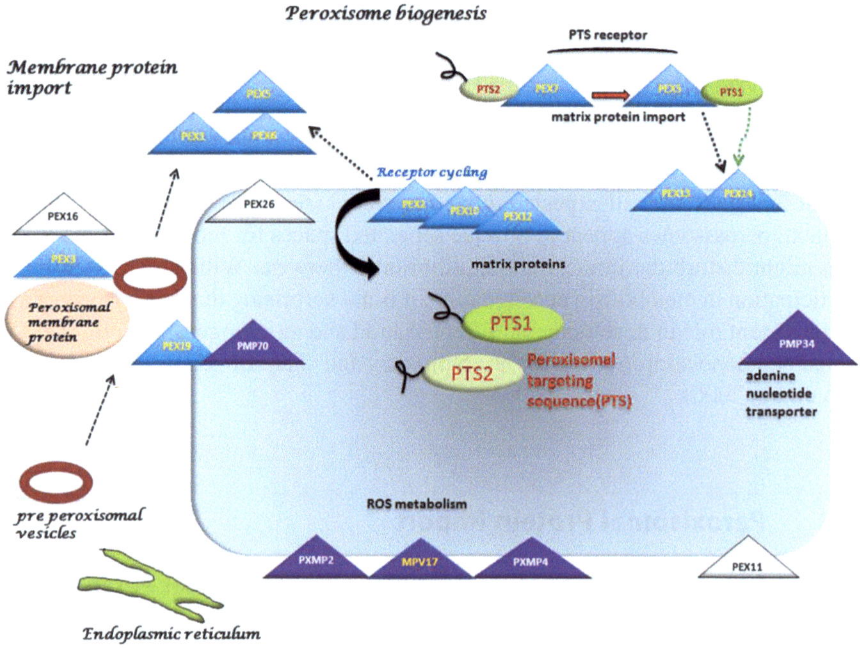

Fig. 9.1 Peroxisomal biogenesis and protein import in yeast. Proteins known to participate in peroxisomal biogenesis and peroxisomal import in yeast and the respective pathways are illustrated schematically following information from the KEGG database (Kanehisa 2013). The PEX genes with functions described in the KEGG database are shown in *blue triangles*. For abbreviations, see legend of Table 9.1

9.2.2 Peroxisomal Targeting Signal 1 (PTS1) Motif Description

PTS1 was originally discovered as a conserved C-terminal tripeptide sequence that triggers peroxisomal import posttranslationally (Distel et al. 1987; Gould et al. 1987, 1988, 1989). After the initial characterization of several proteins with peroxisomal and glycosomal targeting signals (Distel et al. 1987; Gould et al. 1987; Osinga et al. 1985) in the late 1980s, it was clear that there had to be also specific cytosolic import receptors (Subramani 1992) which were later identified to be orthologues from the PEX5 protein family that bind the peroxisomal targeting signal 1 (PTS1) motif sequence in targeted proteins (Dodt et al. 1995; Fransen et al. 1999; McCollum et al. 1993; Van der Leij et al. 1993).

The first protein whose peroxisomal targeting signal was linked to the currently known PTS1 C-terminal sequence motif was firefly luciferase (Gould et al. 1987). The discoverers quickly noted that a similar motif was also conserved in several other peroxisomal proteins (Gould et al. 1988; Keller et al. 1991). Early characterizations were focused on the last three residues which appeared to be conservative variations of the tripeptide -SKL (Gould et al. 1989). In this context, two observations were important for later research efforts. (1) Originally, it was thought

Table 9.1 Abbreviations to Fig. 9.3

Protein family	Description
ABCD	ATP-binding cassette sub-family D.
ACAA1	3-Ketoacyl-CoA thiolase, peroxisomal precursor (EC 2.3.1.16) (Beta-ketothiolase) (Acetyl-CoA acyltransferase) (Peroxisomal 3-oxoacyl-CoA thiolase).
ACOX	Acyl-coenzyme A oxidase , peroxisomal (EC 1.3.3.6).
ACSL	Encodes a peroxisomal protein with acetyl-CoA synthetase activity that is responsible for the activation of acetate for entry into the glyoxylate cycle.
AGPS	Alkyldihydroxyacetonephosphate synthase, peroxisomal precursor (EC 2.5.1.26) (Alkyl-DHAP synthase) (Alkylglycerone-phosphate synthase).
AGXT	Serine-pyruvate aminotransferase (EC 2.6.1.51) (SPT) (Alanine-glyoxylate aminotransferase) (EC 2.6.1.44) (AGT).
AMACR	Alpha-methylacyl-CoA racemase (EC 5.1.99.4) (2-methylacyl-CoA racemase).
BAAT	Bile acid Coenzyme A: amino acid *N*-acyltransferase.
CAT	Catalase (EC 1.11.1.6).
CRAT	Carnitine *O*-acetyltransferase (EC 2.3.1.7) (Carnitine acetylase) (CAT) (Carnitine acetyltransferase) (CrAT).
CROT	Peroxisomal carnitine *O*-octanoyltransferase (EC 2.3.1.137) (COT).
DHRS4	Dehydrogenase/reductase SDR family (NADPH- dependent carbonyl reductase/ NADP-retinol dehydrogenase) (CR) (PHCR) (Peroxisomal short-chain alcohol dehydrogenase) (NADPH-dependent retinol dehydrogenase/reductase) (NDRD).
DHPAT	Glyceronephosphate *O*-acyltransferase (EC:2.3.1.42).
DAO	D-amino-acid oxidase (EC 1.4.3.3) (DAMOX) (DAO) (DAAO).
DDO	D-aspartate oxidase (EC 1.4.3.1) (DASOX) (DDO).
ECH	Peroxisomal 3,2-*trans*-enoyl-CoA isomerase (EC 5.3.3.8) (Dodecenoyl-CoA isomerase) (Delta(3),delta(2)-enoyl-CoA isomerase) (D3,D2-enoyl-CoA isomerase) (DBI-related protein 1) (DRS-1).
EPHX2	Epoxide hydrolase 2 (EC 3.3.2.3) (Soluble epoxide hydrolase) (SEH) (Epoxide hydratase).
FAR	Fatty acyl CoA reductase that reduces fatty acids to fatty alcohols. Male sterility domain containing.
GSTK1	Glutathione *S*-transferase kappa 1 (EC 2.5.1.18) (GST 13-13) (Glutathione *S*-transferase subunit 13) (GST class-kappa) (GSTK1-1).
HAO	Hydroxyacid oxidase. ((S)-2-hydroxy-acid oxidase, peroxisomal) (Long chain alpha-hydroxy acid oxidase) (Long-chain L-2-hydroxy acid oxidase).
HMGCL	Hydroxymethylglutaryl-CoA lyase, (HMG-CoA lyase) (HL) (3-hydroxy-3-methylglutarate-CoA lyase).
HPCL2	2-Hydroxyphytanoyl-CoA lyase (EC 4.1.-.-) (2-HPCL).
IDH	Isocitrate dehydrogenase [NADP] (EC 1.1.1.42) (Oxalosuccinate decarboxylase) (IDH) (NADP(+)-specific ICDH) (IDP).
INOS	Nitric oxide synthase, inducible (EC 1.14.13.39) (NOS type II) (Inducible NOS) (iNOS).
MLYCD	Malonyl-CoA decarboxylase, (EC 4.1.1.9) (MCD).
MPV17	MPV17-domain bearing proteins.
MVK	Mevalonate kinase (EC 2.7.1.36) (MK).

(continued)

Table 9.1 (continued)

Protein family	Description
NUDT12	Peroxisomal NADH pyrophosphatase NUDT12 (EC 3.6.1.22) (Nucleoside diphosphate-linked moiety X motif 12) (Nudix motif 12).
NUDT19	Nudix (nucleoside diphosphate linked moiety X)-type motif 19.
PAOX	Peroxisomal N1-acetyl-spermine/spermidine oxidase (EC 1.5.3.11) (Polyamine oxidase).
PDCR	Peroxisomal 2,4-dienoyl-CoA reductase (EC 1.3.1.34) (2,4-dienoyl-CoA reductase 2) (pDCR).
PECR	Peroxisomal *trans*-2-enoyl-CoA reductase (EC 1.3.1.38) (TERP) (HPDHase) (pVI-ARL) (2,4-dienoyl-CoA reductase-related protein) (DCR-RP).
PECI	Peroxisomal 3,2-*trans*-enoyl-CoA isomerase [EC:5.3.3.8].
PEX1	Peroxisome biogenesis factor 1 (Peroxin-1) (Peroxisome biogenesis disorder protein 1).
PEX2	Peroxisomal membrane protein 2. Peroxisome assembly factor 1 (PAF-1) (Peroxin-2) (35 kDa peroxisomal membrane protein) (RING finger protein 72).
PEX3	Peroxisomal biogenesis factor 3 (Peroxin-3) (Peroxisomal assembly protein PEX3).
PEX5	Peroxisome membrane protein 5 and PEX5-related protein.
PEX6	Peroxisome membrane protein. Peroxisome assembly factor 2 (PAF-2) (Peroxisomal-type ATPase 1) (Peroxin-6) (Peroxisomal biogenesis factor 6).
PEX7	Peroxisomal targeting signal 2 receptor (PTS2 receptor) (Peroxin-7).
PEX10	Peroxisome assembly protein 10 (Peroxin-10) (RING finger protein 69).
PEX11	Peroxisomal membrane protein 11 (Peroxin-11) (Peroxisomal biogenesis factor 11).
PEX12	Peroxisome assembly protein 12 (Peroxin-12) (Peroxisome assembly factor 3) (PAF-3).
PEX13	Peroxisomal membrane protein PEX13 (Peroxin-13).
PEX14	Peroxisomal membrane protein PEX14 (Peroxin-14) (Peroxisomal membrane anchor protein PEX14) (PTS1 receptor docking protein).
PEX16	Peroxisomal membrane protein PEX16 (Peroxin-16) (Peroxisomal biogenesis factor 16).
PEX19	Peroxisomal biogenesis factor 19 (Peroxin-19) (Peroxisomal farnesylated protein) (33 kDa housekeeping protein).
PEX26	Peroxisomal membran protein 26. Peroxisome assembly protein 26 (Peroxin-26). [Source:Uniprot/SWISSPROT;Acc:Q7Z412].
PHYH	Phytanoyl-CoA dioxygenase, peroxisomal precursor (EC 1.14.11.18) (phytanoyl-CoA alpha-hydroxylase) (PhyH) (Phytanic acid oxidase).
PIPOX	Peroxisomal sarcosine oxidase (EC 1.5.3.1) (PSO) (L-pipecolate oxidase) (EC 1.5.3.7) (L-pipecolic acid oxidase).
PMP34	Peroxisomal membrane protein *PMP34*.
PMVK	Phosphomevalonate kinase (EC 2.7.4.2) (PMKase).
PRDX1	Peroxiredoxin 1 (EC 1.11.1.15) (thioredoxin peroxidase 2) (thioredoxin-dependent peroxide reductase 2) (Proliferation-associated protein PAG) (Natural killer cell enhancing factor A) (NKEF-A).
PRDX5	Peroxiredoxin 5.
PTE1	Peroxisomal acyl-coenzyme A thioester hydrolase 1 (EC 3.1.2.2) (Peroxisomal long-chain acyl-coA thioesterase 1) (HIV-Nef associated acyl coA thioesterase) (Thioesterase II) (hTE) (hACTEIII) (hACTE-III) (PTE-2).

(continued)

Table 9.1 (continued)

Protein family	Description
PTE2	Peroxisomal acyl-coenzyme A thioester hydrolase 2a (EC 3.1.2.2) (Peroxisomal long-chain acyl-coA thioesterase 2) (ZAP128) (CTE-Ia).
PXMP2	Peroxisomal membrane protein (PMP22) is a major component of peroxisomal membranes. PMP22 seems to be involved in pore forming activity and may contribute to the unspecific permeability of the organelle membrane.
PXMP4	Peroxisomal membrane protein 4 (24 kDa peroxisomal intrinsic membrane protein).
SCPX	Sterol carrier protein 2 (SCP-2), Sterol carrier protein X (SCP-X) (SCPX).
SOD	Superoxide dismutase.
VLACS	Solute carrier family 27 (fatty acid transporter).
XDH	Xanthine dehydrogenase/oxidase [includes: Xanthine dehydrogenase (EC 1.17.1.4) (XD); Xanthine oxidase (EC 1.17.3.2) (XO) (Xanthine oxidoreductase)].

that cytosolic proteins getting C-terminally appended -SKL become always redirected into peroxisomes. Yet, mouse dihydrofolate reductase (DHFR), however, is remaining cytoplasmic after deletion of the last two C-terminal amino acids and subsequent attaching of an -SKL tripeptide (Distel et al. 1987; Kragler et al. 1993). Thus, the adjacent, N-terminally upstream sequence has an influence on the import. (2) Besides -SKL, there are other C-terminal tripeptides that import proteins into peroxisomes, although their efficiency might not be as good. Subsequent large-scale studies led to the classification into major (canonical) and minor (non-canonical) PTS1 tripeptides (Boden and Hawkins 2005; Emanuelsson et al. 2003; Hawkins et al. 2007; Neuberger et al. 2003a, b; Reumann 2004; Schluter et al. 2010). Major PTS1 C-terminal tripeptides (e.g., SKL, ARL, and PRL) are the prime signals of high-abundance proteins and are present in most eukaryotes, indicating that these signals are sufficient for peroxisome targeting. Minor PTS1 tripeptides such as -SSI, -ASL, and -SLM which are generally limited to a few, more rarely expressed peroxisomal proteins tend to have a taxon-specific role, for example, in plants (Reumann et al. 2007, 2009). Minor targeting tripeptides are generally weak signals that require secondary targeting-enhancing patterns (e.g., basic residues) immediately upstream of the tripeptide for functionality. But such enhancer patterns have been incompletely defined for metazoa (Neuberger et al. 2003a) and they apparently differ among taxonomic groups.

Later studies complemented the motif characterization through various approaches including site-directed mutagenesis of known motifs, peptide library approaches using two-hybrid (Lametschwandtner et al. 1998) or phage display (Fransen et al. 1999) assays as well as statistical physical property analysis of not only the C-terminal tripeptide of known peroxisomal proteins but also the preceding region (Neuberger et al. 2003a). Today, at least 12 C-terminal residues are considered to be critical for productive complex formation between PEX5-PTS1 (Neuberger et al. 2003a, b). The total motif is considered to consist of two major regions, the original three-residues motif that directly interacts with the PEX5-binding cavity and another N-terminally adjacent segment of 9–10 residues that

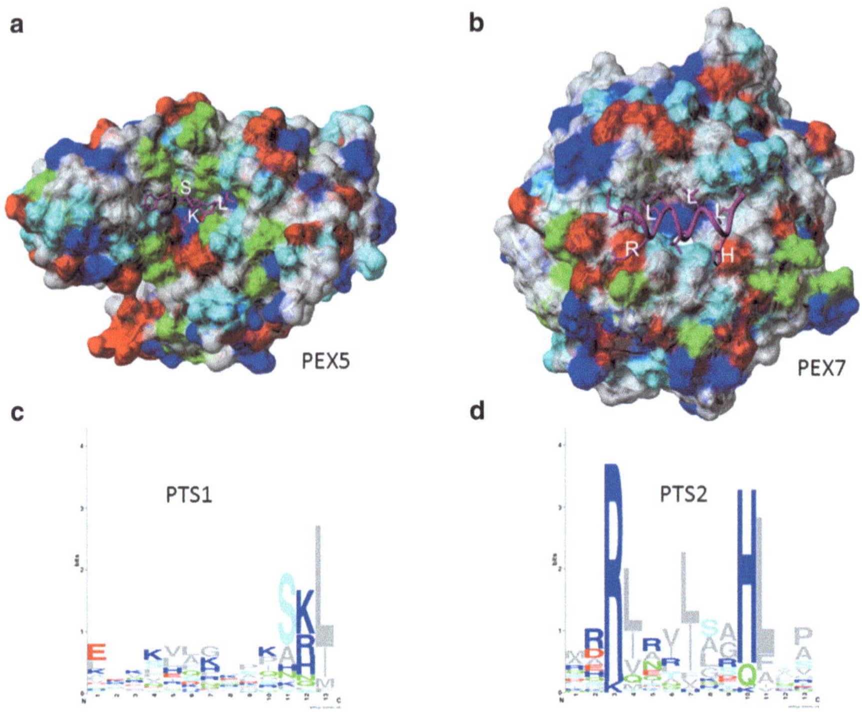

Fig. 9.2 The peroxisomal targeting signal type 1 and type 2 motifs. (**a**) The typical PTS1 peptide SKL is shown to be bound to the structure of the tetratricoppetide domain of PEX5. The surface coloring corresponds to the color of residues in the sequence logo of known PTS1 motifs (see **c**). (**b**) A known PTS2 motif sequence is shown bound to the PEX7 structure. The surface coloring corresponds to the color of residues in the sequence logo of known PTS1 motifs (see **d**). (**c**) Sequence logo of known PTS1 motif sequences. (**d**) Sequence logo of known PTS2 motif sequences

provides flexibility and solubility in the aqueous medium so that the terminal tripeptide is mechanically enabled to enter the binding cavity (Eisenhaber et al. 2004; Eisenhaber and Eisenhaber 2007).

In the crystal structure of the complex of PEX5 with a canonical PTS1 peptide (Gatto et al. 2000), it was confirmed that the terminal tripeptide is deeply buried in the TPR-repeat binding pocket and, hence, comprises the strongest specificity constraints (Fig. 9.2a). At the same time, a weaker binding tripeptide could be compensated by specific additional upstream positions (numbering starts with -1 for the residue at the N-terminal side of the C-terminal tripeptide) especially -4 and -5 (Maynard et al. 2004) and also -2 for species-specific differences (Lametschwandtner et al. 1998). Although not buried, these residues are still in contact with the structural surface of the PEX5 receptor. A sequence entropy logo of the C-termini of a non-redundant set of experimentally verified PTS1-containing proteins allows simple visualization of position-specific amino acid requirements in the PTS1 motif (Fig. 9.2c). At the same time, significant correlations have been found with physical properties describing flexible linker regions with preferentially small and polar residues extending at least until position -9 and -10 beyond the classical tripeptide motif (Neuberger et al. 2003a). Such linker regions are commonly found in distinct but conceptually similar peptide–protein recognition scenarios (Eisenhaber et al. 1998; Maurer-Stroh et al. 2002; Maurer-Stroh and Eisenhaber 2005; Van et al. 2009, 2010) and are understood to be necessary for keeping the peptide extended and the specific binding motif accessible for structural recognition by the import receptor (Eisenhaber et al. 2004; Eisenhaber and Eisenhaber 2007).

9.2.3 Peroxisomal Targeting Signal 1 (PTS1) Prediction

Starting from the first identified canonical motif -SKL, the pattern of allowed amino acids per position was soon extended to [SAGCN]-[RKH]-[LIVMAF] with the discovery of additional peroxisomal proteins. Such pattern can be used to predict whether a protein sequence contains a PTS1 motif. This pattern is applied, for example, within the PSORT2 subcellular localization prediction program (Nakai and Horton 1999). However, for direct pattern searches, every additional allowed amino acid at certain pattern positions makes the motif more ambiguous (thus, increasing the false-positive prediction rate to devastating proportions) and there is the risk of losing specificity and increasing the chance of random hits (e.g., increase the number of false positive predictions). Furthermore, only the terminal three residues were considered in these early motif predictors.

In 2003, the PTS1 Predictor (alternatively called PeroPS = peroxisomal localization prediction suite) by Neuberger et al. became available with a specially designed scoring function utilizing the full-length PTS1 motif weighted by the importance of the known regions and considering position-specific scoring matrices together with physical property constraints (Neuberger et al. 2003b). Compared to the straight-forward pattern search, this approach allowed high sensitivity (e.g., finding more than of the 90 % known PTS1 proteins or true positive predictions) while not over-predicting (e.g., less than 0.4 % false positives in proteomes of prokaryotes which do not have peroxisomes). These results were a dramatic improvement reaching a new order of magnitude for lowering the number of false positive predictions compared to previous permissive pattern methods.

Afterwards, a neural network-based predictor called PeroxiP by Emanuelsson et al. was published (Emanuelsson et al. 2003). Its performance relative to the PTS1 Predictor as well as a newer SVM-based machine learning method called Pts1Prowler by Hawkins et al. was compared subsequently (Hawkins et al. 2007). In the benchmark on a set of newly experimentally verified peroxisomal proteins excluded from the respective learning procedures, Pts1Prowler showed a much better specificity and sensitivity than PeroxiP but appeared on par with the PTS1 Predictor's sensitivity while improving slightly in specificity. An interesting observation was that the respective true positive predictions were not always shared between Pts1Prowler and the PTS1 Predictor indicating potential to use them for complementary predictions.

Although the PTS1 Predictor already had some taxon-specific scoring options (Metazoa, Fungi, General), it did not provide one dedicated to plants since the learning sets for them were insufficient at those times. Reumann et al. studied low-abundance plant proteins with non-canonical PTS1 patterns which were difficult to predict by previous methods and they recently developed PredPlantPTS1 for the prediction of plant peroxisomal proteins (Reumann et al. 2012). Interestingly, on a small challenging set of ten plant proteins they found that both PredPlantPTS1 and the PTS1 Predictor have reasonably good prediction sensitivity while Pts1Prowler appeared too restrictive and missed all proteins.

It should be noted that, in predictor development, the greater goal is not just high sensitivity to already known examples and a low rate of false-positive predictions among known non-relevant protein entries. Most importantly, a prediction tool needs to have the potential to filter out potentially new targets that are useful for experimental study. In the overwhelming majority of sequence-based prediction scenarios, the number of known examples is much too low for automated learning algorithms to extract the correct pattern in an unsupervised mode. In the PTS1/PeroPS effort, the authors wished to model the recognition process between the C-terminal of the target-protein and the receptor (Eisenhaber et al. 2004; Eisenhaber and Eisenhaber 2007) and, thus, formulated a number of physical conditions based on the analysis of various experimental data that finally entered the score functions (Neuberger et al. 2003a, b). Not surprisingly, this prediction tool performs quite well in the recognition of example sequences that were not seen at the time of predictor development and it also reports how the various residues of the C-terminal in the query sequences contribute to the prediction outcome.

9.2.4 The PTS1 Signal in the Hierarchy of Translocation Signals

The generally low false positive prediction rate in the range of ~0.5 % of the PTS1/PeroPS predictor allows the tool to be applied for full proteome runs. Such an analysis generated predictions systematically among some protein families that were generally not known in the peroxisomal context (Neuberger et al. 2003b). To note, these peroxisomal motif detection tools test a query sequence only for the possible occurrence of a productive C-terminal PTS1 motif. Of course, a PTS1 signal can come about by chance during evolution and, thus, provide material for selection. Yet, systematic occurrences tend to indicate deeper reasons. In vivo, the PTS1 signal has to compete with other translocation signals. For example, we obtained a set of lysozymes with consensus -CRL tripeptides at their C termini. The C-terminal tripeptide might be functional as PTS1 signal but, possibly, cannot act as such because the protein gets exported into the endoplasmatic reticulum immediately after translation and never comes into vicinity of a Pex5 receptor molecule. Several experimentally verified protein examples (Holbrook et al. 2000; Oatey et al. 1996a, b; Oda et al. 2000) show that the PTS1 signal is lower in hierarchy compared with N-terminal signals such as the mitochondrial targeting signal if they are present in the same protein, e.g., as a result of alternative transcription and translation initiation sites.

Subsequently, the functionality of several predicted peroxisomal translocation signals of the PTS1 type in out-of context proteins (in chicken lysozyme, human tyrosinase, yeast mitochondrial ribosomal protein L2, and bacterial glutamate-1-semialdehyde 2,1-aminomutase) has been successfully tested experimentally (Neuberger et al. 2004). The lysozyme case was especially striking. As soon as the N-terminal signal peptide for extracellular translocation was rendered dysfunctional by mutation, the lysozyme ended up in the peroxisomes.

Facultative PTS1-type targeting signals appear to be a widespread tool for alternatively allocating several core enzymes of glycolysis such as glyceraldehyde-3-phosphate dehydrogenase (GAPDH) and 3-phosphoglycerate kinase (PGK) to peroxisomes in numerous fungal species (Freitag et al. 2012). These cryptic PTS1 signals were discovered with the PTS1/PeroPS prediction tool and subsequently tested experimentally. In vivo, they are activated by various posttranscriptional processes. In the basidiomycete plant pathogen Ustilago maydis, peroxisomal targeting of Pgk1 results from ribosomal read-through. Alternative splicing is responsible for generating the PTS1 of Gapdh. In the filamentous ascomycete *Aspergillus nidulans*, peroxisomal targeting of the same enzymes Pgk1 and Gapdh is achieved by exactly the opposite mechanisms. PTS1 motifs were also discovered for the glycolytic enzymes triose-phosphate isomerase and fructose-bisphosphate aldolase. It was shown for Ustlilago maydis mutants that the absence of peroxisomal isoforms of Gapdh or Pgk1 goes hand in hand with reduced virulence.

Taken together, these results support the view that functional localization signals can evolve in unrelated protein sequences as a result of neutral mutations. Translocation signals and subcellular targeting are hierarchically organized and the signal accessibility in the given biological context has the decisive role whether the originally silent signals get actively used in vivo. Apparently, silent functional signals have the potential to acquire importance in future evolutionary scenarios and in pathological conditions (see also Danpure et al. 1996).

9.2.5 Peroxisomal Targeting Signal 2 (PTS2) Motif Description

Concurrent with the discovery of the C-terminal PTS1 motif, it became clear that not all peroxisomal proteins have their targeting signal in the C-terminus and it did not take long for the identification of the first peroxisomal targeting signal 2 (PTS2) in 3-ketoacyl-CoA thiolase (Swinkels et al. 1991) at the opposite end of the protein sequence. There are also cases where proteins such as catalase A of yeast (Kragler et al. 1993) contain two independent peroxisomal targeting signals at either end of the amino acid sequence. Thus, the type 2 PTS (PTS2) is located in the N-terminus of another set of matrix proteins that are proteolytically processed after import into peroxisomes (Elgersma et al. 1997; Kurochkin et al. 2007).

The importance of specifically conserved positively charged residues was also recognized (Glover et al. 1994) as well as some similarities to mitochondrial targeting signals (Osumi et al. 1992). At first, the PTS2 signal was characterized as nonapeptide with the pattern (-R/K-X6-H/Q-A/L/F-; X indicates any amino acid) embedded in the N-terminal part of the target protein (Petriv et al. 2004). The PTS2 import receptor Pex7p was soon discovered thereafter (Braverman et al. 1997; Marzioch et al. 1994) but the number of in vivo PTS2-dependent peroxisomal proteins was only growing slowly. Possibly because it was overshadowed in importance by the more abundantly used PTS1-dependent mechanism, it took a while to shed more light on further motif details and the molecular recognition between PTS2 and Pex7p.

Through statistical physical property analysis of known PTS2 motifs, structural modeling and experimental binding assays with site-directed mutations in PTS2 peptides and the Pex7p receptor, Kunze et al. recently proposed that the PTS2 motif could be forming an amphipathic helix (Kunze et al. 2011). In a helical conformation, the three constrained hydrophobic residues in the motif are lined up and would fit into an evolutionary conserved hydrophobic groove at the Pex7p beta-propeller surface (Fig. 9.2b). They also identified complementary negative charge patterns in the Pex7p binding interface that is critical for recognition of the conserved positive charges of the PTS2 motif as confirmed by charge disruption and charge exchange mutation experiments (Kunze et al. 2011). The sequence entropy logo of a non-redundant set of experimentally verified N-terminal PTS2 peptides summarizes currently known position-specific amino acid requirements in the PTS2 motif (Fig. 9.2d). As was known for PTS1 and other motifs, the surrounding region of the sequence-constrained helical motif also shows the tendency to be structurally flexible in many cases which could help to have the PTS2 helix accessible for Pex7p binding as well as possible cleavage after import (Kunze et al. 2011).

This model of PTS2 interaction with the WD40 domain of Pex7p was fully confirmed by a recent crystallographic study (Pan et al. 2013) of Pex7p–Pex21p with the PTS2 segment from 3-ketoacyl-CoA thiolase. Not only was the beta-propeller structure of the Pex7p domain, the helical arrangement of the PTS2 motif segment on top of it and their relative orientation was correctly predicted (Fig. 9.2). The authors of (Kunze et al. 2011) also got the interacting residues at the receptor surface right (Fig. 9.3). No state-of-the-art automated docking method can achieve this precision; it shows the power of the careful analysis of the surface conservation during evolution and of the complementary charge patterns that gave the critical clues to inform the docking simulation published in 2011.

It is notable that Pex5p and Pex7p are present in yeast, mammals and plants but the mechanism of protein transport by means of these components is somewhat different. In yeast, the PTS1 and PTS2 pathways are independent as PTS2 pathway is active in yeasts mutants with the defects of PTS1 pathway like *pex5* and vice versa (Elgersma et al. 1998; Marzioch et al. 1994; McCollum et al. 1993; Rehling et al. 1996; Van der Leij et al. 1993). The PTS2 pathway is completely dependent on the PTS1 pathway in plants as well as in mammals. After establishing the

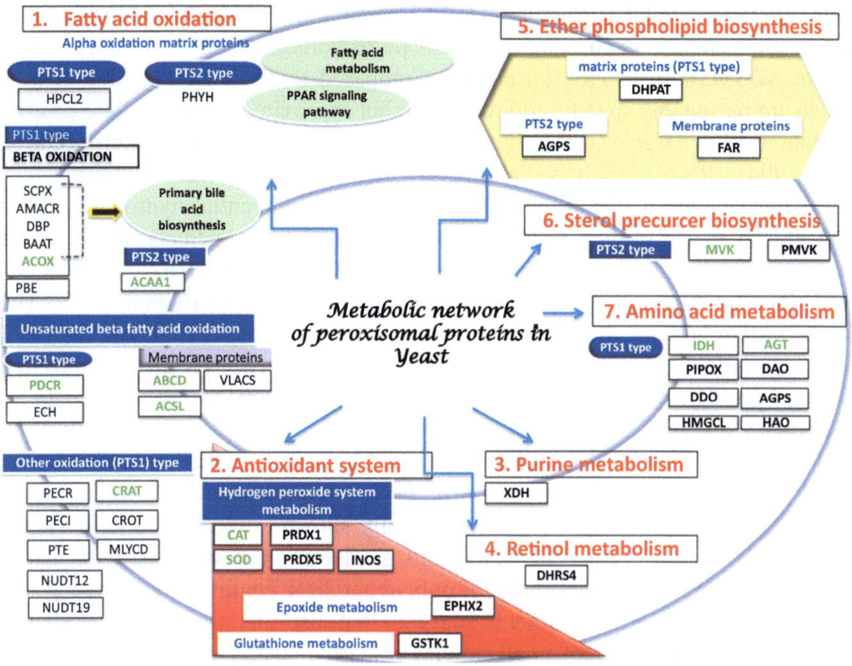

Fig. 9.3 The network of peroxisomal proteins. This figure together with Fig. 9.1 shows the network of peroxisomal protein families with special emphasis in yeast and the type of signals (PTS1 or PTS2) involved for their import based on information from the KEGG database (Kanehisa 2013). Out of different pathways like fatty acid oxidation, antioxidant system, purine metabolism, retinol metabolism, ether phospholipid biosynthesis, sterol precursor biosynthesis, and amino acid metabolism, families shown in *green* have been identified in yeast. The abbreviations used are summarized in Table 9.1

interaction with Pex5p, the complex of PTS2 -containing proteins and Pex7p is transported to the machinery that consists of Pex14p and other unknown proteins at the peroxisomal membrane (Nito et al. 2002).

9.2.6 Peroxisomal Targeting Signal 2 (PTS2) Prediction

Due to the small number of known natural PTS2-dependent peroxisomal proteins, it has been difficult to construct a robust prediction tool for PTS2 motifs so far. However, since there are more positions constrained for specific amino acids than for the PTS1 motif, consensus patterns were at least initially useful for quick screens to find highly similar motifs in other peroxisomal proteins. However, the more PTS2 sequences become verified, the more permissive will be the associated patterns, giving rise to false positive predictions.

To estimate the amount of false positive predictions that can be expected, we applied a similar specificity test as in the benchmarks for Pts1Prowler (Hawkins et al. 2007) and the PTS1 Predictor (Neuberger et al. 2003b) by searching against a prokaryote genome which is known to lack peroxisomes (e.g., *Escherichia coli* K12 substr. WG3110 from NCBI RefSeq). Compared to the PTS1 prediction tools which are reasonably specific and would predict approximately 0.3 % false positive random hits in large screens, the first derived PTS2 consensus motif (Glover et al. 1994), [RK]-[LVI]-X-X-X-X-X-[QH]-[LA], would predict about 10 % false positives. This motif still did not include the conserved central hydrophobic position which can add further stringency in searches and an updated consensus pattern was derived later (Petriv et al. 2004) as [RK]-[LVIQ]-X-X-[LVIHQ]-[LSGAK]-X-[QH]-[LAF] which results in a false positive prediction rate of about 1.7 %. This can be further brought down to 0.3 % false positives at comparable sensitivity to the pattern approach (70 %) when applying the same sequence profile plus physical property penalty scoring function approach as used in the PTS1 Predictor to the set of known PTS2 motifs (Kunze et al. 2011).

This preliminary PTS2 Predictor was used to screen the human proteome for novel peroxisomal proteins (Kunze et al. 2011). While 4 out of the top 14 newly predicted peptides were indeed experimentally shown to localize to peroxisomes when fused to EGFP, only 2 of them did so also in context of their full-length protein. This modest "success" rate of only about 28 % confirmation of new PTS2-targeting capability in experiments suggests that also this preliminary prediction tool is limited by the current small amount of learning data for PTS2 prediction. At the same time, it is not known if there would indeed be many more unknown PTS2-dependent peroxisomal proteins left to be discovered in the human proteome. However, new motif sequence information that would be needed for more robust predictions could come from experimental tests of homologues of currently known PTS2-dependent peroxisomal proteins in other species. The availability of the crystal structure of the Pex7p/PTS2 signal complex (PDB entry 3W15; Pan et al. 2013) could also serve as a new starting point for in silico testing of variable peptide sequences to further explore the amino acid sequence restrictions for productive binding and, subsequently, for using this information for deriving a better prediction tool.

9.2.7 The mPTS Pathway

Peroxisomal membrane proteins are imported posttranslational in peroxisomal membrane. They are encoded by nuclear genes and translated on cytoplasmic ribosomes. The mechanism of import of PMPs is different from peroxisomal matrix proteins (Chang et al. 1999; Erdmann and Blobel 1996; Gould et al. 1996). There are two import pathways which has been stated for targeting of PMPs (Van and Fransen 2006). In the first pathway, hydrophobic proteins are directly inserted into existing peroxisomal membranes after being synthesized in the cytosol (Sparkes et al. 2005). In the second pathway, PMPs are synthesized on the rough endoplasmic reticulum where they concentrate in pre-peroxisomal vesicles (Tabak et al. 2003, 2008a). The vesicles formed act as initiator for de novo formation of peroxisomes or can fuse with existing peroxisome membranes. Hence, PMPs can be classified into two types named as class I and class II (Rucktaschel et al. 2011a). Class I corresponds to PMPs that are imported to peroxisomal membranes via a Pex19p-dependent pathway whereas class II PMPs are targeted independently of Pex19p to peroxisome membranes (Jones et al. 2004).

The class I mPTS motif is the canonical motif which is destined for the import of target proteins to the peroxisomal membrane. It is ill defined and, apparently, comprises of few discontinuous subdomains. One of these subdomains is a bunch of basic amino acids such as arginines and lysines within a loop of the protein (i.e., between membrane spans) which face toward the peroxisomal matrix. mPTS motifs have been identified for several PMPs that have a basic amino acid sequence in arrangement with at least one transmembrane region (Dyer et al. 1996; Honsho and Fujiki 2001).

The distinctiveness of the mPTS receptor is somewhat unclear and the best candidate to infer is the product obtained from Pex19p (Neuberger et al. 2003a; Schueller et al. 2010). Targeting peroxisomal membrane proteins (PMPs) is a complex phenomenon as targeting involves a number of processes like recognition of PMPs in the cytosol and their penetration into the peroxisomal membrane. Pex19p, a farnesylated protein associated with the peroxisomal membrane and the prospective candidate protein for the PMP recognition event, interacts with most PMPs. It can function as import receptor and/or chaperone required for PMPs stability at the peroxisomal membrane (Matsuzono et al. 2006). This Pex19p–PMP complex is directed to the peroxisomal membrane by docking to its membrane-anchored binding partner Pex3p (Fang et al. 2004). Finally, with the aid of Pex19p and Pex3p, PMPs are inserted into the peroxisomal membrane (Rucktaschel et al. 2011b; Theodoulou et al. 2013b).

Only a small part of class II PMPs is sorted indirectly to peroxisomes through vesicles from the endoplasmic reticulum. The potentially alternative mPTS of class II is located in the N-terminal region which contains a transmembrane region but not a Pex19p-binding site (Halbach et al. 2009a; Soukupova et al. 1999). Pex3p, Pex16p, and Pex22p belong to the class II PMPs (Fujiki et al. 2006; Halbach et al. 2009b).

The mechanisms accountable for targeting to the peroxisomal membrane and integration into the endoplasmic reticulum are still inadequately understood and further critical studies need to be done in this area (Halbach et al. 2009c; Perry et al. 2009; Tabak et al. 2008b; Titorenko and Rachubinski 1998). The knowledge about the mPTS motifs is still too fragmentary to attempt sequence pattern formalization and its application for a sequence-based prediction.

9.3 Peroxisomal Proteome and the Related Challenges for Theoretical Research

In the following, we review the literature and electronic databases with regard to the size of the peroxisome proteome and to which extent the various import pathways appear to contribute to it. We also consider some proteins and pathways associated with the peroxisome before we turn to the case of peroxisomal proteases (see the following section) with great detail.

The peroxisomal proteome (or the peroxiproteome as coined by Kurochkin et al. 2003) is the key (similar to the proteome of other cellular substructures) for getting the complete list of functions that the peroxisome can contribute to the total capacities of the cell. This would also lead to a better understanding of peroxisome biogenesis, a process under the control of *PEX* gene-encoded peroxins and, possibly, several other proteins that guide peroxisome assembly, division and inheritance (Ma et al. 2011). Finally, we know now that the genome is ubiquitously transcribed resulting in the production of a variety of non-coding RNAs and it is still a mystery whether some of them influence peroxisome function.

Despite many decades of peroxisome research, the list of peroxisomal functions and of interaction (see Fig. 9.3 for a summary of the pathways in yeast based on the information in KEGG; Kanehisa 2013) with other subcellular organelles (Andrade-Navarro et al. 2009) is still growing. The list of recently discovered peroxisomal functions in plants, animals, and fungi include jasmonate synthesis (Kienow et al. 2008; Schneider et al. 2005; Schrader and Fahimi 2008), auxin synthesis (Wiszniewski et al. 2009; Zolman et al. 2000), pheromone production (Joo et al. 2009, 2010; Spiegel et al. 2011), isoprenoid biosynthesis (Clastre et al. 2011; Hogenboom et al. 2004a, b, c; Kovacs et al. 2007; Sapir-Mir et al. 2008; Simkin et al. 2011; Thabet et al. 2011), biotin synthesis (Tanabe et al. 2011), toxin synthesis (Saikia and Scott 2009; Imazaki et al. 2010), polyamine metabolism (Kamada-Nobusada et al. 2008; Nishikawa et al. 2000a, b; Ono et al. 2012; Wu et al. 2003), phyloquinone (vitamin K1) biosynthesis (Babujee et al. 2010), glycine betaine synthesis (Fujiwara et al. 2008; Mitsuya et al. 2009), viral innate immune defense (Dixit et al. 2010), contributions to the GPI-anchor biosynthesis (Kanzawa et al. 2009), degradation of polyunsaturated dicarboxylic long-chain fatty acids (Nguyen et al. 2008), and H_2O_2 signaling in hypothalamic neurons (Diano et al. 2011).

Mitochondria and peroxisomes are closely associated organelles with co-dependent relationship and, hence, are vital for human health and development which can help in understanding their impact on disease states and functions associated with them (Andrade-Navarro et al. 2009; Schrader and Yoon 2007). Evidences state that these organelles share components of their division machinery that led to improved understanding of the dynamics of compartmentalized chemical reactions and of the import of proteins. It also became clear that this is a general strategy which is observed in plants, mammals, and fungi (Delille et al. 2009). Recent peroxisome studies support a broader definition, thus, the peroxisome is a cellular compartment involved in several metabolic, as well as signaling and developmental processes (Hoepfner et al. 2005). Recent studies also support for the ER's involvement in peroxisome biogenesis which somehow contradicts the earlier belief that peroxisomes only replicate autonomously. Also the ER is necessary both for de novo biogenesis of peroxisomes (van der Zand et al. 2010), for normal peroxisome growth and division (Motley and Hettema 2007; Yan et al. 2008).

9.3.1 Assessing the Peroxisomal Proteome

There are good reasons to believe that the currently known list of peroxisomal functions and their involvement in human pathology is still considerably incomplete (Eisenhaber 2012; Kuznetsov et al. 2013) since about half of the human genes is hardly functionally characterized at the molecular and cellular level. It is known or at least predicted that many of them are potentially targeted to the peroxisome (Neuberger et al. 2003b). For example, among the 485 confidently predicted PTS1-targeted proteins in *Arabidopsis thaliana*, there is absolutely no functional description for 57 of them and many others are only faintly understood. To note, many matrix proteins are relatively small, well-studied enzymes; hence, this evaluation provides just a lower bound since the rate of uncharacterized genes among those coding for membrane proteins is considerably larger.

Complete genome sequencing of human and other eukaryotes opens opportunities to assess the complete proteome of subcellular structures including peroxisomes. The straightforward experimental approach is via purification of the peroxisomal fraction with subsequent mass-spectrometric analysis (Bussell et al. 2013). At the computational side, it is possible to derive the full proteome from the fully sequenced genome and to use subcellular localization prediction (Emanuelsson et al. 2003; Neuberger et al. 2003b).

Besides of many other methodical vagaries (such as membrane proteins being a notoriously difficult target for mass-spectrometry), both approaches depend on the completeness of the organism's protein list and it has been shown that these reference proteomes fluctuated spectacularly over the time of the last decade (Sirota et al. 2012). It influences the protein list determination of organelles dramatically. There are also biological reasons for the possible proteome differences derived from cells of differing origin. Peroxisomes are dynamic in their form and, undeniably, in their functions too (Theodoulou et al. 2013a) depending on the environmental and cellular status. For example, malate synthetase 1 is targeted to peroxisomes if yeast is grown on oleic acid medium but not in the case of glucose (Kunze et al. 2002). These organelles are pleomorphic and to a large extent differ among species with respect to their enzyme content and definite metabolic functions (Schluter et al. 2010). Apparently, plants have the highest number of predicted peroxisomal proteins and fungi have the smallest number (Emanuelsson et al. 2003). Further, the peroxisomes are multifunctional organelles that are capable of adapting to various cell types by adding or deleting enzymes involved in a variety of pathways (Sparkes et al. 2005).

Despite the reservations, it can be stated with confidence that the peroxisomes contain many proteins with yet unknown functions, many pathways await their discovery and some of the better studied proteins might provide future surprises with additional functions and roles to be found.

9.3.2 The Proteome of Peroxisomes of Plants and Other Model Organisms: Current Status of Annotation

Considerable effort of the community has been directed to elucidate the proteome of *A. thaliana* as the example of a plant species. Various experimental and computational screening protocols were applied to find genes that encode proteins targeted to peroxisomes via the canonical import pathways (Hayashi et al. 1997; Kamada et al. 2003; Reumann 2004). The *A. thaliana* proteome was especially intensively studied with mass-spectrometry approaches (Eubel et al. 2008; Fukao et al. 2002, 2003; Reumann et al. 2007, 2009). The purification process of peroxisomes is difficult under the conditions of avoiding contamination by other major cellular organelles and, hence, the analytical depth of the investigation of the *A. thaliana* peroxisome has not yet reached that of other organelles (Bussell et al. 2013).

In the past 10 years, there has been considerable expansion seen in the total number of potential peroxisomal proteins. In silico findings indicate that, apparently, putative C-terminal PTS1 sequences of plants may have the highest number in comparison with animal and fungal genomes (Emanuelsson et al. 2003). About 542 proteins have been predicted to contain a PTS1 sequence as reported in the version 3 of the database SUBA (as of November 23rd, 2012; using PredPlantPTS1 search algorithm of Reumann) and around 110 additional proteins were found to be potentially PTS2 targeted and transported to peroxisomes, mPTS or other mechanisms (Kaur and Hu 2011; Reumann 2004). In the AraPerox database, about 280 genes have been implicated to encode proteins containing putative PTS1 and PTS2 peptides in 2004 (Kamada et al. 2003; Reumann 2004), and 440 in 2008 (Reumann et al. 2007, 2009). These lists include PEX proteins, demonstrated membrane proteins and proteins that are imported using non-standard targeting peptides like catalase and sarcosine oxidase (Goyer et al. 2004; Oshima et al. 2008). Except for the few exceptions listed above, all known matrix proteins from plant peroxisomes are thought to carry either a C-terminal PTS1 or an N-terminal PTS2. About 110 entries have the newly demonstrated PTS1 signals such as SSL, SSI, ASL, and AKI.

In Table 9.2, illustrative data for the sizes of the peroxisomal proteome as described in currently available versions of sequence databases are provided. In addition, the size of the proteome as estimated by the prediction tool PeroxiP is listed and we also added the number of targets predicted by PTS1/PeroPS. To note, the latter two are essentially estimates of the number of proteins targeted to peroxisomes via the PTS1 pathway.

As can be seen, we are far from achieving a consensus with regard to the size and exact list of proteins with peroxisomal localization. Apparently, a number of databases, most importantly UniProt (Dimmer et al. 2012), do not reflect the current state of the art in the field and do not report many experimentally known and confidently predicted examples. Whereas PTS1 prediction tools such as PTS1/PeroPS appear to generate quite comprehensive lists useful for experimental

Table 9.2 Status of peroxisomal localization annotation as of July 2013 in databases and with prediction tools

	S. cerevisiae	*Arabidopsis thaliana*	*Mus musculus*	*Homo sapiens*
Total proteome in UniProt	6,629	31,851	42,893	71,401
Peroxisomal proteins in UniProt	58	109	104	95
Proteins in PeroxisomeDB	75	77	105	100
Araperox (2004)	n/a	220 (PTS1) 60 (PTS2)	n/a	n/a
Araperox (2008)	n/a	440	n/a	n/a
SUBA	n/a	133 (experimentally confirmed)	n/a	n/a
SUBA—predictions	n/a	542 (PTS1) 110 (PTS2, etc.)	n/a	n/a
Predicted by *PTS1/PeroPS*	41 (99)	485 (1089)	254 (612)	414 (967)
Predicted by PeroxiP	27/64/77/277	61/198/337/1146	59/198/217/947	44/240/243/1427

For yeast, thale cress, mouse, and human, illustrative data from sequence databases and prediction tools are summarized that characterize the current status of peroxisomal sequence annotation (as of the time of writing—July 2013). First, the total proteome as extracted from UniProt (Dimmer et al. 2012) is provided (after redundancy reduction due to obvious doublets and minor sequence variations). Then, we list the number of peroxisomal proteins annotated with the keyword "peroxisome" in UniProt. Next, the size of the peroxisomal proteome as in PeroxisomeDB is listed (Schluter et al. 2010). For thale cress, also the data from the SUBA (Tanz et al. 2013) and the Araperox (Reumann et al. 2007) databases are given. Finally, the PTS1/PeroPS predictor (Neuberger et al. 2003b) was rerun over the current version of the taxon-specific proteome and we provide the numbers for confidently predicted and, in parentheses, for twilight zone hits. In the last row, the predicted sizes of the peroxiproteome from Table 4 of (Emanuelsson et al. 2003) in accordance with their four different methodologies and filtering criteria are listed

verification studies when the twilight zone hits are also included, it is much less clear what the approximate numbers of PTS2 or mPTS targeted proteins are. The available data support the impression that peroxisomes contain up to a few hundred proteins out of which some may be obligatory components and, possibly, a larger list of others resides in the peroxisome only under certain physiological conditions.

At present, the best sources for the network organization of peroxisomal proteins are the more general KEGG database (Kanehisa 2013) and the specialized Peroxisome Knowledgebase (Willemsen et al. 2008). In most cases, PTS1 targeted proteins are in included into the pathway maps. Not all of the recently discovered peroxisomal functions are presented.

9.3.3 Peroxisomal Disorders in Homo Sapiens and the Role on Import Pathways

The deficiency of peroxisomal function can lead to severe and fatal peroxisomal disorders in humans. Some of the well-known examples are the Zellweger syndrome spectrum (ZSS) disorders and rhizomelic chondrodysplasia punctata (RCDP) type I (Ebberink et al. 2010; Steinberg et al. 2006; Wanders and Waterham 2006). Since many pathways and biomolecular mechanisms in peroxisomes are studied very well, a number of peroxisomal disorders known in humans can be classified under specific peroxisome biogenesis processes and peroxisomal enzyme/protein deficiencies. The list of diseases includes Zellweger syndrome, infantile refsum disease, Acyl-CoA oxidase deficiency, D-bifunctional protein deficiency, etc. Glutaric acidemia type3 (GA3) disorder is a case that is still poorly understood because of the lack of complete understanding about the type of proteins and genes involved.

Several disease mutations affect proteins imported via the PTS1 pathway into the peroxisome. Since the responsible sequence motif is quite well understood (see Sect. 9.2), prediction tools such as PTS1/PeroPS can be used to extract proteins from the human proteome that are peroxisomally localized and independently known to be involved in pathological states. We applied this approach to information contained in the KEGG database (Kanehisa 2013) and summarized these cases in Table 9.3. Some PTS1-dependent imported, pathology related proteins are in the following families:

- DHAPAT family (Glycerophosphate *O*-acteyltranferase, ID: O15228) involved in rhizomelic chondrodysplasia
- CAT family (catalase, ID: P04040, pdb structures known) involved in acatalasia
- AGT (serine-glyoxylate transaminase, ID: P21549, pdb structures known, isoform-peroxisome, and mitochondria except in some HP1 patients where AGT is found in the mitochondrial matrix) in primary hyperoxaluria
- AMACR family (alpha-methylacyl-CoA racemase, ID: Q9UHK6, isoform-peroxisome and mitochondria) in Peroxisomal beta-oxidation enzyme deficiency Congenital bile acid synthesis defect (CBAS)
- SCPX family (Sterol carrier protein 2, ID: P22307, isoform-peroxisome and mitochondria) in Peroxisomal beta-oxidation enzyme deficiency Leukoencephalopathy with dystonia and motor neuropathy
- HSD17B4, DBP family (hydroxysteroid (17-beta) dehydrogenase 4, ID: P51659, pdb structures known) in peroxisomal beta-oxidation enzyme deficiency

Some further families, which are PTS1 targeted and are involved in peroxisomal disorders, are AMACR, ACOX, DBP, SCPX, BAAT, MLYCD, DHAPAT, AGT, HMGCL, IDH, CAT, SOD, INOS, and XDH (the latter two are also known as drug target; all protein family codes as used in KEGG, see Table 9.3 for further information). A few others are known to be imported via PTS2-dependent pathway: PHYH, AGPS, and MVK families are involved.

Table 9.3 Human peroxisomal protein families known to be involved in pathologies

Family	Enzyme	EC no.	Uniprot ID/PDB entry	Subcellular localization	Pathways	Disease
DHAPAT	Glycero-nephosphate *O*-acyltransferase	2.3.1.42	O15228/no	Peroxisome, Membrane; peripheral membrane protein; Matrix side	Glycero-phospholipid metabolism	Rhizomelic Chondrodysplasia Punctata
INOS	Nitric oxide synthase 2, inducible	1.14.13.39	P35228/yes 3E7G 4NOS 3HR4 3EJ8 1NSI 2NSI	Peroxisome	Arginine and proline metabolism, Calcium signaling pathway, HIF-1 signaling pathway	Salmonella infection Pertussis Leishmaniasis Chagas disease Toxoplasmosis Amoebiasis Tuberculosis Pathways in cancer Small cell lung cancer Esophageal cancer (drug targets)
CAT	Catalase	1.11.1.6	P04040/yes 1DGF 1DGG 1DGH 1DGB 1F4J 1QQW	Peroxisome	Tryptophan metabolism, Glyoxylate and dicarboxylate metabolism, Metabolic pathways	Acatalasia Amyotrophic lateral sclerosis (ALS)
SOD	Superoxide dismutase 1, soluble	1.15.1.1	P00441/yes 4A7U and many more	Cytoplasm The pathogenic variant ALS1 Arg-86 and Ala-94 gradually aggregates and accumulates in mitochondria.	Antioxidant system	Amyotrophic lateral sclerosis (ALS) Huntington's disease Prion diseases
XDH	Xanthine dehydrogenase	1.17.1.4 1.17.3.2	P47989/yes 2E1Q 2CKJ	Cytoplasm, peroxisome, secreted	Purine metabolism, Caffeine metabolism, Drug metabolism—other enzymes	Xanthinuria, drug targets

(continued)

Table 9.3 (continued)

Family	Enzyme	EC no.	Uniprot ID/PDB entry	Subcellular localization	Pathways	Disease
AGT	Serine-glyoxylate transaminase	2.6.1.44 2.6.1.51	P21549/yes 3R9A 1J04 1H0C	Peroxisome. mitochondrion matrix. Note: Except in some HP1 patients where AGT is found in the mitochondrial matrix.	Alanine, aspartate and glutamate metabolism, Glycine, serine and threonine metabolism	Primary hyperoxaluria
IDH	Isocitrate dehydrogenase 1	1.1.1.42	P48735/yes 4JA8	Cytoplasm, peroxisome	Citrate cycle (TCA cycle) Glutathione metabolism Metabolic pathways 2-Oxocarboxylic acid metabolism Biosynthesis of amino acids	D-2-hydroxyglutaric aciduria
HMGCL	3-hydroxymethyl-3-methylglutaryl-CoA lyase	4.1.3.4	P35914/yes 2CW6 3MP4 3MP5 3MP3	Mitochondrion Matrix. peroxisome. Note: Unprocessed form is peroxisomal	Synthesis and degradation of ketone bodies, Valine, leucine and isoleucine degradation, Butanoate metabolism	3-Hydroxy-3-methylglutaryl-CoA lyase deficiency
MLYCD	Malonyl-CoA decarboxylase	4.1.1.9	O95822/yes 2YGW 4F0X	Mitochondrion. cytoplasm. peroxisome	Beta-alanine metabolism, Propanoate metabolism	Malonyl-CoA decarboxylase deficiency
AMACR	Alpha-methylacyl-CoA racemase	5.1.99.4	Q9UHK6/ No	Peroxisome mitochondrion	Primary bile acid biosynthesis	Peroxisomal beta-oxidation enzyme deficiency Congenital bile acid synthesis defect (CBAS)

ACOX	Acyl-CoA oxidase 1, palmitoyl	1.3.3.6	Q15067/no	Peroxisome	Fatty acid metabolism, alpha-linolenic acid metabolism, biosynthesis of unsaturated fatty acids, PPAR pathway	Peroxisomal beta-oxidation enzyme deficiency
HSD17B4, DBP,	Hydroxysteroid (17-beta) dehydrogenase 4	4.2.1.107 4.2.1.119 1.1.1.n12	P51659/yes (isoforms) 1IKT 1ZBQ 1S9C	Peroxisome	Primary bile acid biosynthesis	Peroxisomal beta-oxidation enzyme deficiency
SCPX	Sterol carrier protein 2	2.3.1.176	P22307/no isoforms	Mitochondrion. peroxisome	Primary bile acid biosynthesis, PPAR pathway	Peroxisomal beta-oxidation enzyme deficiency Leukoencephalopathy with dystonia and motor neuropathy
BAAT	Bile acid CoA: amino acid *N*-acyltransferase (glycine *N*-choloyltransferase	2.3.1.65 3.1.2.2	Q14032/no	Cytoplasm	Primary bile acid biosynthesis, taurine and hypotaurine metabolism, biosynthesis of unsaturated fatty acids, bile secretion	Peroxisomal beta-oxidation enzyme deficiency

Enzyme EC number, UniProt ids, localization, metabolic pathways and the relevant peroxisomal disorders are listed. UniProt identifiers shown in cyan were identified as carrying a PTS1 motif with prediction tools as applied in our human proteome-wide analysis

9.4 Peroxisomal Proteases

As an example to illustrate how theoretical and experimental approaches interdigitate in the functional characterization of peroxisomal proteins, the case of peroxisomal proteases is reviewed in greater detail. Proteolytic enzymes perform a variety of important functions in cells and proteases present in peroxisomes are not an exception from this rule. Discovered relatively recently, peroxisomal proteases have been quickly recognized as essential components of the organelle providing the maintenance of the peroxisomal homeostasis through proper management of protein quality control, processing of peroxisome-targeting sequence and regulation of various biochemical pathways. Deficiency of individual peroxisomal proteases caused severe phenotypes in mice. Further studies are expected to implicate malfunctioning peroxisomal proteases in various human disorders. Three proteases have been identified in mammalian peroxisomes so far, including insulin-degrading enzyme (IDE), PsLon and Tysnd1. Computational approaches have been instrumental in the discovery of peroxisomal proteases. The regulatory mechanisms, structural and functional aspects of peroxisomal proteases are far from being fully understood and await further investigation.

9.4.1 Insulin Degrading Enzyme (IDE; EC 3.4.24.56)

IDE was the first protease to be identified in peroxisomes. Cloning and subsequent computational analysis of IDE established that it is a highly conserved Zn^{2+}-metalloprotease found in bacteria, fungi, plants, and animals (Duckworth et al. 1998). IDE contains an inverted zinc-metalloprotease core motif (HXXEH) that places it in clan ME of the M16A family of metalloproteases (Rawlings et al. 2012).

The existence of an enzyme responsible for degradation of insulin has been postulated in 1947 when Mirsky and Broh-Kahn described the presence of insulin inactivating activity in various tissue extracts (Mirsky and Broh-Kahn 1949). IDE was considered a cytosolic enzyme until 1993 when Baumeister et al. have sequenced cDNA for rat IDE and found that the deduced amino acid sequence of the enzyme exhibits 95 % identity to the human IDE and 47 % identity to the Drosophila IDE and all of the sequences contain a conserved carboxyl-terminal peroxisomal targeting sequence (PTS1) S/AKL (Baumeister et al. 1993). Localization of IDE in peroxisomes has been confirmed in a number of studies using various approaches including subcellular fractionation experiments (Authier et al. 1994), immunofluorescence and immunocryoelectron microscopy (Kuo et al. 1994). The presence of IDE in peroxisomes seemed to be at odds with the absence of insulin in this organelle and with the fact that cytosol, endosomes and lysosomes are the major sites of insulin degradation (Duckworth et al. 1998). Site-directed mutagenesis to destroy PTS1 in human IDE abolished its import into peroxisomes but did not affect the ability of IDE to degrade insulin by intact cells (Chesneau et al. 1997).

Moreover, IDE is expressed not only in insulin-sensitive cells but almost in all tissues and cell types suggesting a more general role for this enzyme.

As the consequence of their metabolic function, peroxisomes generate large levels of H_2O_2 creating a highly oxidizing environment leading to a continuous oxidative damage to the organelle constituent proteins. Using oxidized lysozyme as a model substrate, Morita et al. demonstrated that it was degraded by peroxisomal extracts much faster than the non-oxidized form of the protein. IDE was identified as the major peroxisomal protease responsible for this effect (Morita et al. 2000). However, it seems unlikely for IDE to cleave large proteins, even if they are extensively damaged, considering that IDE is a peptidase whose catalytic chamber could accommodate peptides that are shorter than 50 amino acid residues (Malito et al. 2008). A possible explanation for the observed effect is that lysozyme in that study was oxidized by a metal-catalyzed oxidation system that produces H_2O_2 and induces protein fragmentation (Wolff and Dean 1986). Oxidation-generated protein fragments rather than oxidized proteins per se seem to be the likely IDE substrates.

IDE, in addition to insulin, was found to degrade other growth factors including insulin-like growth factor (IGF)-II and transforming growth factor-α (TGF-α). However, it does not act or acts very slowly on highly related peptides such as proinsulin, IGF-I, and EGF. Analysis of the cleavage sites on three-dimensional models of insulin and TGF-α suggested that IDE recognizes tertiary structure of the substrates rather than their primary sequence (Duckworth et al. 1998). But exact sequence/structural features recognized by IDE remained an enigma.

It was noticed that IDE substrates share an ability to form, under certain conditions, amyloid fibrils (Kurochkin and Goto 1994). For example, insulin heated in the presence of diluted acid forms amyloid fibrils (Waugh 1946). Insulin also deposits in a form of amyloid fibrils at the sites of repeated injections of the hormone (Dische et al. 1988). Another IDE substrate atrial natriuretic peptide (ANP) is known to be a major component of amyloid deposits in the atria (Kaye et al. 1986). To test the idea that IDE recognizes amyloid-forming peptides, the protease has been incubated with the amyloid β peptide (Aβ), the chief component of amyloid deposits found in Alzheimer's disease brain (Kurochkin and Goto 1994). IDE not only efficiently degraded Aβ but was the only protein that could be cross-linked to the peptide in crude brain and liver extracts (Kaye et al. 1986). Analysis of the sequences of IDE substrates revealed that they contain a common structural motif, HNHHHPSH, where H is wholly or partly hydrophobic character, N is small and neutral, P is polar, and S is polar and/or small amino acid residue (Kurochkin 1998), which was previously suggested to confer on the peptides amyloid-forming properties (Turnell and Finch 1992). This result immediately suggested that IDE would act on other amyloid fibril precursors (Kurochkin 1998). Indeed, the following studies demonstrated that IDE was able to degrade amylin that accumulates as amyloid deposits in the pancreatic islets of Type 2 diabetes (Bennett et al. 2000). The ABri and ADan amyloid peptides are associated with British and Danish familial dementia (Morelli et al. 2005). Amyloid-forming peptides do not share any sequence or structural homology in their monomeric state but assume highly similar β-pleated sheet conformation when in amyloid fibrils.

Therefore, it has been hypothesized that IDE forms a surface platform on which the protease tests the peptide's ability to adopt a β-sheet structure before making a decision on whether to digest it (Kurochkin 2001). The comparison of crystal structures of IDE-free and IDE-bound substrates confirmed this prediction. Upon binding to IDE, insulin B chain, Aβ, amylin, and glucagon were found to undergo substantial conformational changes to form β-strands (Shen et al. 2006). Recently, a large number of proteins and short peptides have been shown to be converted from their soluble states into amyloid-like fibrils suggesting that amyloid formation is a generic property of polypeptides (Chiti and Dobson 2006). The fact that amyloid aggregates are not frequently found in biological systems is an indication that a variety of mechanisms exist to control these aggregation-prone species.

Consistent with probable house-keeping role of IDE as a protein quality control mechanism, the protease was detected in multiple subcellular locations and in all tissues tested so far. In addition to peroxisomes and cytosol, IDE was found in mitochondria (Leissring et al. 2004). Mitochondrial isoform of IDE is generated by an alternative translation start that adds an N-terminal segment containing mitochondrial targeting sequence (Leissring et al. 2004). A fraction of the enzyme is found at the plasma membrane (Seta and Roth 1997), although its sequence has no recognizable membrane-spanning regions. A small but detectable IDE is present in nuclear fraction (Akiyama et al. 1988; Authier et al. 1994) consistent with the presence of nuclear localization sequence in the sequence of the enzyme (Glebov et al. 2011). Despite the lack of a canonical signal peptide sequence, IDE is also found in the extracellular fluids including blood and cerebrospinal fluid. It is also detected in conditioned media of cultured cells suggesting involvement of non-conventional transport pathways. Recently, IDE was detected in exosomes (Bulloj et al. 2010), small (30–100 nm) membrane vesicles of endocytic origin secreted by most cell types. Using bioinformatics approach, Glebov et al. identified a novel amino acid motif EKPPHY close to the C-terminus of IDE that is identical with the motif within the C terminus of the bacterial SlyX protein (InterPro Database number, IPR007236) and demonstrated that this motif sequence plays a crucial role in IDE secretion (Glebov et al. 2011). Thus IDE represents an example of a single protein targeted to multiple subcellular locations and performing diverse functions. Cells utilize various strategies to achieve redistribution of the enzyme between their different subcompartments. The ratio of cytoplasmic to peroxisomal IDE, for example, is regulated by the levels of protein p70, the cytosolic protein physically associated with IDE (Authier et al. 1996). The expression of long IDE isoform that is targeted to mitochondria is regulated by mitochondrial biogenesis pathway (PGC-1α/NRF-1) (Leal et al. 2013). IDE sorting to exosomes is stimulated by cholesterol lowering drugs statins (Tamboli et al. 2010).

Consistent with the in vitro activity, loss-of-function mutations of the IDE gene in mice resulted in elevated brain Aβ and plasma insulin levels (Farris et al. 2003). Transgenic overexpression of IDE in neurons effectively reduced Aβ accumulation and rescued premature lethality present in Alzheimer's disease mouse model

(Leissring et al. 2003). A full spectrum of IDE substrates is yet unknown but is very likely to include a large number of amyloid-forming peptides. The deposition of peptides as amyloid fibrils is associated with a number of medical disorders that affect various organs and tissues. Enhancement of IDE activity could provide a useful therapeutic approach for prevention and treatment of these diseases. IDE exhibits allosteric kinetic behavior, with small peptide substrates increasing the activity of the enzyme toward the same or other small peptides including Aβ (Song et al. 2003). In addition, alkylation of a single cysteine residue in IDE was found to significantly activate hydrolysis of Aβ but not of other substrates (Neant-Fery et al. 2008). These finding suggest that synthetic small molecules enhancing IDE activity in a substrate-specific manner can be eventually developed.

9.4.2 Tysnd1 (EC 3.4.21.-)

Computational approaches have been of crucial importance in the discovery of peroxisomal processing protease. The existence of a protease responsible for the removal of the N-terminal part of PTS2 targeted proteins has been known since 1982 when Furuta et al. have demonstrated that 3-ketoacyl-CoA thiolase (Acaa1) was translated in a form that is 3 kDa larger than the mature enzyme (Furuta et al. 1982). Since 1980 peroxisomes also have been known to possess catalytic activity responsible for processing of PTS1-targeted acyl-CoA oxidase (Acox1) from a full-length form of 75.5 kDa (form A) to 50.1 kDa (form B) and 19 kDa (form C) forms (Osumi and Hashimoto 1980). Biochemical methods to purify protease(s) responsible for these activities have been unsuccessful probably because of instability of the protease(s) when removed from its natural environment or presence of a co-purified inhibitor. In 2002, the RIKEN Mouse Gene Encyclopedia Project released the sequence of 60,770 mouse cDNA clones (Okazaki et al. 2002). This data set provided the most comprehensive compilation of the mammalian full-length cDNAs at the time. In the search for novel peroxisomal candidate proteins, a conceptually translated protein database derived from the FANTOM2 dataset was scanned for PTS1 (Kurochkin et al. 2003). Among several novel peroxisomal candidate proteins, this search revealed a clone 1300019 N10 encoding a protein of 568 amino acids containing two protease-related domains, glutamyl endopeptidase I (IPR008256), trypsin-like serine and cysteine proteases (IPR009003). The amino acid sequence of a clone 1300019 N10 is weakly similar to a protease-related protein F3H9.3 from *A. thaliana*, which also contains SKL at its C-terminus. This putative novel protease was later designated Tysnd1 for trypsin domain containing 1. Tysnd1 was a good candidate for a peroxisomal processing protease and subsequently has been experimentally characterized. The peroxisomal localization of Tysnd1 was confirmed by subcellular fractionation and confocal microscopy (Kurochkin et al. 2007). Co-transfection of cells with the expression constructs for Tysnd1 and PTS2 protein pre-thiolase (Acaa1) led to a conversion of the 44-kDa Acaa1 precursor to the mature 41-kDa form. Unexpectedly, Tysnd1 was also found to be responsible for the processing of three PTS1 proteins Acox1, Scp2 and Hsd17b4 (Kurochkin et al. 2007). Tysnd1 acted on its substrates not only when transfected to cells but also in tube experiments when recombinant Tysnd1 cleaved the recombinant substrates to fragments found in vivo proving that it is a bona fide peroxisomal processing protease (Kurochkin et al. 2007).

What is the significance of Tysnd1-mediated removal of the PTS2 pre-sequence? Nair et al. demonstrated that the PTS2 receptor Pex7p follows the "extended shuttle" model of peroxisome import receptor function whereby Pex7p goes all the way inside peroxisomes and then is reexported back to cytosol (Nair et al. 2004). It can be imagined then that retention of the PTS2 presequence would cause translocation of PTS2 proteins in and out of peroxisomes in a complex with Pex7p leading to mislocalization of PTS proteins to cytosol. Investigation of localization of PTS2 proteins in Tysnd1 knock-out mice has nicely confirmed this hypothesis. Tysnd1 deficiency interfered with peroxisomal localization of PTS2 proteins Acaa1, Phyh, and Agps (Mizuno et al. 2013). As a consequence, Tysnd1 knock-out mice had decreased plasmalogen and increased phytanic acid levels and had a phenotype somewhat resembling RCDP type 1 disease (RCDP1) (Mizuno et al. 2013) which is caused by nonsense mutations in PST2 receptor PEX7. The changes in the molecular species composition of choline and ethanolamine plasmalogens could be responsible for the fragility of the sperm acrosomal membrane and are likely cause of male infertility in Tynsd1−/− mice (Mizuno et al. 2013).

Interestingly, Tysnd1 functional homologue is absent in *Saccharomyces cerevisiae* and the yeast's PTS2-presequence is not cleaved off. However, this has no adverse effect on the yeast's peroxisomal function. One important difference between yeast and mammalian peroxisomes is the pH in the organelles. While yeast peroxisomal matrix is acidic (Nicolay et al. 1987) or alkaline when yeast are grown on oleate (van Roermund et al. 2004), the mammalian peroxisome pH (6.9–7.1) resembles that of cytosolic compartment (Jankowski et al. 2001). We can imagine that Pex7p forms with a PTS2 protein an acid/alkaline-sensitive complex. In the yeast peroxisomal matrix, PTS2 protein then would be released from the bound Pex7p, while neutral pH conditions of mammalian peroxisomes would require proteolytic removal of the PTS2 sequence to achieve this effect.

The physiological significance of the intra-peroxisomal processing of PTS1 proteins is more difficult to comprehend since proteolytic processing of Acox1, Scp2, and Hsd17b4 does not affect their catalytic properties or the extent of their accumulation in peroxisomes. Noteworthy, all these three Tysnd1 substrates are components of the β-oxidation pathway of fatty acids and their processed forms are known to assemble into a large multienzyme complex. It was proposed that the protein fragments produced through Tysnd1-mediated site-specific proteolysis would acquire enhanced ability to form the complex (Kurochkin et al. 2007). In this scenario, Tysnd1 would function as a positive regulator of peroxisomal β-oxidation of fatty acids because the complex would allow a more efficient transfer of lipid intermediates between the enzymes involved in the same pathway. Indeed, Tysnd1 downregulation in a cell-based assay (Okumoto et al. 2011) and in the knock-out mice model (Mizuno et al. 2013) led to significantly lower activity of the fatty acid β-oxidation pathway.

The exact substrate recognition mechanism of Tysnd1 is still unclear. Sequences surrounding the cleavage sites in the PTS2-targeted protein Acaa1 (AAPC*SAGF), the PTS1-targeted proteins Hsd17b4 (AAPA*ATSG) and Scp2 (AAPT*SSAG) share the Ala–Ala–Pro motif. However, PTS2 protein Agps (TNEC*KARR) and PTS1 protein Acox1 (PQQV*AVWP) are processed at distinct sites. It is therefore possible that Tysnd1 recognizes not only a primary sequence but also some elements of tertiary structure in its substrate. Specificity of Tysnd1 might be regulated by its oligomerization status. Tysnd1 ortholog in higher plants Deg15, for example, functions as PTS2 presequence processing protease in the dimeric form and as a general degrading protease in the monomeric form (Helm et al. 2007). Human Tysnd1 was shown to exist as a dimer, trimer and a higher mass complex (Okumoto et al. 2011). Self-cleavage of Tysnd1 may also potentially produce a form of the enzyme with distinct substrate specificity (Kurochkin et al. 2007; Okumoto et al. 2011). Interestingly, the processing efficiency of Acox1 and Scp2 by recombinant Tysnd1 is very inefficient as compared with that of Acaa1 (Kurochkin et al. 2007). It is possible that auxiliary protein factors or metabolites present in peroxisomes could regulate specificity and protease activity of Tysnd1. Regarding the catalytic mechanism of Tysnd1, it is likely a serine protease. Both mammalian and plant forms of the enzyme contain the catalytic triad His–Asp–Ser (His-372, Asp-408, and Ser-481 in human Tysnd1 and His-392, Asp-491, and Ser-580 in Arabidopsis Deg15). Mutational analysis of Tysnd1 revealed that conserved Ser-481 is absolutely necessary for the proteolytic activity of the enzyme in vitro (Okumoto et al. 2011). Structural analysis of substrate-free and substrate-bound forms of Tysnd1 will be essential to establish mechanisms governing the substrate specificity of the enzyme.

9.4.3 Lon Peptidase 2, Peroxisomal (Lonp2; EC 3.4.21.53)

Peroxisomal Lon (pLon) protease is the most recently identified member of the highly conserved family of ATP-stimulated serine proteases, which belong to the superfamily of AAA-ATPases. Lon proteases are found in prokaryotic as well as in eukaryotic cells where they are involved in elimination of misfolded, damaged and short-lived proteins (Gur and Sauer 2008; Venkatesh et al. 2012).The first member of the Lon protease family discovered was *E. coli* protease La. It was found to be responsible for ~ 50 % of the turnover of abnormal proteins arising from premature translational termination (Kowit and Goldberg 1977) and is thought to eliminate misfolded proteins produced under diverse stress conditions.

Mitochondrial Lon proteases are found in a variety of eukaryotes. They function both as ATP-dependent proteases, degrading non-assembled and damaged proteins, and as chaperones, assisting protein complex assemblies (Lee and Suzuki 2008). Peroxisomal forms of Lon protease have been identified using proteomics and computational approaches. The first pLon protease has been identified by mass-spectrometry analysis of proteins present in highly purified rat liver peroxisomes (Kikuchi et al. 2004). Rat pLon protease is 852 amino acids in length and contains three domains characteristic of the Lon family of proteases: an N-terminal domain (LonNdomain), an AAA (ATPases associated with diverse cellular activities) module and a protease domain (P domain with conserved catalytic dyad). Human pLon shares 95 % identity with the rat and 96 % identity with the mouse orthologs, respectively. Human, rat and mouse pLon proteases possess the classical PTS1 signal SKL at their C-termini. Immunoelectron microscopy demonstrated that the enzyme is indeed localized to the matrix of peroxisome (Kikuchi et al. 2004).

By analogy to the mitochondrial Lon protease, the peroxisomal form of Lon protease was proposed to be involved in the protein quality control through degradation of misfolded proteins or acting as chaperone in the folding process of peroxisomal matrix proteins. The first confirmation that this might be the case came from the yeast studies. A search in the genome database of the methylotrophic yeast Hansenula polymorpha revealed that this organism contains two putative Lon proteases, one of them harboring a degenerate PTS1 signal ARI (Aksam et al. 2007). The putative pLon protease shows 39 % sequence identity with the mouse pLon protease (Aksam et al. 2007). Its subcellular localization to peroxisomes has been confirmed experimentally (Aksam et al. 2007).

To examine the role of the pLon protease in the degradation of misfolded proteins the authors analyzed the fate of a peroxisome-targeted mutant form of dihydrofolate reductase (DHFR) that contains amino acid substitutions destabilizing the protein structure. In a strain with the deleted pLon gene the mutant DHFR accumulated, while in the wild-type strain it was degraded supporting the role of pLon protease in elimination of misfolded proteins (Aksam et al. 2007). Electron microscopy revealed that peroxisomes in the pLon-deleted strain contained electron dense aggregates (Aksam et al. 2007) resembling those found in *S. cerevisiae* mutants defective in the mitochondrial Lon protease and representing accumulated protein aggregates (Wagner et al. 1994). The study also found that matrix proteins from the wild-type cells, but not those from the pLon protease lacking cells, efficiently degraded in vitro synthesized non-assembled alcohol oxidase suggesting that the pLon protease is involved in clearance of non-assembled peroxisomal matrix proteins (Aksam et al. 2007). Deletion of a gene for pLon protease resulted in enhanced oxidative stress and decreased cell viability (Aksam et al. 2007). Mitochondrial form of Lon protease selectively recognizes and degrades mildly oxidized forms of proteins (Bota and Davies 2002).

To test whether pLon protease is able to selectively degrade oxidatively damaged proteins, the fungal pLon protease was incubated with the native and H_2O_2-pretreated catalase-peroxidase. Only oxidatively modified form of catalase-peroxidase was digested by the pLon protease (Bartoszewska et al. 2012). In vitro assays revealed that pLon protease displayed proteolytic activity toward alpha-casein and beta-casein, both natively disordered proteins (Bartoszewska et al. 2012). This activity was not detected in the absence of ATP and was fully abolished by the single amino acid substitution in the conserved catalytic dyad (Bartoszewska et al. 2012). The significant homology with its mitochondrial counterpart suggested that pLon protease might also possess chaperone activity. Purified pLon protease was found to significantly decrease aggregation of guanidine hydrochloride-denatured citrate synthase diluted into a buffer containing pLon protease (Bartoszewska et al. 2012). The proteolytically inactive mutant of pLon protease facilitated refolding of citrate synthase with the same efficiency as the wild-type active protein suggesting that the protease chaperone-like activity is independent of the proteolytic catalytic dyad (Bartoszewska et al. 2012). Thus pLon protease might perform housekeeping function in peroxisomal matrix by its involvement in protein quality control due to its ability to degrade or refold oxidized and otherwise damaged proteins. In addition to this function, mammalian pLon protease was shown to bind and degrade self-processed forms of Tysnd1 (Okumoto et al. 2011). Because Tysnd1 plays a key role in the regulation of peroxisomal beta-oxidation activity (Mizuno et al. 2013; Okumoto et al. 2011), pLon protease was proposed to modulate this process (Okumoto et al. 2011).

Mammalian (Omi et al. 2008) and *A. thaliana* (Lingard and Bartel 2009) pLon proteases were also shown to facilitate sorting of matrix proteins into peroxisomes. In the fungi, however, pLon protease is not involved in sustaining matrix protein import (Aksam et al. 2007; Bartoszewska et al. 2012) suggesting that pLon proteases might display species-specific functionalities. In this connection, Lingard and Bartel noticed that plant pLon proteases are more similar to chordate pLon proteases than either are to their yeast relatives (Lingard and Bartel 2009). The phylogenetic analysis revealed that a subset of yeasts have acquired a pLon isoform in an evolutionary event distinct from the event in which the plants and animals pLon isoforms were evolved (Lingard and Bartel 2009). The authors proposed that this might be the reason for the observed functional differences between these lineages (Lingard and Bartel 2009).

Although pLon proteases selectively degrade several misfolded and oxidized proteins in vitro, the full spectrum of their physiological substrates is still unknown. The molecular basis of pLon protease substrate recognition has also not been defined. Lon proteases from bacteria and mitochondria in eukaryotes have been studied more extensively. It was established that they recognize in their substrates clusters of hydrophobic residues enriched in aromatic residues that are accessible in unfolded polypeptides but hidden in most native structures (Gur and Sauer 2008; Venkatesh et al. 2012). The first attempts have been made to establish macromolecular structure of pLon protease. The biophysical data suggest that the fungal pLon protease forms a heptamer (Bartoszewska et al. 2012), an oligomeric state reported previously for mitochondrial Lon from *S. cerevisiae* (Stahlberg et al. 1999).

Despite their potential versatile activities and important roles in supporting peroxisomal functions pLon proteases are not found in several organisms. The analysis of available databases revealed that pLon proteases are very common in yeast and filamentous fungi, yet they are not found in the yeast species *S. cerevisiae* and *Candida glabrata*, for which the mitochondrial isoforms of Lon protease were identified (Aksam et al. 2007). Surprisingly, other eukaryotes, including *Drosophila melanogaster*, also appear to lack peroxisomal Lon protease isoform (Lingard and Bartel 2009) suggesting a possibility that pLon protease functions might be compensated by other cellular components.

9.5 About Database and Prediction Tool Resources for Peroxisome Research

9.5.1 Peroxisome Databases

The peroxisome database (PeroxisomeDB) (http://www.peroxisomedb.org/) offers a comprehensive glance of different peroxisomal genes, their encoded proteins, metabolic pathways, disorders associated and evolutionary relationships in peroxisomal proteins. The new release PeroxisomeDB 2.0 has added peroxisomal proteins from plants, fungi, yeasts, and other lower eukaryotes. This database contains 2,819 proteins in 38 organisms (Schluter et al. 2010). It gives detailed information about global peroxisome metabolome from different organisms, about PTS1- and PTS2-directed enzymes and their corresponding metabolic pathways.

The peroxisome Knowledge Base (PxKB) provides information about ~60 key peroxisomal pathways specific relevant for human biology (http://www.peroxisomekb.nl/). It provides an overview of peroxisomal functions and disorders including fatty acid oxidation (Willemsen et al. 2008). PxKB integrates information about human peroxisomes by developing so-called concept maps (~160 concept maps in version 1.0) that were curated by leading experts in the field.

AraPerox Database (available at http://www.araperox.uni-goettingen.de/ and also at www3.uis.no/AraPeroxV1) describes Arabidopsis peroxisomal proteins (Reumann 2004). It contains also a huge number of unknown predicted Arabidopsis proteins which carry a potential plant peroxisome targeting signal type 1 (PTS1) or type 2 (PTS2).

The SUBA database (SUBcellular location database for Arabidopsis proteins, http://www.suba.bcs.uwa.edu.au) is a database for finding subcellular location of Arabidopsis proteins (SUBA3, http://suba.plantenergy.uwa.edu.au). It is a repository of manual literature curation of large-scale subcellular proteomics, fluorescent protein visualization, and protein–protein interaction (PPI) datasets with subcellular targeting calls from 22 prediction programs. The localisation data in SUBA encompasses 13 distinct subcellular locations, 6,743 non- redundant proteins and represents the proteins encoded in the transcripts responsible for 51 % of Arabidopsis expressed sequence tags.

Further Databases with Relevance for Peroxisome Research are:

- dbPEX, PEX Gene Database, a catalog of gene variations in peroxisome assembly diseases (http://www.dbpex.org/home.php)
- Excellent Electron Microscopic Images of Peroxisomes (from the "Electron Microscopic Atlas" by Dr. med. H. Jastrow) at http://www.uni-mainz.de/FB/Medizin/Anatomie/workshop/EM/EMPeroxisomE.html
- The illuminated Plant Cell—Peroxisome (*A. thaliana*) at http://www.illuminatedcell.com/Peroxisome.html

- The Arabidopsis Information Resource—TAIR (Lamesch et al. 2012) at http://www.peroxisome.msu.edu/database.html

9.5.2 Prediction Tools for Peroxisomal Targeting

There is still much more to making a good prediction tool for a small pattern in a generally non-globular segment beyond collecting a learning set of sequences (maybe, consisting of verified positive and negative examples) and running some machine learning algorithm over it (Eisenhaber et al. 2003a, b, 2004; Eisenhaber and Eisenhaber 2007, 2010). In the following lists, the attempt was more on getting a more complete picture of available prediction tools when, at the same time, many have insufficient behavior with regard to recognizing known peroxisomally targeted or non-targeted proteins or with regard to finding new examples among those sequences that were not experimentally studied when the prediction tool was developed. Thus, the reader is advised to take precautions and to consider comments provided in Sect. 9.2 of this chapter and other relevant scientific literature (Eisenhaber and Eisenhaber 2010).

The PTS1/PeroPS predictor finds proteins with a C-terminus and predicts for peroxisomal import via the PTS1 pathway (with a metazoan-, fungi-specialized and with a taxon-independent predictor; http://mendel.imp.univie.ac.at/myristate/) (Eisenhaber et al. 2003a, b; Neuberger et al. 2003b). Reliably predicted targets should have a non-negative total score; queries with a score larger than –10 are considered as twilight zone hits. In all other cases, the protein is predicted not to have a PTS1 signal. The sensitivity of this predictor in recognizing documented PTS1 motif containing proteins was found to be close to 90 % for reliable prediction. The predictor distinguishes even SKL-appended non-peroxisomal proteins (Neuberger et al. 2003a).

The tool PredPlantPTS1 is the first ever web server specifically developed for the prediction of plant PTS1 proteins and it is freely accessible at ppp.gobics.de. This web server gives information about the peroxisomal targeting probability of the given sequence, information whether a particular non-canonical PTS1 tripeptide has already been experimentally verified and the prediction scores for the single C-terminal 14 amino acid residues (Reumann et al. 2012). It also provides information about predicted residues that inhibit peroxisome targeting and about optimization by site-directed mutagenesis to raise the peroxisome targeting efficiency. This server identifies low-abundance and stress-inducible peroxisomal proteins and informs about the entire peroxisomal proteome of Arabidopsis and other agronomically important crop plants.

PeroxiP (Emanuelsson et al. 2003) generates in silico prediction of the peroxisomal proteome in fungi, plants and animals (http://www.bioinfo.se/PeroxiP/). Machine-learning techniques have been utilized for recognizing peroxisomal targeting signals (PTS1) with PFAM domain-based cross-species comparisons of 8 eukaryotic genomes. The predicted proteins were organized into 29 families encompassing most of the known steps in peroxisomal pathways known at that time.

WoLF PSORT (http://wolfpsort.org/) predicts the subcellular localization sites of proteins based on their amino acid sequences. This predictor uses simple k-nearest neighbor classifier for prediction after the protein amino acid sequences have been converted into numerical localization features which are based on amino acid composition, sorting signals, and functional motifs (Horton et al. 2007). It provides information about list of proteins of known localization with the most similar localization features to the query and individual localization features. Uniprot and Gene Ontology links are provided for the alignment of query sequence to find similar proteins.

Further tools for subcellular localization with relevance for predicting peroxisomal localization are:

- AdaBoost at http://chemdata.shu.edu.cn/subcell/ (Niu et al. 2008)
- ATP (ambiguous targeting predictor) for the prediction of dual protein targeting to plant organelles (Mitschke et al. 2009)
- EpiLoc (http://epiloc.cs.queensu.ca) a text-based system for predicting protein subcellular location (Brady and Shatkay 2008)
- Euk-mPLoc 2.0—A New Method for Predicting the Subcellular Localization of Eukaryotic Proteins with Both Single and Multiple Sites (Chou and Shen 2010a)
- iPSORT (http://hypothesiscreator.net/iPSORT/) – an extensive feature detection of N-terminal protein sorting signals. (Bannai et al. 2002)
- MultiLoc2 (http://www-bs.informatik.uni-tuebingen.de/Services/MultiLoc2) – integrating phylogeny and Gene Ontology terms for subcellular protein localization prediction (Blum et al. 2009).
- Plant-mPLoc (http://www.csbio.sjtu.edu.cn/bioinf/plant-multi/) – (Chou and Shen 2010b)
- PProwler 1.2: detecting and sorting targeting peptides with neural networks and support vector machines. (Hawkins and Boden 2006)
- Predotar: A tool for rapidly screening proteomes for N-terminal targeting sequences (Small et al. 2004)
- PredSL (http://bioinformatics.biol.uoa.gr/PredSL/) N-terminal sequence-based prediction of protein subcellular localization (Petsalaki et al. 2006)
- SLPFA (http://sunflower.kuicr.kyoto-u.ac.jp/~tamura/slpfa.html) subcellular location prediction of proteins using support vector machines with alignment of block sequences utilizing amino acid composition (Tamura and Akutsu 2007)
- SLP-Local: representation of protein sequences for prediction of subcellular location using support vector machines (Matsuda et al. 2005)
- SubLoc: support vector machine-based method for subcellular localization of human proteins using amino acid compositions, their order, and similarity search (www.imtech.res.in/raghava/hslpred/; bioinformatics.uams.edu/raghava/hslpred/) (Garg et al. 2005)
- TargetP (http://www.cbs.dtu.dk/services/TargetP/) (Emanuelsson et al. 2007)
- Yloc—Interpretable Subcellular Localization Prediction at http://abi.inf.uni-tuebingen.de/Services/YLoc/webloc.cgi (Briesemeister et al. 2010)

References

Akiyama H, Shii K, Yokono K, Yonezawa K, Sato S, Watanabe K, Baba S (1988) Cellular localization of insulin-degrading enzyme in rat liver using monoclonal antibodies specific for this enzyme. Biochem Biophys Res Commun 155(2):914–922

Aksam EB, Koek A, Kiel JA, Jourdan S, Veenhuis M, van dKI (2007) A peroxisomal lon protease and peroxisome degradation by autophagy play key roles in vitality of Hansenula polymorpha cells. Autophagy 3(2):96–105

Andrade-Navarro MA, Sanchez-Pulido L, McBride HM (2009) Mitochondrial vesicles: an ancient process providing new links to peroxisomes. Curr Opin Cell Biol 21(4):560–567

Authier F, Rachubinski RA, Posner BI, Bergeron JJ (1994) Endosomal proteolysis of insulin by an acidic thiol metalloprotease unrelated to insulin degrading enzyme. J Biol Chem 269(4):3010–3016

Authier F, Cameron PH, Taupin V (1996) Association of insulin-degrading enzyme with a 70 kDa cytosolic protein in hepatoma cells. Biochem J 319(Pt 1):149–158

Babujee L, Wurtz V, Ma C, Lueder F, Soni P, van DA, Reumann S (2010) The proteome map of spinach leaf peroxisomes indicates partial compartmentalization of phylloquinone (vitamin K1) biosynthesis in plant peroxisomes. J Exp Bot 61(5):1441–1453

Baker A, Sparkes IA (2005) Peroxisome protein import: some answers, more questions. Curr Opin Plant Biol 8(6):640–647

Bannai H, Tamada Y, Maruyama O, Nakai K, Miyano S (2002) Extensive feature detection of N-terminal protein sorting signals. Bioinformatics 18(2):298–305

Bartoszewska M, Williams C, Kikhney A, Opalinski L, van Roermund CW, de Boer R, Veenhuis M, van der Klei IJ (2012) Peroxisomal proteostasis involves a Lon family protein that functions as protease and chaperone. J Biol Chem 287(33):27380–27395

Baumeister H, Muller D, Rehbein M, Richter D (1993) The rat insulin-degrading enzyme. Molecular cloning and characterization of tissue-specific transcripts. FEBS Lett 317(3):250–254

Bennett RG, Duckworth WC, Hamel FG (2000) Degradation of amylin by insulin-degrading enzyme. J Biol Chem 275(47):36621–36625

Blum T, Briesemeister S, Kohlbacher O (2009) MultiLoc2: integrating phylogeny and gene ontology terms improves subcellular protein localization prediction. BMC Bioinformatics 10:274

Boden M, Hawkins J (2005) Prediction of subcellular localization using sequence-biased recurrent networks. Bioinformatics 21(10):2279–2286

Bonekamp NA, Volkl A, Fahimi HD, Schrader M (2009) Reactive oxygen species and peroxisomes: struggling for balance. Biofactors 35(4):346–355

Bork P, Koonin EV (1998) Predicting functions from protein sequences–where are the bottlenecks? Nat Genet 18(4):313–318

Bork P, Dandekar T, Diaz-Lazcoz Y, Eisenhaber F, Huynen M, Yuan Y (1998) Predicting function: from genes to genomes and back. J Mol Biol 283(4):707–725

Bota DA, Davies KJ (2002) Lon protease preferentially degrades oxidized mitochondrial aconitase by an ATP-stimulated mechanism. Nat Cell Biol 4(9):674–680

Brady S, Shatkay H (2008) EpiLoc: a (working) text-based system for predicting protein subcellular location. Pac Symp Biocomput 604–615

Braverman N, Steel G, Obie C, Moser A, Moser H, Gould SJ, Valle D (1997) Human PEX7 encodes the peroxisomal PTS2 receptor and is responsible for rhizomelic chondrodysplasia punctata. Nat Genet 15(4):369–376

Briesemeister S, Rahnenfuhrer J, Kohlbacher O (2010) YLoc–an interpretable web server for predicting subcellular localization. Nucleic Acids Res 38(Web Server Issue):W497–W502

Bulloj A, Leal MC, Xu H, Castano EM, Morelli L (2010) Insulin-degrading enzyme sorting in exosomes: a secretory pathway for a key brain amyloid-beta degrading protease. J Alzheimers Dis 19(1):79–95

Bussell JD, Behrens C, Ecke W, Eubel H (2013) Arabidopsis peroxisome proteomics". Front Plant Sci 4:101
Chang CC, Warren DS, Sacksteder KA, Gould SJ (1999) PEX12 interacts with PEX5 and PEX10 and acts downstream of receptor docking in peroxisomal matrix protein import. J Cell Biol 147 (4):761–774
Chesneau V, Perlman RK, Li W, Keller GA, Rosner MR (1997) Insulin-degrading enzyme does not require peroxisomal localization for insulin degradation. Endocrinology 138(8):3444–3451
Chiti F, Dobson CM (2006) Protein misfolding functional amyloid, and human disease. Annu Rev Biochem 75:333–366
Chou KC, Shen HB (2010a) A new method for predicting the subcellular localization of eukaryotic proteins with both single and multiple sites: Euk-mPLoc 2.0. PLoS One 5(4):e9931
Chou KC, Shen HB (2010b) Plant-mPLoc: a top-down strategy to augment the power for predicting plant protein subcellular localization. PLoSOne 5(6):e11335
Clastre M, Papon N, Courdavault V, Giglioli-Guivarc'h N, St-Pierre B, Simkin AJ (2011) Subcellular evidence for the involvement of peroxisomes in plant isoprenoid biosynthesis. Plant Signal Behav 6(12):2044–2046
Danpure CJ, Jennings PR, Leiper JM, Lumb MJ, Oatey PB (1996) Targeting of alanine: glyoxylate aminotransferase in normal individuals and its mistargeting in patients with primary hyperoxaluria type 1. Ann N Y Acad Sci 804:477–490
De DC (1969) The peroxisome: a new cytoplasmic organelle. Proc R Soc Lond B Biol Sci 173 (30):71–83
Delille HK, Alves R, Schrader M (2009) Biogenesis of peroxisomes and mitochondria: linked by division. Histochem Cell Biol 131(4):441–446
Diano S, Liu ZW, Jeong JK, Dietrich MO, Ruan HB, Kim E, Suyama S, Kelly K, Gyengesi E, Arbiser JL, Belsham DD, Sarruf DA, Schwartz MW, Bennett AM, Shanabrough M, Mobbs CV, Yang X, Gao XB, Horvath TL (2011) Peroxisome proliferation-associated control of reactive oxygen species sets melanocortin tone and feeding in diet-induced obesity. Nat Med 17(9):1121–1127
Dimmer EC, Huntley RP, Faruque Y, Sawford T, O'Donovan C, Martin MJ, Bely B, Browne P, Mun CW, Eberhardt R, Gardner M, Laiho K, Legge D, Magrane M, Pichler K, Poggioli D, Sehra H, Auchincloss A, Axelsen K, Blatter MC, Boutet E, Braconi-Quintaje S, Breuza L, Bridge A, Coudert E, Estreicher A, Famiglietti L, Ferro-Rojas S, Feuermann M, Gos A, Gruaz-Gumowski N, Hinz U, Hulo C, James J, Jimenez S, Jungo F, Keller G, Lemercier P, Lieberherr D, Masson P, Moinat M, Pedruzzi I, Poux S, Rivoire C, Roechert B, Schneider M, Stutz A, Sundaram S, Tognolli M, Bougueleret L, Rgoud-Puy G, Cusin I, Duek-Roggli P, Xenarios I, Apweiler R (2012) The UniProt-GO annotation database in 2011. Nucleic Acids Res 40(Database issue):D565–D570
Dische FE, Wernstedt C, Westermark GT, Westermark P, Pepys MB, Rennie JA, Gilbey SG, Watkins PJ (1988) Insulin as an amyloid-fibril protein at sites of repeated insulin injections in a diabetic patient. Diabetologia 31(3):158–161
Distel B, Veenhuis M, Tabak HF (1987) Import of alcohol oxidase into peroxisomes of Saccharomyces cerevisiae. EMBO J 6(10):3111–3116
Dixit E, Boulant S, Zhang Y, Lee AS, Odendall C, Shum B, Hacohen N, Chen ZJ, Whelan SP, Fransen M, Nibert ML, Superti-Furga G, Kagan JC (2010) Peroxisomes are signaling platforms for antiviral innate immunity. Cell 141(4):668–681
Dodt G, Braverman N, Wong C, Moser A, Moser HW, Watkins P, Valle D, Gould SJ (1995) Mutations in the PTS1 receptor gene, PXR1, define complementation group 2 of the peroxisome biogenesis disorders. Nat Genet 9(2):115–125
Duckworth WC, Bennett RG, Hamel FG (1998) Insulin degradation: progress and potential. Endocr Rev 19(5):608–624
Dyer JM, McNew JA, Goodman JM (1996) The sorting sequence of the peroxisomal integral membrane protein PMP47 is contained within a short hydrophilic loop. J Cell Biol 133(2):269–280

Ebberink MS, Kofster J, Wanders RJ, Waterham HR (2010) Spectrum of PEX6 mutations in Zellweger syndrome spectrum patients. Hum Mutat 31(1):E1058–E1070

Eisenhaber F (2012) A decade after the first full human genome sequencing: when will we understand our own genome? J Bioinform Comput Biol 10(5):1271001

Eisenhaber B, Eisenhaber F (2007) Posttranslational modifications and subcellular localization signals: indicators of sequence regions without inherent 3D structure? Curr Protein Pept Sci 8 (2):197–203

Eisenhaber B, Eisenhaber F (2010) Prediction of posttranslational modification of proteins from their amino acid sequence. Methods Mol Biol 609:365–384

Eisenhaber B, Bork P, Eisenhaber F (1998) Sequence properties of GPI-anchored proteins near the omega-site: constraints for the polypeptide binding site of the putative transamidase. Protein Eng 11(12):1155–1161

Eisenhaber F, Eisenhaber B, Kubina W, Maurer-Stroh S, Neuberger G, Schneider G, Wildpaner M (2003a) Prediction of lipid posttranslational modifications and localization signals from protein sequences: big-Pi, NMT and PTS1. Nucleic Acids Res 31(13):3631–3634

Eisenhaber F, Eisenhaber B, Maurer-Stroh S (2003b) Prediction of post-translational modifications from amino acid sequence: problems, pitfalls, methodological hints. In: Andrade MM (ed) Bioinformatics and genomes: current perspectives, 1st edn. Horizon Scientific Press, Wymondham, pp 81–105

Eisenhaber B, Eisenhaber F, Maurer-Stroh S, Neuberger G (2004) Prediction of sequence signals for lipid post-translational modifications: insights from case studies. Proteomics 4(6):1614–1625

Elgersma Y, Kwast L, van den Berg M, Snyder WB, Distel B, Subramani S, Tabak HF (1997) Overexpression of Pex15p, a phosphorylated peroxisomal integral membrane protein required for peroxisome assembly in S.cerevisiae, causes proliferation of the endoplasmic reticulum membrane. EMBO J 16(24):7326–7341

Elgersma Y, Elgersma-Hooisma M, Wenzel T, McCaffery JM, Farquhar MG, Subramani S (1998) A mobile PTS2 receptor for peroxisomal protein import in Pichia pastoris. J Cell Biol 140 (4):807–820

Emanuelsson O, Elofsson A, von Heijne G, Cristobal S (2003) In silico prediction of the peroxisomal proteome in fungi, plants and animals. J Mol Biol 330(2):443–456

Emanuelsson O, Brunak S, von Heijne G, Nielsen H (2007) Locating proteins in the cell using TargetP, SignalP and related tools. Nat Protoc 2(4):953–971

Erdmann R, Blobel G (1996) Identification of Pex13p a peroxisomal membrane receptor for the PTS1 recognition factor. J Cell Biol 135(1):111–121

Eubel H, Meyer EH, Taylor NL, Bussell JD, O'Toole N, Heazlewood JL, Castleden I, Small ID, Smith SM, Millar AH (2008) Novel proteins, putative membrane transporters, and an integrated metabolic network are revealed by quantitative proteomic analysis of Arabidopsis cell culture peroxisomes. Plant Physiol 148(4):1809–1829

Fang Y, Morrell JC, Jones JM, Gould SJ (2004) PEX3 functions as a PEX19 docking factor in the import of class I peroxisomal membrane proteins. J Cell Biol 164(6):863–875

Farris W, Mansourian S, Chang Y, Lindsley L, Eckman EA, Frosch MP, Eckman CB, Tanzi RE, Selkoe DJ, Guenette S (2003) Insulin-degrading enzyme regulates the levels of insulin, amyloid beta-protein, and the beta-amyloid precursor protein intracellular domain in vivo. Proc Natl Acad Sci U S A 100(7):4162–4167

Fransen M, Van Veldhoven PP, Subramani S (1999) Identification of peroxisomal proteins by using M13 phage protein VI phage display: molecular evidence that mammalian peroxisomes contain a 2,4-dienoyl-CoA reductase. Biochem J 340(Pt 2):561–568

Freitag J, Ast J, Bolker M (2012) Cryptic peroxisomal targeting via alternative splicing and stop codon read-through in fungi. Nature 485(7399):522–525

Fujiki Y, Matsuzono Y, Matsuzaki T, Fransen M (2006) Import of peroxisomal membrane proteins: the interplay of Pex3p- and Pex19p-mediated interactions. Biochim Biophys Acta 1763(12):1639–1646

Fujiki Y, Nashiro C, Miyata N, Tamura S, Okumoto K (2012) New insights into dynamic and functional assembly of the AAA peroxins, Pex1p and Pex6p, and their membrane receptor Pex26p in shuttling of PTS1-receptor Pex5p during peroxisome biogenesis. Biochim Biophys Acta 1823(1):145–149

Fujiwara T, Hori K, Ozaki K, Yokota Y, Mitsuya S, Ichiyanagi T, Hattori T, Takabe T (2008) Enzymatic characterization of peroxisomal and cytosolic betaine aldehyde dehydrogenases in barley. Physiol Plant 134(1):22–30

Fukao Y, Hayashi M, Nishimura M (2002) Proteomic analysis of leaf peroxisomal proteins in greening cotyledons of Arabidopsis thaliana. Plant Cell Physiol 43(7):689–696

Fukao Y, Hayashi M, Hara-Nishimura I, Nishimura M (2003) Novel glyoxysomal protein kinase, GPK1, identified by proteomic analysis of glyoxysomes in etiolated cotyledons of Arabidopsis thaliana. Plant Cell Physiol 44(10):1002–1012

Furuta S, Hashimoto T, Miura S, Mori M, Tatibana M (1982) Cell-free synthesis of the enzymes of peroxisomal beta-oxidation. Biochem Biophys Res Commun 105(2):639–646

Garg A, Bhasin M, Raghava GP (2005) Support vector machine-based method for subcellular localization of human proteins using amino acid compositions, their order, and similarity search. J Biol Chem 280(15):14427–14432

Gatto GJ Jr, Geisbrecht BV, Gould SJ, Berg JM (2000) Peroxisomal targeting signal-1 recognition by the TPR domains of human PEX5. Nat Struct Biol 7(12):1091–1095

Girzalsky W, Platta HW, Erdmann R (2009) Protein transport across the peroxisomal membrane. Biol Chem 390(8):745–751

Glebov K, Schutze S, Walter J (2011) Functional relevance of a novel SlyX motif in non-conventional secretion of insulin-degrading enzyme. J Biol Chem 286(26):22711–22715

Glover JR, Andrews DW, Subramani S, Rachubinski RA (1994) Mutagenesis of the amino targeting signal of Saccharomyces cerevisiae 3-ketoacyl-CoA thiolase reveals conserved amino acids required for import into peroxisomes in vivo. J Biol Chem 269(10):7558–7563

Gould SG, Keller GA, Subramani S (1987) Identification of a peroxisomal targeting signal at the carboxy terminus of firefly luciferase. J Cell Biol 105(6 Pt 2):2923–2931

Gould SJ, Keller GA, Subramani S (1988) Identification of peroxisomal targeting signals located at the carboxy terminus of four peroxisomal proteins. J Cell Biol 107(3):897–905

Gould SJ, Keller GA, Hosken N, Wilkinson J, Subramani S (1989) A conserved tripeptide sorts proteins to peroxisomes. J Cell Biol 108(5):1657–1664

Gould SJ, Kalish JE, Morrell JC, Bjorkman J, Urquhart AJ, Crane DI (1996) Pex13p is an SH3 protein of the peroxisome membrane and a docking factor for the predominantly cytoplasmic PTs1 receptor. J Cell Biol 135(1):85–95

Goyer A, Johnson TL, Collakova E, Shachar-Hill Y, Rhodes D, Hanson AD (2004) Characterization and metabolic function of a peroxisomal sarcosine and pipecolate oxidase from Arabidopsis. J Biol Chem 279(17):16947–16953

Gur E, Sauer RT (2008) Recognition of misfolded proteins by Lon, a AAA(+) protease. Genes Dev 22(16):2267–2277

Halbach A, Rucktaschel R, Rottensteiner H, Erdmann R (2009a) The N-domain of Pex22p can functionally replace the Pex3p N-domain in targeting and peroxisome formation. J Biol Chem 284(6):3906–3916

Halbach A, Rucktaschel R, Rottensteiner H, Erdmann R (2009b) The N-domain of Pex22p can functionally replace the Pex3p N-domain in targeting and peroxisome formation. J Biol Chem 284(6):3906–3916

Halbach A, Rucktaschel R, Rottensteiner H, Erdmann R (2009c) The N-domain of Pex22p can functionally replace the Pex3p N-domain in targeting and peroxisome formation. J Biol Chem 284(6):3906–3916

Hawkins J, Boden M (2006) Detecting and sorting targeting peptides with neural networks and support vector machines. J Bioinform Comput Biol 4(1):1–18

Hawkins J, Mahony D, Maetschke S, Wakabayashi M, Teasdale RD, Boden M (2007) Identifying novel peroxisomal proteins. Proteins 69(3):606–616

Hayashi M, Aoki M, Kondo M, Nishimura M (1997) Changes in targeting efficiencies of proteins to plant microbodies caused by amino acid substitutions in the carboxy-terminal tripeptide. Plant Cell Physiol 38(6):759–768

Helm M, Luck C, Prestele J, Hierl G, Huesgen PF, Frohlich T, Arnold GJ, Adamska I, Gorg A, Lottspeich F, Gietl C (2007) Dual specificities of the glyoxysomal/peroxisomal processing protease Deg15 in higher plants. Proc Natl Acad Sci U S A 104(27):11501–11506

Hoepfner S, Severin F, Cabezas A, Habermann B, Runge A, Gillooly D, Stenmark H, Zerial M (2005) Modulation of receptor recycling and degradation by the endosomal kinesin KIF16B. Cell 121(3):437–450

Hogenboom S, Tuyp JJ, Espeel M, Koster J, Wanders RJ, Waterham HR (2004a) Human mevalonate pyrophosphate decarboxylase is localized in the cytosol. Mol Genet Metab 81 (3):216–224

Hogenboom S, Tuyp JJ, Espeel M, Koster J, Wanders RJ, Waterham HR (2004b) Mevalonate kinase is a cytosolic enzyme in humans. J Cell Sci 117(Pt 4):631–639

Hogenboom S, Tuyp JJ, Espeel M, Koster J, Wanders RJ, Waterham HR (2004c) Phosphomevalonate kinase is a cytosolic protein in humans. J Lipid Res 45(4):697–705

Holbrook JD, Birdsey GM, Yang Z, Bruford MW, Danpure CJ (2000) Molecular adaptation of alanine:glyoxylate aminotransferase targeting in primates. Mol Biol Evol 17(3):387–400

Honsho M, Fujiki Y (2001) Topogenesis of peroxisomal membrane protein requires a short, positively charged intervening-loop sequence and flanking hydrophobic segments. study using human membrane protein PMP34. J Biol Chem 276(12):9375–9382

Horton P, Park KJ, Obayashi T, Fujita N, Harada H, Ms-Collier CJ, Nakai K (2007) WoLF PSORT: protein localization predictor. Nucleic Acids Res 35(Web Server issue):W585–W587

Imazaki A, Tanaka A, Harimoto Y, Yamamoto M, Akimitsu K, Park P, Tsuge T (2010) Contribution of peroxisomes to secondary metabolism and pathogenicity in the fungal plant pathogen Alternaria alternata. Eukaryot Cell 9(5):682–694

Jankowski A, Kim JH, Collins RF, Daneman R, Walton P, Grinstein S (2001) In situ measurements of the pH of mammalian peroxisomes using the fluorescent protein pHluorin. J Biol Chem 276(52):48748–48753

Jones JM, Morrell JC, Gould SJ (2004) PEX19 is a predominantly cytosolic chaperone and import receptor for class 1 peroxisomal membrane proteins. J Cell Biol 164(1):57–67

Joo HJ, Yim YH, Jeong PY, Jin YX, Lee JE, Kim H, Jeong SK, Chitwood DJ, Paik YK (2009) Caenorhabditis elegans utilizes dauer pheromone biosynthesis to dispose of toxic peroxisomal fatty acids for cellular homoeostasis. Biochem J 422(1):61–71

Joo HJ, Kim KY, Yim YH, Jin YX, Kim H, Kim MY, Paik YK (2010) Contribution of the peroxisomal acox gene to the dynamic balance of daumone production in Caenorhabditis elegans. J Biol Chem 285(38):29319–29325

Kamada T, Nito K, Hayashi H, Mano S, Hayashi M, Nishimura M (2003) Functional differentiation of peroxisomes revealed by expression profiles of peroxisomal genes in Arabidopsis thaliana. Plant Cell Physiol 44(12):1275–1289

Kamada-Nobusada T, Hayashi M, Fukazawa M, Sakakibara H, Nishimura M (2008) A putative peroxisomal polyamine oxidase, AtPAO4, is involved in polyamine catabolism in Arabidopsis thaliana. Plant Cell Physiol 49(9):1272–1282

Kanehisa M (2013) Molecular network analysis of diseases and drugs in KEGG. Methods Mol Biol 939:263–275

Kanzawa N, Maeda Y, Ogiso H, Murakami Y, Taguchi R, Kinoshita T (2009) Peroxisome dependency of alkyl-containing GPI-anchor biosynthesis in the endoplasmic reticulum. Proc Natl Acad Sci U S A 106(42):17711–17716

Kaur N, Hu J (2011) Defining the plant peroxisomal proteome: from Arabidopsis to rice. Front Plant Sci 2:103

Kaye GC, Butler MG, Ardenne AJ, Edmondson SJ, Camm AJ, Slavin G (1986) Identification of immunoreactive atrial natriuretic peptide in atrial amyloid. J Clin Pathol 39(5):581–582

Keller GA, Krisans S, Gould SJ, Sommer JM, Wang CC, Schliebs W, Kunau W, Brody S, Subramani S (1991) Evolutionary conservation of a microbody targeting signal that targets proteins to peroxisomes, glyoxysomes, and glycosomes. J Cell Biol 114(5):893–904

Kienow L, Schneider K, Bartsch M, Stuible HP, Weng H, Miersch O, Wasternack C, Kombrink E (2008) Jasmonates meet fatty acids: functional analysis of a new acyl-coenzyme A synthetase family from Arabidopsis thaliana. J Exp Bot 59(2):403–419

Kikuchi M, Hatano N, Yokota S, Shimozawa N, Imanaka T, Taniguchi H (2004) Proteomic analysis of rat liver peroxisome: presence of peroxisome-specific isozyme of Lon protease. J Biol Chem 279(1):421–428

Kovacs WJ, Tape KN, Shackelford JE, Duan X, Kasumov T, Kelleher JK, Brunengraber H, Krisans SK (2007) Localization of the pre-squalene segment of the isoprenoid biosynthetic pathway in mammalian peroxisomes. Histochem Cell Biol 127(3):273–290

Kowit JD, Goldberg AL (1977) Intermediate steps in the degradation of a specific abnormal protein in Escherichia coli. J Biol Chem 252(23):8350–8357

Kragler F, Langeder A, Raupachova J, Binder M, Hartig A (1993) Two independent peroxisomal targeting signals in catalase A of Saccharomyces cerevisiae. J Cell Biol 120(3):665–673

Kunze M, Kragler F, Binder M, Hartig A, Gurvitz A (2002) Targeting of malate synthase 1 to the peroxisomes of Saccharomyces cerevisiae cells depends on growth on oleic acid medium. Eur J Biochem 269(3):915–922

Kunze M, Neuberger G, Maurer-Stroh S, Ma J, Eck T, Braverman N, Schmid JA, Eisenhaber F, Berger J (2011) Structural requirements for interaction of peroxisomal targeting signal 2 and its receptor PEX7. J Biol Chem 286(52):45048–45062

Kuo WL, Gehm BD, Rosner MR, Li W, Keller G (1994) Inducible expression and cellular localization of insulin-degrading enzyme in a stably transfected cell line. J Biol Chem 269 (36):22599–22606

Kurochkin IV (1998) Amyloidogenic determinant as a substrate recognition motif of insulin-degrading enzyme. FEBS Lett 427(2):153–156

Kurochkin IV (2001) Insulin-degrading enzyme: embarking on amyloid destruction. Trends Biochem Sci 26(7):421–425

Kurochkin IV, Goto S (1994) Alzheimer's beta-amyloid peptide specifically interacts with and is degraded by insulin degrading enzyme. FEBS Lett 345(1):33–37

Kurochkin IV, Konagaya A, Nagashima T, Schonbach C, RIKEN GER Group (2003) Identification of potential peroxisomal proteins in mouse. Genome Res 13(6B):1560

Kurochkin IV, Mizuno Y, Konagaya A, Sakaki Y, Schonbach C, Okazaki Y (2007) Novel peroxisomal protease Tysnd1 processes PTS1- and PTS2-containing enzymes involved in beta-oxidation of fatty acids. EMBO J 26(3):835–845

Kuznetsov V, Lee HK, Maurer-Stroh S, Molnar MJ, Pongor S, Eisenhaber B, Eisenhaber F (2013) How bioinformatics influences health informatics: usage of biomolecular sequences, expression profiles and automated microscopic image analyses for clinical needs and public health. Health Inform Sci Syst 1:2

Lamesch P, Berardini TZ, Li D, Swarbreck D, Wilks C, Sasidharan R, Muller R, Dreher K, Alexander DL, Garcia-Hernandez M, Karthikeyan AS, Lee CH, Nelson WD, Ploetz L, Singh S, Wensel A, Huala E (2012) The Arabidopsis Information Resource (TAIR): improved gene annotation and new tools. Nucleic Acids Res 40(Database issue):D1202–D1210

Lametschwandtner G, Brocard C, Fransen M, Van VP, Berger J, Hartig A (1998) The difference in recognition of terminal tripeptides as peroxisomal targeting signal 1 between yeast and human is due to different affinities of their receptor Pex5p to the cognate signal and to residues adjacent to it. J Biol Chem 273(50):33635–33643

Lanyon-Hogg T, Warriner SL, Baker A (2010) Getting a camel through the eye of a needle: the import of folded proteins by peroxisomes. Biol Cell 102(4):245–263

Leal MC, Magnani N, Villordo S, Buslje CM, Evelson P, Castano EM, Morelli L (2013) Transcriptional regulation of insulin-degrading enzyme modulates mitochondrial amyloid beta (abeta) peptide catabolism and functionality. J Biol Chem 288(18):12920–12931

Lee I, Suzuki CK (2008) Functional mechanics of the ATP-dependent Lon protease- lessons from endogenous protein and synthetic peptide substrates. Biochim Biophys Acta 1784(5):727–735

Leissring MA, Farris W, Chang AY, Walsh DM, Wu X, Sun X, Frosch MP, Selkoe DJ (2003) Enhanced proteolysis of beta-amyloid in APP transgenic mice prevents plaque formation, secondary pathology, and premature death. Neuron 40(6):1087–1093

Leissring MA, Farris W, Wu X, Christodoulou DC, Haigis MC, Guarente L, Selkoe DJ (2004) Alternative translation initiation generates a novel isoform of insulin-degrading enzyme targeted to mitochondria. Biochem J 383(Pt. 3):439–446

Lingard MJ, Bartel B (2009) Arabidopsis LON2 is necessary for peroxisomal function and sustained matrix protein import. Plant Physiol 151(3):1354–1365

Ma C, Agrawal G, Subramani S (2011) Peroxisome assembly: matrix and membrane protein biogenesis. J Cell Biol 193(1):7–16

Malito E, Hulse RE, Tang WJ (2008) Amyloid beta-degrading cryptidases: insulin degrading enzyme, presequence peptidase, and neprilysin. Cell Mol Life Sci 65(16):2574–2585

Marzioch M, Erdmann R, Veenhuis M, Kunau WH (1994) PAS7 encodes a novel yeast member of the WD-40 protein family essential for import of 3-oxoacyl-CoA thiolase, a PTS2-containing protein, into peroxisomes. EMBO J 13(20):4908–4918

Mast FD, Fagarasanu A, Rachubinski R (2010) The peroxisomal protein importomer: a bunch of transients with expanding waistlines. Nat Cell Biol 12(3):203–205

Matsuda S, Vert JP, Saigo H, Ueda N, Toh H, Akutsu T (2005) A novel representation of protein sequences for prediction of subcellular location using support vector machines. Protein Sci 14 (11):2804–2813

Matsuzono Y, Matsuzaki T, Fujiki Y (2006) Functional domain mapping of peroxin Pex19p: interaction with Pex3p is essential for function and translocation. J Cell Sci 119(Pt 17):3539–3550

Maurer-Stroh S, Eisenhaber F (2005) Refinement and prediction of protein prenylation motifs. Genome Biol 6(6):R55

Maurer-Stroh S, Eisenhaber B, Eisenhaber F (2002) N-terminal N-myristoylation of proteins: refinement of the sequence motif and its taxon-specific differences. J Mol Biol 317(4):523–540

Maynard EL, Gatto GJ Jr, Berg JM (2004) Pex5p binding affinities for canonical and noncanonical PTS1 peptides. Proteins 55(4):856–861

McCollum D, Monosov E, Subramani S (1993) The pas8 mutant of Pichia pastoris exhibits the peroxisomal protein import deficiencies of Zellweger syndrome cells–the PAS8 protein binds to the COOH-terminal tripeptide peroxisomal targeting signal, and is a member of the TPR protein family. J Cell Biol 121(4):761–774

McNew JA, Goodman JM (1996) The targeting and assembly of peroxisomal proteins: some old rules do not apply. Trends Biochem Sci 21(2):54–58

Meinecke M, Cizmowski C, Schliebs W, Kruger V, Beck S, Wagner R, Erdmann R (2010) The peroxisomal importomer constitutes a large and highly dynamic pore. Nat Cell Biol 12(3):273–277

Mirsky IA, Broh-Kahn RH (1949) The inactivation of insulin by tissue extracts; the distribution and properties of insulin inactivating extracts. Arch Biochem 20(1):1–9

Mitsuya S, Yokota Y, Fujiwara T, Mori N, Takabe T (2009) OsBADH1 is possibly involved in acetaldehyde oxidation in rice plant peroxisomes. FEBS Lett 583(22):3625–3629

Mizuno Y, Ninomiya Y, Nakachi Y, Iseki M, Iwasa H, Akita M, Tsukui T, Shimozawa N, Ito C, Toshimori K, Nishimukai M, Hara H, Maeba R, Okazaki T, Alodaib AN, Al AM, Jacob M, Alkuraya FS, Horai Y, Watanabe M, Motegi H, Wakana S, Noda T, Kurochkin IV, Mizuno Y, Schonbach C, Okazaki Y (2013) Tysnd1 deficiency in mice interferes with the peroxisomal localization of PTS2 enzymes, causing lipid metabolic abnormalities and male infertility. PLoS Genet 9(2):e1003286

Morelli L, Llovera RE, Alonso LG, Frangione B, de Prat-Gay G, Ghiso J, Castano EM (2005) Insulin-degrading enzyme degrades amyloid peptides associated with British and Danish familial dementia. Biochem Biophys Res Commun 332(3):808–816

Morita M, Kurochkin IV, Motojima K, Goto S, Takano T, Okamura S, Sato R, Yokota S, Imanaka T (2000) Insulin-degrading enzyme exists inside of rat liver peroxisomes and degrades oxidized proteins. Cell Struct Funct 25(5):309–315

Motley AM, Hettema EH (2007) Yeast peroxisomes multiply by growth and division. J Cell Biol 178(3):399–410

Nair DM, Purdue PE, Lazarow PB (2004) Pex7p translocates in and out of peroxisomes in Saccharomyces cerevisiae. J Cell Biol 167(4):599–604

Nakai K, Horton P (1999) PSORT: a program for detecting sorting signals in proteins and predicting their subcellular localization. Trends Biochem Sci 24(1):34–36

Neant-Fery M, Garcia-Ordonez RD, Logan TP, Selkoe DJ, Li L, Reinstatler L, Leissring MA (2008) Molecular basis for the thiol sensitivity of insulin-degrading enzyme. Proc Natl Acad Sci U S A 105(28):9582–9587

Neuberger G, Maurer-Stroh S, Eisenhaber B, Hartig A, Eisenhaber F (2003a) Motif refinement of the peroxisomal targeting signal 1 and evaluation of taxon-specific differences. J Mol Biol 328 (3):567–579

Neuberger G, Maurer-Stroh S, Eisenhaber B, Hartig A, Eisenhaber F (2003b) Prediction of peroxisomal targeting signal 1 containing proteins from amino acid sequence. J Mol Biol 328(3):581–592

Neuberger G, Kunze M, Eisenhaber F, Berger J, Hartig A, Brocard C (2004) Hidden localization motifs: naturally occurring peroxisomal targeting signals in non-peroxisomal proteins. Genome Biol 5(12):R97

Nguyen SD, Baes M, Van Veldhoven PP (2008) Degradation of very long chain dicarboxylic polyunsaturated fatty acids in mouse hepatocytes, a peroxisomal process. Biochim Biophys Acta 1781(8):400–405

Nicolay K, Veenhuis M, Douma AC, Harder W (1987) A 31P NMR study of the internal pH of yeast peroxisomes. Arch Microbiol 147(1):37–41

Nishikawa M, Hagishita T, Yurimoto H, Kato N, Sakai Y, Hatanaka T (2000a) Primary structure and expression of peroxisomal acetylspermidine oxidase in the methylotrophic yeast Candida boidinii. FEBS Lett 476(3):150–154

Nishikawa M, Hagishita T, Yurimoto H, Kato N, Sakai Y, Hatanaka T (2000b) Primary structure and expression of peroxisomal acetylspermidine oxidase in the methylotrophic yeast Candida boidinii. FEBS Lett 476(3):150–154

Nito K, Hayashi M, Nishimura M (2002) Direct interaction and determination of binding domains among peroxisomal import factors in Arabidopsis thaliana. Plant Cell Physiol 43(4):355–366

Niu B, Jin YH, Feng KY, Lu WC, Cai YD, Li GZ (2008) Using AdaBoost for the prediction of subcellular location of prokaryotic and eukaryotic proteins. Mol Divers 12(1):41–45

Oatey PB, Lumb MJ, Danpure CJ (1996a) Molecular basis of the variable mitochondrial and peroxisomal localisation of alanine-glyoxylate aminotransferase. Eur J Biochem 241(2):374–385

Oatey PB, Lumb MJ, Jennings PR, Danpure CJ (1996b) Context dependency of the PTS1 motif in human alanine: glyoxylate aminotransferase 1. Ann Y Acad Sci 804:652–653

Oda T, Uchida C, Miura S (2000) Mitochondrial targeting signal-induced conformational change and repression of the peroxisomal targeting signal of the precursor for rat liver serine:pyruvate/alanine:glyoxylate aminotransferase. J Biochem 127(4):665–671

Okazaki Y, Furuno M, Kasukawa T, Adachi J, Bono H, Kondo S, Nikaido I, Osato N, Saito R, Suzuki H, Yamanaka I, Kiyosawa H, Yagi K, Tomaru Y, Hasegawa Y, Nogami A, Schonbach C, Gojobori T, Baldarelli R, Hill DP, Bult C, Hume DA, Quackenbush J, Schriml LM, Kanapin A, Matsuda H, Batalov S, Beisel KW, Blake JA, Bradt D, Brusic V, Chothia C, Corbani LE, Cousins S, Dalla E, Dragani TA, Fletcher CF, Forrest A, Frazer KS, Gaasterland T, Gariboldi M, Gissi C, Godzik A, Gough J, Grimmond S, Gustincich S, Hirokawa N, Jackson IJ, Jarvis ED, Kanai A, Kawaji H, Kawasawa Y, Kedzierski RM, King BL, Konagaya A, Kurochkin IV, Lee Y, Lenhard B, Lyons PA, Maglott DR, Maltais L, Marchionni L, McKenzie L, Miki H, Nagashima T, Numata K, Okido T, Pavan WJ,

Pertea G, Pesole G, Petrovsky N, Pillai R, Pontius JU, Qi D, Ramachandran S, Ravasi T, Reed JC, Reed DJ, Reid J, Ring BZ, Ringwald M, Sandelin A, Schneider C, Semple CA, Setou M, Shimada K, Sultana R, Takenaka Y, Taylor MS, Teasdale RD, Tomita M, Verardo R, Wagner L, Wahlestedt C, Wang Y, Watanabe Y, Wells C, Wilming LG, Wynshaw-Boris A, Yanagisawa M, Yang I, Yang L, Yuan Z, Zavolan M, Zhu Y, Zimmer A, Carninci P, Hayatsu N, Hirozane-Kishikawa T, Konno H, Nakamura M, Sakazume N, Sato K, Shiraki T, Waki K, Kawai J, Aizawa K, Arakawa T, Fukuda S, Hara A, Hashizume W, Imotani K, Ishii Y, Itoh M, Kagawa I, Miyazaki A, Sakai K, Sasaki D, Shibata K, Shinagawa A, Yasunishi A, Yoshino M, Waterston R, Lander ES, Rogers J, Birney E, Hayashizaki Y (2002) Analysis of the mouse transcriptome based on functional annotation of 60,770 full-length cDNAs. Nature 420(6915):563–573

Okumoto K, Kametani Y, Fujiki Y (2011) Two proteases, trypsin domain-containing 1 (Tysnd1) and peroxisomal lon protease (PsLon), cooperatively regulate fatty acid beta-oxidation in peroxisomal matrix. J Biol Chem 286(52):44367–44379

Omi S, Nakata R, Okamura-Ikeda K, Konishi H, Taniguchi H (2008) Contribution of peroxisome-specific isoform of Lon protease in sorting PTS1 proteins to peroxisomes. J Biochem 143 (5):649–660

Ono Y, Kim DW, Watanabe K, Sasaki A, Niitsu M, Berberich T, Kusano T, Takahashi Y (2012) Constitutively and highly expressed Oryza sativa polyamine oxidases localize in peroxisomes and catalyze polyamine back conversion. Amino Acids 42(2–3):867–876

Oshima Y, Kamigaki A, Nakamori C, Mano S, Hayashi M, Nishimura M, Esaka M (2008) Plant catalase is imported into peroxisomes by Pex5p but is distinct from typical PTS1 import. Plant Cell Physiol 49(4):671–677

Osinga KA, Swinkels BW, Gibson WC, Borst P, Veeneman GH, Van Boom JH, Michels PA, Opperdoes FR (1985) Topogenesis of microbody enzymes: a sequence comparison of the genes for the glycosomal (microbody) and cytosolic phosphoglycerate kinases of Trypanosoma brucei. EMBO J 4(13B):3811–3817

Osumi T, Hashimoto T (1980) Purification and properties of mitochondrial and peroxisomal 3-hydroxyacyl-CoA dehydrogenase from rat liver. Arch Biochem Biophys 203(1):372–383

Osumi T, Tsukamoto T, Hata S (1992) Signal peptide for peroxisomal targeting: replacement of an essential histidine residue by certain amino acids converts the amino-terminal presequence of peroxisomal 3-ketoacyl-CoA thiolase to a mitochondrial signal peptide. Biochem Biophys Res Commun 186(2):811–818

Pan D, Nakatsu T, Kato H (2013) Crystal structure of peroxisomal targeting signal-2 bound to its receptor complex Pex7p-Pex21p. Nat Struct Mol Biol 20(8):987–93

Perry RJ, Mast FD, Rachubinski RA (2009) Endoplasmic reticulum-associated secretory proteins Sec20p, Sec39p, and Dsl1p are involved in peroxisome biogenesis. Eukaryot Cell 8(6):830–843

Petriv OI, Tang L, Titorenko VI, Rachubinski RA (2004) A new definition for the consensus sequence of the peroxisome targeting signal type 2. J Mol Biol 341(1):119–134

Petsalaki EI, Bagos PG, Litou ZI, Hamodrakas SJ (2006) PredSL: a tool for the N-terminal sequence-based prediction of protein subcellular localization. Genomics Proteomics Bioinformatics 4(1):48–55

Rawlings ND, Barrett AJ, Bateman A (2012) MEROPS: the database of proteolytic enzymes, their substrates and inhibitors. Nucleic Acids Res 40(Database issue):D343–D350

Rayapuram N, Subramani S (2006) The importomer–a peroxisomal membrane complex involved in protein translocation into the peroxisome matrix. Biochim Biophys Acta 1763(12):1613–1619

Rehling P, Marzioch M, Niesen F, Wittke E, Veenhuis M, Kunau WH (1996) The import receptor for the peroxisomal targeting signal 2 (PTS2) in Saccharomyces cerevisiae is encoded by the PAS7 gene. EMBO J 15(12):2901–2913

Reumann S (2004) Specification of the peroxisome targeting signals type 1 and type 2 of plant peroxisomes by bioinformatics analyses. Plant Physiol 135(2):783–800

Reumann S, Babujee L, Ma C, Wienkoop S, Siemsen T, Antonicelli GE, Rasche N, Luder F, Weckwerth W, Jahn O (2007) Proteome analysis of Arabidopsis leaf peroxisomes reveals novel targeting peptides, metabolic pathways, and defense mechanisms. Plant Cell 19 (10):3170–3193

Reumann S, Quan S, Aung K, Yang P, Manandhar-Shrestha K, Holbrook D, Linka N, Switzenberg R, Wilkerson CG, Weber AP, Olsen LJ, Hu J (2009) In-depth proteome analysis of Arabidopsis leaf peroxisomes combined with in vivo subcellular targeting verification indicates novel metabolic and regulatory functions of peroxisomes. Plant Physiol 150 (1):125–143

Reumann S, Buchwald D, Lingner T (2012) PredPlantPTS1: a web server for the prediction of plant peroxisomal proteins. Front Plant Sci 3:194

Rucktaschel R, Girzalsky W, Erdmann R (2011a) Protein import machineries of peroxisomes. Biochim Biophys Acta 1808(3):892–900

Rucktaschel R, Girzalsky W, Erdmann R (2011b) Protein import machineries of peroxisomes. Biochim Biophys Acta 1808(3):892–900

Saikia S, Scott B (2009) Functional analysis and subcellular localization of two geranylgeranyl diphosphate synthases from Penicillium paxilli. Mol Genet Genomics 282(3):257–271

Sapir-Mir M, Mett A, Belausov E, Tal-Meshulam S, Frydman A, Gidoni D, Eyal Y (2008) Peroxisomal localization of Arabidopsis isopentenyl diphosphate isomerases suggests that part of the plant isoprenoid mevalonic acid pathway is compartmentalized to peroxisomes. Plant Physiol 148(3):1219–1228

Schluter A, Real-Chicharro A, Gabaldon T, Sanchez-Jimenez F, Pujol A (2010) PeroxisomeDB 2.0: an integrative view of the global peroxisomal metabolome. Nucleic Acids Res 38(Database issue):D800–D805

Schneider K, Kienow L, Schmelzer E, Colby T, Bartsch M, Miersch O, Wasternack C, Kombrink E, Stuible HP (2005) A new type of peroxisomal acyl-coenzyme A synthetase from Arabidopsis thaliana has the catalytic capacity to activate biosynthetic precursors of jasmonic acid. J Biol Chem 280(14):13962–13972

Schrader M, Fahimi HD (2008) The peroxisome: still a mysterious organelle. Histochem Cell Biol 129(4):421–440

Schrader M, Yoon Y (2007) Mitochondria and peroxisomes: are the 'big brother' and the 'little sister' closer than assumed? Bioessays 29(11):1105–1114

Schueller N, Holton SJ, Fodor K, Milewski M, Konarev P, Stanley WA, Wolf J, Erdmann R, Schliebs W, Song YH, Wilmanns M (2010) The peroxisomal receptor Pex19p forms a helical mPTS recognition domain. EMBO J 29(15):2491–2500

Seta KA, Roth RA (1997) Overexpression of insulin degrading enzyme: cellular localization and effects on insulin signaling. Biochem Biophys Res Commun 231(1):167–171

Shen Y, Joachimiak A, Rosner MR, Tang WJ (2006) Structures of human insulin-degrading enzyme reveal a new substrate recognition mechanism. Nature 443(7113):870–874

Simkin AJ, Guirimand G, Papon N, Courdavault V, Thabet I, Ginis O, Bouzid S, Giglioli-Guivarc'h N, Clastre M (2011) Peroxisomal localisation of the final steps of the mevalonic acid pathway in planta. Planta 234(5):903–914

Sirota FL, Batagov A, Schneider G, Eisenhaber B, Eisenhaber F, Maurer-Stroh S (2012) Beware of moving targets: reference proteome content fluctuates substantially over the years. J Bioinform Comput Biol 10(6):1250020

Small I, Peeters N, Legeai F, Lurin C (2004) Predotar: a tool for rapidly screening proteomes for N-terminal targeting sequences. Proteomics 4(6):1581–1590

Song ES, Juliano MA, Juliano L, Hersh LB (2003) Substrate activation of insulin-degrading enzyme (insulysin). A potential target for drug development. J Biol Chem 278(50):49789–49794

Soukupova M, Sprenger C, Gorgas K, Kunau WH, Dodt G (1999) Identification and characterization of the human peroxin PEX3. Eur J Cell Biol 78(6):357–374

Sparkes IA, Hawes C, Baker A (2005) AtPEX2 and AtPEX10 are targeted to peroxisomes independently of known endoplasmic reticulum trafficking routes. Plant Physiol 139(2):690–700

Spiegel CN, Batista-Pereira LG, Bretas JA, Eiras AE, Hooper AM, Peixoto AA, Soares MJ (2011) Pheromone gland development and pheromone production in lutzomyia longipalpis (Diptera: Psychodidae: Phlebotominae). J Med Entomol 48(3):489–495

Stahlberg H, Kutejova E, Suda K, Wolpensinger B, Lustig A, Schatz G, Engel A, Suzuki CK (1999) Mitochondrial Lon of Saccharomyces cerevisiae is a ring-shaped protease with seven flexible subunits. Proc Natl Acad Sci U S A 96(12):6787–6790

Steinberg SJ, Dodt G, Raymond GV, Braverman NE, Moser AB, Moser HW (2006) Peroxisome biogenesis disorders. Biochim Biophys Acta 1763(12):1733–1748

Subramani S (1992) Targeting of proteins into the peroxisomal matrix. J Membr Biol 125(2):99–106

Swinkels BW, Gould SJ, Bodnar AG, Rachubinski RA, Subramani S (1991) A novel, cleavable peroxisomal targeting signal at the amino-terminus of the rat 3-ketoacyl-CoA thiolase. EMBO J 10(11):3255–3262

Tabak HF, Murk JL, Braakman I, Geuze HJ (2003) Peroxisomes start their life in the endoplasmic reticulum. Traffic 4(8):512–518

Tabak HF, van der Zand A, Braakman I (2008a) Peroxisomes: minted by the ER. Curr Opin Cell Biol 20(4):393–400

Tabak HF, van der Zand A, Braakman I (2008b) Peroxisomes: minted by the ER. Curr Opin Cell Biol 20(4):393–400

Tamboli IY, Barth E, Christian L, Siepmann M, Kumar S, Singh S, Tolksdorf K, Heneka MT, Lutjohann D, Wunderlich P, Walter J (2010) Statins promote the degradation of extracellular amyloid {beta}-peptide by microglia via stimulation of exosome-associated insulin-degrading enzyme (IDE) secretion. J Biol Chem 285(48):37405–37414

Tamura T, Akutsu T (2007) Subcellular location prediction of proteins using support vector machines with alignment of block sequences utilizing amino acid composition. BMC Bioinformatics 8:466

Tanabe Y, Maruyama J, Yamaoka S, Yahagi D, Matsuo I, Tsutsumi N, Kitamoto K (2011) Peroxisomes are involved in biotin biosynthesis in Aspergillus and Arabidopsis. J Biol Chem 286(35):30455–30461

Tanz SK, Castleden I, Hooper CM, Vacher M, Small I, Millar HA (2013) SUBA3: a database for integrating experimentation and prediction to define the SUBcellular location of proteins in Arabidopsis. Nucleic Acids Res 41(Database issue):D1185–D1191

Thabet I, Guirimand G, Courdavault V, Papon N, Godet S, Dutilleul C, Bouzid S, Giglioli-Guivarc'h N, Clastre M, Simkin AJ (2011) "The subcellular localization of periwinkle farnesyl diphosphate synthase provides insight into the role of peroxisome in isoprenoid biosynthesis", J. Plant Physiol 168(17):2110–2116

Theodoulou FL, Bernhardt K, Linka N, Baker A (2013a) Peroxisome membrane proteins: multiple trafficking routes and multiple functions? Biochem J 451(3):345–352

Theodoulou FL, Bernhardt K, Linka N, Baker A (2013b) Peroxisome membrane proteins: multiple trafficking routes and multiple functions? Biochem J 451(3):345–352

Titorenko VI, Rachubinski RA (1998) Mutants of the yeast Yarrowia lipolytica defective in protein exit from the endoplasmic reticulum are also defective in peroxisome biogenesis. Mol Cell Biol 18(5):2789–2803

Tolbert NE, Yamazaki RK (1969) Leaf peroxisomes and their relation to photorespiration and photosynthesis. Ann N Y Acad Sci 168(2):325–341

Tolbert NE, Oeser A, Kisaki T, Hageman RH, Yamazaki RK (1968) Peroxisomes from spinach leaves containing enzymes related to glycolate metabolism. J Biol Chem 243(19):5179–5184

Turnell WG, Finch JT (1992) Binding of the dye congo red to the amyloid protein pig insulin reveals a novel homology amongst amyloid-forming peptide sequences. J Mol Biol 227 (4):1205–1223

Van der Leij I, Franse MM, Elgersma Y, Distel B, Tabak HF (1993) PAS10 is a tetratricopeptide-repeat protein that is essential for the import of most matrix proteins into peroxisomes of Saccharomyces cerevisiae. Proc Natl Acad Sci U S A 90(24):11782–11786

van der Zand A, Braakman I, Tabak HF (2010) Peroxisomal membrane proteins insert into the endoplasmic reticulum. Mol Biol Cell 21(12):2057–2065

van Roermund CW, de Jong M, IJlst L, van Marle J, Dansen TB, Wanders RJ, Waterham HR (2004) The peroxisomal lumen in Saccharomyces cerevisiae is alkaline. JCell Sci 117(no. Pt 18):4231–4237

Van AE, Fransen M (2006) Targeting signals in peroxisomal membrane proteins. Biochim Biophys Acta 1763(12):1629–1638

Van DP, Maurer-Stroh S, Plasman K, Van DJ, Colaert N, Timmerman E, De Bock PJ, Goethals M, Rousseau F, Schymkowitz J, Vandekerckhove J, Gevaert K (2009) Analysis of protein processing by N-terminal proteomics reveals novel species-specific substrate determinants of granzyme B orthologs. Mol Cell Proteomics 8(2):258–272

Van DP, Maurer-Stroh S, Hao H, Colaert N, Timmerman E, Eisenhaber F, Vandekerckhove J, Gevaert K (2010) The substrate specificity profile of human granzyme A. Biol Chem 391 (8):983–997

Venkatesh S, Lee J, Singh K, Lee I, Suzuki CK (2012) Multitasking in the mitochondrion by the ATP-dependent Lon protease. Biochim Biophys Acta 1823(1):56–66

Wagner I, Arlt H, van Dyck L, Langer T, Neupert W (1994) Molecular chaperones cooperate with PIM1 protease in the degradation of misfolded proteins in mitochondria. EMBO J 13 (21):5135–5145

Wanders RJ, Waterham HR (2006) Peroxisomal disorders: the single peroxisomal enzyme deficiencies. Biochim Biophys Acta 1763(12):1707–1720

Waugh DF (1946) Reactions involved in insulin fibril formation. Fed Proc 5(1 Pt 2):111

Willemsen AM, Jansen GA, Komen JC, van Hoff S, Waterham HR, Brites PM, Wanders RJ, van Kampen AH (2008) Organization and integration of biomedical knowledge with concept maps for key peroxisomal pathways. Bioinformatics 24(16):i21–i27

Wiszniewski AA, Zhou W, Smith SM, Bussell JD (2009) Identification of two Arabidopsis genes encoding a peroxisomal oxidoreductase-like protein and an acyl-CoA synthetase-like protein that are required for responses to pro-auxins. Plant Mol Biol 69(5):503–515

Wolff SP, Dean RT (1986) Fragmentation of proteins by free radicals and its effect on their susceptibility to enzymic hydrolysis. Biochem J 234(2):399–403

Wong WC, Maurer-Stroh S, Eisenhaber F (2010) More than 1,001 problems with protein domain databases: transmembrane regions, signal peptides and the issue of sequence homology. PLoS Comput Biol 6(7):e1000867

Wu T, Yankovskaya V, McIntire WS (2003) Cloning, sequencing, and heterologous expression of the murine peroxisomal flavoprotein, N1-acetylated polyamine oxidase. J Biol Chem 278 (23):20514–20525

Yan M, Rachubinski DA, Joshi S, Rachubinski RA, Subramani S (2008) Dysferlin domain-containing proteins, Pex30p and Pex31p, localized to two compartments, control the number and size of oleate-induced peroxisomes in Pichia pastoris. Mol Biol Cell 19(3):885–898

Zolman BK, Yoder A, Bartel B (2000) Genetic analysis of indole-3-butyric acid responses in Arabidopsis thaliana reveals four mutant classes. Genetics 156(3):1323–1337

Live Imaging of Peroxisomes and Peroxules in Plants

10

Kiah A. Barton, Erica-Ashley Jaipargas, Nigel Griffiths, and Jaideep Mathur

Abstract

Peroxisomes are single membrane-bound organelles found in all eukaryotic cells. Many different fluorescent proteins, including photo-convertible proteins have been used to highlight peroxisomes and observe their actin-dependent motility in living plant cells. One of the insights resulting from live imaging of peroxisomes is their rapid response to changes in subcellular oxidative stress whereby thin tubules that have been named peroxules are extended and retracted. Peroxules result from transient changes in the peroxisomal membrane and disappear quickly once the redox equilibrium in the cell is re-established. However, under high stress intensity, or when the period of stress gets prolonged, peroxisomes elongate into 3–7 μm long tubules that progress into a beaded morphology, and finally undergo fission. Live imaging also shows that peroxules and elongated peroxisomes align with contiguous tubules of the endoplasmic reticulum and allows new insights on the peroxisome–ER relationship in plants.

Keywords

Peroxisome • Peroxule • Endoplasmic reticulum • Live imaging

K.A. Barton • E.-A. Jaipargas • N. Griffiths • J. Mathur (✉)
Laboratory of Plant Development and Interactions, Department of Molecular and Cellular Biology, University of Guelph, 50 Stone Road, Guelph, ON, Canada N1G2W1
e-mail: jmathur@uoguelph.ca

C. Brocard and A. Hartig (eds.), *Molecular Machines Involved in Peroxisome Biogenesis and Maintenance*, DOI 10.1007/978-3-7091-1788-0_10,

10.1 Introduction

10.1.1 The Visualization of Peroxisomes

Peroxisomes are small, single membrane-bound eukaryotic organelles first reported at the ultrastructural level by Rhodin (1954) in the proximal convoluted tubules of mouse kidneys and subsequently by Rouiller and Bernhard (1956) in hepatic parenchymal cells. Soon after, these microbodies were reported in plants (Porter and Caulfield 1958; Mollenhauer et al. 1966; Frederick et al. 1968; Frederick and Newcomb 1969). The discovery of catalase and other hydrogen peroxide-producing oxidases in microbodies led to their being named peroxisomes (de Duve and Baudhuin 1966; de Duve 1969), and their recognition as major producers as well as scavengers of reactive oxygen and nitrogen species (del Río et al. 2003; Palma et al. 2009). By the mid-1960s several cytochemical procedures had been developed for the localization of catalase in peroxisomes using 3,3′ diaminobenzidine (DAB), and these formed the basis for observing peroxisomes in different cell types and organisms as well as their relationship with other organelles (Graham and Karnovsky 1966; Novikoff and Goldfischer 1969; Hirai 1969; Fahimi 1969; Frederick and Newcomb 1969; Vigil 1970). However, following the discovery and cloning of the Green Fluorescent Protein (GFP) by Prasher et al. (1992), it became possible to target GFP specifically to peroxisomes and observe them in living mammalian, yeast (Table 10.1) and plant cells (Table 10.2). Whereas fluorescent proteins of many different colors have often been targeted to peroxisomes by adding a tripeptide peroxisome targeting sequence (PTS1) at their C-terminus, many other fusion proteins have been created using PTS2 sequences as well as full-length proteins. This review deals mainly with live imaging-based studies on peroxisomes in plants but uses contextual comparisons with other systems.

10.1.2 Live Cell Imaging-Based Observations on Peroxisome Motility

The most eye-catching aspects of live imaging of peroxisomes are their bright, punctate nature and their rapid, erratic motility. Closer observations show that peroxisome motility involves oscillations, short-range movements, and long saltations. Movement can be bidirectional, and the actions of an individual peroxisome appear independent from those around it. Further, peroxisomes can travel in opposite directions along the same path simultaneously, and this movement can be against the flow of the surrounding cytosol. Plant peroxisomes, and those of yeast, move along the actin cytoskeleton and depend on myosin motors for long distance movement. In contrast, peroxisomes in animal systems have developed a microtubule-based mechanism for transport. Despite this difference the patterns of motility, and achievable velocities and displacements are similar, and in both systems oscillatory motion appears independent of the cytoskeletal system.

Table 10.1 Various fusion proteins available for the microscopic visualization of peroxisomes in a range of mammals and yeasts

PTS1/PTS2/ Pex targeted	Organism[a]	Colour[b]	Probe	Accession No.	References
Pex3p	Y (*Hansenula polymorpha*)	G	N50.Pex3.GFP	U37763	van Zutphen et al. (2008)
	Y (*Pichia pastoris*)	R	Pex3p-mRFP	N/A	Zhang et al. (2005)
	Y (*Saccharomyces cerevisiae*)	Y	pGAL1-PEX3-mYFP	NM_001180637	Hoepfner et al. (2001)
Pex8p	Y (*Pichia pastoris*)	G	GFP-Pex8p	N/A	Zhang et al. (2005)
Pex10p	Y (*Hansenula polymorpha*)	C	Pex10-CFP	N/A	Haan et al. (2006)
Pex11pα	M	Y	Pex11pα-YFP	NM_003847	Delille and Schrader (2008)
Pex11pβ	M	Y	Pex11pβ-YFP_m	N/A	Delille et al. (2010)
Pex14p	Y (*Hansenula polymorpha*)	G	Pex14-GFP	N/A	Haan et al. (2006)
Pex16p	M	G[c]	pPEX16-PAGFP	Q9Y5Y5	Kim et al. (2006)
Pex19p	M (Human)	Y	YFP-Pex19	NM_002857	Delille and Schrader (2008)
PPARα –γ	M (Monkey; Human; Mouse)	Y	EYFP-PPARα, -β, and -γ1	N/A	Feige et al. (2005)
PTS1	M	B	BFP-PTS1	–	Ito et al. (2000)
	M	G	GFP-PTS1	–	Ito et al. (2000)
	M (Human; Rat)	R	SNAFL2-PTS1	–	Dansen et al. (2001)
	Y (*Hansenula polymorpha*)	G	P_{AOX}.*eGFP.SKL*	–	Leão-Helder et al. (2003)
	Y (*Pichia pastoris*)	B	BFP-SKL	–	Farré et al. (2008)
	Y (*Pichia pastoris*)	G	GFP-SKL	–	Monosov et al. (1996)
	Y (*Saccharomyces cerevisiae*)	C	CFP-PTS1	–	Hoepfner et al. (2001)
PTS2	M	B	PTS2-BFP	–	Ito et al. (2000)
	M	G	PTS2-GFP	–	Ghaedi et al. (1999), Ito et al. (2000)
PTS1 & PTS2	M	B	PTS2-BFP-PTS1	–	Ito et al. (2000)
	M	G	PTS2-GFP-PTS1	–	Ito et al. (2000)

This list is by no means exhaustive. These are a mere select few of the currently available fluorescent proteins fused to peroxisomal targets to allow the subcellular visualization of peroxisomes in mammals and yeasts

[a]*M* mammal, *Y* yeast

[b]*B* blue, *C* cyan, *G* green, *R* red, *Y* yellow

[c]Irreversible photoconversion, off to on

Table 10.2 Various fusion proteins available for the microscopic visualization of peroxisomes in plants

PTS1/PTS2/Pex targeted	Organism	Colour[a]	Probe	Accession No.	References
Pex1p	Onion	R	TagRFP-PEX1	N/A	Goto et al. (2011)
Pex2p	Arabidopsis	Y	*At*PEX2-eYFP	AT1G79810.1	Sparkes et al. (2005)
Pex6p	Onion	R	TagRFP-PEX6	N/A	Goto et al. (2011)
Pex10p	Arabidopsis	Y	AtPEX10-YFP	AT2G26350	Sparkes et al. (2003)
Pex11pa	Arabidopsis	C	CFP-PEX11a	AT1G47750	Orth et al. (2007)
Pex11pb	Arabidopsis	C	CFP-PEX11b	AT3G47430	Orth et al. (2007)
Pex11pc	Arabidopsis	C	CFP-PEX11c	AT1G01820	Orth et al. (2007)
Pex11pd	Arabidopsis	C	CFP-PEX11d	AT2G45740	Orth et al. (2007)
Pex11pe	Arabidopsis	C	CFP-PEX11e	AT3G61070	Orth et al. (2007)
PTS1	Tobacco	C	CFP-PTS1	–	Reumann et al. (2009)
	Arabidopsis	E[b]	EosFP-PTS1	–	Sinclair et al. (2009)
	Arabidopsis	G	GFP-PTS1	–	Mano et al. (2002)
	Arabidopsis	G	GFP-SKL	–	Rodríguez-Serrano et al. (2009)
	Arabidopsis	R	RFP-PTS1	–	Lin et al. (2004)
	Arabidopsis, Onion	Y	EYFP-PTS1	–	Mathur et al. (2002)
	Arabidopsis	Y	YFP-SKL	–	Orth et al. (2007)
PTS2	Arabidopsis	G	GFP-PTS2	–	Mano et al. (2002)
	Arabidopsis	R	PTS2-RFP	–	Lin et al. (2004)

This list is by no means exhaustive. These are a mere select few of the currently available fluorescent proteins fused to peroxisomal targets to allow the subcellular visualization of peroxisomes in plants

[a]*B* blue, *C* cyan, *G* green, *R* red, *Y* yellow

[b]Photoconversion, wavelength shift

10.1.3 Actin–Myosin-Based Motility

The role of the actin cytoskeleton in plant peroxisome movement was elucidated through several key studies that performed both co-localization and drug-based experiments (Collings et al. 2002; Jedd and Chua 2002; Mano et al. 2002; Mathur et al. 2002). It was shown simultaneously by several groups that actin-disrupting drugs such as Latrunculin B and Cytochalasin D caused peroxisomes to exhibit only actin-independent oscillatory movement, while microtubule depolymerizing drugs had no apparent effect. Immunolocalization and live-cell fluorescent protein-based visualization showed that peroxisomes appeared to both associate with microfilaments and to move along them (Collings et al. 2002; Jedd and Chua 2002; Mano et al. 2002; Mathur et al. 2002). Similar experiments in yeast showed the same patterns (Chang et al. 2007; Hoepfner et al. 2001) and reinforced the view that in both systems intact actin microfilaments are required for peroxisome saltations.

Myosin motors are responsible for the transport of many organelles along actin and were thus the most likely candidate for assisting in peroxisome motility. Initial results on the effect of myosin inhibitors such as 2,3-butanedione monoxime (BDM) and *N*-ethylmaleimide (NEM) were inconclusive; one study found that BDM did not abolish peroxisome movement but NEM resulted in complete arrest (Mathur et al. 2002), while another showed that BDM did cause cessation of movement (Jedd and Chua 2002). Continued work has suggested, however, that several myosins known to be involved in the movement of other organelles may have partially redundant functions in moving peroxisomes and that this process involves calcium and ATP (Rodríguez-Serrano et al. 2009; Sparkes 2010). Fluorescent fusions with myosin sequences including MYA1 (XI-1), MYA2 (XI-2), XI-K and XI-I have been shown to at least partially localize to peroxisomes (Avisar et al. 2008; Hashimoto et al. 2005; Li and Nebenfuhr 2007; Reisen and Hanson 2007; Sparkes 2011). Mutation of either XI-2 or XI-K causes a marked decrease in peroxisome velocity but not a complete cessation of movement, while mutation of XI-1 results in only a slight decrease in average velocity (Peremyslov et al. 2008). Additionally, expression of dominant negative forms of myosins XI-K, XI-2, XI-E, or XI-F (Avisar et al. 2008; Peremyslov et al. 2008; Sparkes et al. 2008) has detrimental effects on peroxisome movement. Interestingly, though individual mutations of XI-2 and XI-K are more detrimental than XI-1, a xi-1/xi-k double mutant shows a more dramatic decrease in velocity than the xi-2/xi-k double mutant (Prokhnevsky et al. 2008). A quadruple mutant, xi-k xi-1 xi-2 xi-i, shows a complete lack of actin-dependent peroxisome movement (Peremyslov et al. 2010).

It therefore appears that in plants, peroxisome movement occurs through the orchestrated effort of many different Class XI myosins, which contrasts from yeast systems where only single peroxisome-associated myosins have been described. In the yeast *Saccharomyces cerevisiae*, this association appears to involve the myosin Myo2p (Hoepfner et al. 2001), a peroxisomal membrane protein, Inp2p, that mediates peroxisome attachment to myosin (Fagarasanu et al. 2006), and a second peroxisomal membrane protein, Inp1p, that can bind to an unknown element on the cell surface and acts to anchor a peroxisome in place (Fagarasanu et al. 2005; Chang et al. 2007). In *Yarrowia lipolytica*, the peroxisome biogenesis protein Pex3p and a paralogue, Pex3pB, may play partly redundant roles in the peroxisome–myosin interaction (Chang et al. 2009). A peroxisomal membrane protein that performs a similar function in plants remains to be identified.

The motility of peroxisomes along the actin cytoskeleton can be seen to change depending upon cell type, developmental stage, and environmental stress. A correlation has been drawn between cell size and speed of peroxisome movement within plant cells, wherein elongated cells exhibit more rapid motility. Thus peroxisomes are seen to move most quickly in root phloem, hypocotyl, trichome, root, or epidermal cells, while those in the mesophyll or guard cells move relatively slowly (Jedd and Chua 2002; Mano et al. 2002; Muench and Mullen 2003). The developmental stage of the plant also appears to affect the rate of peroxisome movement, as it has been shown that in leaf epidermal cells peroxisomes move more slowly in young plants and increase speed with maturity (Rodríguez-Serrano et al. 2009). The opposite trend can be seen in a growing root, wherein young elongated cells toward the tip show more rapid peroxisomal movement than older cells further up the root (Jedd and Chua 2002). Additionally, various stress conditions have noticeable effects on plant peroxisome movement. In response to exposure to hydrogen peroxide or high light, a stress known to induce hydrogen peroxide production, peroxisomes become less motile in terms of both velocity and average displacement. Prolonged exposure results in near complete cessation of movement, but upon removal of the stressor, peroxisomes can be seen to slowly recover. They exhibit first a wriggling motion before slowly returning to normal motility, a process that is often accompanied by peroxisome fission (Sinclair et al. 2009). Though hydrogen peroxide exposure appears to cause a cessation in peroxisome movement, exposure to superoxide radicals in this experiment for up to 2 h had no noticeable effect. A separate study however showed that superoxide exposure induced by treatment with cadmium causes peroxisomes to increase in speed over the course of several days, until the rate of motility has more than doubled (Rodríguez-Serrano et al. 2009). Though these stresses both result from ROS exposure, they appear to have very different effects on the peroxisome. It is thus possible that peroxisomes respond with unique behavior to different ROS types; however, it is important to note the difference in time frame, as one experiment focused on changes in the range of minutes (Sinclair et al. 2009) and the second in days (Rodríguez-Serrano et al. 2009).

Peroxisome movement in plants thus depends on the coordinated effort of the actin cytoskeleton, several myosin motors, and potentially at least one unidentified linker protein; this behavior can be modified depending on internal or external factors, presumably through modification of the movement system at one of these key points. Though microtubules are not involved in peroxisome motility in plants, correlations have nonetheless been drawn between them and peroxisome activity. Both the pausing and oscillatory behavior that is often exhibited by peroxisomes has been found to occur a significant portion of the time at junctions between microtubules and microfilaments. However despite this correlation, microtubule-disrupting drugs have no apparent effect on the frequency of pausing (Chuong et al. 2005; Hamada et al. 2012). It has moreover been shown that several peroxisomal matrix proteins possess tubulin-binding activities and that the import of at least one of these, the peroxisomal multifunctional protein, is less efficient when microtubules are disrupted (Chuong et al. 2005).

10.1.4 Microtubule-Based Motility

In contrast to the results found in plant and yeast cells, peroxisome movement in animal cells is believed to be microtubule based. Treatment of cells with microtubule disrupting drugs such as nocodazole, coclemid, vinblastine, demecocline, and oryzalin caused peroxisomes to stop long-range saltations and also demonstrated that, like in plants and yeast, oscillatory motion is independent of the cytoskeleton (Huber et al. 1997; Rapp et al. 1996; Schrader et al. 2000; Schrader 2001; Wiemer et al. 1997). Co-visualization in both live and fixed cells showed the close association between peroxisomes and microtubules (Kulic et al. 2008; Rapp et al. 1996). Peroxisome movement in animal cells thus occurs along microtubule-based tracks and additionally is not dependent on microtubule dynamics, as microtubule stabilizing drugs have no effect (Wiemer et al. 1997).

Kinesin and dynein are motor proteins required for microtubule-based movement within the cell, where kinesin controls movement to the positive end of microtubules, typically oriented toward the cell periphery, and dynein controls movement in the minus direction, toward the center of the cell (Kural et al. 2005). Work in both mammalian and Drosophila cells have shown that these motors are both involved in peroxisome motility (Ally et al. 2009; Kaur et al. 2009; Kulic et al. 2008; Kural et al. 2005; Schrader et al. 2000). Though they control movement along microtubules in different directions, the motors appear to act in concert and require each other for successful transport. Consequently, knockdown of either protein results in a cessation of peroxisome movement (Ally et al. 2009; Kaur et al. 2009; Kural et al. 2005) while neither can induce motility on their own in in vitro conditions (Schrader et al. 2000). Additionally, knockdown of a subunit of the dynactin complex that mediates dynein activity disrupted peroxisome motility (Kaur et al. 2009). The process of peroxisome movement along microtubules has additionally been shown to be Ca2+ and ATP dependent and involve regulation by the lysophosphatidic acid (LPA) receptor and G proteins (Huber et al. 1997, 1999; Rapp et al. 1996; Wiemer et al. 1997). The bidirectional movement exhibited by many peroxisomes likely results from the transfer of a peroxisome from one motor to another (Kural et al. 2005).

Similar to the plant system, a peroxisomal membrane protein that serves as a kinesin, dynein, or microtubule anchor has not been identified to date. It has however been found that the peroxisomal import protein Pex14p may at least partially play this role, as it has been shown to bind tubulin directly and peroxisomal remnants in Pex14p-deficient cells lack microtubule-based motility (Bharti et al. 2011). A CLIP-like linker protein that mediates direct peroxisome binding to microtubules has been proposed to exist (Schrader et al. 2003). The Rho type protein, RhoA, may play a role in mediating this microtubule binding, as inactive RhoA allows peroxisome movement, while upon exposure to dominant active RhoA, peroxisomes cease long distance saltations and exhibit only oscillatory movement (Schollenberger et al. 2010).

Interestingly, though the actin cytoskeleton does not play a significant role in peroxisome movement in animal cells, Cytochalasin D does cause a 25 % drop in peroxisome speed in treated cells (Rapp et al. 1996). It has been suggested that actin may play a role in controlling peroxisome features such as size, shape, and number (Schollenberger et al. 2010).

10.1.5 Other Views on Peroxisomal Motility

Whereas peroxisome motility is typically attributed to their interactions with the cytoskeleton and associated motors, the fission yeast, *Schizosaccharomyces pombe*, suggests an exception. Drug treatments showed that peroxisome movement in fission yeast is independent of microtubules or microfilaments and instead utilizes mitochondria for transportation (Jourdain et al. 2008). In plants, the possibility had been suggested that small peroxisomal vesicles might be moved around through an actin polymerization-based mechanism (Mathur et al. 2002). Whereas the direct visualization of comet like actin tails as observed in the "rocketing motility" of bacteria and vesicles (Machesky 1999) has not been provided, peroxisome movement is aberrant in mutants in different subunits of the actin-related protein (ARP) 2/3 complex (Mathur et al. 2003). Through more recent observations aimed at understanding the erratic path taken by peroxisomes in cortical areas of the plant cell, it is suggested that peroxisome movement occurs in conjunction with the movement of the ER, possibly due to peroxisome–ER membrane contact sites (Barton et al. 2013).

10.2 Live Cell Imaging-Based Observations on Peroxisome Morphology

Fluorescent protein-aided highlighting of peroxisomes in different organisms shows them predominantly as spherical organelles with di-ameters ranging from 0.4 to 1.5 μm (Fig. 10.1a). However, other shapes have been published including tubular peroxisomes whose aggregations can suggest a reticulum (Schrader et al. 1999; Fig. 10.1d), individual peroxisomes with a torus shape (Cutler et al. 2000), and peroxisomes with extended tails (Delille et al. 2010; Mano et al. 2004; Fig. 10.1d). Most often spherical peroxisomes can change into tubules (Fig. 10.1b, c) with lengths ranging from 3 to 7 μm and diameters ranging from 0.3 to 0.7 μm. The elongation of a peroxisome generally indicates the addition of extramembrane proteins and lipids to the existing organelle in preparation for its fission (Koch and Brocard 2011). In plants, peroxin 11 isoforms (Lingard and

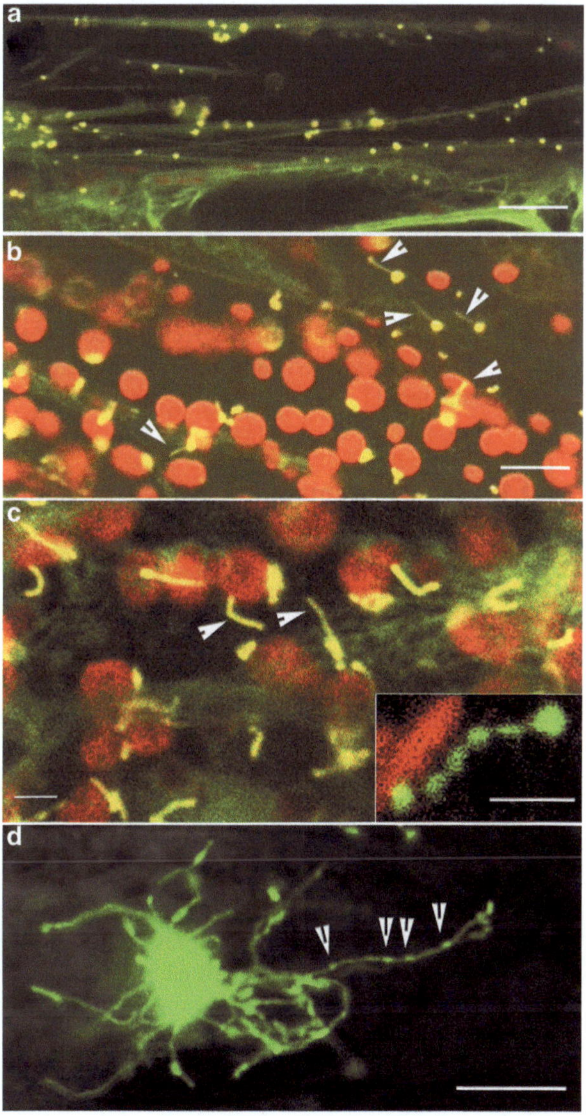

Fig. 10.1 Live imaging of peroxisomes using targeted fluorescent proteins shows the morphological range that they may display. (**a**) Small peroxisomes of 0.4–1.5 μm diameter (*yellow* due to YFP-PTS1) align and move along actin filaments (*green* due to GFP:mTalin). (**b**) A single plant cell showing peroxisomes (*yellow*) that extend and retract thin tubular extensions called peroxules (*arrowheads*) sporadically in response to increased subcellular hydroxyl stress. (**c**) A plant cell exhibiting elongated peroxisomes (*yellow*; *arrowheads*) upon exposure to 0.8 M H_2O_2 for 60 s. The inset "**c**" shows the subsequent beading and bead separation in an elongated peroxisome. (**d**) A cluster of tubular peroxisomes observed in the apm1 mutant of Arabidopsis. One of the tubules exhibits fluorescent bulges (*arrowheads*) that appear sporadically as the tubule stretches out. Tubular peroxisomes morph constantly through contiguity with neighboring organelles. Chloroplasts are in *red* due to chlorophyll autofluorescence. Size bars in **a**, **d** = 10; **b**, **c**, *c* = 5 μm

Trelease 2006; Desai and Hu 2008) are implicated in this process since their overexpression results in long, smooth peroxisomes (Lingard and Trelease 2006; Kobayashi et al. 2007; Nito et al. 2007). Under normal conditions the tubular peroxisomes become progressively constricted through the activity of protein complexes including dynamin-related proteins (DRP3A and DRP3B) and associated anchor proteins (FIS1A and FIS1B) (Zhang and Hu 2009; Aung et al. 2010). At this stage the tubular peroxisome appears beaded (Fig. 10.1c). It contorts repeatedly and finally fragments into smaller peroxisomes (Fig. 10.1c-inset) that are separated over time.

Live imaging of peroxisomes in plants has drawn attention to an elastic tubular peroxisomal extension that appears sporadically soon after a cell is exposed to conditions that cause oxidative stress (Sinclair et al. 2009; Fig. 10.1b). This transient state suggests that fresh membrane components are being imported into an existing spherical peroxisome and thus causing it to elongate. Based purely on its morphological resemblance to stroma-filled tubules called "stromules" from plastids (Kolher and Hanson 2000), this transient peroxisomal form was given the name "peroxule" (Scott et al. 2007). Since the introduction of "peroxules" into peroxisomal literature is relatively recent (Scott et al. 2007; Sinclair et al. 2009), here we provide its detailed description and suggest where we consider it appropriate to use the term.

10.2.1 Peroxules: Imaging Artifacts or Actual Extensions of Peroxisomes?

A mild increase in subcellular stress such as a short exposure to H_2O_2 or to high intensity light leads to the extension of peroxules (Sinclair et al. 2009). These elastic extensions retract quickly to the parent peroxisome body if the causal stress is short lived, but the whole peroxisome progresses into a tubule of uniform diameter if the stress persists (Fig. 10.2a). However, many peroxisomes are irregularly shaped and not completely spherical. Therefore when viewed at certain angles they give the impression of having small projections (Fig. 10.2b–e). Similarly a peroxisome undergoing fission can have its numerous parts connected by narrow necks, rotation around which can present views of different faces of the organelle (Fig. 10.2f). We have found while visualizing peroxisomes and other small motile organelles through optical sections in an *x*,*y* plane over time (*x*,*y*,*t*), that their rotation can convey the impression of small projections where none exist. The rotation of the organelles can often be discerned by following several consecutive frames after the projection first appears. These artifacts result from the limited ability of most older laser scanning confocal microscopes to perform rapid four-dimensional (*x*,*y*,*z*,*t*) scans that would allow the rapid creation of a complete 3D image of an organelle. The power of confocal microscopy lies in its ability to eliminate out-of-focus blur and provide sharp images. However, the flip-side of achieving confocality is that it limits the plane of view to a very shallow *z*-axis resolution. In practice, this does not pose a problem in observing large or nearly static organelles since gradual, sequential changes in the focal plane allow the instrument to be operated as though it has an infinite depth of field. Nevertheless when the size of a motile organelle is nearly equal to or lies within the *z*-step size being applied during optical sectioning, the view of the object changes rapidly from one scan to another (Fig. 10.2g). While visualizing peroxisomes and peroxules we considered all these possible artifacts before concluding that transient, tubular extensions from peroxisomes are real. Our most convincing evidence is based on

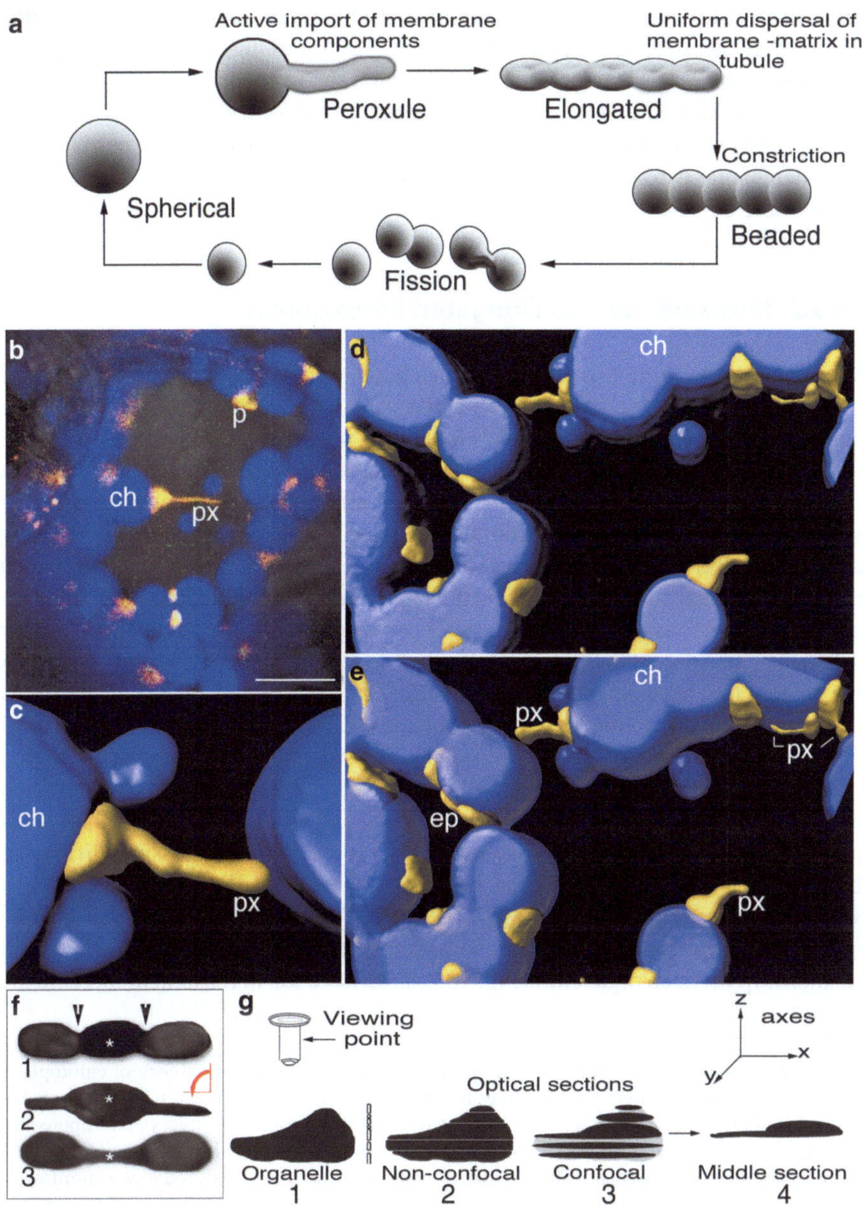

Fig. 10.2 Peroxules in plant cells and possible imaging artifacts. (**a**) Diagrammatic depiction of the intermediate state of peroxule extension and retraction as a spherical-appearing peroxisome accumulates fresh membranes and undergoes elongation into a tubular form which in turn undergoes constriction (beading) and fission to produce more peroxisomes, each capable of importing matrix proteins and growing in size. (**b**) A confocal section showing peroxisomes (p) associated with chloroplasts (ch) following a 2-min irradiation with white light of 1,500 μMol m^{-2} s^{-2} intensity. A peroxule (px) extended by a peroxisome is shown (*arrowhead*). Size bar = 10 μm. (**c**) An isosurface volume rendered portion of the image shown in "**b**" using Imaris software shows a tubular peroxule (px) extending away from a chloroplast (ch). (**d–e**) Isosurface volume rendering of an image stack shows peroxi-somes in different stages. Whereas

observation of peroxisomes in the vicinity of chloroplasts (Fig. 10.2b–e). A thin tubule may be stretched out to more than 5–8 times the length of its parent peroxisome (Fig. 10.2b, c). The peroxule may be smooth tipped or display a small beaded structure at its end. Such a peroxisomal form described by Jedd and Chua (2002) was considered as a budding peroxisome. Whereas most peroxisomes display a single peroxule, occasionally we have observed peroxules from a single peroxisome being extended in different directions simultaneously (Fig. 10.1b).

10.2.2 Peroxules Versus Elongated Peroxisomes

While using the term peroxule, we considered whether it is actually useful for describing a particular stage in the peroxisomal cycle or would merely create confusion with the generally used "elongated" peroxisome. We suggest that use of the term "peroxule" be limited to the specific, transient state during which a peroxisome is actively acquiring membrane components and elongating progressively (Fig. 10.2a). By comparison an elongated peroxisome would have already acquired membrane constituents and adjusted the shape into a tubule within which matrix components are regularly distributed. Compared to the peroxule, the elongated peroxisome is relatively inelastic; it can bend and distort but does not appear to retract into a parent body. Indeed, there is no separate parent body any longer as the matrix has become redistributed within the tubule. The elongated tubule is susceptible to constriction by mechanochemical GTPases like DRP3A and DRP3B which create a beaded tubule phenotype in preparation for subsequent fission (Fig. 10.2a). The term peroxules thus gives a specific morphological interpretation to the preparatory steps of membrane protrusion and elongation described by Koch and Brocard (2011) for peroxisome proliferation. The progressive difference between elongating and elongated becomes apparent in 3D reconstructions and volume rendered image stacks (Fig. 10.2d, e). While maintaining the full opacity during image rendering allows an appreciation of overall morphology and spatial relationship between organelles (Fig. 10.2d), we have found it useful to increase the

Fig. 10.2 (continued) "**d**" maintains full opacity, "**e**" has increased transparency of chloroplasts which allows an impression of peroxisomes and their extension behind the plastid. It is possible to distinguish between peroxules (px) and elongated-beaded peroxisomes (ep) as well as appreciate that many peroxisomes are not perfectly spherical. (**f**) Artifact depicting how a single organelle "1" with constrictions (*arrowheads*) along its length that create narrow necks around which different portions of the organelle can freely rotate, can be viewed in the *xyz* axes and be misinterpreted as showing profiles that differ considerably from its actual shape. The ends of a single tubular clay model "1" were twisted by 90° for image "2" and the entire model rotated by 90° for obtaining image "3." The *asterisk* in 1,2,3 shows the same area which appears convex-lens shaped or thin tubular depending upon the view. (**g**) Optical sectioning using a confocal microscope limits the depth of field in the *z*-axis. The wedge shaped organelle shown in "1" gives a different profile in optical section "3." Whereas under normal epifluorescence most profiles would appear spherical due to the blooming effect of fluorescent objects the confocal optical sections limit the depth of field and the mid-section "4" could suggest a tubule being extended by a larger main body

transparency of chloroplasts to distinguish between the simple tubular peroxule and the beaded elongated peroxisome ready for fission (Fig. 10.2e).

The progressive relationship between a peroxule and a completely elongated peroxisome was also investigated indirectly through another approach. It is known that prolonged stress or increased ROS levels results in peroxisomes appearing as narrow, beaded tubules rather than single spheres (Schrader et al. 1999). Two possibilities exist for achieving the tubular form: the fusion of many individual peroxisomes to form a single elongated organelle or the stretching out of a single peroxisome into a tubule. Sinclair et al. (2009) used differential coloring of peroxisomes that were targeted by a green to red photoconvertible monomeric EosFP for discriminating between the two possibilities. Photoconversion of a few individual peroxisomes in a cell converted them to red fluorescent bodies while the rest remained green. Individual, differentially colored peroxisomes exhibited peroxules of the same color as the parent body. Subsequent treatment with H_2O_2 resulted in the cell displaying elongated peroxisomes. The experiment relied on the observation that if the resultant tubule consisted of beads of different color then fusion of multiple peroxisomes could be concluded. Alternatively tubules consisting of beads of a single color only would suggest that a single peroxisome elongated into a tubule and became beaded subsequently. Sinclair et al. (2009) observed the latter event consistently and concluded that a single peroxisome elongates considerably.

10.2.3 Should There Be Medial and Terminal Peroxules?

The tubular peroxisomes in the apm1 mutant of Arabidopsis (Mano et al. 2004) display an interesting phenomenon wherein a GFP-labelled tubule sporadically displays small bulges along its length (Fig. 10.1d arrowheads). Similar transient bodies of GFP have often been observed in plastid stromules, where they have been described as "giving the appearance of a thin snake that has swallowed a small rodent" (Hanson and Sattarzadeh 2011). Like peroxisomes, mitochondria are also abnormally elongated in the apm1/drp3A mutant (Mano et al. 2004). According to Logan et al. (2004) elongated mitochondria exhibit a similar feature where narrow regions connect two broader regions in a tubular mitochondrion. The narrow region of the mitochondrion has been called a "matrixule" and based on the location, a discrimination is made between a terminal matrixule and a medial matrixule (Logan et al. 2004; Scott et al. 2007; Movie 3: http://www.plantmitochondria.net/Plant_Mitochondria/Movies.html). The mechanism for the sporadic narrowing and broadening of tubules has not been elucidated so far in any of the publications reporting them (Hanson and Sattarzadeh 2011; Logan et al. 2004; Scott et al. 2007). We speculate that the observation might either be an artifact of fluorescence microscopy or a result of molding of a tubule through its close alignment with other organelles around them (Mathur et al. 2012). However in peroxisomes, since the transient bulges in a tubule appear to be connected by the thinner areas of the same tubule, unlike the name "medial matrixule" given by Logan et al. (2004) we do not consider them as medial peroxules. As mentioned earlier we reserve use of the term to clearly tubular, elastic extensions from individual peroxisomes rather than portions of a tubule within a single elongated peroxisome.

10.3 Simultaneous Visualization of Peroxisomes and the ER

A large number of publications based on transmission electron microscopy have suggested the association between peroxisomes and the endoplasmic reticulum (ER; Novikoff and Shin 1964; Svoboda et al. 1967; Magalhaes and Magalhaes 1971; Novikoff et al. 1973, Frederick et al. 1975; Hirai et al. 1983; Tabak et al. 2003; Geuze et al. 2003; van der Zand et al. 2006). The peroxisome field has debated the role of the ER in the peroxisome life-cycle for over 50 years and appears to have concluded that peroxisomes are endomembrane derivatives (Kunau 2005; Titorenko and Rachubinski 2009; Hu et al. 2012). Certain peroxisomal proteins have been shown to transit through the ER (Mullen et al. 1999; Kim et al. 2006; Karnik and Trelease 2007) and lend support to their ER biogenesis (Mullen and Trelease 2006; Titorenko and Rachubinski 2009; Hu et al. 2012).

While trying to understand the dynamic behavior of peroxules and elongated peroxisomes, Sinclair et al. (2009) reported that peroxules extend along paths defined by the ER. Recently we have undertaken a detailed investigation through the simultaneous visualization of peroxisomes and the ER in Arabidopsis plants co-expressing fluorescent probes for both organelles (Barton et al. 2013). The co-visualization suggests contiguity between peroxisomes and the ER but provides no sign of their luminal continuity. While the overexpression of some peroxisomal proteins like peroxisomal ascorbate peroxidase (Mullen et al. 1999) and Pex16p (Karnik and Trelease 2005, 2007) to produce thin ER-like tubules has been considered as indicative of pre-peroxisomal ER domains involved in peroxisome biogenesis, there is no actual evidence in plants of peroxisomes budding from the ER (Trelease and Lingard 2006). By contrast, observations by Sinclair et al. (2009) suggest that the actual behaviour of peroxisomes can be attributed to the possible presence of membrane contact sites (MCS) between the ER and peroxisomes. Notably the dynamic behaviour of peroxisomes is very similar to mitochondria for which MCS with the ER involving specific proteins have been established (English and Voeltz 2013).

10.4 Visualization of Peroxisomes and Their Relationship to Other Organelles and Vesicles

Peroxisomes, as major components of the cellular ROS scavenging machinery, are affected by the general activity in a living cell. It is also known that a majority of peroxisomal proteins are imported directly from the cytosol while some proteins may be routed via the ER. Recent evidence favors the idea that peroxisomes can be assembled through the fusion of biochemically distinct vesicles from the ER

(Titorenko et al. 2000; van der Zand et al. 2010, 2012; van der Zand and Tabak 2013). Alternatively cell free sorting of peroxisomal membrane proteins such as pex11p, and pex3p budding from the ER has been demonstrated in vitro (Lam et al. 2010; Agrawal et al. 2011). It would be interesting to see if peroxisome assembly through vesicle fusion can be observed in plant cells where there is no evidence for peroxisome formation directly from the ER (Trelease and Lingard 2006).

The peroxisome–mitochondria relationship has been explored in the drp3a/apm1 mutants and it has been shown that both organelles are affected by the same mechanochemical dynamin-related protein (Fujimoto et al. 2009; Aung et al. 2010). In addition it has been shown in animal cells that mitochondria derived vesicles (MDVs) deliver specific cargo to peroxisomes (Andrade-Navarro et al. 2009; Braschi et al. 2010). Whether a similar phenomenon involving MDVs occurs in plants has not been explored. Moreover, in plants, plastids provide an extra source of ROS in addition to mitochondria and endomembranes. Excellent electron micrographs supported thorough biochemical studies (Tolbert and Yamazaki 1969) have shown peroxisome activity to be intimately linked with light mediated activity of chloroplasts (Desai and Hu 2008). Live imaging has already revealed the rapid extension of peroxules from peroxisomes in response to illumination (Sinclair et al. 2009). The precise relationship between plastids, peroxisomes, mitochondria, and the ER remains unexplored but should provide interesting information on the hierarchy of subcellular response and the maintenance of homeostasis in plant cells.

Conclusions

Live imaging of peroxisomes has revealed details about the cytoskeleton mediated motility of these organelles and has further potential for identifying specific motor molecules involved in the process. It is still unclear how the same organelle might be able to utilize two completely different cytoskeletal tracks and motors, or other speculated mechanisms for motility. Similarly, visualization of the relationship between peroxisomes and other organelles, especially the ER, is already providing fresh insights for understanding these organelles. Whereas much of the cell biological research being contemplated now requires the combined efforts of live imaging and high-resolution electron microscopy, for peroxisomes specifically, the identification of proteins involved in their membrane contact sites with other organelles as well as mechanisms of their reconstitution from assorted vesicles remain exciting prospects that can still be explored through live imaging.

Acknowledgments JM acknowledges funding from the Natural Sciences and Engineering Research Council of Canada (NSERC), the Canada Foundation for Innovation (CFI), and the Ministry of Research and Innovation, Ontario, Canada.

References

Agrawal G, Joshi S, Subramani S (2011) Cell-free sorting of peroxi-somal membrane proteins from the endoplasmic reticulum. Proc Natl Acad Sci U S A 108:9113–9118

Ally S, Larson AG, Barlan K et al (2009) Opposite-polarity motors activate one another to trigger cargo transport in live cells. J Cell Biol 187:1071–1082

Andrade-Navarro MA, Sanchez-Pulido L, McBride HM (2009) Mi-tochondrial vesicles: an ancient process providing new links to peroxi-somes. Curr Opin Cell Biol 21:560–567

Aung K, Zhang X, Hu J (2010) Peroxisome division and prolifera-tion in plants. Biochem Soc Trans 38:817–822

Avisar D, Prokhnevsky AI, Makarova KS et al (2008) Myosin XI-K is required for rapid trafficking of Golgi stacks, peroxisomes, and mitochondria in leaf cells of Nicotiana benthamiana. Plant Physiol 146:1098–1108

Barton K, Mathur N, Mathur J (2013) Simultaneous live-imaging of peroxisomes and the ER in plant cells suggests contiguity but no luminal continuity between the two organelles. Front Physiol 4:196

Bharti P, Schliebs W, Schievelbusch T et al (2011) PEX14 is re-quired for microtubule-based peroxisome motility in human cells. J Cell Sci 124:1759–1768

Braschi E, Goyon V, Zunino R et al (2010) Vps35 mediates vesicle transport between the mitochondria and peroxisomes. Curr Biol 20:1310–1315

Chang J, Fagarasanu A, Rachubinski RA (2007) Peroxisomal pe-ripheral membrane protein YlInp1p is required for peroxisome in-heritance and influences the dimorphic transition in the yeast Yar-rowia lipolytica. Eukaryot Cell 6:1528–1537

Chang J, Mast FD, Fagarasanu A et al (2009) Pex3 peroxisome bio-genesis proteins function in peroxisome inheritance as class V myosin receptors. J Cell Biol 187:233–246

Chuong SDX, Park N-I, Freeman MC et al (2005) The peroxisomal multifunctional protein interacts with cortical microtubules in plant cells. BMC Cell Biol 6:40

Collings DA, Harper JDI, Marc J et al (2002) Life in the fast lane: actin-based motility of plant peroxisomes. Can J Bot 80:430–441

Cutler SR, Ehrhardt DW, Griffitts JS et al (2000) Random GFP::cDNA fusions enable visualization of subcellular structures in cells of Arabidopsis at a high frequency. Proc Natl Acad Sci U S A 97:3718–3723

Dansen TB, Pap EHW, Wanders RJA et al (2001) Targeted fluores-cent probes in peroxisome function. Histochem J 33:65–69

De Duve C (1969) The peroxisome: a new cytoplasmic organelle. Proc R Soc Lond B Biol Sci 173:71–83

De Duve C, Baudhuin P (1966) Peroxisomes (microbodies and re-lated particles). Physiol Rev 46:323–357

del Río LA, Corpas FJ, Sandalio LM et al (2003) Plant peroxisomes, reactive oxygen metabolism and nitric oxide. IUBMB Life 55:71–81

Delille HK, Schrader M (2008) Targeting of hFis1 to peroxisomes is mediated by Pex19p. J Biol Chem 283:31107–31115

Delille HK, Agricola B, Guimaraes SC et al (2010) Pex11pβ-mediated growth and division of mammalian peroxisomes follows a maturation pathway. J Cell Sci 123:2750–2762

Desai M, Hu J (2008) Light induces peroxisome proliferation in Ar-abidopsis seedlings through the photoreceptor phytochrome A, the transcription factor HY5 HOMOLOG, and the peroxisomal protein PEROXIN11b. Plant Physiol 146:1117–1127

English AR, Voeltz GK (2013) Endoplasmic reticulum structure and interconnections with other organelles. Cold Spring Harb Perspect Biol 5:a013227

Fagarasanu A, Fagarasanu M, Eitzen GA et al (2006) The peroxiso-mal membrane protein Inp2p is the peroxisome-specific receptor for the myosin V motor Myo2p of Saccharomyces cerevisiae. Dev Cell 10:587–600

Fagarasanu M, Fagarasanu A, Tam YYC et al (2005) Inp1p is a pe-roxisomal membrane protein required for peroxisome inheritance in Saccharomyces cerevisiae. J Cell Biol 169:765–775

Fahimi HD (1969) Cytochemical localization of peroxidatic activity of catalase in rat hepatic microbodies (peroxisomes). J Cell Biol 43:275–288

Farré JC, Manjithaya R, Mathewson RD et al (2008) PpAtg30 tags peroxisomes for turnover by selective autophagy. Dev Cell 14:365–376

Feige JN, Gelman L, Tudor C et al (2005) Fluorescence imaging re-veals the nuclear behaviour of peroxisome proliferator-activated re-ceptor/retinoid X receptor heterodimers in the absence and presence of ligand. J Biol Chem 280:17880–17890

Frederick SE, Newcomb EH (1969) Microbody-like organelles in leaf cells. Science 163:1353–1355

Frederick SE, Newcomb EH, Vigil EL et al (1968) Fine-structural characterization of plant microbodies. Planta 81:229–252

Frederick SE, Gruber PJ, Newcomb EH (1975) Plant microbodies. Protoplasma 84:1–29

Fujimoto M, Arimura S, Mano S et al (2009) Arabidopsis dynamin-related proteins DRP3A and DRP3B are functionally redundant in mi-tochondrial fission, but have distinct roles in peroxisomal fission. Plant J 58:388–400

Geuze HJ, Murk JL, Stroobants AK et al (2003) Involvement of the endoplasmic reticulum in peroxisome formation. Mol Biol Cell 14:2900–2907

Ghaedi K, Kawai A, Okumoto K et al (1999) Isolation and charac-terization of novel ovary cell mutants using Green Fluorescent Protein. Exp Cell Res 248:489–497

Goto S, Mano S, Nakamori C et al (2011) Arabidopsis ABERRANT PEROXISOME MORPHOLOGY9 is a peroxin that recruits the PEX1-PEX6 complex to peroxisomes. Plant Cell 23:1573–1587

Graham RC, Karnovsky MJ (1966) The early stages of absorption of injected horseradish peroxidase in the proximal tubules of mouse kidney: ultrastructural cytochemistry by a new technique. J Histochem Cytochem 14:291–302

Haan GJ, Baerends RJS, Krikken AM et al (2006) Reassembly of peroxisomes in Hansenula polymorph pex3 cells on reintroduction of Pex3p involves the nuclear envelope. FEMS Yeast Res 6:186–194

Hamada T, Tominaga M, Fukaya T et al (2012) RNA processing bodies, peroxisomes, Golgi bodies, mitochondria, and endoplasmic reticulum tubule junctions frequently pause at cortical microtubules. Plant Cell Physiol 53:699–708

Hanson MR, Sattarzadeh A (2011) Stromules: Recent insights into a long neglected feature of plastid morphology and function. Plant Physiol 155:1486–1492

Hashimoto K, Igarashi H, Mano S et al (2005) Peroxisomal localization of a myosin XI isoform in Arabidopsis thaliana. Plant Cell Physiol 46:782–789

Hirai KI (1969) Light microscopic study of the peroxidatic activity of catalase in formaldehyde-fixed rat liver. J Histochem Cytochem 17:585–590

Hirai K, Yamauchi M, Witschi H et al (1983) Disintegration of lung peroxisomes during differentiation of type II cells to type I cells in bu-tylated hydroxytoluene-administered mice. Exp Mol Pathol 39:129–138

Hoepfner D, van den Berg M, Philippsen P et al (2001) A role for Vps1p, actin, and the Myo2p motor in peroxisome abundance and in-heritance in Saccharomyces cerevisiae. J Cell Biol 155:979–990

Hu J, Baker A, Bartel B et al (2012) Plant peroxisomes: biogenesis and function. Plant Cell 24:2279–2303

Huber CM, Saffrich R, Anton M et al (1997) A heterotrimeric G protein-phospholipase A2 signaling cascade is involved in the regu-lation of peroxisomal motility in CHO cells. J Cell Sci 110:2955–2968

Huber CM, Saffrich R, Ansorge W et al (1999) Receptor-mediated regulation of peroxisomal motility in CHO and endothelial cells. EMBO J 18:5476–5485

Ito M, Ito R, Huang Y et al (2000) Rapid isolation and characteriza-tion of CHO mutants deficient in peroxisome biogenesis using the pe-roxisomal forms of fluorescent proteins. Mol Cell Res 1496:232–242

Jedd G, Chua NH (2002) Visualization of peroxisomes in living plant cells reveals acto-myosin-dependent cytoplasmic streaming and peroxisome budding. Plant Cell Physiol 43:384–392

Jourdain I, Sontam D, Johnson C et al (2008) Dynamin-dependent biogenesis, cell cycle regulation and mitochondrial association of peroxisomes in fission yeast. Traffic 9:353–365

Karnik SK, Trelease RN (2005) Arabidopsis peroxin 16 coexists at steady state in peroxisomes and endoplasmic reticulum. Plant Physiol 138:1967–1981

Karnik SK, Trelease RN (2007) Arabidopsis peroxin 16 trafficks through the ER and an intermediate compartment to pre-existing pe-roxisomes via overlapping molecular targeting signals. J Exp Bot 58:1677–1693

Kaur N, Reumann S, Hu J (2009) Peroxisome biogenesis and func-tion. Arabidopsis Book 7:e0123

Kim PK, Mullen RT, Schumann U et al (2006) The origin and maintenance of mammalian peroxisomes involves a de novo PEX16-dependent pathway from the ER. J Cell Biol 173:521–532

Kobayashi S, Tanaka A, Fujiki Y (2007) Fis1, DLP1, and Pex11p coordinately regulate peroxisome morphogenesis. Exp Cell Res 313:1675–1686

Koch J, Brocard C (2011) Membrane elongation factors in organelle maintenance: the case of peroxisome proliferation. Biomol Concept 2:353–364

Kolher RH, Hanson MR (2000) Plastid tubules of higher plants are tissue-specific and developmentally regulated. J Cell Sci 113:81–89

Kulic IM, Brown AEX, Kim H et al (2008) The role of microtubule movement in bidirectional organelle transport. Proc Natl Acad Sci 105:10011–10016

Kunau WH (2005) Peroxisome biogenesis: end of the debate. Curr Biol 15:R774–R776

Kural C, Kim H, Syed S et al (2005) Kinesin and dynein move a pe-roxisome in vivo: a tug-of-war or coordinated movement? Science 308:1469–1472

Lam SK, Yoda N, Schekman R (2010) A vesicle carrier that medi-ates peroxisome protein traffic from the endoplasmic reticulum. Proc Natl Acad Sci U S A 107:21523–21528

Leão-Helder AN, Krikken AM, van der Klei IJ et al (2003) Tran-scriptional down-regulation of peroxisome numbers affects selective peroxisome degradation in Hansenula polymorpha. J Biol Chem 278:40749–40756

Li J, Nebenfuhr A (2007) Organelle targeting of myosin XI is medi-ated by two globular tail domains with separate cargo binding sites. J Biol Chem 282:20593–20602

Lin Y, Cluette-Brown JE, Goodman HM (2004) The peroxisome de-ficient Arabidopsis mutant sse1 exhibits impaired fatty acid synthesis. Plant Physiol 135:814–827

Lingard MJ, Trelease RN (2006) Five Arabidopsis peroxin 11 hom-ologs individually promote peroxisome elongation, duplication or ag-gregation. J Cell Sci 119:1961–1972

Logan DC, Scott I, Tobin AK (2004) ADL2a, like ADL2b, is in-volved in the control of higher plant mitochondrial morphology. J Exp Bot 55:783–785

Machesky LM (1999) Rocket-based motility: a universal mecha-nism? Nat Cell Biol 1:E29–E31

Magalhaes MM, Magalhaes MC (1971) Microbodies (peroxisomes) in rat adrenal cortex. J Ultrastruct Res 37:563–573

Mano S, Nakamori C, Hayashi M et al (2002) Distribution and characterization of peroxisomes in Arabidopsis by visualization with GFP: dynamic morphology and actin-dependent movement. Plant Cell Physiol 43:331–341

Mano S, Nakamori C, Kondo M et al (2004) An Arabidopsis dy-namin-related protein, DRP3A, controls both peroxisomal and mito-chondrial division. Plant J 38:487–498

Mathur J, Mathur N, Hulskamp M (2002) Simultaneous visualiza-tion of peroxisomes and cytoskeletal elements reveals actin and not microtubule-based peroxisome motility in plants. Plant Physiol 128:1031–1045

Mathur J, Mathur N, Kirik V et al (2003) Arabidopsis CROOKED encodes for the smallest subunit of the ARP2/3 complex and controls cell shape by region specific fine F-actin formation. Development 130:3137–3146

Mathur J, Mammone A, Barton K (2012) Organelle extensions in plant cells. J Int Plant Biol 54:851–867

Mollenhauer HH, Morre DJ, Kelley AG (1966) The widespread occurrence of plant cytosomes resembling animal microbodies. Proto-Plasma 62:44–52

Monosov EZ, Wenzel TJ, Lüers GH et al (1996) Labeling of peroxi-somes with green fluorescent protein in living P. pastoris cells. J Histochem Cytochem 44:581–589

Muench DG, Mullen RT (2003) Peroxisome dynamics in plant cells: a role for the cytoskeleton. Plant Sci 164:307–315

Mullen RT, Lisenbee CS, Miernyk JA et al (1999) Peroxisomal membrane ascorbate peroxidase is sorted to a membranous network that resembles a subdomain of the endoplasmic reticulum. Plant Cell 11:2167–2185

Mullen RT, Trelease RN (2006) The ER-peroxisome connection in plants: Development of the "ER semi-autonomous peroxisome matu-ration and replication" model for plant peroxisome biogenesis. Biochim Biophys Acta Mol Cell Res 1763:1655–1668

Nito K, Kamigaki A, Kondo M et al (2007) Functional classification of Arabidopsis peroxisome biogenesis factors proposed from analyses of knockdown mutants. Plant Cell Physiol 48:763–774

Novikoff AB, Shin WY (1964) The endoplasmic reticulum in the Golgi zone and its relations to microbodies, Golgi apparatus and au-tophagic vacuoles in rat liver cells. J Microsc 3:187–206

Novikoff AB, Goldfischer S (1969) Visualization of peroxisomes (microbodies) and mitochondria with diaminobenzidine. J Histochem Cytochem 17:675–680

Novikoff AB, Phyllis M, Novikoff PM et al (1973) Studies on mi-croperoxisomes IV. Interrelations of microperoxisomes, endoplasmic reticulum and lipofuscin granules. J Histochem Cytochem 21:1010–1020

Orth T, Reumann S, Zhang X et al (2007) The PEROXIN11 protein family controls peroxisome proliferation in Arabidopsis. Plant Cell 19:333–350

Palma JM, Corpas FJ, del Río LA (2009) Proteome of plant peroxi-somes: new perspectives on the role of these organelles in cell biology. Proteomics 9:2301–2312

Peremyslov VV, Prokhnevsky AI, Avisar D et al (2008) Two class XI myosins function in organelle trafficking and root hair develop-ment in Arabidopsis. Plant Physiol 146:1109–1116

Peremyslov VV, Prokhnevsky AI, Dolja VV (2010) Class XI myo-sins are required for development, cell expansion, and F-actin organization in Arabidopsis. Plant Cell 22:1883–1897

Porter KR, Caulfield JB (1958) The formation of the cell plate dur-ing cytokinesis in Allium cepa L. In: Mollenstedt H (ed) Proceedings of the 4th international conference of electron microscopy, vol 2. Springer, Berlin, pp 503–509

Prasher DC, Eckenrode VK, Ward WW et al (1992) Primary struc-ture of the Aequorea victoria green fluorescent protein. Gene 111:229–233

Prokhnevsky AI, Peremyslov VV, Dolja VV (2008) Overlapping functions of the four class XI myosins in Arabidopsis growth, root hair elongation, and organelle motility. Proc Natl Acad Sci U S A 105:19744–19749

Rapp S, Saffrich R, Anton M et al (1996) Microtubule-based peroxisome movement. J Cell Sci 109:837–849

Reisen D, Hanson MR (2007) Association of six YFP-myosin XI-tail fusions with mobile plant cell organelles. BMC Plant Biol 7:6

Reumann S, Quan S, Aung K et al (2009) In-depth proteome analy-sis of Arabidopsis leaf peroxisomes combined with in vivo subcellular targeting verification indicates novel metabolic and regulatory func-tions of peroxisomes. Plant Physiol 150:125–143

Rhodin J (1954) Correlation of ultrastructural organization and function in normal and experimentally changed proximal convoluted tubule cells of the mouse kidney. Aktiebolaget Godvil, Stockholm

Rodríguez-Serrano M, Romero-Puertas MC, Sparkes I et al (2009) Peroxisome dynamics in Arabidopsis plants under oxidative stress in-duced by cadmium. Free Rad Biol Med 47:1632–1639
Rouiller C, Bernhard W (1956) "Microbodies" and the problem of mitochondrial regeneration in liver cells. J Biophys Biochem Cytol 2:355–360
Schollenberger L, Gronemeyer T, Huber CM et al (2010) RhoA reg-ulates peroxisome association to microtubules and the actin cyto-skeleton. PLoS One 5:e13886
Schrader M (2001) Tubulo-reticular clusters of peroxisomes in liv-ing COS-7 cells: dynamic behavior and association with lipid drop-lets. J Histochem Cytochem 49:1421–1429
Schrader M, Wodopia R, Fahimi DH (1999) Induction of tubular peroxisomes by UV irradiation and reactive oxygen species in hepg2 cells. J Histochem Cytochem 47:1141–1148
Schrader M, King SJ, Stroh TA, Schroer TA (2000) Real time imaging reveals a peroxisomal reticulum in living cells. J Cell Sci 113:3663–3671
Schrader M, Thiemann M, Fahimi HD (2003) Peroxisomal motility and interaction with microtubules. Microsc Res Techniq 61:171–178
Scott I, Sparkes IA, Logan DC (2007) The missing link: inter-organellar connections in mitochondria and peroxisomes? Trends Plant Sci 12:380–381
Sinclair AM, Trobacher CP, Mathur N et al (2009) Peroxule exten-sion over ER-defined paths constitutes a rapid subcellular response to hydroxyl stress. Plant J 59:231–242
Sparkes IA (2010) Motoring around the plant cell: insights from plant myosins. Biochem Soc Trans 38:833–838
Sparkes I (2011) Recent advances in understanding plant myosin function: life in the fast lane. Mol Plant 4:805–812
Sparkes IA, Brandizzi F, Slocombe SP et al (2003) An Arabidopsis pex10 null mutant is embryo lethal, implicating peroxisomes in an es-sential role during plant embryogenesis. Plant Physiol 133:1809–1819
Sparkes IA, Hawes C, Baker A (2005) AtPEX2 and AtPEX10 are targeted to peroxisomes independently of known endoplasmic reticu-lum trafficking routes. Plant Physiol 139:690–700
Sparkes IA, Teanby NA, Hawes C (2008) Truncated myosin XI tail fusions inhibit peroxisome, Golgi, and mitochondrial movement in tobacco leaf epidermal cells: a genetic tool for the next generation. J Exp Bot 59:2499–2512
Svoboda D, Grady H, Azarnoff D (1967) Microbodies in experimen-tally altered cells. J Cell Biol 35:127–152
Tabak HF, Murk JL, Braakman I et al (2003) Peroxisomes start their life in the endoplasmic reticulum. Traffic 4:512–518
Titorenko VI, Rachubinski RA (2009) Spatiotemporal dynamics of the ER-derived peroxisomal endomembrane system. Int Rev Cell Mol Biol 272:191–244
Titorenko VI, Chan H, Rachubinski RA (2000) Fusion of small pe-roxisomal vesicles in vitro reconstructs an early step in the in vivo multistep peroxisome assembly pathway of Yarrowia lipolytica. J Cell Biol 148:29–44
Tolbert NE, Yamazaki RK (1969) Leaf peroxisomes and their rela-tion to photorespiration and photosynthesis. Ann N Y Acad Sci 168:325–341
Trelease RN, Lingard MJ (2006) Participation of the plant ER in peroxisomal biogenesis. In: Robinson DG (ed) The plant endoplasmic reticulum, plant cell monographs 4. Springer, Heidelberg, pp 205–232
van der Zand A, Tabak HF (2013) Peroxisomes: offshoots of the ER. Curr Opin Cell Biol 24:1–6
van der Zand A, Braakman I, Geuze HJ et al (2006) The return of the peroxisome. J Cell Sci 119:989–994
van der Zand A, Braakman I, Tabak HF (2010) Peroxisomal mem-brane proteins insert into the endoplasmic reticulum. Mol Biol Cell 21:2057–2065
van der Zand A, Gent J, Braakman I, Tabak HF (2012) Biochemi-cally distinct vesicles from the endoplasmic reticulum fuse to form pe-roxisomes. Cell 149:397–409

van Zutphen T, Veenhuis M, van der Klei IJ (2008) Pex14 is the sole component of the peroxisomal translocon that is required for pexoph-agy. Autophagy 4:63–66

Vigil EL (1970) Cytochemical and developmental changes in micro-bodies (glyoxysomes) and related organelles of Castor bean endosperm. J Cell Biol 46:435–454

Wiemer EAC, Wenzel T, Deerinck TJ et al (1997) Visualization of the peroxisomal compartment in living mammalian cells: dynamic behavior and association with microtubules. J Cell Biol 136:71–80

Zhang X, Hu J (2009) Two small protein families, DYNAMIN-RELATED PROTEIN3 and FISSION1, are required for peroxisome fission in Arabidopsis. Plant J 57:146–159

Zhang L, Léon S, Subramani S (2005) Two independent pathways traffic the intraperoxisomal peroxin PpPex8p into peroxisomes: mechanism and evolutionary implications. Mol Biol Cell 17:690–699

Dissecting Peroxisome-Mediated Signaling Pathways: A New and Exciting Research Field

11

Bo Wang, Oksana Apanasets, Marcus Nordgren, and Marc Fransen

Abstract

Peroxisomes are multifunctional organelles that play an important role in the metabolism of lipids and reactive oxygen species. As many cellular signaling functions are regulated via lipids, lipid second messengers, and oxidative stress-related factors, it is not surprising to see that these organelles are increasingly recognized as critical regulators of cellular signaling events. To fulfill these signaling functions, peroxisomes physically and functionally interact with other cell organelles, including mitochondria. Recent progress in the development of tools to visualize and modulate molecular processes at the subcellular level has made it possible to gain a better insight into the potential mechanisms governing peroxisomal signaling. This chapter is intended to provide a comprehensive overview of the tools and strategies that are currently available to study peroxisome-mediated signaling pathways in living cells. To provide the reader with relevant background information, we also highlight key studies that have contributed to our understanding of how peroxisomes may function as important sites of redox-, lipid-, inflammatory-, and viral-mediated signal transduction.

Keywords

Peroxisomes • Live-cell imaging • Oxidative stress • Intracellular signaling

11.1 Introduction

Peroxisomes are functionally complex organelles that play a central role in diverse metabolic processes. These processes can vary widely between species, cell types, and physiological and developmental stages (Fransen 2012). For a long time,

B. Wang • O. Apanasets • M. Nordgren • M. Fransen (✉)
Laboratory of Lipid Biochemistry and Protein Interactions, Department of Cellular and Molecular Medicine, KU Leuven, Leuven, Belgium
e-mail: marc.fransen@med.kuleuven.be

C. Brocard and A. Hartig (eds.), *Molecular Machines Involved in Peroxisome Biogenesis and Maintenance*, DOI 10.1007/978-3-7091-1788-0_11, © Springer-Verlag Wien 2014

peroxisomes were thought to function solely as metabolic organelles. However, over the last years, it has become increasingly clear that peroxisome-derived metabolites can exert cytoprotective effects by activating pro-survival pathways or initiate signaling cascades that ultimately induce pathophysiological responses (Titorenko and Terlecky 2011). In addition, peroxisomes are increasingly recognized as signaling platforms in the battle between viruses and their hosts (Dixit et al. 2010; Lazarow 2011). In the following sections, we first review the emerging concepts highlighting the potential role of peroxisomes in diverse signaling pathways, with an emphasis on mammals. Afterwards, we provide an overview of tools and approaches that are currently available to explore peroxisome-mediated signaling events.

11.2 Peroxisomes as Signaling Platforms

It is currently a common belief that most, if not all, intracellular signaling pathways are controlled by redox regulation and/or lipid second messengers. In addition, it is well known that the specific cellular responses to signaling molecules depend on their identity, concentration, and spatial distribution. Given that peroxisomes actively contribute to the bioavailability of various reactive oxygen and nitrogen species (ROS/RNS), bioactive lipids, and inflammatory factors (see below), it is not surprising to see that these organelles are increasingly recognized as important intracellular signaling platforms (Fig. 11.1; Beach et al. 2012). The following sections are intended to summarize key findings in this rapidly moving and exciting research area. Importantly, the exploration of this research field has just begun, and as such, there is still a large gap in our understanding of how peroxisomes are incorporated into subcellular communication networks.

11.2.1 Redox Signaling

For a long time, it was thought that ROS/RNS were only toxic byproducts of aerobic metabolism. However, in the meantime, it has become clear that some ROS/RNS, e.g., hydrogen peroxide (H_2O_2) and nitric oxide ($NO^•$) can act as intracellular messengers at subtoxic concentrations (Fransen et al. 2012). In general, this signaling occurs through the oxidative modification of reactive cysteine residues. In this context, it should be mentioned that many signaling components like kinases, phosphatases, and transcription factors contain cysteine residues that can be reversibly modified in a redox-responsive manner (Barford 2004). Examples of such modifications include, but are not limited to, disulfide formation, *S*-nitrosylation, and *S*-glutathionylation. As these modifications often influence protein activity, a chronic or acute disturbance in redox homeostasis can be expected to deregulate vital cellular signaling pathways. This may influence cell fate by, for example, promoting cell growth or death.

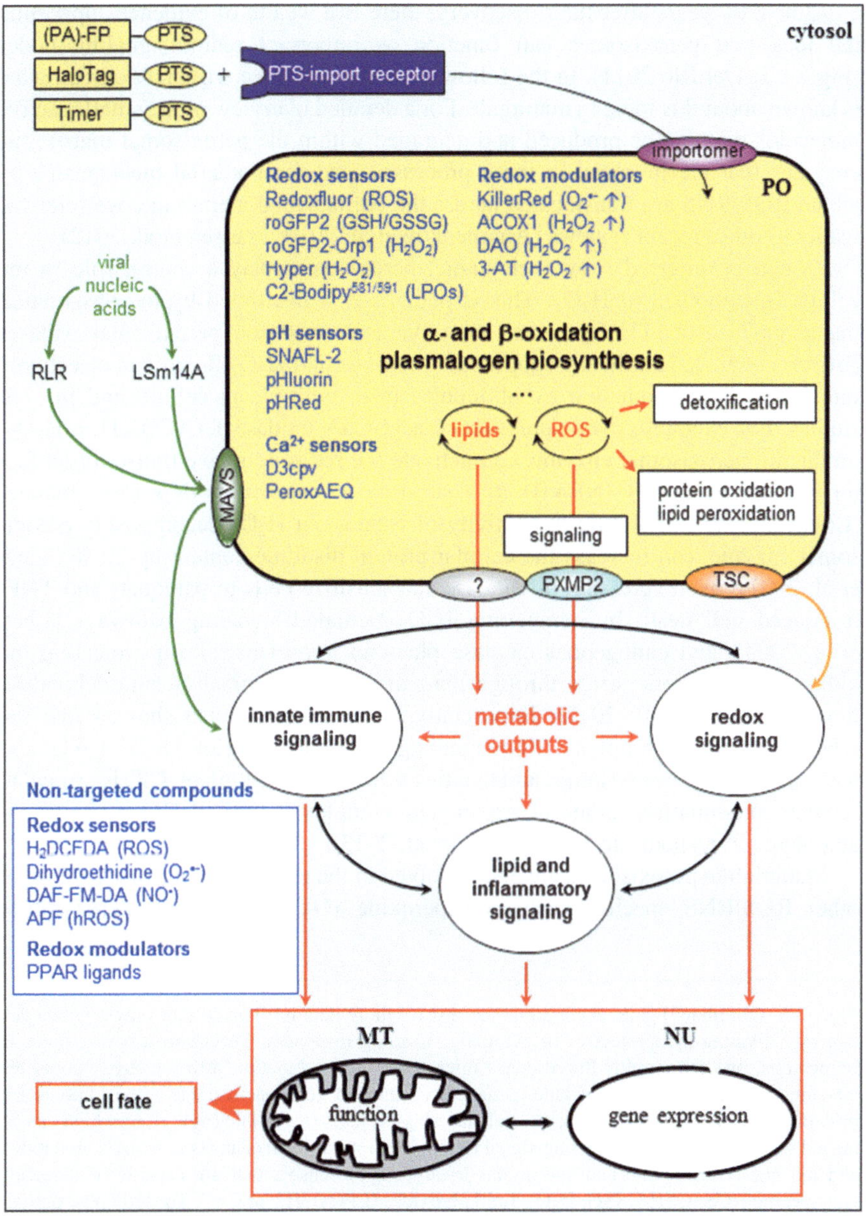

Fig. 11.1 Model depicting the role of peroxisomes as intracellular signaling platforms, and tools to study them. Peroxisomes play a central role in cellular lipid and ROS metabolism (the *rotating arrows* represent anabolic and catabolic pathways), and substantial evidence supports the view that the functional state of these organelles can influence redox-, lipid-, and inflammatory-mediated signaling pathways. Peroxisomes can also activate and propagate antiviral innate immune responses (via peroxisomal MAVS, an adaptor protein for pattern recognition receptors such as RIG-I-like receptor (RLR) and LSm14A) and regulate autophagy (via a tuberous sclerosis complex (TSC) at the peroxisomal membrane). Taken together, alterations in peroxisome activity may—directly or indirectly—modulate mitochondrial function and trigger changes in nuclear

Almost 60 years after their discovery, there is a wealth of evidence supporting the idea that peroxisomes can function as important redox signaling nodes (Fig. 11.1; Del Río 2011). In the following paragraphs, we will elaborate on what is known about this topic in mammals. For a detailed overview of the small reactive molecules that can be produced and degraded within the peroxisomal matrix, the enzymes that are involved in these processes, and the potential mechanisms by which ROS/RNS are translocated across the peroxisomal membrane, we refer the reader to other recent reviews (Antonenkov et al. 2010; Fransen et al. 2012).

As can be inferred from their name, peroxisomes play a central role in the cellular metabolism of H_2O_2. This is perhaps best illustrated by the observation that about 35 % of all H_2O_2 formed in rat liver is derived from peroxisomal oxidases (Boveris et al. 1972). Strong support that peroxisomal H_2O_2 fluxes can effectively influence cellular signaling events comes from multiple in cellulo and in vivo studies. For example, overexpression of acyl-CoA oxidase 1 (ACOX1), a H_2O_2-producing peroxisomal enzyme, can activate the redox-sensitive transcription factor nuclear factor κB (NF-κB) in a substrate concentration-dependent manner (Li et al. 2000); inhibiting the activity of catalase, a H_2O_2-decomposing peroxisomal enzyme, can increase the cellular protein disulfide content by 20 % (Yang et al. 2007); overexpression of catalase can sensitize cells to paraquat- and TNF-α-induced cell death by dampening H_2O_2-mediated signaling pathways (Chen et al. 2004); and endogenous catalase plays an important role in protecting the kidney from diabetic stress through maintaining peroxisomal and mitochondrial fitness (Hwang et al. 2012). Importantly, a recent study also showed that the tuberous sclerosis complex (TSC) signaling node consisting of TSC1, TSC2, and Rheb localizes to peroxisomes and that this subcellular location of TSC is critical to regulate mammalian target of rapamycin complex 1 (mTORC1) activity and autophagy in response to ROS (Zhang et al. 2013).

Mammalian peroxisomes are also involved in the production and degradation of other ROS/RNS species, such as superoxide ($O_2^{\bullet -}$), $NO^{\bullet}$, and peroxynitrite

---

Fig. 11.1 (continued) gene expression (see text for more details). This in turn may regulate cell fate by activating cytoprotective or cytotoxic signaling responses. To visualize peroxisomes in living cells, one can employ fluorescence microscopy in combination with selective fluorescent labeling techniques. These include genetically encoded (photoactivatable (PA)) fluorescent proteins (FPs), fluorescent timers, and self-labeling proteins (e.g., HaloTag), all of which can be tagged with a peroxisomal targeting signal (PTS). To monitor local changes in peroxisomal redox and ion homeostasis, one can use organelle-targeted biosensors that are capable of detecting non-specific ROS, GSH/GSSG, H_2O_2, lipid peroxides (LPOs), H^+, or Ca^{2+}. To study how peroxisomal metabolism affects the global cellular redox state, one can also use non-targetable redox sensors, such as H_2DCFDA, dihydroethidine, DAF-FM-DA, APF, and C11-Bodipy$^{581/591}$. Finally, to modulate the peroxisomal redox state, one can expose cells to peroxisome proliferators, alter the expression levels and/or activity of peroxisomal ROS-modulating enzymes, or employ peroxisomal KillerRed (see text for more details). The organelles are not drawn to scale. *3-AT* 3-amino-1,2,4-triazole, *ACOX1* acyl-CoA oxidase 1, *DAO* D-amino acid oxidase, *hROS* highly reactive oxygen species, *MT* mitochondrion, *NU* nucleus, *PO* peroxisome, *PPAR* peroxisome proliferator-activated receptor

($ONOO^-$) (Fransen et al. 2012). Unfortunately, with few exceptions, virtually nothing is known regarding the specific effects and physiological functions of these peroxisomal ROS/RNS species. However, it should be pointed out that there are some indications that the peroxisomal pool of NOS2, the inducible nitric oxide synthase, may function as a local enzyme activity-modulating factor (Stolz et al. 2002). This hypothesis is in agreement with the findings that the appearance of NOS2 inside peroxisomes has been associated with a decrease in catalase activity (Stolz et al. 2002) and various peroxisomal proteins, including catalase, can be *S*-nitrosylated (Doulias et al. 2013).

Other strong but indirect evidence that peroxisomes can indeed function as redox signaling platforms comes from studies in mice. For example, it has been shown that (1) preservation of peroxisome function is essential to reduce renal ROS levels and alleviate kidney injury after ischemia/reperfusion or cisplatin treatment (Hasegawa et al. 2010), (2) peroxisomal ROS metabolism plays a key role in the regulation of the hypothalamic melanocortin tone and food intake in diet-induced obesity (Diano et al. 2011), and (3) the accumulation of functionally compromised peroxisomes alters the cellular redox equilibrium and attenuates organ injury induced by lipopolysaccharides (Vasko et al. 2013). Again, the molecular mechanisms of how peroxisomes exactly contribute to these physiological and pathological redox signaling processes are poorly understood. However, here it is essential to mention that peroxisomes share an intricate redox-sensitive relationship with mitochondria (Ivashchenko et al. 2011; Walton and Pizzitelli 2012). This in turn suggests that mitochondria may act as dynamic receivers, integrators, and transmitters of peroxisome-derived mediators of oxidative stress. For more information regarding how peroxisomes and mitochondria can communicate with each other, we refer the reader to another recent review (Fransen et al. 2013).

11.2.2 Lipid and Inflammatory Signaling

Many cellular signaling functions are regulated via lipids and lipid second messengers. Remarkably, despite the fact that peroxisomes are actively involved in the metabolism of many of these compounds, very little is known about how these organelles contribute to cross-compartmental lipid signaling. However, as (1) peroxisomes harbor enzymes that are involved in the biosynthesis of plasmalogens and docosahexaenoic acid (DHA) (Van Veldhoven 2010), (2) these lipids function as potential reservoirs for bioactive molecules such as prostaglandins, thromboxanes, leukotrienes, resolvins, docosatrienes, and neuroprotectins (Braverman and Moser 2012), and (3) some of these molecules (e.g., the eicosanoids) can also be degraded through peroxisomal β-oxidation (Van Veldhoven 2010), it comes as no surprise that these organelles are garnering increasing attention as signaling platforms in inflammation and immunoregulation (Fig. 11.1). This is evidenced by the following observations: a variety of the bioactive molecules listed above (e.g., DHA and eicosanoids) are natural ligands of peroxisome proliferator-activated receptors (PPARs), who have emerged as key players at the crossroads of lipid signaling and inflammation (Wahli and Michalik 2012); peroxisomes provide oligodendrocytes with an essential protective function against axon degeneration and neuroinflammation (Kassmann et al. 2007); abnormal accumulation of very long-chain fatty acids, a condition associated with peroxisome dysfunction, can lead to an increase in inflammatory cytokine expression (Singh et al. 2009); the interleukin-1 signaling pathway is significantly activated in skin fibroblasts from patients lacking functional ACOX1 (El Hajj et al. 2012); and inactivation of neuronal multifunctional protein 2, a central enzyme of peroxisomal β-oxidation, causes neuroinflammation (Verheijden et al. 2013).

Over the years, it has become increasingly clear that lipid signaling, inflammation, and oxidative stress are inextricably linked processes. In addition, it is well known that virtually all stress stimuli trigger changes in lipid composition. For example, exposure of cells to oxidative stress can induce sphingolipid metabolism and lipid peroxidation (Bikman and Summers 2011). This in turn may lead to the accumulation of ceramides, ceramide metabolites, and highly reactive electrophilic aldehydes. Importantly, as (1) many of these compounds can act as important messengers in signaling events that lead to cell proliferation, differentiation, and senescence (Chen and Niki 2006; Bikman and Summers 2011) and (2) brains and fibroblasts of patients and mice with peroxisomal disorders display a significant increase in the level of C26:1/0-ceramide (Pettus et al. 2004) and lipid peroxidation products (Fourcade et al. 2008; Baarine et al. 2012), it is very likely that changes in peroxisomal metabolism can directly or indirectly modulate cytoprotective and cytotoxic signal transduction pathways (Titorenko and Terlecky 2011). This idea is further corroborated by the observation that cells lacking GNPAT, a peroxisomal enzyme catalyzing the first step in ether phospholipid biosynthesis, are more vulnerable to oxidative stress (Brodde et al. 2012). Note that plasmalogens, among other functions, have been found to possess antioxidant capacities (Braverman and Moser 2012).

11.2.3 Innate Immune Signaling

Innate immunity provides the first line of defense against pathogen invasion. The initiation of innate immune responses relies on the recognition of pathogen components by pattern recognition receptors (PRRs). Activated PRRs deliver signals to specific adaptor proteins that, in turn, orchestrate complex host defense mechanisms. These include the activation of transcription factors, e.g. NF-κB, activator protein 1, and interferon regulatory factors (IRFs) and the subsequent production of type 1 interferons (IFNs), inflammatory cytokines, and chemokines (Chen and Jiang 2013). Interestingly, a few years ago, it was reported that a small portion of mitochondrial antiviral signaling protein (MAVS), a RIG-I-like receptor (RLR) adaptor protein, is located on peroxisomes (Fig. 11.1) (Dixit et al. 2010). RLRs are a family of cytosolic PRRs that function as innate immune receptors for specific RNA virus ligands. Kagan and coworkers also found that (1) RNA viruses can activate MAVS-dependent signaling from peroxisomes, (2) this signaling process is temporally and qualitatively different from that of mitochondrial MAVS (peroxisomal MAVS elicits rapid but transient, type I IFN-independent innate immune responses; mitochondrial MAVS responses are delayed but sustained and type I IFN-mediated), and (3) MAVS responses from both peroxisomes and mitochondria are needed for maximal antiviral activity (Dixit et al. 2010). Recently, these findings have been extended by the observations that MAVS is also localized to mitochondria-associated ER membranes (MAMs), and that dynamic MAM tethering to mitochondria and peroxisomes coordinates MAVS localization to form a signaling synapse between membranes (Horner et al. 2011). Interestingly, the hepatitis C virus NS3-4A protease complex specifically targets MAVS at these sites to ablate RIG-I signaling of immune defenses (Horner et al. 2011). Finally, also LSm14A, a processing body-associated sensor of viral RNA and DNA, is translocated to peroxisomes upon viral infection (Li et al. 2012). Importantly, this process is also MAVS dependent and mediates IRF3 activation and IFN-β production (Li et al. 2012).

Over the last years, evidence has accumulated suggesting that some viruses exploit peroxisomes for their replication. Unfortunately, the precise role of these organelles in viral replication is yet to be determined. For a state-of-the-art overview of research in this field, we refer the reader to another review (Lazarow 2011). Finally, there are indications that peroxisomes may also modulate inflammatory immune responses. For example, it has been shown that a peroxisome deficiency in Drosophila larvae causes an increase in the expression of genes involved in innate immunity and humoral responses (Mast et al. 2011) and that the immune system is activated in *nestin-Pex5* knockout mice (Bottelbergs et al. 2012). Again, the molecular mechanisms underlying these phenomena remain to be fully elucidated. However, it has been hypothesized that the upregulation of innate immunity genes may be a response to increased levels of unused peroxisomal metabolites (Mast et al. 2011).

11.3 Tools to Study Peroxisome-Mediated Signaling Pathways

Signaling pathways can be considered as cascades of biochemical reaction-diffusion processes that govern cellular functions in a spatiotemporal manner. Many of these processes are triggered by transient changes in the internal or external milieu (e.g., nutrient status, Ca^{2+} levels, pH, redox state, etc.) and ultimately lead to transcriptional reprogramming. As a dysregulation of these signaling pathways often contributes to the etiology and progression of disease, it is critical to understand these processes at the molecular and (sub)cellular level. During recent years, the development of new methods to visualize proteins and biochemical reactions in function of time and space in living cells has revolutionized signaling research. In the next sections, we further elaborate on the tools and strategies that are currently available to study peroxisome-mediated signaling pathways in living cells (Fig. 11.1). We first briefly outline different strategies that can be used to visualize (distinct populations of) peroxisomes by direct fluorescence microscopy. Next, we discuss a number of tools that can be used to detect and modulate peroxisomal signaling activity in a spatially and temporally controlled manner. Note that, as (1) the continuous development of new and improved synthetic and genetically encoded probes offers an increasing array of possibilities for imaging spatiotemporal processes in living cells, and (2) it is impossible to outline all these possibilities and variations, we limit ourselves to an overview of probes that can be used to track peroxisomal redox changes and ion fluctuations.

11.3.1 Fluorescence Microscopic Methods to Visualize Peroxisomes in Living Cells

About half a century ago, peroxisomes were first visualized in rat liver by electron microscopy (Rhodin 1954). This technique, which is still widely used today, was for a long time the most prominent—if not only—approach to get a visual image of these organelles. However, as (1) the use of electron microscopy to image biological systems has some intrinsic limitations (e.g., no live samples, high cost and time expenses, etc.), and (2) during the last decade, rapid progress has been made in the area of live-cell imaging by fluorescence microscopy (e.g., development of sensitive and specific fluorescent probes for functional analysis), the latter technology has often become the first method of choice to study the subcellular localization and dynamics of ions, metabolites, signaling molecules, and proteins (Wang et al. 2012). In the next paragraph, we highlight how some of these advances have been applied to study peroxisome dynamics. Here it is important to note that peroxisomes can rapidly adapt their morphology and number in response to changes in the cellular environment (Ribeiro et al. 2012).

Most strategies that are currently employed to visualize peroxisomes in living cells make use of fluorescent proteins (FPs) that are fused to peroxisomal signal peptides (e.g., PTS1, PTS2, or mPTS) via recombinant DNA techniques. Although these probes are excellent tools for monitoring the import capacity, appearance, and trafficking of the total population of peroxisomes, they do not allow researchers to optically distinguish pools of proteins and organelles that are synthesized at different time points. To deal with this problem, two strategies have already been adopted from other research fields. The first one utilized a photoactivatable green FP (PAGFP) to investigate peroxisome maturation from the ER (Kim et al. 2006). Photoactivatable FPs are capable to change their spectral properties in response to irradiation with light of specific wavelengths, thereby enabling the spatial and temporal visualization of specific structures and tracking of their signal in time. A potential disadvantage of many of these proteins is that optimal photoactivation requires intense irradiation with UV light, a condition known to cause oxidative stress. The second strategy exploited the HaloTag technology to study peroxisome growth, division, and degradation (Huybrechts et al. 2009; Delille et al. 2010). This technology is a chemistry-based method for protein labeling in which synthetic chloroalkane ligands covalently bind to a dehalogenase-based protein fusion tag. As

(1) the chloroalkane linker can be attached to different membrane-permeable functional groups (e.g., fluorophores, affinity handles, etc.), (2) these probes spontaneously react with the HaloTag fusion protein, (3) unbound probes can be easily washed away to reduce the signal-to-noise ratio, and (4) labeling of newly-synthesized proteins can be completely turned off by using the HaloTag blocking ligand, this strategy is very well suited to conduct pulse-chase experiments in cells expressing a HaloTag-fusion protein (Los et al. 2008). This technology is also perfectly suited to visualize distinct pools of proteins, generated from a single genetic HaloTag-encoding construct, by sequentially incubating the cells with different fluorescent ligands. In the context of this chapter, it is worth mentioning that the HaloTag technology has been used to show that a disturbance in peroxisomal redox balance may function as a trigger for peroxisome degradation (Ivashchenko et al. 2011). As for PAGFP, a potential drawback of the HaloTag technology is the relative large size (±33 kDa) of the protein tag (Los et al. 2008). However, alternative strategies employing fluorogenic probes in combination with small protein tags (e.g., the 1 kDa tetracysteine tag) are currently available and have already been applied successfully in related research fields (Hori and Kikuchi 2013). Another attractive approach that can potentially be adopted to simultaneously track the localization and age of individual peroxisomes is the use of fluorescent timers, which change their color with time due to a chemical conversion of the chromophore (Chudakov et al. 2010).

11.3.2 Indicators and Modulators of Peroxisomal Signal Transduction

Peroxisomes are actively involved in cellular ROS metabolism and alterations in the peroxisomal redox state are likely to have important consequences on cellular physiology. As such, it is essential to gain a better understanding of how changes in peroxisomal redox metabolism impact cell signaling pathways. A requirement to carry out such studies is to have access to tools that are suitable to detect and modulate the peroxisomal redox state. In the sections 11.3.3 and 11.3.4, we discuss approaches that are currently used for these purposes.

As hydrogen (H^+) and calcium (Ca^{2+}) ions are potent modulators of essentially all biological processes (Swietach et al. 2013), it is also of great scientific value to gain a better understanding of how changes in peroxisomal metabolism influence subcellular homeostasis of these molecules. In Sect. 11.3.5, we discuss the biosensors that have already been used to monitor these signaling ions in peroxisomes in living cells.

11.3.3 Redox Sensors

H_2DCFDA The cell-permeable indicator dye 2′,7′-dichlorodihydrofluorescein diacetate (H_2DCFDA) is one of the most popular probes to assess oxidative stress in living cells. After passing through the membrane, this non-fluorescent lipophilic compound is de-esterified to a hydrophilic alcohol (H_2DCF) that can be oxidized to 2′,7′-dichlorofluorescein (DCF), a highly fluorescent molecule ($E_{ex} = 502$ nm; $E_{em} = 523$ nm) (Karlsson et al. 2010). H_2DCFDA, which is thought to function as a general oxidative stress indicator, is also commonly used in the peroxisome field. For example, this compound has been employed to demonstrate that ROS levels are markedly increased in fibroblasts from X-linked adrenoleukodystrophy patients (Fourcade et al. 2008). Importantly, the use of H_2DCFDA to evaluate cellular oxidative stress faces several limitations: oxidation of H_2DCF to DCF is an irreversible process; DCF fluorescence may not always be a result of exposure to ROS, but may also reflect the relocation of lysosomal iron and/or mitochondrial cytochrome c to the cytosol; DCF itself can also generate ROS upon visible light illumination; and the compound cannot be targeted to specific subcellular compartments (Karlsson et al. 2010).

Dihydroethidine Another compound that can be used to visualize ROS production in living cells is dihydroethidine (DHE). DHE is a cell-permeable non-targetable redox sensor that reacts preferentially with $O_2^{\bullet -}$ to form 2-hydroxyethidium (Dikalov et al. 2007). This molecule rapidly intercalates into double-stranded DNA and this in turn results in a marked increase in nuclear fluorescence ($E_{ex} = 510$ nm; $E_{em} = 590$ nm). DHE has already been successfully used to demonstrate that $O_2^{\bullet -}$ production in peroxisome-deficient cerebellar neurons was almost twice that of littermate controls (Müller et al. 2011). Note that, as (1) DHE is light-sensitive, (2) in cells and tissues, DHE can be oxidized to ethidium in an $O_2^{\bullet -}$independent manner, and (3) ethidium is difficult to distinguish from 2-hydroxyethidium by conventional fluorescence techniques upon excitation at 510 nm, care should be taken when interpreting results (Dikalov et al. 2007).

DAF-FM DA The cell-permeable indicator dye 4-amino-5-methylamino-2′, 7′-difluorofluorescein diacetate (DAF-FM DA) can be used to detect $NO^{\bullet}$ (Kojima et al. 1999). Once inside the cell, this compound is quickly deacetylated by intracellular esterases to DAF-FM, which upon reaction with $NO^{\bullet}$ becomes fluorescent ($E_{ex} = 495$ nm; $E_{em} = 515$ nm). As the fluorescence intensity of the $NO^{\bullet}$ adduct of DAF-FM is stable above pH 5.8, the probe can be used to visualize the temporal and spatial distribution of $NO^{\bullet}$ in living cells (Nagano 2009). Potential disadvantages of DAF-FM include photobleaching and the fact that the probe can also react with $ONOO^-$ (Nagano 2009).

APF Aminophenyl fluorescein (APF) is a cell-permeable compound that is selective for the detection of highly reactive oxygen species (hROS), such as $ONOO^-$, hydroxyl radicals ($^{\bullet}OH$), and hypochlorite (^-OCl) (Nagano 2009). Upon oxidation, APF exhibits a bright green fluorescence ($E_{ex} = 490$ nm; $E_{em} = 515$ nm). A major disadvantage of APF is that this probe cannot detect low levels of hROS because it can easily leak out of cells (Nagano 2009).

Redox-Sensitive Green Fluorescent Proteins Approximately a decade ago, Remington and colleagues developed novel GFP probes suitable for monitoring redox changes in individual cells (Hanson et al. 2004). These probes, called roGFPs, contain two engineered cysteine residues on adjacent surface-exposed β-strands close to the chromophore. Disulfide formation between these residues promotes protonation of the chromophore and this in turn increases the protein's excitation peak near 400 nm at the expense of the peak near 480 nm. As oxidation of the thiol pair causes reciprocal changes in the emission intensities (around 510 nm) when excited at these two different wavelengths, the fluorescence ratios can provide accurate insight into the redox environment of the chromophore. Importantly, the use of genetically encoded ratiometric fluorescent probes alleviates several shortcomings of chemical redox sensors. For example, the results will not depend on sensor concentration, cellular thickness, and the sensor distribution pattern; these probes show reversible changes in fluorescence, and they can be targeted to specific subcellular locations (Schwarzländer et al. 2008). Importantly, a variety of roGFP biosensors has been developed and characterized (Lukyanov and Belousov 2013). These include, among others, roGFP2, Grx1-roGFP2, and roGFP2-Orp1. RoGFP2 preferentially interacts with glutaredoxins (Grxs) and is therefore particularly suited to monitor changes in the glutathione redox couple (GSH/GSSG) (Hanson et al. 2004). However, as the availability of Grxs is often a rate-limiting factor in roGFP2-equilibration with intracellular thiols, the response of this probe to changes in redox potential is rather slow. To solve this problem, Dick and colleagues fused roGFP2 to human Grx1 and demonstrated that the oxidation of this fusion protein by GSSG is at least 100,000 times faster as compared to uncoupled roGFP2 (Gutscher et al. 2009). Next, the same research group demonstrated that fusing roGFP2 to the yeast peroxidase Orp1 resulted in a H_2O_2-specific probe with the spectral properties of roGFP2. In the meantime, we and others have shown that a peroxisome-targeted variant of roGFP2 is sufficiently sensitive to detect changes in the peroxisomal redox environment in response to altered growth conditions or upon challenge with cell-permeant oxidants and reductants, toxicologically relevant metal ions, or green light illumination after peroxisomal KillerRed expression (Schwarzländer et al. 2008; Ivashchenko et al. 2011).

Redoxfluor Redoxfluor is a fluorescence resonance energy transfer (FRET)-based redox sensor, which consists of three parts: cerulean, a variant of enhanced cyan fluorescent protein (CFP); a tandemly repeated fragment of a partial region within the carboxy-terminal cysteine-rich domain of Yap1, a yeast transcription factor which activates expression of antioxidant genes in response to oxidative stress; and citrine, a variant of enhanced yellow fluorescent protein (YFP) (Yano et al. 2010). Upon oxidation and excitation at 434 nm, CFP emission (near 476 nm) is increased at the expense of YFP emission (near 527 nm), thereby decreasing the yellow-to-cyan emission ratio. Importantly, Redoxfluor does not measure a specific type of ROS. It rather detects quantitative changes of various redox-related compounds such as H_2O_2, glutathione, and thioredoxin. Redoxfluor has already been successfully used to compare the redox state within peroxisomes and the cytosol in yeast and Chinese hamster ovary cells harboring functional or dysfunctional peroxisomes (Yano et al. 2010). One interesting observation of this study was that the cytosolic redox state of cell mutants lacking functional peroxisomes was more reductive than that of control cells.

Hyper Hyper is a genetically encoded fluorescent biosensor that has a high sensitivity and specificity for H_2O_2 (Belousov et al. 2006). The probe consists of the regulatory domain of the *Escherichia coli* transcription factor OxyR (OxyR-RD), a positive regulator of H_2O_2-inducible genes that is inserted into a circularly permuted YFP (cpYFP). Upon exposure of OxyR-RD to H_2O_2, a disulfide bond is formed between two cysteine residues and this in turn induces a conformational change that alters cpYFP sufficiently to shift its excitation maximum from 420 to 500 nm. As these H_2O_2-dependent changes are fully reversible, changes in the fluorescence ratios of Hyper (emission peak: 516 nm) upon excitation at 420 and 500 nm reflect alterations in local H_2O_2 levels (Belousov et al. 2006). However, in this context, it is essential to point out that Hyper measurements are strongly influenced by pH and therefore it is advised to simultaneously monitor local changes in pH (see below). By employing targeted variants of Hyper, it has been shown that (1) in plants, peroxisomal catalase activity and H_2O_2 levels are inversely correlated (Costa et al. 2010), and (2) in mammals, the toxicity of excess long-chain nonesterified fatty acids in insulin-producing β-cells is mediated by H_2O_2 derived from peroxisomal, but not mitochondrial β-oxidation (Elsner et al. 2011).

Bodipy-PTS1 To detect lipid peroxidation in peroxisomes in living cells, Wirtz and coworkers covalently linked a C-terminal peroxisomal matrix protein targeting signal (PTS1) to C2-Bodipy$^{581/591}$, a membrane-permeable oxidation-sensitive fluorescent lipid peroxidation probe (Dansen and Wirtz 2001). As the red fluorescence of this fatty acid analogue (emission peak: 595 nm) shifts to green fluorescence (emission peak: 520 nm) upon oxidation, it can be used as a ratiometric indicator of free radical processes that have the potential to oxidize lipids in the peroxisomal membrane. For reasons that are not clear, this probe has—until now—only been used to show that the intraperoxisomal redox state in rat fibroblasts becomes more oxidized upon incubation of the cells with phytanic acid or arachidonic acid, which are substrates for peroxisomal α- and β-oxidation, respectively.

11.3.4 Redox Modulators

Activators and Inhibitors of H_2O_2 Production Peroxisomes contain copious amounts of enzymes that can produce or degrade H_2O_2 (Antonenkov et al. 2010). Over the years, it has become clear that the expression levels and activities of these enzymes can be influenced by several factors. Examples include the PPAR-mediated induction of peroxisomal enzymes in rodents fed a diet containing fibrate drugs, industrial phthalate ester plasticizers, xenobiotics, or high fat (Kozawa et al. 2011). Importantly, long-term treatment of rats with these compounds results in disproportionate increases in H_2O_2-generating enzymes and catalase and this in turn has been linked to increased oxidative stress and DNA damage (Kasai et al. 1989). These findings have been further substantiated by other studies, which showed that overexpression of ACOX1 can activate NF-κB DNA binding activity in a substrate concentration-dependent manner (Li et al. 2000), and the toxicity of long chain non-esterified fatty acids in insulin-producing β-cells is caused by H_2O_2 derived from peroxisomal β-oxidation (Elsner et al. 2011). Note that the results of the first study indirectly indicate that it is most likely possible to modulate the intraperoxisomal redox state by altering the activities of other bona fide H_2O_2-producing peroxisomal enzymes. One attractive candidate is D-amino acid oxidase (DAO). Indeed, a nuclear-targeted variant of this enzyme has already been successfully used to study H_2O_2-dependent signaling events in the nucleus (Halvey et al. 2007). Note that exposure of the cells to DAO substrates (e.g., D-proline and *N*-acetyl-D-alanine) is most likely to result in less non-peroxisomal oxidative stress than exposure of the cells to free fatty acids (Soardo et al. 2011). Another attractive approach to modulate the peroxisomal redox state involves the up- or downregulation of peroxisomal antioxidant enzyme activities. Such experiments have already been successfully performed for catalase. For example, it has been shown that restoration of peroxisomal catalase import (e.g., by protein transduction of catalase-SKL, a catalase analogue with enhanced peroxisome targeting efficiency) can restore a normal oxidative state in aging cells (Titorenko and Terlecky 2011).

KillerRed-PTS1 KillerRed is a red fluorescent photosensitizer that efficiently generates ROS upon green light illumination (Bulina et al. 2006). As such, targeted variants of this protein can be used as powerful tools to study the downstream effects of local ROS production in a spatiotemporally controlled manner. We recently employed a peroxisomal variant of KillerRed to study the downstream effects of peroxisome-derived ROS (Ivashchenko et al. 2011). This study revealed that generation of excess ROS inside peroxisomes quickly perturbs the mitochondrial redox balance and leads to mitochondrial fragmentation.

11.3.5 Ion Signaling

SNAFL-2-PTS1 SNAFL-2-PTS1 is a cell-permeable fluorescent reporter molecule that can be used to monitor the intraperoxisomal pH in living cells (Dansen et al. 2000). It consists of a seminaphthofluorescein (SNAFL-2) moiety that is covalently linked to a PTS1 sequence. As the probe can be present in a protonated or deprotonated state (pKa ~7.7), and the acidic (excitation peak: 488 nm; emission peak: 546 nm) and basic (excitation peak: 568 nm; emission peak: 625 nm) forms have different spectral characteristics, the ratio of the red and green fluorescence intensities can be used to calculate the pH by comparing these values with a calibration curve. By using this biosensor, these authors have demonstrated that peroxisomes in human fibroblasts have an internal pH of ±8.2. However, peroxisomes in fibroblasts from patients lacking Pex7p, the import receptor for peroxisomal matrix proteins containing an N-terminal peroxisomal targeting signal (PTS2), have a pH of ±6.5 (Dansen et al. 2000). The pathophysiological relevance of this observation remains to be established.

pHluorin-PTS1 An alternative method to measure the intraperoxisomal pH in a non-invasive manner is by using pHluorin-PTS1, a peroxisomally located pH-sensitive mutant of GFP (Jankowski et al. 2001). This pH-sensor has a bimodal excitation spectrum (with peaks at 395 nm and 475 nm) and an emission maximum at 509 nm. As, upon acidification, the protein's excitation peak near 395 nm increases at the expense of the peak near 475 nm, the corresponding emission ratio can—after calibration—provide an accurate estimate of the probe's local pH environment. Interestingly, by employing this biosensor, these authors found that mammalian peroxisomes do not regulate their own pH, but that their pH resembles that of the cytosol (pH 6.9–7.1). Note that this finding is in conflict with that obtained by SNAFL-2-PTS1 (Dansen et al. 2000). Here it is worth mentioning that the p*K*a of pHluorin is 6.5 (Miesenböck et al. 1998), which is most likely too low to enable reliable measurements in case of high pH. To bring more clarity to this issue, the use of a peroxisome-targeted variant of pHRed (p*K*a ~7.8), a recently developed ratiometric red fluorescent pH sensor, may offer an alternative (Tantama et al. 2011).

D3cpv-PTS1 D3cpv is a computationally designed genetically encoded high-affinity FRET Ca^{2+} sensor that has been optimized to monitor calcium even in the presence of a large excess of native calmodulin (Palmer et al. 2006). The probe has one excitation peak near 425 nm and two emission maxima near 480 and 540 nm. Upon binding to Ca^{2+}, the protein's fluorescence emission peak near 540 nm is increased at the expense of the peak near 480 nm. Therefore, the 540/480 nm ratio can be used to monitor Ca^{2+}. By appending a PTS1 sequence to this probe, Pozzan and colleagues could demonstrate that (1) the Ca^{2+} concentration of peroxisomes in living mammalian cells at rest is similar to that of the cytosol, and (2) peroxisomes do not act as Ca^{2+} stores from which Ca^{2+} can be mobilized upon stimulation (Drago et al. 2008).

PeroxAEQ Aequorin is a Ca^{2+}-sensitive photoprotein that is composed of two distinct units: the apoprotein apoaequorin, which has three high-affinity binding sites for Ca^{2+}; and the prosthetic group coelenterazine, a blue light-emitting membrane-permeable molecule found in many aquatic organisms (Rizzuto et al. 1992). Both components reconstitute spontaneously to form a functional protein that, upon Ca^{2+}-binding, undergoes a conformational change (Chudakov et al. 2010). This change triggers the conversion of the protein into an oxygenase, which oxidizes coelenterazine to coelenteramide. The subsequent relaxation of coelenteramide to its ground state results in the emission of blue light (near 469 nm), which can be detected by a luminometer. Palmieri and colleagues generated a PTS1-tagged variant of apoaequorin, termed PeroxAEQ, to study the dynamics of free Ca^{2+} in the peroxisomal matrix (Lasorsa et al. 2008). In contrast to the findings reported by Pozzan and colleagues (see above), these researchers found that (1) the Ca^{2+} concentration of peroxisomes in living mammalian cells at rest is approximately 20 times higher than that of the cytosol, and (2) peroxisomes transiently take up calcium upon stimulation with agonists that induce Ca^{2+} release from intracellular stores. Further work needs to be done to reconcile these apparently conflicting studies.

Conclusions and Perspectives

Hidden in the shadow of mitochondria, peroxisomes have only recently begun to emerge as potentially important signaling platforms in various physiological and pathological processes (Titorenko and Terlecky 2011). As illustrated in Sect. 11.2, an increasing amount of evidence supports the idea that these organelles are at the crossroads between lipid metabolism, redox signaling, and inflammation. This is further emphasized by the observation that dysfunctional peroxisomes sensitize cells and organs for oxidative stress, a condition associated with the initiation and progression of age-related disorders such as type 2 diabetes, cancer, and neurodegenerative diseases (Fransen et al. 2013). In this context, deciphering the role of peroxisomes in various cellular signaling processes is of paramount importance, both for basic and translational research. Over the past years, through the development of novel and improved methods and technologies, this research field has fundamentally changed. In this chapter, we have mainly focused on the use of a wide range of fluorescent probes that are suitable for live cell imaging (see Sect. 11.3). However, as (1) each probe has its intrinsic limitations (e.g., Hyper measurements are strongly influenced by pH), and (2) different probes sometimes yield different results (e.g., see Sect. 11.3.5: compare SNAFL-2-PTS1 with pHluorin-PTS1, and D3cpv-PTS1 with PeroxAEQ), results obtained from live cell microscopy should be confirmed by independent methods (e.g., biochemical assays, redox proteomics, microarray analysis, etc.) to avoid misinterpretation. Finally, it is needless to say that we are still far from understanding the mechanistic details of how peroxisomes are embedded in cellular signaling networks. For example, the proximal targets of peroxisomal ROS are largely unknown; the molecular mechanisms underlying stress-related communication events between peroxisomes and mitochondria remain to be discovered; and it is even not yet clear under which conditions peroxisomes act as net sources or sinks of ROS. Finding answers to these and many other questions will certainly advance our understanding of how peroxisomes are integrated into developmental programs, and how dysregulation of peroxisome homeostasis may lead to pathological conditions.

Acknowledgments MF is supported by grants from the "Fonds voor Wetenschappelijk Onderzoek- Vlaanderen (Onderzoeksproject G.0754.09)" and the "Bijzonder Onderzoeksfonds van de KU Leuven (OT/09/045)." MN is supported by a FLOF fellowship from the Department of Cellular and Molecular Medicine, KU Leuven. BW is a recipient of a DBOF fellowship (DBOF/10/059) from the KU Leuven.

References

Antonenkov VD, Grunau S, Ohlmeier S et al (2010) Peroxisomes are oxidative organelles. Antioxid Redox Signal 13:525–537

Baarine M, Andréoletti P, Athias A et al (2012) Evidence of oxidative stress in very long chain fatty acid-treated oligodendrocytes and potentialization of ROS production using RNA interference-directed knockdown of ABCD1 and ACOX1 peroxisomal proteins. Neuroscience 213:1–18

Barford D (2004) The role of cysteine residues as redox-sensitive regulatory switches. Curr Opin Struct Biol 14:679–686

Beach A, Burstein MT, Richard VR et al (2012) Integration of peroxisomes into an endomembrane system that governs cellular aging. Front Physiol 3:283. doi:10.3389/fphys.2012.00283

Belousov VV, Fradkov AF, Lukyanov KA et al (2006) Genetically encoded fluorescent indicator for intracellular hydrogen peroxide. Nat Methods 3:281–286

Bikman BT, Summers SA (2011) Ceramides as modulators of cellular and whole-body metabolism. J Clin Invest 121:4222–4230

Bottelbergs A, Verheijden S, Van Veldhoven PP et al (2012) Peroxisome deficiency but not the defect in ether lipid synthesis causes activation of the innate immune system and axonal loss in the central nervous system. J Neuroinflammation 9:61. doi:10.1186/1742-2094-9-61

Boveris A, Oshino N, Chance B (1972) The cellular production of hydrogen peroxide. Biochem J 128:617–630

Braverman NE, Moser AB (2012) Functions of plasmalogen lipids in health and disease. Biochim Biophys Acta 1822:1442–1452

Brodde A, Teigler A, Brugger B et al (2012) Impaired neurotransmission in ether lipid-deficient nerve terminals. Hum Mol Genet 21:2713–2724

Bulina ME, Chudakov DM, Britanova OV et al (2006) A genetically encoded photosensitizer. Nat Biotechnol 24:95–99

Chen H, Jiang Z (2013) The essential adaptors of innate immune signaling. Protein Cell 4:27–39

Chen ZH, Niki E (2006) 4-hydroxynonenal (4-HNE) has been widely accepted as an inducer of oxidative stress. Is this the whole truth about it or can 4-HNE also exert protective effects? IUBMB Life 58:372–373

Chen X, Liang H, Van Remmen H et al (2004) Catalase transgenic mice: characterization and sensitivity to oxidative stress. Arch Biochem Biophys 422:197–210

Chudakov DM, Matz MV, Lukyanov S et al (2010) Fluorescent proteins and their applications in imaging living cells and tissues. Physiol Rev 90:1103–1163

Costa A, Drago I, Behera S et al (2010) H_2O_2 in plant peroxisomes: an in vivo analysis uncovers a Ca^{2+}-dependent scavenging system. Plant J 62:760–772

Dansen TB, Wirtz KW (2001) The peroxisome in oxidative stress. IUBMB Life 51:223–230

Dansen TB, Wirtz KW, Wanders RJ et al (2000) Peroxisomes in human fibroblasts have a basic pH. Nat Cell Biol 2:51–53

Del Río LA (2011) Peroxisomes as a cellular source of reactive nitrogen species signal molecules. Arch Biochem Biophys 506:1–11

Delille HK, Agricola B, Guimaraes SC et al (2010) Pex11pbeta-mediated growth and division of mammalian peroxisomes follows a maturation pathway. J Cell Sci 123:2750–2762

Diano S, Liu ZW, Jeong JK et al (2011) Peroxisome proliferation-associated control of reactive oxygen species sets melanocortin tone and feeding in diet-induced obesity. Nat Med 17:1121–1127

Dikalov S, Griendling KK, Harrison DG (2007) Measurement of reactive oxygen species in cardiovascular studies. Hypertension 49:717–727

Dixit E, Boulant S, Zhang Y et al (2010) Peroxisomes are signaling platforms for antiviral innate immunity. Cell 141:668–681

Doulias PT, Tenopoulou M, Greene JL et al (2013) Nitric oxide regulates mitochondrial fatty acid metabolism through reversible protein S-nitrosylation. Sci Signal 6(256):rs1. doi:10.1126/scisignal.2003252

Drago I, Giacomello M, Pizzo P et al (2008) Calcium dynamics in the peroxisomal lumen of living cells. J Biol Chem 283:14384–14390

El Hajj HI, Vluggens A, Andreoletti P et al (2012) The inflammatory response in acyl-CoA oxidase 1 deficiency (pseudoneonatal adrenoleukodystrophy). Endocrinology 153:2568–2575

Elsner M, Gehrmann W, Lenzen S (2011) Peroxisome-generated hydrogen peroxide as important mediator of lipotoxicity in insulin-producing cells. Diabetes 60:200–208

Fourcade S, López-Erauskin J, Galino J et al (2008) Early oxidative damage underlying neurodegeneration in X-adrenoleukodystrophy. Hum Mol Genet 17:1762–1773

Fransen M (2012) Peroxisome dynamics: molecular players, mechanisms, and (dys)functions. ISRN Cell Biol article ID 714192. doi:10.5402/2012/714192

Fransen M, Nordgren M, Wang B et al (2012) Role of peroxisomes in ROS/RNS-metabolism: implications for human disease. Biochim Biophys Acta 1822:1363–1373

Fransen M, Nordgren M, Wang B et al (2013) Aging, age-related diseases and peroxisomes. Subcell Biochem 69:45–65

Gutscher M, Sobotta MC, Wabnitz GH et al (2009) Proximity-based protein thiol oxidation by H_2O_2-scavenging peroxidases. J Biol Chem 284:31532–31540

Halvey PJ, Hansen JM, Johnson JM et al (2007) Selective oxidative stress in cell nuclei by nuclear-targeted D-amino acid oxidase. Antioxid Redox Signal 9:807–816

Hanson GT, Aggeler R, Oglesbee D et al (2004) Investigating mitochondrial redox potential with redox-sensitive green fluorescent protein indicators. J Biol Chem 279:13044–13053

Hasegawa K, Wakino S, Yoshioka K et al (2010) Kidney-specific overexpression of Sirt1 protects against acute kidney injury by retaining peroxisome function. J Biol Chem 285:13045–13056

Hori Y, Kikuchi K (2013) Protein labeling with fluorogenic probes for no-wash live-cell imaging of proteins. Curr Opin Chem Biol 17:644–650

Horner SM, Liu HM, Park HS et al (2011) Mitochondrial-associated endoplasmic reticulum membranes (MAM) form innate immune synapses and are targeted by hepatitis C virus. Proc Natl Acad Sci USA 108:14590–14595

Huybrechts SJ, Van Veldhoven PP, Brees C et al (2009) Peroxisome dynamics in cultured mammalian cells. Traffic 10:1722–1733

Hwang I, Lee J, Huh JY et al (2012) Catalase deficiency accelerates diabetic renal injury through peroxisomal dysfunction. Diabetes 61:728–738

Ivashchenko O, Van Veldhoven PP, Brees C et al (2011) Intraperoxisomal redox balance in mammalian cells: oxidative stress and interorganellar cross-talk. Mol Biol Cell 22:1440–1451

Jankowski A, Kim JH, Collins RF et al (2001) In situ measurements of the pH of mammalian peroxisomes using the fluorescent protein pHluorin. J Biol Chem 276:48748–48753

Karlsson M, Kurz T, Brunk UT et al (2010) What does the commonly used DCF test for oxidative stress really show? Biochem J 428:183–190

Kasai H, Okada Y, Nishimura S et al (1989) Formation of 8-hydroxydeoxyguanosine in liver DNA of rats following long-term exposure to a peroxisome proliferator. Cancer Res 49:2603–2605

Kassmann CM, Lappe-Siefke C, Baes M et al (2007) Axonal loss and neuroinflammation caused by peroxisome-deficient oligodendrocytes. Nat Genet 39:969–976

Kim PK, Mullen RT, Schumann U et al (2006) The origin and maintenance of mammalian peroxisomes involves a de novo PEX16-dependent pathway from the ER. J Cell Biol 173:521–532

Kojima H, Urano Y, Kikuchi K et al (1999) Fluorescent indicators for imaging nitric oxide. Angew Chem Int Ed Eng 38:3209–3212

Kozawa S, Honda A, Kajiwara N et al (2011) Induction of peroxisomal lipid metabolism in mice fed a high-fat diet. Mol Med Rep 4:1157–1162

Lasorsa FM, Pinton P, Palmieri L et al (2008) Peroxisomes as novel players in cell calcium homeostasis. J Biol Chem 283:15300–15308

Lazarow PB (2011) Viruses exploiting peroxisomes. Curr Opin Microbiol 14:458–469

Li Y, Tharappel JC, Cooper S et al (2000) Expression of the hydrogen peroxide-generating enzyme fatty acyl CoA oxidase activates NF-kappaB. DNA Cell Biol 19:113–120

Li Y, Chen R, Zhou Q et al (2012) LSm14A is a processing body-associated sensor of viral nucleic acids that initiates cellular antiviral response in the early phase of viral infection. Proc Natl Acad Sci USA 109:11770–11775

Los GV, Encell LP, McDougall MG et al (2008) HaloTag: a novel protein labeling technology for cell imaging and protein analysis. ACS Chem Biol 3:373–382

Lukyanov KA, Belousov VV (2013) Genetically encoded fluorescent redox sensors. Biochim Biophys Acta. doi:10.1016/j.bbagen.2013.05.030

Mast FD, Li J, Virk MK et al (2011) A Drosophila model for the Zellweger spectrum of peroxisome biogenesis disorders. Dis Model Mech 4:659–672

Miesenböck G, De Angelis DA, Rothman JE (1998) Visualizing secretion and synaptic transmission with pH-sensitive green fluorescent proteins. Nature 394:192–195

Müller CC, Nguyen TH, Ahlemeyer B et al (2011) PEX13 deficiency in mouse brain as a model of Zellweger syndrome: abnormal cerebellum formation, reactive gliosis and oxidative stress. Dis Model Mech 4:104–119

Nagano T (2009) Bioimaging probes for reactive oxygen species and reactive nitrogen species. J Clin Biochem Nutr 45:111–124

Palmer AE, Giacomello M, Kortemme T et al (2006) Ca^{2+} indicators based on computationally redesigned calmodulin-peptide pairs. Chem Biol 13:521–530

Pettus BJ, Baes M, Busman M et al (2004) Mass spectrometric analysis of ceramide perturbations in brain and fibroblasts of mice and human patients with peroxisomal disorders. Rapid Commun Mass Spectrom 18:1569–1574

Rhodin J (1954) Correlation of ultrastructural organization and function in normal and experimentally changed proximal tubule cells of the mouse kidney. Dissertation, Karolinska Institutet

Ribeiro D, Castro I, Fahimi HD et al (2012) Peroxisome morphology in pathology. Histol Histopathol 27:661–676

Rizzuto R, Simpson AW, Brini M et al (1992) Rapid changes of mitochondrial Ca^{2+} revealed by specifically targeted recombinant aequorin. Nature 358:325–327

Schwarzländer M, Fricker MD, Müller C et al (2008) Confocal imaging of glutathione redox potential in living plant cells. J Microsc 231:299–316

Singh J, Khan M, Singh I (2009) Silencing of Abcd1 and Abcd2 genes sensitizes astrocytes for inflammation: implication for X-adrenoleukodystrophy. J Lipid Res 50:135–147

Soardo G, Donnini D, Domenis L et al (2011) Oxidative stress is activated by free fatty acids in cultured human hepatocytes. Metab Syndr Relat Disord 9:397–401

Stolz DB, Zamora R, Vodovotz Y et al (2002) Peroxisomal localization of inducible nitric oxide synthase in hepatocytes. Hepatology 36:81–93

Swietach P, Youm JB, Saegusa N et al (2013) Coupled Ca^{2+}/H^{+} transport by cytoplasmic buffers regulates local Ca^{2+} and H^{+} ion signaling. Proc Natl Acad Sci USA 110:E2064–E2073

Tantama M, Hung YP, Yellen G (2011) Imaging intracellular pH in live cells with a genetically encoded red fluorescent protein sensor. J Am Chem Soc 133:10034–10037

Titorenko VI, Terlecky SR (2011) Peroxisome metabolism and cellular aging. Traffic 12:252–259

Van Veldhoven PP (2010) Biochemistry and genetics of inherited disorders of peroxisomal fatty acid metabolism. J Lipid Res 51:2863–2895

Vasko R, Ratliff BB, Bohr S et al (2013) Endothelial peroxisomal dysfunction and impaired pexophagy promotes oxidative damage in lipopolysaccharide-induced acute kidney injury. Antioxid Redox Signal 19:211–230

Verheijden S, Bottelbergs A, Krysko O et al (2013) Peroxisomal multifunctional protein-2-deficiency causes neuroinflammation and degeneration of Purkinje cells independent of very long chain fatty acid accumulation. Neurobiol Dis 58:258–269

Wahli W, Michalik L (2012) PPARs at the crossroads of lipid signaling and inflammation. Trends Endocrinol Metab 23:351–363

Walton PA, Pizzitelli M (2012) Effects of peroxisomal catalase inhibition on mitochondrial function. Front Physiol 3:108. doi:10.3389/fphys.2012.00108

Wang S, Li N, Pan W et al (2012) Advances in functional fluorescent and luminescent probes for imaging intracellular small-molecule reactive species. Trend Anal Chem 39:3–37

Yang Y, Song Y, Loscalzo J (2007) Regulation of the protein disulfide proteome by mitochondria in mammalian cells. Proc Natl Acad Sci USA 104:10813–10817

Yano T, Oku M, Akeyama N et al (2010) A novel fluorescent sensor protein for visualization of redox states in the cytoplasm and in peroxisomes. Mol Cell Biol 30:3758–3766

Zhang J, Kim J, Alexander A et al (2013) A tuberous sclerosis complex signalling node at the peroxisome regulates mTORC1 and autophagy in response to ROS. Nat Cell Biol 15:1186–1196

The Proteomics Toolbox Applied to Peroxisomes

12

Silke Oeljeklaus, Andreas Schummer, and Bettina Warscheid

Abstract

The full understanding of a cell's functionality and activity requires comprehensive knowledge about the properties and the individual contribution of all players involved, both as single entities and as parts of functional units such as membrane-enclosed organelles or larger multi-protein complexes. This comprises, among others, the gathering of information about the accurate subcellular localization of proteins and the interaction networks they form as well as dynamic alterations thereof upon metabolic, developmental, or environmental changes. The proteomics toolbox provides us here with a powerful means for the systematic, discovery-driven analysis of protein properties on a proteome-wide scale. In clever combination with classical biochemical or cell biological methods, state-of-the-art mass spectrometry (MS)-based proteomics has boosted the functional and quantitative analysis of proteins beyond the sheer generation of identification lists.

In this chapter, we will highlight the potential of modern MS-based proteomics research for the in-depth analysis of different aspects of peroxisome biology. The focus will be on quantitative MS strategies based on label-free (e.g., protein correlation profiling) or stable isotope-labeling techniques. We will review the applicability of these powerful approaches to the virtually complete delineation of the peroxisomal proteome including the discovery of new peroxisomal constituents as well as the characterization of peroxisomal membrane protein complexes providing new insights into distinct aspects of biogenesis and functioning of peroxisomes.

S. Oeljeklaus • A. Schummer • B. Warscheid (✉)
Institute of Biology II and BIOSS—Centre for Biological Signalling Studies, University of Freiburg, Schänzlestr. 1, 79104 Freiburg, Germany
e-mail: bettina.warscheid@biologie.uni-freiburg.de

C. Brocard and A. Hartig (eds.), *Molecular Machines Involved in Peroxisome Biogenesis and Maintenance*, DOI 10.1007/978-3-7091-1788-0_12, © Springer-Verlag Wien 2014

Keywords
Peroxisomes • Subcellular fractionation • Protein complexes • Affinity purification • Mass spectrometry • Stable isotope labeling • Organellar proteomics • Protein correlation profiling • Quantitative proteomics

Abbreviations

AP	Affinity purification
AP-AM	Affinity purification after mixing
AP-MS	Affinity purification-mass spectrometry
AP-PM	Affinity purification prior to mixing
ER	Endoplasmic reticulum
ESI	Electrospray ionization
GFP	Green fluorescence protein
GPF	Gas phase fractionation
ICAT	Isotope-coded affinity tag
iTRAQ	Isobaric tags for absolute and relative quantification
LC	Liquid chromatography
MALDI	Matrix-assisted laser desorption/ionization
MS	Mass spectrometry
MS/MS	Tandem mass spectrometry
PA	Protein A
POI	Protein of interest
PPI	Protein–protein interactions
PTM	Posttranslational modification
PTS	Peroxisomal targeting signal
QTOF	Quadrupole time of flight
SILAC	Stable isotope labeling with amino acids in cell culture
TAP	Tandem affinity purification

12.1 Introduction

The ultimate goal of cell biological endeavors is to decipher the fundamental molecular mechanisms driving and regulating development, survival, or pathology of cells on a system's wide level. To this end, it is essential to reveal the contribution of individual cellular components, their interplay, localization within the cell and dynamic behavior upon environmental changes. The basic information for life and survival are provided by the genome defining the general biological potential of organisms. However, the genome is rather static, remaining largely unaltered under varying internal and external stimuli or perturbations. Cellular functions and activities are predominantly carried out by proteins, which in their entirety—the proteome—exceed the genome's complexity by several orders of magnitude. The proteome is highly dynamic basically constantly adapting to the environment, which is reflected in spatiotemporal changes in expression, abundance, subcellular localization, interactions, and/or posttranslational modification (PTM) of individual proteins.

Genome and proteome are linked by the transcriptome reflecting the subset of genes expressed in a cell under distinct conditions at a given time. Quantitative information about mRNA levels is frequently used to estimate the abundance of corresponding proteins. However, studies comparing mRNA and protein levels showed that these often only partially correlate (Schwanhausser et al. 2011; Gygi et al. 1999b). Moreover, transcriptome data lack important information about the spatial organization of proteins and the dynamics of their distribution within the cell. Nuclear-encoded organellar proteins, for example, are posttranslationally imported into the organelle of destination. In addition, proteins may exhibit dual or multiple localizations, they may shuttle between different organelles and differ in their abundance in different subcellular compartments. Therefore, it is necessary to directly analyze the abundance levels and further properties of proteins.

Fuelled by the availability of genome sequences of various organisms ranging from bacteria, yeast, the model plant *Arabidopsis thaliana*, mice and rat to man as well as significant advancements in sample preparation, mass spectrometric systems, and computational data analysis, MS-based proteomics has established itself as key technology for the comprehensive study of protein properties on a proteome-wide scale (Aebersold and Mann 2003). At its best, latest instrumentation allows for the identification of thousands of proteins in a single liquid chromatography (LC)/MS run (Hebert et al. 2014). More importantly, however, the availability of sophisticated quantitative MS approaches facilitates the functional characterization of proteins beyond the mere generation of protein identification lists.

Quantitative MS strategies rely on stable isotope labeling or label-free methods allowing for the simultaneous identification of proteins in different samples. Stable, non-radioactive isotope-coded mass tags can be introduced into proteins of samples subjected to different experimental conditions either metabolically during growth of cells and organisms or chemically on the level of extracted proteins or proteolytic peptides. The mass tags convey fixed, predictable mass shifts to proteins or peptides allowing for mixing and joint processing of differentially labeled samples for MS analysis. Differences in protein composition and abundance between the samples are then revealed by the comparison of MS signal intensities or peak areas extracted from mass spectra of isotope-labeled peptide pairs. An advantage of stable isotope labeling approaches is the fact that an internal standard is generated for each peptide present in a proteolytic mixture, which facilitates the accurate relative quantification of proteins. In label-free quantification approaches, samples to be compared are analyzed in separate LC/MS runs. The comparison of protein abundances is generally based on signal intensities or peak areas of peptide ions on the level of MS scans or the frequency of peptide fragmentation events attributed to peptides of a distinct protein, an approach referred to as "spectral counting" (Liu et al. 2004).

For more detailed information on principles, pros and cons as well as applications of different quantitative MS strategies, please refer to recent reviews (e.g., Beynon and Pratt 2005; Mallick and Kuster 2010; Bantscheff et al. 2012).

Combined with adequate biochemical techniques for subcellular fractionation and/or enrichment of functional units, quantitative proteomics has proven to provide valuable means for the analysis of changes in protein abundance and PTMs, for the assignment of proteins to distinct subcellular organelles as well as the composition and structure of protein complexes and larger protein interaction networks (Ong and Mann 2005; Walther and Mann 2010). In this chapter, we will review and discuss the applicability of advanced quantitative MS-based strategies for the identification of new peroxisomal proteins in global organellar proteomics studies and for the thorough characterization of peroxisomal membrane protein complexes. We will highlight the strong potential of quantitative proteomics research to provide new insight into the physiological role of distinct peroxisomal proteins as well as the functionality of the entire organelle.

12.2 Organellar Proteomics

Detailed functional understanding of organelles requires in-depth knowledge about the molecular components residing in these lipid-bilayer surrounded entities of eukaryotic cells. In cell biology, classical approaches for the characterization of organelles comprise optical techniques allowing for the study of their morphological features inside cells and for the localization of proteins up to a genome-wide scale, for example by expressing proteins fused to a generic tag (e.g., GFP) in living cells. In the field of biochemistry, organelles are isolated based on their physical and antigenic properties applying different established cell fractionation methods such as differential centrifugation, equilibrium density-gradient centrifugation as well as affinity-based purification techniques. In both strategic tracks, data interpretation is facilitated by the specific detection of distinct organellar marker proteins using fluorescence- and immuno-based methodologies or enzyme activity assays. These classical approaches in cell biology and biochemistry are typically hypothesis-driven focusing on specific components and their functions rather than providing a holistic and unbiased view of the composition, dynamics, and functions of organelles. Here, the introduction and rapid advancement of proteomics technologies combined with powerful hypothesis-generating concepts will pave the way for obtaining a more comprehensive understanding of the molecular, functional and dynamic properties of organelles in a whole-cell context as highlighted in various recent reviews (Yates et al. 2005; Andersen and Mann 2006; Walther and Mann 2010; Wiederhold et al. 2010; Drissi et al. 2013). In the following, we will exclusively focus on organellar proteomics studies for the characterization of peroxisomes in the model system *Saccharomyces cerevisiae* as well as in rodents and humans. For the proteomics study of plant peroxisomes, we would like to refer to a recent in-depth review by Bussell et al. (2013).

12.2.1 Proteomics Applied to the Study of Yeast Peroxisomes

In the yeast *S. cerevisiae*, the number and size of peroxisomes depend on the growth conditions with 1–4 peroxisomes in cells cultured with glucose as inoculum to up to 14 peroxisomes when cells are shifted to medium containing oleic acid as sole carbon source (Veenhuis et al. 1987). Proliferation of peroxisomes induced by oleate is accompanied by an increase in the activities of enzymes involved in the degradation of long-chain fatty acids by β-oxidation. Under these conditions, peroxisomes were found to constitute approx. 10 % of the cytoplasmic volume of *S. cerevisiae* cells (Veenhuis et al. 1987), making them amenable for biochemical purification. In a pioneering study from 1995, Erdmann and Blobel succeeded in the purification of peroxisomal membranes from *S. cerevisiae* cells cultured for 9 h in oleate-containing medium to form mature peroxisomes with a density of 1.21 g/cm^3 (Erdmann and Blobel 1995). Using differential centrifugation followed by successive sucrose and Nycodenz density gradient centrifugation, peroxisomes of high density could be well separated from both mitochondria and the endoplasmic reticulum (ER) followed by purification of peroxisomal membranes by consecutive extraction under low salt, high salt, and alkaline conditions. However, it is of interest to note here that due to their fragileness, the majority of peroxisomes did not remain intact during the extensive isolation procedure leading to a loss of matrix protein content even before membrane extraction. For analysis of the composition, a total of 1 mg of purified peroxisomal membranes were solubilized in SDS, separated by reversed-phase LC followed by SDS-PAGE and visualized by Coomassie staining revealing the presence of about 30 distinct proteins, the majority of which were of unknown identity. Through classical NH_2-terminal protein sequencing combined with DNA cloning and gene sequencing, Erdmann and Blobel could finally determine the identity of a highly prominent protein band with an apparent molecular weight of 27 kDa as PMP27. Today, PMP27 is known as Pex11p, an integral peroxisomal membrane protein conserved from yeast to humans with essential functions in peroxisome division, proliferation, and maintenance (reviewed in Thoms and Erdmann 2005).

Despite these substantial data already outlining the molecular blue-print of yeast peroxisomal membranes in the mid-1990s, protein-centered research was clearly limited by the availability of technologies for polypeptide sequencing at both high sensitivity and high speed. During the nineties, this bottleneck had been overcome by the development of new ionization methods allowing for the gentle transfer of intact polypeptides into the gas phase for mass spectrometric analysis (for a brief history on of MS technologies, see Yates 2011). These seminal innovations were finally recognized by the Nobel Prize in Chemistry awarded to John Bennet Fenn for the development of electrospray ionization (ESI) and to Koichi Tanaka for innovations in soft desorption ionization of large biomolecules in 2002. It therefore came as no surprise that with the beginning of the twenty-first century N-terminal sequencing was progressively replaced by innovative tandem MS (MS/MS) technologies for peptide and protein analysis at unforeseen speed and sensitivity

(Peng and Gygi 2001; Aebersold and Mann 2003; Han et al. 2008; Mann and Kelleher 2008; Yates 2011).

Schäfer et al. (2001) were the first to make use of innovative MS-based sequencing methods for the proteomics characterization of mature peroxisomes from *S. cerevisiae* cells grown in oleate-containing medium. Peroxisomal membranes were purified following the protocol established by Erdmann and Blobel (1995), separated by SDS-PAGE and analyzed by matrix-assisted laser desorption/ionization time-of-flight (MALDI-TOF) MS and ESI-MS/MS on an ion trap system following in-gel protein digestion with trypsin. For the analysis, the total amount of protein used for SDS-PAGE was approx. 10 μg, demonstrating the high potential of sensitive MS-based peptide sequencing technologies. Yet, application of MALDI-TOF-MS was impeded by both the low efficiency of peptide fragmentation by post-source decay and the limited resolving power of SDS-PAGE. By contrast, direct coupling of nano-capillary LC employing analytical columns of 75 μm inner diameter and operating at flow rates of 300 nl/min with ESI/ion trap MS technology (Peng and Gygi 2001) allowed for the identification of 25 peroxisomal proteins including Pex11p, Pex3p, and several components central to the peroxisomal matrix protein import machine (i.e., Pex12p, Pex13p, Pex14p, and Pex8p) (Schäfer et al. 2001). However, as a consequence of increased sensitivity with peptide detection and sequencing limits in the low femtomole range, nano-LC/ESI-MS/MS analyses also led to the identification of a high number of low abundant contaminants indicating that "pure" peroxisomal membrane preparations still contained fractions of mitochondria, the ER and the cytoplasm.

In successive work, Yi et al. (2002) sought to obviate SDS-PAGE for protein separation by devising an MS approach applicable to the analysis of complex peptide mixtures. Tryptic peptides were generated from purified peroxisomal membrane fractions obtained by discontinuous Nycodenz gradient centrifugation and subsequent hypotonic lysis of peroxisomes from *S. cerevisiae*. To increase the effective dynamic range of LC/MS and, thus, the number of sequenced peptides per analysis, gas phase fractionation in the mass-to-charge dimension ($GPF_{m/z}$) of the mass analyzer was performed using 13 overlapping 100 *m*/*z*-windows covering the full mass range (*m*/*z* 400–1,800) for the selection of precursor ions in MS/MS experiments. In this approach, $GPF_{m/z}$ requiring multiple LC/MS runs allowed to compensate for limited MS/MS scan speed to most comprehensively fragment all peptide ions detected in complex mixture. This yielded, with the detection of 38 peroxisomal proteins, a virtually complete core proteome of yeast peroxisomes; however, this was accompanied by 143 co-purified components of which the majority is known to reside in other cellular compartments. Thus, for the characterization of organelles, biochemists and mass spectrometrists were facing a dilemma: opting for high sensitivity, while at the same time loosing specificity as organelles cannot be purified to homogeneity.

In order to address this issue, Aitchison and coworkers were the first to make use of chemical protein labeling with stable isotopes in conjunction with quantitative MS analysis to reveal those components that specifically enrich during the purification procedure indicative for their association with peroxisomes (Marelli et al. 2004). To quantitatively compare the abundance levels of proteins present in organellar fractions of different purity and composition, they used the commercially available isotope-coded affinity tagging (ICAT) technology (Gygi et al. 1999a). The ICAT reagent consists of three parts: (1) a reactive iodoacetyl group for site-directed derivatization of the nucleophilic sulfhydryl group of cysteine residues in proteins, (2) a linker region for stable isotope coding of the mass tag, and (3) a biotin tag allowing for the specific enrichment of stable isotope-coded peptides generated by tryptic digestion for subsequent LC/MS analysis. Use of different "light" ($^{1}H_8$ or $^{12}C_9$) and "heavy" ($^{2}H_8$ or $^{13}C_9$) versions of the ICAT reagent bestow fixed mass differences (8 Da or 9 Da) between differentially labeled peptides which are readily detectable by MS. However, an inherent limitation of the ICAT approach is that only cysteine-containing proteins can be identified and quantified. To accurately delineate resident components of peroxisomes as well as to identify new peroxisomal candidate proteins, two different ICAT experiments were performed. In the first, membrane-enriched fractions of peroxisomes were quantitatively compared with mitochondrial membranes and, in the second, with peroxisomal membranes of higher purity obtained from a yeast strain expressing Pex11p-PA (Protein A of *Staphylococcus aureus*) amenable for additional affinity purification on IgG Sepharose (Marelli et al. 2004). In total, LC/MS analyses on a low-resolution ion trap instrument resulted in the identification of approx. 350 proteins of which 71 (20 %) were selected as peroxisomal candidate proteins based on their enrichment factors and mathematical data modeling. The high potential of this elegant quantitative proteomics approach was demonstrated by the identification of 38 proteins with known functions in peroxisome biology as well as the discovery of three new peroxisome-associated proteins, namely the cytosolic enzyme Gpd1p, COPII-like Emp24p and the small GTPase Rho1p with a potential role in the biogenesis and movement of peroxisomes. Gdp1p comprising a peroxisomal targeting signal type 2 (PTS2) in its N-terminal region has recently been confirmed to dynamically relocalize from the cytosol to peroxisomes via the Pex7p-dependent route of matrix protein import in response to oleic acid exposure (Jung et al. 2010). Interestingly, localization of Gdp1p to peroxisomes occurs in a stress-dependent manner and is modulated by reversible phosphorylation of two distinct serine residues adjacent to the PTS2 signal sequence. However, while Gpd1p has been suggested to play a role in redox potential regulation in cells, its precise function in peroxisome biochemistry has remained elusive so far. Similarly, Pnc1p, a nicotinamidase, was found to distribute between the cytosol, nucleus and peroxisomes in a stress-dependent manner, indicating that enzyme localization and molecular function are closely linked in order to allow cells to dynamically respond to different external and internal cues (Jung et al. 2010).

12.2.2 MS-Driven Studies of Rodent and Human Peroxisomes

Number and size of peroxisomes are dynamically modulated in response to the organism's cellular and metabolic state. For example, an increase in the number of peroxisomes is observed as a consequence of a lipid-rich diet or the intake of lipid-lowering drugs, whereas conditions of fatty liver disease or treatment with catalase inhibitors are accompanied with a significant decrease in the number of peroxisomes. Since functions of peroxisomes are manifold and further depend on the organ as well as vary between species (Islinger et al. 2010), well-adjusted alterations in their enzymatic apparatus are required. This is exemplified by human peroxisomes lacking several proteins known to reside in peroxisomes of *Mus musculus* and the yeast *Yarrowia lipolytica* (Schluter et al. 2007). In humans, defects in peroxisome biogenesis or deficiencies in the function of single peroxisomal enzymes ultimately manifest themselves in severe inherited diseases such as Zellweger syndrome or X-linked adrenoleukodystrophy (Wanders and Waterham 2006), demonstrating the biochemical, physiological and clinical importance of the organelle. Today, more than 15 peroxisomal diseases are known; however, it can be anticipated that additional links between peroxisomal dysfunctions and human disorders exist. Therefore, complete knowledge of the proteome of mammalian peroxisomes is highly desirable as it provides the unique possibility to get new clues about its biogenesis, biochemistry and function, which will eventually lead to a better understanding of its important role in human health and disease.

Challenges in the study of mammalian peroxisomes primarily rely on the isolation of this highly dynamic organelle diminutive in size and number from tissue sections. Proteomics research has mainly focused on the characterization of liver and kidney peroxisomes from rat or mouse (Yi et al. 2002; Kikuchi et al. 2004; Islinger et al. 2006, 2007; Mi et al. 2007; Wiese et al. 2007, 2012), while only a single report on human liver peroxisomes exists today (Gronemeyer et al. 2013b). Common to the majority of studies was the analysis of peroxisome-enriched fractions by 2D or 1D gel electrophoresis followed by MS-based protein identification. In two independent studies, the applicability of benzyldimethyl-*n*-hexadecylammonium chloride (16-BAC)/SDS-PAGE for the analysis of mammalian peroxisomes was shown, facilitating the identification of microsomal glutathione-*S*-transferase (mGST) and the nudrix hydrolase 19, referred to as Rp2, as new peroxisomal proteins in rat liver (Islinger et al. 2006) and in mouse kidney (Ofman et al. 2006), respectively. However, while this 2D gel electrophoretic system generally allows for the analysis of integral membrane proteins using 16-BAC as cationic and SDS as anionic detergent (Hartinger et al. 1996; Zahedi et al. 2007; Braun et al. 2007), it provides only moderate resolution, hampering the quantification of dye-stained proteins by densitometry. Thus, traditional 2D PAGE with isoelectric focusing in the first dimension was employed for a comparative analysis of the matrix proteome of liver and kidney peroxisomes, revealing a higher expression of proteins involved in fatty acid α- and β-oxidation, amino acid and nucleotide metabolism in liver (Mi et al. 2007). Nevertheless, shortcomings of 2D PAGE such as limited sensitivity, resolution, reproducibility, and speed as well as strong bias against proteins with an extreme isoelectric point, molecular weight, or hydropathy remain inherent to this classical technique. Also as a consequence thereof, proteomics strategies exploiting the power of quantitative MS have come to the fore. Despite this paradigm shift, the generation of highly purified organelle fractions has remained a key factor in identifying also minor components potentially important for the proper functioning of these cellular machines. To this end, Kikuchi et al. (2004) prepared peroxisomes from rat liver by sequentially performing isopycnic centrifugation on a Nycodenz gradient and immunoaffinity chromatography employing an antibody against Pmp70, an abundant peroxisomal membrane protein. Highly purified rat liver peroxisomes were then examined by gel-enhanced LC/MS analysis. To also yield a high coverage of the membrane-bound subproteome including low abundant peroxisomal biogenesis factors, affinity-purified peroxisomes were further subjected to mild alkaline treatment prior to SDS-PAGE and LC/MS analysis. This comprehensive organellar proteomics approach led to the identification of approx. 50 genuine peroxisomal proteins including virtually all membrane-bound components of peroxisomes. Only the peroxisomal receptor proteins Pex7p and Pex19p remained elusive, indicative of their mainly cytoplasmic localization. Assessment of the relative abundance of proteins detected in highly purified peroxisomal fractions further pointed to the association of distinct components of other compartments (e.g., ER and mitochondria) as well as proteins of the Rab family with peroxisomes. The latter finding was confirmed in successive studies showing that distinct Rab proteins locate to the peroxisomal membrane where Rab8 and Rab18 present in their GDP-bound state promote peroxisome proliferation (Schollenberger et al. 2010; Gronemeyer et al. 2013a).

During the recent years, implementation of quantitative MS methodology has further expedited the quest for new constituents of mammalian peroxisomes by enabling the identification of genuine peroxisomal proteins against a large background of co-purified contaminants originating from other cellular compartments. In a fashion analogous to the subcellular localization of proteins by immunoblotting across density gradient fractions, quantitative high-resolution MS can be exploited to simultaneously establish organelle-characteristic abundance profiles of hundreds of proteins present in a density gradient (reviewed in Andersen and Mann 2006). The potential of this method, termed protein correlation profiling (PCP), for the accurate subcellular localization of proteins was demonstrated for the first time by the identification of new constituents of human centrosomes (Andersen et al. 2003) and, further, by the delineation of a mammalian organelle map comprising ten subcellular structures only partially separated across a mouse liver density gradient (Foster et al. 2006). Yet, low abundant organelles such as peroxisomes remained poorly described in this subcellular-specific proteome map.

To address this issue, our group was the first to report a PCP approach tailored to a most complete description of the proteome of mouse kidney peroxisomes with the specific objective of identifying new peroxisomal proteins (Wiese et al. 2007). Following tissue homogenization, a light mitochondrial fraction was subjected to Nycodenz density gradient centrifugation for separation of peroxisomes from mitochondria, lysosomes, and microsomes (Fig. 12.1). Comprehensive MS-based proteomics analyses of both the peroxisomal peak fraction and a purified membrane fraction thereof allowed for virtually complete coverage of the established peroxisomal proteome comprising 42 matrix and 22 membrane proteins. Yet, due to the

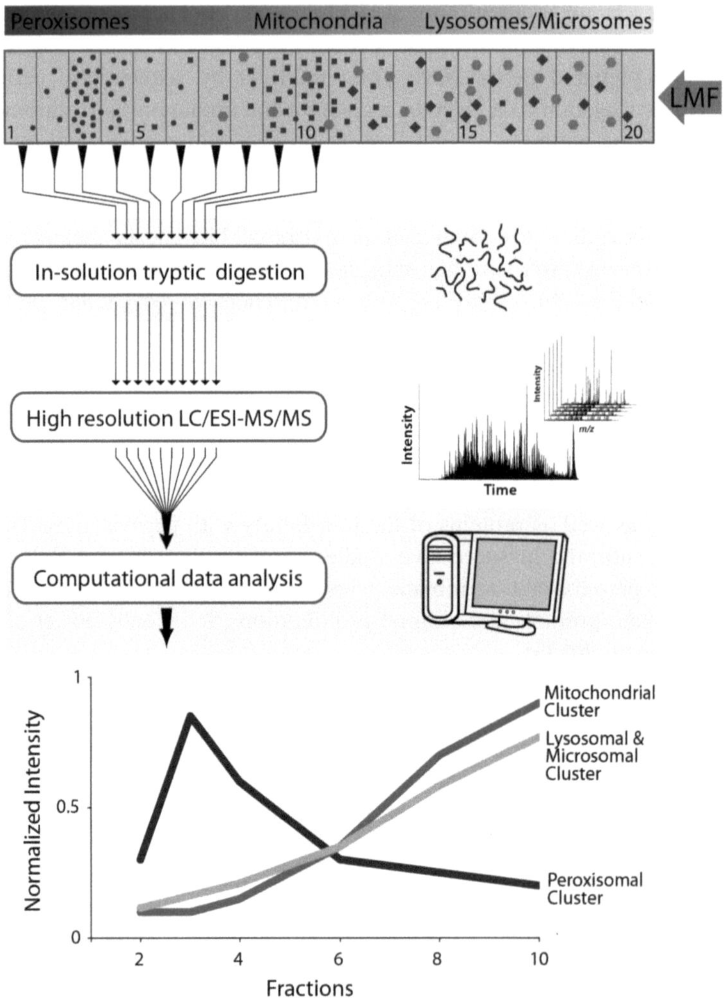

Fig. 12.1 Study of mammalian peroxisomes by protein correlation profiling. Organelles of a light mitochondrial fraction (LMF) obtained from a kidney or liver homogenate are separated by Nycodenz density gradient centrifugation. Following fractionation of the gradient, selected fractions including the peroxisomal peak fractions are subjected to tryptic in-solution digestion and high-resolution LC/ESI-MS/MS for protein identification. In order to differentiate between true peroxisomal proteins and co-purified proteins from other subcellular compartments, protein profiles based on the abundance of each protein in distinct fractions are established. Marker proteins for different organelles define the respective consensus profiles and proteins following the characteristic consensus profile for an organelle are considered as resident proteins of this organelle, whereas deviations from the profile indicate contaminants. Further details about PCP are given in the main text

high sensitivity of mass analyses, these 64 genuine peroxisome constituents accounted only for ¼ of all the proteins (including 65 proteins of unknown localization) identified.

To further obtain information about the subcellular localization of proteins, several consecutive fractions of the dense part of the density gradient containing peroxisomes were directly subjected to tryptic digestion followed by LC/MS measurements on a high resolution LTQ-FTICR instrument applying $GPF_{m/z}$ in combination with single ion monitoring scans. Computational data analysis then allowed for the establishment of protein abundance profiles by plotting normalized intensities of proteins obtained against the respective gradient fractions (Fig. 12.1). In this label-free approach, quantification of peptides and, thus, proteins is based on measured peptide ion signal intensities detected in high-resolution MS scans, while sequence information is retrieved from MS/MS scans. Finally, organellar consensus profiles are established for clustering analysis allowing for sorting individual components to distinct subcellular compartments in a global manner. Since protein profiles can also feature characteristics of consensus profiles of two or more distinct subcellular compartments, PCP analysis generally provides the potential to also reveal multiple localization sites of proteins.

As a result of this advanced quantitative organellar proteomics approach, 15 new candidate proteins of mouse kidney peroxisomes were identified (Wiese et al. 2007). Of these, six candidate proteins (i.e., Acad11, Acbd5, Mdh1, Cyb5a, Dia1, and Aldh3a2) had also been detected in preparations of rat liver peroxisomes in previous work (Kikuchi et al. 2004; Islinger et al. 2006) and five proteins (Acad11, Zadh2, Acbd5, Pmp52, and Mosc2) were further studied by immunofluorescence microscopy (Wiese et al. 2007), confirming their peroxisomal localization as listed in Table 12.1. Of note, the membrane-bound iron–sulfur protein Mosc2 (Marc2) was demonstrated to exhibit a dual localization in mitochondria and peroxisomes. Pmp52 (Tmem135) represent a new component of peroxisomal membranes which is related to Pmp24 (Pxmp4) (Reguenga et al. 1999); however, the function of both proteins has remained elusive so far. A further, yet not confirmed peroxisomal membrane protein is Atad1, a member of the AAA-superfamily of ATPases involved in protein unfolding or disassembly of protein complexes and aggregates, suggesting its functioning in peroxisomal homeostasis (Wiese et al. 2007). Among the newly identified peroxisomal matrix candidate proteins, the fatty aldehyde dehydrogenase variant encoded by the gene Aldh3a2 is of particular interest as mutations in the gene cause Sjogren–Larsson syndrome, an inherited human neurocutaneous disorder associated with conditions of abnormal lipid accumulation, defective eicosanoid metabolism, and increased formation of aldehyde adducts with lipids or proteins (Rizzo and Carney 2005).

The high applicability of high-resolution MS-based protein abundance profiling to the in-depth characterization of "hard to grasp" peroxisomes was also demonstrated by two successive studies. In the first, label-free protein profiling was used to determine alterations in the proteome of kidney peroxisomes isolated from wild-type and Pex7 knockout mice, showing not only the near absence of all known PTS2 proteins, namely acetyl-Coenzyme A acyltransferase (Acaa1),

Table 12.1 New proteins associated with mammalian peroxisomes identified by quantitative proteomics approaches

Uniprot accession no.	Gene name	Protein name	Predicted PTS sequence	Localization by Uniprot	Tissue and organism	Reported in study
Q80XL6, mouse B3DMA2, rat	Acad11	Acyl-Coenzyme A dehydrogenase family, member 11	RLTARM[a], mouse QLKARM[a], rat	Peroxisome, mitochondrion	Mouse kidney, rat liver	Wiese et al. (2007), Islinger et al. (2007)
Q8BGC4, mouseD4A264, rat	Zadh2	Zinc-binding alcohol dehydrogenase domain containing protein 2	PVSSKL[a]	Peroxisome	MOUSE kidney, rat liver	Wiese et al. (2007), Islinger et al. (2007)
Q5CZX6, mouse A0FKI7, rat	Acbd5	Acyl-CoA-binding domain-containing protein 5	None	Peroxisome membrane	Mouse kidney, rat liver	Wiese et al. (2007), Islinger et al. (2007)
Q8BSY5, mouse Q5U4F4, rat	Tmem135	Transmembrane protein 135 (peroxisomal membrane protein 52, PMP52)	RIPSAFRHL[b]	Peroxisome membrane	Mouse kidney, rat liver	Wiese et al. (2007), Islinger et al. (2007)
Q922Q1, mouse O88994, rat	Marc2 (Mosc2)	MOSC domain-containing protein 2, mitochondrial (MOCO sulphurase C-terminal domain containing 2)	None	Mitochondrion outer membrane, peroxisome	Mouse kidney, rat liver	Wiese et al. (2007), Islinger et al. (2007)
Q96CN7, humanQ6I7R3, rat	ISOC1	Isochorismatase domain-containing protein 1	GLLSKV[a] RLVPLQIQL[b]	Peroxisome	Human liver, rat liver	Gronemeyer et al. (2013a, b), Islinger et al. (2007)
Q6YN16, human	HSDL2	Hydroxysteroid dehydrogenase-like protein 2	QMNARL[a]	Peroxisome, mitochondrion	Human liver	Gronemeyer et al. (2013a, b)
P00338, human	LDHA	Isoform 1 of L-lactate dehydrogenase A chain	None	Cytoplasm	Human liver	Gronemeyer et al. (2013a, b)
P07327, human	ADH1A	Alcohol dehydrogenase 1A	None	Cytoplasm	Human liver	Gronemeyer et al. (2013a, b)
P40925, human	MDH1	Malate dehydrogenase, cytoplasmic	None	Cytoplasm	Human liver	Gronemeyer et al. (2013a, b)

Q5M884, rat Q78JN3, mouse	Eci3	Protein Eci3 (δ3,δ2-enoyl-CoA isomerase)	SRKAKL[a], rat KAKAKL[a], mouse	–	Rat liver, mouse kidney	Islinger et al. (2007), Wiese et al. (2007)
B1H261, rat	Tysnd1	Trypsin domain containing 1	APRSKL[a]	–	Rat liver	Islinger et al. (2007)
Q5RJY4, rat	Dhrs7b	Dehydrogenase/reductase SDR family member 7B (short chain dehydrogenase 7b)	None	Peroxisome membrane	Rat liver	Islinger et al. (2007)
Q9JJ19, rat	Slc9a3r1 (Nherf1)	Na(+)/H(+) exchange regulatory cofactor NHE-RF1 (Ezrin-radixin-moesin-binding phosphoprotein 50)	None	Cell membrane, cell projection, cytoplasm	Rat liver	Islinger et al. (2007)

For all proteins, localization to peroxisomes was confirmed by fluorescence microscopy studies. Alternative gene and protein names are listed in parentheses
[a]PTS1
[b]PTS

alkylglycerone phosphate synthase (Agps), and phytanoyl-CoA hydroxylase (Phyh), but also the drastic up-regulation of several PTS1 proteins essential for very long-chain fatty acid β-oxidation or ether-phospholipid biosynthesis (Wiese et al. 2012). In the second study, elaborate PCP analysis facilitated a most comprehensive characterization of peroxisomes from human liver, leading to the identification of 59 known peroxisomal constituents and additional five new proteins (i.e., ISOC1, HSDL2, MDH1, LDHA, and ADH1A) at least partially associated with human peroxisomes (Gronemeyer et al. 2013b) (Table 12.1). Supported by immunofluorescence studies and enzyme activity profiles, association of fractions of MDH1 and LDHA with human peroxisomes further points to the existence of alternative pathways for the regeneration of NAD^+ consumed during fatty acid α- and β-oxidation in peroxisomes. Of note, the existence of a lactate shuttle mechanism in mammalian peroxisomes has been suggested before (Baumgart et al. 1996; McClelland et al. 2003).

Following a stable isotope labeling approach, Islinger et al. (2007) examined dynamic changes in the proteome of rat liver peroxisomes in response to treatment with bezafibrate, a drug known to induce peroxisome proliferation and peroxisomal β-oxidation (Fahimi et al. 1982). Figure 12.2 depicts the overall experimental strategy beginning with the isolation of peroxisomes from control and bezafibrate-treated rats by standard density gradient centrifugation and further suborganellar fractionation yielding distinct peroxisomal fractions enriched in integral membrane, peripheral membrane or matrix proteins. Following protein digestion using trypsin, the iTRAQ (isobaric tag for the relative and absolute quantitation) method was employed to derivatize primary amino groups in peptides by stable isotope-containing tags (Ross et al. 2004). Following chemical labeling, peptide samples can be mixed in equal ratio to facilitate MS-based protein identification and relative quantification in the same experiment.

A unique feature of the commercially available iTRAQ is that it enables sample multiplexing by providing up to eight reagents distinct in their isotopic forms but identical in their main composition. The isobaric reagents consist of a reporter group (i.e., N-methylpiperazine), a carbonyl balance group and a peptide-reactive group (i.e., N-hydroxysuccinimide ester). Since reporter and balance group feature a nominal mass of 145.1 Da, iTRAQ-labeled peptides are detected as single peaks in MS survey spectra. In MS/MS experiment, collisional activation of iTRAQ-labeled peptides induces cleavage of the carbonyl balance group allowing for the generation of distinguishable reporter ions detected in the low mass-to-charge (m/z) range (Fig. 12.2). Here, reporter signal intensities reflect the relative quantities of peptides and, thus, respective proteins among the samples (for in depth-reviews on pros and cons of isobaric tagging for relative protein quantification, see Evans et al. 2012; Christoforou and Lilley 2012; Treumann and Thiede 2010; Aggarwal et al. 2006).

In their study, Islinger et al. (2007) quantitatively compared integral and peripheral peroxisomal membrane fractions as well as peroxisomal matrix fractions in two repeats from control and bezafribate-treated rats (Fig. 12.2). To obtain a most comprehensive dataset, iTRAQ-labeled sample mixtures were subjected to 2D

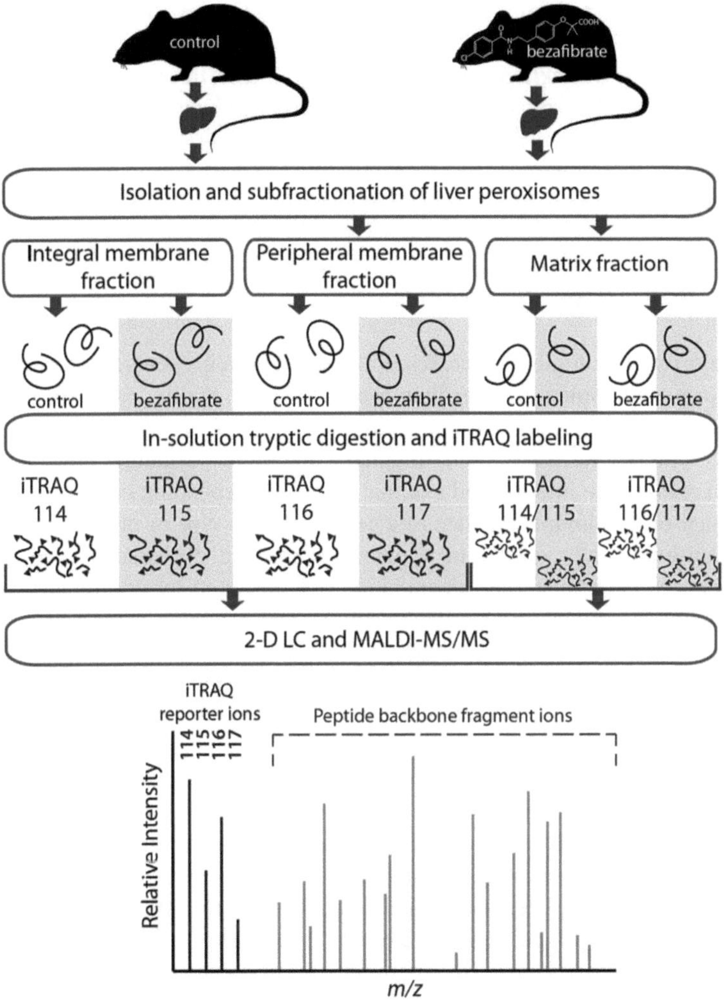

Fig. 12.2 Quantitative proteomics strategy for the comparative analysis of rat liver peroxisomes using iTRAQ. Peroxisomes were purified from livers of control and bezafibrate-treated rats using density gradient centrifugation and further subfractionated to obtain samples enriched in integral membrane, peripheral membrane or matrix proteins. Proteins were tryptically digested, the resulting peptides of the subfractions differentially labeled using iTRAQ and mixed as depicted for subsequent 2D LC separation prior to MALDI-TOF/TOF analysis. This approach enables MS-based protein identification and relative quantification in the same experiment

LC followed by MALDI-TOF/TOF analyses for protein identification and quantification. Applying stringent criteria, they identified 57 bona fide peroxisomal proteins affected differently in their expression levels by bezafibrate, leading them to the conclusion that liver peroxisomes may contribute to a higher extent to the degradation of long-chain fatty acids as anticipated before. From their dataset, they selected ten candidate proteins not known to localize to rat liver peroxisomes before and verified their peroxisomal residency by immunofluorescence microscopy. Strikingly, as outlined in Table 12.1, this set of new peroxisomal proteins identified in rat liver was highly consistent with data obtained by PCP analysis of mouse kidney peroxisomes (Wiese et al. 2007), demonstrating the high potential and accuracy of advanced quantitative MS-based proteomics strategies for the study of peroxisomal proteomes in mammals.

12.3 Interaction Proteomics

The majority of biological processes is driven by proteins generally forming larger (multi)protein complexes rather than acting as single entities (Alberts 1998). The composition of individual protein complexes as well as the intricate networks proteins typically form are tightly regulated and highly dynamic; they change and adapt in response to internal and external stimuli or depending on, for example, the nutritional or developmental state of a cell, the tissue type or organism they occur in. Accordingly, key events in a peroxisome's life cycle such as biogenesis, maintenance, inheritance and degradation are mediated by coordinated protein–protein interactions and large protein machines (for example, see Hasan et al. 2013; Platta et al. 2013; Platta and Erdmann 2007; Nuttall et al. 2011). To fully comprehend the molecular mechanisms underlying the functionality of peroxisomes—as well as dysfunctions—it is therefore of crucial importance to identify individual protein–protein interactions (PPIs) as well as to delineate the composition of multiprotein complexes. Furthermore, it is necessary to reveal the dynamics of their composition including assembly and disassembly as well as to eventually reveal changes in peroxisome-centered interaction networks in response to different stimuli. As a result, proteins so far not related to peroxisomes—with either unknown function or different functional annotation—may be found to be essential for certain aspects of peroxisome biology. The study of PPIs further holds the potential to discover crosstalk between peroxisomes and other subcellular structures.

Classical biochemical strategies for the analysis of PPIs are largely based on affinity purification of a protein of interest (POI) by co-immunoprecipitation or epitope tagging of this protein using it as bait for the isolation of interaction partners followed by SDS-PAGE and detection of co-purified proteins by immunoblot analysis. Specific binding partners are revealed by comparing the immunosignals of the isolated complex with those of an adequate negative control processed in parallel. Although this generic approach has facilitated the characterization of countless protein complexes covering virtually all aspects of cellular function and activity, the success of such studies generally depends on prior knowledge, hypotheses about potential interaction partners and the availability of suitable antibodies. Thus, the quest for individual interaction partners, not to mention the establishment of entire protein networks, often proves to be a tedious task. In contrast, MS-based strategies allows for the systematic analysis of protein complexes in an unbiased, discovery-driven way with the potential to obtain results beyond preexisting assumptions and expectations. In the following, we will discuss achievements as well as pros and cons of several, often ground-breaking interaction proteomics studies in the context of peroxisome biology.

12.3.1 Study of Membrane-Bound Peroxisomal Protein Complexes by MS

About a decade ago, a number of pioneering large-scale protein interaction studies based on the combination of affinity purification and MS (AP-MS) were published (Gavin et al. 2002, 2006; Ho et al. 2002; Krogan et al. 2006). Attempting at systematically characterizing protein complexes and establishing a global protein interaction map of the yeast *S. cerevisiae*, thousands of proteins were epitope-tagged and protein complexes purified either following a two-step protocol using the tandem affinity purification (TAP) tag (Gavin et al. 2002, 2006; Krogan et al. 2006) or via the FLAG tag in a single purification step (Ho et al. 2002). Affinity-purified complexes were individually analyzed following a standard proteomics workflow consisting of protein separation by SDS-PAGE, proteolytic in-gel digestion, and LC/MS analysis for protein identification. These studies impressively demonstrated the high potential of AP-MS: they considerably contributed to an improved understanding of numerous cellular processes and facilitated deeper insights into the modular organization of the entire yeast proteome into protein complexes built up of core components and transient binders. However, the majority of "newly" identified interactions reported have not been validated—a general drawback of proteome-wide interaction studies. In addition, the gain of knowledge they provided with respect to peroxisome biology remained marginal.

Peroxisomal protein complexes are generally underrepresented in large-scale studies reflecting the challenges inherent to the analysis of constituents of this low abundant organelle and emphasizing the necessity to carefully choose experimental conditions. Gavin et al. (2002, 2006), for instance, used yeast cells cultured in the presence of glucose, which is known to result in the repression of genes coding for peroxisomal proteins (Gurvitz and Rottensteiner 2006; Veenhuis et al. 1987). In addition, numerous peroxisomal proteins, among these the peroxins, belong to the class of integral membrane proteins or are membrane-associated. Due to their hydrophobic nature, these proteins are still difficult to analyze by MS (Whitelegge 2013). Furthermore, the analysis of membrane protein complexes requires specifically tailored protocols for the detergent-assisted extraction from lipid bilayers without compromising their integrity, which is generally beyond the scope of large-scale studies. However, in a recent study, Babu et al. (2012) used a set of three different non-denaturing detergents, i.e., Triton X-100, *n*-dodecyl β-D-maltopyranoside (DDM), and octaethylene glycol monododecyl ether, to purify a total of 1,228 TAP-tagged putative integral, peripheral and lipid-anchored membrane proteins of *S. cerevisiae*. Among these were 35 peroxisomal membrane-bound proteins. Except for the association of Pex1p with Pex6p and Pex14p with Pex17p, however, none of the known peroxin–peroxin interactions were detected in this work.

In contrast, targeted AP-MS tailored to the analysis of individual peroxisomal membrane proteins facilitated several important findings regarding the biology of peroxisomes in different organisms thereby proving its potential for advancing our understanding of processes essential for the functionality of peroxisomes. In a seminal work, Kunau and coworkers used *S. cerevisiae* cells grown under peroxisome-proliferating conditions and expressing PA-tagged Pex14p or Pex2p for AP-MS analysis (Agne et al. 2003). Protein complexes were affinity-purified from crude cellular membrane fractions treated with the mild non-ionic detergent digitonin, which allows for membrane extraction of proteins under preservation of the complex as opposed to Triton X-100 and DDM. AP-MS data were confirmed by immunoblot analyses. The approach enabled to delineate the peroxisomal matrix protein import machinery of *S. cerevisiae* and, ultimately, the dissection of this "importomer" into docking complex (Pex14p/Pex17p/Pex13p) and RING finger complex (Pex2p/Pex10p/Pex12p) connected by Pex8p at the *trans* side of the peroxisomal membrane.

AP-MS studies further showed that Fox3p is the major PTS2 protein in *S. cerevisiae* under oleate growth conditions (Grunau et al. 2009) and revealed that Pex14p is required for microtubule-based peroxisome motility in human cells by acting as membrane anchor for microtubules, thus assigning an additional, novel function to human Pex14p (Bharti et al. 2011). In addition, they enabled the identification of a novel peroxin, Pex33p, as component of the peroxisomal docking complex indispensable for the biogenesis of glyoxysomes and Woronin bodies in the filamentous fungus *Neurospora crassa* (Managadze et al. 2010).

12.3.2 Quantitative AP-MS for the Identification of Transient Interaction Partners

An inherent challenge of AP-MS experiments is the reliable discrimination between true interaction partners and copurified, nonspecific background proteins (reviewed in Ong and Mann 2005; Oeljeklaus et al. 2009; Kaake et al. 2010), which is reinforced by continuous advancements in modern LC/MS technologies. Attempts to increase the purity of protein complexes to get rid of nonspecifically binding proteins thereby increasing the reliability of protein interaction data often employ multistep purification methods—such as the TAP-tag technology used for most of the large-scale studies discussed above—and/or high stringent washing conditions. While this will decrease the number of background proteins still present in a purified complex, it will also increase the likelihood of losing specific low abundant and transiently associated interaction partners. Here, quantitative AP-MS exploiting the advantages of metabolic labeling has proven to be a powerful strategy for a most comprehensive characterization of protein complexes not only facilitating the reliable identification of binding partners but also enabling the differentiation between stable core components and transiently associated interaction partners (Kaake et al. 2010; Oeljeklaus et al. 2009; Ong and Mann 2005).

First applied to the analysis of protein complexes in human cells (Wang and Huang 2008; Mousson et al. 2008), an advanced dual-track quantitative AP-MS strategy employing stable isotope labeling with amino acids in cell culture (SILAC) has been applied to the in-depth characterization of membrane-bound Pex14p and Pex30p complexes of *S. cerevisiae* (David et al. 2013; Oeljeklaus et al. 2012) as illustrated in Fig. 12.3a. Yeast cells expressing PA-tagged Pex14p or Pex30p were grown in the presence of unlabeled "light" arginine and lysine, while control cells expressing the wild-type versions were metabolically labeled by growth in medium containing the corresponding "heavy" amino acids. Following the classical SILAC approach (Blagoev et al. 2003; Ong et al. 2002, 2003), differentially labeled cells were mixed in equal ratios immediately after harvesting. Protein complexes were affinity-purified from digitonin extracts of crude membrane fractions and subjected to SDS-PAGE followed by LC/MS and computational data analysis for protein identification and quantification. True components of a protein complex are in general specifically enriched with the tagged POI from light labeled cells and, thus, typically exhibit high abundance ratios whereas background proteins show abundance ratios of approximately one. Since SILAC thereby allows for the reliable discrimination between specific interaction partners and co-purified contaminants, protein complexes can be purified employing a simple one-step purification protocol and/or mild washing conditions. However, although this classical approach of affinity purification after mixing (AP-AM) exploits the advantage of SILAC, i.e., mixing of differentially labeled samples at the earliest time-point possible resulting in a minimum of experimental variations due to uneven sample handling (Fig. 12.3a, AP-AM track), it facilitates the exchange of labeled and unlabeled interaction partners only transiently associated with the bait or the complex during the process of affinity purification. This kind of interactors may therefore be misclassified as co-purified contaminants. For this reason, Pex14p and Pex30p affinity purifications were additionally performed prior to mixing (AP-PM), i.e., separately from cells expressing the tagged POI and from control cells (Fig. 12.3a, AP-PM track). Differentially SILAC-labeled samples were combined just prior to SDS-PAGE and LC/MS analysis, thereby impeding the exchange of transient binding partners. Integration of interaction data derived from both AP-AM and AP-PM experiments ultimately allowed for the definition of stable core components and proteins transiently interacting with the complex, as exemplified by sections of peptide mass spectra of the stable Pex14p interactor Pex17p, the transient interactor Pex5p and Por1p, a mitochondrial contaminant in Fig. 12.3b.

The application of this advanced SILAC AP-MS approach to the analysis of the peroxisomal importomer using its central component Pex14p as bait led to the establishment of the so far most exhaustive Pex14p interactome composed of nine stable core components and 12 transiently interacting proteins. In addition to the well-established constituents of the peroxisomal importomer, the core complex further contains Pex11p and Dyn2p, the cytoplasmic light chain dynein of *S. cerevisiae* (Oeljeklaus et al. 2012). In accordance with these data, the interaction

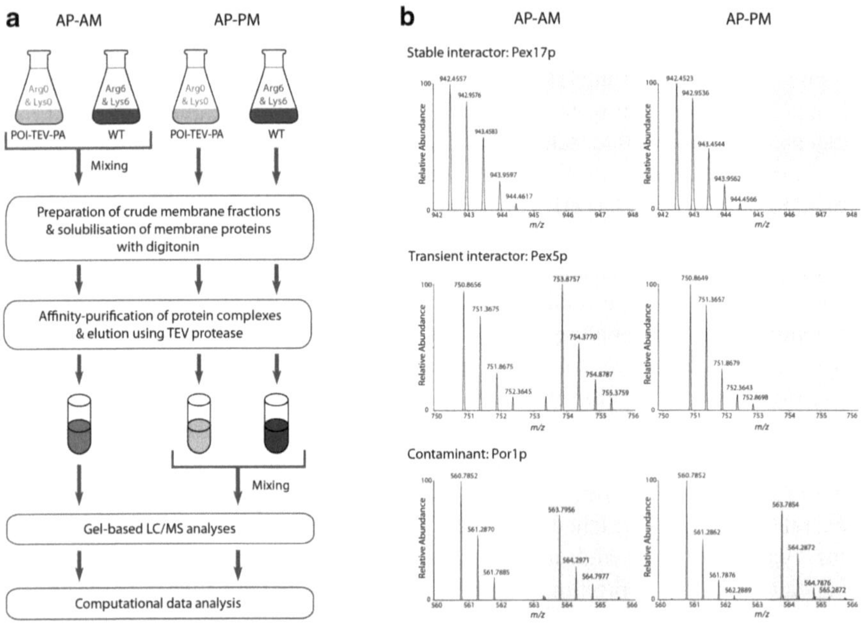

Fig. 12.3 SILAC-based quantitative AP-MS for identification of stable and transient interactors of peroxisomal membrane protein complexes. (**a**) Workflow for the dual-track SILAC approach employed for the characterization of Pex14p and Pex30p complexes (Oeljeklaus et al. 2012; David et al. 2013). *S. cerevisiae* cells expressing Pex14p or Pex30p (POI, protein of interest) fused to a cleavage site for the tobacco etch virus protease (TEV) and the Protein A (PA) tag or the wild-type (WT) variant were grown under peroxisome-proliferating conditions in the presence of "light" $^{12}C_6$-containing arginine and lysine (Arg0, Lys0) or in medium containing the respective $^{13}C_6$-coded "heavy" counterparts (Arg6, Lys6). Protein complexes were affinity-purified using IgG Sepharose from crude membrane fractions treated with digitonin either after mixing of differentially SILAC-labeled cells (AP-AM) or prior to mixing (AP-PM). Following TEV protease digest, eluted proteins were subjected to SDS-PAGE and tryptic in-gel digestion. The resulting peptide mixtures were analyzed by high-resolution LC/MS/MS followed by computational data analyses for protein identification and relative quantification. (**b**) Discrimination between core components and transient Pex14p interactors by integration of data obtained in AP-AM and AP PM experiments. Shown are sections of mass spectra of tryptic peptides containing $^{12}C_6$/$^{13}C_6$-arginine or -lysine identified in Pex14p complexes following the AP-AM or AP-PM track. For the peptide derived from Pex17p (SSGQPSESIDDFVFQIK), the heavy isotopic version from the wild-type strain is absent in both AP-AM and AP-PM experiments as indicated by the shaded areas outlining the *m*/*z* range of the respective heavy peptides, which is characteristic for stable core components. In contrast, the Pex5p peptide GFTHIDMoxNAHITK was detected in both light and heavy form at a ratio of approx. 1 in AP-PM data while in AP-AM data, only the light species was present specifying Pex5p as specific transient Pex14p interaction partner. The peptide SPPVYSDIR of Por1p, a mitochondrial protein, is observed as peptide pair with equal abundance of light and heavy version in both AP-AM and AP-PM experiments and represents peptides of copurified contaminants. *m*/*z*, mass-to-charge

between Dyn2p and the peroxisomal docking complex has recently been verified in the yeast *Y. lipolytica* where Dyn2p is involved in modulating peroxisome biogenesis. Cells lacking Dyn2p exhibited impaired peroxisome function and biogenesis, abnormal peroxisome morphology and showed mislocalization of peroxisomal matrix proteins (Chang et al. 2013). Among the proteins identified as transient interaction partners of Pex14p complexes were, among others, Pex5p and the PTS1-containing peroxisomal matrix proteins Cat2p, Idp3p, and Mdh3p, which reflects the biological role of Pex5p as receptor for PTS1 proteins cycling between a cytosolic and a membrane-bound form (Dodt and Gould 1996; Kerssen et al. 2006).

Applied to Pex30p, the dual-track SILAC-AP-MS strategy, combined with live-cell imaging, facilitated the discovery of ER-to-peroxisome contact sites (EPCONS) (David et al. 2013). These macromolecular membrane protein complexes contain, among others, Pex30p and the reticulon homology proteins Rtn1p, Rtn2p, and Yop1p residing in the membrane of the ER. They act as hubs for the regulation of peroxisome proliferation and movement with Pex30p trafficking through the ER and establishing the contact between ER tubules and peroxisomes. Interestingly, Pex30p was found to transiently interact with all subunits of the COPI coatomer. Although further studies are necessary to ascertain the functional significance of this transient interaction, it is conceivable that vesicular trafficking is involved in the retrograde transport of Pex30p. In support of this hypothesis, Pex30p contains a dilysine motif (KKXX) at its C-terminus, a sorting signal for COPI vesicles (Cosson and Letourneur 1994; David et al. 2013).

These intriguing new findings demonstrate the high value of quantitative AP-MS for the study of peroxisomal membrane protein complexes allowing for the identification of new players in peroxisome biology and for uncovering so far unknown physical connections to other neighboring subcellular organelles.

Conclusions and Future Perspectives

Organellar proteomics approaches applied to peroxisomes have advanced from the mere identification of proteins in peroxisomal preparations to the relative quantitative analysis of proteins allowing, for example, to systematically localize proteins to peroxisomes or to reveal dynamic changes in the peroxisomal proteome in response to a distinct stimulus. Large-scale quantitative MS studies allowed here to establish a most complete inventory of the proteome of liver and kidney peroxisomes including the discovery of numerous new proteins residing in or associated with the organelle. Based on the currently collected datasets, it can be anticipated that the proteome of mammalian peroxisomes comprises at least 100 different constituents of which approx. ⅔ are located in the matrix and ⅓ in the membrane compartment of the organelle. We would like to encourage scientists active in the field of peroxisome research to pursue functional studies on these new peroxisomal constituents (e.g., Pmp52, Acad11, Acbd5) and further candidate proteins such as Aldh3a2 or Atad1 identified by quantitative proteomics endeavors. Such research effort would hold the great potential for gaining new insights into biogenesis, biochemistry and functions of mammalian peroxisomes and, moreover, for discovering new human diseases caused by the deficiency or malfunctioning of peroxisomal proteins.

In this chapter, we have further highlighted the applicability of quantitative AP-MS methodologies to comprehensively characterize peroxisomal membrane protein complexes leading to the identification of new players involved in biogenesis and proliferation of peroxisomes. Since the approach also allows for identifying proteins that only transiently associate with peroxisomes, it provides an effective tool to explore further how peroxisomes communicate with other parts of the cell. In this context, we still need to improve our understanding of peroxisomes being an integral part of complex metabolic and signaling networks in eukaryotic cells. We would like to refer to two recent reports highlighting new functions of mammalian peroxisomes in antiviral immune response (Dixit et al. 2010) and in blocking mTORC1 (mammalian target of rapamycin complex 1)-dependent signaling pathways leading to the induction of autophagy in response to oxidative stress (Zhang et al. 2013). In future research, the proper application of MS-based proteomics approaches tailored to the analysis of posttranslational modifications of proteins will provide us with powerful means to investigate signaling functions of peroxisomes as well as to discover new regulatory processes involved in the biogenesis and homeostasis of peroxisomes in response to metabolic or environmental changes.

Acknowledgments We would like to apologize to all the scientists whose work was not cited. This work was supported by grants of the Deutsche Forschungsgemeinschaft (FOR1905) and the Excellence Initiative of the German Federal & State Governments (Grant EXC 294 BIOSS Centre for Biological Signalling Studies).

References

Aebersold R, Mann M (2003) Mass spectrometry-based proteomics. Nature 422(6928):198–207

Aggarwal K, Choe LH, Lee KH (2006) Shotgun proteomics using the iTRAQ isobaric tags. Brief Funct Genomic Proteomic 5(2):112–120

Agne B, Meindl NM, Niederhoff K, Einwachter H, Rehling P, Sickmann A, Meyer HE, Girzalsky W, Kunau WH (2003) Pex8p: an intraperoxisomal organizer of the peroxisomal import machinery. Mol Cell 11(3):635–646

Alberts B (1998) The cell as a collection of protein machines: preparing the next generation of molecular biologists. Cell 92(3):291–294

Andersen JS, Mann M (2006) Organellar proteomics: turning inventories into insights. EMBO Rep 7(9):874–879

Andersen JS, Wilkinson CJ, Mayor T, Mortensen P, Nigg EA, Mann M (2003) Proteomic characterization of the human centrosome by protein correlation profiling. Nature 426 (6966):570–574

Babu M, Vlasblom J, Pu S, Guo X, Graham C, Bean BD, Burston HE, Vizeacoumar FJ, Snider J, Phanse S, Fong V, Tam YY, Davey M, Hnatshak O, Bajaj N, Chandran S, Punna T, Christopolous C, Wong V, Yu A, Zhong G, Li J, Stagljar I, Conibear E, Wodak SJ, Emili A, Greenblatt JF (2012) Interaction landscape of membrane-protein complexes in Saccharomyces cerevisiae. Nature 489(7417):585–589

Bantscheff M, Lemeer S, Savitski MM, Kuster B (2012) Quantitative mass spectrometry in proteomics: critical review update from 2007 to the present. Anal Bioanal Chem 404 (4):939–965

Baumgart E, Fahimi HD, Stich A, Volkl A (1996) L-lactate dehydrogenase A4- and A3B isoforms are bona fide peroxisomal enzymes in rat liver. Evidence for involvement in intraperoxisomal NADH reoxidation. J Biol Chem 271(7):3846–3855

Beynon RJ, Pratt JM (2005) Metabolic labeling of proteins for proteomics. Mol Cell Proteomics 4 (7):857–872

Bharti P, Schliebs W, Schievelbusch T, Neuhaus A, David C, Kock K, Herrmann C, Meyer HE, Wiese S, Warscheid B, Theiss C, Erdmann R (2011) PEX14 is required for microtubule-based peroxisome motility in human cells. J Cell Sci 124(Pt 10):1759–1768

Blagoev B, Kratchmarova I, Ong SE, Nielsen M, Foster LJ, Mann M (2003) A proteomics strategy to elucidate functional protein-protein interactions applied to EGF signaling. Nat Biotechnol 21(3):315–318

Braun RJ, Kinkl N, Beer M, Ueffing M (2007) Two-dimensional electrophoresis of membrane proteins. Anal Bioanal Chem 389(4):1033–1045

Bussell JD, Behrens C, Ecke W, Eubel H (2013) Arabidopsis peroxisome proteomics. Front Plant Sci 4:101

Chang J, Tower RJ, Lancaster DL, Rachubinski RA (2013) Dynein light chain interaction with the peroxisomal import docking complex modulates peroxisome biogenesis in yeast. J Cell Sci 126 (Pt 20):4698–4706

Christoforou AL, Lilley KS (2012) Isobaric tagging approaches in quantitative proteomics: the ups and downs. Anal Bioanal Chem 404(4):1029–1037

Cosson P, Letourneur F (1994) Coatomer interaction with di-lysine endoplasmic reticulum retention motifs. Science 263(5153):1629–1631

David C, Koch J, Oeljeklaus S, Laernsack A, Melchior S, Wiese S, Schummer A, Erdmann R, Warscheid B, Brocard C (2013) A combined approach of quantitative interaction proteomics and live-cell imaging reveals a regulatory role for ER reticulon homology proteins in peroxisome biogenesis. Mol Cell Proteomics 12(9):2408–2425

Dixit E, Boulant S, Zhang Y, Lee AS, Odendall C, Shum B, Hacohen N, Chen ZJ, Whelan SP, Fransen M, Nibert ML, Superti-Furga G, Kagan JC (2010) Peroxisomes are signaling platforms for antiviral innate immunity. Cell 141(4):668–681

Dodt G, Gould SJ (1996) Multiple PEX genes are required for proper subcellular distribution and stability of Pex5p, the PTS1 receptor: evidence that PTS1 protein import is mediated by a cycling receptor. J Cell Biol 135(6 Pt 2):1763–1774

Drissi R, Dubois ML, Boisvert FM (2013) Proteomics methods for subcellular proteome analysis. FEBS J 280(22):5626–5634

Erdmann R, Blobel G (1995) Giant peroxisomes in oleic acid-induced Saccharomyces cerevisiae lacking the peroxisomal membrane protein Pmp27p. J Cell Biol 128(4):509–523

Evans C, Noirel J, Ow SY, Salim M, Pereira-Medrano AG, Couto N, Pandhal J, Smith D, Pham TK, Karunakaran E, Zou X, Biggs CA, Wright PC (2012) An insight into iTRAQ: where do we stand now? Anal Bioanal Chem 404(4):1011–1027

Fahimi HD, Reinicke A, Sujatta M, Yokota S, Ozel M, Hartig F, Stegmeier K (1982) The short- and long-term effects of bezafibrate in the rat. Ann N Y Acad Sci 386:111–135

Foster LJ, de Hoog CL, Zhang Y, Xie X, Mootha VK, Mann M (2006) A mammalian organelle map by protein correlation profiling. Cell 125(1):187–199

Gavin AC, Aloy P, Grandi P, Krause R, Boesche M, Marzioch M, Rau C, Jensen LJ, Bastuck S, Dumpelfeld B, Edelmann A, Heurtier MA, Hoffman V, Hoefert C, Klein K, Hudak M, Michon AM, Schelder M, Schirle M, Remor M, Rudi T, Hooper S, Bauer A, Bouwmeester T, Casari G, Drewes G, Neubauer G, Rick JM, Kuster B, Bork P, Russell RB, Superti-Furga G (2006) Proteome survey reveals modularity of the yeast cell machinery. Nature 440(7084):631–636

Gavin AC, Bosche M, Krause R, Grandi P, Marzioch M, Bauer A, Schultz J, Rick JM, Michon AM, Cruciat CM, Remor M, Hofert C, Schelder M, Brajenovic M, Ruffner H, Merino A,

Klein K, Hudak M, Dickson D, Rudi T, Gnau V, Bauch A, Bastuck S, Huhse B, Leutwein C, Heurtier MA, Copley RR, Edelmann A, Querfurth E, Rybin V, Drewes G, Raida M, Bouwmeester T, Bork P, Seraphin B, Kuster B, Neubauer G, Superti-Furga G (2002) Functional organization of the yeast proteome by systematic analysis of protein complexes. Nature 415(6868):141–147

Gronemeyer T, Wiese S, Grinhagens S, Schollenberger L, Satyagraha A, Huber LA, Meyer HE, Warscheid B, Just WW (2013a) Localization of Rab proteins to peroxisomes: a proteomics and immunofluorescence study. FEBS Lett 587(4):328–338

Gronemeyer T, Wiese S, Ofman R, Bunse C, Pawlas M, Hayen H, Eisenacher M, Stephan C, Meyer HE, Waterham HR, Erdmann R, Wanders RJ, Warscheid B (2013b) The proteome of human liver peroxisomes: identification of five new peroxisomal constituents by a label-free quantitative proteomics survey. PLoS One 8(2):e57395

Grunau S, Schliebs W, Linnepe R, Neufeld C, Cizmowski C, Reinartz B, Meyer HE, Warscheid B, Girzalsky W, Erdmann R (2009) Peroxisomal targeting of PTS2 pre-import complexes in the yeast Saccharomyces cerevisiae. Traffic 10(4):451–460

Gurvitz A, Rottensteiner H (2006) The biochemistry of oleate induction: transcriptional upregulation and peroxisome proliferation. Biochim Biophys Acta 1763(12):1392–1402

Gygi SP, Rist B, Gerber SA, Turecek F, Gelb MH, Aebersold R (1999a) Quantitative analysis of complex protein mixtures using isotope-coded affinity tags. Nat Biotechnol 17(10):994–999

Gygi SP, Rochon Y, Franza BR, Aebersold R (1999b) Correlation between protein and mRNA abundance in yeast. Mol Cell Biol 19(3):1720–1730

Han X, Aslanian A, Yates JR 3rd (2008) Mass spectrometry for proteomics. Curr Opin Chem Biol 12(5):483–490

Hartinger J, Stenius K, Hogemann D, Jahn R (1996) 16-BAC/SDS-PAGE: a two-dimensional gel electrophoresis system suitable for the separation of integral membrane proteins. Anal Biochem 240(1):126–133

Hasan S, Platta HW, Erdmann R (2013) Import of proteins into the peroxisomal matrix. Front Physiol 4:261

Hebert AS, Richards AL, Bailey DJ, Ulbrich A, Coughlin EE, Westphall MS, Coon JJ (2014) The one hour yeast proteome. Mol Cell Proteomics 13(1):339–47

Ho Y, Gruhler A, Heilbut A, Bader GD, Moore L, Adams SL, Millar A, Taylor P, Bennett K, Boutilier K, Yang L, Wolting C, Donaldson I, Schandorff S, Shewnarane J, Vo M, Taggart J, Goudreault M, Muskat B, Alfarano C, Dewar D, Lin Z, Michalickova K, Willems AR, Sassi H, Nielsen PA, Rasmussen KJ, Andersen JR, Johansen LE, Hansen LH, Jespersen H, Podtelejnikov A, Nielsen E, Crawford J, Poulsen V, Sorensen BD, Matthiesen J, Hendrickson RC, Gleeson F, Pawson T, Moran MF, Durocher D, Mann M, Hogue CW, Figeys D, Tyers M (2002) Systematic identification of protein complexes in Saccharomyces cerevisiae by mass spectrometry. Nature 415(6868):180–183

Islinger M, Cardoso MJ, Schrader M (2010) Be different–the diversity of peroxisomes in the animal kingdom. Biochim Biophys Acta 1803(8):881–897

Islinger M, Luers GH, Li KW, Loos M, Volkl A (2007) Rat liver peroxisomes after fibrate treatment. A survey using quantitative mass spectrometry. J Biol Chem 282(32):23055–23069

Islinger M, Luers GH, Zischka H, Ueffing M, Volkl A (2006) Insights into the membrane proteome of rat liver peroxisomes: microsomal glutathione-S-transferase is shared by both subcellular compartments. Proteomics 6(3):804–816

Jung S, Marelli M, Rachubinski RA, Goodlett DR, Aitchison JD (2010) Dynamic changes in the subcellular distribution of Gpd1p in response to cell stress. J Biol Chem 285(9):6739–6749

Kaake RM, Wang X, Huang L (2010) Profiling of protein interaction networks of protein complexes using affinity purification and quantitative mass spectrometry. Mol Cell Proteomics 9(8):1650–1665

Kerssen D, Hambruch E, Klaas W, Platta HW, de Kruijff B, Erdmann R, Kunau WH, Schliebs W, Kerssen A, de Valk HW, Visser GH (2006) Membrane association of the cycling peroxisome import receptor Pex5p. J Biol Chem 281(37):27003–27015

Kikuchi M, Hatano N, Yokota S, Shimozawa N, Imanaka T, Taniguchi H (2004) Proteomic analysis of rat liver peroxisome: presence of peroxisome-specific isozyme of Lon protease. J Biol Chem 279(1):421–428

Krogan NJ, Cagney G, Yu H, Zhong G, Guo X, Ignatchenko A, Li J, Pu S, Datta N, Tikuisis AP, Punna T, Peregrin-Alvarez JM, Shales M, Zhang X, Davey M, Robinson MD, Paccanaro A, Bray JE, Sheung A, Beattie B, Richards DP, Canadien V, Lalev A, Mena F, Wong P, Starostine A, Canete MM, Vlasblom J, Wu S, Orsi C, Collins SR, Chandran S, Haw R, Rilstone JJ, Gandi K, Thompson NJ, Musso G, St Onge P, Ghanny S, Lam MH, Butland G, Altaf-Ul AM, Kanaya S, Shilatifard A, O'Shea E, Weissman JS, Ingles CJ, Hughes TR, Parkinson J, Gerstein M, Wodak SJ, Emili A, Greenblatt JF (2006) Global landscape of protein complexes in the yeast Saccharomyces cerevisiae. Nature 440(7084):637–643

Liu H, Sadygov RG, Yates JR 3rd (2004) A model for random sampling and estimation of relative protein abundance in shotgun proteomics. Anal Chem 76(14):4193–4201

Mallick P, Kuster B (2010) Proteomics: a pragmatic perspective. Nat Biotechnol 28(7):695–709

Managadze D, Wurtz C, Wiese S, Schneider M, Girzalsky W, Meyer HE, Erdmann R, Warscheid B, Rottensteiner H (2010) Identification of PEX33, a novel component of the peroxisomal docking complex in the filamentous fungus Neurospora crassa. Eur J Cell Biol 89(12):955–964

Mann M, Kelleher NL (2008) Precision proteomics: the case for high resolution and high mass accuracy. Proc Natl Acad Sci U S A 105(47):18132–18138

Marelli M, Smith JJ, Jung S, Yi E, Nesvizhskii AI, Christmas RH, Saleem RA, Tam YY, Fagarasanu A, Goodlett DR, Aebersold R, Rachubinski RA, Aitchison JD (2004) Quantitative mass spectrometry reveals a role for the GTPase Rho1p in actin organization on the peroxisome membrane. J Cell Biol 167(6):1099–1112

McClelland GB, Khanna S, Gonzalez GF, Butz CE, Brooks GA (2003) Peroxisomal membrane monocarboxylate transporters: evidence for a redox shuttle system? Biochem Biophys Res Commun 304(1):130–135

Mi J, Kirchner E, Cristobal S (2007) Quantitative proteomic comparison of mouse peroxisomes from liver and kidney. Proteomics 7(11):1916–1928

Mousson F, Kolkman A, Pijnappel WW, Timmers HT, Heck AJ (2008) Quantitative proteomics reveals regulation of dynamic components within TATA-binding protein (TBP) transcription complexes. Mol Cell Proteomics 7(5):845–852

Nuttall JM, Motley A, Hettema EH (2011) Peroxisome biogenesis: recent advances. Curr Opin Cell Biol 23(4):421–426

Oeljeklaus S, Meyer HE, Warscheid B (2009) New dimensions in the study of protein complexes using quantitative mass spectrometry. FEBS Lett 583(11):1674–1683

Oeljeklaus S, Reinartz BS, Wolf J, Wiese S, Tonillo J, Podwojski K, Kuhlmann K, Stephan C, Meyer HE, Schliebs W, Brocard C, Erdmann R, Warscheid B (2012) Identification of core components and transient interactors of the peroxisomal importomer by dual-track stable isotope labeling with amino acids in cell culture analysis. J Proteome Res 11(4):2567–2580

Ofman R, Speijer D, Leen R, Wanders RJ (2006) Proteomic analysis of mouse kidney peroxisomes: identification of RP2p as a peroxisomal nudix hydrolase with acyl-CoA diphosphatase activity. Biochem J 393(Pt 2):537–543

Ong SE, Blagoev B, Kratchmarova I, Kristensen DB, Steen H, Pandey A, Mann M (2002) Stable isotope labeling by amino acids in cell culture, SILAC, as a simple and accurate approach to expression proteomics. Mol Cell Proteomics 1(5):376–386

Ong SE, Kratchmarova I, Mann M (2003) Properties of 13C-substituted arginine in stable isotope labeling by amino acids in cell culture (SILAC). J Proteome Res 2(2):173–181

Ong SE, Mann M (2005) Mass spectrometry-based proteomics turns quantitative. Nat Chem Biol 1 (5):252–262

Peng J, Gygi SP (2001) Proteomics: the move to mixtures. J Mass Spectrom 36(10):1083–1091

Platta HW, Erdmann R (2007) Peroxisomal dynamics. Trends Cell Biol 17(10):474–484

Platta HW, Hagen S, Erdmann R (2013) The exportomer: the peroxisomal receptor export machinery. Cell Mol Life Sci 70(8):1393–1411

Reguenga C, Oliveira ME, Gouveia AM, Eckerskorn C, Sa-Miranda C, Azevedo JE (1999) Identification of a 24 kDa intrinsic membrane protein from mammalian peroxisomes. Biochim Biophys Acta 1445(3):337–341

Rizzo WB, Carney G (2005) Sjogren-Larsson syndrome: diversity of mutations and polymorphisms in the fatty aldehyde dehydrogenase gene (ALDH3A2). Hum Mutat 26(1):1–10

Ross PL, Huang YN, Marchese JN, Williamson B, Parker K, Hattan S, Khainovski N, Pillai S, Dey S, Daniels S, Purkayastha S, Juhasz P, Martin S, Bartlet-Jones M, He F, Jacobson A, Pappin DJ (2004) Multiplexed protein quantitation in Saccharomyces cerevisiae using amine-reactive isobaric tagging reagents. Mol Cell Proteomics 3(12):1154–1169

Schäfer H, Nau K, Sickmann A, Erdmann R, Meyer HE (2001) Identification of peroxisomal membrane proteins of Saccharomyces cerevisiae by mass spectrometry. Electrophoresis 22 (14):2955–2968

Schluter A, Fourcade S, Domenech-Estevez E, Gabaldon T, Huerta-Cepas J, Berthommier G, Ripp R, Wanders RJ, Poch O, Pujol A (2007) PeroxisomeDB: a database for the peroxisomal proteome, functional genomics and disease. Nucleic Acids Res 35(Database issue):D815–822

Schollenberger L, Gronemeyer T, Huber CM, Lay D, Wiese S, Meyer HE, Warscheid B, Saffrich R, Peranen J, Gorgas K, Just WW (2010) RhoA regulates peroxisome association to microtubules and the actin cytoskeleton. PLoS One 5(11):e13886

Schwanhausser B, Busse D, Li N, Dittmar G, Schuchhardt J, Wolf J, Chen W, Selbach M (2011) Global quantification of mammalian gene expression control. Nature 473(7347):337–342

Thoms S, Erdmann R (2005) Dynamin-related proteins and Pex11 proteins in peroxisome division and proliferation. FEBS J 272(20):5169–5181

Treumann A, Thiede B (2010) Isobaric protein and peptide quantification: perspectives and issues. Expert Rev Proteomics 7(5):647–653

Veenhuis M, Mateblowski M, Kunau WH, Harder W (1987) Proliferation of microbodies in Saccharomyces cerevisiae. Yeast 3(2):77–84

Walther TC, Mann M (2010) Mass spectrometry-based proteomics in cell biology. J Cell Biol 190 (4):491–500

Wanders RJ, Waterham HR (2006) Biochemistry of mammalian peroxisomes revisited. Annu Rev Biochem 75:295–332

Wang X, Huang L (2008) Identifying dynamic interactors of protein complexes by quantitative mass spectrometry. Mol Cell Proteomics 7(1):46–57

Whitelegge JP (2013) Integral membrane proteins and bilayer proteomics. Anal Chem 85 (5):2558–2568

Wiederhold E, Veenhoff LM, Poolman B, Slotboom DJ (2010) Proteomics of Saccharomyces cerevisiae organelles. Mol Cell Proteomics 9(3):431–445

Wiese S, Gronemeyer T, Brites P, Ofman R, Bunse C, Renz C, Meyer HE, Wanders RJA, Warscheid B (2012) Comparative profiling of the peroxisomal proteome of wildtype and Pex7 knockout mice by quantitative mass spectrometry. Int J Mass Spectrometry 312:30–40

Wiese S, Gronemeyer T, Ofman R, Kunze M, Grou CP, Almeida JA, Eisenacher M, Stephan C, Hayen H, Schollenberger L, Korosec T, Waterham HR, Schliebs W, Erdmann R, Berger J, Meyer HE, Just W, Azevedo JE, Wanders RJ, Warscheid B (2007) Proteomics characterization of mouse kidney peroxisomes by tandem mass spectrometry and protein correlation profiling. Mol Cell Proteomics 6(12):2045–2057

Yates JR 3rd (2011) A century of mass spectrometry: from atoms to proteomes. Nat Methods 8:633–637

Yates JR 3rd, Gilchrist A, Howell KE, Bergeron JJ (2005) Proteomics of organelles and large cellular structures. Nat Rev Mol Cell Biol 6(9):702–714

Yi EC, Marelli M, Lee H, Purvine SO, Aebersold R, Aitchison JD, Goodlett DR (2002) Approaching complete peroxisome characterization by gas-phase fractionation. Electrophoresis 23(18):3205–3216

Zahedi RP, Moebius J, Sickmann A (2007) Two-dimensional BAC/SDS-PAGE for membrane proteomics. Subcell Biochem 43:13–20

Zhang J, Kim J, Alexander A, Cai S, Tripathi DN, Dere R, Tee AR, Tait-Mulder J, Di Nardo A, Han JM, Kwiatkowski E, Dunlop EA, Dodd KM, Folkerth RD, Faust PL, Kastan MB, Sahin M, Walker CL (2013) A tuberous sclerosis complex signalling node at the peroxisome regulates mTORC1 and autophagy in response to ROS. Nat Cell Biol 15(10):1186–1196

Part IV

Protein Transport Across the Peroxisomal Membrane

The Matrix Protein Import Complex in Yeast

13

Daniel Effelsberg, Ralf Erdmann, and Wolfgang Schliebs

Abstract

Nearly all peroxisomal matrix proteins contain specific peroxisomal targeting signals (PTS1 or PTS2) that are posttranslationally recognized and bound in the cytosol by soluble import receptors. Whereas the PTS1 receptor Pex5p targets its cargo autonomously to the peroxisomal membrane, the PTS2-receptor Pex7p requires auxiliary proteins, so-called co-receptors. Docking of receptor–cargo complexes at the peroxisomal membrane is facilitated by a membrane subcomplex consisting of Pex13p, Pex14p, and Pex17p. Pex5p together with its docking partner Pex14p forms a transient pore in the peroxisomal membrane, which allows the translocation of folded and even oligomerized PTS1 proteins. After release of the matrix enzymes into the peroxisomal lumen, the PTS receptors are dislocated from the membrane to the cytosol and thus made available for the next import cycle. In this review, we will report most recent advances in understanding the formation and function of receptor complexes in the cytosol and at the peroxisomal membrane.

Keywords

Matrix protein import • PTS receptor • Docking complex • Translocation pore

13.1 Introduction

Peroxisomal matrix proteins are synthesized on free ribosomes in the cytosol and are posttranslationally transported across the peroxisomal membrane. Matrix protein import is a unidirectional process starting with receptor recognition in the cytosol (I), docking of receptor–cargo complexes at the peroxisomal membrane (II), pore formation (III) and translocation of cargo (Fig. 13.1). At least three

D. Effelsberg • R. Erdmann • W. Schliebs (✉)
Institute for Physiological Chemistry, Ruhr-University Bochum, 44780 Bochum, Germany
e-mail: wolfgang.schliebs@rub.de

C. Brocard and A. Hartig (eds.), *Molecular Machines Involved in Peroxisome Biogenesis and Maintenance*, DOI 10.1007/978-3-7091-1788-0_13, © Springer-Verlag Wien 2014

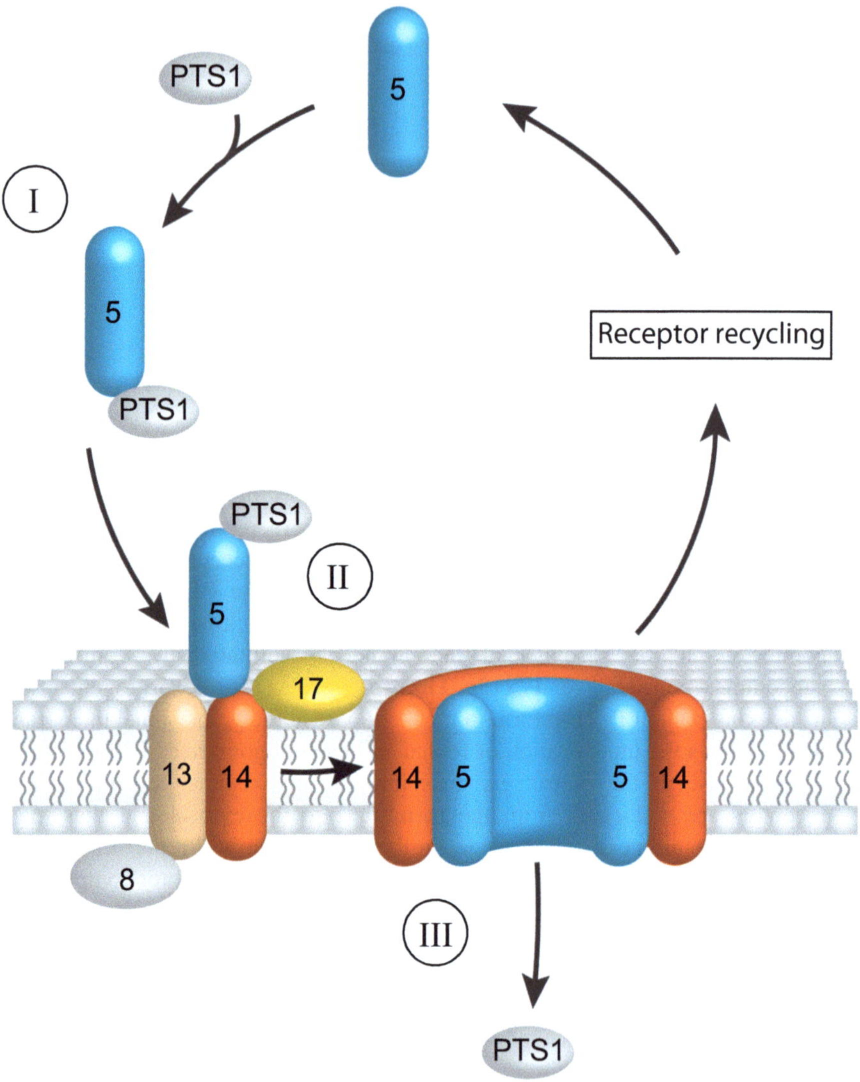

Fig. 13.1 The peroxisomal PTS1-dependent importomer under oleate-induced conditions. Peroxisomal matrix protein import starts with the recognition of PTS1 containing enzymes by the cycling receptors Pex5p in the cytosol (I). The PTS1 receptor–cargo complex is targeted to the peroxisomal membrane and initially interacts with Pex14p, a constituent of the docking complex (II). The next step of the import cascade is pore formation and the release of the cargo into the peroxisomal lumen (III) followed by the recycling of the receptor for a new round of import. Please note that there might be differences to the views how Pex5p is positioned with respect to the translocation pore (Chap. 16 in this book)

different matrix protein import complexes (receptor–cargo complex, docking complex and a transient pore complex) are sequentially formed along the import route in the classical model organism, the yeast *Saccharomyces cerevisiae*. Here, we will review the requirements and regulatory mechanisms for the formation of matrix protein import complexes in the cytosol and at the peroxisomal membrane.

13.2 Formation of the Cytosolic Receptor/Cargo Preimport Complexes

13.2.1 Matrix Protein Import Signals

The import of matrix proteins is initiated by receptor recognition of specific targeting signals which are located at the enzymes destined for peroxisomes. Hitherto, two classes of well-characterized peroxisomal targeting signals (PTS) are known which are highly conserved among all organisms from yeast to humans. The first class, PTS1, consists of a C-terminal tripeptide with the consensus sequence (S/A/C)(K/H/R)(L/M). The targeting sequence -SKL was first identified by mutation analysis of firefly luciferase, which was imported into mammalian peroxisomes (Gould et al. 1989; Keller et al. 1987). Although only the extreme carboxy-terminal tripeptide is specifically bound by the receptor, the nine amino acids further upstream also seem to contribute to receptor recognition, e.g. by providing an unstructured linker (Neuberger et al. 2003). Almost all known yeast peroxisomal matrix proteins contain a PTS1.

The second class of PTS consists of the nonapeptide PTS2 with the consensus sequence (R/K)(L/I/V)X5(H/Q)(L/A/F) (Lazarow 2006), which is located near the N-terminus. The targeting signal was first identified within the sequence of rat liver 3-ketoacyl-CoA thiolase (Swinkels et al. 1991). In *S. cerevisiae*, only two proteins, the β-oxidation enzyme 3-ketoacyl-CoA thiolase (Fox3p) and the glycerol-3-phosphate dehydrogenase (Gpd1p), were shown to contain a functional PTS2 motif, which is required and sufficient to target these proteins into peroxisomes (Purdue et al. 1998; Jung et al. 2009). The yeast nudix hydrolase Pcd1p bears a potential PTS2 motif, but the Pex7p-dependent import into peroxisomes has not yet been confirmed experimentally (Cartwright et al. 2000). In the yeast *Pichia pastoris*, the intraperoxisomal Pex8p is targeted by either PTS1 or PTS2 (Wolf et al. 2010; Ma et al. 2009; Zhang et al. 2006). In *S. cerevisiae* Pex8p, sequences resembling PTS1 and PTS2 exist but neither is required for import (Rehling et al. 2000). The PTS2 motif of 3-ketoacyl-CoA thiolase is conserved from yeast to human. Two exceptions are known: thiolases from *Caenorhabditis elegans* and the diatom *Phaeodactylum tricornutum* do not contain PTS2 but instead harbour a PTS1 (Swinkels et al. 1991; Bun-Ya et al. 1997; Gonzalez et al. 2011). However, both organisms in general seem to be unable to import PTS2 proteins indicated by the lack of PTS2 receptor Pex7p (Gonzalez et al. 2011). On the contrary to the small number of PTS2 cargoes known in yeast and mammals, almost one-third of the peroxisomal matrix proteins in plants are targeted to peroxisomes via a PTS2 (Lazarow 2006).

A third type of peroxisomal targeting signal, sometimes termed as PTS3, is an internal sequence, which is also recognized by the PTS1 receptor but is not yet clearly defined by a consensus sequence (Schäfer et al. 2004). In *P. pastoris* acyl-CoA oxidase is a PTS1 protein, though in other yeasts like *Y. lipolytica* and *S. cerevisiae* the enzyme lacks a PTS1 but still interacts with the PTS1 import receptor Pex5p. Mutations in the C-terminal Pex5p-binding region for PTS1 proteins (TPR domains) do not result in an import defect for the acyl-CoA oxidase. Instead, the N-terminal half of Pex5p is sufficient to target acyl-CoA oxidase into peroxisomes (van der Klei and Veenhuis 2006; Schäfer et al. 2004). Another PTS3 candidate is carnitine acetyl-transferase (Cat2p) of *S. cerevisiae*, which exhibits a dual localization in mitochondria as well in peroxisomes and contains both an N-terminal mitochondrial targeting signal and a PTS1 at the C-terminus. After deletion of both targeting signals, Cat2p is still imported into peroxisomes by binding to the N-terminal region of Pex5p (Elgersma et al. 1995; Klein et al. 2002).

The latter example strongly suggests that cryptic peroxisomal targeting signals might exist, which become accessible by regulatory terms. This has been shown for the glycolytic enzymes glyceraldehyde-3-phosphate dehydrogenase (GAPDH) and 3-phosphoglycerate kinase (PGK) in *Ustilago maydis* and *Aspergillus nidulans* (Freitag et al. 2012). In the plant pathogenic fungus *U. maydis*, translational read-through results in the peroxisomal isoform of PGK, while in the filamentous fungus *A. nidulans* alternative splicing leads to a peroxisomal localization of GAPDH and vice versa. Also in yeasts, cryptic PTS1 motifs in GAPDH and PGK of *Y. lipolytica* and GAPDH of *C. albicans* are presumed on basis of sequence similarities. However, functionality of these potential targeting signals has still to be confirmed experimentally (Freitag et al. 2012).

Another way of how proteins can enter peroxisomes is the so-called piggy-back transport (Wolf et al. 2010). This allows proteins lacking a PTS to hitchhike onto a PTS-containing protein for peroxisomal transport. This was discovered by the co-expression of an artificially truncated version of the 3-ketoacyl thiolase, lacking the first 16 amino acids, together with the full-length protein, which did result in the transport of the truncated thiolase into peroxisomes. This observation demonstrated that a PTS-lacking protein could be imported into peroxisomes by forming a stable complex with another PTS protein. Furthermore, this was the first example for import of oligomeric proteins into peroxisomes (Glover et al. 1994). Another example for piggy-back mechanism concerns the β-oxidation enzyme Δ3, Δ2-enoyl-CoA isomerase (Eci1p), which contains a functional PTS1-like motif and a PTS2-like sequence. However, both targeting signals are dispensable for import into peroxisomes. Instead, the presence of Dci1p (dodecenoyl-CoA isomerase) is necessary for the PTS1-dependent translocation of Eci1p as a hetero-dimer (Yang et al. 2001). In mammals, Cu/Zn superoxide dismutase 1 (SOD1) lacks an endogenous PTS, but is transported to the peroxisomal matrix by its physiological interaction partner "copper chaperone of SOD1" (CCS). This pathway was the first proven example for a natural piggy-back import mechanism (Islinger et al. 2009).

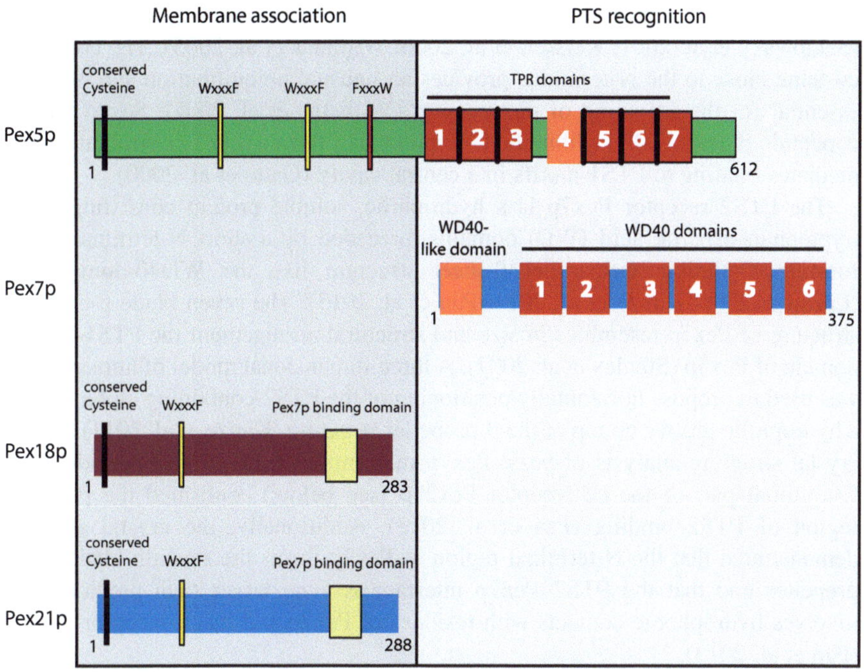

Fig. 13.2 Structure of the receptors contributing to the oleate-induced PTS import. The PTS1 receptor Pex5p can be structurally and functionally separated into two halves. Within its unstructured N-terminal half, Pex5p bears sequence motifs necessary for membrane association (WxxxF, FxxxW) and a conserved cysteine for monoubiquitination required for receptor recycling. The C-terminal half contains seven TPR domains forming a ring-like structure enabling PTS1 binding. The fourth TPR domain is less conserved and therefore depicted in orange. PTS2 receptor Pex7p contains six WD40 motifs and one WD40-like domain, arranged in a seven blade propeller fold enabling PTS2 binding. The PTS2 co-receptors Pex18p and Pex21p interact via a C-terminal Pex7p-binding domain with the cargo-laden PTS2 receptor. Structurally, the co-receptors resemble the N-terminal half of Pex5p, harboring the same set of protein-binding motifs

13.2.2 PTS-Receptors

PTS1 and PTS2 sequences are recognized in the cytosol by their cognate well-conserved receptors Pex5p and Pex7p, respectively. Both receptors are mainly soluble cytosolic proteins, but undergo a cycle in which they associate with peroxisomal membranes and form subsequently subcomplexes with various membrane peroxins. After the release of cargo protein into the peroxisomal matrix, PTS1 and PTS2 receptors are transported back to the cytosol. In addition to numerous protein–protein interactions, Pex5p shows intrinsic lipid binding activity and is able to insert spontaneously into membranes in vitro (Kerssen et al. 2006). From a structural and functional point of view, the PTS1 receptor Pex5p can be divided into two halves (Schliebs and Kunau 2006). The structurally disordered N-terminal regions bear known binding sites for membrane peroxins (Fig. 13.2). The WxxxF motifs were identified as high affinity binding sites for Pex14p and Pex13p (Saidowsky et al. 2001; Kerssen et al. 2006; Williams et al. 2005). The conserved cysteine close to the N-terminus provides an unusual ubiquitination site, which is essential for the recycling of the receptor (Williams et al. 2007). Seven tetratricopeptide repeats (TPR) within the C-terminal half form a ring-like structure which mediates binding to PTS1 motifs in a central cavity (Gatto et al. 2000).

The PTS2 receptor Pex7p is a hydrophilic, soluble protein consisting of six tryptophan–aspartic acid (WD) domains, preceded by a short N-terminal region forming a similar antiparallel β-sheet structure like the WD40-domains do (Lazarow 2006; Kunze et al. 2011; Pan et al. 2013). The seven blade β-propeller structure of Pex7p resembles in size and structural arrangement the PTS1-binding domain of Pex5p (Stanley et al. 2007). A three-dimensional model of human Pex7p was used to propose horizontally positioning of the PTS2-containing alpha helix in a hydrophilic groove on top of the β-propeller structure (Kunze et al. 2011). Recent crystal structure analysis of yeast Pex7p in complex with a PTS2 peptide and a C-terminal part of the co-receptor Pex21p (see below) confirmed the proposed region of PTS2 binding (Pan et al. 2013). Additionally, the crystal structure demonstrated that the N-terminal region of Pex7p forms the seventh blade of the propeller and that the PTS2–Pex7p interface is even larger than predicted and involves hydrophobic contacts with residues of Pex7p and its co-receptor Pex21p (Pan et al. 2013).

Pex7p recognizes the cargo proteins in the cytosol, but for the targeting of the receptor–cargo complex to the peroxisomal membrane a co-receptor is required. In *S. cerevisiae* the redundant peroxins Pex18p and Pex21p interact with the Pex7p–cargo complex and the ternary complex of cargo, receptor and co-receptor is directed to the docking complex at the peroxisomal membrane (Grunau et al. 2009). The order of events is that first the PTS2 is recognized by Pex7p and then the cargo loaded receptor interacts with its co-receptor, which mediates the association with the docking complex (Grunau et al. 2009). The recently resolved crystal structure revealed that co-receptor binding also strengthens the Pex7p–PTS2 interaction (Pan et al. 2013).

Besides the Pex7p-binding domain, both PTS2 co-receptors also bear a conserved cysteine and a WxxxF-motif. In other organisms, different co-receptors can be found, e.g., the long isoform of PEX5 (PEX5L) in mammals or Pex20p in *Pichia pastoris* and *Neurospora crassa* (Schliebs and Kunau 2006).

13.2.3 PTS Recognition

Efficiency of cargo binding of the import receptors depends from several factors. The most critical parameters include (1) the abundance of binding partners, (2) the affinity of complex formation, and (3) the contribution of auxiliary proteins.

The abundance of receptor and peroxisomal matrix enzymes in yeast depends on gene expression as a response to varying environmental conditions. While Pex5p and Pex7p are constitutively expressed (Kal et al. 1999), several peroxisomal enzymes, including those of the β-oxidation are induced by activation of a promoter element called oleate response element (ORE) and therefore abundant in the cytosol of oleate-grown cells. ORE-inducible proteins are, e.g., Fox2p (PTS1), Fox3p (PTS2), Mdh3p (PTS1) and Cta1p (PTS1) (Karpichev and Small 1998).

Furthermore, the affinity of the receptor for the substrate plays an important role for preimport complex formation. Binding studies with the C-terminus of human Pex5p and peptides, containing a PTS1 sequence revealed equilibrium binding constants (KD values) ranging between 1 nM and 4 μM (Maynard et al. 2004; Gatto et al. 2003; Ghosh and Berg 2010). A more recent study on human PTS1–Pex5p affinities revealed a range of four orders of magnitude for affinities (Ghosh and Berg 2010). Interestingly, the authors reported a correlation between high PTS1–Pex5p affinities and low expression level of corresponding proteins. On the other hand, highly abundant proteins often possess PTS1 with weaker affinity. This relation may provide a rather uniform population of Pex5p–cargo complexes and similar probabilities of initial targeting for most peroxisomal matrix proteins, while still allowing a broad variety of expression levels for some PTS proteins (Ghosh and Berg 2010).

The affinities of PTS2–Pex7p have not been quantitatively determined. However, the stability of purified Pex7p–Fox3p complexes (Grunau et al. 2009) and interaction studies by two-hybrid analyses (Rehling et al. 1996) suggest that the binding properties are comparable with Pex5p–PTS1 complexes.

As a third critical factor for cargo–receptor complex formation, auxiliary proteins have been considered. It has been suggested that chaperones play an important role for receptor binding and peroxisomal matrix protein import. The mostly cytosolic J-domain containing protein Djp1p, a protein homologous to *E. coli* DnaJ, is specifically required for peroxisomal protein import and essential for both PTS1 and PTS2 pathway. Proteins bearing a J-domain recruit members of the Hsp70 family to their site of action. Djp1p and its cognate chaperone may act as a trigger for PTS affinity or even regulate the PTS receptor docking onto the peroxisomal membrane (Hettema et al. 1998). Related to this, two highly abundant chaperons of the Hsp70 family, Ssa1p and Ssa2p, were co-purified with PTS2 preimport complexes (Grunau et al. 2009).

In higher eukaryotes, heat shock proteins are also involved in the formation of the PTS1 receptor–cargo complex. Two possible roles have been suggested, either that Hsp70p binding increases the affinity of Pex5p to the PTS1 protein (Harano et al. 2001) or that Hsp70p functions as a chaperone for folding of matrix proteins (Harper et al. 2003).

The recently solved crystal structure of Pex7p/PTS2/Pex21p revealed that the PTS2 cofactors specifically stabilize the Pex7p–PTS2 binding. In this case, the PTS2 co-receptor directly covers the hydrophobic surfaces of both receptor and PTS2 (Pan et al. 2013).

13.3 Formation of the Docking Complex

After recognition of cargo by the appropriate receptor in the cytosol, the receptor–cargo complex docks at the cytosolic side of the peroxisomal membrane. It is generally assumed that cargo recognition leads to conformational changes in the N-terminal half of the receptor resulting in the accessibility of primary docking sites (Stanley et al. 2006; Wolf et al. 2010). Such mechanism provides directionality of receptor targeting and in addition allows the docking machinery to distinguish between cargo-loaded and empty receptor as a load-control (Wolf et al. 2010).

The N-terminal half of Pex5p (Pex5p-N) contains all known binding sites for other peroxins and is supposed to mediate membrane binding and multiple sequential interactions with import machinery constituents during the receptor cycle (Fig. 13.1; Girzalsky et al. 2010; Stanley et al. 2006). In *S. cerevisiae*, it was shown that the N-terminal half of Pex5p, lacking the C-terminal PTS1-binding domain, can still facilitate the import of the matrix protein Fox1p. This finding suggests that Pex5p-N is sufficient for receptor docking and cargo transport into peroxisomes (Schäfer et al. 2004; Grou et al. 2012). Moreover, it was shown that Pex5p-N could be functionally replaced by Pex18p as Pex18p fused to Pex5p-C is able to partially restore the PTS1 import in Pex5p deficient yeast cells. Based on these results, it is considered that the PTS2 co-receptors might fulfill roles similar to those of Pex5p-N in the PTS1 import (Schäfer et al. 2004).

13.3.1 Composition, Topology and Stoichiometry of the Docking Subcomplex

The docking to the peroxisomal membrane in the yeast *S. cerevisiae* is facilitated by the so-called docking subcomplex consisting of Pex14p, Pex13p and Pex17p (Erdmann and Schliebs 2005).

The integral membrane protein Pex13p contains two transmembrane segments with both N- and C-terminal ends exposed into the cytosol. With one exception in trypanosomes (Brennand et al. 2012), Pex13p contains an SH3 domain at the C-terminus. Pex14p behaves as an integral membrane protein in most species, but in baker's yeast the topology is uncertain (Albertini et al. 1997; Brocard et al. 1997). Several results suggest that different Pex14 subpopulations exist, which differ in alkaline extractability (Niederhoff et al. 2005). Interestingly, two distinct interacting sites between Pex14p and Pex13p have been identified. The SH3 domain is exposed to the cytosol and binds a PxxP-like motif of Pex14p (Douangamath et al. 2002; Pires et al. 2003). Additionally, the intraluminal fragment, which is located between the two transmembrane segments of yeast Pex13p, contains a distinct Pex14p-binding site (Schell-Steven et al. 2005). This suggests that Pex14p at least partially is translocated across the membrane. However, the functional role of such an intraperoxisomal interaction of Pex13p is not known. Pex17p is a peripheral membrane protein attached to Pex14p at the outside of peroxisomes (Huhse et al. 1998).

The stoichiometry of the central constituents of the docking subcomplex is not clear and seems to differ from species to species. Interestingly, the stoichiometric relationship between Pex13p and Pex14p is critical for protein import. Overexpression of one of these proteins impairs protein import whereas co-overexpression has no effect (Bottger et al. 2000).

S. cerevisiae Pex14p is able to form homo-oligomers (Albertini et al. 1997). In mammalian cells, a homodimeric interaction through the conserved N-terminus of Pex14p, which is also the binding site for Pex5p, was shown. Further, binding of the PTS1-receptor leads to the complete dissociation of Pex14p homodimers (Su et al. 2010). Additionally, Pex14p homo-oligomers of larger sizes were identified by cross-linking assays (Itoh and Fujiki 2006).

A recent study reported that also human PEX13 homo-oligomerizes in peroxisomes of living cells. This requires a highly conserved W313 residue of the SH3 domain of Pex13p, which is not required for its interaction with PEX14. Interestingly, impairing self-association of Pex13p leads to disruption of PTS1 protein import (Krause et al. 2013). Homo-oligomerization has not yet been observed for yeast Pex13p.

The unclear situation of the oligomerization grade of peroxins of the docking complex and their topology might reflect not only species-specific differences but also suggests the existence of subpopulations of these proteins, which act dynamically at different states of the receptor cycle.

13.3.2 Docking of the Receptors

Although detailed structural information about the docking complex is still scarce, the recent view suggests a very complex hierarchical scenario during receptor cycling in yeast: Before docking, Pex14p, itself required as a membrane anchor for Pex17p, is bound by the integral membrane protein Pex13p via its SH3 domain. Probably, all Pex5p-binding sites of Pex14p are fully accessible for the arriving receptor–cargo complex. In yeast, the receptor interacts with a C-terminal Pex5p-binding site of Pex14p (Niederhoff et al. 2005). At a later step during receptor processing, Pex5p binds the conserved N-terminal domain of Pex14p via a short N-terminal region comprising amino acid positions 246 and 267 (Kerssen et al. 2006). One striking feature within this stretch of amino acids is a di-aromatic pentapeptide that represents an inverted WxxxF motif.

The docking event seems to be different in human peroxisomes. The receptor contains at least seven WxxxF motifs which all bind PEX14 with high affinity and could thereby serve as initial contact sites (Saidowsky et al. 2001). Recent results suggest that the N-terminally located Pex14p binding site of Pex5p acts as an initial docking site from where Pex14p can slide to other WxxxF motifs with different function (Neuhaus et al. unpublished).

The function of the Pex5p–Pex13p interaction is not yet clear. Pex13p was originally defined as docking protein for the PTS1-receptor (Erdmann and Blobel 1996; Elgersma et al. 1996; Gould et al. 1996). More recent studies revealed that

Pex14p and Pex13p have redundant function with respect to PTS1-receptor docking (Bottger et al. 2000). In yeast, one of the two typical WxxxF motifs interacts with its C-terminal Src-homology-3 (SH3) domain of Pex13p which is exposed at the cytosolic site (Douangamath et al. 2002; Pires et al. 2003).

Beside this redundant function with respect to the docking step, Pex13p plays an unknown role later in the PTS1 receptor cycle. This is strongly suggested by the observation that Pex13p has a higher affinity to unloaded receptor than to cargo-loaded (Urquhart et al. 2000; Otera et al. 2002).

Recent data on the early events of PTS2 pathway indicate that the PTS2–preimport complex initially binds to Pex13p and then assembles into high-molecular weight complexes containing both Pex14p and Pex13p (Grunau et al. 2009; Stein et al. 2002; Girzalsky et al. 2010). The distinct role of co-receptors and Pex7p for docking is still unclear. Whereas Pex18p and Pex21p are supposed to direct the complex to the peroxisomal membrane, Pex7p provides high-affinity binding sites for both Pex13p and Pex14p (Schell-Steven et al. 2005). However, a slight two-hybrid based interaction between Pex21p and both docking proteins could be observed (Einwächter et al. 2001), which might be stronger in the presence of the peroxisomal membrane. A membrane complex consisting of Pex18p–Pex7p–Fox3p with a molecular mass of 150 kDa was co-isolated together with Pex13p probably representing the PTS2 preimport complex after docking (Grunau et al. 2009).

The peripheral membrane protein Pex17p, which does not directly interact with either Pex5p or Pex7p, turned out to be a binding partner of Pex14p and is thus considered as the third docking-complex constituent (Huhse et al. 1998; Snyder et al. 1999; Girzalsky et al. 2010). In some filamentous fungi, a chimeric protein consisting of Pex14p-like N-terminal domain and a Pex17p-like C-terminal domain has been described, called Pex14/17p or Pex33p (Managadze et al. 2010; Opalinski et al. 2010). A homolog of these Pex17p-like proteins has not yet been identified in higher eukaryotes. The function of the Pex17p-like proteins is not clear. Interestingly, Pex17p-like proteins are not essential for receptor docking but still essential for cargo import into the matrix (Platta et al. 2012).

Although the discussed initial interactions with docking peroxins are multivalent and species specific, at least one general principle is emerging. WxxxF motifs are always required to establish contacts with the peroxisomal membrane. So far, multiple functions have been assigned to these diaromatic pentapeptide sequences. The human WxxxF motifs are involved in Pex14p and Pex13p binding as well as in cargo release (Freitas et al. 2011). Yeast Pex5p contains two typical motifs and one inversed WxxxF motif within the N-terminal region of the protein (Fig. 13.2). Whereas the two internal motifs seem to establish the association with Pex13p and Pex14p, the function of the N-terminal WxxxF motif is unknown. Beside their functional heterogeneity, another question regards the varying number of these motifs in distinct species. Whereas yeast Pex5p only possesses one WxxxF motif with binding capacity for Pex14p, Pex5p of higher eukaryotes contain up to 9 of these motifs (Schliebs et al. 1999).

13.4 Formation of the Translocation Complex

In contrast to the translocons of the ER, mitochondria and chloroplasts, a characteristic feature of peroxisomes is their capability to import folded and even oligomeric proteins. The identification of the nature of the translocon required for the transport of proteins across the peroxisomal membrane, for a long time was subject of speculations. A model for peroxisomal import of PTS1 proteins proposed the existence of a transient protein conducting channel, which is formed also by membrane integrated receptors (Erdmann and Schliebs 2005).

13.4.1 Structure of the Peroxisomal Translocon

The PTS1-receptor Pex5p together with the docking protein Pex14p forms the minimal import unit for the PTS1- and PTS2-containing Pex8p in the yeast *P. pastoris* (Ma et al. 2009). The functional role of these peroxins in *S. cerevisiae* was further substantiated by electro-physiological characterization of an affinity-purified membrane-bound Pex5p/Pex14p complex. These studies revealed that the complex exhibits the features of a regulated pore with a diameter of up to 9 nm (Meinecke et al. 2010). Gating of the Pex5p/Pex14p channel could be induced by incubation with cytosolic receptor–cargo complexes. The enormous size of the water-filled channel and the receptor–cargo-dependent opening gives a first glimpse of how large oligomerized proteins are transferred across the membrane.

One obvious structural question concerns the nature of the inner pore-forming unit, either Pex5p or Pex14p? The primary suspect would be Pex14p due to its known transmembrane topology and its ability to form homo-oligomers. As discussed before, controversial results concerning the topology of Pex14p could be explained by the dynamic switch between a peripheral and integral state (Azevedo and Schliebs 2006). It can be assumed that the Pex14p population which is stably associated with the membrane might represent the pore constituent, whereas the peripheral Pex14p is required for docking.

However, several in vivo data suggest that Pex5p, and not Pex14p, forms the central core of the PTS1 import channel. In *S. cerevisiae*, a Pex5p mutant lacking functional binding sites for both Pex13p and the N-terminal domain of Pex14p can mediate PTS1 protein import (Kerssen et al. 2006). In the yeast *H. polymorpha*, overproduction of Pex5p rescues the import defect of a Pex14p-deficient mutant (Salomons et al. 2000) and *Arabidopsis* Pex14p-deficient mutant still imports protein (Monroe-Augustus et al. 2011).

In contrast to Pex14p, the predominantly soluble Pex5p does not contain any predictable transmembrane segment. Nonetheless, recombinant human and yeast PTS1 receptors were shown to insert spontaneously into artificial phospholipid membranes (Kerssen et al. 2006). It had been suggested that Pex5p mechanistically might resemble pore forming toxins, which form a pore by the assembly of amphipathic helices (Erdmann and Schliebs 2005). Along this line, membrane insertion of Pex5p could be facilitated by five amphipathic helices within the

N-terminal half (Meinecke et al. 2010). An almost complete insertion of Pex5p into the peroxisomal membrane is in accordance with protease protection assays in human cells, which show that only a small N-terminal region remains accessible from the cytosolic site (Gouveia et al. 2003). This region of Pex5p contains the conserved cysteine, which is ubiquitinated to facilitate relocation into the cytosol.

Size and stoichiometry of the membrane-associated Pex5p/Pex14p complex are difficult to determine due to the transient nature of the pore. Soluble complexes formed by human Pex5p cargo and the Pex5p-binding domain of Pex14p were analyzed by small angle X-ray scattering, revealing a Pex5p–Pex14p stoichiometry of 1:6 (Shiozawa et al. 2009). However, under in vivo conditions the stoichiometry could be affected by steric hindrances and the phospholipid environment. The purification and identification of Pex5p subcomplexes in yeast indicated that the pore-forming subcomplex contains Pex5p and Pex14p in a ratio of 1:1 (Meinecke et al. 2010).

The size of the pore forming Pex5p/Pex14p complex as determined by size-exclusion chromatography (Meinecke et al. 2010) was approximately 600 kDa. Accordingly, six molecules of each Pex5p and Pex14p seem to constitute the pore, which had been isolated in a closed state. The size of the pore in the open state, which is induced by the receptor–cargo complex, is still elusive.

In yeast, both PTS1 and PTS2 pathways converge at the level of Pex14p at the peroxisomal membrane. The PTS1 import is functional in the absence of the PTS2-receptor and vice versa (Zhang and Lazarow 1996; Marzioch et al. 1994), opening the possibility that both import pathways use distinct translocation channels. In the PTS2 pathway, the function of the N-domain of Pex5p is performed by the co-receptor Pex18p. Affinity-purification of Pex18p-ProteinA from solubilized membranes of *S. cerevisiae* revealed subcomplexes of different sizes and composition (Grunau et al. 2009). The bulk of Pex18p at the membrane was found in a 150 kDa complex, containing Pex7p and cargo-protein and most likely representing the docking state of the PTS2–preimport complex (as discussed in Sect. 11.2.3). A second high molecular weight complex of Pex18p did contain Pex14p but neither Pex7p nor cargo-protein, which might indicate that this subcomplex represents the translocation pore after receptor–cargo release.

13.4.2 Gating of the Transient Translocation Pore

For the PTS1 pathway, the elementary role of Pex5p and Pex14p in the formation of the highly dynamic translocation pore is consolidated. However, there is general agreement that other peroxins are required to maintain and to regulate efficient transport of cargo proteins. For instance, rapid formation and breakdown of the pore by removal of receptor molecules are tightly coupled processes. The insertion of the receptor–cargo complex into the membrane takes place in an ATP-independent manner (Schäfer et al. 2004; Schliebs and Kunau 2006). According to the recently postulated "Export-driven Import Model," the ATP-dependent dislocation of the membrane-bound receptor might provide the driving force for cargo translocation and release (Schliebs et al. 2010). For further details of the receptor recycling, see the accompanying review on the exportomer (Platta et al. 2012). Here, we will briefly address possible role of Pex5p- and Pex14p-interacting proteins that are not directly involved in receptor recycling. In baker's yeast, these are peroxins Pex13p, Pex8p, and Pex17p.

Pex13p contains multiple binding sites for Pex5p and Pex14p (Schell-Steven et al. 2005) but is not an integral constituent of the transient PTS1 pore (Meinecke et al. 2010; Gouveia et al. 1999). It is generally accepted that Pex13p can serve as docking protein for the PTS-receptors (Elgersma et al. 1996; Grunau et al. 2009). Interestingly, Pex13p has a higher affinity to unloaded receptor than to the cargo-loaded receptor (Urquhart et al. 2000; Otera et al. 2002). This might indicate that Pex13p might perform a function in receptor binding after receptor–cargo dissociation. Interestingly, a Pex5p mutant disrupted in Pex13p-binding site is still able to facilitate protein import suggesting that at least assembly of the pore and translocation should be intact (Bottger et al. 2000; Kerssen et al. 2006). Possibly, Pex13p could be involved in disassembly of Pex5p/Pex14p complexes.

Pex8p is supposed to initiate receptor recycling by linking the docking complex/transient pore to the ubiquitination machinery (Agne et al. 2003). However, Pex8p has also been considered as a cargo-releasing factor (Wang et al. 2003). This assumption is based on in vitro assays, which show that the presence of Pex8p causes dissociation of a Pex5p–PTS1–peptide complex. However, Pex8p has been identified only in yeast and whether a functional orthologue exists in higher eukaryotes is still unclear. Subramani and coworkers could show that Pex8p import into peroxisomes requires Pex5p and Pex14p (Ma et al. 2009).

The Pex14p-interacting Pex17p has only been found in fungi so far. Pex17p does not seem to be a constitutive member of the Pex5p-dependent import pore (Meinecke et al. 2010). This finding was somewhat surprising since Pex14p and Pex17p form a stable heterodimer. The simplest explanation is that different subpopulations of Pex14p exist: a heterodimeric Pex14p–Pex17p complex with unknown function and the Pex14p–Pex5p complex, which constitutes the import pore. This assumption is further supported by the finding that two homologous Pex14p genes exist in *Neurospora crassa*. The longer isoform shares significant sequence similarity with Pex17p (Managadze et al. 2010).

Taken together, our knowledge on the functional role of the three pore-interacting peroxins Pex13p, Pex8p and Pex17p in peroxisomal protein import is still scarce. In this respect, it will be of great interest to study assembly and activity of the import pore in mutants that are affected in binding of pore-interacting peroxins.

13.5 Alternative Translocation Complexes

The function of peroxisomes seems to be determined by the set of enzymes, which are expressed under definite environmental conditions. So far, biogenesis of peroxisomes in the yeast *S. cerevisiae* has been explored extensively under oleate-inducing conditions. Thus, the screening for *pex*-mutants for identification of *PEX* genes was based on the ability of yeast to degrade fatty acids in intact peroxisomes (Erdmann et al. 1989). It can be assumed that the major constituents of the matrix import machinery, in particular the receptors, are evolutionary optimized. In this context, the question arises whether yeast cells can adapt their import machineries to other growth conditions.

Here, the possible role of the second PTS2 co-receptor Pex21p comes into focus. Pex18p and Pex21p have redundant function with respect to targeting Pex7p–cargo complex to peroxisomes and most likely, by forming the transient PTS2 import channel. Purdue et al. claimed that Pex18p is the predominant PTS2 co-receptor for Fox3p under oleate-inducing conditions (Purdue et al. 1998). In accordance, the promoter region of Pex18p contains an oleate-responsive element (ore) (Smith et al. 2002). It is tempting to speculate that under different growth conditions e.g. glucose repression, Pex18p becomes downregulated, while Pex21p could be present at higher amounts. The possible "backup" function of a second PTS2 co-receptor could be important for yeast cells due to the demand of fast changing environmental conditions.

The yeast *S. cerevisiae* contains numerous duplicated genes due to whole genome duplication 100 million years ago (Wolfe and Shields 1997). Gene duplication is often followed by the acquisition of novel gene functions or expression diversity (Hittinger and Carroll 2007). Noteworthy, a Pex5p paralog has been identified in the genome of *S. cerevisiae*. The TPR-containing protein, named Ymr018wp shares 27 % sequence identity with Pex5p (Amery et al. 2001). Most strikingly, sequence motifs required for Pex5p interaction with other proteins are also conserved in Ymr018wp. Accordingly, the paralog contains a conserved N-terminal cysteine, which in Pex5p is a targeting site for ubiquitination, two WxxxF motifs as possible interaction sites for docking proteins, and six TPR domains, which are arranged in a similar manner as found in the PTS1-binding domain of Pex5p. Despite this striking conservation of typical receptor structural features (Amery et al. 2001), gene disruption of YMR018w did not affect matrix protein import (Amery et al. 2001). However, the mutants were tested under growth conditions in which Pex5p is abundant. Therefore, it is still possible that the paralog acts as a PTS1-receptor under different growth conditions, maybe for only a specific subset of peroxisomal enzymes.

In human cells, another Pex5p-related protein, Pex5Rp, was identified and shown to bind PTS1 proteins (Amery et al. 2001). However, no receptor function has yet been assigned to this protein. Pex5Rp is preferentially expressed in brain and does neither bind Pex14p nor Pex12p; it could not restore PTS1 import in Pex5p-deficient mouse fibroblasts (Amery et al. 2001). Moreover, all known binding motifs, which are typical for the PTS1-receptor (conserved cysteine and WxxxF-motifs) are missing in the sequence of Pex5Rp.

Final Conclusion

Challenging features of peroxisomal matrix protein import include (1) the transport of folded and oligomerized proteins, (2) to import proteins against higher concentration in an unidirectional manner, (3) to maintain impermeability of the peroxisomal membrane for small compounds, and (4) to adapt to various environmental conditions. During the last years, we gained some insight into molecular background required for the PTS-driven import process. Structural studies of receptor complexes with cargo proteins or other peroxins evoke a complex hierarchical organization of the initial steps of protein import. Pex5p and Pex14p were identified as core constituents of a transient import channel for large peroxisomal proteins. Other interacting peroxins like Pex13p, Pex17p, Pex8p or those involved in Pex5p recycling are supposed to play a regulatory role in pore assembly and gating. The elucidation of the concerted function of peroxins in PTS1- and PTS2-dependent protein import and the characterization of alternative import pathways will contribute to our understanding of the basic mechanistic principles of the unique peroxisomal protein import.

References

Agne B, Meindl NM, Niederhoff K, Einwachter H, Rehling P, Sickmann A, Meyer HE, Girzalsky W, Kunau WH (2003) Pex8p: an intraperoxisomal organizer of the peroxisomal import machinery. Mol Cell 11(3):635–646

Albertini M, Rehling P, Erdmann R, Girzalsky W, Kiel JAKW, Veenhuis M, Kunau W-H (1997) Pex14p, a peroxisomal membrane protein binding both receptors of the two PTS-dependent import pathways. Cell 89:83–92

Amery L, Sano H, Mannaerts GP, Snider J, Van Looy J, Fransen M, Van Veldhoven PP (2001) Identification of PEX5p-related novel peroxisome-targeting signal 1 (PTS1)-binding proteins in mammals. Biochem J 357(Pt 3):635–646

Azevedo JE, Schliebs W (2006) Pex14p, more than just a docking protein. Biochim Biophys Acta 1763(12):1574–1584

Bottger G, Barnett P, Klein AT, Kragt A, Tabak HF, Distel B (2000) Saccharomyces cerevisiae PTS1 receptor Pex5p interacts with the SH3 domain of the peroxisomal membrane protein Pex13p in an unconventional, non-PXXP-related manner. Mol Biol Cell 11(11):3963–3976

Brennand A, Rigden DJ, Michels PA (2012) Trypanosomes contain two highly different isoforms of peroxin PEX13 involved in glycosome biogenesis. FEBS Lett 586(13):1765–1771. doi:10.1016/j.febslet.2012.05.019

Brocard C, Lametschwandtner G, Koudelka R, Hartig A (1997) Pex14p is a member of the protein linkage map of Pex5p. EMBO J 16(18):5491–5500

Bun-Ya M, Maebuchi M, Hashimoto T, Yokota S, Kamiryo T (1997) A second isoform of 3-ketoacyl-CoA thiolase found in Caenorhabditis elegans, which is similar to sterol carrier protein x but lacks the sequence of sterol carrier protein 2. Eur J Biochem 245(2):252–259

Cartwright JL, Gasmi L, Spiller DG, McLennan AG (2000) The Saccharomyces cerevisiae PCD1 gene encodes a peroxisomal nudix hydrolase active towards coenzyme A and its derivatives. J Biol Chem 275:32925–30

Douangamath A, Filipp FV, Klein AT, Barnett P, Zou P, Voorn-Brouwer T, Vega MC, Mayans OM, Sattler M, Distel B, Wilmanns M (2002) Topography for independent binding of alpha-helical and PPII-helical ligands to a peroxisomal SH3 domain. Mol Cell 10(5):1007–1017

Einwächter H, Sowinski S, Kunau WH, Schliebs W (2001) Yarrowia lipolytica Pex20p, Saccharomyces cerevisiae Pex18p/Pex21p and mammalian Pex5pL fulfil a common function in the early steps of the peroxisomal PTS2 import pathway. EMBO Rep 2:1035–1039

Elgersma Y, van Roermund CW, Wanders RJ, Tabak HF (1995) Peroxisomal and mitochondrial carnitine acetyltransferases of Sac-charomyces cerevisiae are encoded by a single gene. Embo J 14(14):3472–3479
Elgersma Y, Kwast L, Klein A, Voorn-Brouwer T, van den Berg M, Metzig B, America T, Tabak HF, Distel B (1996) The SH3 domain of the Saccharomyces cerevisiae peroxisomal membrane protein Pex13p functions as a docking site for Pex5p, a mobile receptor for the import of PTS1 containing proteins. J Cell Biol 135:97–109
Erdmann R, Blobel G (1996) Identification of Pex13p, a peroxisomal membrane receptor for the PTS1 recognition factor. J Cell Biol 135:111–121
Erdmann R, Schliebs W (2005) Peroxisomal matrix protein import: the transient pore model. Nat Rev Mol Cell Biol 6(9):738–742
Erdmann R, Veenhuis M, Mertens D, Kunau W-H (1989) Isolation of peroxisome-deficient mutants of Saccharomyces cerevisiae. Proc Natl Acad Sci USA 86:2432–2436
Freitag J, Ast J, Bolker M (2012) Cryptic peroxisomal targeting via alternative splicing and stop codon read-through in fungi. Nature 485(7399):522–525. doi:10.1038/nature11051
Freitas MO, Francisco T, Rodrigues TA, Alencastre IS, Pinto MP, Grou CP, Carvalho AF, Fransen M, Sa-Miranda C, Azevedo JE (2011) PEX5 protein binds monomeric catalase blocking its tetramerization and releases it upon binding the N-terminal domain of PEX14. J Biol Chem 286(47):40509–40519
Gatto GJ, Geisbrecht BV, Gould SJ, Berg JM (2000) Peroxisomal targeting signal-1 recognition by the TPR domains of human PEX5. Nat Struct Biol 7(12):1091–1095
Gatto GJ Jr, Maynard EL, Guerrerio AL, Geisbrecht BV, Gould SJ, Berg JM (2003) Correlating structure and affinity for PEX5:PTS1 complexes. Biochemistry 42(6):1660–1666
Ghosh D, Berg JM (2010) A proteome-wide perspective on peroxi-some targeting signal 1(PTS1)-Pex5p affinities. J Am Chem Soc 132(11):3973–3979
Girzalsky W, Saffian D, Erdmann R (2010) Peroxisomal protein translocation. Biochim Biophys Acta 1803(6):724–731
Glover JR, Andrews DW, Rachubinski RA (1994) Saccharomyces cerevisiae peroxisomal thiolase is imported as a dimer. Proc Natl Acad Sci USA 91(22):10541–10545
Gonzalez NH, Felsner G, Schramm FD, Klingl A, Maier UG, Bolte K (2011) A single peroxisomal targeting signal mediates matrix protein import in diatoms. PLoS One 6(9):e25316
Gould SJ, Keller GA, Hosken N, Wilkinson J, Subramani S (1989) A conserved tripeptide sorts proteins to peroxisomes. J Cell Biol 108:1657–1664
Gould SJ, Kalish JE, Morrell JC, Bjorkman J, Urquhart AJ, Crane DI (1996) Pex13p is an SH3 protein of the peroxisome membrane and a docking factor for the predominantly cytoplasmic PTS1 receptor. J Cell Biol 135:85–95
Gouveia AM, Reguenga C, Oliveira ME, Eckerskorn C, Sa-Miranda C, Azevedo JE (1999) Alkaline density gradient floatation of membranes: polypeptide composition of the mammalian peroxisomal membrane. Anal Biochem 274(2):270–277
Gouveia AM, Guimaraes CP, Oliveira ME, Reguenga C, Sa-Miranda C, Azevedo JE (2003) Characterization of the peroxisomal cycling receptor Pex5p import pathway. Adv Exp Med Biol 544:219–220
Grou CP, Francisco T, Rodrigues TA, Freitas MO, Pinto MP, Car-valho AF, Domingues P, Wood SA, Rodriguez-Borges JE, Sa-Miranda C, Fransen M, Azevedo JE (2012) Identification of ubiquitin-specific protease 9X (USP9X) as a deubiquitinase acting on ubiquitin-peroxin 5 (PEX5) thioester conjugate. J Biol Chem 287(16):12815–12827
Grunau S, Schliebs W, Linnepe R, Neufeld C, Cizmowski C, Reinartz B, Meyer HE, Warscheid B, Girzalsky W, Erdmann R (2009) Peroxisomal targeting of PTS2 pre-import complexes in the yeast Sac-charomyces cerevisiae. Traffic 10(4):451–460
Harano T, Nose S, Uezu R, Shimizu N, Fujiki Y (2001) Hsp70 regu-lates the interaction between the peroxisome targeting signal type 1 (PTS1)-receptor Pex5p and PTS1. Biochem J 357:157–165

Harper CC, Berg JM, Gould SJ (2003) PEX5 binds the PTS1 inde-pendently of Hsp70 and the peroxin PEX12. J Biol Chem 278(10):7897–7901

Hettema EH, Ruigrok CCM, Koerkamp MG, van den Berg M, Tabak HF, Distel B, Braakman I (1998) The cytosolic DnaJ-like pro-tein Djp1p is involved specifically in peroxisomal protein import. J Cell Biol 142(2):421–434

Hittinger CT, Carroll SB (2007) Gene duplication and the adaptive evolution of a classic genetic switch. Nature 449(7163):677–681

Huhse B, Rehling P, Albertini M, Blank L, Meller K, Kunau WH (1998) Pex17p of Saccharomyces cerevisiae is a novel peroxin and component of the peroxisomal protein translocation machinery. J Cell Biol 140(1):49–60

Islinger M, Li KW, Seitz J, Volkl A, Luers GH (2009) Hitchhiking of Cu/Zn superoxide dismutase to peroxisomes–evidence for a natural piggyback import mechanism in mammals. Traffic 10 (11):1711–1721

Itoh R, Fujiki Y (2006) Functional domains and dynamic assembly of the peroxin Pex14p, the entry site of matrix proteins. J Biol Chem 281(15):10196–10205

Jung S, Marelli M, Rachubinski RA, Goodlett DR, Aitchison JD (2009) Dynamic changes in the subcellular distribution of Gpd1p in response to cell stress. J Biol Chem 285:6739–6749

Kal AJ, van Zonneveld AJ, Benes V, van den Berg M, Koerkamp MG, Albermann K, Strack N, Ruijter JM, Richter A, Dujon B, An-sorge W, Tabak HF (1999) Dynamics of gene expression revealed by comparison of serial analysis of gene expression transcript profiles from yeast grown on two different carbon sources. Mol Biol Cell 10(6):1859–1872

Karpichev IV, Small GM (1998) Global regulatory functions of Oaf1p and Pip2p (Oaf2p), transcription factors that regulate genes encoding peroxisomal proteins in Saccharomyces cerevisiae. Mol Cell Biol 18(11):6560–6570

Keller GA, Gould S, DeLuca M, Subramani S (1987) Firefly lucif-erase is targeted to peroxisomes in mammalian cells. Proc Natl Acad Sci USA 84:3264–3268

Kerssen D, Hambruch E, Klaas W, Platta HW, de Kruijff B, Erd-mann R, Kunau WH, Schliebs W (2006) Membrane association of the cycling peroxisome import receptor Pex5p. J Biol Chem 281(37):27003–27015

Klein AT, van Den Berg M, Bottger G, Tabak HF, Distel B (2002) Saccharomyces cerevisiae acyl-CoA oxidase follows a novel, non-PTS1, import pathway into peroxisomes that is dependent on Pex5p. J Biol Chem 277:25011–25019

Krause C, Rosewich H, Woehler A, Gartner J (2013) Functional analysis of PEX13 mutation in a Zellweger syndrome spectrum patient reveals novel homooligomerization of PEX13 and its role in human peroxisome biogenesis. Hum Mol Genet 22(19):3844–57

Kunze M, Neuberger G, Maurer-Stroh S, Ma J, Eck T, Braverman N, Schmid JA, Eisenhaber F, Berger J (2011) Structural requirements for interaction of peroxisomal targeting signal 2 and its receptor PEX7. J Biol Chem 286(52):45048–45062

Lazarow PB (2006) The import receptor Pex7p and the PTS2 tar-geting sequence. Biochim Biophys Acta 1763(12):1599–1604

Ma C, Schumann U, Rayapuram N, Subramani S (2009) The perox-isomal matrix import of Pex8p requires only PTS receptors and Pex14p. Mol Biol Cell 20(16):3680–3689

Managadze D, Wurtz C, Wiese S, Schneider M, Girzalsky W, Meyer HE, Erdmann R, Warscheid B, Rottensteiner H (2010) Identification of PEX33, a novel component of the peroxisomal docking complex in the filamentous fungus Neurospora crassa. Eur J Cell Biol 89(12):955–964

Marzioch M, Erdmann R, Veenhuis M, Kunau W-H (1994) PAS7 encodes a novel yeast member of the WD-40 protein family essential for import of 3-oxoacyl-CoA thiolase, a PTS2-containing protein, into peroxisomes. EMBO J 13:4908–4918

Maynard EL, Gatto GJ Jr, Berg JM (2004) Pex5p binding affini-ties for canonical and noncanonical PTS1 peptides. Proteins 55(4):856–861

Meinecke M, Cizmowski C, Schliebs W, Kruger V, Beck S, Wagner R, Erdmann R (2010) The peroxisomal importomer constitutes a large and highly dynamic pore. Nat Cell Biol 12(3):273–277

Monroe-Augustus M, Ramon NM, Ratzel SE, Lingard MJ, Chris-tensen SE, Murali C, Bartel B (2011) Matrix proteins are inefficiently imported into Arabidopsis peroxisomes lacking the receptor-docking peroxin PEX14. Plant Mol Biol 77(1–2):1–15

Neuberger G, Maurer-Stroh S, Eisenhaber B, Hartig A, Eisenhaber F (2003) Motif refinement of the peroxisomal targeting signal 1 and evaluation of taxon-specific differences. J Mol Biol 328 (3):567–579

Niederhoff K, Meindl-Beinker NM, Kerssen D, Perband U, Schafer A, Schliebs W, Kunau WH (2005) Yeast Pex14p possesses two func-tionally distinct Pex5p and one Pex7p binding sites. J Biol Chem 280(42):35571–35578

Opalinski L, Kiel JA, Homan TG, Veenhuis M, van der Klei IJ (2010) Penicillium chrysogenum Pex14/17p–a novel component of the peroxisomal membrane that is important for penicillin production. FEBS J 277(15):3203–3218

Otera H, Setoguchi K, Hamasaki M, Kumashiro T, Shimizu N, Fu-jiki Y (2002) Peroxisomal targeting signal receptor Pex5p interacts with cargoes and import machinery components in a spatiotemporally differentiated manner: Conserved Pex5p WXXXF/Y motifs are critical for matrix protein import. Mol Cell Biol 22(6):1639–1655

Pan D, Nakatsu T, Kato H (2013) Crystal structure of peroxisomal targeting signal-2 bound to its receptor complex Pex7p-Pex21p. Nat Struct Mol Biol 20(8):987–93

Pires JR, Hong X, Brockmann C, Volkmer-Engert R, Schneider-Mergener J, Oschkinat H, Erdmann R (2003) The ScPex13p SH3 do-main exposes two distinct binding sites for Pex5p and Pex14p. J Mol Biol 326(5):1427–1435

Platta HW, Hagen S, Erdmann R (2012) The exportomer: the peroxisomal receptor export machinery. Cell Mol Life Sci 70(8):1393–411

Purdue PE, Yang X, Lazarow PB (1998) Pex18p and Pex21p, a nov-el pair of related peroxins essential for peroxisomal targeting by the PTS2 pathway. J Cell Biol 143(7):1859–1869

Rehling P, Marzioch M, Niesen F, Wittke E, Veenhuis M, Kunau W-H (1996) The import receptor for the peroxisomal targeting signal 2 (PTS2) in Saccharomyces cerevisiae is encoded by the PAS7 gene. EMBO J 15(12):2901–2913

Rehling P, Skaletz-Rorowski A, Girzalsky W, Voorn-Brouwer T, Franse MM, Distel B, Veenhuis M, Kunau WH, Erdmann R (2000) Pex8p, an intraperoxisomal peroxin of Saccharomyces cerevisiae re-quired for protein transport into peroxisomes binds the PTS1 receptor Pex5p. J Biol Chem 275(5):3593–3602

Saidowsky J, Dodt G, Kirchberg K, Wegner A, Nastainczyk W, Kunau WH, Schliebs W (2001) The di-aromatic pentapeptide repeats of the human peroxisome import receptor PEX5 are separate high af-finity binding sites for the peroxisomal membrane protein PEX14. J Biol Chem 276(37):34524–34529

Salomons FA, Kiel JA, Faber KN, Veenhuis M, van der Klei IJ (2000) Overproduction of Pex5p stimulates import of alcohol oxidase and dihydroxyacetone synthase in a Hansenula polymorpha Pex14 null mutant. J Biol Chem 275(17):12603–12611

Schäfer A, Kerssen D, Veenhuis M, Kunau WH, Schliebs W (2004) Functional similarity between the peroxisomal PTS2 receptor binding protein Pex18p and the N-terminal half of the PTS1 receptor Pex5p. Mol Cell Biol 24(20):8895–8906

Schell-Steven A, Stein K, Amoros M, Landgraf C, Volkmer-Engert R, Rottensteiner H, Erdmann R (2005) Identification of a novel, in-traperoxisomal Pex14-binding site in Pex13: association of Pex13 with the docking complex is essential for peroxisomal matrix protein im-port. Mol Cell Biol 25(8):3007–3018

Schliebs W, Kunau WH (2006) PTS2 co-receptors: diverse proteins with common features. Biochim Biophys Acta 1763(12):1605–1612

Schliebs W, Saidowsky J, Agianian B, Dodt G, Herberg FW, Kunau WH (1999) Recombinant human peroxisomal targeting signal receptor PEX5. Structural basis for interaction of PEX5 with PEX14. J Biol Chem 274(9):5666–5673

Schliebs W, Girzalsky W, Erdmann R (2010) Peroxisomal protein import and ERAD: variations on a common theme. Nat Rev Mol Cell Biol 11(12):885–890

Shiozawa K, Konarev PV, Neufeld C, Wilmanns M, Svergun DI (2009) Solution structure of human Pex5.Pex14.PTS1 protein com-plexes obtained by small angle X-ray scattering. J Biol Chem 284(37):25334–25342

Smith JJ, Marelli M, Christmas RH, Vizeacoumar FJ, Dilworth DJ, Ideker T, Galitski T, Dimitrov K, Rachubinski RA, Aitchison JD (2002) Transcriptome profiling to identify genes involved in peroxi-some assembly and function. J Cell Biol 158(2):259–271

Snyder WB, Koller A, Choy AJ, Johnson MA, Cregg JM, Rangell L, Keller GA, Subramani S (1999) Pex17p is required for import of both peroxisome membrane and lumenal proteins and interacts with Pex19p and the peroxisome targeting signal-receptor docking com-plex in pichia pastoris. Mol Biol Cell 10(12):4005–4019

Stanley WA, Filipp FV, Kursula P, Schuller N, Erdmann R, Schliebs W, Sattler M, Wilmanns M (2006) Recognition of a functional peroxisome type 1 target by the dynamic import receptor Pex5p. Mol Cell 24(5):653–663

Stanley WA, Fodor K, Marti-Renom MA, Schliebs W, Wilmanns M (2007) Protein translocation into peroxisomes by ring-shaped import receptors. FEBS Lett 581(25):4795–4802

Stein K, Schell-Steven A, Erdmann R, Rottensteiner H (2002) In-teractions of Pex7p and Pex18p/Pex21p with the peroxisomal docking machinery: implications for the first steps in PTS2 protein import. Mol Cell Biol 22(17):6056–6069

Su JR, Takeda K, Tamura S, Fujiki Y, Miki K (2010) Monomer-dimer transition of the conserved N-terminal domain of the mammali-an peroxisomal matrix protein import receptor, Pex14p. Biochem Bi-ophys Res Commun 394(1):217–221

Swinkels BW, Gould SJ, Bodnar AG, Rachubinski RA, Subramani S (1991) A novel, cleavable peroxisomal targeting signal at the amino-terminus of the rat 3-ketoacyl-CoA thiolase. EMBO J 10:3255–3262

Urquhart AJ, Kennedy D, Gould SJ, Crane DI (2000) Interaction of Pex5p, the type 1 peroxisome targeting signal receptor, with the pe-roxisomal membrane proteins Pex14p and Pex13p. J Biol Chem 275(6):4127–4136

van der Klei IJ, Veenhuis M (2006) PTS1-independent sorting of peroxisomal matrix proteins by Pex5p. Biochim Biophys Acta 1763(12):1794–1800

Wang D, Visser NV, Veenhuis M, van der Klei IJ (2003) Physical interactions of the peroxisomal targeting signal 1 receptor pex5p, studied by fluorescence correlation spectroscopy. J Biol Chem 278(44):43340–43345

Williams C, van den Berg M, Distel B (2005) Saccharomyces cere-visiae Pex14p contains two independent Pex5p binding sites, which are both essential for PTS1 protein import. FEBS Lett 579(16):3416–3420

Williams C, van den Berg M, Sprenger RR, Distel B (2007) A con-served cysteine is essential for Pex4p-dependent ubiquitination of the peroxisomal import receptor Pex5p. J Biol Chem 282 (31):22534–22543

Wolf J, Schliebs W, Erdmann R (2010) Peroxisomes as dynamic or-ganelles: peroxisomal matrix protein import. FEBS J 277(16):3268–3278

Wolfe KH, Shields DC (1997) Molecular evidence for an ancient duplication of the entire yeast genome. Nature 387(6634):708–713

Yang X, Purdue PE, Lazarow PB (2001) Eci1p uses a PTS1 to enter peroxisomes: either its own or that of a partner, Dci1p. Eur J Cell Biol 80(2):126–138

Zhang JW, Lazarow PB (1996) PEB1(PAS7) is an intraperoxisomal receptor for the NH2-terminal, Type 2, peroxisomal targeting sequence of thiolase: Peb1p itself is targeted to peroxisomes by an NH2-terminal peptide. J Cell Biol 132(3):325–334

Zhang L, Leon S, Subramani S (2006) Two independent pathways traffic the intraperoxisomal peroxin PpPex8p into peroxisomes: mechanism and evolutionary implications. Mol Biol Cell 17(2):690–699

Protein Transport In and Out of Plant Peroxisomes

14

Bonnie Bartel, Sarah E. Burkhart, and Wendell A. Fleming

Abstract

Plant peroxisomes house conserved functions such as β-oxidation and hydrogen peroxide decomposition along with specialized tasks including hormone metabolism and photorespiration. Phenotypes stemming from defects in these pathways have been exploited to isolate and characterize peroxisome-defective mutants in the reference plant *Arabidopsis thaliana*. Because peroxisome function is essential for plant viability, partial loss-of-function alleles have been frequently recovered from forward genetic screens. Analysis of these mutants has revealed the broad outlines of matrix protein import in plants and that these processes may be more similar between plants and mammals than between plants and yeast. Here we review matrix protein import into plant peroxisomes and the emerging understanding of how these matrix proteins may be degraded when they are damaged or no longer needed.

Keywords

Arabidopsis thaliana • organelle biogenesis • peroxin • peroxisome • protein targeting

14.1 Functions of Plant Peroxisomes

Many metabolic pathways sequestered in peroxisomes produce hydrogen peroxide (H_2O_2), which can damage cellular constituents. To prevent this damage, plant peroxisomes detoxify H_2O_2 using systems that include catalase in the matrix (Fig. 1; reviewed in Mhamdi et al. 2012) and an ascorbate-dependent system at the membrane (Lisenbee et al. 2005; Eastmond 2007). One oxidative process

B. Bartel (✉) • S.E. Burkhart • W.A. Fleming
Department of Biochemistry and Cell Biology, Rice University, 6100 Main Street, Houston, TX 77005, USA
e-mail: bartel@rice.edu

C. Brocard and A. Hartig (eds.), *Molecular Machines Involved in Peroxisome Biogenesis and Maintenance*, DOI 10.1007/978-3-7091-1788-0_14, © Springer-Verlag Wien 2014

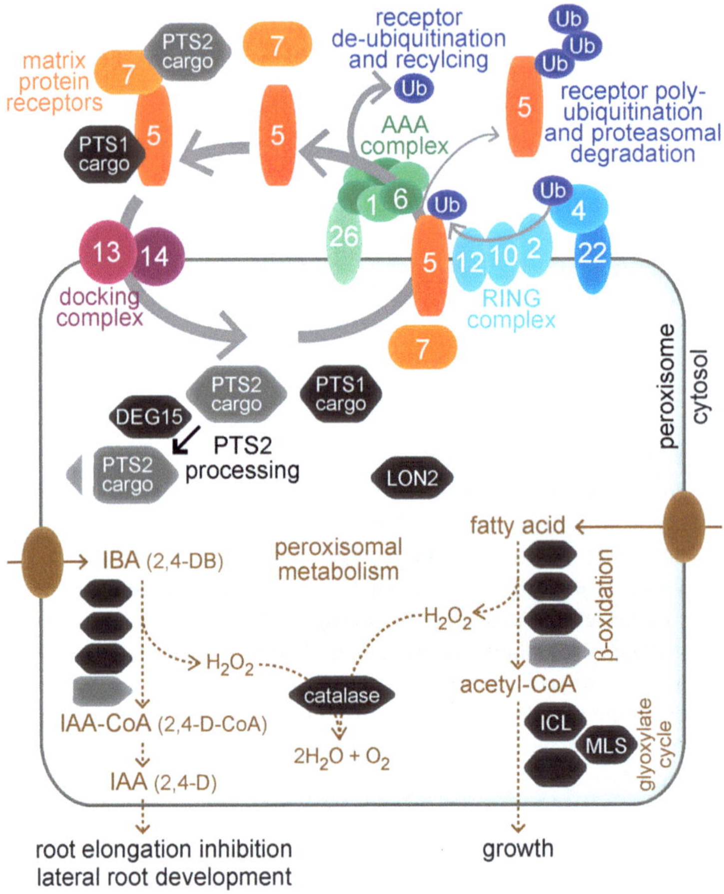

Fig. 14.1 *Plant peroxisome metabolism and peroxins implicated in matrix protein import.* Proteins destined for the peroxisomal matrix contain an N- or C-terminal peroxisome-targeting signal (PTS). Various peroxins (numbered ovals) function to bring these proteins into the peroxisome or to return the matrix protein receptors (PEX5 and PEX7) back to the cytosol. PEX5 is degraded when not efficiently recycled, as when the AAA ATPase PEX6 is mutated. Once in the peroxisome, the PTS2-containing region is cleaved by the protease DEG15. Among the pathways housed in plant peroxisomes are H_2O_2-inactivating systems (e.g., catalase), the glyoxylate cycle enzymes isocitrate lyase (ICL) and malate synthase (MLS), and various oxidative pathways, including those that β-oxidize IBA and fatty acids

housed in peroxisomes is fatty acid β-oxidation. Unlike in mammals, where peroxisomes divide β-oxidation tasks with mitochondria, fatty acid β-oxidation is solely peroxisomal in plants (reviewed in Graham and Eastmond 2002). In young seedlings, glyoxylate cycle enzymes process acetyl-CoA freed by fatty acid β-oxidation, which allows sucrose production prior to the onset of photosynthesis (Fig. 14.1). As triacylglycerol stores are depleted and seedlings develop photosynthetic capability, glyoxylate cycle enzymes are degraded and peroxisomes acquire several photorespiration enzymes. Photorespiration is a salvage pathway needed when chloroplastic RuBisCO fixes O_2 instead of CO_2. Photorespiration enzymes are present in the chloroplasts, mitochondria, and peroxisomes, requiring shuttling of intermediates among these organelles.

Plant peroxisomes also β-oxidize precursors of the critical hormones auxin and jasmonic acid. Auxin influences apical dominance, tropisms, shoot and root elongation, and root branching (reviewed in Woodward and Bartel 2005b); jasmonic acid functions in plant defense and reproductive development (reviewed in Hu et al. 2012). Indole-3-butyric acid (IBA) is β-oxidized to active auxin, indole-3-acetic acid (IAA), in peroxisomes (Zolman et al. 2000; Strader et al. 2010). Some peroxisomal enzymes may be dedicated to IBA β-oxidation, whereas others also act in fatty acid β-oxidation (reviewed in Strader and Bartel 2011). Like IBA, the side chain of a jasmonic acid precursor is shortened via β-oxidation in the peroxisome (reviewed in Hu et al. 2012), a necessary step in active jasmonate production. Beyond fatty acid utilization, photorespiration, and hormone metabolism, plant peroxisomes participate in additional diverse metabolic processes (reviewed in Hu et al. 2012).

14.2 Genetic Approaches to Uncovering Plant Peroxins

Various forward genetic screens have exploited the physiological consequences of disrupting peroxisomal β-oxidation to uncover *Arabidopsis* peroxin (*pex*) mutants. For example, screening for mutants resistant to inhibitory effects of IBA on root or hypocotyl (seedling stem) elongation that remain responsive to IAA identified mutants defective in not only a transporter importing IBA (and fatty acids) into the peroxisome (Zolman et al. 2001), fatty acid β-oxidation enzymes (Adham et al. 2005), and IBA β-oxidation enzymes (Zolman et al. 2007, 2008; Strader et al. 2011) but also several peroxins (Zolman et al. 2000, 2005; Monroe-Augustus 2004; Zolman and Bartel 2004; Ramón and Bartel 2010). A subset of these *IBA-response* (*ibr*) mutants also require an exogenous fixed carbon source such as sucrose for growth, suggesting reduced β-oxidation of both IBA and fatty acids (Zolman et al. 2000). As expected, *pex* mutants often display both IBA resistance and sucrose dependence (Fig. 14.2a; Tables 14.1 and 14.2), whereas IBA-resistant but sucrose-independent mutants may disrupt enzymes specifically acting in IBA to IAA conversion (Zolman et al. 2007, 2008; Strader et al. 2011). However, sucrose dependence is not a defining criterion of *pex* mutants, as weak *pex* alleles can display IBA resistance without sucrose dependence (Woodward and Bartel 2005a; Burkhart et al. 2013).

Like IBA, 2,4-dichlorophenoxybutyric acid (2,4-DB) is β-oxidized to the synthetic auxin, 2,4-dichlorophenoxyacetic acid (Wain and Wightman 1954), which activates auxin signaling to inhibit root elongation (reviewed in Woodward and Bartel 2005b). The *peroxisome defective* (*ped*) mutants emerged from screens for resistance to the inhibitory effects of 2,4-DB on root elongation accompanied by sucrose dependence (Hayashi et al. 1998). *ped* mutants are defective in a

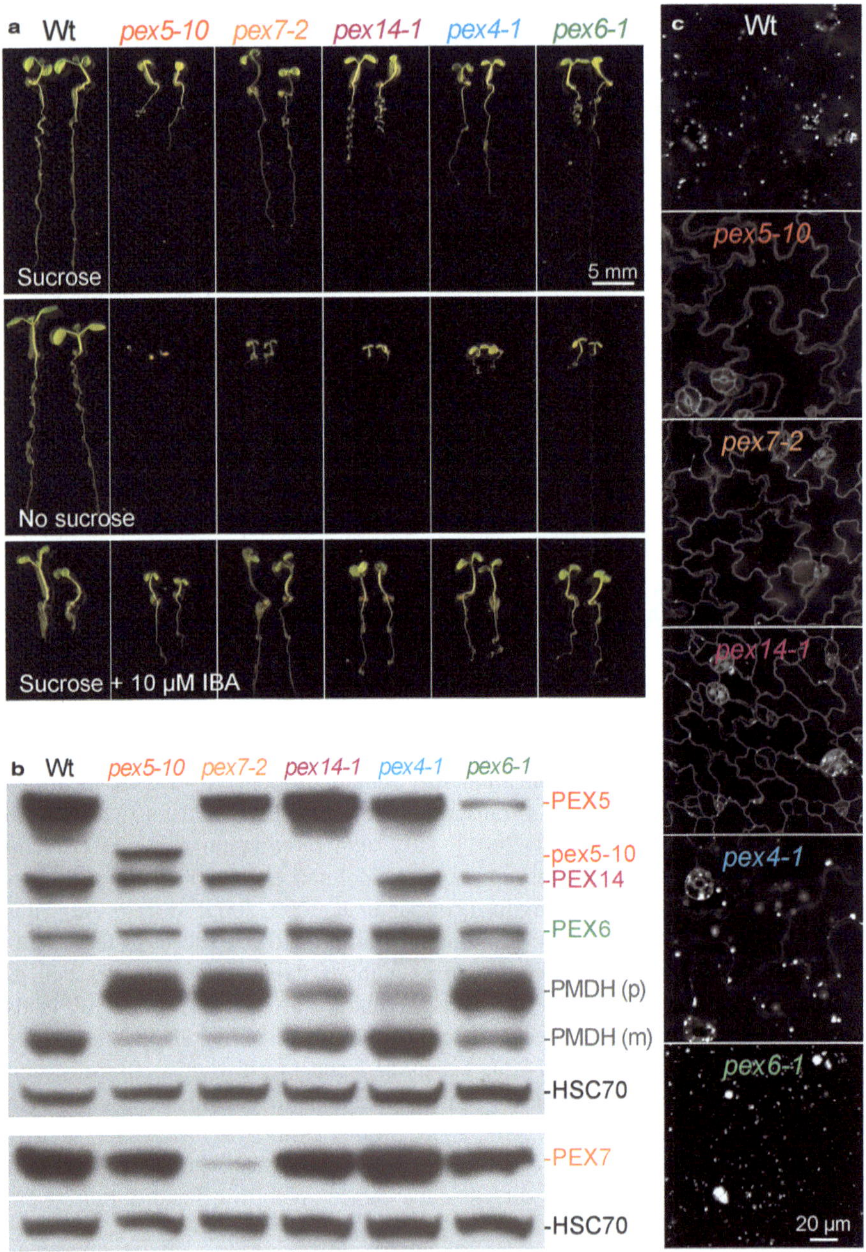

Fig. 14.2 *Phenotypes of typical Arabidopsis peroxin mutants*. (**a**) Plant *pex* mutants often display sucrose dependence and IBA resistance. Seedlings grown for 8 days on the indicated media are shown. Wild type (Wt) does not require sucrose and is sensitive to the inhibitory effects of IBA on root elongation; *pex* mutant growth typically is impaired without sucrose and less inhibited by IBA. (**b**) Plant *pex* mutants often display incomplete PTS2 processing and may show reduced peroxin levels. Extracts from 8-day-old seedlings grown on sucrose-supplemented medium were processed for immunoblotting on duplicate membranes serially probed with antibodies to the indicated proteins in the top four or bottom two panels. PMDH is synthesized as a precursor

peroxisomal fatty acid transporter (Hayashi et al. 2002), a fatty acid β-oxidization enzyme (Hayashi et al. 1998), and a peroxin (Hayashi et al. 2000a).

Screening directly for mutants that are *sugar dependent* (*sdp*) during germination yielded genes encoding several peroxins and fatty acid β-oxidation enzymes (Eastmond 2006). In addition, the *sdp* screen revealed several mutations that would not confer IBA- or 2,4-DB-resistance, including those in glyoxylate cycle enzymes, a H_2O_2 detoxification enzyme, and a lipase involved in oil breakdown (Eastmond 2006, 2007).

Peroxin alleles also emerged from microscopy-based screens for mislocalized peroxisomally targeted GFP derivatives. *aberrant peroxisome morphology* (*apm* or *apem*) mutants have revealed proteins involved in peroxisome division (Mano et al. 2004), metabolite transport (Mano et al. 2011), and several peroxins (Mano et al. 2006; Goto et al. 2011); a *persistent fluorescence* (*pfl*) screen yielded a fatty acid β-oxidation enzyme and several peroxins (Burkhart et al. 2013). Interestingly, microscopy-based approaches have recovered weak *pex* alleles that do not confer sucrose dependence or 2,4-DB resistance (Mano et al. 2006; Goto et al. 2011).

In addition to following peroxisomally targeted GFP derivatives microscopically (Fig. 14.2c), monitoring matrix protein processing via immunoblotting (Fig. 14.2b) provides an indirect measure of protein import and thus peroxin function. Proteins are targeted to the matrix following recognition of a C-terminal type 1 or an N-terminal type 2 peroxisome-targeting signal (PTS). Unlike the noncleaved PTS1 tripeptide present on most peroxisome-targeted proteins (Gould et al. 1988), the N-terminal region that includes the PTS2 nonapeptide is cleaved after cargo import in plants (Gietl et al. 1994) and mammals (Hijikata et al. 1987; Swinkels et al. 1991). In plants, the DEG15 peroxisomal protease performs this cleavage (Helm et al. 2007; Schumann et al. 2008). A *deg15* null allele fails to process PTS2-containing proteins (Helm et al. 2007) and is slightly resistant to IBA (Lingard and Bartel 2009) and 2,4-DB (Schumann et al. 2008). Because DEG15 is targeted to peroxisomes via a PTS1, PTS2 processing requires both PTS1 and PTS2 pathways, and many *pex* mutants display PTS2-processing defects (Fig. 14.2b; Tables 14.1 and 14.2).

Fig. 14.2 (continued) (p) containing the PTS2 signal that is processed to a mature (m) protein in the peroxisome. *pex* mutants typically display PMDH processing defects, suggesting inefficient protein import. *pex5-10* accumulates reduced levels of a truncated PEX5 variant and *pex14-1* lacks detectable full-length PEX14 protein. *pex6-1* has reduced PEX5 levels, supporting the hypothesis that PEX5 is degraded rather than recycled in this mutant. *pex7-2* has low levels of not only PEX7 but also PEX5. HSC70 is a loading control. (**c**). Plant *pex* mutants display varying extents of cytosolic mislocalization of peroxisomally-targeted GFP reporters. Cotyledon epidermal cells from 8-day-old light-grown wild-type (Wt) and *pex* seedlings constitutively expressing *PTS2-GFP* were imaged for GFP fluorescence using confocal microscopy with identical settings. Wt seedlings display peroxisomal fluorescence (puncta) whereas *pex* mutants often display fluorescence in the cytosol, which is concentrated around the cell periphery by the large central vacuole. Because *Arabidopsis* mutants completely lacking peroxisome function are not viable, even severe *pex* mutants retain some puncta

Table 14.1 *Arabidopsis* mutants defective in receptor and docking peroxins

Mutant	Mutation (mutagen)	Phenotypes[a]	References
PEX5 (At5g56290)—PTS1 receptor			
pex5-1	Missense (EMS)	IBA^R, suc^D, reduced PTS2 import, reduced PTS2 processing	Zolman et al. (2000), Woodward and Bartel (2005a)
pex5-10	Insertion in coding sequence (T-DNA)	IBA^R, suc^D, reduced PTS1 and PTS2 import, reduced PTS2 processing	Zolman et al. (2005), Khan and Zolman (2010)
pex5i	Reduced function (RNAi)	2,4-DB^R, suc^D, reduced PTS1 and PTS2 import	Hayashi et al. (2005)
PEX7 (At1g29260)—PTS2 receptor			
pex7-1	Insertion 5′ of ATG (T-DNA)	IBA^R, *pex5* enhancer, reduced PTS1 and PTS2 import, reduced PEX5, reduced PTS2 processing	Woodward and Bartel (2005a), Ramón and Bartel (2010)
pex7-2	Missense (EMS)	IBA^R, suc^D, *pex5* enhancer, reduced PTS1 and PTS2 import, reduced PEX5, reduced PTS2 processing	Ramón and Bartel (2010)
pex7i	Reduced function (RNAi)	2,4-DB^R, suc^D, reduced PTS2 import	Hayashi et al. (2005)
PEX13 (At3g07560)—membrane peroxin, PEX5/PEX7 docking			
pex13-1	Insertion 5′ of ATG (T-DNA)	*pex* modifier: enhancer of *pex5* and *pex14*; suppressor of *pex4* and *pex6*	Ratzel et al. (2011)
abstinence by mutual consent (amc)	Insertion in coding sequence (T-DNA)	Lethal; reduced PTS1 import in pollen	Boisson-Dernier et al. (2008)
aberrant peroxisome morphology2 (apm2)	Nonsense in last exon (EMS)	2,4-DB^R, slightly reduced PTS1 and PTS2 import, *apm4 (pex12)* enhancer, reduced PTS2 processing	Mano et al. (2006)
pex13i	Reduced function (RNAi)	2,4-DB^R, suc^D, reduced PTS1 and PTS2 import	Nito et al. (2007)
pex13-4	Missense (EMS)	IBA^R, suc^D, reduced PTS1 and PTS2 import	Woodward et al. (2014)
PEX14 (At5g62810)—membrane peroxin, PEX5/PEX7 docking			
peroxisome defective2 (ped2)	Nonsense (EMS)	2,4-DB^R, suc^D, reduced PTS1 and PTS2 import, absent PEX14 protein, reduced PTS2 processing	Hayashi et al. (2000a)
pex14-1, *pex14-4*	Nonsense; splicing (EMS)	IBA^R, suc^D, reduced PTS1 and PTS2 import (*pex14-1*), absent or altered PEX14 protein, reduced PTS2 processing	Monroe-Augustus et al. (2011)

(continued)

Table 14.1 (continued)

Mutant	Mutation (mutagen)	Phenotypes[a]	References
pex14-2, *pex14-3*	Insertion in coding sequence (T-DNA)	IBA^R, suc^D, reduced PTS1 import (*pex14-2*), absent or altered PEX14 protein, reduced PTS2 processing	Monroe-Augustus et al. (2011)
pex14-5	Nonsense (EMS)	IBA^R, suc^D, reduced PTS1 import, absent PEX14 protein, reduced PTS2 processing	Burkhart et al. (2013)
pex14-6	Splicing (EMS)	IBA^R, reduced PTS1 import, reduced PEX14, reduced PTS2 processing	Burkhart et al. (2013)

[a]IBA^R, IBA resistant; 2,4-DB^R, 2,4-DB resistant; suc^D, sucrose dependent

Reverse genetic techniques such as gene silencing using RNA interference (RNAi; Sharp 1999) and gene disruption via transfer-DNA (T-DNA) insertions (Krysan et al. 1999; Alonso et al. 2003) allow investigation of the importance of putative plant peroxins. Null alleles with insertions in peroxin coding sequences often confer lethality (Hu et al. 2002; Schumann et al. 2003; Fan et al. 2005; Boisson-Dernier et al. 2008), but mutants with T-DNAs inserted upstream of coding regions can be informative (Woodward and Bartel 2005a; Zolman et al. 2005; Ratzel et al. 2011), and a few *pex* mutants harboring T-DNAs within coding sequences are viable (Khan and Zolman 2010; Monroe-Augustus et al. 2011). In addition, partially reducing function by RNAi avoids complications caused by lethality; RNAi lines have been used to characterize most of the well-conserved *Arabidopsis* peroxins (Fan et al. 2005; Hayashi et al. 2005; Nito et al. 2007).

14.3 PEX5 and PEX7: Interdependent Matrix Protein Receptors

Most peroxins are involved in importing proteins destined for the peroxisomal matrix (Fig. 14.1). PEX5 and PEX7 are receptors for PTS1- and PTS2-containing proteins, respectively (reviewed in Hu et al. 2012). *Arabidopsis* PEX7 interacts with PEX5 (Nito et al. 2002) and requires PEX5 to deliver PTS2 cargo into peroxisomes (Hayashi et al. 2005; Woodward and Bartel 2005a). Moreover, *in vitro* import of PTS2 cargo into plant peroxisomes is enhanced by addition of PTS1 cargo (Johnson and Olsen 2003).

The PEX5 N-terminal region contains a PEX7-binding domain and pentapeptide repeats that aid in docking with peroxisomal membrane peroxins (Nito et al. 2002); the C-terminal tetratricopeptide repeats recognize the PTS1 (Gatto et al. 2000). Null alleles of *PEX5* have not been reported, but reducing *PEX5* levels by RNAi confers resistance to 2,4-DB root elongation inhibition, dependence on sucrose for normal development, and reduced PTS1 and PTS2 import (Hayashi et al. 2005). Similarly, the *pex5-10* insertion allele (Fig. 14.2), which accumulates reduced levels of a

Table 14.2 *Arabidopsis* mutants defective in receptor-recycling peroxins

Mutant	Mutation (mutagen)	Phenotypes[a]	References
PEX4 (At5g25760)—ubiquitin-conjugating enzyme			
pex4-1	Missense (EMS)	IBA^R, suc^D, reduced PTS2 processing	Zolman et al. (2005)
pex4i	Reduced function (RNAi)	2,4-DB^R, suc^D, reduced PTS1 and PTS2 import	Nito et al. (2007)
PEX22 (At3g21865)—membrane peroxin tethering PEX4			
pex22-1	Insertion 5′ of ATG (T-DNA)	Enhancer of *pex4-1*	Zolman et al. (2005)
PEX2 (At1g79810)—membrane peroxin, RING-finger Ub-protein ligase			
ted3	Missense (EMS)	Dominant suppressor of *det1-1*	Hu et al. (2002)
pex2-T-DNA	Insertion in coding sequence (T-DNA)	Lethal	Hu et al. (2002)
pex2i	Reduced function (RNAi)	2,4-DB^R, suc^D, reduced PTS1 and PTS2 import	Nito et al. (2007)
PEX10 (At2g26350)—membrane peroxin, RING-finger ubiquitin-protein ligase			
pex10-1	Insertion in coding sequence (T-DNA)	Lethal	Schumann et al. (2003), Sparkes et al. (2003)
pex10i	Reduced function (RNAi)	2,4-DB^R, suc^D, reduced PTS1 and PTS2 import, reduced growth, variegated leaves, reduced fertility	Nito et al. (2007)
PEX12 (At3g04460)—membrane peroxin, RING-finger ubiquitin-protein ligase			
aberrant peroxisome morphology4 (apm4)	Missense (EMS)	2,4-DB^R, suc^D, *apm2 (pex13)* enhancer, reduced PTS1 and PTS2 import, reduced PTS2 processing	Mano et al. (2006)
pex12-T-DNA	Insertion in coding sequence (T-DNA)	Lethal	Fan et al. (2005)
pex12i	Reduced function (RNAi)	IBA^R, 2,4-DB^R, suc^D, reduced PTS1 and PTS2 import	Fan et al. (2005), Nito et al. (2007)
PEX1 (At5g08470)—AAA-family ATPase			
pex1i	Reduced function (RNAi)	2,4-DB^R, suc^D, reduced PTS1 and PTS2 import	Nito et al. (2007)
PEX6 (At1g03000)—AAA-family ATPase			
pex6-1	Missense (EMS)	IBA^R, suc^D, low PEX5, reduced PTS2 processing	Zolman and Bartel (2004), Burkhart et al. (2013)

(continued)

Table 14.2 (continued)

Mutant	Mutation (mutagen)	Phenotypes[a]	References
pex6-2	Missense (EMS)	IBA^R, normal PEX5, reduced PTS2 processing	Burkhart et al. (2013)
pex6i	Reduced function (RNAi)	2,4-DB^R, suc^D, reduced PTS1 and PTS2 import	Nito et al. (2007)
PEX26/PEX15/APEM9 (*At3g10572*)—membrane peroxin tethering PEX1/PEX6			
apem9-1	Missense (EMS)	Reduced PTS1 and PTS2 import	Goto et al. (2011)
apem9-2, *apem9-3*	Insertion in coding sequence (T-DNA)	Lethal	Goto et al. (2011)
apem9i	Reduced function (RNAi)	2,4-DB^R, suc^D, reduced PTS1 and PTS2 import, reduced PTS2 processing	Goto et al. (2011)

[a]IBA^R, IBA resistant; 2,4-DB^R, 2,4-DB resistant; suc^D, sucrose dependent

truncated pex5 protein, displays PTS1- and PTS2-import defects, IBA resistance, and sucrose dependence (Zolman et al. 2005; Khan and Zolman 2010; Ramón and Bartel 2010). The *pex5-1* missense allele was isolated from an IBA-resistant root elongation screen and is sucrose dependent (Zolman et al. 2000). *pex5-1* displays defects in importing PTS2-GFP but not GFP-PTS1, suggesting that the lesion disrupts PEX5–PEX7 association (Woodward and Bartel 2005a). In addition, PTS2 (but not PTS1) import can be restored in *pex5-10* by overexpressing the PEX5 N-terminal region that includes the pentapeptide repeats and the PEX7-binding domain (Khan and Zolman 2010). These results indicate that PTS1 and PTS2 import both require PEX5 function in *Arabidopsis*.

The *Arabidopsis pex5-1* missense allele (Zolman et al. 2000) alters an amino acid analogous to one altered in a mammalian *PEX5* mutant that also disrupts PTS2 but not PTS1 import (Matsumura et al. 2000). In humans, this region is part of a "long" PEX5 isoform. Alternative splicing of human *PEX5* results in a "short" PEX5 isoform that is more similar to yeast Pex5p and does not bind PEX7 (Braverman et al. 1998; Otera et al. 1998, 2000). Although only "long" *PEX5* transcripts are detected in *Arabidopsis* (Zolman et al. 2000; Lee et al. 2006), rice *PEX5* (like mammalian *PEX5*) has two splice variants (Lee et al. 2006). Whereas expressing the short rice PEX5 restores PTS1 import to *Arabidopsis pex5-10* protoplasts, only the long isoform binds PEX7 and restores PTS2 import to *pex5-10* (Lee et al. 2006). Because yeast have only "short" Pex5p the additional functions of the longer isoform are performed by the Pex18p and Pex21p co-receptors, which are essential for PTS2 import (Purdue et al. 1998) and contain a Pex7p-binding region similar to the long Pex5p isoform (Dodt et al. 2001). It is intriguing that plant PTS2 import is more similar to the mammalian system than is the yeast co-receptor system or the metazoans *Caenorhabditis elegans* (Gurvitz et al. 2000; Motley et al. 2000) and *Drosophila melanogaster* (Faust et al. 2012), which appear to lack PTS2 proteins entirely.

PEX7 is a soluble protein comprised of WD repeats (Woodward and Bartel 2005a). Null alleles of *PEX7* have not been reported in plants, but *Arabidopsis PEX7* RNAi lines display reduced PTS2 import and are 2,4-DB resistant and sucrose dependent (Hayashi et al. 2005). Two *Arabidopsis pex7* alleles have been characterized. The *pex7-1* mutant harbors a T-DNA upstream of the *PEX7* start codon (Woodward and Bartel 2005a) and *pex7-2* (Fig. 14.2) emerged from an *ibr* screen (Ramón and Bartel 2010). Both alleles display IBA resistance and reduced PEX7 protein levels, but only *pex7-2* is sucrose dependent (Woodward and Bartel 2005a; Ramón and Bartel 2010). *pex7* mutants display not only the expected reduced import of PTS2-GFP (Woodward and Bartel 2005a) but also a surprising reduction in GFP-PTS1 import (Ramón and Bartel 2010). Reduced PEX5 levels accompany this import defect (Ramón and Bartel 2010), demonstrating that PTS1- and PTS2-import pathways are interdependent in *Arabidopsis*.

14.4 PEX13 and PEX14: Receptor Docking

After cargo binding, the PEX5–PEX7 complex docks with the peroxisomal membrane proteins PEX13 and PEX14 (Fig. 14.1). The C-terminal domain of PEX13 interacts with the N-terminal region of PEX7 but not PEX5 (Mano et al. 2006), whereas the N-terminal region of PEX14 binds the WXXXF motifs in the N-terminal region of PEX5 (Nito et al. 2002). Although mammalian and yeast Pex14p directly interact with Pex7p (Albertini et al. 1997; Shimizu et al. 1999), *Arabidopsis* PEX14 does not appear to interact with PEX7 in yeast two-hybrid assays (Nito et al. 2002).

An allelic series of *Arabidopsis pex13* mutants has been characterized. A *pex13* T-DNA allele, *abstinence by mutual consent* (*amc*), is defective in paternal–maternal gametophyte recognition, resulting in lethality (Boisson-Dernier et al. 2008). Reducing *PEX13* levels by RNAi confers 2,4-DB resistance, sucrose dependence, and impaired PTS1 and PTS2 import (Nito et al. 2007). The *pex13-4* missense allele displays similarly severe peroxisomal defects (Woodward et al. 2014). The *apm2* nonsense mutation in the last exon of *PEX13* emerged from a GFP-PTS1 mislocalization screen and displays slightly reduced PTS1 and PTS2 import along with slight 2,4-DB resistance (Mano et al. 2006). The weakest reported *pex13* allele is caused by a T-DNA inserted in the *PEX13* 5′-UTR. This *pex13-1* mutation does not cause notable physiological or molecular defects, but enhances the physiological and molecular defects in *pex5* and *pex14* mutants (Ratzel et al. 2011), indicating that peroxisome function is impaired.

PEX14 is the *Arabidopsis* peroxin most frequently recovered from forward genetic screens for peroxisome defects, perhaps because *PEX14* is the only reported

Arabidopsis PEX gene for which null alleles are viable (Tables 14.1 and 14.2). For example, the *ped2* mutant isolated via 2,4-DB resistance contains a premature stop codon and lacks detectable PEX14 protein (Hayashi et al. 2000a). *ped2* is sucrose dependent and displays photorespiration defects and partially impaired PTS1 and PTS2 import (Hayashi et al. 2000a). In addition, a *pex14* nonsense allele was recovered from an *ibr* screen (Fig. 14.2) (Monroe-Augustus et al. 2011) and two *pex14* alleles emerged from a matrix protein stabilization screen (Burkhart et al. 2013). Although PEX14 clearly contributes to import efficiency, analysis of multiple *pex14* nonsense, splicing, and insertion alleles (Table 14.1) suggests that PEX14 is not absolutely required for matrix protein import in *Arabidopsis* (Monroe-Augustus et al. 2011). PEX14 dispensability contrasts with PEX13, which is essential for *Arabidopsis* viability (Boisson-Dernier et al. 2008), and is surprising because yeast Pex5p appears to deliver cargo through a transient pore comprised of Pex5p and Pex14p (Meinecke et al. 2010). The interdependence of PEX5 and PEX7 in plants (Ramón and Bartel 2010), coupled with the ability of *Arabidopsis* PEX7 to directly bind PEX13 (Mano et al. 2006), suggests that *Arabidopsis* PEX13 may have acquired partial docking responsibility carried by PEX14 in other systems.

14.5 PEX5 Recycling: Ubiquitination and Retrotranslocation

After cargo delivery, PEX5 is ubiquitinated, removed from the peroxisomal membrane, de-ubiquitinated, and used in further import cycles. PEX4 is an ubiquitin-conjugating (UBC) enzyme tethered to the peroxisome by PEX22 (Fig. 14.1). In yeast, Pex4p is necessary for Pex5p monoubiquitination by the Pex12p ubiquitin-protein ligase as it exits the peroxisome (Platta et al. 2009). When Pex4p is absent, yeast Pex5p can be polyubiquitinated by Pex2p in a Ubc4-dependent manner and targeted to the proteasome for degradation (Thoms and Erdmann 2006; Platta et al. 2007). Although null alleles of *PEX4* or *PEX22* have not been described in plants, a *PEX4* RNAi line displays typical peroxisome defects: 2,4-DB resistance, sucrose dependence, and PTS1- and PTS2-import defects (Nito et al. 2007). *pex4-1*, an *Arabidopsis* missense allele that emerged from an *ibr* screen, displays sucrose dependence and slight PTS2-processing defects (Fig. 14.2; Zolman et al. 2005). *pex22-1*, which carries a T-DNA insertion upstream of the *PEX22* start codon, does not display notable peroxisome defects on its own but enhances *pex4-1* defects (Zolman et al. 2005). Although PEX5 levels are normal in *Arabidopsis pex4-1* and *pex4-1 pex22-1* mutants (Zolman et al. 2005), a greater fraction of PEX5 is organelle-associated in *pex4-1* than in wild type (Ratzel et al. 2011), consistent with a role for PEX4 in ubiquitin-dependent PEX5 recycling. The obvious IBA resistance and sucrose dependence of *pex4-1* appears disproportionate to the slight defect in matrix protein import of this mutant (Fig. 14.2c), suggesting that PEX4 may have substrates in addition to PEX5 or that PEX5 stuck in the peroxisomal membrane impairs peroxisomal function.

Peroxisomal matrix protein import in yeast and humans requires the RING finger ubiquitin-protein ligases Pex2p, Pex10p, and Pex12p, which assist in Pex5p ubiquitination (reviewed in Schliebs et al. 2010). *Arabidopsis* PEX2, PEX10 and PEX12 display zinc-dependent monoubiquitination activity *in vitro*, and the RING-finger regions of PEX2 and PEX12 interact with DSK2a and DSK2b, ubiquitin-binding proteins implicated in shuttling proteins to the proteasome (Kaur et al. 2013). However, in vivo targets of RING-finger peroxin ubiquitination have not been verified in plants.

Null alleles of the *Arabidopsis* RING-finger peroxin genes confer embryo lethality (Hu et al. 2002; Schumann et al. 2003; Sparkes et al. 2003; Fan et al. 2005; Prestele et al. 2010), implying that peroxisome function is essential for embryogenesis. Targeting *PEX2*, *PEX10*, or *PEX12* by RNAi confers 2,4-DB resistance, sucrose dependence, and matrix protein import defects (Fan et al. 2005; Nito et al. 2007). A weak *pex12* missense allele, *apm4*, was isolated from a GFP-PTS1 mislocalization screen and displays PTS2-processing defects, developmental delays, and 2,4-DB resistance (Mano et al. 2006). Yeast RING-finger peroxins form a complex that requires all three members for stability (Agne et al. 2003). Because PEX2, PEX10, and PEX12 protein levels were not monitored in the plant studies, it is not known whether defects result directly from loss of the targeted peroxin or from secondary effects of destabilizing an interacting RING-finger peroxin.

In an effort to assign specific functions to individual RING-finger peroxins, versions of PEX2, PEX10, or PEX12 carrying missense mutations in the RING-finger domain (ΔZn) were expressed in wild-type *Arabidopsis* (Schumann et al. 2007; Prestele et al. 2010). Plants expressing *PEX12-ΔZn* resemble wild type. Expressing *PEX2-ΔZn* reduces GFP-PTS1 import into peroxisomes (Prestele et al. 2010), consistent with RNAi results indicating that *PEX2* is required for matrix protein import (Nito et al. 2007). Plants expressing *PEX10-ΔZn* import GFP-PTS1 normally and are not IBA resistant or sucrose dependent but display photorespiration defects, abnormal peroxisome size and shape, and reduced peroxisome-chloroplast association (Schumann et al. 2007; Prestele et al. 2010), suggesting a plant-specific role for PEX10 in mediating inter-organellar interactions.

PEX2 also may have plant-specific roles. The *ted3* allele of *pex2* was recovered as a suppressor of the *de-etiolated1* (*det1*) mutant, which displays photomorphogenesis defects (Hu et al. 2002). The *ted3* missense mutation alters a residue adjacent to the PEX2 RING-finger domain. Because the *ted3* allele is dominant, however, it is not known whether *det1* suppression reflects a normal peroxisomal function in photomorphogenesis or whether this allele acquired a new function that ameliorates *det1* defects. In either case, the *ted3* mutant does not display typical peroxisome-defective phenotypes (Hu et al. 2002).

Pex1p and Pex6p are interacting AAA ATPases that, in yeast, are needed to retrotranslocate ubiquitinated Pex5p from the peroxisomal membrane back to the cytosol (Thoms and Erdmann 2006). Null alleles of these peroxins have not been reported in plants, but reducing *PEX1* or *PEX6* by RNAi in *Arabidopsis* confers

2,4-DB resistance, sucrose dependence, and PTS1 and PTS2 import defects (Nito et al. 2007). Although *pex1* alleles have not been reported from forward-genetic screens, *pex6* mutants have been recovered (Zolman and Bartel 2004; Burkhart et al. 2013). The *pex6-1* missense mutant (Fig. 14.2) was isolated in an *ibr* screen and also displays sucrose dependence, defects in processing PTS2 proteins, and reduced pigmentation typical of photorespiration defects (Zolman and Bartel 2004). PEX5 levels are reduced in *pex6-1*, suggesting that PEX5 is not efficiently retrotranslocated and remains in the peroxisomal membrane until degraded (Zolman and Bartel 2004). Indeed, PEX5 in *pex6-1* is more organelle-associated rather than distributed between the organelle and cytosol (Ratzel et al. 2011), and *PEX5* overexpression partially rescues *pex6-1* defects (Zolman and Bartel 2004). Moreover, *pex4-1* restores PEX5 levels to the *pex6-1* mutant (Ratzel et al. 2011), implying that reduced PEX5 levels in *pex6-1* result from ubiquitin-dependent degradation. A weaker missense allele, *pex6-2*, displays some IBA and 2,4-DB resistance but is sucrose independent and maintains normal PEX5 levels (Burkhart et al. 2013).

Interestingly, reduced *PEX13* expression in the *pex13-1* mutant partially suppresses the sucrose dependence of both *pex4-1* and *pex6-1* (Ratzel et al. 2011). This suppression suggests that failure to efficiently recycle PEX5 in *pex4-1* and *pex6-1* mutants can be offset by reducing PEX5 docking at the peroxisome and supports the notion that PEX5 lingering in the peroxisomal membrane impedes peroxisomal metabolism.

PEX26 (in mammals) and Pex15p (in yeast) are membrane anchors that recruit the Pex1p–Pex6p complex to the peroxisome (reviewed in Fujiki et al. 2012; Grimm et al. 2012). Like the PEX22 membrane anchor for PEX4 (Zolman et al. 2005), PEX26/Pex15p homologs are not apparent in plant genomes. However, the *apem9-1* missense allele identified in a GFP-PTS1 mislocalization screen is defective in the *Arabidopsis* PEX26/Pex15p functional equivalent (Goto et al. 2011). APEM9 binds PEX6 and recruits PEX6 and PEX1 to peroxisomes (Goto et al. 2011). Although the mild *apem9-1* allele does not display PTS2-processing defects, *APEM9* RNAi lines have more severe defects, and *apem9* null mutations confer embryo lethality (Goto et al. 2011).

14.6 Do Peroxins Function in Matrix Protein Degradation?

In addition to roles in matrix protein import, plant peroxins may assist in the disposal of matrix proteins that are damaged or no longer needed. The glyoxylate cycle enzymes isocitrate lyase and malate synthase are degraded shortly after germination as seedlings begin deriving fixed carbon from photosynthesis instead of fatty acids (Hayashi et al. 2000b). These proteins thus provide model substrates to study peroxisomal matrix protein degradation in plants (Lingard et al. 2009; Burkhart et al. 2013). Peroxisomal import is required for efficient degradation. For example, glyoxylate cycle enzymes are stabilized in *pex5-10* and *pex14* mutants

(Lingard et al. 2009; Burkhart et al. 2013), which have marked matrix protein import defects (Khan and Zolman 2010; Monroe-Augustus et al. 2011).

Endoplasmic reticulum-associated protein degradation (ERAD) is the ubiquitin-dependent process exporting misfolded ER proteins to the 26S proteasome for degradation (reviewed in Hampton and Sommer 2012). The similarity of the Pex5p-retrotranslocating peroxins to ERAD components (Schluter et al. 2006; Schliebs et al. 2010) suggests a mechanism for peroxisomal matrix protein degradation in which obsolete or damaged matrix proteins are retrotranslocated from the peroxisome for proteasomal degradation (Lingard et al. 2009; Burkhart et al. 2013). This idea is supported by the stabilization of glyoxylate cycle enzymes in *pex4-1 pex22-1* and *pex6-2* mutants (Zolman et al. 2005; Lingard et al. 2009; Burkhart et al. 2013). Because PEX4 and PEX22 function in ubiquitination and PEX6 is implicated in peroxisomal protein export (Fig. 14.1), glyoxylate cycle enzyme stabilization in these mutants suggests that damaged or unnecessary proteins might be ubiquitinated and sent to the proteasome for degradation.

Alternatively or in addition, matrix proteins may be degraded by resident peroxisomal proteases. LON2 is one of several proteases in *Arabidopsis* peroxisomes (Ostersetzer et al. 2007; Reumann et al. 2007; Eubel et al. 2008; Lingard and Bartel 2009). *lon2* mutants are IBA resistant, exhibit age-dependent PTS1 and PTS2 import defects, but do not appear to degrade matrix proteins more slowly than wild type (Lingard and Bartel 2009; Burkhart et al. 2013). A *lon2* suppressor screen revealed that mutants defective in autophagy (*atg2*, *atg3*, and *atg7*) can restore peroxisome function to *lon2* (Farmer et al. 2013). The observation that blocking autophagy ameliorates *lon2* defects suggests that targeting of peroxisomes for autophagy is accelerated when LON2 is deficient, slowing matrix protein import and conferring physiological defects. Interestingly, *lon2 atg* double mutants markedly stabilize glyoxylate cycle enzymes compared to wild type or the single mutants (Farmer et al. 2013), suggesting a LON2 role in matrix protein degradation that is masked by excessive autophagy in *lon2* single mutants.

14.7 Open Questions and Future Directions

As in other organisms, plant peroxisomes can grow and divide by fission, and pre-peroxisomes can bud *de novo* from the ER and mature via fusion to form functional peroxisomes (reviewed in Hu et al. 2012). Determining the importance of *de novo* biogenesis versus growth and division in different tissues, during various developmental stages, and in response to environmental challenges are important areas for future investigation.

Compared to peroxins involved in peroxisomal matrix protein import, we know less about the earliest steps in peroxisome biogenesis from the ER in plants. In mammals and certain yeasts, three peroxins are key to this biogenesis: Pex3p, Pex16p, and Pex19p (Heiland and Erdmann 2005; Toro et al. 2009). Membrane ER-localized Pex16p recruits Pex3p; Pex3p docks Pex19p; and Pex19p recruits peroxisomal membrane proteins to the ER or pre-peroxisomes (Heiland and

Erdmann 2005; Toro et al. 2009). *PEX3*, *PEX16*, and *PEX19* homologs are found in plants, but the corresponding mutants have not emerged from forward-genetic screens, perhaps because *Arabidopsis PEX3* and *PEX19* are duplicated. A *pex16* mutant, *shrunken seed1* (*sse1*), was discovered in a transgenic line due to its seed phenotype (Lin et al. 1999). *sse1* seeds store starch instead of the normal triacylglycerol storage form (Lin et al. 2006), but the reason this allele displays phenotypes unlike other *Arabidopsis pex* mutants is mysterious. Reducing *PEX16* via RNAi confers enlarged peroxisomes and 2,4-DB resistance, but the seedlings remain sucrose independent and appear to import matrix proteins normally (Nito et al. 2007). Similarly, RNAi-mediated reduction of *PEX3* or *PEX19* results in enlarged peroxisomes that lack obvious matrix protein import defects, but these lines are sucrose independent and show wild-type 2,4-DB sensitivity, suggesting normal β-oxidation (Nito et al. 2007). It remains to be determined if these RNAi lines are too weak to substantially impair peroxisome biogenesis, if to-be-discovered peroxins are needed for pre-peroxisome budding from the ER, or if heightened peroxisome fission can compensate for *de novo* biogenesis defects in plants.

In addition to early peroxisome biogenesis, there are processes about which little is known. Plant PEX5 ubiquitination and deubiquitination is inferred from mutant phenotypes and analogy to other organisms, but ubiquitinated PEX5 derivatives have not been reported in plants, and a PEX5 deubiquitinating enzyme has not been identified. Moreover, the mechanism through which PEX7 returns to the cytosol after cargo delivery is unknown. GFP-PEX7 accumulates in the peroxisomal membrane (Singh et al. 2009) and leads to the degradation of untagged PEX7 (Cui et al. 2013), suggesting that an exposed N-terminus is needed for PEX7 recycling.

Although the frameworks and tools discussed above are in place to understand how plant matrix proteins are imported and degraded, the signals triggering shifts in peroxisome contents during seedling development are not identified, and little is known about how plant peroxisomes are degraded. Peroxisome-specific autophagy (pexophagy) is characterized in yeast and mammals (reviewed in Till et al. 2012). General autophagy is well studied in plants (Li and Vierstra 2012), but yeast pexophagy-specific components are not conserved in plants, and the first reports of plant pexophagy are just emerging (Farmer et al. 2013; Kim et al. 2013; Shibata et al. 2013). It will be interesting to learn whether specific peroxins are used to target plant peroxisomes for pexophagy and how this process is induced and regulated in plants.

Are there more peroxins to be discovered in plants? *Arabidopsis* homologs of several yeast peroxins have not been identified, and forward genetic screens can identify plant peroxins not recognizable by homology (Goto et al. 2011). These screens are not saturated, as numerous *PEX* genes are represented by only one viable allele (Tables 14.1 and 14.2), and mutant alleles of several known peroxin genes (*PEX1*, *PEX2*, *PEX3*, *PEX10*, *PEX11*, *PEX19*, *PEX22*) have not been reported. This deficiency can be explained by duplicated genes in only a few cases (*PEX3*, *PEX11*, *PEX19*). Because essential genes require viable partial loss-of-function alleles displaying an observable phenotype, extensive screening will be required to recover mutants in all *Arabidopsis* peroxins.

What would be an ideal screen for new *Arabidopsis* peroxin mutants? Combining a metabolic screen (IBA or 2,4-DB resistance, sucrose dependence) with PTS2-processing defects is expected to focus a screen on *pex* mutants, as defects in metabolic enzymes seem unlikely to impair PTS2 processing. However, this assumption is contradicted by the observation that disrupting a thiolase acting in β-oxidation impairs processing of a PTS2 protein (Burkhart et al. 2013). The exquisite sensitivity of microscopy-based screens to detect weak alleles (Mano et al. 2004; Goto et al. 2011; Burkhart et al. 2013) suggests that this approach will continue to be fruitful. Moreover, the application of genetic modifier screens starting with existing *pex*-mutants remains unexplored in plants and is likely to yield new insights into peroxisome biogenesis and function.

Acknowledgments We apologize to those whose work we were unable to discuss due to space limitations. We thank Kim Gonzalez, Yun-Ting Kao, Mauro Rinaldi, Andrew Woodward, and Pierce Young for critical comments on the manuscript. Confocal microscopy was performed on equipment obtained through a Shared Instrumentation Grant from the National Institutes of Health (S10RR026399). The authors' research is supported by the National Institutes of Health (R01-GM079177), the National Science Foundation (MCB-0745122 and MCB-1244182), and the Robert A. Welch Foundation (C-1309).

References

Adham AR, Zolman BK, Millius A, Bartel B (2005) Mutations in *Arabidopsis* acyl-CoA oxidase genes reveal distinct and overlapping roles in β-oxidation. Plant J 41:859–874

Agne B, Meindl NM, Niederhoff K, Einwächter H, Rehling P, Sickmann A, Meyer HE, Girzalsky W, Kunau W-H (2003) Pex8p: an intraperoxisomal organizer of the peroxisomal import machinery. Mol Cell 11:635–646

Albertini M, Rehling P, Erdmann R, Girzalsky W, Kiel JAKW, Veenhuis M, Kunau WH (1997) Pex14p, a peroxisomal membrane protein binding both receptors of the two PTS-dependent import pathways. Cell 89:83–92

Alonso JM, Stepanova AN, Leisse TJ, Kim CJ, Chen H, Shinn P, Stevenson DK, Zimmerman J, Barajas P, Cheuk R, Gadrinab C, Heller C, Jeske A, Koesema E, Meyers CC, Parker H, Prednis L, Ansari Y, Choy N, Deen H, Geralt M, Hazari N, Hom E, Karnes M, Mulholland C, Ndubaku R, Schmidt I, Guzman P, Aguilar-Henonin L, Schmid M, Weigel D, Carter DE, Marchand T, Risseeuw E, Brogden D, Zeko A, Crosby WL, Berry CC, Ecker JR (2003) Genome-wide insertional mutagenesis of *Arabidopsis thaliana*. Science 301:653–657

Boisson-Dernier A, Frietsch S, Kim T-H, Dizon MB, Schroeder JI (2008) The peroxin loss-of-function mutation *abstinence by mutual consent* disrupts recognition between male and female gametophytes. Curr Biol 18:63–68

Braverman N, Dodt G, Gould SJ, Valle D (1998) An isoform of Pex5p, the human PTS1 receptor, is required for the import of PTS2 proteins into peroxisomes. Hum Mol Genet 7:1195–1205

Burkhart SE, Lingard MJ, Bartel B (2013) Genetic dissection of peroxisome-associated matrix protein degradation in *Arabidopsis thaliana*. Genetics 193:125–141

Cui S, Fukao Y, Mano S, Yamada K, Hayashi M, Nishimura M (2013) Proteomic analysis reveals that the Rab GTPase RabE1c is involved in the degradation of the peroxisomal protein receptor PEX7 (peroxin 7). J Biol Chem 288:6014–6023

Dodt G, Warren D, Becker E, Rehling P, Gould SJ (2001) Domain mapping of human PEX5 reveals functional and structural similarities to *Saccharomyces cerevisiae* Pex18p and Pex21p. J Biol Chem 276:41769–41781

Eastmond PJ (2006) *SUGAR-DEPENDENT1* encodes a patatin domain triacylglycerol lipase that initiates storage oil breakdown in germinating *Arabidopsis* seeds. Plant Cell 18:665–675

Eastmond PJ (2007) *MONODEHYROASCORBATE REDUCTASE4* is required for seed storage oil hydrolysis and postgerminative growth in *Arabidopsis*. Plant Cell 19:1376–1387

Eubel H, Meyer EH, Taylor NL, Bussell JD, O'Toole N, Heazlewood JL, Castleden I, Small ID, Smith SM, Millar AH (2008) Novel proteins, putative membrane transporters, and an integrated metabolic network are revealed by quantitative proteomic analysis of *Arabidopsis* cell culture peroxisomes. Plant Physiol 148:1809–1829

Fan J, Quan S, Orth T, Awai C, Chory J, Hu J (2005) The *Arabidopsis PEX12* gene is required for peroxisome biogenesis and is essential for development. Plant Physiol 139:231–239

Farmer LM, Rinaldi MA, Young PG, Danan CH, Burkhart SE, Bartel B (2013) Disrupting autophagy restores peroxisome function to an *Arabidopsis lon2* mutant and reveals a role for the LON2 protease in peroxisomal matrix protein degradation. Plant Cell 25:4085–4100

Faust JE, Verma A, Peng C, McNew JA (2012) An inventory of peroxisomal proteins and pathways in *Drosophila melanogaster*. Traffic 13:1378–1392

Fujiki Y, Nashiro C, Miyata N, Tamura S, Okumoto K (2012) New insights into dynamic and functional assembly of the AAA peroxins, Pex1p and Pex6p, and their membrane receptor Pex26p in shuttling of PTS1-receptor Pex5p during peroxisome biogenesis. Biochim Biophys Acta 1823:145–149

Gatto GJ Jr, Geisbrecht BV, Gould SJ, Berg JM (2000) Peroxisomal targeting signal-1 recognition by the TPR domains of human PEX5. Nat Struct Biol 7:1091–1095

Gietl C, Faber KN, van der Klei IJ, Veenhuis M (1994) Mutational analysis of the N-terminal topogenic signal of watermelon glyoxysomal malate dehydrogenase using the heterologous host *Hansenula polymorpha*. Proc Natl Acad Sci USA 91:3151–3155

Goto S, Mano S, Nakamori C, Nishimura M (2011) *Arabidopsis* ABERRANT PEROXISOME MORPHOLOGY9 is a peroxin that recruits the PEX1-PEX6 complex to peroxisomes. Plant Cell 23:1573–1587

Gould SJ, Keller GA, Subramani S (1988) Identification of peroxisomal targeting signals located at the carboxy terminus of four peroxisomal proteins. J Cell Biol 107:897–905

Graham IA, Eastmond PJ (2002) Pathways of straight and branched chain fatty acid catabolism in higher plants. Prog Lipid Res 41:156–181

Grimm I, Saffian D, Platta HW, Erdmann R (2012) The AAA-type ATPases Pex1p and Pex6p and their role in peroxisomal matrix protein import in *Saccharomyces cerevisiae*. Biochim Biophys Acta 1823:150–158

Gurvitz A, Langer S, Piskacek M, Hamilton B, Ruis H, Hartig A (2000) Predicting the function and subcellular location of *Caenorhabditis elegans* proteins similar to *Saccharomyces cerevisiae* β-oxidation enzymes. Yeast 17:188–200

Hampton RY, Sommer T (2012) Finding the will and the way of ERAD substrate retrotranslocation. Curr Opin Cell Biol 24:460–466

Hayashi M, Toriyama K, Kondo M, Nishimura M (1998) 2,4-dichlorophenoxybutyric acid-resistant mutants of *Arabidopsis* have defects in glyoxysomal fatty acid β-oxidation. Plant Cell 10:183–195

Hayashi M, Nito K, Toriyama-Kato K, Kondo M, Yamaya T, Nishimura M (2000a) *At*Pex14p maintains peroxisomal functions by determining protein targeting to three kinds of plant peroxisomes. EMBO J 19:5701–5710

Hayashi M, Toriyama K, Kondo K, Kato A, Mano S, De Bellis L, Hayashi-Ishimaru Y, Yamaguchi K, Hayashi H, Nishimura M (2000b) Functional transformation of plant peroxisomes. Cell Biochem Biophys 32:295–304

Hayashi H, Nito K, Takei-Hoshi R, Yagi M, Kondo M, Suenaga A, Yamaya T, Nishimura M (2002) Ped3p is a peroxisomal ATP-binding cassette transporter that might supply substrates for fatty acid β-oxidation. Plant Cell Physiol 43:1–11

Hayashi M, Yagi M, Nito K, Kamada T, Nishimura M (2005) Differential contribution of two peroxisomal protein receptors to the maintenance of peroxisomal functions in *Arabidopsis*. J Biol Chem 280:14829–14835

Heiland I, Erdmann R (2005) Biogenesis of peroxisomes. Topogenesis of the peroxisomal membrane and matrix proteins. FEBS J 272:2362–2372

Helm M, Lück C, Prestele J, Hierl G, Huesgen PF, Frohlich T, Arnold GJ, Adamska I, Görg A, Lottspeich F, Gietl C (2007) Dual specificities of the glyoxysomal/peroxisomal processing protease DEG15 in higher plants. Proc Natl Acad Sci USA 104:11501–11506

Hijikata M, Ishii N, Kagamiyama H, Osumi T, Hashimoto T (1987) Structural analysis of cDNA for rat peroxisomal 3-ketoacyl-CoA thiolase. J Biol Chem 262:8151–8158

Hu J, Aguirre M, Peto C, Alonso J, Ecker J, Chory J (2002) A role for peroxisomes in photomorphogenesis and development of *Arabidopsis*. Science 297:405–409

Hu J, Baker A, Bartel B, Linka N, Mullen RT, Reumann S, Zolman BK (2012) Plant peroxisomes: biogenesis and function. Plant Cell 24:2279–2303

Johnson TL, Olsen LJ (2003) Import of the peroxisomal targeting signal type 2 protein 3-ketoacyl-Coenzyme A thiolase into glyoxysomes. Plant Physiol 133:1991–1999

Kaur N, Zhao Q, Xie Q, Hu J (2013) *Arabidopsis* RING peroxins are E3 ubiquitin ligases that interact with two homologous ubiquitin receptor proteins. J Integr Plant Biol 55:108–120

Khan BR, Zolman BK (2010) *pex5* mutants that differentially disrupt PTS1 and PTS2 peroxisomal matrix protein import in *Arabidopsis*. Plant Physiol 154:1602–1615

Kim J, Lee H, Lee HN, Kim SH, Shin KD, Chung T (2013) Autophagy-related proteins are required for degradation of peroxisomes in *Arabidopsis* hypocotyls during seedling growth. Plant Cell 25:4956–4966

Krysan PJ, Young JC, Sussman MR (1999) T-DNA as an insertional mutagen in *Arabidopsis*. Plant Cell 11:2283–2290

Lee JR, Jang HH, Park JH, Jung JH, Lee SS, Park SK, Chi YH, Moon JC, Lee YM, Kim SY, Kim JY, Yun DJ, Cho MJ, Lee KO, Lee SY (2006) Cloning of two splice variants of the rice PTS1 receptor, *Os*Pex5pL and *Os*Pex5pS, and their functional characterization using *pex5*-deficient yeast and *Arabidopsis*. Plant J 47:457–466

Li F, Vierstra RD (2012) Autophagy: a multifaceted intracellular system for bulk and selective recycling. Trends Plant Sci 17:526–537

Lin Y, Sun L, Nguyen LV, Rachubinski RA, Goodman HM (1999) The Pex16p homolog SSE1 and storage organelle formation in *Arabidopsis* seeds. Science 284:328–330

Lin Y, Ulanov AV, Lozovaya V, Widholm J, Zhang G, Guo J, Goodman HM (2006) Genetic and transgenic perturbations of carbon reserve production in *Arabidopsis* seeds reveal metabolic interactions of biochemical pathways. Planta 225:153–164

Lingard MJ, Bartel B (2009) *Arabidopsis* LON2 is necessary for peroxisomal function and sustained matrix protein import. Plant Physiol 151:1354–1365

Lingard MJ, Monroe-Augustus M, Bartel B (2009) Peroxisome-associated matrix protein degradation in *Arabidopsis*. Proc Natl Acad Sci USA 106:4561–4566

Lisenbee CS, Lingard MJ, Trelease RN (2005) *Arabidopsis* peroxisomes possess functionally redundant membrane and matrix isoforms of monodehydroascorbate reductase. Plant J 43:900–914

Mano S, Nakamori C, Kondo M, Hayashi M, Nishimura M (2004) An *Arabidopsis* dynamin-related protein, DRP3A, controls both peroxisomal and mitochondrial division. Plant J 38:487–498

Mano S, Nakamori C, Nito K, Kondo M, Nishimura M (2006) The *Arabidopsis pex12* and *pex13* mutants are defective in both PTS1- and PTS2-dependent protein transport to peroxisomes. Plant J 47:604–618

Mano S, Nakamori C, Fukao Y, Araki M, Matsuda A, Kondo M, Nishimura M (2011) A defect of peroxisomal membrane protein 38 causes enlargement of peroxisomes. Plant Cell Physiol 52:2157–2172

Matsumura T, Otera H, Fujiki Y (2000) Disruption of the interaction of the longer isoform of Pex5p, Pex5pL, with Pex7p abolishes peroxisome targeting signal type 2 protein import in mammals. Study with a novel *PEX5*-impaired Chinese hamster ovary cell mutant. J Biol Chem 275:21715–21721

Meinecke M, Cizmowski C, Schliebs W, Kruger V, Beck S, Wagner R, Erdmann R (2010) The peroxisomal importomer constitutes a large and highly dynamic pore. Nat Cell Biol 3:273–277

Mhamdi A, Noctor G, Baker A (2012) Plant catalases: peroxisomal redox guardians. Arch Biochem Biophys 525:181–194

Monroe-Augustus M (2004) Genetic approaches to elucidating the mechanisms of indole-3-acetic acid and indole-3-butyric acid function in *Arabidopsis thaliana*. Ph.D. Rice University, Houston

Monroe-Augustus M, Ramón NM, Ratzel SE, Lingard MJ, Christensen SE, Murali C, Bartel B (2011) Matrix proteins are inefficiently imported into *Arabidopsis* peroxisomes lacking the receptor-docking peroxin PEX14. Plant Mol Biol 77:1–15

Motley AM, Hettema EH, Ketting R, Plasterk R, Tabak HF (2000) *Caenorhabditis elegans* has a single pathway to target matrix proteins to peroxisomes. EMBO Rep 1:40–46

Nito K, Hayashi M, Nishimura M (2002) Direct interaction and determination of binding domains among peroxisomal import factors in *Arabidopsis thaliana*. Plant Cell Physiol 43:355–366

Nito K, Kamigaki A, Kondo M, Hayashi M, Nishimura M (2007) Functional classification of *Arabidopsis* peroxisome biogenesis factors proposed from analyses of knockdown mutants. Plant Cell Physiol 48:763–774

Ostersetzer O, Kato Y, Adam Z, Sakamoto W (2007) Multiple intracellular locations of Lon protease in *Arabidopsis*: evidence for the localization of AtLon4 to chloroplasts. Plant Cell Physiol 48:881–885

Otera H, Okumoto K, Tateishi K, Ikoma Y, Matsuda E, Nishimura M, Tsukamoto T, Osumi T, Ohashi K, Higuchi O, Fujiki Y (1998) Peroxisome targeting signal type 1 (PTS1) receptor is involved in import of both PTS1 and PTS2: Studies with PEX5-defective CHO cell mutants. Mol Cell Biol 18:388–399

Otera H, Harano T, Honsho M, Ghaedi K, Mukai S, Tanaka A, Kawai A, Shimizu N, Fujiki Y (2000) The mammalian peroxin Pex5pL, the longer isoform of the mobile peroxisome targeting signal (PTS) Type 1 transporter, translocates the Pex7p-PTS2 protein complex into peroxisomes via its initial docking site, Pex14p. J Biol Chem 275:21703–21714

Platta HW, El Magraoui F, Schlee D, Grunau S, Girzalsky W, Erdmann R (2007) Ubiquitination of the peroxisomal import receptor Pex5p is required for its recycling. J Cell Biol 177:197–204

Platta HW, Magraoui FE, Baumer BE, Schlee D, Girzalsky W, Erdmann R (2009) Pex2 and Pex12 function as protein-ubiquitin ligases in peroxisomal protein import. Mol Cell Biol 29:5505–5516

Prestele J, Hierl G, Scherling C, Hetkamp S, Schwechheimer C, Isono E, Weckwerth W, Wanner G, Gietl C (2010) Different functions of the C3HC4 zinc RING finger peroxins PEX10, PEX2, and PEX12 in peroxisome formation and matrix protein import. Proc Natl Acad Sci USA 107:14915–14920

Purdue PE, Yang X, Lazarow PB (1998) Pex18p and Pex21p, a novel pair of related peroxins essential for peroxisomal targeting by the PTS2 pathway. J Cell Biol 143:1859–1869

Ramón NM, Bartel B (2010) Interdependence of the peroxisome-targeting receptors in *Arabidopsis thaliana*: PEX7 facilitates PEX5 accumulation and import of PTS1 cargo into peroxisomes. Mol Biol Cell 21:1263–1271

Ratzel SE, Lingard MJ, Woodward AW, Bartel B (2011) Reducing *PEX13* expression ameliorates physiological defects of late-acting peroxin mutants. Traffic 12:121–134

Reumann S, Babujee L, Ma C, Wienkoop S, Siemsen T, Antonicelli GE, Rasche N, Luder F, Weckwerth W, Jahn O (2007) Proteome analysis of *Arabidopsis* leaf peroxisomes reveals novel targeting peptides, metabolic pathways, and defense mechanisms. Plant Cell 19:3170–3193

Schliebs W, Girzalsky W, Erdmann R (2010) Peroxisomal protein import and ERAD: variations on a common theme. Nat Rev Mol Cell Biol 11:885–890

Schluter A, Fourcade S, Ripp R, Mandel JL, Poch O, Pujol A (2006) The evolutionary origin of peroxisomes: an ER-peroxisome connection. Mol Biol Evol 23:838–845

Schumann U, Wanner G, Veenhuis M, Schmid M, Gietl C (2003) *AthPEX10*, a nuclear gene essential for peroxisome and storage organelle formation during *Arabidopsis* embryogenesis. Proc Natl Acad Sci USA 100:9626–9631

Schumann U, Prestele J, O'Geen H, Brueggeman R, Wanner G, Gietl C (2007) Requirement of the C3HC4 zinc RING finger of the *Arabidopsis PEX10* for photorespiration and leaf peroxisome contact with chloroplasts. Proc Natl Acad Sci USA 104:1069–1074

Schumann H, Huesgen PF, Gietl C, Adamska I (2008) The DEG15 serine protease cleaves peroxisomal targeting signal 2-containing proteins in *Arabidopsis*. Plant J 148:1847–1856

Sharp PA (1999) RNAi and double-stranded RNA. Genes Dev 13:139–141

Shibata M, Oikawa K, Yoshimoto K, Kondo M, Mano S, Yamada K, Hayashi M, Sakamoto W, Ohsumi Y, Nishimura M (2013) Highly oxidized peroxisomes are selectively degraded via autophagy in *Arabidopsis*. Plant Cell 25(12):4967–4983

Shimizu N, Itoh R, Hirono Y, Otera H, Ghaedi K, Tateishi K, Tamura S, Okumoto K, Harano T, Mukai S, Fujiki Y (1999) The peroxin Pex14p. cDNA cloning by functional complementation on a Chinese hamster ovary cell mutant, characterization, and functional analysis. J Biol Chem 274:12593–12604

Singh T, Hayashi M, Mano S, Arai Y, Goto S, Nishimura M (2009) Molecular components required for the targeting of PEX7 to peroxisomes in *Arabidopsis thaliana*. Plant J 60:488–498

Sparkes IA, Brandizzi F, Slocombe SP, El-Shami M, Hawes C, Baker A (2003) An *Arabidopsis pex10* null mutant is embryo lethal, implicating peroxisomes in an essential role during plant embryogenesis. Plant Physiol 133:1809–1819

Strader LC, Bartel B (2011) Transport and metabolism of the endogenous auxin precursor indole-3-butyric acid. Mol Plant 4:477–486

Strader L, Culler Hendrickson A, Cohen J, Bartel B (2010) Conversion of endogenous indole-3-butyric acid to indole-3-acetic acid drives cell expansion in *Arabidopsis* seedlings. Plant Physiol 153:1577–1586

Strader LC, Wheeler DL, Christensen SE, Berens JC, Cohen JD, Rampey RA, Bartel B (2011) Multiple facets of Arabidopsis seedling development require indole-3-butyric acid-derived auxin. Plant Cell 23:984–999

Swinkels BW, Gould SJ, Bodnar AG, Rachubinski RA, Subramani S (1991) A novel, cleavable peroxisomal targeting signal at the amino-terminus of the rat 3-ketoacyl-CoA thiolase. EMBO J 10:3255–3262

Thoms S, Erdmann R (2006) Peroxisomal matrix protein receptor ubiquitination and recycling. Biochim Biophys Acta 1763:1620–1628

Till A, Lakhani R, Burnett SF, Subramani S (2012) Pexophagy: the selective degradation of peroxisomes. Int J Cell Biol 2012:512721

Toro AA, Araya CA, Cordova GJ, Arredondo CA, Cardenas HG, Moreno RE, Venegas A, Koenig CS, Cancino J, Gonzalez A, Santos MJ (2009) Pex3p-dependent peroxisomal biogenesis initiates in the endoplasmic reticulum of human fibroblasts. J Cell Biochem 107:1083–1096

Wain RL, Wightman F (1954) The growth-regulating activity of certain ω-substituted alkyl carboxylic acids in relation to their β-oxidation within the plant. Proc R Soc London Ser B 142:525–536

Woodward AW, Bartel B (2005a) The *Arabidopsis* peroxisomal targeting signal type 2 receptor PEX7 is necessary for peroxisome function and dependent on PEX5. Mol Biol Cell 16:573–583

Woodward AW, Bartel B (2005b) Auxin: regulation, action, and interaction. Ann Bot 95:707–735

Woodward AW, Fleming WA, Ratzel SE, Burkhart SE, Bjornson M, Bartel B (2014) A viable *Arabidopsis pex13* missense allele confers severe peroxisomal defects and decreases PEX5 association with peroxisomes. Plant Mol Biol (in revision)

Zolman BK, Bartel B (2004) An *Arabidopsis* indole-3-butyric acid-response mutant defective in PEROXIN6, an apparent ATPase implicated in peroxisomal function. Proc Natl Acad Sci USA 101:1786–1791

Zolman BK, Yoder A, Bartel B (2000) Genetic analysis of indole-3-butyric acid responses in *Arabidopsis thaliana* reveals four mutant classes. Genetics 156:1323–1337

Zolman BK, Silva ID, Bartel B (2001) The *Arabidopsis pxa1* mutant is defective in an ATP-binding cassette transporter-like protein required for peroxisomal fatty acid b-oxidation. Plant Physiol 127:1266–1278

Zolman BK, Monroe-Augustus M, Silva ID, Bartel B (2005) Identification and functional characterization of *Arabidopsis* PEROXIN4 and the interacting protein PEROXIN22. Plant Cell 17:3422–3435

Zolman BK, Nyberg M, Bartel B (2007) IBR3, a novel peroxisomal acyl-CoA dehydrogenase-like protein required for indole-3-butyric acid response. Plant Mol Biol 64:59–72

Zolman BK, Martinez N, Millius A, Adham AR, Bartel B (2008) Identification and characterization of *Arabidopsis* indole-3-butyric acid response mutants defective in novel peroxisomal enzymes. Genetics 180:237–251

The Peroxisomal Exportomer 15

Harald W. Platta, Stefanie Hagen, and Ralf Erdmann

Abstract

Peroxisomes are ubiquitous cell organelles of eukaryotic cells. Depending on environmental changes and cellular demands, peroxisomes display a high plasticity in metabolic functions. A prerequisite to carry out their physiological tasks is compartmentalization of peroxisomal enzymes in the lumen of this organelle, the peroxisomal matrix. The matrix proteins are synthesized on free polyribosomes in the cytosol and harbor a peroxisomal targeting sequence (PTS). They are targeted to the peroxisomal membrane by soluble PTS-receptors. Following the release of the cargo enzyme into the peroxisomal matrix, the PTS-receptor is ubiquitinated and exported back to the cytosol to facilitate further rounds of matrix protein import. The retrotranslocation of the receptor is facilitated by a molecular machinery that comprises enzymes required for the ubiquitination as well as for the ATP-dependent extraction of the receptor from the membrane. Furthermore, recent evidence indicates that the export machinery of the receptors might function as molecular motor not only for the retrotranslocation of the receptors themselves but also for the import of peroxisomal matrix proteins. This is thought to be achieved by coupling the ATP-dependent removal of the PTS-receptor with the cargo protein translocation into the organelle. In this review, we will discuss the combined data on the architecture and molecular function of the peroxisomal receptor export machinery, the peroxisomal exportomer.

H.W. Platta (✉)
Biochemie Intrazellulärer Transportprozesse, Ruhr-Universität Bochum, 44780 Bochum, Germany
e-mail: harald.platta@rub.de

S. Hagen • R. Erdmann (✉)
Systembiochemie, Ruhr-Universität Bochum, 44780 Bochum, Germany
e-mail: ralf.erdmann@rub.de

C. Brocard and A. Hartig (eds.), *Molecular Machines Involved in Peroxisome Biogenesis and Maintenance*, DOI 10.1007/978-3-7091-1788-0_15, © Springer-Verlag Wien 2014

Keywords
Peroxisome • Exportomer • Peroxin • Protein import • Ubiquitination • AAA-type ATPases

15.1 Introduction to the Cellular Functions of Peroxisomes

Peroxisomes are single membrane-bound organelles that can be found in all eukaryotic cells with the exception of spermatozoa and mature erythrocytes (Novikoff et al. 1973). Peroxisomes display a high variability in their enzyme content and metabolic tasks that can be adjusted according to cellular needs. The enzymes in the lumen of peroxisomes are often highly concentrated and tightly packed to form crystalline inclusions that are visible as electron-dense structures. The beta-oxidation of fatty acids and the detoxification of the hydrogen peroxide are regarded as the central and most conserved functions of peroxisomes (Cooper and Beevers 1969; Lazarow and DeDuve 1976). Furthermore, the beta-oxidation pathway is linked to the synthesis of signaling molecules, like phytohormones in plants (Baker et al. 2006; Kienow et al. 2008) and pheromones in *Caenorhabditis elegans* and insects (Joo et al. 2010; Spiegel et al. 2011). Mammalian peroxisomes have a key function in the biosynthesis of ether lipids and bile acids (Wanders and Waterham 2006a). Peroxisomes house important steps of penicillin biosynthesis in some filamentous fungi (Meijer et al. 2010; Müller et al. 1991), but also certain enzymes required for the biosynthesis of Vitamin K1 in plants (Widhalm et al. 2012) or the synthesis of siderophores required for iron uptake and virulence of *Aspergillus* species (Gründlinger et al. 2013). Depending on the metabolic state of the cell, the number of peroxisomes can be dynamically regulated either by the proliferation of peroxisomes or their selective autophagic degradation via pexophagy (Grunau et al. 2011; Opaliński et al. 2011; Till et al. 2012; Tower et al. 2011).

Defects in peroxisome function are the molecular cause for human inborn errors that are caused by mutation of single metabolic enzymes (Wanders and Waterham 2006b) or genes coding for proteins that are required for the biogenesis of the organelles (Steinberg et al. 2006). The peroxisomal biogenesis disorders (PBDs) form a spectrum of autosomal recessive metabolic disorders that are collectively characterized by abnormal peroxisome assembly and result in multisystemic disorders that often lead to death in early infancy (Baes and Van Veldhoven 2012; Nagotu et al. 2012; Waterham and Ebberink 2012). Furthermore, the physiological function of peroxisomes contributes to the cellular protection mechanism against the progressive brain damage and cognitive decline caused by Alzheimer's disease (Fanelli et al. 2013; Kou et al. 2011; Lizard et al. 2012).

The formation of peroxisomes depends on specific biogenesis factors, the peroxins (Distel et al. 1996). To date, 34 different peroxins have been described. In general, they are involved in the six key stages of peroxisomal biogenesis which comprise the (1) *de novo* formation and (2) proliferation of peroxisomes, (3) their inheritance and (4) regulated degradation by an authophagic process called pexophagy as well as the import of (5) peroxisomal membrane and (6) matrix proteins (Fagarasanu et al. 2010; Islinger et al. 2012; Liu et al. 2012; Platta and Erdmann 2007b; Theodoulou et al. 2013; Hasan et al. 2013).

In this review, we will discuss the peroxisomal matrix protein import with emphasis on the function of the peroxisomal membrane complexes that are involved in the ubiquitination and energy-consuming dislocation of the dynamic import receptors and describe their concerted function as receptor export machinery, the peroxisomal exportomer (Platta et al. 2013).

15.2 Peroxisomal Matrix Proteins Are Imported by Cycling Receptors

All peroxisomal proteins are encoded in the nucleus, synthesized on free ribosomes in the cytosol and imported posttranslationally. Most remarkably, peroxisomes are capable to accommodate fully folded proteins. Frequently, even oligomeric or cofactor-bound proteins are imported (Leon et al. 2006a; Girzalsky et al. 2009). The import of folded proteins distinguishes peroxisomes from other organelles like chloroplasts or mitochondria, which all import unfolded proteins, but it makes them comparable to the Tat (Twin-arginine translocation) pathways of bacteria and thylakoid membranes (Albiniak et al. 2012; Palmer and Berks 2012). However, in contrast to these translocation systems, peroxisomal matrix protein import is facilitated by dynamic receptors that cycle between a soluble state in the cytosol and a membrane-associated state at the peroxisomal membrane (Hasan et al. 2013; Liu et al. 2012; Platta and Erdmann 2007a). The import cycle can conceptually be divided into five steps, comprising (1) cargo recognition in the cytosol, (2) receptor–cargo docking at the peroxisome, (3) cargo translocation across the membrane, (4) cargo release into the matrix, and (5) receptor ubiquitination and export back to the cytosol (Fig. 15.1).

The events concerning the cargo transport from the cytosol to the peroxisomal lumen are discussed in detail in Chaps. 13 (Effelsberg et al.) and 14 (Bartel et al.) of this book. In brief, proteins destined for the peroxisomal matrix usually harbor a peroxisomal targeting sequence (PTS). Most peroxisomal matrix proteins carry a C-terminal PTS1-sequence, which is recognized by the PTS1-receptor Pex5p, while a subset of proteins displays an N-terminal PTS2-sequence via which they are ferried to the peroxisomal membrane by the PTS2-receptor Pex7p and its co-receptors, like *S. cerevisiae* Pex18p or *P. pastoris* Pex20p. Here, the cargo-bound PTS-receptors interact with constituents of the docking complex (Pex13p, Pex14p), which results in the formation of a transient import pore and finally the transloction and release of the cargo.

Subsequent to the liberation of the cargo, the PTS-receptors return to the cytosol for further rounds of matrix protein import (Fig. 15.2). This retrotranslocation is facilitated by the peroxisomal receptor export machinery, the exportomer (Platta et al. 2013). The monoubiquitination of the PTS-receptors is regarded as a central

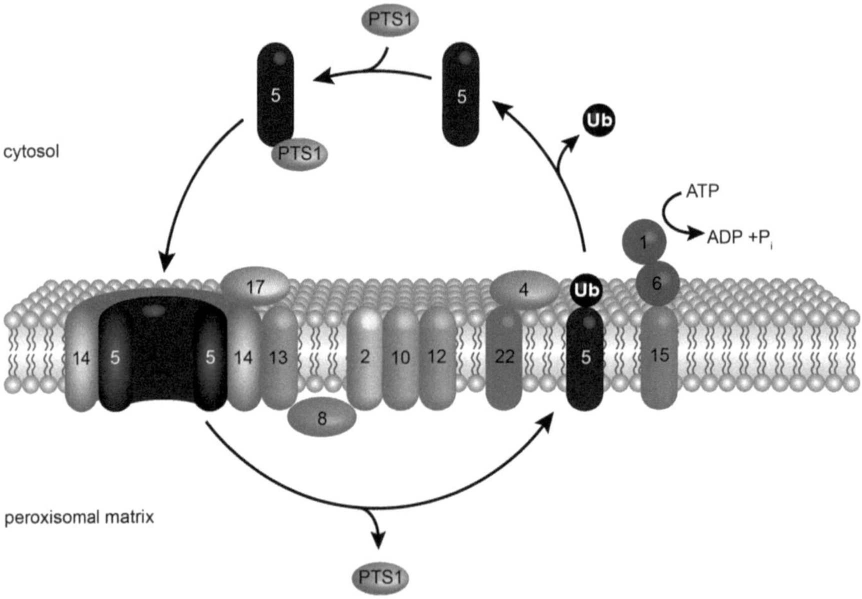

Fig. 15.1 Peroxisomal PTS1-protein import in *Saccharomyces cerevisiae*. Peroxisomal matrix proteins are imported into peroxisomes via soluble receptors, which shuttle between the cytosol and the peroxisomal membrane. The matrix proteins are recognized by their peroxisomal targeting signal 1 (PTS1) in the cytosol via the receptor Pex5p, forming a receptor–cargo complex. At the peroxisomal membrane this complex binds to the docking complex (Pex13p, Pex14p and Pex17p), and this binding leads to the formation of a transient pore, whose exact molecular constitution is still under discussion but at least contains Pex5p and Pex14p. In the next step, the cargo is translocated into the peroxisomal lumen in an unknown manner; Pex8p might be involved in the receptor-cargo dissociation. At the end of the receptor cycle, the receptor is removed from the membrane and transported back to the cytosol for another round of import. To this end, Pex5p is monoubiquitinated by the ubiquitin-conjugating Pex4p (E2), which is anchored at the membrane by Pex22p, and by the ubiquitin ligase Pex12p (E3), which forms the RING-finger complex together with the other ubiquitin ligases Pex2p and Pex10p. The ubiquitin signal leads to an ATP-dependent dislocation of Pex5p from the peroxisomal membrane, performed by the Pex15p-anchored AAA peroxins Pex1p and Pex6p. Prior to a new round of import, the ubiquitin moiety is removed from the receptor

event in this process and has been shown to depend on the ubiquitin-conjugating enzyme Pex4p, its membrane anchor Pex22p and the presence of the peroxisomal RING–peroxin complex (Pex2p, Pex10p, Pex12p). The ubiquitination is thought to prime the PTS-receptors for the recognition by the AAA-type ATPase complex (Pex1p, Pex6p), which functions as dislocase by extracting the modified PTS-receptors from the membrane. Furthermore, the recent "export-driven-import model" postulates that the ATP-dependent export of the PTS-receptors may be directly linked to the translocation of the cargo proteins into the peroxisomal matrix. In case the monoubiquitination-dependent receptor recycling pathway is

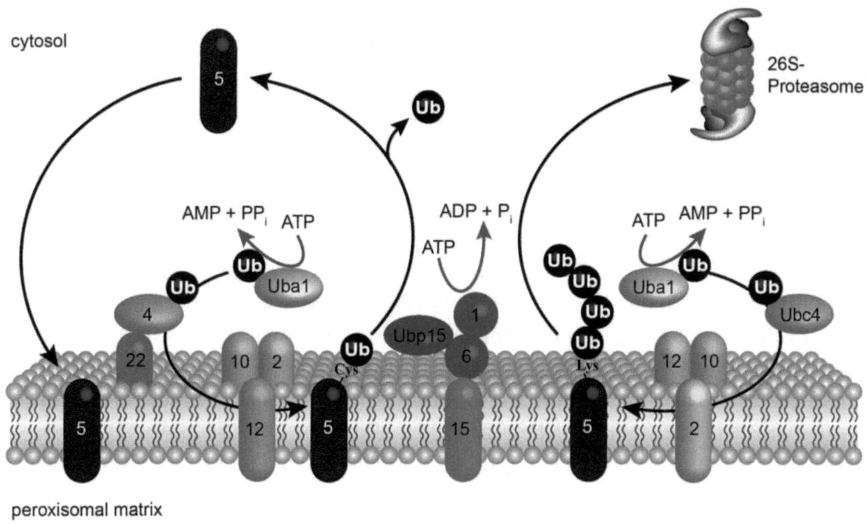

Fig. 15.2 Ubiquitination and export of the PTS1-receptor at the peroxisomal membrane in *Saccharomyces cerevisiae*. At the peroxisomal membrane, the PTS1-receptor Pex5p is either mono- or polyubiquitinated. Both ubiquitination cascades are initiated by the ATP-dependent ubiquitin-activating enzyme Uba1p (E1). For monoubiquitination of Pex5p as part of the typical receptor cycle (*left*), the activated ubiquitin is transferred to the ubiquitin-conjugating enzyme Pex4p (E2) and then attached to a conserved cysteine of the receptor by assistance of the RING-ligase Pex12p (E3). The ubiquitinated receptor is ATP-dependent export of Pex5p catalyzed by the Pex15p-anchored AAA-peroxins Pex1p and Pex6p. For a new round of import the ubiquitin is cleaved off by the deubiquitinating enzyme Ubp15p. For polyubiquitination of Pex5p as part of a quality control pathway (*right*), the activated ubiquitin is transferred to the ubiquitin-conjugating enzyme Ubc4p (E2) and then delivered to conserved lysines of the receptor by the RING-ligase Pex2p (E3). The following export of Pex5p is also performed by the AAA peroxins Pex1p and Pex6p and the polyubiquitination signal leads to a degradation of Pex5p by the 26S proteasome

blocked, the PTS-receptors become substrates of a polyubiquitination-dependent proteolytic pathway, which promotes their degradation by the 26S proteasome.

15.3 The Peroxisomal Receptor Export Machinery: The Exportomer

The import of matrix proteins depends on the cycle of the PTS-receptors, which itself relies on the function of each constituent of the membrane-bound peroxins. Previous studies elucidated the composition of the peroxisomal membrane-bound subcomplexes, the docking- and the RING complex (Agne et al. 2003; Hazra et al. 2002). These two complexes were referred to as the "importomer" because both are physically connected by Pex8p in *S. cerevisiae* and both are required for matrix protein import (Agne et al. 2003). However, later work revealed that also constituents of the AAA complex and Pex4p complex could be co-purified with the importomer components, which strongly suggests that all membrane-associated peroxin complexes required for matrix protein import are dynamically interconnected (Oeljeklaus et al. 2012; Rosenkranz et al. 2006; Platta et al. 2009). Therefore, an alternative approach to define functionally related subcomplexes could be based on the steps of the PTS-receptor cycle at the membrane in general and on the energy dependence in particular. The current view is that the association of the PTS-receptors with the peroxisomal membrane at the site of the importomer is ATP independent (Miyata and Fujiki 2005; Miyata et al. 2009; Oliveira et al. 2003; Platta et al. 2005). However, the RING complex (Pex2p, Pex10p, Pex12p) as well as the Ubc components (Pex22p, Pex4p, Ubc4p family, UbcH5 family) belong to the ATP-dependent ubiquitination cascade (El Magraoui et al. 2012; Grou et al. 2008; Platta et al. 2007a, b, 2009; Williams et al. 2007, 2008, 2012; Liu and Subramani 2013; Kaur et al. 2013) and form together with the AAA-type ATPase complex (Pex1p, Pex6p, Pex15p, Ubp15p, AWP1) (Debelyy et al. 2011; Hensel et al. 2011; Leon et al. 2006b; Miyata and Fujiki 2005; Miyata et al. 2012) the receptor export machinery, or alternatively, peroxisomal exportomer (Table 15.1).

15.3.1 The Ubiquitin-Conjugating Enzymes Required for Monoubiquitination of the Receptors

Ubiquitination is a posttranslational protein modification that is mediated by a three-step enzyme cascade. The ubiquitin-activating enzyme (E1) activates ubiquitin via an AMP-bound intermediate and transfers it to an ubiquitin-conjugating enzyme (E2). Finally, an ubiquitin-protein ligase (E3) binds the ubiquitin-charged E2 as well as the substrate protein, thereby enabling the transfer of the ubiquitin moiety to the target amino acid residue of the substrate (Kerscher et al. 2006; Ravid and Hochstrasser 2008). Regularly, the epsilon-amino group of a lysine within the target protein is covalently linked to ubiquitin via an isopeptide bond. Interestingly, ubiquitin can also be attached via a peptide bond to the alpha-amino group to the N-terminal amino acid, or via an oxyester bond to a threonine or serine, or even via a thioester bond to a cysteine (Wang et al. 2012).

The peroxisomal matrix protein import depends on the unusual ubiquitination of a conserved cysteine of the PTS1-receptor Pex5p (Carvalho et al. 2007; Okumoto et al. 2011; Williams et al. 2007) and of the PTS2-co-receptors Pex18p (Hensel et al. 2011) or Pex20p (Liu and Subramani 2013). The E2-enzyme that has been demonstrated to catalyze the cysteine-dependent monoubiquitination of *S. cerevisiae* Pex5p both in vivo and in vitro is Pex4p (Ubc10p; Platta et al. 2007a; Williams et al. 2007). Recently, the monoubiquitination of *S. cerevisiae* Pex18p (El Magraoui et al. 2013) and *P. pastoris* Pex20p (Liu and Subramani 2013) has been demonstrated to depend on the presence of Pex4p as well.

The soluble E2-enzyme Pex4p is essential for the import of both PTS1 and PTS2 proteins and therefore was the first E2-enzyme shown to be essential for the

Table 15.1 Peroxisomal import factors

Classification	Role	Yeast peroxin	Human peroxin	Plant peroxin
Import receptors	PTS1 receptor	Pex5p	Pex5pS	Pex5pS
	PTS2 receptor	Pex7p	Pex7p	Pex7p
	PTS2 co-receptor	Pex18p/ Pex21p	Pex5pL	Pex5pL
		(*Pp* Pex20p)		
Membrane association of the receptors	Docking complex	Pex13p	Pex13p	Pex13p
		Pex14p	Pex14p	Pex14p
		Pex17p		
	Importomer assembly	Pex8p (*Pp*: Pex3p)		
Export of the receptors	RING finger ligase complex	Pex2p	Pex2p	Pex2p
		Pex10p	Pex10p	Pex10p
		Pex12p	Pex12p	Pex12p
				DSK2a/2b
	Receptor ubiquitin conjugation	Pex4p	UbcH5a/b/c	Pex4p
		Pex22p		Pex22p
		Ubc1p/4p/ 5p		
	Receptor deubiquitination	Ubp15p		
			USP9X	
	AAA export complex	Pex1p	Pex1p	Pex1p
		Pex6p	Pex6p	Pex6p
		Pex15p	Pex26p	APEM9
			AWP1	

The table lists the known yeast, human and plant peroxisomal proteins required for peroxisomal matrix protein import
Abbreviations: *Pp Pichia pastoris*

biogenesis of an organelle (Crane et al. 1994; van der Klei et al. 1998; Wiebel and Kunau 1992; Zolman et al. 2005). Pex4p is anchored to peroxisomes via the membrane protein Pex22p (Koller et al. 1999; Zolman et al. 2005). The crystal structure of *S. cerevisiae* Pex4p complexed to Pex22p (without its membrane domain) revealed that the Pex22p-binding site in Pex4p does not resemble a common substrate-binding motif and therefore it has been suggested that Pex22p may act as a co-activator of this E2-enzyme (Williams et al. 2012).

The molecular function of the Pex4p-catalyzed monoubiquitination of the membrane-bound Pex5p is to prime the PTS1-receptor for export (Platta et al. 2007a). While Pex4p and Pex22p are well conserved in yeasts and plants, they are absent in the genomes of mammals (Kiel et al. 2006). Instead, members of the E2D family of E2-enzymes (UbcH5a, UbcH5b and UbcH5c) fulfill the function of Pex4p in mammals (Grou et al. 2008). They catalyze the monoubiquitination of mammalian Pex5p on the conserved cysteine and therefore are required for the receptor export in vitro (Grou et al. 2008). Even though the three UbcH5 proteins carry out a central task in peroxisome biogenesis, their cellular targets are not restricted to this organelle (Brzovic and Klevit 2006; Gonen et al. 1999; Saville et al. 2004). Future work may reveal why the monoubiquitination of Pex5p has been transferred to the promiscuous UbcH5 proteins. One possible explanation could be that they control cellular events that are interconnected with peroxisome function in a concerted manner.

15.3.2 The RING–Peroxin Complex

The import of peroxisomal matrix proteins requires the presence of the three RING-finger proteins Pex2p, Pex10p, and Pex12p (Albertini et al. 2001; Chang et al. 1999; Eckert and Johnsson 2003; Okumoto et al. 2000; Berteaux-Lecellier et al. 1995; Krazy and Michels 2006; Peraza-Reyes et al. 2008; Sparkes et al. 2003). They have been found to form a distinct subcomplex at the peroxisomal membrane (Agne et al. 2003; Hazra et al. 2002). Defects in the assembly of the human RING complex are the second most common cause of peroxisomal biogenesis disorders (Ebberink et al. 2011; Steinberg et al. 2006). The mammalian Pex2p (formerly PAF-1) was the first gene that could be linked to PBDs (Shimozawa et al. 1992; Tsukamoto et al. 1991). Work from *S. cerevisiae* and *A. thaliana* in recent years has uncovered that all three peroxins display ubiquitin-protein ligases activity (Kaur et al. 2013; Platta et al. 2009; Williams et al. 2008). The RING–peroxins are directly involved in the ubiquitination of the PTS1-receptor Pex5p in *S. cerevisiae* (Platta et al. 2009; Williams et al. 2008). Their activity is also required for the formation of ubiquitinated Pex20p in *P. pastoris* (Liu and Subramani 2013).

In general, E3-enzymes determine the substrate specificity of ubiquitination reactions because they bind the ubiquitin-charged E2-enzyme and the substrate, thereby insuring a specific transfer of ubiquitin to the target amino acid. RING-type E3-enzymes catalyze the direct transfer of ubiquitin from the E2-enzyme to the substrate (Deshaies and Joazeiro 2009). They belong to the superfamily of Treble-Clef fold-containing proteins. This scaffold structure, which is stabilized by a Zn^{2+}-ion, functions as an interaction motif in diverse proteins even outside the ubiquitin system (Burroughs et al. 2011). The canonical RING-finger domain (Freemont et al. 1991) binds two Zn^{2+}-ions through its conserved Cys and His residues in a "cross-brace" arranged manner (Deshaies and Joazeiro 2009). The RING domains of Pex2p and Pex10p coordinate two Zn^{2+}-ions, whereas the RING-finger of Pex12p binds only one Zn^{2+}-ion (Koellensperger et al. 2007). It is interesting to note that also several members of the RBR-(RING-between-RING) family of E3-enzymes, e.g., the Parkin-like Ariadne, contain an active RING domain at their carboxy-terminus containing a single Zn^{2+}-ion (Eisenhaber et al. 2007).

The RING–peroxins Pex2p, Pex10p, and Pex12p assemble to a distinct complex and stabilize each other in vivo (Agne et al. 2003; Hazra et al. 2002). Based on earlier binary interaction studies (Albertini et al. 2001; Chang et al. 1999; Eckert and Johnsson 2003; Okumoto et al. 2000) and recent in vitro interaction data on all three RING domains (El Magraoui et al. 2012), the RING–peroxins are thought to form a heterotrimeric complex. Pex10p (RING) functions as central component of the ternary complex as it directly binds to Pex2p(RING) and Pex12p(RING) thereby bridging the indirect interaction between these two RING domains (El Magraoui et al. 2012). The heteromeric architecture of the RING complex has a direct influence on the E3-ligase activity of the RING–peroxins because the ubiquitination activity of the combined Pex10p/Pex12p RING-domains is enhanced in presence of Pex4p in vitro (El Magraoui et al. 2012).

Pex10p also fulfills additional tasks that are distinct from Pex2p and Pex12p. A systematic functional screen of all peroxins in *A. thaliana* uncovered that only Pex10p has a pleiotropic growth phenotype (Nito et al. 2007). Furthermore, overexpression experiments of proteins with mutated RING domain in wild-type background suggested that *A. thaliana* Pex10p but not Pex2p or Pex12p are required for the contact of peroxisomes to chloroplasts during photorespiration (Prestele et al. 2010; Schumann et al. 2007). However, whether this association is due to a physical interaction or due to a functional interaction via ubiquitination events remains to be investigated.

Early studies already linked the function of the RING–peroxins to the recycling of Pex5p (Chang et al. 1999; Dodt and Gould 1996) and Pex20p (Leon et al. 2006b) as these receptors accumulate at the peroxisomal membrane in cells with disrupted RING complex. Because the monoubiquitination of Pex5p is reported to be essential for the export of Pex5p (Grou et al. 2008; Platta et al. 2007a; Okumoto et al. 2011) and as Pex12p (RING) cooperates with Pex10p(RING) in vitro (El Magraoui et al. 2012) and catalyzes this Pex4p-dependent modification in vivo (Platta et al. 2009), the Pex10p/Pex12p unit may function as the physiologic active ligase complex dedicated to the monoubiquitination-mediated export of the PTS1-receptor.

15.3.3 The Peroxisomal AAA-Type ATPase Complex

The ubiquitinated PTS1-receptor Pex5p is substrate for the peroxisomal AAA-type ATPase complex, which functions as dislocase that extracts Pex5p from the membrane and thereby exports it back to the cytosol (Fujiki et al. 2012; Grimm et al. 2012; Miyata and Fujiki 2005; Platta et al. 2005, 2008). The two peroxisomal AAA proteins Pex1p and Pex6p display a non-redundant and essential function in this process (Birschmann et al. 2005; Kiel et al. 1999, 2000; Tamura et al. 1998; Tamura et al. 2006).

The AAA peroxins associate with peroxisomes via an interaction of Pex6p to the tail-anchored membrane protein Pex15p in yeast and the orthologous Pex26p in mammals as well as APEM9 in plants (Birschmann et al. 2003; Furuki et al. 2006; Goto et al. 2011; Matsumoto et al. 2003a, b). An impaired assembly of the human AAA complex is the most common cause of Zellweger syndrome spectrum disorders (Geisbrecht et al. 1998; Steinberg et al. 2006). It is interesting to point out that Pex1p (formerly PAS1) was the first peroxin to be identified and also one of the founding members of the AAA family (Beyer 1997; Erdmann et al. 1991; Kunau et al. 1993).

In general, AAA proteins are characterized by a conserved modular architecture. They can be classified as P-loop NTPases, which are characterized by conserved motifs for NTP binding (Walker A motif) and hydrolysis (Walker B motif; Walker et al. 1982). AAA proteins in particular are defined by the evolutionary conserved AAA domain that contains the Walker A and B motifs as well as other conserved regions like the Second Region of Homology (SRH; Beyer 1997; Neuwald et al. 1999; Wendler et al. 2012). Pex1p and Pex6p harbor two AAA domains (AAA-D1 and AAA-D2) as well as an N-terminal domain (NTD). The binding and hydrolysis of ATP by the AAA peroxins are thought to result in conformational changes, as shown for p97 (Beuron et al. 2003), ClpX (Stinson et al. 2013) or NSF (Cipriano et al. 2013).

Most AAA proteins form active oligomers with predominantly hexameric constitution (Iyer et al. 2004). However, the current knowledge on the structural assembly of the AAA peroxins Pex1p and Pex6p is still scarce and even though they are thought to form a hetero-oligomeric complex, the stoichiometry has not yet been solved. Distinct ATP-binding and hydrolysis sites contribute to the assembly of the AAA complex (Birschmann et al. 2003, 2005; Nashiro et al. 2011; Tamura et al. 2006; Saffian et al. 2012). In yeast, ATP binding and hydrolysis in Pex6p regulate the assembly and disassembly with Pex15p (Birschmann et al. 2003), while the Pex1p–Pex6p interaction is influenced by ATP binding in D2 of Pex1p (Birschmann et al. 2005). Furthermore, the release of the AAA peroxins from the peroxisomal membrane might be regulated by the E2-enzyme Pex4p, because Pex1p and Pex6p accumulate at the peroxisome in Pex4p-deficient yeast cells. This might indicate that the ubiquitin-dependent PTS1-receptor cycle and the dynamic ATPase cycle of the AAA peroxins are interconnected (Rosenkranz et al. 2006).

In addition to their involvement in matrix protein import, the AAA peroxins have been suggested to function in the fusion of pre-peroxisomal vesicles in yeasts (Titorenko and Rachubinski 2000; van der Zand et al. 2012), while Pex6p seems to be involved in the suppression of different cell death mechanisms (Jungwirth et al. 2008; Seo et al. 2007; Warner et al. 2003). However, the best analyzed function of Pex1p and Pex6p to date is their role in peroxisomal matrix protein import (Miyata and Fujiki 2005; Platta et al. 2005; Grimm et al. 2012; Fujiki et al. 2012).

While accumulating evidence strongly indicates that the purpose of monoubiquitination is to prime Pex5p for AAA complex-mediated dislocation, the direct mechanistic purpose of this modification remains elusive. In this context, it is interesting to note that the X-ray structure of the N-domain of murine Pex1p contains a double-psi-beta-barrel fold (Shiozawa et al. 2004). This fold is also present in the N-domain of p97, where it functions as binding module for ubiquitin (Park et al. 2005). However, if the domain found in Pex1p carries out a similar function still has to be investigated. AWP1 (Associated with PRK1) has been identified as a novel binding protein of human Pex6p (Miyata et al. 2012) and is supposed to contribute to linking of the AAA peroxins to the ubiquitinated Pex5p. Accordingly, AWP1 is required for peroxisomal biogenesis in vivo and the protein interacts with both Pex6p as well as with monoubiquitinated Pex5p (Miyata et al. 2012). Thus, AWP1 might function as specific adaptor, which links the modified Pex5p to the AAA peroxins and enables them to transfer their suggested pulling force to the monoubiquitinated PTS1-receptor. Interestingly, AWP1 has also been described as an ubiquitin-binding modulator of NF-kappaB (Fenner et al. 2009).

15.3.4 Deubiquitination of the Receptor

The ubiquitin moiety is removed from the PTS1-receptor during or shortly after the export step but certainly prior to a new round of matrix protein import. In general, the cleavage of ubiquitin from a substrate protein is catalyzed by ubiquitin hydrolases that are called deubiquitinating enzymes (Amerik and Hochstrasser 2004). The ubiquitin hydrolase Ubp15p has been identified as a binding partner of Pex6p in *S. cerevisiae* (Debelyy et al. 2011). Ubp15p functions as deubiquitinating enzyme acting on Pex5p, which represents the first characterized target of this enzyme (Debelyy et al. 2011). Work based on an in vitro system with mammalian proteins suggests that the thioester bond between ubiquitin and Pex5p can be cleaved either non-enzymatically via a nucleophilic attack of glutathione or, as the major pathway, enzyme-catalyzed by ubiquitin hydrolases (Grou et al. 2009b). USP9X has been described as the main deubiquitinating enzyme acting on mammalian Pex5p (Grou et al. 2012). USP9X is a cytosolic protein whose function is not restricted to peroxisomal protein import because it has been described to take part in the regulation of the transforming growth factor beta (TGFbeta) pathway (Dupont et al. 2009).

15.4 Functional Link Between Receptor Export and Cargo Release

Early work has defined that the import of peroxisomal matrix proteins requires the hydrolysis of ATP (Imanaka et al. 1987). Later studies identified the export of the receptor back to the cytosol as the energy-dependent step (Oliveira et al. 2003; Gouveia et al. 2003). In recent years, it has become evident that the ubiquitination machinery (Carvalho et al. 2007; Grou et al. 2008, 2009b; Okumoto et al. 2011; Platta et al. 2007a) as well as the AAA complex (Leon et al. 2006b; Miyata and Fujiki 2005; Miyata et al. 2012; Platta et al. 2005, 2007a; Kerssen et al. 2006) can be regarded as the only ATP-consuming factors of the peroxisomal protein import machinery. This indicates that energy consumption, matrix-protein import, and PTS-receptor export merge at the exportomer.

In this respect, it is interesting to note that the protein composition of the exportomer is functionally and evolutionary related to the proteins of the endoplasmic reticulum associated degradation (ERAD) machinery (Gabaldon et al. 2006; Schluter et al. 2006). ERAD can be defined as a mechanism by which misfolded proteins are polyubiquitinated and extracted from the ER in order to be disposed by the 26S proteasome in the cytosol (Hampton and Sommer 2012). Translocation systems that are in many aspects comparable to the exportomer and ERAD are the mitochondria associated degradation (MAD) for proteins of the outer mitochondrial membrane (Taylor and Rutter 2011) as well as the pre-protein translocator of complex plastids called symbiont-derived ERAD-like machinery (SELMA; Bolte et al. 2011). Therefore, a mechanistic parallel can be drawn between the exportomer, ERAD-, MAD-, and SELMA substrates because all are extracted by mechanoenzymes of the AAA-type ATPase family in an ubiquitination-dependent manner (Bolte et al. 2011; Platta et al. 2007b; Schliebs et al. 2010).

Based on this similarity, a model has been proposed that draws a direct interconnection of receptor export and the translocation of matrix proteins across the peroxisomal membrane (Schliebs et al. 2010). This "export-driven import model" is supported by the fact that the presence of a functional exportomer is a prerequisite for the import of matrix proteins. This, ATP is required for the ubiquitin- and AAA-driven extraction of the receptor and might be mechanically coupled to the translocation of the cargo proteins over the membrane.

Accordingly, the import defects observed in mutants of the exportomer can be explained in two ways. First, the binding capacity for functional PTS-receptors at the peroxisomal membrane seems to be limited. In fact, a decreased rate of receptor export caused by the functional impairment of the export machinery leads to an accumulation of PTS-receptors at the membrane (Leon et al. 2006b; Platta et al. 2004) and therefore would block the docking of new receptor–cargo complexes from the cytosol. In *A. thaliana*, the physiological defects of mutated and only insufficiently active Pex6p could be partially overcome when it was co-expressed with a weak allele of the docking protein Pex13p (Ratzel et al. 2011). This finding strongly indicates that the import and export rates of the PTS-receptors need to be balanced. Second, this model suggests that export of the receptor and the release of the cargo might be directly linked by a concerted mechanism. Work on the ubiquitination of the *S. cerevisiae* PTS2-co-receptor Pex18p delivered first direct evidence for such a connection (Hensel et al. 2011). Based on protease-protection assays, it was revealed that Pex7p is partially protease protected in wild-type cells, while Pex18p remains accessible. This topology is reversed when the cysteine of Pex18p is mutated or the AAA peroxins Pex1p/Pex6p are deleted (Hensel et al. 2011). This finding strongly indicates that monoubiquitination of Pex18p as well as AAA complex governs the import of cargo-loaded Pex7p. However, in the mammalian system, it is not yet clear whether the cargo release step itself requires ATP hydrolysis (Miyata et al. 2009) or does not (Alencastre et al. 2009).

In conclusion, the receptor export machinery is thought to function as the energy-consuming import motor for matrix proteins, either indirectly via balanced receptor import/export rates and/or directly via an interconnection of receptor export and cargo translocation.

15.5 Polyubiquitination of the PTS-Receptors

Under certain conditions, the PTS-receptors Pex5p, Pex18p, and Pex20p are polyubiquitinated on lysine residues in order to mark them for the degradation by the 26S proteasome (Hensel et al. 2011; Kiel et al. 2005b; Leon et al. 2006b; Platta et al. 2007a; Williams et al. 2007). This proteolytic pathway is induced when the normal monoubiquitination-dependent recycling pathway is blocked, as it is the case when constituents of the Pex4p or AAA complexes are deleted or the conserved cysteine of the PTS-receptor is mutated (Kiel et al. 2005a; Kragt et al. 2005; Platta et al. 2004; Leon and Subramani 2007; Hensel et al. 2011). The polyubiquitination of the *S. cerevisiae* PTS1-receptor Pex5p is predominantly catalyzed by Ubc4p and to a minor portion by the partial redundant enzymes Ubc5p and Ubc1p (Kiel et al. 2005a; Kragt et al. 2005; Platta et al. 2004). These three ubiquitin-conjugating enzymes display a high sequence similarity and are involved in diverse other cellular processes as well (Seufert and Jentsch 1990; Seufert et al. 1990). Both Pex10p (Williams et al. 2008) as well as Pex2p (Platta et al. 2009) have been suggested to function as E3 enzymes for the polyubiquitination of Pex5p. In this respect, it is interesting to note that a recent in vitro study demonstrates that Pex10p (RING) can synergistically enhance the ubiquitination activity of the Ubc4p–Pex2p (RING) enzyme pair (El Magraoui et al. 2012). This result suggests that both RING–peroxins may act together in the Ubc4p-dependent generation of K48-linked polyubiquitin chains on Pex5p.

Receptor polyubiquitination is enhanced when the export machinery is affected in its function and therefore the purpose of this modification is likely to remove the aberrant receptor molecules from the membrane when the normal extraction and recycling reaction is hampered. However, mutagenesis of the lysine residues required for polyubiquitination of *S. cerevisiae* Pex5p does not lead to a growth defect on oleate medium (Platta et al. 2007a; Williams et al. 2007). Interestingly, polyubiquitination of Pex5p can also be regarded as an alternative export signal. In vitro export assays demonstrated that a fraction of Pex5p is still exported even in a Pex4p-deficient system when the two conserved lysine residues required for polyubiquitination were still present (Platta et al. 2007a). Moreover, mutation of the conserved cysteine in *P. pastoris* Pex20p (Leon and Subramani 2007) induces polyubiquitination of Pex20p but still retains a partial functional receptor molecule that displays partial complementation in growth tests. Interestingly, only both the non-essential lysine targets for polyubiquitination of Pex20p as well as to the typically monoubiquitinated cysteine are mutated, the receptor completely loses its functionality (Leon and Subramani 2007). These data demonstrate that the enhanced degradation of Pex20p can restore the matrix protein import to a certain extent, supposedly because the receptors are removed efficiently enough to allow the docking of further cargo-bound receptors. This mechanism has been described as RADAR (receptor accumulation and degradation in the absence of recycling; Leon et al. 2006a, b) in order to distinguish it from the non-essential quality control. However, it should be noted that the mutation of the conserved cysteine of *S. cerevisiae* Pex5p and Pex18p alone is already sufficient to fully abolish the function of these receptors (Hensel et al. 2011; Williams et al. 2007). In this respect, it is interesting to note that degradation of Pex5p occurs much slower in *S. cerevisiae* than in most other species (Collins et al. 2000; Dodt and Gould 1996; van der Klei et al. 1998; Zolman and Bartel 2004; Zolman et al. 2005). Therefore, the observed instability of Pex5p in exportomer mutants in these species is most likely due to rapid degradation via K48-linked polyubiquitination as described for the PTS1-receptor of *H. polymorpha* (Kiel et al. 2005b).

The *S. cerevisiae* PTS2-co-receptor Pex18p behaves somewhat different from the PTS2-co-receptor Pex20p in *P. pastoris* and *H. polymorpha* because Pex18p shows a constitutive turnover already under wild-type conditions (Hensel et al. 2011; Leon et al. 2006b; Otzen et al. 2005; Purdue and Lazarow 2001). Currently, the functional impact of this instability is not known. In contrast to Pex18p, the PTS2-receptor Pex7p of *S. cerevisiae* is a stable protein (Hensel et al. 2011). So far, no indications for an ubiquitination of yeast Pex7p have been found. Interestingly, a recent report describes the polyubiquitination and degradation of *Arabidopsis* Pex7p when the dominant-negative GFP-Pex7p species is expressed in the cell (Cui et al. 2013). However, it is not clear if this mechanism is conserved in other organisms.

In general, the removal of the PTS-receptors via polyubiquitination is initiated when the monoubiquitination-dependent recycling pathway is blocked and therefore may function as alternative export signal.

Concluding Remarks

The combined work of several laboratories on the ubiquitination and recycling of the PTS-receptors has helped to uncover the functional contribution of distinct peroxisomal subcomplexes to the dislocation step and therefore enabled the definition of the peroxisomal receptor export machinery, the exportomer (Platta et al. 2013).

Certainly, many open questions remain to be answered and one of the most intriguing ones concerns the finding that Pex5p, Pex18p, and Pex20p are monoubiquitinated on a cysteine via a thioester bond and not by a more common isopeptide bond to a lysine. The first evidence that ubiquitin can be attached to cysteine, serine or threonine residues came from studies of viral MARCH (Membrane-associated RING-CH) E3 ligases that ubiquitinate MHC I (Major Histocompatibility Complex I) molecules (Cadwell and Coscoy 2005; Wang et al. 2007), and recent work demonstrates that this uncommon ubiquitination can also take place during ERAD (Ishikura et al. 2010; Shimizu et al. 2010). However, it is unclear how the specificity for these non-lysine ubiquitination reactions is ensured because the E2 and E3 enzymes involved are not restricted to this kind of modification and can also modify lysine residues (Wang et al. 2012). Interestingly, the cysteine of mammalian Pex5p (Grou et al. 2009b, 2012) and *P. pastoris* Pex20p (Leon and Subramani 2007) can be replaced by a lysine, which results in a still largely functional protein. Thus, even though the cysteine and the thioester-bond mediated ubiquitination of the PTS-receptors are evolutionary conserved, they are not essential for the principle export mechanisms and therefore may mainly represent an important regulatory device.

There are several different possibilities to explain the function of the conserved cysteine of the peroxisomal receptors. (1) The first concept is based on the fact that thioester bonds are less stable in comparison to isopeptide bonds. Therefore, the duration of the ubiquitin moiety at the PTS-receptor might be restricted in order to disable the formation of a polyubiquitin chain or to prevent the recognition by proteasomal adaptors. The rapid non-enzymatic disruption of the thioester bond of Ub-Pex5p in a mammalian in vitro system supports the idea that the cysteine-ubiquitination protects the PTS-receptors against degradation (Grou et al. 2009a, b). (2) Another concept takes into account that that certain E3 enzymes, like HECT-type ligases (Kee and Huibregtse 2007) or the RBR-type ligases (Wenzel et al. 2011) form an ubiquitin-thioester intermediate on a cysteine before this ubiquitin molecule is finally transferred to the substrate protein. One hypothetical model could be that once Pex12p/Pex10p have modified one of the receptor molecules of the oligomeric pore, Pex5p itself could catalyze an intra-oligomeric ubiquitin transfer in a relay-like system in order to accelerate the decomposition of the pore (Erdmann and Schliebs 2005; Platta et al. 2013). Interestingly, the E2-enzyme E2-230 K represents an example of an intramolecular ubiquitin transfer (Berleth and Pickart 1996), where ubiquitin is transferred from the first cysteine to a second cysteine of E2-230 K prior to attachment of the ubiquitin to the target protein. (3) A third concept is related to recent work on the regulation of the peroxisomal redox balance (Ivashchenko et al. 2011), which contributes to the general functional role of peroxisomes in the control of the cellular levels of reactive oxygen species (Bonekamp et al. 2009). One possibility is that the cysteine required for monoubiquitination might be accessible for redox changes. This might have a direct impact on the availability of this residue for the monoubiquitination and therefore could control the import/export rates of the receptor (Fransen et al. 2012).

In conclusion, the understanding of the late acting peroxins as concerted acting components of the exportomer will be instrumental to uncover the molecular mechanism underlying peroxisomal matrix protein import.

Acknowledgements We apologize to all the scientists whose work could not be cited due to space limitations. This work was supported by grants of the Deutsche Forschungsgemeinschaft (SFB 642 and FOR 1905) to RE and HWP.

References

Agne B, Meindl NM, Niederhoff K, Einwächter H, Rehling P, Sickmann A, Meyer HE, Girzalsky W, Kunau WH (2003) Pex8p. An intraperoxisomal organizer of the peroxisomal import machinery. Mol Cell 11(3):635–646

Albertini M, Girzalsky W, Veenhuis M, Kunau W-H (2001) Pex12p of *Saccharomyces cerevisiae* is a component of a multi-protein complex essential for peroxisomal matrix protein import. Eur J Cell Biol 80(4):257–270

Albiniak AM, Baglieri J, Robinson C (2012) Targeting of lumenal proteins across the thylakoid membrane. J Exp Bot 63(4):1689–1698

Alencastre IS, Rodrigues TA, Grou CP, Fransen M, Sá-Miranda C, Azevedo JE (2009) Mapping the Cargo Protein Membrane Translocation Step into the PEX5 Cycling Pathway. J Biol Chem 284(40):27243–27251

Amerik AY, Hochstrasser M (2004) Mechanism and function of deubiquitinating enzymes. Biochim Biophys Acta 1695:189–207

Baes M, Van Veldhoven PP (2012) Mouse models for peroxisome biogenesis defects and β-oxidation enzyme deficiencies. Biochim Biophys Acta 1822(9):1489–1500

Baker A, Graham IA, Holdsworth M, Smith SM, Theodoulou FL (2006) Chewing the fat: beta-oxidation in signalling and development. Trends Plant Sci 11(3):124–132

Berleth ES, Pickart CM (1996) Mechanism of ubiquitin conjugating enzyme E2-230 K: catalysis involving a thiol relay? Biochemistry 35(5):1664–1671

Berteaux-Lecellier V, Picard M, Thompson-Coffe C, Zickler D, Panvier-Adoutte A, Simonet JM (1995) A nonmammalian homolog of the *PAF1* gene (Zellweger syndrome) discovered as a gene involved in caryogamy in the fungus Podospora anserina. Cell 81(7):1043–1051

Beuron F, Flynn TC, Ma J, Kondo H, Zhang X, Freemont PS (2003) Motions and negative cooperativity between p97 domains revealed by cryo-electron microscopy and quantised elastic deformational model. J Mol Biol 327(3):619–629

Beyer A (1997) Sequence analysis of the AAA protein family. Protein Sci 6(10):2043–2058

Birschmann I, Stroobants AK, Van Den Berg M, Schäfer A, Rosenkranz K, Kunau WH, Tabak HF (2003) Pex15p of *Saccharomyces cerevisiae* Provides a Molecular Basis for Recruitment of the AAA Peroxin Pex6p to Peroxisomal Membranes. Mol Biol Cell 14(6):2226–2236

Birschmann I, Rosenkranz K, Erdmann R, Kunau WH (2005) Structural and functional analysis of the interaction of the AAA-peroxins Pex1p and Pex6p. FEBS J 272(1):47–58

Bolte K, Gruenheit N, Felsner G, Sommer MS, Maier UG, Hempel F (2011) Making new of old: recycling and modification of an ancient protein translocation system during eukaryotic evolution. Mechanistic comparison and phylogenetic analysis of ERAD, SELMA and the peroxisomal importomer. Bioessays 33(5):368–376

Bonekamp NA, Völkl A, Fahimi HD, Schrader M (2009) Reactive oxygen species and peroxisomes: struggling for balance. Biofactors 35(4):346–355

Brzovic PS, Klevit RE (2006) Ubiquitin transfer from the E2 perspective: why is UbcH5 so promiscuous? Cell Cycle 5:2867–2873

Burroughs AM, Iyer LM, Aravind L (2011) Functional diversification of the RING finger and other binuclear treble clef domains in prokaryotes and the early evolution of the ubiquitin system. Mol Biosyst 7(7):2261–2277

Cadwell K, Coscoy L (2005) Ubiquitination on nonlysine residues by a viral E3 ubiquitin ligase. Science 309:127–130

Carvalho AF, Pinto MP, Grou CP, Alencastre IS, Fransen M, Sa-Miranda C, Azevedo JE (2007) Ubiquitination of mammalian Pex5p, the peroxisomal import receptor. J Biol Chem 282 (43):31267–31272

Chang CC, Warren DS, Sacksteder KA, Gould SJ (1999) PEX12 Interacts with PEX5 and PEX10 and Acts Downstream of Receptor Docking in Peroxisomal Matrix Protein Import. J Cell Biol 147(4):761–774

Cipriano DJ, Jung J, Vivona S, Fenn TD, Brunger AT, Bryant Z (2013) Processive ATP-driven disassembly of SNARE complexes by the N-ethylmaleimide sensitive factor molecular machine. J Biol Chem. doi:10.1074/jbc.M1113.476705

Collins CS, Kalish JE, Morrell JC, McCaffery JM, Gould SJ (2000) The peroxisome biogenesis factors Pex4p, Pex22p, Pex1p, and Pex6p Act in the terminal steps of peroxisomal matrix protein import. Mol Cell Biol 20(20):7516–7526

Cooper TG, Beevers H (1969) Beta oxidation in glyoxysomes from castor bean endosperm. J Biol Chem 244(13):3514–3520

Crane DI, Kalish JE, Gould SJ (1994) The *Pichia pastoris PAS4* gene encodes a ubiquitin-conjugation enzyme required for peroxisome assembly. J Biol Chem 269(34):21835–21844

Cui S, Fukao Y, Mano S, Yamada K, Hayashi M, Nishimura M (2013) Proteomic analysis reveals that the Rab GTPase RabE1c is involved in the degradation of the peroxisomal protein receptor PEX7 (peroxin 7). J Biol Chem 288(8):doi: 1.1074/jbc.M1112.438143

Debelyy MO, Platta HW, Saffian D, Hensel A, Thoms S, Meyer HE, Warscheid B, Girzalsky W, Erdmann R (2011) Ubp15p, a ubiquitin hydrolase associated with the peroxisomal export machinery. J Biol Chem 286(32):28223–28234

Deshaies RJ, Joazeiro CA (2009) RING domain E3 ubiquitin ligases. Annu Rev Biochem 78:399–434

Distel B, Erdmann R, Gould SJ, Blobel G, Crane DI, Cregg JM, Dodt G, Fujiki Y, Goodman JM, Just WW, Kiel JA, Kunau WH, Lazarow PB, Mannaerts GP, Moser HW, Osumi T, Rachubinski RA, Roscher A, Subramani S, Tabak HF, Tsukamoto T, Valle D, van der Klei I, van Veldhoven PP, Veenhuis M (1996) A unified nomenclature for peroxisome biogenesis factors. J Cell Biol 135(1):1–3

Dodt G, Gould SJ (1996) Multiple *PEX* genes are required for proper subcellular distribution and stability of Pex5p, the PTS1 receptor: Evidence that PTS1 protein import is mediated by a cycling receptor. J Cell Biol 135:1763–1774

Dupont S, Mamidi A, Cordenonsi M, Montagner M, Zacchigna L, Adorno M, Martello G, Stinchfield MJ, Soligo S, Morsut L, Inui M, Moro S, Modena N, Argenton F, Newfeld SJ, Piccolo S (2009) FAM/USP9x, a deubiquitinating enzyme essential for TGFbeta signaling, controls Smad4 monoubiquitination. Cell 136(1):123–135

Ebberink MS, Mooijer PA, Gootjes J, Koster J, Wanders RJ, Waterham HR (2011) Genetic classification and mutational spectrum of more than 600 patients with a Zellweger syndrome spectrum disorder. Hum Mutat 32(1):59–69

Eckert JH, Johnsson N (2003) Pex10p links the ubiquitin conjugating enzyme Pex4p to the protein import machinery of the peroxisome. J Cell Sci 116(17):3623–3634

Eisenhaber B, Chumak N, Eisenhaber F, Hauser MT (2007) The ring between ring fingers (RBR) protein family. Genome Biol 8(3):209

El Magraoui F, Bäumer BE, Platta HW, Baumann JS, Girzalsky W, Erdmann R (2012) The RING-type ubiquitin ligases Pex2p, Pex10p and Pex12p form a heteromeric complex that displays enhanced activity in an ubiquitin conjugating enzyme-selective manner. FEBS J 279 (11):2060–2070

El Magraoui F, Brinkmeier R, Schrötter A, Girzalsky W, Müller T, Marcus K, Meyer HE, Erdmann R, Platta HW (2013) Distinct ubiquitination cascades act on the peroxisomal targeting signal type 2 co-receptor Pex18p. Traffic 14(12):1290–1301

Erdmann R, Schliebs W (2005) Peroxisomal matrix protein import: the transient pore model. Nat Rev Mol Cell Biol 6(9):738–742

Erdmann R, Wiebel FF, Flessau A, Rytka J, Beyer A, Fröhlich KU, Kunau W-H (1991) *PAS1*, a yeast gene required for peroxisome biogenesis, encodes a member of a novel family of putative ATPases. Cell 64(3):499–510

Fagarasanu A, Mast FD, Knoblach B, Rachubinski RA (2010) Molecular mechanism of organelle inheritance: lessons from peroxisomes in yeast. Nat Rev Mol Cell Biol 11(9):644–654

Fanelli F, Sepe S, D'Amelio M, Bernardi C, Cristiano L, Cimini A, Cecconi F, Ceru' MP, Moreno S (2013) Age-dependent roles of peroxisomes in the hippocampus of a transgenic mouse model of Alzheimer's disease. Mol Neurodegener 8 (8):doi:10.1186/1750-1326-1188-1188

Fenner BJ, Scannell M, Prehn JH (2009) Identification of polyubiquitin binding proteins involved in NF-kappaB signaling using protein arrays. Biochim Biophys Acta 1794(7):1010–1016

Fransen M, Nordgren M, Wang B, Apanasets O (2012) Role of peroxisomes in ROS/RNS-metabolism: Implications for human disease. Biochim Biophys Acta 1822(9):1363–1373

Freemont PS, Hanson IM, Trowsdale J (1991) A novel cysteine-rich sequence motif. Cell 64 (3):483–484

Fujiki Y, Nashiro C, Miyata N, Tamura S, Okumoto K (2012) New insights into dynamic and functional assembly of the AAA peroxins, Pex1p and Pex6p, and their membrane receptor Pex26p in shuttling of PTS1-receptor Pex5p during peroxisome biogenesis. Biochim Biophys Acta 1823(1):145–149

Furuki S, Tamura S, Matsumoto N, Miyata N, Moser A, Moser HW, Fujiki Y (2006) Mutations in the peroxin Pex26p responsible for peroxisome biogenesis disorders of complementation group 8 impair its stability, peroxisomal localization, and interaction with the Pex1p x Pex6p complex. J Biol Chem 281(3):1317–1323

Gabaldon T, Snel B, van Zimmeren F, Hemrika W, Tabak H, Huynen MA (2006) Origin and evolution of the peroxisomal proteome. Biol Direct 1:8

Geisbrecht BV, Collins CS, Reuber BE, Gould SJ (1998) Disruption of a PEX1-PEX6 interaction is the most common cause of the neurologic disorders Zellweger syndrome, neonatal adrenoleukodystrophy, and infantile Refsum disease. Proc Natl Acad Sci U S A 95 (15):8630–8635

Girzalsky W, Platta HW, Erdmann R (2009) Protein transport across the peroxisomal membrane. Biol Chem 390(8):745–751

Gonen H, Bercovich B, Orian A, Carrano A, Takizawa C, Yamanaka K, Pagano M, Iwai K, Ciechanover A (1999) Identification of the ubiquitin carrier proteins, E2s, involved in signal-induced conjugation and subsequent degradation of IkappaBalpha. J Biol Chem 274 (21):14823–14830

Goto S, Mano S, Nakamori C, Nishimura M (2011) Arabidopsis ABERRANT PEROXISOME MORPHOLOGY9 is a peroxin that recruits the PEX1-PEX6 complex to peroxisomes. Plant Cell 23(4):1573–1587

Gouveia AM, Guimaraes CP, Oliveira ME, Reguenga C, Sa-Miranda C, Azevedo JE (2003) Characterization of the peroxisomal cycling receptor Pex5p import pathway. Adv Exp Med Biol 544:213–220

Grimm I, Saffian D, Platta HW, Erdmann R (2012) The AAA-type ATPases Pex1p and Pex6p and their role in peroxisomal matrix protein import in Saccharomyces cerevisiae. Biochim Biophys Acta 1823(1):150–158

Grou CP, Carvalho AF, Pinto MP, Wiese S, Piechura H, Meyer HE, Warscheid B, Sa-Miranda C, Azevedo JE (2008) Members of the E2D (UbcH5) family mediate the ubiquitination of the conserved cysteine of Pex5p, the peroxisomal import receptor. J Biol Chem 283(21):14190–14197

Grou CP, Carvalho AF, Pinto MP, Alencastre IS, Rodrigues TA, Freitas MO, Francisco T, Sa-Miranda C, Azevedo JE (2009a) The peroxisomal protein import machinery–a case report of transient ubiquitination with a new flavor. Cell Mol Life Sci 66(2):254–262

Grou CP, Carvalho AF, Pinto MP, Huybrechts SJ, Sa-Miranda C, Fransen M, Azevedo JE (2009b) Properties of the ubiquitin-Pex5p thiol ester conjugate. J Biol Chem 284(16):10504–10513

Grou CP, Francisco T, Rodrigues TA, Freitas MO, Pinto MP, Carvalho AF, Domingues P, Wood SA, Rodríguez-Borges JE, Sá-Miranda C, Fransen M, Azevedo JE (2012) Identification of ubiquitin-specific protease 9X (USP9X) as a deubiquitinase acting on the ubiquitin-peroxin 5 (PEX5) thioester conjugate. J Biol Chem 287(16):12815–12827

Grunau S, Lay D, Mindthoff S, Platta HW, Girzalsky W, Just WW, Erdmann R (2011) The Phosphoinositide-3-kinase Vps34p is required for pexophagy in Saccharomyces cerevisiae. Biochem J 434:161–170

Gründlinger M, Yasmin S, Lechner BE, Geley S, Schrettl M, Hynes M, Haas H (2013) Fungal siderophore biosynthesis is partially localized in peroxisomes. Mol Microbiol 88(5):862–875

Hampton RY, Sommer T (2012) Finding the will and the way of ERAD substrate retrotranslocation. Curr Opin Cell Biol 24(4):460–466

Hasan S, Platta HW, Erdmann R (2013) Import of proteins into the peroxisomal matrix. Front Physiol 4:261

Hazra PP, Suriapranata I, Snyder WB, Subramani S (2002) Peroxisome remnants in pex3Delta cells and the requirement of Pex3p for interactions between the peroxisomal docking and translocation subcomplexes. Traffic 3(8):560–574

Hensel A, Beck S, El Magraoui F, Platta HW, Girzalsky W, Erdmann R (2011) Cysteine-dependent ubiquitination of Pex18p is linked to cargo translocation across the peroxisomal membrane. J Biol Chem 286:43495–43505

Imanaka T, Small GM, Lazarow PB (1987) Translocation of acyl-CoA oxidase into peroxisomes requires ATP hydrolysis but not a membrane potential. J Cell Biol 105:2915–2922

Ishikura S, Weissman AM, Bonifacino JS (2010) Serine residues in the cytosolic tail of the T-cell antigen receptor alpha-chain mediate ubiquitination and endoplasmic reticulum-associated degradation of the unassembled protein. J Biol Chem 285(31):23916–23924

Islinger M, Grille S, Fahimi HD, Schrader M (2012) The peroxisome: an update on mysteries. Histochem Cell Biol 137(5):547–574

Ivashchenko O, Van Veldhoven PP, Brees C, Ho YS, Terlecky SR, Fransen M (2011) Intraperoxisomal redox balance in mammalian cells: oxidative stress and interorganellar cross-talk. Mol Biol Cell 22(9):1440–1451

Iyer LM, Leipe DD, Koonin EV, Aravind L (2004) Evolutionary history and higher order classification of AAA + ATPases. J Struct Biol 146(1–2):11–31

Joo HJ, Kim KY, Yim YH, Jin YX, Kim H, Kim MY, Paik YK (2010) Contribution of the peroxisomal acox gene to the dynamic balance of daumone production in Caenorhabditis elegans. J Biol Chem 285(38):29319–29325

Jungwirth H, Ring J, Mayer T, Schauer A, Buttner S, Eisenberg T, Carmona-Gutierrez D, Kuchler K, Madeo F (2008) Loss of peroxisome function triggers necrosis. FEBS Lett 582 (19):2882–2886

Kaur N, Zhao Q, Xie Q, Hu J (2013) Arabidopsis RING peroxins are E3 ubiquitin ligases that interact with two homologous ubiquitin receptor proteins(F). J Integr Plant Biol 55(1):108–120

Kee Y, Huibregtse JM (2007) Regulation of catalytic activities of HECT ubiquitin ligases. Biochem Biophys Res Commun 354(2):329–333

Kerscher O, Felberbaum R, Hochstrasser M (2006) Modification of proteins by ubiquitin and ubiquitin-like proteins. Annu Rev Cell Dev Biol 22:159–180

Kerssen D, Hambruch E, Klaas W, Platta HW, de Kruijff B, Erdmann R, Kunau WH, Schliebs W (2006) Membrane association of the cycling peroxisome import receptor Pex5p. J Biol Chem 281(37):27003–27015

Kiel JA, Hilbrands RE, van der Klei IJ, Rasmussen SW, Salomons FA, van der Heide M, Faber KN, Cregg JM, Veenhuis M (1999) *Hansenula polymorpha* Pex1p and Pex6p are peroxisome-associated AAA proteins that functionally and physically interact. Yeast 15(11):1059–1078

Kiel JA, Hilbrands RE, Bovenberg RA, Veenhuis M (2000) Isolation of Penicillium chrysogenum *PEX1* and *PEX6* encoding AAA proteins involved in peroxisome biogenesis. Appl Microbiol Biotechnol 54(2):238–242

Kiel JA, Emmrich K, Meyer HE, Kunau WH (2005a) Ubiquitination of the peroxisomal targeting signal type 1 receptor, Pex5p, suggests the presence of a quality control mechanism during peroxisomal matrix protein import. J Biol Chem 280(3):1921–1930

Kiel JA, Otzen M, Veenhuis M, van der Klei IJ (2005b) Obstruction of polyubiquitination affects PTS1 peroxisomal matrix protein import. Biochim Biophys Acta 1745(2):176–186

Kiel JA, Veenhuis M, van der Klei IJ (2006) PEX genes in fungal genomes: common, rare or redundant. Traffic 7(10):1291–1303
Kienow L, Schneider K, Bartsch M, Stuible HP, Weng H, Miersch O, Wasternack C, Kombrink E (2008) Jasmonates meet fatty acids: functional analysis of a new acyl-coenzyme A synthetase family from Arabidopsis thaliana. J Exp Bot 59(2):403–419
Koellensperger G, Daubert S, Erdmann R, Hann S, Rottensteiner H (2007) Characterisation of zinc-binding domains of peroxisomal RING finger proteins using size exclusion chromatography/inductively coupled plasma-mass spectrometry. Biol Chem 388(11):1209–1214
Koller A, Snyder WB, Faber KN, Wenzel TJ, Rangell L, Keller GA, Subramani S (1999) Pex22p of *Pichia pastoris*, essential for peroxisomal matrix protein import, anchors the ubiquitin-conjugating enzyme, Pex4p, on the peroxisomal membrane. J Cell Biol 146(1):99–112
Kou J, Kovacs GG, Höftberger R, Kulik W, Brodde A, Forss-Petter S, Hönigschnabl S, Gleiss A, Brügger B, Wanders R, Just W, Budka H, Jungwirth S, Fischer P, Berger J (2011) Peroxisomal alterations in Alzheimer's disease. Acta Neuropathol 122(3):271–283
Kragt A, Voorn-Brouwer T, van den Berg M, Distel B (2005) The Saccharomyces cerevisiae peroxisomal import receptor Pex5p is monoubiquitinated in wild type cells. J Biol Chem 280 (9):7867–7874
Krazy H, Michels PA (2006) Identification and characterization of three peroxins–PEX6, PEX10 and PEX12–involved in glycosome biogenesis in Trypanosoma brucei. Biochim Biophys Acta 1763(1):6–17
Kunau W-H, Beyer A, Franken T, Götte K, Marzioch M, Saidowsky J, Skaletz-Rorowski A, Wiebel FF (1993) Two complementary approaches to study peroxisome biogenesis in *Saccharomyces cerevisiae*: Forward and reversed genetics. Biochimie 75:209–224
Lazarow PB, DeDuve C (1976) A fatty acyl-CoA oxidazing system in rat liver peroxisomes; enhancement by clofibrate, a hypolipidemic drug. Proc Natl Acad Sci U S A 73:2043–2046
Leon S, Subramani S (2007) A conserved cysteine residue of Pichia pastoris Pex20p is essential for its recycling from the peroxisome to the cytosol. J Biol Chem 282(10):7424–7430
Leon S, Goodman JM, Subramani S (2006a) Uniqueness of the mechanism of protein import into the peroxisome matrix: transport of folded, co-factor-bound and oligomeric proteins by shuttling receptors. Biochim Biophys Acta 1763(12):1552–1564
Leon S, Zhang L, McDonald WH, Yates J 3rd, Cregg JM, Subramani S (2006b) Dynamics of the peroxisomal import cycle of PpPex20p: ubiquitin-dependent localization and regulation. J Cell Biol 172(1):67–78
Liu X, Subramani S (2013) Unique requirements for mono- and polyubiquitination of the peroxisomal targeting signal co-receptor, Pex20. J Biol Chem 288(10):7230–7240
Liu X, Ma C, Subramani S (2012) Recent advances in peroxisomal matrix protein import. Curr Opin Cell Biol 24(4):484–489
Lizard G, Rouaud O, Demarquoy J, Cherkaoui-Malki M, Iuliano L (2012) Potential roles of peroxisomes in Alzheimer's disease and in dementia of the Alzheimer's type. J Alzheimers Dis 29(2):241–254
Matsumoto N, Tamura S, Fujiki Y (2003a) The pathogenic peroxin Pex26p recruits the Pex1p-Pex6p AAA ATPase complexes to peroxisomes. Nat Cell Biol 5(5):454–460
Matsumoto N, Tamura S, Furuki S, Miyata N, Moser A, Shimozawa N, Moser HW, Suzuki Y, Kondo N, Fujiki Y (2003b) Mutations in novel peroxin gene PEX26 that cause peroxisome-biogenesis disorders of complementation group 8 provide a genotype-phenotype correlation. Am J Hum Genet 73(2):233–246
Meijer WH, Gidijala L, Fekken S, Kiel JA, van den Berg MA, Lascaris R, Bovenberg RA, van der Klei IJ (2010) Peroxisomes are required for efficient penicillin biosynthesis in Penicillium chrysogenum. Appl Environ Micobiol 76(17):5702–5709
Miyata N, Fujiki Y (2005) Shuttling mechanism of peroxisome targeting signal type 1 receptor Pex5: ATP-independent import and ATP-dependent export. Mol Cell Biol 25(24):10822–10832

Miyata N, Hosoi K, Mukai S, Fujiki Y (2009) In vitro import of peroxisome-targeting signal type 2 (PTS2) receptor Pex7p into peroxisomes. Biochim Biophys Acta 1793(5):860–870

Miyata N, Okumoto K, Mukai S, Noguchi M, Fujiki Y (2012) AWP1/ZFAND6 Functions in Pex5 Export by Interacting with Cys-monoubiquitinated Pex5 and Pex6 AAA ATPase. Traffic 13 (1):168–183

Müller WH, van der Krift TP, Krouwer AJ, Wösten HA, van der Voort LH, Smaal EB, Verkleij AJ (1991) Localization of the pathway of the penicillin biosynthesis in Penicillium chrysogenum. EMBO J 10(2):489–495

Nagotu S, Kalel VC, Erdmann R, Platta HW (2012) Molecular basis of peroxisomal biogenesis disorders caused by defects in peroxisomal matrix protein import. Biochim Biophys Acta 1822 (9):1326–1336

Nashiro C, Kashiwagi A, Matsuzaki T, Tamura S, Fujiki Y (2011) Recruiting Mechanism of the AAA peroxins, Pex1p and Pex6p, to Pex26p on Peroxisome Membrane. Traffic 12(6):774–788

Neuwald AF, Aravind L, Spouge JL, Koonin EV (1999) AAA+: A class of chaperone-like ATPases associated with the assembly, operation, and disassembly of protein complexes. Genome Res 9(1):27–43

Nito K, Kamigaki A, Kondo M, Hayashi M, Nishimura M (2007) Functional classification of Arabidopsis peroxisome biogenesis factors proposed from analyses of knockdown mutants. Plant Cell Physiol 48(6):763–774

Novikoff AB, Novikoff PM, Davis C, Quintana N (1973) Studies on microperoxisomes. V are microperoxisomes ubiquitous in mammalian cells? J Histochem Cytochem 21(8):737–755

Oeljeklaus S, Reinartz BS, Wolf J, Wiese S, Tonillo J, Podwojski K, Kuhlmann K, Stephan C, Meyer HE, Schliebs W, Brocard C, Erdmann R, Warscheid B (2012) Identification of core components and transient interactors of the peroxisomal importomer by dual-track stable isotope labeling with amino acids in cell culture analysis. J Proteome Res 11(4):2567–2580

Okumoto K, Abe I, Fujiki Y (2000) Molecular anatomy of the peroxin Pex12p: RING finger domain is essential for the Pex12p function and interacts with the peroxisome targeting signal type 1-receptor Pex5p and a RING peroxin, Pex10p. J Biol Chem 275(33):25700–25710

Okumoto K, Misono S, Miyata N, Matsumoto Y, Mukai S, Fujiki Y (2011) Cysteine ubiquitination of PTS1 receptor Pex5p regulates Pex5p recycling. Traffic 12(8):1067–1083

Oliveira ME, Gouveia AM, Pinto RA, Sa-Miranda C, Azevedo JE (2003) The energetics of Pex5p-mediated peroxisomal protein import. J Biol Chem 278(41):39483–39488

Opaliński L, Veenhuis M, van der Klei IJ (2011) Peroxisomes: membrane events accompanying peroxisome proliferation. Int J Biochem Cell Biol 43(6):847–851

Otzen M, Wang D, Lunenborg MG, van der Klei IJ (2005) Hansenula polymorpha Pex20p is an oligomer that binds the peroxisomal targeting signal 2 (PTS2). J Cell Sci 118(Pt 15):3409–3418

Palmer T, Berks BC (2012) The twin-arginine translocation (Tat) protein export pathway. Nat Rev Microbiol 10(7):483–964

Park S, Isaacson R, Kim HT, Silver PA, Wagner G (2005) Ufd1 exhibits the AAA-ATPase fold with two distinct ubiquitin interaction sites. Structure 13(7):995–1005

Peraza-Reyes L, Zickler D, Berteaux-Lecellier V (2008) The peroxisome RING-finger complex is required for meiocyte formation in the fungus Podospora anserina. Traffic 9(11):1998–2009

Platta HW, Erdmann R (2007a) Peroxisomal dynamics. Trends Cell Biol 17(10):474–484

Platta HW, Erdmann R (2007b) The peroxisomal protein import machinery. FEBS Lett 581 (15):2811–2819

Platta HW, Girzalsky W, Erdmann R (2004) Ubiquitination of the peroxisomal import receptor Pex5p. Biochem J 384(Pt 1):37–45

Platta HW, Grunau S, Rosenkranz K, Girzalsky W, Erdmann R (2005) Functional role of the AAA peroxins in dislocation of the cycling PTS1 receptor back to the cytosol. Nat Cell Biol 7 (8):817–822

Platta HW, El Magraoui F, Schlee D, Grunau S, Girzalsky W, Erdmann R (2007a) Ubiquitination of the peroxisomal import receptor Pex5p is required for its recycling. J Cell Biol 177(2):197–204
Platta HW, Thoms S, Kunau WH, Erdmann R (2007b) Function of the ubiquitin-conjugating enzyme Pex4p and the AAA peroxins Pex1p and Pex6p in peroxisomal protein transport, vol 25. The Enzymes Elsevier Academic Press, Amsterdam
Platta HW, Debelyy MO, El Magraoui F, Erdmann R (2008) The AAA peroxins Pex1p and Pex6p function as dislocases for the ubiquitinated peroxisomal import receptor Pex5p. Biochem Soc Trans 36:99–104
Platta HW, El Magraoui F, Baumer BE, Schlee D, Girzalsky W, Erdmann R (2009) Pex2 and Pex12 function as protein-ubiquitin ligases in peroxisomal protein import. Mol Cell Biol 29 (20):5505–5516
Platta HW, Hagen S, Erdmann R (2013) The exportomer: the peroxisomal receptor export machinery. Cell Mol Life Sci 70(8):1393–1411
Prestele J, Hierl G, Scherling C, Hetkamp S, Schwechheimer C, Isono E, Weckwerth W, Wanner G, Gietl C (2010) Different functions of the C3HC4 zinc RING finger peroxins PEX10, PEX2, and PEX12 in peroxisome formation and matrix protein import. Proc Natl Acad Sci U S A 107(33):14915–14920
Purdue PE, Lazarow PB (2001) Pex18p is constitutively degraded during peroxisome biogenesis. J Biol Chem 276(50):47684–47689
Ratzel SE, Lingard MJ, Woodward AW, Bartel B (2011) Reducing PEX13 expression ameliorates physiological defects of late-acting peroxin mutants. Traffic 12(1):121–134
Ravid T, Hochstrasser M (2008) Diversity of degradation signals in the ubiquitin-proteasome system. Nat Rev Mol Cell Biol 9(9):679–690
Rosenkranz K, Birschmann I, Grunau S, Girzalsky W, Kunau WH, Erdmann R (2006) Functional association of the AAA complex and the peroxisomal importomer. FEBS J 273(16):3804–3815
Saffian D, Grimm I, Girzalsky W, Erdmann R (2012) ATP-dependent assembly of the heteromeric Pex1p-Pex6p-complex of the peroxisomal matrix protein import machinery. J Struct Biol 179 (2):126–132
Saville MK, Sparks A, Xirodimas DP, Wardrop J, Stevenson LF, Bourdon JC, Woods YL, Lane DP (2004) Regulation of p53 by the ubiquitin-conjugating enzymes UbcH5B/C in vivo. J Biol Chem 279(40):42169–42181
Schliebs W, Girzalsky W, Erdmann R (2010) Peroxisomal protein import and ERAD: variations on a common theme. Nat Rev Mol Cell Biol 11(12):885–890
Schluter A, Fourcade S, Ripp R, Mandel JL, Poch O, Pujol A (2006) The evolutionary origin of peroxisomes: an ER-peroxisome connection. Mol Biol Evol 23(4):838–845
Schumann U, Prestele J, O'Geen H, Brueggeman R, Wanner G, Gietl C (2007) Requirement of the C3HC4 zinc RING finger of the Arabidopsis PEX10 for photorespiration and leaf peroxisome contact with chloroplasts. Proc Natl Acad Sci U S A 104(3):1069–1074
Seo JG, Lai CY, Miceli MV, Jazwinski SM (2007) A novel role of peroxin PEX6: suppression of aging defects in mitochondria. Aging Cell 6(3):405–413
Seufert W, Jentsch S (1990) Ubiquitin-conjugating enzymes UBC4 and UBC5 mediate selective degradation of short-lived and abnormal proteins. EMBO J 9(2):543–550
Seufert W, McGrath JP, Jentsch S (1990) UBC1 encodes a novel member of an essential subfamily of yeast ubiquitin-conjugating enzymes involved in protein degradation. EMBO J 9(13):4535–3541
Shimizu Y, Okuda-Shimizu Y, Hendershot LM (2010) Ubiquitylation of an ERAD substrate occurs on multiple types of amino acids. Mol Cell 40(6):917–926
Shimozawa N, Tsukamoto T, Suzuki Y, Orii T, Shirayoshi Y, Mori T, Fujiki Y (1992) A human gene responsible for Zellweger syndrome that affects peroxisome assembly. Science 255 (5048):1132–1134

Shiozawa K, Maita N, Tomii K, Seto A, Goda N, Akiyama Y, Shimizu T, Shirakawa M, Hiroaki H (2004) Structure of the N-terminal domain of PEX1 AAA-ATPase. Characterization of a putative adaptor-binding domain. J Biol Chem 279(48):50060–50068

Sparkes IA, Brandizzi F, Slocombe SP, El-Shami M, Hawes C, Baker A (2003) An arabidopsis pex10 null mutant is embryo lethal, implicating peroxisomes in an essential role during plant embryogenesis. Plant Physiol 133(4):1809–1819

Spiegel CN, Batista-Pereira LG, Bretas JA, Eiras AE, Hooper AM, Peixoto AA, Soares MJ (2011) Pheromone gland development and pheromone production in lutzomyia longipalpis (Diptera: Psychodidae: Phlebotominae). J Med Entomol 48(3):489–495

Steinberg SJ, Dodt G, Raymond GV, Braverman NE, Moser AB, Moser HW (2006) Peroxisome biogenesis disorders. Biochim Biophys Acta 1763(12):1733–1748

Stinson BM, Nager AR, Glynn SE, Schmitz KR, Baker TA, Sauer RT (2013) Nucleotide binding and conformational switching in the hexameric ring of a AAA + machine. Cell 153(3):628–639

Tamura S, Shimozawa N, Suzuki Y, Tsukamoto T, Osumi T, Fujiki Y (1998) A cytoplasmic AAA family peroxin, Pex1p, interacts with Pex6p. Biochem Biophys Res Commun 245(3):883–886

Tamura S, Yasutake S, Matsumoto N, Fujiki Y (2006) Dynamic and functional assembly of the AAA peroxins, Pex1p and Pex6p, and their membrane receptor Pex26p. J Biol Chem 281 (38):27693–27704

Taylor EB, Rutter J (2011) Mitochondrial quality control by the ubiquitin-proteasome system. Biochem Soc Trans 39(5):1509–1513

Theodoulou FL, Bernhardt K, Linka N, Baker A (2013) Peroxisome membrane proteins: multiple trafficking routes and multiple functions? Biochem J 451(3):345–352

Till A, Lakhani R, Burnett SF, Subramani S (2012) Pexophagy: the selective degradation of peroxisomes. Int J Cell Biol 2012:512721

Titorenko VI, Rachubinski RA (2000) Peroxisomal membrane fusion requires two AAA family ATPases, Pex1p and Pex6p. J Cell Biol 150(4):881–886

Tower RJ, Fagarasanu A, Aitchison JD, Rachubinski RA (2011) The Peroxin Pex34p functions with the Pex11 family of peroxisomal divisional proteins to regulate the peroxisome population in yeast. Mol Biol Cell 22(10):1727–1738

Tsukamoto T, Miura S, Fujiki Y (1991) Restoration by a 35 K membrane protein of peroxisome assembly in a peroxisome-deficient mammalian cell mutant. Nature 350(6313):77–81

van der Klei IJ, Hilbrands RE, Kiel JAKW, Rasmussen SW, Cregg JM, Veenhuis M (1998) The ubiquitin-conjugating enzyme Pex4p of *Hansenula polymorpha* is required for efficient functioning of the PTS1 import machinery. EMBO J 17(13):3608–3618

van der Zand A, Gent J, Braakman I, Tabak HF (2012) Biochemically distinct vesicles from the endoplasmic reticulum fuse to form peroxisomes. Cell 149(2):397–409

Walker JE, Saraste M, Runswick MJ, Gay NJ (1982) Distantly related sequences in the alpha- and beta-subunits of ATP synthase, myosin, kinases and other ATP-requiring enzymes and a common nucleotide binding fold. Embo J 1(8):945–951

Wanders RJ, Waterham HR (2006a) Biochemistry of mammalian peroxisomes revisited. Annu Rev Biochem 75:295–332

Wanders RJ, Waterham HR (2006b) Peroxisomal disorders: the single peroxisomal enzyme deficiencies. Biochim Biophys Acta 1763(12):1707–1720

Wang X, Herr RA, Chua WJ, Lybarger L, Wiertz EJ, Hansen TH (2007) Ubiquitination of serine, threonine, or lysine residues on the cytoplasmic tail can induce ERAD of MHC-I by viral E3 ligase mK3. J Cell Biol 177(4):613–624

Wang X, Herr RA, Hansen TH (2012) Ubiquitination of substrates by esterification. Traffic 13 (1):19–24

Warner DR, Roberts EA, Greene RM, Pisano MM (2003) Identification of novel Smad binding proteins. Biochem Biophys Res Commun 312(4):1185–1190

Waterham HR, Ebberink MS (2012) Genetics and molecular basis of human peroxisome biogenesis disorders. Biochim Biophys Acta 1822(9):1430–1441

Wendler P, Ciniawsky S, Kock M, Kube S (2012) Structure and function of the AAA + nucleotide binding pocket. Biochim Biophys Acta 1823(1):2–14
Wenzel DM, Lissounov A, Brzovic PS, Klevit RE (2011) UBCH7 reactivity profile reveals parkin and HHARI to be RING/HECT hybrids. Nature 474:105–108
Widhalm JR, Ducluzeau AL, Buller NE, Elowsky CG, Olsen LJ, Basset GJ (2012) Phylloquinone (vitamin K(1)) biosynthesis in plants: two peroxisomal thioesterases of lactobacillales origin hydrolyze 1,4-dihydroxy-2-naphthoyl-Coa. Plant J 71(2):205–215
Wiebel FF, Kunau W-H (1992) The PAS2 protein essential for peroxisome biogenesis is related to ubiquitin-conjugating enzymes. Nature 359(6390):73–76
Williams C, van den Berg M, Sprenger RR, Distel B (2007) A conserved cysteine is essential for Pex4p-dependent ubiquitination of the peroxisomal import receptor Pex5p. J Biol Chem 282 (31):22534–22543
Williams C, van den Berg M, Geers E, Distel B (2008) Pex10p functions as an E3 ligase for the Ubc4p-dependent ubiquitination of Pex5p. Biochem Biophys Res Commun 374(4):620–624
Williams C, van den Berg M, Panjikar S, Stanley WA, Distel B, Wilmanns M (2012) Insights into ubiquitin-conjugating enzyme/ co-activator interactions from the structure of the Pex4p: Pex22p complex. EMBO J 31(2):391–402
Zolman BK, Bartel B (2004) An arabidopsis indole-3-butyric acid-response mutant defective in PEROXIN6, an apparent ATPase implicated in peroxisomal function. Proc Natl Acad Sci U S A 101(6):1786–1791
Zolman BK, Monroe-Augustus M, Silva ID, Bartel B (2005) Identification and functional characterization of arabidopsis PEROXIN4 and the interacting protein PEROXIN22. Plant Cell 17 (12):3422–3435

Factors Involved in Ubiquitination and Deubiquitination of PEX5, the Peroxisomal Shuttling Receptor

16

Tony A. Rodrigues, Tânia Francisco, Andreia F. Carvalho, Manuel P. Pinto, Cláudia P. Grou, and Jorge E. Azevedo

Abstract

Peroxisomal matrix proteins are synthesized on cytosolic ribosomes and post-translationally targeted to the organelle by the soluble factor PEX5. Besides a role as a receptor, and probably as a chaperone, PEX5 also holds the key to the matrix of the organelle. Indeed, the available data suggest that PEX5 itself pushes these proteins across the peroxisomal membrane using as driving force the strong protein–protein interactions that it establishes with components of the peroxisomal membrane docking/translocation module (DTM). In recent years, much has been learned on how this transport system is reset and kept fine-tuned. Notably, this involves covalent modification of PEX5 with ubiquitin. Two types of PEX5 ubiquitination have been characterized: monoubiquitination at a conserved cysteine, a mandatory event for the extraction of PEX5 from the DTM; and polyubiquitination, probably the result of a quality control mechanism aiming at clearing the DTM from entangled PEX5 molecules. Monoubiquitination of PEX5 is transient in nature and the factors that reverse this modification have recently been identified.

Keywords

Protein trafficking • PEX5 • Transient ubiquitination • Monoubiquitination • Polyubiquitination • Deubiquitination

---

T.A. Rodrigues • T. Francisco • J.E. Azevedo (✉)
Organelle Biogenesis and Function Group, Instituto de Biologia Celular e Molecular (IBMC), Universidade do Porto, R. do Campo Alegre, 823, 4150-180 Porto, Portugal

Instituto de Ciências Biomédicas Abel Salazar (ICBAS), Universidade do Porto, R. de Jorge Viterbo Ferreira, 228, 4050-313 Porto, Portugal
e-mail: jazevedo@ibmc.up.pt

A.F. Carvalho • M.P. Pinto • C.P. Grou
Organelle Biogenesis and Function Group, Instituto de Biologia Celular e Molecular (IBMC), Universidade do Porto, R. do Campo Alegre, 823, 4150-180 Porto, Portugal

C. Brocard and A. Hartig (eds.), *Molecular Machines Involved in Peroxisome Biogenesis and Maintenance*, DOI 10.1007/978-3-7091-1788-0_16, © Springer-Verlag Wien 2014

Abbreviations

AAA	ATPases associated with diverse cellular activities
DTM	Docking/translocation module
DUB	Deubiquitinase
E1	Ubiquitin-activating enzyme
E2	Ubiquitin-conjugating enzyme
E3	Ubiquitin ligase
PEX	Peroxin
PIM	Peroxisomal import machinery
PTS	Peroxisome targeting signal
REM	Receptor export module
RING	Really Interesting New Gene
TPRs	Tetratricopeptide repeats
Ub	Ubiquitin

16.1 Introduction

Peroxisomes are single membrane-bound organelles involved in numerous metabolic pathways (Hu et al. 2012; Islinger et al. 2010; Michels et al. 2006; Wanders and Waterham 2006). Their importance for human health and development is dramatically illustrated by a group of genetic diseases, the peroxisomal biogenesis disorders, in which peroxisome functions are partially or even completely impaired (Waterham and Ebberink 2012). Peroxisomal biogenesis disorders are caused by mutations in *PEX* genes, which encode proteins specifically involved in peroxisome maintenance and inheritance, the so-called peroxins (Distel et al. 1996). The majority of mammalian peroxins are components of the peroxisomal protein import machinery (PIM; Fig. 16.1). Collectively these proteins ensure the correct delivery of newly synthesized proteins to the organelle lumen. Most PIM peroxins are part of one of two functional/structural units: PEX13, PEX14, and the "Really Interesting New Gene" (RING) peroxins PEX2, PEX10, and PEX12 compose the membrane-embedded docking/translocation module (DTM; Agne et al. 2003; Oeljeklaus et al. 2012; Reguenga et al. 2001); the two peroxisomal "ATPases associated with diverse cellular activities" (AAA ATPases), PEX1 and PEX6, together with their peroxisomal membrane anchor, PEX26, comprise the receptor export module (REM; Matsumoto et al. 2003; Birschmann et al. 2003). The peroxisomal protein shuttling receptor PEX5 and its adaptor protein PEX7 complete the list of mammalian peroxins that integrate the PIM (Dodt et al. 1995; Fransen et al. 1995; Braverman et al. 1997). In addition to peroxins, the mammalian PIM also comprises other proteins, which are mostly involved in ubiquitination/deubiquitination events. Due to the fact that their function is not restricted to the PIM they are not classified as peroxins.

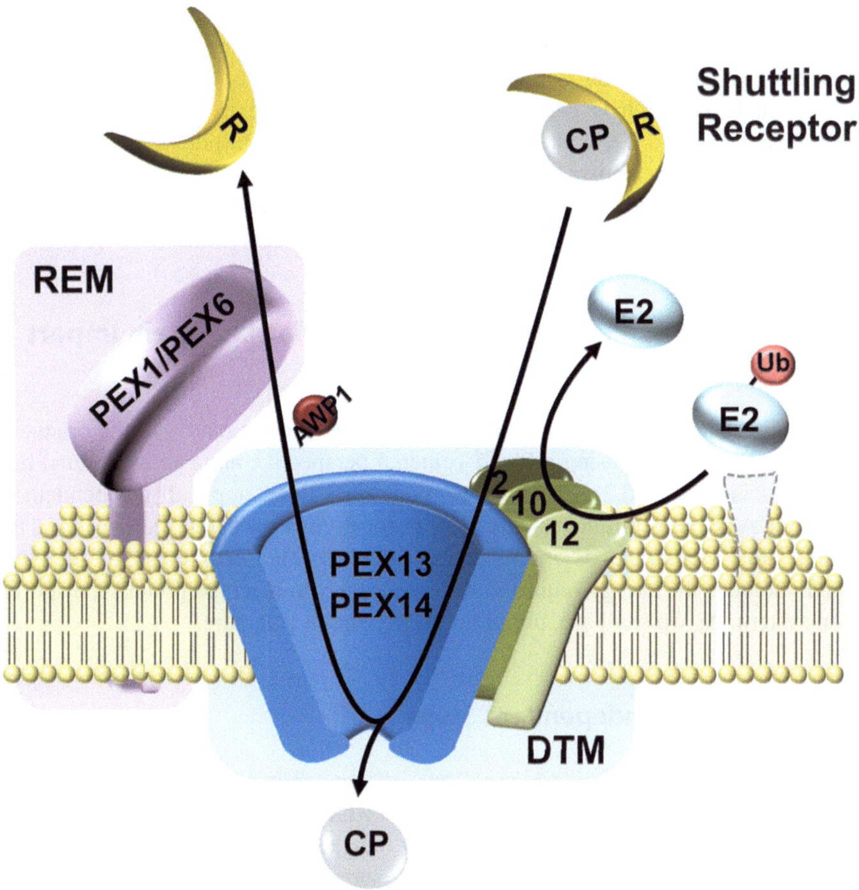

Fig. 16.1 Components of the peroxisomal protein import machinery (PIM). The components of the PIM are organized into structural/functional units. The docking/translocation module (DTM), a membrane-embedded protein complex comprises: PEX13, PEX14, and the RING peroxins PEX2, PEX10, and PEX12, whereas PEX1 and PEX6 and their membrane anchor, PEX26, constitute the receptor export module (REM). The shuttling receptor (R), a cargo protein (CP), ubiquitin (Ub), AWP1 (a soluble protein involved in receptor recycling), and the ubiquitin-conjugating enzyme E2 are also depicted. In some organisms the E2 is bound to the peroxisomal membrane via a membrane anchor (*dashed-line shape*)

The mammalian peroxins referred to above have orthologs in all peroxisome-containing organisms characterized so far from yeasts and fungi to plants (Kiel et al. 2006; Schluter et al. 2006). Strikingly, however, the reverse is not true. Indeed, several peroxins found in plants and lower eukaryotes do not exist in mammals. Apparently, evolution led to simpler PIMs. At least two different mechanisms seem to be behind this simplification. In one case, the function of two yeast/fungi/plant peroxins, PEX4 and PEX22, ended up being carried out by a family of mammalian ubiquitin-conjugating enzymes (E2D1/2/3) involved in many other pathways. In another case, a peroxin (PEX5) acquired the capacity to perform two different tasks (import of both PTS1- and PTS2-containing proteins; see below; Braverman et al. 1998; Galland et al. 2007; Otera et al. 1998; Woodward and Bartel 2005), each of which is performed by a different peroxin in yeasts/fungi (see Schliebs and Kunau 2006, and references cited therein). Despite these differences, the basics of the mechanism of protein import into the peroxisomal matrix remained relatively well conserved during evolution (Galland and Michels 2010; Grou et al. 2009a; Hu et al. 2012; Platta et al. 2013; Ma and Subramani 2009).

16.2 The PEX5-Mediated Peroxisomal Matrix Protein Import Pathway

The import pathway of newly synthesized proteins into the matrix of the organelle is generally described using a PEX5-centered perspective and can be divided into two parts (see Fig. 16.2). The first part comprises three steps, all independent of cytosolic ATP, and culminates with the translocation of the cargo protein across the peroxisomal membrane, as briefly explained below. The second part, on the other hand, includes the energy-requiring steps necessary to reset the protein transportation system. As we shall see, ubiquitin plays a major role here.

16.2.1 The ATP-Independent Steps

Peroxisomal matrix proteins possess one of two types of peroxisomal targeting signals (PTSs). The majority of them harbor a PTS type 1 (PTS1), a tripeptide with the sequence S-K-L or a variant present at their extreme C-termini (Brocard and Hartig 2006; Gould et al. 1989). A few peroxisomal matrix proteins possess instead a PTS2. This is an N-terminal-degenerated nonapeptide with the sequence (R/K)-(L/V/I)-X5-(H/Q)-(L/A) (Kunze et al. 2011; Lazarow 2006; Swinkels et al. 1991). In contrast to the PTS1, which remains intact upon import, the PTS2 is generally cleaved in the peroxisomal matrix by a serine protease (Kurochkin et al. 2007; Schuhmann et al. 2008). In mammals, plants, and many other organisms, all peroxisomal matrix proteins are transported to the peroxisome by PEX5 (Otera et al. 1998; Braverman et al. 1998; Galland et al. 2007; Woodward and Bartel 2005). This is a 70-kDa monomeric protein which in vivo displays a dual subcellular localization, peroxisomal and cytosolic, reflecting its role as a shuttling receptor (Costa-Rodrigues et al. 2005; Dodt and Gould 1996; Shiozawa et al. 2009). Interestingly, binding of PEX5 to newly synthesized matrix proteins that are oligomeric in their native state strongly inhibits their oligomerization, suggesting that PEX5 is also a chaperone/holdase (Freitas et al. 2011). This property is probably crucial to avoid premature oligomerization in the cytosol of proteins that no longer expose their PTS1 upon oligomerization (Luo et al. 2008; Tanaka et al. 2008). Structurally, PEX5 comprises two main domains: (1) a natively unfolded N-terminal half that interacts with the peroxins PEX7, PEX13, and PEX14 (Braverman et al. 1998;

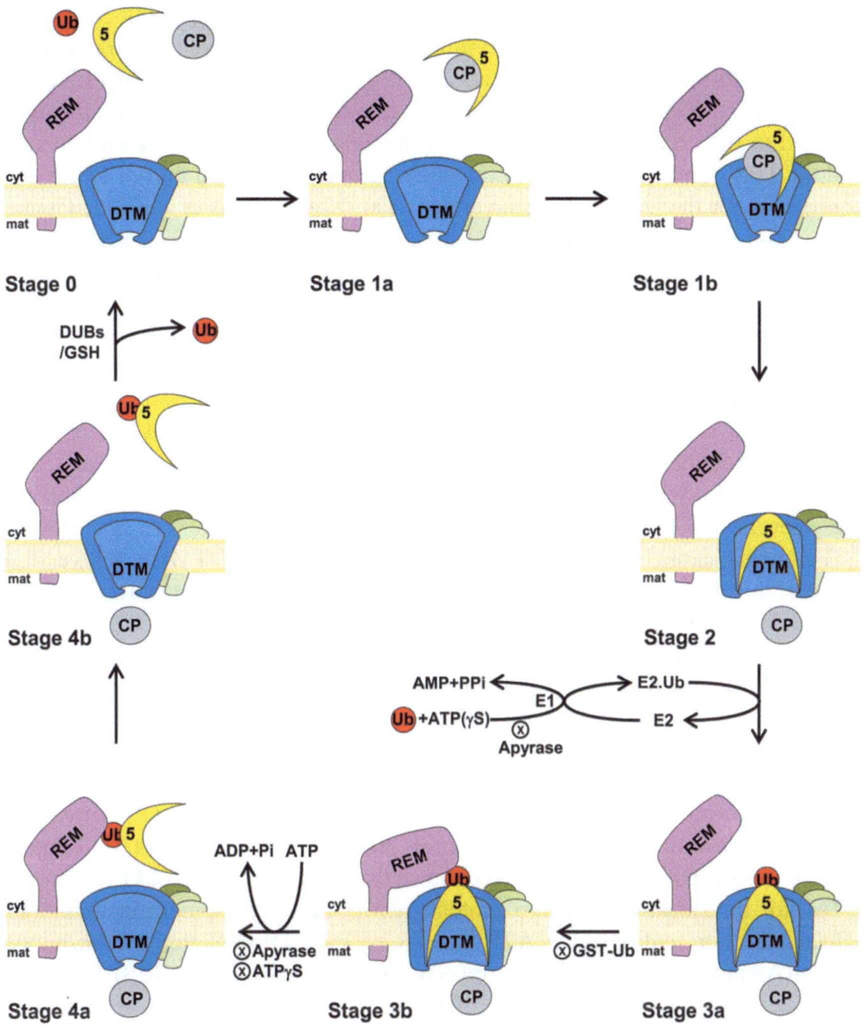

Fig. 16.2 The PEX5-mediated protein import pathway. The PEX5-mediated protein import pathway comprises eight steps. The PIM intermediates are referred to as stages, which are numbered 0 to 4 (some substages "a" and "b" are of conceptual nature). PEX5 (stage 0) binds a cargo protein (CP) in the cytosol (stage 1a). It then docks at the docking/translocation module (DTM; stage 1b), where it is inserted with the concomitant translocation of the cargo protein into the organelle matrix (stage 2). PEX5 is then monoubiquitinated (stage 3a), so that it can be recognized by the Receptor Export Module (REM) and dislocated into the cytosol, in an ATP-dependent manner (stage 3b to stage 4b). Finally, cytosolic Ub-PEX5 is deubiquitinated probably by a combination of enzymatic (DUBs) and non-enzymatic (GSH) mechanisms regenerating free PEX5 (stage 0). Strategies/reagents that have been used to block this pathway at different steps are indicated (*circled x*). Note that ATPγS, a non-hydrolyzable ATP analogue is efficiently used by the ubiquitin-activating enzyme (E1), but not by the REM. *Ub* ubiquitin, *E2* ubiquitin-conjugating enzyme, *DUBs* deubiquitinases, *GSH* glutathione, *GST-Ub* GST-tagged ubiquitin, *cyt* cytosol, *mat* peroxisomal matrix

Carvalho et al. 2006; Costa-Rodrigues et al. 2005; Otera et al. 2000; Saidowsky et al. 2001); and (2) a structured C-terminal half possessing seven tetratricopeptide repeats (TPRs) domains (Gatto et al. 2000). The interaction between PEX5 and PTS1 proteins is direct and sufficient to ensure that these proteins are efficiently targeted to the organelle. The interaction involves the PTS1 peptide on one side, and the TPR domains of PEX5 on the other, but the N-terminal half of PEX5 also contributes for the interaction (Braverman et al. 1998; Freitas et al. 2011; Gunkel et al. 2004; Klein et al. 2001, 2002; Oshima et al. 2008). The PTS2–PEX5 interaction requires the adaptor protein PEX7 (Galland et al. 2007; Lazarow 2006; Otera et al. 1998; Woodward and Bartel 2005; Braverman et al. 1998). In lower eukaryotes, PEX5 does not interact with PEX7, and therefore the receptor function of PEX5 is restricted to PTS1 proteins. In these organisms, targeting of PTS2 proteins is ensured by a species-specific receptor (PEX20, PEX18 or PEX21) which displays structural/functional similarities with the N-terminal half of mammalian PEX5, including the capacity to interact with PEX7 (Dodt et al. 2001; Lazarow 2006; Schliebs and Kunau 2006; Einwachter et al. 2001). We refer to these peroxins as PEX5-like proteins.

Following cargo recognition, PEX5 interacts with the DTM in a reversible manner; this docking step is then followed by the temperature-dependent insertion of PEX5 into the DTM (Costa-Rodrigues et al. 2004; Francisco et al. 2013). PEX5 at this stage displays a transmembrane topology having most of its polypeptide chain facing the peroxisomal matrix, whereas a 2-kDa N-terminal portion remains exposed to the cytosol (Gouveia et al. 2000, 2003a). Importantly, insertion of PEX5 into the DTM is a cargo protein-dependent process (Gouveia et al. 2003b). These observations are at the basis of the current model proposing that PEX5 pushes cargo proteins across the peroxisomal membrane as it gets inserted into the DTM (Azevedo et al. 2004; Grou et al. 2009a; Oliveira et al. 2003). Remarkably, in vitro import experiments have shown that neither insertion of PEX5 into the DTM nor translocation of cargo proteins across the peroxisomal membrane are affected by non-hydrolyzable ATP analogs or by ATP depletion of the import assays (Alencastre et al. 2009; Oliveira et al. 2003; Francisco et al. 2013). Likewise, ionophores have no effect on any of these events (Alencastre et al. 2009), in agreement with the fact that the peroxisomal membrane is readily permeable to small ions/molecules (Antonenkov and Hiltunen 2012; Rokka et al. 2009). Apparently, the PIM uses neither the energy of ATP hydrolysis nor a membrane potential to transport proteins from the cytosol into the organelle matrix. Altogether, these findings led us to propose that the driving force for the cargo protein translocation step resides in the strong protein–protein interactions that PEX5 establishes with components of the DTM (Oliveira et al. 2003).

16.3 Resetting the Peroxisomal Protein Import Machinery

The interaction of DTM-embedded PEX5 with components of this membrane module is essentially irreversible (Costa-Rodrigues et al. 2004). Therefore, it is not surprising that extraction of the receptor back into the cytosol requires energy input. Understanding the details of this extraction step was a difficult but particularly interesting task because it ended up revealing that the DTM is also an ubiquitin ligase. Indeed, PEX5 is monoubiquitinated every time it passes through the DTM, a mandatory modification for its subsequent ATP-dependent extraction back into the cytosol.

16.3.1 PEX5 Monoubiquitination

As stated above, DTM-embedded PEX5 exposes approximately 2 kDa of its N terminus to the cytosol. This small N-terminal domain includes a cysteine-containing motif that is conserved not only in PEX5 proteins from all organisms but also in PEX5-like proteins. Interestingly, deletion of the first 17 amino acids of human PEX5, which contains this motif, does not affect its capacity to become inserted into the DTM in a cargo-dependent manner, but renders it completely incompetent in the export step (Costa-Rodrigues et al. 2004). Likewise, deletion of the first 19 N-terminal amino acids of *Pichia pastoris* PEX20 also interferes with the normal recycling step (Leon et al. 2006). Similar results were observed when the conserved cysteine of both *P. pastoris* PEX20 and human PEX5 was replaced by a serine (Carvalho et al. 2007a; Leon and Subramani 2007). Clearly, this conserved cysteine residue has a determinant role in the receptor recycling step but the reason why it is so important became apparent only when this residue was found to be monoubiquitinated (Carvalho et al. 2007b; Williams et al. 2007). This modification is absolutely required for the next step of the PEX5-mediated protein import pathway, the extraction of monoubiquitinated PEX5 back into the cytosol (Carvalho et al. 2007b; Platta et al. 2007). Recent data confirmed that the PEX5-like proteins, PEX20 and PEX18, are also ubiquitinated at the conserved cysteine residue (Hensel et al. 2011; Liu and Subramani 2013).

The reason why a cysteine, and not the classical lysine residue, is the acceptor of ubiquitin in this reaction remains unknown. This is even more puzzling when we take into consideration the fact that substitution of the conserved cysteine residue in PEX5 by a lysine results in a seemingly normal protein that enters the DTM receives a single ubiquitin molecule and is exported back into the cytosol as efficiently as the wild-type protein in in vitro assays. Furthermore, when expressed in embryonic fibroblasts from a *PEX5* knockout mouse, this *pex5* mutant protein is capable of restoring peroxisomal protein import, again, as efficiently as the wild-type PEX5 protein (Grou et al. 2009b). Nevertheless, some hypotheses regarding the conserved cysteine have been formulated. These include the possibility to deubiquitinate cytosolic Ub-PEX5 using a non-enzymatic mechanism (see below) or the potential to block the DTM under some conditions (e.g., oxidative stress) through chemical modification of the conserved cysteine residue (e.g., oxidation, glutathiolation, nitrosylation, etc.). Obstruction of the DTM by export-incompetent PEX5 molecules would result in a cytosolic localization for newly synthesized peroxisomal enzymes (e.g., catalase, epoxide hydrolase, and glutathione S-transferase κ), a situation that might be advantageous under some stress conditions (see also Fransen et al. 2012 and Grou et al. 2009b).

16.3.2 PEX5 Dislocation

It is presently believed that monoubiquitination of DTM-embedded PEX5 serves no purpose other than preparing the receptor for the export step. Several arguments support this idea. First, as stated above, insertion of cargo-loaded PEX5 into the DTM in an in vitro import system is not affected when ATP is removed from the assays (Oliveira et al. 2003). Under these conditions, monoubiquitination of PEX5 at the DTM is no longer possible because the ubiquitin-activating enzyme (E1) uses ATP to activate ubiquitin. Likewise, PEX5 proteins lacking the conserved cysteine are still able to enter the DTM where they acquire the expected transmembrane topology (Carvalho et al. 2007a, b). Furthermore, and in agreement with these findings, PEX5-mediated import of pre-thiolase, a PTS2 protein, and its processing in the peroxisomal matrix are also not affected by removal of ATP from import reactions and similar results were obtained recently in our laboratory when studying the import pathway of a PTS1 protein (Alencastre et al. 2009; Francisco et al. 2013). Thus, monoubiquitination is necessary neither for the docking/insertion steps of PEX5 into the DTM nor for cargo protein translocation and release into the peroxisomal matrix. On the other hand, as stated above, PEX5 mutant proteins that cannot be monoubiquitinated are not substrates for the REM and accumulate at the DTM. Furthermore, monoubiquitination of PEX5 in in vitro import assays using a GST-ubiquitin fusion protein leads to the same outcome (Carvalho et al. 2007b). Altogether, these findings suggest that it is not the covalent modification of PEX5 per se that prepares the receptor for the export step (e.g., by inducing a conformational alteration of PEX5), but rather that the ubiquitin moiety in the DTM-embedded Ub-PEX5 conjugate provides a context-specific "handle" for the REM.

How the REM peroxins, PEX1 and PEX6, recognize DTM-embedded Ub-PEX5 remains largely unknown. Nevertheless, recent data suggest that the interaction between Ub-PEX5 and the REM may not be direct (Miyata et al. 2012). Indeed, using a mammalian in vitro import/export assay the authors found a cytosolic protein that stimulated export of PEX5 from the DTM. The protein was identified as AWP1, an ubiquitin-binding protein best known for its participation in the NF-κB signaling pathway (Chang et al. 2011). Further biochemical characterization of this protein led the authors to propose that AWP1 mediates the interaction between monoubiquitinated PEX5 and the REM (Miyata et al. 2012).

16.3.3 PEX5 Deubiquitination

Export of monoubiquitinated PEX5 from the DTM can be easily observed using a mammalian peroxisomal in vitro assay, particularly if the export reaction is made in the presence of a general deubiquitinase (DUB) inhibitor (e.g., ubiquitin aldehyde; Grou et al. 2009b). In contrast, all attempts to detect the mammalian or yeast Ub-PEX5 thioester conjugate in cytosolic fractions obtained from cells/organs yielded negative results; Ub-PEX5 could only be detected in organelle fractions (Grou et al. 2009b; Williams et al. 2007). Apparently, in vivo Ub-PEX5 is deubiquitinated very rapidly after export from the DTM. The factors that catalyze this deubiquitination step have been recently identified. Using biochemical approaches, UBP15 in *Saccharomyces cerevisiae* and USP9X in mammals have been identified as the DUBs acting on Ub-PEX5. Interestingly, however, knockout and knockdown of UBP15 and USP9X genes, respectively, did not result in the cytosolic accumulation of Ub-PEX5 (Debelyy et al. 2011; Grou et al. 2012). Obviously, there are other ways to deubiquitinate PEX5. These may include other less specific/active DUBs (Debelyy et al. 2011; Grou et al. 2012) or, as proposed previously, even a non-enzymatic mechanism because the thioester bond linking ubiquitin to PEX5 is much more labile than the typical isopeptide bond found in most ubiquitin conjugates (Grou et al. 2009b). Indeed, soluble Ub-PEX5 (but not DTM-embedded Ub-PEX5) is easily disrupted in the presence of 5 mM glutathione (a physiological concentration) displaying a half-life of just 2.3 min (Grou et al. 2009b).

16.4 The Ubiquitin-Conjugating Enzymes (E2s) and the Ubiquitin Ligase(s) (E3) of the PEX5-Mediated Protein Import Pathway

In order to react with a protein, ubiquitin has first to be activated by the ATP-dependent ubiquitin-activating enzyme (E1) yielding a thioester intermediate. This activated form of ubiquitin is then transferred to the catalytic cysteine of an ubiquitin-conjugating enzyme (E2). Finally, the Ub-E2 thioester conjugate is recruited by an ubiquitin ligase (E3) to the vicinity of the protein substrate which will then react with the activated ubiquitin (Spasser and Brik 2012). The specificity of a given protein ubiquitination reaction is thus imposed by the particular E2/E3 pair that catalyses that reaction (Metzger et al. 2013).

Most eukaryotic organisms have 1–2 E1s, dozens of E2s and an even larger number of E3s (e.g., more than 600 in mammals; Hutchins et al. 2013). The latter can be grouped into several different classes (Metzger et al. 2012), but of relevance here are the RING family of ubiquitin ligases (Deshaies and Joazeiro 2009). These E3s are characterized by a small protein domain (the RING domain) that binds two Zn^{2+} ions and adopts a typical cross-braced folding (Budhidarmo et al. 2012). RING domains have two distinct functions. First, they interact directly with E2s acting as recruitment platforms for Ub-E2 conjugates; second, they increase the reactivity of the E2-bound ubiquitin by repositioning the ubiquitin molecule further exposing the thioester bond to nucleophile attack (Metzger et al. 2013). This activation phenomenon is frequently explored in in vitro ubiquitination assays to test whether a RING domain really has E3 activity and also to define the E2(s) that are accepted by a given E3. Several experimental readouts are used in these assays, including monitoring auto-ubiquitination of the E3, ubiquitination of model substrates or even E3-dependent reactivity of the E2-Ub conjugate with single amino acids (e.g., lysine and cysteine; Wenzel et al. 2011; Swanson et al. 2001). In many of these assays, proximity of the nucleophilic protein/amino acid substrate to the RING-E2-Ub, more than its identity, is the crucial factor. This is achieved either by simply increasing the concentration of the substrate or by fusing it to the RING domain.

E3s can also be classified according to the mechanisms they use to recruit substrates. Some E3s possess substrate recruiting domains in other regions of their polypeptide chain (Metzger et al. 2013). Many, however, are subunits of larger proteins complexes and depend on other partners of the complex to recruit the substrate (e.g., Cullin RING ligases; Sarikas et al. 2011). These are frequently referred to as multi-subunit E3s (see also Metzger et al. 2013).

Which E2s and E3s participate in the PEX5-mediated protein import pathway? The first answer to this question was provided by two independent studies on yeast PEX4, an E2 long-known for its involvement in peroxisomal protein import (Wiebel and Kunau 1992). Indeed, it was shown that yeast strains lacking PEX4 do not monoubiquitinate PEX5 at the conserved cysteine residue (Williams et al. 2007), and are unable to recycle peroxisomal PEX5 back into the cytosol (Platta et al. 2007). Interestingly, and in contrast to the majority of E2s which are soluble proteins, PEX4 is stable and active only when bound to the peroxisomal membrane by another protein, PEX22 (Koller et al. 1999). Orthologs of both PEX4 and PEX22 have been found using bioinformatic analyses in several yeasts/fungi and plants, suggesting that all these organisms have a peroxisomal E2 dedicated to the peroxisomal protein import pathway (Kiel et al. 2006; Schluter et al. 2006; Zolman et al. 2005). Strikingly, however, no orthologs could be found in mammals and many other organisms. Proteomic studies aiming at identifying new mammalian peroxisomal proteins also failed to reveal the existence of a peroxisome-bound E2 (Islinger et al. 2007; Kikuchi et al. 2004; Wiese et al. 2007). An explanation for these negative findings was provided by a biochemical characterization of the mammalian E2. Using an in vitro system that recapitulates all steps of the peroxisomal protein import pathway, it was found that a low-speed centrifugation of post-nuclear supernatants was sufficient to separate peroxisomes from the E2 activity involved in PEX5 monoubiquitination. Purification of this activity led to the identification of three almost identical cytosolic E2s, E2D1/2/3 (UbcH5a/b/c in humans) (Grou et al. 2008), a group of E2s involved in many other biological pathways (Gonen et al. 1999; Saville et al. 2004).

As stated above, monoubiquitination of PEX5 at the conserved cysteine occurs at the DTM after cargo protein-dependent insertion of the receptor into this membrane module. Importantly, monoubiquitination of PEX5 can also be observed in an in vitro system comprising highly pure rat liver peroxisomes and recombinant E1 and UbcH5c (Grou et al. 2008), a finding strongly suggesting that the DTM itself is the E3 ligase catalyzing this unconventional ubiquitination. Three of the five core components of this module are the RING peroxins, PEX2, PEX10, and PEX12 and thus they are the most obvious candidates to perform this function. Interestingly, several studies suggest that PEX5 can still enter the DTM in cells lacking RING peroxins (Agne et al. 2003; Chang et al. 1999; Collins et al. 2000; Dodt and Gould 1996). Apparently, and similarly to multi-subunit E3s, the substrate-recruiting function of the DTM/E3 resides not in the RING peroxins but rather in other subunits of the complex.

Detailed mechanistic data on how the DTM monoubiquitinates PEX5 are still scarce. In vitro ubiquitination assays using recombinant RING domains from yeast and plant peroxins have shown that they all display E3 activity when assayed with human UbcH5 (Kaur et al. 2013; Williams et al. 2008) or with yeast PEX4 (Platta et al. 2009). Thus, in principle, any of the three RING peroxins could promote monoubiquitination of PEX5 at the conserved cysteine. However, data supporting this possibility are not yet available, despite some attempts (Platta et al. 2009). A definite answer to this question will probably require reconstituting a major part of the DTM/E3 using purified components and determine which of the RING peroxins, if any individually, is capable of promoting the correct type of ubiquitination, at the correct amino acid residue of PEX5. Such experiments may turn out to be quite demanding specially if we take into consideration very recent in vivo data suggesting that RING peroxins are not redundant and that all three together are required for receptor mono- and polyubiquitination (Liu and Subramani 2013, and see below).

16.5 Polyubiquitination of PEX5

Monoubiquitination of PEX5 is not the only type of ubiquitination occurring at the DTM. In yeast mutant strains lacking PIM components that act at late steps of the pathway (i.e., PEX5 monoubiquitination and its ATP-dependent dislocation from the DTM), small amounts of polyubiquitinated PEX5 are detected in peroxisomes (Kiel et al. 2005a; Platta et al. 2004; Kragt et al. 2005). Furthermore, the steady-state levels of PEX5 are diminished in some of these mutant strains (Collins et al. 2000; Koller et al. 1999) suggesting that this polyubiquitination event targets PEX5 for proteasomal degradation. A similar decrease in the steady-state levels of PEX5 in human cell lines from some patients with Peroxisome Biogenesis Disorders was also reported (Yahraus et al. 1996; Dodt and Gould 1996). Further characterization of this phenomenon in yeasts revealed that polyubiquitination of PEX5 is mediated by the multipurpose E2s Ubc1/Ubc4/Ubc5 (Kiel et al. 2005a; Kragt et al. 2005; Platta et al. 2004). Polyubiquitination of PEX5 does not seem to occur at the conserved cysteine residue; rather, one or two lysines located near the conserved cysteine have been identified as the ubiquitination sites (Kiel et al. 2005b; Williams et al. 2007). Although polyubiquitination of PEX5 has not yet been detected in wild-type strains, and substitution of those two PEX5 lysines by arginines has no phenotypic effects (Platta et al. 2007), it is possible, nevertheless, that this alternative way to remove PEX5 from the DTM is important whenever the normal recycling mechanism cannot be used, e.g., if PEX5 becomes entangled in the DTM.

Conclusions

The first clue that ubiquitin should play some role in the peroxisomal protein import pathway dates back to 1992 when one of the yeast genes involved in this pathway was found to encode the ubiquitin-conjugating enzyme, PEX4 (Wiebel and Kunau 1992). The awareness, a few years later, that the three RING peroxins present in all peroxisome-containing organisms might well be members of a vast family of ubiquitin ligases E3s (Joazeiro and Weissman 2000), fed this suspicion. However, the main mechanistic connection between ubiquitin and the PIM remained elusive for many years, and only in 2007 did we understand that ubiquitination at the PIM is not simply a manifestation of the quality control ubiquitin-proteasome system. Rather, ubiquitination is a mandatory step of this protein sorting pathway, occurring every single time a PEX5 molecule delivers a cargo protein into the matrix of the organelle. Clearly, the disruption of the Ub-PEX5 thioester conjugate by the thiol reagents commonly used in SDS-PAGE analyses tricked many researchers in the field, us included, for too long.

Any new finding ends up raising more questions than those it solved. One of the numerous questions still waiting for an answer is why a cysteine, and not, a lysine residue is used as the ubiquitin acceptor in PEX5. Also, the enzymology of the RING peroxins remains vastly unknown. How do they function? How are they regulated? In principle, either a single RING domain or a dimer of RINGs (Metzger et al. 2013) should be sufficient to catalyze monoubiquitination of PEX5. Why then are the three RING peroxins non-redundant and all necessary for receptor monoubiquitination (Liu and Subramani 2013)? Is it possible that the three RING peroxins are actually modules of a multi-RING E3 ligase? Obviously there are still many other new findings to be made.

Please note that there might be differences to the views expressed in the article by Effelsberg et al. in Chap. 13.

Acknowledgments This work was funded by FEDER funds through the Operational Competitiveness Programme—COMPETE and by National Funds through FCT—Fundação para a Ciência e a Tecnologia under the project FCOMP-01-0124-FEDER-019731 (PTDC/BIA-BCM/118577/2010). T. A. R., T. F., M. P. P. and C. P. G. are supported by Fundação para a Ciência e a Tecnologia, Programa Operacional Potencial Humano do QREN, and Fundo Social Europeu. A. F. C. is supported by Programa Ciência, funded by Programa Operacional Potencial Humano do QREN, Tipologia 4.2, Promoção do Emprego Científico, by Fundo Social Europeu and by national funds from Ministério da Ciência, Tecnologia e Ensino Superior.

References

Agne B, Meindl NM, Niederhoff K, Einwachter H, Rehling P, Sickmann A, Meyer HE, Girzalsky W, Kunau WH (2003) Pex8p: an intraperoxisomal organizer of the peroxisomal import machinery. Mol Cell 11(3):635–646

Alencastre IS, Rodrigues TA, Grou CP, Fransen M, Sa-Miranda C, Azevedo JE (2009) Mapping the cargo protein membrane translocation step into the PEX5 cycling pathway. J Biol Chem 284(40):27243–27251

Antonenkov VD, Hiltunen JK (2012) Transfer of metabolites across the peroxisomal membrane. Biochim Biophys Acta 1822(9):1374–1386

Azevedo JE, Costa-Rodrigues J, Guimaraes CP, Oliveira ME, Sa-Miranda C (2004) Protein translocation across the peroxisomal membrane. Cell Biochem Biophys 41(3):451–468

Birschmann I, Stroobants AK, van den Berg M, Schafer A, Rosenkranz K, Kunau WH, Tabak HF (2003) Pex15p of Saccharomyces cerevisiae provides a molecular basis for recruitment of the AAA peroxin Pex6p to peroxisomal membranes. Mol Biol Cell 14(6):2226–2236

Braverman N, Steel G, Obie C, Moser A, Moser H, Gould SJ, Valle D (1997) Human PEX7 encodes the peroxisomal PTS2 receptor and is responsible for rhizomelic chondrodysplasia punctata. Nat Genet 15(4):369–376

Braverman N, Dodt G, Gould SJ, Valle D (1998) An isoform of pex5p, the human PTS1 receptor, is required for the import of PTS2 proteins into peroxisomes. Hum Mol Genet 7(8):1195–1205

Brocard C, Hartig A (2006) Peroxisome targeting signal 1: is it really a simple tripeptide? Biochim Biophys Acta 1763(12):1565–1573

Budhidarmo R, Nakatani Y, Day CL (2012) RINGs hold the key to ubiquitin transfer. Trends Biochem Sci 37(2):58–65

Carvalho AF, Costa-Rodrigues J, Correia I, Costa Pessoa J, Faria TQ, Martins CL, Fransen M, Sa-Miranda C, Azevedo JE (2006) The N-terminal half of the peroxisomal cycling receptor Pex5p is a natively unfolded domain. J Mol Biol 356(4):864–875

Carvalho AF, Grou CP, Pinto MP, Alencastre IS, Costa-Rodrigues J, Fransen M, Sa-Miranda C, Azevedo JE (2007a) Functional characterization of two missense mutations in Pex5p–C11S and N526K. Biochim Biophys Acta 1773(7):1141–1148

Carvalho AF, Pinto MP, Grou CP, Alencastre IS, Fransen M, Sa-Miranda C, Azevedo JE (2007b) Ubiquitination of mammalian Pex5p, the peroxisomal import receptor. J Biol Chem 282 (43):31267–31272

Chang CC, Warren DS, Sacksteder KA, Gould SJ (1999) PEX12 interacts with PEX5 and PEX10 and acts downstream of receptor docking in peroxisomal matrix protein import. J Cell Biol 147 (4):761–774

Chang EJ, Ha J, Kang SS, Lee ZH, Kim HH (2011) AWP1 binds to tumor necrosis factor receptor-associated factor 2 (TRAF2) and is involved in TRAF2-mediated nuclear factor-kappaB signaling. Int J Biochem Cell Biol 43(11):1612–1620

Collins CS, Kalish JE, Morrell JC, McCaffery JM, Gould SJ (2000) The peroxisome biogenesis factors pex4p, pex22p, pex1p, and pex6p act in the terminal steps of peroxisomal matrix protein import. Mol Cell Biol 20(20):7516–7526

Costa-Rodrigues J, Carvalho AF, Gouveia AM, Fransen M, Sa-Miranda C, Azevedo JE (2004) The N terminus of the peroxisomal cycling receptor, Pex5p, is required for redirecting the peroxisome-associated peroxin back to the cytosol. J Biol Chem 279(45):46573–46579

Costa-Rodrigues J, Carvalho AF, Fransen M, Hambruch E, Schliebs W, Sa-Miranda C, Azevedo JE (2005) Pex5p, the peroxisomal cycling receptor, is a monomeric non-globular protein. J Biol Chem 280(26):24404–24411

Debelyy MO, Platta HW, Saffian D, Hensel A, Thoms S, Meyer HE, Warscheid B, Girzalsky W, Erdmann R (2011) Ubp15p, a ubiquitin hydrolase associated with the peroxisomal export machinery. J Biol Chem 286(32):28223–28234

Deshaies RJ, Joazeiro CA (2009) RING domain E3 ubiquitin ligases. Annu Rev Biochem 78:399–434

Distel B, Erdmann R, Gould SJ, Blobel G, Crane DI, Cregg JM, Dodt G, Fujiki Y, Goodman JM, Just WW, Kiel JA, Kunau WH, Lazarow PB, Mannaerts GP, Moser HW, Osumi T, Rachubinski RA, Roscher A, Subramani S, Tabak HF, Tsukamoto T, Valle D, van der Klei I, van Veldhoven PP, Veenhuis M (1996) A unified nomenclature for peroxisome biogenesis factors. J Cell Biol 135(1):1–3

Dodt G, Gould SJ (1996) Multiple PEX genes are required for proper subcellular distribution and stability of Pex5p, the PTS1 receptor: evidence that PTS1 protein import is mediated by a cycling receptor. J Cell Biol 135(6 Pt 2):1763–1774

Dodt G, Braverman N, Wong C, Moser A, Moser HW, Watkins P, Valle D, Gould SJ (1995) Mutations in the PTS1 receptor gene, PXR1, define complementation group 2 of the peroxisome biogenesis disorders. Nat Genet 9(2):115–125

Dodt G, Warren D, Becker E, Rehling P, Gould SJ (2001) Domain mapping of human PEX5 reveals functional and structural similarities to Saccharomyces cerevisiae Pex18p and Pex21p. J Biol Chem 276(45):41769–41781

Einwachter H, Sowinski S, Kunau WH, Schliebs W (2001) Yarrowia lipolytica Pex20p, Saccharomyces cerevisiae Pex18p/Pex21p and mammalian Pex5pL fulfil a common function in the early steps of the peroxisomal PTS2 import pathway. EMBO Rep 2(11):1035–1039

Francisco T, Rodrigues TA, Freitas MO, Grou CP, Carvalho AF, Sá-Miranda C, Pinto MP, Azevedo JE (2013) A cargo-centered perspective on the PEX5 receptor-mediated peroxisomal protein import pathway. J Biol Chem 288(40):29151–29159

Fransen M, Brees C, Baumgart E, Vanhooren JC, Baes M, Mannaerts GP, Van Veldhoven PP (1995) Identification and characterization of the putative human peroxisomal C-terminal targeting signal import receptor. J Biol Chem 270(13):7731–7736

Fransen M, Nordgren M, Wang B, Apanasets O (2012) Role of peroxisomes in ROS/RNS-metabolism: implications for human disease. Biochim Biophys Acta 1822(9):1363–1373

Freitas MO, Francisco T, Rodrigues TA, Alencastre IS, Pinto MP, Grou CP, Carvalho AF, Fransen M, Sa-Miranda C, Azevedo JE (2011) PEX5 protein binds monomeric catalase blocking its tetramerization and releases it upon binding the N-terminal domain of PEX14. J Biol Chem 286(47):40509–40519

Galland N, Michels PA (2010) Comparison of the peroxisomal matrix protein import system of different organisms. Exploration of possibilities for developing inhibitors of the import system of trypanosomatids for anti-parasite chemotherapy. Eur J Cell Biol 89(9):621–637

Galland N, Demeure F, Hannaert V, Verplaetse E, Vertommen D, Van der Smissen P, Courtoy PJ, Michels PA (2007) Characterization of the role of the receptors PEX5 and PEX7 in the import of proteins into glycosomes of Trypanosoma brucei. Biochim Biophys Acta 1773(4):521–535

Gatto GJ Jr, Geisbrecht BV, Gould SJ, Berg JM (2000) A proposed model for the PEX5-peroxisomal targeting signal-1 recognition complex. Proteins 38(3):241–246

Gonen H, Bercovich B, Orian A, Carrano A, Takizawa C, Yamanaka K, Pagano M, Iwai K, Ciechanover A (1999) Identification of the ubiquitin carrier proteins, E2s, involved in signal-induced conjugation and subsequent degradation of IkappaBalpha. J Biol Chem 274 (21):14823–14830

Gould SJ, Keller GA, Hosken N, Wilkinson J, Subramani S (1989) A conserved tripeptide sorts proteins to peroxisomes. J Cell Biol 108(5):1657–1664

Gouveia AM, Reguenga C, Oliveira ME, Sa-Miranda C, Azevedo JE (2000) Characterization of peroxisomal Pex5p from rat liver. Pex5p in the Pex5p-Pex14p membrane complex is a transmembrane protein. J Biol Chem 275(42):32444–32451

Gouveia AM, Guimaraes CP, Oliveira ME, Reguenga C, Sa-Miranda C, Azevedo JE (2003a) Characterization of the peroxisomal cycling receptor, Pex5p, using a cell-free in vitro import system. J Biol Chem 278(1):226–232

Gouveia AM, Guimaraes CP, Oliveira ME, Sa-Miranda C, Azevedo JE (2003b) Insertion of Pex5p into the peroxisomal membrane is cargo protein-dependent. J Biol Chem 278(7):4389–4392

Grou CP, Carvalho AF, Pinto MP, Wiese S, Piechura H, Meyer HE, Warscheid B, Sa-Miranda C, Azevedo JE (2008) Members of the E2D (UbcH5) family mediate the ubiquitination of the conserved cysteine of Pex5p, the peroxisomal import receptor. J Biol Chem 283(21):14190–14197

Grou CP, Carvalho AF, Pinto MP, Alencastre IS, Rodrigues TA, Freitas MO, Francisco T, Sa-Miranda C, Azevedo JE (2009a) The peroxisomal protein import machinery–a case report of transient ubiquitination with a new flavor. Cell Mol Life Sci 66(2):254–262

Grou CP, Carvalho AF, Pinto MP, Huybrechts SJ, Sa-Miranda C, Fransen M, Azevedo JE (2009b) Properties of the ubiquitin-pex5p thiol ester conjugate. J Biol Chem 284(16):10504–10513

Grou CP, Francisco T, Rodrigues TA, Freitas MO, Pinto MP, Carvalho AF, Domingues P, Wood SA, Rodriguez-Borges JE, Sa-Miranda C, Fransen M, Azevedo JE (2012) Identification of ubiquitin-specific protease 9X (USP9X) as a deubiquitinase acting on ubiquitin-peroxin 5 (PEX5) thioester conjugate. J Biol Chem 287(16):12815–12827

Gunkel K, van Dijk R, Veenhuis M, van der Klei IJ (2004) Routing of Hansenula polymorpha alcohol oxidase: an alternative peroxisomal protein-sorting machinery. Mol Biol Cell 15 (3):1347–1355

Hensel A, Beck S, El Magraoui F, Platta HW, Girzalsky W, Erdmann R (2011) Cysteine-dependent ubiquitination of Pex18p is linked to cargo translocation across the peroxisomal membrane. J Biol Chem 286(50):43495–43505

Hu J, Baker A, Bartel B, Linka N, Mullen RT, Reumann S, Zolman BK (2012) Plant peroxisomes: biogenesis and function. Plant Cell 24(6):2279–2303

Hutchins AP, Liu S, Diez D, Miranda-Saavedra D (2013) The repertoires of ubiquitinating and deubiquitinating enzymes in eukaryotic genomes. Mol Biol Evol 30(5):1172–1187

Islinger M, Luers GH, Li KW, Loos M, Volkl A (2007) Rat liver peroxisomes after fibrate treatment. A survey using quantitative mass spectrometry. J Biol Chem 282(32):23055–23069

Islinger M, Cardoso MJ, Schrader M (2010) Be different–the diversity of peroxisomes in the animal kingdom. Biochim Biophys Acta 1803(8):881–897

Joazeiro CA, Weissman AM (2000) RING finger proteins: mediators of ubiquitin ligase activity. Cell 102(5):549–552

Kaur N, Zhao Q, Xie Q, Hu J (2013) Arabidopsis RING peroxins are E3 ubiquitin ligases that interact with two homologous ubiquitin receptor proteins(F). J Integr Plant Biol 55(1):108–120

Kiel JA, Emmrich K, Meyer HE, Kunau WH (2005a) Ubiquitination of the peroxisomal targeting signal type 1 receptor, Pex5p, suggests the presence of a quality control mechanism during peroxisomal matrix protein import. J Biol Chem 280(3):1921–1930

Kiel JA, Otzen M, Veenhuis M, van der Klei IJ (2005b) Obstruction of polyubiquitination affects PTS1 peroxisomal matrix protein import. Biochim Biophys Acta 1745(2):176–186

Kiel JA, Veenhuis M, van der Klei IJ (2006) PEX genes in fungal genomes: common, rare or redundant. Traffic 7(10):1291–1303

Kikuchi M, Hatano N, Yokota S, Shimozawa N, Imanaka T, Taniguchi H (2004) Proteomic analysis of rat liver peroxisome: presence of peroxisome-specific isozyme of Lon protease. J Biol Chem 279(1):421–428

Klein AT, Barnett P, Bottger G, Konings D, Tabak HF, Distel B (2001) Recognition of peroxisomal targeting signal type 1 by the import receptor Pex5p. J Biol Chem 276(18):15034–15041

Klein AT, van den Berg M, Bottger G, Tabak HF, Distel B (2002) Saccharomyces cerevisiae acyl-CoA oxidase follows a novel, non-PTS1, import pathway into peroxisomes that is dependent on Pex5p. J Biol Chem 277(28):25011–25019

Koller A, Snyder WB, Faber KN, Wenzel TJ, Rangell L, Keller GA, Subramani S (1999) Pex22p of Pichia pastoris, essential for peroxisomal matrix protein import, anchors the ubiquitin-conjugating enzyme, Pex4p, on the peroxisomal membrane. J Cell Biol 146(1):99–112

Kragt A, Voorn-Brouwer T, van den Berg M, Distel B (2005) The Saccharomyces cerevisiae peroxisomal import receptor Pex5p is monoubiquitinated in wild type cells. J Biol Chem 280 (9):7867–7874

Kunze M, Neuberger G, Maurer-Stroh S, Ma J, Eck T, Braverman N, Schmid JA, Eisenhaber F, Berger J (2011) Structural requirements for interaction of peroxisomal targeting signal 2 and its receptor PEX7. J Biol Chem 286(52):45048–45062

Kurochkin IV, Mizuno Y, Konagaya A, Sakaki Y, Schonbach C, Okazaki Y (2007) Novel peroxisomal protease Tysnd1 processes PTS1- and PTS2-containing enzymes involved in beta-oxidation of fatty acids. EMBO J 26(3):835–845

Lazarow PB (2006) The import receptor Pex7p and the PTS2 targeting sequence. Biochim Biophys Acta 1763(12):1599–1604

Leon S, Subramani S (2007) A conserved cysteine residue of Pichia pastoris Pex20p is essential for its recycling from the peroxisome to the cytosol. J Biol Chem 282(10):7424–7430

Leon S, Zhang L, McDonald WH, Yates J 3rd, Cregg JM, Subramani S (2006) Dynamics of the peroxisomal import cycle of PpPex20p: ubiquitin-dependent localization and regulation. J Cell Biol 172(1):67–78

Liu X, Subramani S (2013) Unique requirements for mono- and polyubiquitination of the peroxisomal targeting signal co-receptor, Pex20. J Biol Chem 288(10):7230–7240

Luo B, Norris C, Bolstad ES, Knecht DA, Grant DF (2008) Protein quaternary structure and expression levels contribute to peroxisomal-targeting-sequence-1-mediated peroxisomal import of human soluble epoxide hydrolase. J Mol Biol 380(1):31–41

Ma C, Subramani S (2009) Peroxisome matrix and membrane protein biogenesis. IUBMB Life 61 (7):713–722

Matsumoto N, Tamura S, Fujiki Y (2003) The pathogenic peroxin Pex26p recruits the Pex1p-Pex6p AAA ATPase complexes to peroxisomes. Nat Cell Biol 5(5):454–460

Metzger MB, Hristova VA, Weissman AM (2012) HECT and RING finger families of E3 ubiquitin ligases at a glance. J Cell Sci 125(Pt 3):531–537

Metzger MB, Pruneda JN, Klevit RE, Weissman AM (2013) RING-type E3 ligases: master manipulators of E2 ubiquitin-conjugating enzymes and ubiquitination. Biochim Biophys Acta. doi:10.1016/j.bbamcr.2013.05.026

Michels PA, Bringaud F, Herman M, Hannaert V (2006) Metabolic functions of glycosomes in trypanosomatids. Biochim Biophys Acta 1763(12):1463–1477

Miyata N, Okumoto K, Mukai S, Noguchi M, Fujiki Y (2012) AWP1/ZFAND6 functions in Pex5 export by interacting with cys-monoubiquitinated Pex5 and Pex6 AAA ATPase. Traffic 13 (1):168–183

Oeljeklaus S, Reinartz BS, Wolf J, Wiese S, Tonillo J, Podwojski K, Kuhlmann K, Stephan C, Meyer HE, Schliebs W, Brocard C, Erdmann R, Warscheid B (2012) Identification of core components and transient interactors of the peroxisomal importomer by dual-track stable isotope labeling with amino acids in cell culture analysis. J Proteome Res 11(4):2567–2580

Oliveira ME, Gouveia AM, Pinto RA, Sa-Miranda C, Azevedo JE (2003) The energetics of Pex5p-mediated peroxisomal protein import. J Biol Chem 278(41):39483–39488

Oshima Y, Kamigaki A, Nakamori C, Mano S, Hayashi M, Nishimura M, Esaka M (2008) Plant catalase is imported into peroxisomes by Pex5p but is distinct from typical PTS1 import. Plant Cell Physiol 49(4):671–677

Otera H, Okumoto K, Tateishi K, Ikoma Y, Matsuda E, Nishimura M, Tsukamoto T, Osumi T, Ohashi K, Higuchi O, Fujiki Y (1998) Peroxisome targeting signal type 1 (PTS1) receptor is involved in import of both PTS1 and PTS2: studies with PEX5-defective CHO cell mutants. Mol Cell Biol 18(1):388–399

Otera H, Harano T, Honsho M, Ghaedi K, Mukai S, Tanaka A, Kawai A, Shimizu N, Fujiki Y (2000) The mammalian peroxin Pex5pL, the longer isoform of the mobile peroxisome targeting signal (PTS) type 1 transporter, translocates the Pex7p.PTS2 protein complex into peroxisomes via its initial docking site, Pex14p. J Biol Chem 275(28):21703–21714

Platta HW, Girzalsky W, Erdmann R (2004) Ubiquitination of the peroxisomal import receptor Pex5p. Biochem J 384(Pt 1):37–45

Platta HW, El Magraoui F, Schlee D, Grunau S, Girzalsky W, Erdmann R (2007) Ubiquitination of the peroxisomal import receptor Pex5p is required for its recycling. J Cell Biol 177(2):197–204

Platta HW, El Magraoui F, Baumer BE, Schlee D, Girzalsky W, Erdmann R (2009) Pex2 and pex12 function as protein-ubiquitin ligases in peroxisomal protein import. Mol Cell Biol 29 (20):5505–5516

Platta HW, Hagen S, Erdmann R (2013) The exportomer: the peroxisomal receptor export machinery. Cell Mol Life Sci 70(8):1393–1411

Reguenga C, Oliveira ME, Gouveia AM, Sa-Miranda C, Azevedo JE (2001) Characterization of the mammalian peroxisomal import machinery: Pex2p, Pex5p, Pex12p, and Pex14p are subunits of the same protein assembly. J Biol Chem 276(32):29935–29942

Rokka A, Antonenkov VD, Soininen R, Immonen HL, Pirila PL, Bergmann U, Sormunen RT, Weckstrom M, Benz R, Hiltunen JK (2009) Pxmp2 is a channel-forming protein in Mammalian peroxisomal membrane. PLoS One 4(4):e5090

Saidowsky J, Dodt G, Kirchberg K, Wegner A, Nastainczyk W, Kunau WH, Schliebs W (2001) The di-aromatic pentapeptide repeats of the human peroxisome import receptor PEX5 are separate high affinity binding sites for the peroxisomal membrane protein PEX14. J Biol Chem 276(37):34524–34529

Sarikas A, Hartmann T, Pan ZQ (2011) The cullin protein family. Genome Biol 12(4):220

Saville MK, Sparks A, Xirodimas DP, Wardrop J, Stevenson LF, Bourdon JC, Woods YL, Lane DP (2004) Regulation of p53 by the ubiquitin-conjugating enzymes UbcH5B/C in vivo. J Biol Chem 279(40):42169–42181

Schliebs W, Kunau WH (2006) PTS2 co-receptors: diverse proteins with common features. Biochim Biophys Acta 1763(12):1605–1612

Schluter A, Fourcade S, Ripp R, Mandel JL, Poch O, Pujol A (2006) The evolutionary origin of peroxisomes: an ER-peroxisome connection. Mol Biol Evol 23(4):838–845

Schuhmann H, Huesgen PF, Gietl C, Adamska I (2008) The DEG15 serine protease cleaves peroxisomal targeting signal 2-containing proteins in Arabidopsis. Plant Physiol 148(4):1847–1856

Shiozawa K, Konarev PV, Neufeld C, Wilmanns M, Svergun DI (2009) Solution structure of human Pex5.Pex14.PTS1 protein complexes obtained by small angle X-ray scattering. J Biol Chem 284(37):25334–25342

Spasser L, Brik A (2012) Chemistry and biology of the ubiquitin signal. Angew Chem Int Ed Engl 51(28):6840–6862

Swanson R, Locher M, Hochstrasser M (2001) A conserved ubiquitin ligase of the nuclear envelope/endoplasmic reticulum that functions in both ER-associated and Matalpha2 repressor degradation. Genes Dev 15(20):2660–2674

Swinkels BW, Gould SJ, Bodnar AG, Rachubinski RA, Subramani S (1991) A novel, cleavable peroxisomal targeting signal at the amino-terminus of the rat 3-ketoacyl-CoA thiolase. EMBO J 10(11):3255–3262

Tanaka N, Aoki K, Ishikura S, Nagano M, Imamura Y, Hara A, Nakamura KT (2008) Molecular basis for peroxisomal localization of tetrameric carbonyl reductase. Structure 16(3):388–397

Wanders RJ, Waterham HR (2006) Biochemistry of mammalian peroxisomes revisited. Annu Rev Biochem 75:295–332

Waterham HR, Ebberink MS (2012) Genetics and molecular basis of human peroxisome biogenesis disorders. Biochim Biophys Acta 1822(9):1430–1441

Wenzel DM, Lissounov A, Brzovic PS, Klevit RE (2011) UBCH7 reactivity profile reveals parkin and HHARI to be RING/HECT hybrids. Nature 474(7349):105–108

Wiebel FF, Kunau WH (1992) The Pas2 protein essential for peroxisome biogenesis is related to ubiquitin-conjugating enzymes. Nature 359(6390):73–76

Wiese S, Gronemeyer T, Ofman R, Kunze M, Grou CP, Almeida JA, Eisenacher M, Stephan C, Hayen H, Schollenberger L, Korosec T, Waterham HR, Schliebs W, Erdmann R, Berger J, Meyer HE, Just W, Azevedo JE, Wanders RJ, Warscheid B (2007) Proteomics characterization of mouse kidney peroxisomes by tandem mass spectrometry and protein correlation profiling. Mol Cell Proteomics 6(12):2045–2057

Williams C, van den Berg M, Sprenger RR, Distel B (2007) A conserved cysteine is essential for Pex4p-dependent ubiquitination of the peroxisomal import receptor Pex5p. J Biol Chem 282 (31):22534–22543

Williams C, van den Berg M, Geers E, Distel B (2008) Pex10p functions as an E3 ligase for the Ubc4p-dependent ubiquitination of Pex5p. Biochem Biophys Res Commun 374(4):620–624

Woodward AW, Bartel B (2005) The Arabidopsis peroxisomal targeting signal type 2 receptor PEX7 is necessary for peroxisome function and dependent on PEX5. Mol Biol Cell 16(2):573–583

Yahraus T, Braverman N, Dodt G, Kalish JE, Morrell JC, Moser HW, Valle D, Gould SJ (1996) The peroxisome biogenesis disorder group 4 gene, PXAAA1, encodes a cytoplasmic ATPase required for stability of the PTS1 receptor. EMBO J 15(12):2914–2923

Zolman BK, Monroe-Augustus M, Silva ID, Bartel B (2005) Identification and functional characterization of Arabidopsis PEROXIN4 and the interacting protein PEROXIN22. Plant Cell 17 (12):3422–3435

Part V

Peroxisome Dynamic

Molecular Complex Coordinating Peroxisome Morphogenesis in Mammalian Cells

17

Y. Fujiki, A. Itoyama, Y. Abe, and M. Honsho

Abstract

Peroxisomal division comprises three stages: elongation, constriction, and fission. Potential candidates thus far studied for the factors involved in these stages include Pex11pβ, dynamin-like protein 1 (DLP1), mitochondrial fission factor (Mff), and Fission 1 (Fis1). A poly-unsaturated fatty acid of peroxisomal β-oxidation metabolites, docosahexaenoic acid (C22:6*n*-3), augments hyper-oligomerization of Pex11pβ that gives rise to peroxisomal elongation, a prerequisite for subsequent fission and peroxisome division. Translocation of DLP1, a member of the large GTPase family, from the cytosol to peroxisomes is a prerequisite for membrane fission. However, the molecular machinery for peroxisomal targeting of DLP1 remains elusive. Mff is also localized to peroxisomes, especially at the membrane-constricted regions of elongated peroxisomes. Knockdown of Mff abrogates the fission stage of peroxisomal division and fails to recruit DLP1 to peroxisomes, while ectopic expression of Mff increases the peroxisomal targeting of DLP1. Co-expression of Mff and Pex11pβ increases peroxisome abundance. Overexpression of Mff also increases the interaction between DLP1 and Pex11pβ, which knockdown of Mff, but not Fis1, abolishes. Moreover, Pex11pβ interacts with Mff in a DLP1-dependent manner. Mff contributes to the peroxisomal targeting of DLP1 and plays a key role in the fission of the peroxisomal membrane by acting in concert with Pex11pβ and DLP1. The investigations performed to date suggest that a functional complex comprising Pex11pβ, Mff, and DLP1 promotes Mff-mediated fission during peroxisomal division. With regard to peroxisome morphogenesis, we address recent issues and findings and propose a model for peroxisome division.

Y. Fujiki (✉) • A. Itoyama • Y. Abe • M. Honsho
Department of Biology, Faculty of Sciences, Kyushu University Graduate School, Fukuoka 812-8581, Japan
e-mail: yfujiki@kyudai.jp

C. Brocard and A. Hartig (eds.), *Molecular Machines Involved in Peroxisome Biogenesis and Maintenance*, DOI 10.1007/978-3-7091-1788-0_17, © Springer-Verlag Wien 2014

Keywords
Peroxisome morphogenesis • Elongation • Fission • Division • Peroxin Pex11p • Dynamin-like protein 1 • Mitochondrial fission factor • Fis1 • Fatty acid β-oxidation

Abbreviations

DHA	Docosahexaenoic acid
DLP1	Dynamin-like protein 1
Fis1	Fission 1
Mff	Mitochondrial fission factor
PPARα	Peroxisome proliferator-activated receptor α

17.1 Introduction

The peroxisome is a single membrane-bounded organelle that is present in almost all eukaryotic cells. Peroxisomes constitutively take up membrane and matrix proteins, and dynamically change their morphology, abundance, and composition of peroxisomal proteins in response to extra- and intracellular stimuli (Fagarasanu et al. 2007; Schrader et al. 1998a, 1999). Peroxisomes are generated by either de novo formation (Honsho et al. 1998; Matsuzono et al. 1999; South and Gould 1999; Ghaedi et al. 2000) or division from pre-existing peroxisomes (Lazarow and Fujiki 1985). The "growth and division" model of peroxisome biogenesis where peroxisomes grow and multiply by taking up newly synthesized proteins from the cytosol is generally accepted, whereas the contribution of de novo formation of peroxisomes under normal conditions is a matter of debate (Kim et al. 2006; Agrawal et al. 2011; Lam et al. 2010; Yonekawa et al. 2011).

Peroxisomes proliferate in response to administration of fibrate agents to rodents (Berger and Moller 2002). Hypolipidemic fibrate drugs bind to peroxisome proliferator-activated receptor α (PPARα), which forms a complex with another nuclear receptor, the 9-cis-retinoic acid receptor (RXR). The PPAR/RXR transcription factor complex binds to PPAR-responsive elements in target DNA and increases the expression of a variety of lipid-metabolizing enzymes, including those involved in the peroxisomal fatty acid β-oxidation pathway (Reddy and Hashimoto 2001), and the 28 kDa peroxisomal membrane protein Pex11pβ (Abe et al. 1998). However, exposure to fibrate drugs does not increase peroxisomal β-oxidation activity or the abundance of peroxisomes in primary cultures of human hepatocytes (Blaauboer et al. 1990).

In addition to peroxisome proliferators that mediate PPARα, several chemicals, such as BM15766 (Baumgart et al. 1990) and 4-phenylbutyrate (PBA) (Gondcaille et al. 2005), promote peroxisome proliferation in human cells. Indeed, PBA increases peroxisome abundance in a PPARα-independent manner (Gondcaille et al. 2005), suggesting that peroxisome proliferation is induced by at least two different types of stimuli.

Peroxisome abundance can be induced by the expression of Pex11pβ independently of extracellular stimuli (Schrader et al. 1998b) and peroxisome metabolism (Li and Gould 2002). Based on morphological changes of peroxisomes upon induction by Pex11pβ expression, peroxisome proliferation is suggested to comprise multiple processes including peroxisomal membrane elongation, constriction, and fission. Such processes are confirmed in animal models (Schrader and Fahimi 2006). Given these data, a peroxisomal growth and division model comprising elongation, constriction, and fission of peroxisomes is now generally accepted (Lazarow and Fujiki 1985; Schrader and Fahimi 2006; Fagarasanu et al. 2007; Thoms and Erdmann 2005; Yan et al. 2005).

17.2 Fission Machinery of Peroxisome

According to the growth and division model of peroxisomes, it is conceivable that dysfunction of factor(s) required for peroxisome division may induce aberrant peroxisome morphology. Indeed, Pex11pβ knockdown reduces peroxisome abundance and increases the elongation of peroxisomes in mouse (Li et al. 2002b). In addition, the morphology of peroxisomes is aberrant in a CHO cell mutant (Tanaka et al. 2006) and human fibroblasts (Waterham et al. 2007) that are defective in dynamin-like protein 1 (DLP1), a member of the large GTPase family, and cells knocked down for DLP1 (Li and Gould 2003; Koch et al. 2003), mitochondrial fission factor (Mff; Gandre-Babbe and van der Bliek 2008; Otera et al. 2010; Itoyama et al. 2013), and Fission 1 (Fis1; Koch et al. 2005; Kobayashi et al. 2007). It is noteworthy that most of these morphogenesis factors, except for Pex11p, are shared with mitochondria (Schrader 2006; Koch and Brocard 2012; Itoyama et al. 2013).

In mammals, three Pex11p isoforms, Pex11pα, Pex11pβ, and Pex11pγ, have been identified by a homology search in which the expressed sequence tag database was screened using yeast *PEX* genes (Abe and Fujiki 1998; Abe et al. 1998; Tanaka et al. 2003; Schrader et al. 1998b; Li et al. 2002a). All the isoform gene products are localized to peroxisomes (Abe et al. 1998; Abe and Fujiki 1998; Tanaka et al. 2003; Schrader et al. 1998b; Itoyama et al. 2012). *PEX11β* is constitutively expressed, whereas *PEX11α* and *PEX11γ* are expressed in a tissue-specific manner (Schrader et al. 1998b; Li et al. 2002a). Ectopic expression of *PEX11α* shows very limited proliferation-promoting activity (Schrader et al. 1998b), and *PEX11γ* expression induces the formation of large peroxisomal membrane stacks (Koch and Brocard 2012; Koch et al. 2010). By contrast, ectopic expression of *PEX11β* prominently induces peroxisome proliferation via the elongation step of peroxisomal membranes (Schrader et al. 1998b; Kobayashi et al. 2007). Moreover, extensively elongated peroxisomes are discernible by the expression of Pex11pβ under conditions where the fission step of peroxisome proliferation is inhibited (Koch et al. 2004). Accordingly, Pex11pβ most likely plays a key role in the process of growth and the elongation step during peroxisome division, although the individual functions of mammalian Pex11p family members are not fully understood.

DLP1, Fis1, and Mff, which are involved in peroxisomal division, were originally identified as fission factors in mitochondrial morphogenesis. Recent studies on these factors led to the findings that endogenous DLP1 (Li and Gould 2003; Tanaka et al. 2006), Fis1 (Kobayashi et al. 2007), and Mff (Itoyama et al. 2013) are also localized to peroxisomes. Moreover, an aberrant peroxisomal structure is observed upon reducing the expression level of respective factor (Koch et al. 2003, 2004, 2005; Li and Gould 2003; Kobayashi et al. 2007; Gandre-Babbe and van der Bliek 2008; Otera et al. 2010; Itoyama et al. 2013).

DLP1 promotes the maintenance of peroxisomal and mitochondrial morphology, especially during membrane fission (Ishihara et al. 2009; Waterham et al. 2007; Tanaka et al. 2006). DLP1 is predicted to mediate the fission of peroxisomes and mitochondria via the formation of large multimeric spirals in a molecular machinery similar to that of dynamin at the site of endocytosis (Danino and Hinshaw 2001; Ford et al. 2011; Zhang and Hinshaw 2001). DLP1 and dynamin have several common multidomains including the GTPase, middle, and GTPase effector domains (Ford et al. 2011). In particular, the middle domain functions in the higher order assembly of both proteins, which is required for the formation of functional multimeric spirals (Ingerman et al. 2005; Ramachandran et al. 2007). Therefore, mutations in the DLP1 middle domain result in the abnormal elongation of peroxisomes and hyper-tubulation of mitochondria (Tanaka et al. 2006; Waterham et al. 2007). Translocation of DLP1 from the cytosol to peroxisomes and mitochondria is a prerequisite for membrane fission.

Fis1 and Mff, which are both tail-anchored proteins, are thought to be involved in the peroxisomal targeting of DLP1 in mammalian cells (Gandre-Babbe and van der Bliek 2008; Kobayashi et al. 2007; Koch et al. 2005; Otera et al. 2010). Fis1 and Mff dissociate from large peroxisomal membrane stacks that are induced by the expression of Pex11pγ, suggesting that Fis1 and Mff function in a fission step of peroxisome division (Koch and Brocard 2012). Expression of Fis1, but not Mff, promoted peroxisome proliferation, and silencing the expression of either of these proteins induced elongation of peroxisomes (Koch et al. 2005; Kobayashi et al. 2007; Itoyama et al. 2013), suggesting that they function in a non-redundant manner.

At least nine splicing variants of Mff are represented by multiple expressed sequence tags (Gandre-Babbe and van der Bliek 2008), and several isoforms of Mff are expressed in HeLa cells, HEK293 cells, and fibroblasts (Gandre-Babbe and van der Bliek 2008; Otera et al. 2010; Itoyama et al. 2013). The shortest form of Mff (isoform 8) is sufficient for the recruitment of DLP1 to mitochondria and peroxisomes (Otera et al. 2010; Itoyama et al. 2013). By contrast, one isoform is predominantly expressed in bovine brain (Gandre-Babbe and van der Bliek 2008). The functional differences among these isoforms, if any, remain to be elucidated.

17.3 Mechanism of Peroxisome Fission

Peroxisome proliferation involves multiple processes including peroxisome membrane elongation, constriction, and fission (Fig. 17.1). However, the regulatory mechanism underlying peroxisomal division remains elusive.

17.3.1 Elongation of Peroxisomes

Ectopic expression of Pex11pβ induces proliferation of peroxisomes in mammalian cells, where Pex11pβ apparently promotes membrane elongation at the constriction sites (Schrader et al. 1998b). However, an increase in *PEX11β* expression is not reported under physiological conditions, suggesting that the expression and function of Pex11pβ is regulated by unidentified mechanisms. Our recent study suggests that docosahexaenoic acid (DHA, C22:6*n*-3), a major product of peroxisomal β-oxidation, is an inducer of peroxisome division (Itoyama et al. 2012). DHA induces the elongation and subsequent fission of peroxisomes in fibroblasts from patients with defective fatty acid β-oxidation in a manner dependent on Pex11pβ (Li and Gould 2002; Nguyen et al. 2006; Itoyama et al. 2012), but independent of microtubules (Schrader et al. 1996; Tanaka et al. 2006; Itoyama et al. 2012) and PPARα (Kemp et al. 1998; Gondcaille et al. 2005; Itoyama et al. 2012). Moreover, DHA promotes multiple Pex11pβ-enriched regions on elongated peroxisomes as well as extensions of Pex11pβ-enriched membranes. DHA augments the oligomerization of Pex11pβ in vivo and DHA-containing phospholipids promote homo-oligomerization of Pex11pβ in vitro (Itoyama et al. 2012). These findings suggest that the elongation and fission of peroxisomes is initiated and augmented by the homo-oligomerization of Pex11pβ induced by DHA-containing phospholipids. DHA-mediated elongation of peroxisomes may not be conserved in yeast because *Saccharomyces cerevisiae* cannot produce polyunsaturated fatty acids with more than two double bonds (Yazawa et al. 2009). The contribution of Pex11pα and/or Pex11pγ to the elongation step of peroxisome proliferation is not clearly defined. Knockdown of *PEX11α* significantly attenuates DHA-mediated peroxisome proliferation, but less severely than knockdown of *PEX11β*. Of note, there are no apparent phenotypic differences in fibroblasts between those knocked down for *PEX11β* and those knocked down for both Pex11pβ and *PEX11β* (Itoyama et al. 2012). Thus, it is conceivable that Pex11pβ plays a key role in peroxisomal division and that Pex11pα plays an auxiliary role to Pex11pβ (Itoyama et al. 2012).

17.3.2 Recruitment of DLP1 to Peroxisome Membranes and Assembly of the Peroxisome Fission Complex

DLP1 is essential for the fission of peroxisomes (Tanaka et al. 2006; Waterham et al. 2007). In mitochondria, Fis1, Mff, MiD49, and MiD51 can each recruit DLP1 at one of the rate-limiting steps of mitochondrial fission (Cereghetti

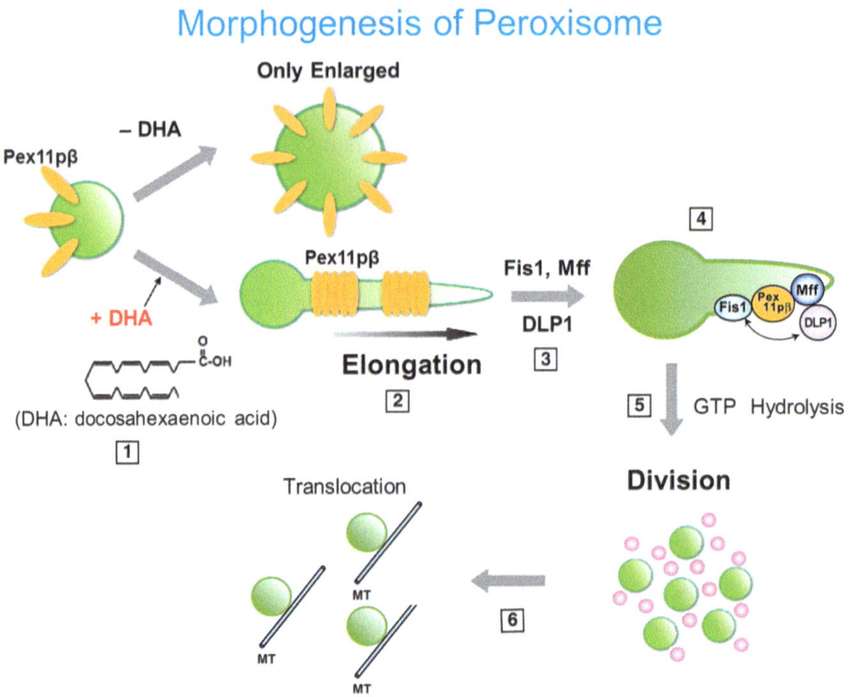

Fig. 17.1 A schematic model of peroxisome morphogenesis. DHA promotes the oligomerization of Pex11pβ, which leads to the formation of Pex11pβ-rich regions and initiates peroxisome elongation (step 1), in which peroxisomes elongate in one direction (step 2). Mff and Fis1 are localized to peroxisomes, especially at the membrane-constricted regions of elongated peroxisomes, where Mff recruits DLP1 (step 3). The functional complex comprising Pex11pβ, Mff, and DLP1 promotes Mff-mediated fission during peroxisomal division (step 4). The complex may include Fis1 that also interacts with DLP1. The activated DLP1 hydrolyzes GTP, leading to cleavage of peroxisomal membranes, thereby giving rise to peroxisomal fission (step 5). After division, daughter peroxisomes translocate via microtubules (MT) (step 6). See the text for details

et al. 2008; Otera and Mihara 2011; Palmer et al. 2011; Zhao et al. 2011; Losón et al. 2013). For instance, overexpression of Mff facilitates the mitochondrial targeting of DLP1, resulting in the fragmentation of mitochondria (Otera et al. 2010). By contrast, peroxisomal proliferation, resulting from peroxisomal fission, is increased only by the overexpression of Mff together with Pex11pβ (Itoyama et al. 2013). However, the proliferation of peroxisomes is significantly suppressed in fibroblasts from patients with defective fatty acid β-oxidation, such as a patient deficient in acyl-CoA oxidase, although DLP1 localizes to peroxisomes (Itoyama et al. 2012). These findings strongly suggest that unlike mitochondrial fission, the recruitment of DLP1 is not a rate-limiting step in peroxisomal division.

Pex11pβ interacts with DLP1 via Mff, suggesting that Pex11pβ forms a ternary complex with Mff and DLP1 during the fission process of peroxisomal division. Mff is sufficient to recruit DLP1 to the target membranes (Otera et al. 2010), suggesting that the interaction between Mff and Pex11pβ is not essential for the recruitment of DLP1 to peroxisomes. Pex11pβ and Mff localize to the constricted regions of elongated peroxisomes, which are devoid of Pex14p (Itoyama et al. 2012, 2013). Therefore, it is likely that the ternary complex comprising Pex11pβ, Mff, and DLP1 promotes fission at the constricted regions of the elongated peroxisomes.

Fis1 interacts with DLP1 and ectopically expressed Fis1 interacts with Pex11pα, Pex11pβ, and Pex11pγ (Koch et al. 2010). Moreover, ectopic expression of Fis1 increases the interplay between Pex11pβ and DLP1 (Kobayashi et al. 2007), suggesting that Fis1 can recruit DLP1 to peroxisomes. However, ectopically expressed Fis1 is not concentrated at the constriction sites (Koch et al. 2005), and the interplay between DLP1 and Pex11pβ is not altered in cells treated with siRNA targeting Fis1 (Itoyama et al. 2013). Taken together, these findings suggest that in mammalian cells, Fis1 contributes less to the morphogenesis of peroxisomes than Mff.

How is the assembly of the fission complex regulated? The interplay between Pex11pβ and Mff is strikingly decreased in cells treated with DLP1 dsRNA, indicating that DLP1 promotes the interaction between Pex11pβ and Mff (Itoyama et al. 2013). Therefore, it is likely that the complex formed by Mff and DLP1 interacts with Pex11pβ, leading to the assembly of large multimeric DLP1 spirals and peroxisome membrane fission (Itoyama et al. 2013). However, how the interplay between Pex11pβ and Mff/DLP1 is regulated remains to be defined. Mff indeed localizes at the membrane-constricted sites of elongated peroxisomes in PEX11$\beta^{-/-}$ mouse embryonic fibroblasts (Itoyama et al. 2013), implying that other factors besides Pex11pβ are involved in the localization of Mff to these sites. Notably, ganglioside-induced differentiation-associated protein 1 (GDAP1) was recently suggested to be required for peroxisome fission downstream of Pex11pβ and upstream of fission steps mediated by Mff and DLP1 (Huber et al. 2013), thereby inferring that GDAP1 mediates the interaction between Pex11pβ and the Mff-DLP1 complex.

Conclusions and Future Perspective

Peroxisomes proliferate by growth and division. We propose a model of peroxisome division as follows: (1) peroxisome division is initiated by the incorporation of DHA into phospholipids in the peroxisomal membrane; (2) Pex11pβ assembles into hyper-oligomers and Pex11pβ-enriched regions on peroxisomal membranes are formed, thereby initiating the elongation of peroxisomes; and (3) existing or newly-recruited DLP1 interacts with Pex11pβ via Mff at the constriction sites of the elongated peroxisomes, giving rise to peroxisomal fission (Fig. 17.1). Understanding the mechanisms underlying peroxisome proliferation has become highly attractive owing to the identification of molecules that are essential for peroxisome division and the elucidation of their functional regulation. However, how peroxisome proliferation is regulated remains far from fully understood. For instance, DHA augments the oligomerization of Pex11pβ and induces the elongation of peroxisomes without proceeding to the fission stage in control fibroblasts, suggesting that a complex mechanism strictly maintains the peroxisome abundance at a constant level under normal conditions (Itoyama et al. 2012). To better appreciate the system controlling peroxisome proliferation under physiological conditions, synchronization of peroxisome proliferation, visualization of peroxisome fission using live cell imaging, and dissection of peroxisome division may be required.

To define such intriguing steps, several new approaches are available. Expression of Pex11pβ carrying a monomeric YFP tag at the C-terminus inhibits the constriction and division of peroxisomes in mammalian cells, which may be useful to assess the properties of the Pex11pβ-enriched region during the elongation of peroxisomes (Delille et al. 2010). Moreover, DHA-induced peroxisome division provides a model experimental system in which peroxisome division can be readily separated into distinct steps of elongation and fission under physiological conditions (Itoyama et al. 2012). These systems could open up new pathways to fully elucidate the mechanisms underlying peroxisome morphogenesis.

Acknowledgments We thank K. Shimizu for figure illustrations. This work was supported in part by a CREST grant (to Y.F.) from the Science and Technology Agency of Japan, Grants-in-Aid for Scientific Research (19058011, 20370039, 24247038, 25112518, and 25116717 to Y.F.; and 23570236 to M.H.), Global Center of Excellence (GCOE) Program, Grants for Excellent Graduate Schools from the Ministry of Education, Culture, Sports, Science and Technology of Japan, Grants from Kyushu University Interdisciplinary Programs in Education and Projects in Research Development (to Y.A. and M.H.), and grants (to Y.F.) from Takeda Science Foundation and the Japan Foundation for Applied Enzymology. A.I. was supported by a Research Fellowship of CREST.

References

Abe I, Fujiki Y (1998) cDNA cloning and characterization of a constitutively expressed isoform of the human peroxin Pex11p. Biochem Biophys Res Commun 252:529–533

Abe I, Okumoto K, Tamura S, Fujiki Y (1998) Clofibrate-inducible, 28-kDa peroxisomal integral membrane protein is encoded by PEX11. FEBS Lett 431:468–472

Agrawal G, Joshi S, Subramani S (2011) Cell-free sorting of peroxisomal membrane proteins from the endoplasmic reticulum. Proc Natl Acad Sci USA 108:9113–9118

Baumgart E, Völkl A, Pill J, Fahimi HD (1990) Proliferation of peroxisomes without simultaneous induction of the peroxisomal fatty acid β-oxidation. FEBS Lett 264:5–9

Berger J, Moller DE (2002) The mechanisms of action of PPARs. Annu Rev Med 53:409–435

Blaauboer BJ, van Holsteijn CW, Bleumink R, Mennes WC, van Pelt FN, Yap SH, van Pelt JF, van Iersel AA, Timmerman A, Schmid BP (1990) The effect of beclobric acid and clofibric acid on peroxisomal β-oxidation and peroxisome proliferation in primary cultures of rat, monkey and human hepatocytes. Biochem Pharmacol 40:521–528

Cereghetti GM, Stangherlin A, Martins de Brito O, Chang CR, Blackstone C, Bernardi P, Scorrano L (2008) Dephosphorylation by calcineurin regulates translocation of Drp1 to mitochondria. Proc Natl Acad Sci USA 105:15803–15808

Danino D, Hinshaw JE (2001) Dynamin family of mechanoenzymes. Curr Opin Cell Biol 13:454–460

Delille HK, Agricola B, Guimaraes SC, Borta H, Lüers GH, Fransen M, Schrader M (2010) Pex11pβ-mediated growth and division of mammalian peroxisomes follows a maturation pathway. J Cell Sci 123:2750–2762

Fagarasanu A, Fagarasanu M, Rachubinski RA (2007) Maintaining peroxisome populations: a story of division and inheritance. Annu Rev Cell Dev Biol 23:321–344
Ford MG, Jenni S, Nunnari J (2011) The crystal structure of dynamin. Nature 477:561–566
Gandre-Babbe S, van der Bliek AM (2008) The novel tail-anchored membrane protein Mff controls mitochondrial and peroxisomal fission in mammalian cells. Mol Biol Cell 19:2402–2412
Ghaedi K, Tamura S, Okumoto K, Matsuzono Y, Fujiki Y (2000) The peroxin Pex3p initiates membrane assembly in peroxisome biogenesis. Mol Biol Cell 11:2085–2102
Gondcaille C, Depreter M, Fourcade S, Lecca MR, Leclercq S, Martin PG, Pineau T, Cadepond F, ElEtr M, Bertrand N, Beley A, Duclos S, De Craemer D, Roels F, Savary S, Bugaut M (2005) Phenylbutyrate up-regulates the adrenoleukodystrophy-related gene as a nonclassical peroxisome proliferator. J Cell Biol 169:93–104
Honsho M, Tamura S, Shimozawa N, Suzuki Y, Kondo N, Fujiki Y (1998) Mutation in PEX16 is causal in the peroxisome-deficient Zellweger syndrome of complementation group D. Am J Hum Genet 63:1622–1630
Huber N, Guimaraes S, Schrader M, Suter U, Niemann A (2013) Charcot-Marie-Tooth disease-associated mutants of GDAP1 dissociate its roles in peroxisomal and mitochondrial fission. EMBO Rep 14:545–552
Ingerman E, Perkins EM, Marino M, Mears JA, McCaffery JM, Hinshaw JE, Nunnari J (2005) Dnm1 forms spirals that are structurally tailored to fit mitochondria. J Cell Biol 170:1021–1027
Ishihara N, Nomura M, Jofuku A, Kato H, Suzuki SO, Masuda K, Otera H, Nakanishi Y, Nonaka I, Goto Y, Taguchi N, Morinaga H, Maeda M, Takayanagi R, Yokota S, Mihara K (2009) Mitochondrial fission factor Drp1 is essential for embryonic development and synapse formation in mice. Nat Cell Biol 11:958–966
Itoyama A, Honsho M, Abe Y, Moser A, Yoshida Y, Fujiki Y (2012) Docosahexaenoic acid mediates peroxisomal elongation, a prerequisite for peroxisome division. J Cell Sci 125:589–602
Itoyama A, Michiyuki S, Honsho M, Yamamoto T, Moser A, Yoshida Y, Fujiki Y (2013) Mff functions with Pex11pβ and DLP1 in peroxisomal fission. Biol Open 2:998–1006
Kemp S, Wei H-M, Lu J-F, Braiterman LT, McGuinness MC, Moser AB, Watkins PA, Smith KD (1998) Gene redundancy and pharmacological gene therapy: implications for X-linked adrenoleukodystrophy. Nat Med 4:1261–1268
Kim PK, Mullen RT, Schumann U, Lippincott-Schwartz J (2006) The origin and maintenance of mammalian peroxisomes involves a de novo PEX16-dependent pathway from the ER. J Cell Biol 173:521–532
Kobayashi S, Tanaka A, Fujiki Y (2007) Fis1, DLP1, and Pex11p coordinately regulate peroxisome morphogenesis. Exp Cell Res 313:1675–1686
Koch J, Brocard C (2012) PEX11 proteins attract Mff and human Fis1 to coordinate peroxisomal fission. J Cell Sci 125:3813–3826
Koch A, Thiemann M, Grabenbauer M, Yoon Y, McNiven MA, Schrader M (2003) Dynamin-like protein 1 is involved in peroxisomal fission. J Biol Chem 278:8597–8605
Koch A, Schneider G, Lüers GH, Schrader M (2004) Peroxisome elongation and constriction but not fission can occur independently of dynamin-like protein 1. J Cell Sci 117:3995–4006
Koch A, Yoon Y, Bonekamp NA, McNiven MA, Schrader M (2005) A role for Fis1 in both mitochondrial and peroxisomal fission in mammalian cells. Mol Biol Cell 16:5077–5086
Koch J, Pranjic K, Huber A, Ellinger A, Hartig A, Kragler F, Brocard C (2010) PEX11 family members are membrane elongation factors that coordinate peroxisome proliferation and maintenance. J Cell Sci 123:3389–3400
Lam SK, Yoda N, Schekman R (2010) A vesicle carrier that mediates peroxisome protein traffic from the endoplasmic reticulum. Proc Natl Acad Sci USA 107:21523–21528
Lazarow PB, Fujiki Y (1985) Biogenesis of peroxisomes. Annu Rev Cell Biol 1:489–530
Li X, Gould SJ (2002) PEX11 promotes peroxisome division independently of peroxisome metabolism. J Cell Biol 156:643–651

Li X, Gould SJ (2003) The dynamin-like GTPase DLP1 is essential for peroxisome division and is recruited to peroxisomes in part by PEX11. J Biol Chem 278:17012–17020

Li X, Baumgart E, Dong G-X, Morrell JC, Jimenez-Sanchez G, Valle D, Smith KD, Gould SJ (2002a) PEX11α is required for peroxisome proliferation in response to 4-phenylbutyrate but is dispensable for peroxisome proliferator-activated receptor alpha-mediated peroxisome proliferation. Mol Cell Biol 22:8226–8240

Li X, Baumgart E, Morrell JC, Jimenez-Sanchez G, Valle D, Gould SJ (2002b) PEX11β deficiency is lethal and impairs neuronal migration but does not abrogate peroxisome function. Mol Cell Biol 22:4358–4365

Losón OC, Song Z, Chen H, Chan DC (2013) Fis1, Mff, MiD49 and MiD51 mediate Drp1 recruitment in mitochondrial fission. Mol Biol Cell 24:659–667

Matsuzono Y, Kinoshita N, Tamura S, Shimozawa N, Hamasaki M, Ghaedi K, Wanders RJA, Suzuki Y, Kondo N, Fujiki Y (1999) Human PEX19: cDNA cloning by functional complementation, mutation analysis in a patient with Zellweger syndrome and potential role in peroxisomal membrane assembly. Proc Natl Acad Sci USA 96:2116–2121

Nguyen T, Bjorkman J, Paton BC, Crane DI (2006) Failure of microtubule-mediated peroxisome division and trafficking in disorders with reduced peroxisome abundance. J Cell Sci 119:636–645

Otera H, Mihara K (2011) Discovery of the membrane receptor for mitochondrial fission GTPase Drp1. Small GTPases 2:167–172

Otera H, Wang C, Cleland MM, Setoguchi K, Yokota S, Youle RJ, Mihara K (2010) Mff is an essential factor for mitochondrial recruitment of Drp1 during mitochondrial fission in mammalian cells. J Cell Biol 191:1141–1158

Palmer CS, Osellame LD, Laine D, Koutsopoulos OS, Frazier AE, Ryan MT (2011) MiD49 and MiD51, new components of the mitochondrial fission machinery. EMBO Rep 12:565–573

Ramachandran R, Surka M, Chappie JS, Fowler DM, Foss TR, Song BD, Schmid SL (2007) The dynamin middle domain is critical for tetramerization and higher-order self-assembly. EMBO J 26:559–566

Reddy JK, Hashimoto T (2001) Peroxisomal β-oxidation and peroxisome proliferator-activated receptor α: an adaptive metabolic system. Annu Rev Nutr 21:193–230

Schrader M (2006) Shared components of mitochondrial and peroxisomal division. Biochim Biophys Acta 1763:531–541

Schrader M, Fahimi HD (2006) Growth and division of peroxisomes. Int Rev Cytol 255:237–290

Schrader M, Burkhardt JK, Baumgart E, Lüers G, Spring H, Völkl A, Fahimi HD (1996) Interaction of microtubules with peroxisomes. Tubular and spherical peroxisomes in HepG2 cells and their alterations induced by microtubule-active drugs. Eur J Cell Biol 69:24–35

Schrader M, Krieglstein K, Fahimi HD (1998a) Tubular peroxisomes in HepG2 cells: selective induction by growth factors and arachidonic acid. Eur J Cell Biol 75:87–96

Schrader M, Reuber BE, Morrell JC, Jimenez-Sanchez G, Obie C, Stroh TA, Valle D, Schroer TA, Gould SJ (1998b) Expression of PEX11β mediates peroxisome proliferation in the absence of extracellular stimuli. J Biol Chem 273:29607–29614

Schrader M, Wodopia R, Fahimi HD (1999) Induction of tubular peroxisomes by UV irradiation and reactive oxygen species in HepG2 cells. J Histochem Cytochem 47:1141–1148

South ST, Gould SJ (1999) Peroxisome synthesis in the absence of preexisting peroxisomes. J Cell Biol 144:255–266

Tanaka A, Okumoto K, Fujiki Y (2003) cDNA cloning and characterization of the third isoform of human peroxin Pex11p. Biochem Biophys Res Commun 300:819–823

Tanaka A, Kobayashi S, Fujiki Y (2006) Peroxisome division is impaired in a CHO cell mutant with an inactivating point-mutation in dynamin-like protein 1 gene. Exp Cell Res 312:1671–1684

Thoms S, Erdmann R (2005) Dynamin-related proteins and Pex11 proteins in peroxisome division and proliferation. FEBS J 272:5169–5181

Waterham HR, Koster J, van Roermund CWT, Mooyer PAW, Wanders RJA, Leonard JV (2007) A lethal defect of mitochondrial and peroxisomal fission. N Engl J Med 356:1736–1741

Yan M, Rayapuram N, Subramani S (2005) The control of peroxisome number and size during division and proliferation. Curr Opin Cell Biol 17:376–383

Yazawa H, Iwahashi H, Kamisaka Y, Kimura K, Uemura H (2009) Production of polyunsaturated fatty acids in yeast Saccharomyces cerevisiae and its relation to alkaline pH tolerance. Yeast 26:167–184

Yonekawa S, Furuno A, Baba T, Fujiki Y, Ogasawara Y, Yamamoto A, Tagaya M, Tani K (2011) Sec16B is involved in the endoplasmic reticulum export of the peroxisomal membrane biogenesis factor peroxin 16 (Pex16) in mammalian cells. Proc Natl Acad Sci USA 108:12746–12751

Zhang P, Hinshaw JE (2001) Three-dimensional reconstruction of dynamin in the constricted state. Nat Cell Biol 3:922–926

Zhao J, Liu T, Jin S, Wang X, Qu M, Uhlen P, Tomilin N, Shupliakov O, Lendahl U, Nister M (2011) Human MIEF1 recruits Drp1 to mitochondrial outer membranes and promotes mitochondrial fusion rather than fission. EMBO J 30:2762–2778

Peroxisome Proliferation: Vesicles, Reticulons and ER-to-Peroxisome Contact Sites

18

Cécile Brocard

Abstract

Peroxisomes originate from the endoplasmic reticulum (ER) and can propagate via a growth and division process. Regulation of peroxisome proliferation has to be stringently controlled by the cell to guarantee that the number of peroxisomes per cell fits the metabolic requirements. Such regulation is achieved through coordination of de novo biogenesis, growth/division, inheritance, and degradation. In this review, I will focus on the role of the ER in the regulation of peroxisome maintenance. I will depict the assembly of components involved in peroxisome proliferation and their intimate interaction with ER resident proteins. Similar to other organelles, peroxisomes are in constant interaction with the rest of the cell. The formation of high molecular weight complexes between ER membrane proteins and proteins regulating peroxisome abundance highlights an important crosstalk between the two organelles. Finally, I propose a role for ER-to-peroxisome contacts sites (EPCONS) in the coordination of peroxisome proliferation.

Keywords

EPCONs • ER • Peroxisome maintenance and proliferation • Reticulon homology proteins

Abbreviations

EPCONS	ER-to-peroxisome contact sites
ER	Endoplasmic reticulum

C. Brocard (✉)
Tissue Med Biosciences, Magnesit-Strasse 1, 3500 Krems, Austria

MFPL, University of Vienna, Dr. Bohr-Gasse 9, 1030 Vienna, Austria
e-mail: CB@tmbiosciences.com; cecile.brocard@univie.ac.at

C. Brocard and A. Hartig (eds.), *Molecular Machines Involved in Peroxisome Biogenesis and Maintenance*, DOI 10.1007/978-3-7091-1788-0_18, © Springer-Verlag Wien 2014

ERES	ER exit sites
MCS	Membrane contact sites
PMPs	Peroxisomal membrane proteins
RHP	Reticulon homology proteins

18.1 Introduction

Peroxisomes represent a dynamic subcellular compartment in constant interaction with the rest of the cell. The number, size, and morphology of peroxisomes can greatly vary suggesting the existence of molecular mechanisms maintaining their abundance. A unique feature of peroxisomes is their stimulation by nutrients, herbicides, xenobiotics, ozone, or during senescence leading to massive proliferation (Lazarow and Fujiki 1985; Pastori and Del Rio 1997) usually associated with increased synthesis of some enzymes predominantly those involved in lipid metabolism (see also Chap. 3).

Early electron microscopic observations suggested the physical association of peroxisomes with the endoplasmic reticulum (Novikoff and Novikoff 1972). However, at this time, no biochemical connection to ER components could be unambiguously identified and ER proteins co-purifying with peroxisomes were typically classified as contaminants. Consequently, a model was adopted in which peroxisomes multiply through a process involving growth and division. mRNAs coding for a specific peroxisomal membrane protein were predominantly found on free polysomes. These results suggested, for the first time, that peroxisomal membrane proteins (PMPs) could be posttranslationally imported (Fujiki et al. 1984). This finding gave rise to a conceptual view of peroxisome biogenesis in which peroxisomes arise through budding and separation from the pre-existing organelles. This concept was entirely in agreement with the observation that most peroxisome biogenesis mutant cells contained empty membrane bags harboring PMPs (Purdue and Lazarow 1995; Santos et al. 1988). These membranous structures called peroxisome ghosts or remnants were believed to serve as template for the reestablishment of peroxisomes upon complementation.

Yet, later findings described the absence of detectable peroxisomes in cells from patients lacking the membrane protein Pex16p. In these cells, peroxisomal matrix proteins mostly remained in the cytosol, while membrane proteins were either degraded or mislocalized to other organelles. Amazingly, restitution of the wild-type gene restored functional peroxisomes in the affected cells (Honsho et al. 1998). In human cells, absence or mutation of any of the three biogenesis factors Pex3p, Pex16p, or Pex19p lead to the complete lack of detectable peroxisomes. In all cases reintroduction of the wild-type copy of the defective gene restores peroxisomes, suggesting that peroxisomes can form de novo. Most studies indicate that all eukaryotic organisms utilize the same structurally conserved set of proteins to generate peroxisomes (Baker and Sparkes 2005; Hayashi and Nishimura 2003; see also Table 1.1). In recent years experimental evidence for de novo peroxisome formation was provided and we now only begin to understand how peroxisomes form and how the ER participates in this process. This review focuses on the role of the ER in peroxisome proliferation.

18.2 A Role for the ER in Peroxisome Biogenesis

The ER is thought to be essential for peroxisome maintenance. Besides protein transport or lipid transfer from the ER to peroxisomes, de novo peroxisome biogenesis was shown to initiate at the ER (Fig. 18.1). Yeasts or mammalian cells devoid the membrane protein Pex3p also lack peroxisomal structures. Peroxisomes are rapidly regenerated upon reintroduction of the *PEX3* gene (Geuze et al. 2003; Haan et al. 2006; Kragt et al. 2005; Toro et al. 2007, 2009). Interestingly, while most organisms contain one Pex3p, the yeast *Yarrowia lipolytica* contains two namely proteins Pex3p and Pex3Bp, seemingly involved in different aspects of peroxisomal maintenance (Chang et al. 2009).

The ER origin of peroxisomes was reconsidered when electron microscopy, immunocytochemistry, and three-dimensional image reconstruction on mouse dendritic cells showed that some PMPs were present at specialized sites of the ER membrane and that those appeared continuous with a peroxisome reticulum (Geuze et al. 2003). Substantial data strengthening this idea came from real-time fluorescence microscopy of living cells using fluorescent versions of the early peroxisome biogenesis factor Pex3p. These studies showed that soon after its synthesis the protein Pex3p localizes to the ER and accumulates in specific subdomains marking sites at which pre-peroxisomal vesicles bud off (Hoepfner et al. 2005; Tam et al. 2005). To mature into fully functional peroxisomes, these vesicles require the function of Pex19p (Gotte et al. 1998), a protein participating in the transport of PMPs (Fang et al. 2004; Jones et al. 2004; Sacksteder and Gould 2000). Similar conclusions were reached from photobleaching experiments using cells from Zellweger patients with defective Pex16p (Kim et al. 2006).

Based on these findings, a new hypothesis developed that peroxisomes originate from the ER, mature, accumulate proteins and divide at some unknown point to give rise to new organelles that yet again accumulate proteins (Kim et al. 2006; Kunau 2005; Motley and Hettema 2007; Schluter et al. 2006; Tabak et al. 2003). Figure 18.1 depicts a model that reconciles all the present findings on peroxisome maintenance.

18.2.1 Peroxisome Membrane Proteins

Because peroxisomes do not enclose DNA, all peroxisomal proteins, including peroxins, translocate to the organelle post-translationally. While import of peroxisomal matrix proteins has been well characterized (see Chap. 13), the mechanism of protein insertion into the peroxisomal membrane is not fully understood.

Two pathways were proposed for the import of PMPs (Jones et al. 2004). Class I PMPs were suggested to predominantly bind Pex19p in the cytosol. Indeed, the

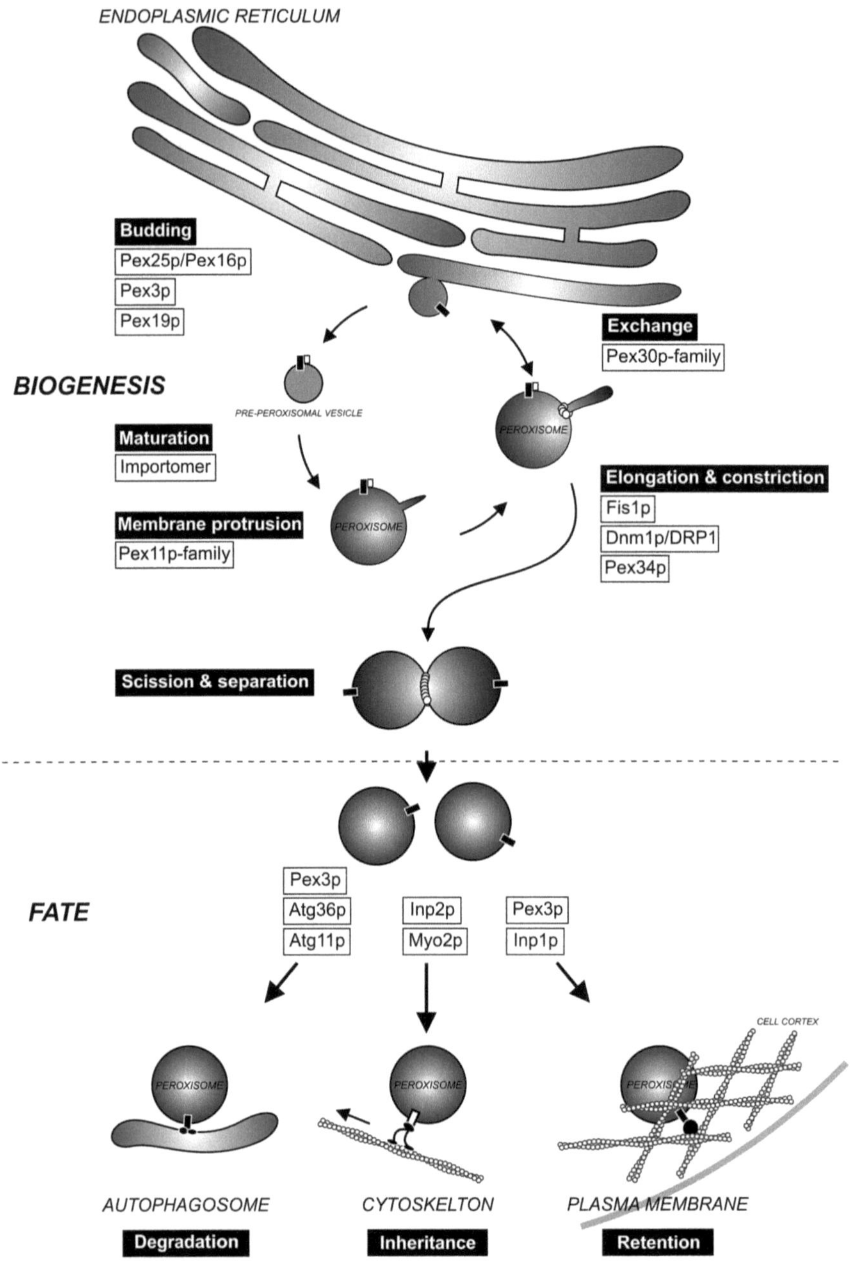

Fig. 18.1 Proposed model for the peroxisome life cycle. Peroxisomes are synthesized de novo through vesicular budding from specialized exit sites of peripheral ER tubules, an event that requires the proteins Pex25p (in yeast) or Pex16p (in mammals) as well as Pex3p and Pex19p. A maturation process allows for import of peroxisomal membrane and matrix proteins through action of Pex19p and the importomer complex, respectively. In the growth/division process, proteins of the Pex11p family allow pre-existing peroxisome to multiply through a multistep process starting with protrusion of the peroxisomal membrane confined at a definite area. At this point,

farnesylated protein, Pex19p, was proposed to act as soluble receptor allowing PMPs to target to the peroxisomal membrane (Fang et al. 2004; Fransen et al. 2001, 2005; Matsuzono and Fujiki 2006). Alternatively, due to its ability to bind to various protein domains it was suggested that Pex19p harbors a chaperone-like activity allowing for PMPs to access the peroxisome membrane (Snyder et al. 2000). The site of Pex19p/interaction of PMPs was thereafter called membrane peroxisomal targeting signal or mPTS (Rottensteiner et al. 2004). The cargo loaded Pex19p interacts with Pex3p at the membrane, thus recruiting PMPs to peroxisomes (Hettema et al. 2000; Muntau et al. 2003). In mammalian cells, the membrane protein Pex16p also participates in the recruitment process and was suggested to act as anchoring factor for the integration of new PMPs into the peroxisomal membrane (Brown and Baker 2003; Hettema et al. 2000; Muntau et al. 2003). A Pex19p-independent import of class II PMPs was proposed for Pex3p. This involves indirect sorting to peroxisome via the ER (Fang et al. 2004). In agreement, the transient knock down of *PEX19* via RNA interference in mammalian cells did not affect the import of Pex3p into peroxisomes (Jones et al. 2004). Together with the observation that cells lacking a functional Pex3p are devoid of peroxisomal structure, these findings suggest a key role for both Pex3p and the ER in the import of peroxisomal membrane proteins.

Several studies exploited the requirement for the presence of Pex3p for peroxisome biogenesis in mutant cells lacking peroxisomes and showed its localization at specialized ER foci. These dynamic ER sub-domains develop into pre-peroxisomal vesicles containing Pex3p (Hoepfner et al. 2005; Kragt et al. 2005; Tam et al. 2005; Toro et al. 2009; Veenhuis et al. 1996). The exact role of the ER in PMP targeting remains to be elucidated but a growing body of evidence suggests that several PMPs traffic through the ER before their localization on the peroxisomal membrane: While overexpression of the protein Pex15p was shown to affect the ER morphology (Elgersma et al. 1997), in the yeast *Y. lipolytica*, the delivery of two PMPs, Pex2p and Pex16p, to pre-existing peroxisomes was shown to occur via the ER (Titorenko and Rachubinski 1998). In agreement with these observations, photobleaching experiments showed that in mammalian cells Pex16p was inserted co-translationally into the ER membrane and served as scaffold to form early peroxisomal vesicles indicating that de novo peroxisome formation from the ER is not solely restricted to engineered model organisms (Kim et al. 2006). To date, no Pex16p was identified in *S. cerevisiae* and it has been postulated that, in this

Fig. 18.1 (continued) peroxisomal matrix proteins are retained in the old membrane and new material is imported into the protrusion thereby expanding into an offspring peroxisome. During elongation and growth, peroxisomes transiently associate with the ER, through a protein complex containing proteins of the Pex30p family, an event that may favor the transfer of lipids and membrane proteins. Membrane scission and separation of the organelles requires the protein proteins Fis1p and Dnm1p/DRP1 both also involved in mitochondrial fission and Pex34p (Tower et al. 2011). Upon cell division, whether peroxisomes are retained in the mother, segregated into the daughter cell or degraded will mainly depend on the proteinaceous interactions of the check-point factor Pex3p

organism, Pex19p assumes the molecular function of Pex16p by recruiting PMPs from the ER thereby facilitating their exit (Ma and Subramani 2009).

18.2.2 Vesicles and Peroxisome Biogenesis

The biogenesis of peroxisomes was proposed to represent a parallel branch of the secretory pathway and occur via vesicular formation from the ER (Schekman 2005). The molecular mechanisms underlying the biogenic pathway of peroxisome formation including vesicle budding and maturation has not been fully clarified so far. Two studies based on in vitro vesicle-budding reactions, however, demonstrated that several peroxisomal proteins trafficked from the ER to peroxisomes. Both strategies aimed at following the pinching of Pex3p-containing vesicles co-packaged with another peroxisomal membrane protein, Pex15p or Pex11p, respectively (Agrawal et al. 2011; Lam et al. 2010). The budding was independent of the COPII coat proteins and Sar1p, suggesting the existence of peroxisome-specific components for the budding event (Lam et al. 2010). Interestingly, even in the absence of Pex3p, vesicles were observed that contained Pex11p questioning the role of Pex3p as sole initiator of peroxisome formation from the ER (Agrawal et al. 2011). The Pex11p-containing vesicles produced in this assay were devoid of peroxisomal matrix proteins. Altogether these observations point to the existence of vesicular events to mediate the transport of PMPs from the ER.

Earlier studies in yeasts using secretory mutants already suggested that part of the ER-associated secretory machinery was involved in peroxisome biogenesis (Perry et al. 2009). A more recent study also showed that the secretory protein Sec16p, known to delineate ER exit sites (ERES), was required for ER export of Pex16p in mammalian cells (Yonekawa et al. 2011). Mammalian cells harbor two Sec16 proteins namely, Sec16Ap and Sec16Bp. These two proteins seem to have developed different functions and the presence of Sec16Bp only was required for the formation of pre-peroxisomal vesicles from the ER. Because in the classical secretory pathway, Sec16p coordinates the assembly of the COPII coat components it is tempting to speculate that peroxisome-specific coat components may exist that require the function of Sec16Bp. The underlying molecular principles of pre-peroxisomal vesicle assembly and budding remain to be discovered. Keeping in mind these findings, it will be interesting to establish which molecular complexes are involved in the assembly of pre-peroxisomal vesicles at the ER and clarify whether already known or unidentified, peroxisome-specific coat components exist.

18.3 Peroxisome Fate

Peroxisome turnover is regulated according to the cellular requirements. Hence, cells rigorously maintain their number of peroxisomes through processes involving proliferation, inheritance, and degradation. During cell division, peroxisomes are inherited to daughter cells. The cytoskeleton has been shown to play a crucial role in regulating peroxisome distribution and fate (see also Chap. 21). Human cells utilize microtubules to move peroxisomes (Rapp et al. 1996; Schrader et al. 1996; Wiemer et al. 1997). In fibroblast from Zellweger patients lacking functional Pex16p dynein motor proteins were shown to be important for events prior to peroxisome formation that may include the budding of vesicles from the ER (Brocard et al. 2005). Microtubules and motor proteins are required for peroxisome biogenesis. However, peroxins other than Pex16p could be involved in microtubule binding. The candidates could enclose the factors involved in regulating peroxisome proliferation. But, peroxisomes also need to acquire membrane lipids to proliferate. Accordingly, communication must exist between peroxisomes and the rest of the cell to strictly adapt the peroxisome number to the metabolic requirements (Guo et al. 2007).

18.3.1 Peroxisome Inheritance

While in mammals, the large number of peroxisomes (typically more than 100 per cell) and symmetric cytokinesis seems to allow for equal distribution of peroxisomes between mother and daughter cells, in budding yeast specific players are necessary. The discovery of *S. cerevisiae* mutants with peroxisome inheritance defects allowed the identification of proteins involved in the anchoring of peroxisomes to the cytoskeleton. Two proteins called Inp1p and Inp2p were identified, which seemingly assume antagonistic function to control inheritance of the peroxisomal population during cell division (Fig. 18.1). While Inp1p anchors peroxisomes to the cortex of the mother cell (Fagarasanu et al. 2005), Inp2p acts by connecting peroxisomes to Myo2p, a myosin motor protein that transports peroxisomes in the daughter cell along actin cables (Fagarasanu et al. 2006). In vivo time-lapse imaging illustrated that peroxisomal movement coincided with the polarity of the actin cables, and the study of cells expressing thermo-sensitive alleles of the class V myosin Moy2p showed that peroxisome movement was stopped at non-permissive temperature suggesting the involvement of myosin (Hoepfner et al. 2001). The protein Inp2p was established as peroxisomal membrane protein, which binds Myo2p thereby directing peroxisomes to the daughter cells during cell division (Fagarasanu et al. 2006). The observation that the levels of Inp2p oscillate during the cell cycle and that overproduction of Inp2 leads to migration of the entire peroxisome population to the daughter cell are in agreement with a role in peroxisome inheritance. While no Inp1p or Inp2p counterpart has been found in mammals, it is interesting to note that in the yeast *Y. lipolytica* peroxisome transport to daughter cells is mediated through Pex3Bp (Chang et al. 2009). A very recent work also reported the participation of *S. cerevisiae* Pex3p in the retention of peroxisomes at specific area of the cell cortex via direct association with Inp1p (Knoblach et al. 2013). The opposite action of Inp1p and Inp2p seems to control the inheritance of the peroxisome population. However, whether preferably newly formed peroxisomes are inherited remains unclear.

18.3.2 Peroxisome Degradation

Elderly, excessive or malfunctioning peroxisomes are selectively degraded through an autophagic process called pexophagy (Iwata et al. 2006; Kiel 2010; Oku and Sakai 2010; Sakai et al. 2006). Two distinct modes are employed, micro- and macropexophagy. In micropexophagy peroxisomes or accumulations thereof are engulfed in a vacuole, whereas in macropexophagy individual peroxisomes are sequestered by membrane layers forming the pexophagosome, which subsequently fuses with the vacuole (see also Chap. 22). It is still matter of debate how peroxisomes are marked for degradation. While the protein Pex14p was shown to be solely required for peroxisomes degradation (Zutphen et al. 2008), removal of Pex3p was reported to initiate peroxisome breakdown (Bellu et al. 2002). As already mentioned Pex3p is necessary for the generation of new peroxisomes. However, it seems to also contribute to peroxisome inheritance (Bellu et al. 2002; Knoblach et al. 2013; Munck et al. 2009). Furthermore, Pex3p can recruit the degradation factor Atg26p to peroxisomes determining autophagic degradation of the organelle (Motley et al. 2012) making this protein a major player in peroxisome fate.

18.4 Peroxisome Proliferation

As already mentioned peroxisome proliferation includes de novo generation from the ER, and multiplication from pre-existing peroxisomes via the growth/division process (Fig. 18.1). These two pathways seem to be intimately linked, however, the exact mechanism for their coordination and cooperation is not fully understood.

18.4.1 Growth/Division

Several proteins are involved in the growth/division process, among which proteins of the Pex11p family play an essential role (Abe and Fujiki 1998; Abe et al. 1998; Erdmann and Blobel 1995; Marshall et al. 1995; Orth et al. 2007; Schrader et al. 1998). In mammals, three genes have been identified, which code for Pex11αp, Pex11βp, and Pex11γp, respectively. Noteworthy, *PEX11α* was shown to be the only mammalian *PEX* gene whose expression was inducible by fibrates. Furthermore, it was reported that 4-phenylbutyrate (4-PBA) induced peroxisome proliferation along with the expression of *PEX11α* in mammalian cell culture (McGuinness et al. 2000; Wei et al. 2000) in a PPARα-independent manner (Li et al. 2002).

In yeast, expression of genes coding for many peroxisomal enzymes and some peroxins is repressed by glucose and induced by fatty acids such as oleic acid (see Chap. 1); among them *PEX11* and *PEX25* (Rottensteiner et al. 2003; Tam et al. 2003). Here, oleate-inducible genes contain an oleate-responsive element (ORE) in their promoter sequence that bind the transcription factor dimer Oaf1p/

Pip2p (Rottensteiner et al. 1997). Disruption of the *PEX11* gene results in the presence of giant peroxisomes in mutant cells whereas, overexpression of this gene leads to the formation of small peroxisomes in larger amounts than in wild type cells (Erdmann and Blobel 1995) indicating a role for Pex11p in peroxisome proliferation. Microarray profiling and proteomic approaches have led to the identification of the genes *PEX25* and *PEX27*. The function of Pex25p (Smith et al. 2002) and Pex27p (Rottensteiner et al. 2003; Tam et al. 2003) together with the dynamin-like/related protein 1 (DLP1/DRP1; see Chap. 20), Vps1p and Dnm1p in yeast (Kuravi et al. 2006; Hoepfner et al. 2001), is required for the maintenance of peroxisome size and number. Cells lacking any of these three proteins exhibit abnormally enlarged peroxisomes.

Pex25p was recently shown to play a crucial role for the reintroduction of peroxisomes in mutant cells lacking this organelle (Huber et al. 2012; Saraya et al. 2011). Although this protein has only been described in yeasts, Pex25p belongs to the conserved protein family of Pex11p-proteins. Interestingly, yeasts lacking any of the three Pex11-proteins contain fewer peroxisomes than wild-type cells. However, among the three proteins only Pex25p rescues the phenotype of mutant cells lacking all three proteins (Huber et al. 2012; Rottensteiner et al. 2003). All Pex11-proteins seem to have evolved from a common ancestor. However, according to their sequence it looks as if Pex25p and Pex27p divided from the tree early in evolution. This may be reflected in the observations that while in yeasts each member of the Pex11 family exerts a specific function, in higher eukaryotes all Pex11 proteins function cooperatively to promote peroxisome proliferation. Indeed, all Pex11p represent membrane elongation factors that remodel the peroxisomal membrane prior to fission. They all contain an amphipathic α-helix as common domain for membrane elongation. The amphipathic α-helix inserts into one leaflet of the lipid bilayer thereby increasing the surface area of one layer with respect to the other hence promoting membrane curvature (Koch and Brocard 2011, 2012; Koch et al. 2010; Opalinski et al. 2010, 2011).

Other factors, Pex28p, Pex29p, and members of the Pex30 family are also involved in growth and division and may participate in the fission event (Vizeacoumar et al. 2003, 2004; Brown et al. 2000). Altogether the coordinated action of these factors contributes to the control of peroxisome shape, number, and size. These PMPs form supramolecular protein complexes and also interact with ER components (David et al. 2013) adding new interrogations on the individuality of the two processes of peroxisome proliferation, de novo biogenesis and growth/division.

18.4.2 Members of the Pex30 Family

Peroxins of the Pex30p family are integral membrane proteins containing dysferlin domains. Dysferlin was identified as a gene mutated in limb-girdle muscular dystrophy (type 2B) and Miyoshi myopathy (Bashir et al. 1998; Liu et al. 1998). Evidence from studies of dysferlin-null mice suggests a function for dysferlin in membrane repair (Bansal et al. 2003). Dysferlin-mediated membrane repair has been suggested to be important to maintain membrane integrity of cardiomyocytes, particularly under conditions of mechanical stress (Han et al. 2007). The mammalian gene coding for dysferlin shows homology to the Fer-1 gene of *Caenorhabditis elegans* (Bashir et al. 1998). Fer-1 is a spermatogenesis factor that is specifically expressed in primary spermatocytes. In spermatids, mutations in Fer-1 cause infertility by impairing fusion of large vesicles called membranous organelles with the plasma membrane (Achanzar and Ward 1997). This fusion event leads to addition of membrane to the plasma membrane at the fusion site a process necessary for the extension of the pseudopodia that cause crawling of the spermatids. Consequently, mutations in Fer-1 lead to immobile spermatids and sterility in *C. elegans* (Achanzar and Ward 1997). Because Dysferlin and Fer-1 contain structural as well as sequence similarities, it was proposed that dysferlin might also be a vesicle-associated membrane protein involved in the docking and fusion of vesicles in skeletal muscle cells.

The only dysferlin-containing proteins known in yeast are the proteins of the Pex30 family. Originally, the protein Pex23p was identified in the yeast *Y. lipolytica* in a screen for mutant cells unable to utilize oleic acid as a sole carbon source (Brown et al. 2000). Homology probing with the Pex23p sequence in *S. cerevisiae* led to the finding of Pex30p, Pex31p, and Pex32p (Vizeacoumar et al. 2004). All members of this protein family contain at least one transmembrane domain and a dysferlin domain as common structural motifs. In line with the role of Fer-1 in membrane fusion it could well be that proteins of the Pex30p family play a role in the attraction of dysferlin-domain protein-containing vesicles to the peroxisomal membrane. In support of this hypothesis, it has been observed that in *pex23* mutants accumulate vesicles containing peroxisomal matrix and membrane proteins (Brown et al. 2000).

Peroxisome proliferation is altered in *S. cerevisiae* cells lacking Pex30p, Pex31p, or Pex32p and the cells harbor an abnormally high number of peroxisomes (Vizeacoumar et al. 2004). From the observation of mutants it has been suggested that while Pex30p would control the peroxisome number, Pex31p and Pex32p would rather regulate their size. Moreover, these three peroxins were proposed to act downstream of Pex28p and Pex29p, two proteins apparently required for the dissociation of peroxisomes after the fission event (Vizeacoumar et al. 2003, 2004).

18.4.3 Reticulon Homology Proteins

Small GTPases of the Rab family are essential for the protein transport machineries of eukaryotic cells. Cycles of nucleotide binding and hydrolysis through Rabs are required for each round of membrane transport. In *S. cerevisiae*, Yip1p is involved in Rab-mediated membrane transport and interacts with Yop1p. While the absence of Yop1 has no apparent effect on cell viability, its overexpression resulted in cell death and accumulation of internal cell membranes. Immunoprecipitation experiments illustrated the association of Yop1p with the Rab-proteins Ypt52p,

Sec4p, Ypt6p, and Ypt7p (Calero et al. 2001). It was then postulated that Yop1p facilitates Rab-mediated events in membrane traffic. Interestingly, a protein called DP1 (deleted in polyposis) or TB2 (Joslyn et al. 1991) was identified as the human homologue of Yop1p. Although the yeast and human proteins only share about 25 % identity, their overall structure is conserved. Both proteins contain extensive hydrophobic domains with their N-termini predicted to be exposed to the cytoplasmic face of the membrane. A study based on in vitro ER network formation reported that Yop1p/DP1 and the reticulon proteins Rtn1p/Rtn4ap generate and maintain the tubular morphology of the cortical ER (Voeltz et al. 2006).

Reticulons are membrane proteins originally referred to as neuronendocrine-specific proteins (NSP) anchored in the ER membrane (van de Velde et al. 1994). They are involved in various cellular processes and interact with proteins as diverse as BACE1 (He et al. 2004), extracellular receptor (Fournier et al. 2001) or fusion and endocytic proteins (Oertle and Schwab 2003). While the *S. cerevisiae* contains two reticulon genes *RTN1* and *RTN2*, the mammalian genome contains four (*RTN1*, *RTN2*, *RTN3,* and *RTN4*). Among those, Rtn4p, also known as Nogo, has been investigated as inhibitory molecule of axonal growth and generation (Klinger et al. 2004). Several isoforms of Rtn4p are expressed namely, Rtn4ap, Rtn4bp and Rtn4cp. Rtn4ap acts as strong inhibitor of neurite outgrowth after spinal cord injury (Bandtlow and Schwab 2000). Reticulon proteins and Yop1p/DP1 contain a structurally conserved reticulon homologous domain that varies from 186 amino acid residues in zebrafish to 276 residues in *S. cerevisiae*. These domains hold a hydrophilic loop of 66 amino acids flanked by two transmembrane domains and a hydrophilic tail. This structure has already been shown to confer common functions upon reticulon homology proteins (RHPs) such as localizing them to the appropriate membrane or mediating protein interactions. The identification of Yop1/DP1 and Rtn4ap as morphogenic proteins has been pinpointed for endomembranes. These proteins are crucially involved in vesicular trafficking, regulation of neurite outgrowth, apoptosis and modulation of axonal junction architecture and their malfunction have been associated with the pathogenesis of Alzheimer's disease and other neurodegenerative diseases. Consequently, cellular homeostasis of RHPs appears as an essential cellular function. For instance, the loop region in Rtn4p has been associated with attractive or repulsive pathways transducing signals to neighboring neurons and probably also non-neuronal cells. This loop region may feature RHPs to function as surface signaling molecules. Because all RHPs are ER membrane proteins and fluctuation in their levels or dysfunction affect intracellular trafficking events they may also participate to peroxisome maintenance.

18.5 Contact with the ER Membrane

The ER consists of a single membrane divided into different dynamic subdomains, the nuclear envelope and the so-called peripheral ER themselves organized into structurally distinct regions. These regions comprise a network of tubules and sheets distributed throughout the whole cytosol. Because these various ER forms

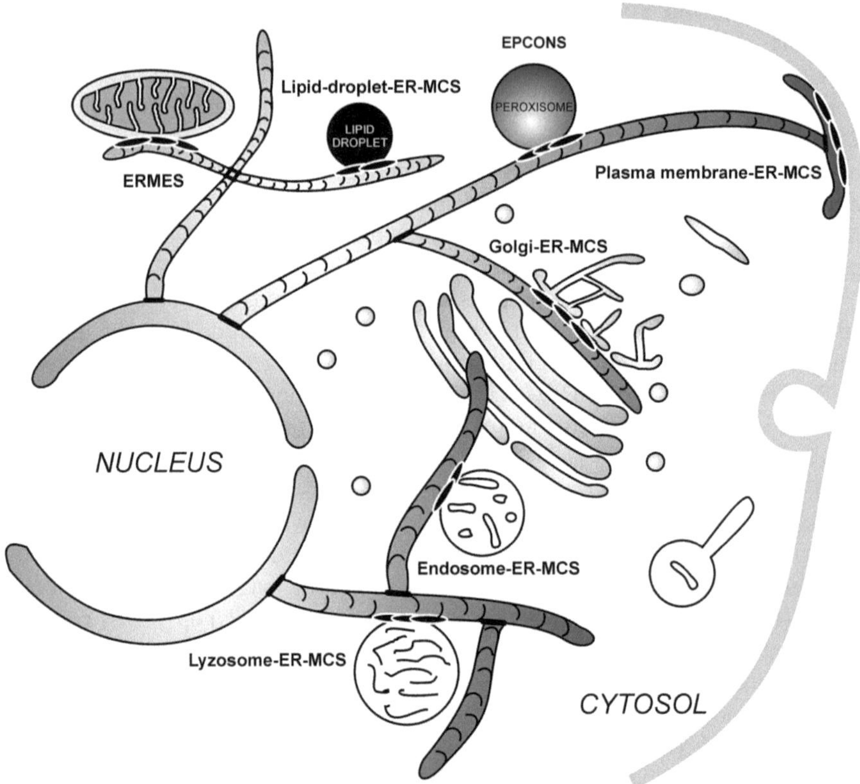

Fig. 18.2 Schematic representation of membranous domains with protein complexes building ER-to-organelle membrane contact sites (MCS). Peripheral ER tubules are represented and show reticulon homology proteins (RHPs; *Black lines* along the ER tubules) shaping the ER by maintaining a high membrane curvature (Voeltz et al. 2006). Tubule branches and membrane interconnections are achieved through the well-balanced action of the dynamin-like GTPase Atlastin/Sey1p (Orso et al. 2009) and the Lunapark family member Lnp1p (Chen et al. 2012) (*Black boxes*). Peripheral ER tubules are in contact with almost every other organelle in the cell through protein scaffolds (*Black ellipses*) established between both membranes, including mitochondria (ERMES), lipid droplets, Golgi (VAP proteins), endosomes, Lysosomes, plasma membrane (Osh-proteins), as well as peroxisomes (EPCONS). The EPCONS could coincide with peroxisome ER exit sites allowing for the synchronization of de novo biogenesis and growth/division of peroxisomes. In yeast, protein of the Pex30 family (Pex28p, Pex29p, Pex30p, Pex31p, Pex32p) regulate this process, and RHPs provide the suitable membrane architecture for enrichment of the peroxisome biogenesis machinery at specific regions of the cortical ER tubules

are achieved within a single continuous phospholipid bilayer, scaffolds are required to maintain functional separation and shape the membrane accordingly. Hence, reticulon homology proteins were shown to organize the tubular network of the peripheral ER (Voeltz et al. 2006; Zurek et al. 2011; Shibata et al. 2010).

18.5.1 Membrane Contact Sites

The ER contacts almost every membrane-bound organelle in the cell, including mitochondria, endosomes, Golgi, as well as the plasma membrane (Fig. 18.2). Multiple MCS were reported between the ER and other organelles. For instance, ER proteins form bridges with mitochondrial proteins through the ERMES protein complex (Kornmann et al. 2009), allowing for phospholipid and calcium exchange between both compartments (Kornmann and Walter 2010). Similarly, the Golgi membrane attaches to the ER membrane through VAP-proteins interacting with lipid transfer binding proteins (Peretti et al. 2008; Spang 2009; Wyles et al. 2002) and the plasma membrane through the Osh protein complex (Schulz and Prinz 2007; Stefan et al. 2011). MCS were also reported to occur with endosomes (Eden et al. 2010), lysosomes (Novikoff et al. 1978; Underwood et al. 1998) and also with lipid droplets (Jacquier et al. 2011).

18.5.2 Contacts Between the ER and Peroxisomes

Contacts between these two organelles have already been proposed in the 70s, and we recently reported that a protein complex assembled around Pex30p is involved in this tethering (David et al. 2013). However, the architecture of the membrane contact sites remains to be investigated. Electron-microscopic studies will shed light on the nature of these contact sites and might even allow for differentiation between pre-existing peroxisomes that dock onto the ER and newly formed peroxisomes that are currently budding off the ER. Once established, the flow of material, including lipids and proteins could be recorded and thus pinpoint the role of the ER in peroxisome maintenance.

ER Retrieval Upon membrane contact between two organelles and vesicular trafficking, proteins are co-transferred with the cargo, which have to return to their original membrane. A well-established Golgi-to-ER retrieval machinery involves components of the COPI coatomer complex for this retrograde transport (Beck et al. 2009; Pinot et al. 2010). We recently showed that the peroxisome proliferation factor Pex30p interacts with all subunits of the coatomer suggesting such a retrieval mechanism to the ER membrane (David et al. 2013). Although it cannot be excluded that a small portion of Pex30p enters the Golgi apparatus and thus need to be transported back to the ER, it is more likely that retrieval occurs between peroxisomes and the ER. Proteins of the Arf family, factors necessary for vesiculation of COPI-coated vesicles, were shown to contribute to peroxisomal formation, maintenance of peroxisomal morphology and membrane protein sorting (Anthonio et al. 2009; Anton et al. 2000; Passreiter et al. 1998). Interestingly, some human and plant Pex11-proteins carry a KKXX-motif, known as ER retrieval signal (Cosson and Letourneur 1994), in their primary sequence. In yeast, Pex30p is the only peroxin that carries such a signal. Understanding the contribution of ER membrane contact and retrograde trafficking to peroxisome maintenance would complete the picture of ER-peroxisome crosstalk and clarify the involvement of the COPI protein complex in this process.

ER-to-Peroxisomes Contact Sites In a recent study, we highlighted the connection of peroxisomes with the tubular ER network and the potential involvement of a large complex of membrane proteins in maintaining this architectural connection that we coined ER-to-peroxisomes contact sites or ER-to-peroxisomes contact sites (EPCONS) (David et al. 2013). These connections are required to maintain suitable peroxisome abundance and it will be interesting to characterize the molecular interchange occurring at these particular sites. Based on our finding that RHPs regulate peroxisome abundance, we proposed a model in which the ER plays a central role in providing lipids and membrane proteins for both, de novo biogenesis and multiplication of pre-existing peroxisomes. The transfer of material is achieved through membrane contact sites, the EPCONS, where RHPs provide the proper membrane architecture, and members of the Pex30-protein family are required for the tethering of peroxisomes onto the ER. Considering that these protein complexes associate at the ER membrane and that components of the COPI-coatomer complex interacted with Pex30p, we also hypothesized that these membrane contact sites may represent ER exit sites during de novo biogenesis of peroxisomes. In this case Pex30p and the RHPs would play an essential role in the regulation and coordination of peroxisome proliferation. Indeed, protein complexes may assemble in certain subdomains of the ER to define either budding site or sites of exchange. Experiments using biogenesis and inheritance mutant yeast cells showed that deletion of RHPs or Pex30p both increased the efficiency of yeast cells to form new peroxisomes (David et al. 2013).

Conclusion and Perspectives

The biogenic pathway leading to peroxisome formation is strongly regulated through connections between peroxisomes and the ER. Although seemingly independent from the ER, during peroxisome growth/division lipids and PMPs are likely to be delivered from this organelle. ER tubules were recently shown to contact and effectively wrap around mitochondrial constriction sites possibly to administer proteins and lipid or even mediate physical constriction (Friedman et al. 2011). In a similar way, the ER could actively participate in peroxisome multiplication. Overall, the two modes of peroxisome proliferation seem to be tightly coordinated in a process requiring the participation of specialized ER subdomains.

Proteins of the Pex30p family through their association with both RHPs and coatomers subunits constitute the first direct molecular link between peroxisomes and the ER. Molecular associations between RHPs and Pex30p proteins determine dynamic ER sub-domains important for the regulation of peroxisome maintenance. It will be interesting to further analyze these contact sites and characterize their detailed function with regard to peroxisome proliferation in de novo biogenesis, growth/division or both. Detailed studies should determine whether EPCONS also embody peroxisome exit sites from the ER.

Another important aspect of peroxisome growth/division relies on the observations that peroxisomes seem to segregate their content during this process. While matrix proteins remain confined to the mother organelle, membrane proteins seem to be differentially positioned along the peroxisomal membrane during division (Delille et al. 2010; Koch and Brocard 2012; Koch et al. 2010). Such partitioning could represent a mechanism by which old and possibly damaged matrix proteins would preferentially remain in the mother peroxisome. Whether a quality control truly exists to differentiate between new and old peroxisome and how this is achieved will be interesting questions to answer in the future.

Until recently, most diseases associated with peroxisomal defects were either due to the lack of function of one particular enzyme or displayed a general peroxisomal biogenesis defect due to mutation in *PEX* genes (see Chaps. 1 and 4). Recently, findings describing the discovery of a *PEX11β* mutation in cells from an affected patient reported the first direct correlation between peroxisome abundance and health (Thoms and Gartner 2012). Future studies will certainly highlight the importance of organelle abundance for health.

Finally, considering its participation in the various aspects of peroxisome maintenance, Pex3p might rather act as check-point for peroxisome fate. In agreement with such role Pex3p was shown to be important for peroxisome inheritance and degradation (see also Chap. 22). Pex3p is definitely required for the generation of functional, mature peroxisomes. Altogether, it appears that the specific array of Pex3p interactions label peroxisomes as new, mature, inheritable or aging. Whether Pex3p folding is influenced by its surrounding membrane and what precisely governs the proteinaceous interactions of this intriguing protein are questions that still remain to be elucidated.

Acknowledgments For findings reported in this work CB was supported by the Elise-Richter-Program grant V39-B09 of the Austrian Science Fund (FWF) and the Austrian Federal Ministry for Science and Research (BMWF) and the FWF grant P-20803.

References

Abe I, Fujiki Y (1998) cDNA cloning and characterization of a constitutively expressed isoform of the human peroxin Pex11p. Biochem Biophys Res Commun 252:529–533

Abe I, Okumoto K, Tamura S, Fujiki Y (1998) Clofibrate-inducible, 28-kDa peroxisomal integral membrane protein is encoded by PEX11. FEBS Lett 431:468–472

Achanzar WE, Ward S (1997) A nematode gene required for sperm vesicle fusion. J Cell Sci 110 (Pt 9):1073–1081

Agrawal G, Joshi S, Subramani S (2011) Cell-free sorting of peroxisomal membrane proteins from the endoplasmic reticulum. Proc Natl Acad Sci USA 108:9113–9118

Anthonio EA, Brees C, Baumgart-Vogt E, Hongu T, Huybrechts SJ, Van Dijck P, Mannaerts GP, Kanaho Y, Van Veldhoven PP, Fransen M (2009) Small G proteins in peroxisome biogenesis: the potential involvement of ADP-ribosylation factor 6. BMC Cell Biol 10:58

Anton M, Passreiter M, Lay D, Thai TP, Gorgas K, Just WW (2000) ARF- and coatomer-mediated peroxisomal vesiculation. Cell Biochem Biophys 32 Spring:27–36

Baker A, Sparkes IA (2005) Peroxisome protein import: some answers, more questions. Curr Opin Plant Biol 8:640–647

Bandtlow CE, Schwab ME (2000) NI-35/250/nogo-a: a neurite growth inhibitor restricting structural plasticity and regeneration of nerve fibers in the adult vertebrate CNS. Glia 29:175–181

Bansal D, Miyake K, Vogel SS, Groh S, Chen CC, Williamson R, McNeil PL, Campbell KP (2003) Defective membrane repair in dysferlin-deficient muscular dystrophy. Nature 423:168–172

Bashir R, Britton S, Strachan T, Keers S, Vafiadaki E, Lako M, Richard I, Marchand S, Bourg N, Argov Z, Sadeh M, Mahjneh I, Marconi G, Passos-Bueno MR, Moreira Ede S, Zatz M, Beckmann JS, Bushby K (1998) A gene related to Caenorhabditis elegans spermatogenesis factor fer-1 is mutated in limb-girdle muscular dystrophy type 2B. Nat Genet 20:37–42

Beck R, Rawet M, Wieland FT, Cassel D (2009) The COPI system: molecular mechanisms and function. FEBS Lett 583:2701–2709

Bellu AR, Salomons FA, Kiel JA, Veenhuis M, Van Der Klei IJ (2002) Removal of Pex3p is an important initial stage in selective peroxisome degradation in Hansenula polymorpha. J Biol Chem 277:42875–42880

Brocard CB, Boucher KK, Jedeszko C, Kim PK, Walton PA (2005) Requirement for microtubules and dynein motors in the earliest stages of peroxisome biogenesis. Traffic 6:386–395

Brown LA, Baker A (2003) Peroxisome biogenesis and the role of protein import. J Cell Mol Med 7:388–400

Brown TW, Titorenko VI, Rachubinski RA (2000) Mutants of the Yarrowia lipolytica PEX23 gene encoding an integral peroxisomal membrane peroxin mislocalize matrix proteins and accumulate vesicles containing peroxisomal matrix and membrane proteins. Mol Biol Cell 11:141–152

Calero M, Whittaker GR, Collins RN (2001) Yop1p, the yeast homolog of the polyposis locus protein 1, interacts with Yip1p and negatively regulates cell growth. J Biol Chem 276:12100–12112

Chang J, Mast FD, Fagarasanu A, Rachubinski DA, Eitzen GA, Dacks JB, Rachubinski RA (2009) Pex3 peroxisome biogenesis proteins function in peroxisome inheritance as class V myosin receptors. J Cell Biol 187:233–246

Chen S, Novick P, Ferro-Novick S (2012) ER network formation requires a balance of the dynamin-like GTPase Sey1p and the Lunapark family member Lnp1p. Nat Cell Biol 14:707–716

Cosson P, Letourneur F (1994) Coatomer interaction with di-lysine endoplasmic reticulum retention motifs. Science 263:1629–1631

David C, Koch J, Oeljeklaus S, Laernsack A, Melchior S, Wiese S, Schummer A, Erdmann R, Warscheid B, Brocard C (2013) A combined approach of quantitative interaction proteomics and live-cell imaging reveals a regulatory role for ER reticulon homology proteins in peroxisome biogenesis. Mol Cell Proteomics 12(9):2408–2425

Delille HK, Agricola B, Guimaraes SC, Borta H, Luers GH, Fransen M, Schrader M (2010) Pex11pbeta-mediated growth and division of mammalian peroxisomes follows a maturation pathway. J Cell Sci 123:2750–2762

Eden ER, White IJ, Tsapara A, Futter CE (2010) Membrane contacts between endosomes and ER provide sites for PTP1B-epidermal growth factor receptor interaction. Nat Cell Biol 12:267–272

Elgersma Y, Kwast L, van den Berg M, Snyder WB, Distel B, Subramani S, Tabak HF (1997) Overexpression of Pex15p, a phosphorylated peroxisomal integral membrane protein required for peroxisome assembly in S.cerevisiae, causes proliferation of the endoplasmic reticulum membrane. EMBO J 16:7326–7341

Erdmann R, Blobel G (1995) Giant peroxisomes in oleic acid-induced Saccharomyces cerevisiae lacking the peroxisomal membrane protein Pmp27p. J Cell Biol 128:509–523

Fagarasanu M, Fagarasanu A, Tam YY, Aitchison JD, Rachubinski RA (2005) Inplp is a peroxisomal membrane protein required for peroxisome inheritance in Saccharomyces cerevisiae. J Cell Biol 169:765–775

Fagarasanu A, Fagarasanu M, Eitzen GA, Aitchison JD, Rachubinski RA (2006) The peroxisomal membrane protein Inp2p is the peroxisome-specific receptor for the myosin V motor Myo2p of Saccharomyces cerevisiae. Dev Cell 10:587–600

Fang Y, Morrell JC, Jones JM, Gould SJ (2004) PEX3 functions as a PEX19 docking factor in the import of class I peroxisomal membrane proteins. J Cell Biol 164:863–875

Fournier AE, GrandPre T, Strittmatter SM (2001) Identification of a receptor mediating Nogo-66 inhibition of axonal regeneration. Nature 409:341–346

Fransen M, Wylin T, Brees C, Mannaerts GP, Van Veldhoven PP (2001) Human pex19p binds peroxisomal integral membrane proteins at regions distinct from their sorting sequences. Mol Cell Biol 21:4413–4424

Fransen M, Vastiau I, Brees C, Brys V, Mannaerts GP, Van Veldhoven PP (2005) Analysis of human Pex19p's domain structure by pentapeptide scanning mutagenesis. J Mol Biol 346:1275–1286

Friedman JR, Lackner LL, West M, DiBenedetto JR, Nunnari J, Voeltz GK (2011) ER tubules mark sites of mitochondrial division. Science 334:358–362

Fujiki Y, Rachubinski RA, Lazarow PB (1984) Synthesis of a major integral membrane polypeptide of rat liver peroxisomes on free polysomes. Proc Natl Acad Sci USA 81:7127–7131

Geuze HJ, Murk JL, Stroobants AK, Griffith JM, Kleijmeer MJ, Koster AJ, Verkleij AJ, Distel B, Tabak HF (2003) Involvement of the endoplasmic reticulum in peroxisome formation. Mol Biol Cell 14:2900–2907

Gotte K, Girzalsky W, Linkert M, Baumgart E, Kammerer S, Kunau WH, Erdmann R (1998) Pex19p, a farnesylated protein essential for peroxisome biogenesis. Mol Cell Biol 18:616–628

Guo T, Gregg C, Boukh-Viner T, Kyryakov P, Goldberg A, Bourque S, Banu F, Haile S, Milijevic S, San KH, Solomon J, Wong V, Titorenko VI (2007) A signal from inside the peroxisome initiates its division by promoting the remodeling of the peroxisomal membrane. J Cell Biol 177:289–303

Haan GJ, Baerends RJ, Krikken AM, Otzen M, Veenhuis M, van der Klei IJ (2006) Reassembly of peroxisomes in Hansenula polymorpha pex3 cells on reintroduction of Pex3p involves the nuclear envelope. FEMS Yeast Res 6:186–194

Han R, Bansal D, Miyake K, Muniz VP, Weiss RM, McNeil PL, Campbell KP (2007) Dysferlin-mediated membrane repair protects the heart from stress-induced left ventricular injury. J Clin Invest 117:1805–1813

Hayashi M, Nishimura M (2003) Entering a new era of research on plant peroxisomes. Curr Opin Plant Biol 6:577–582

He W, Lu Y, Qahwash I, Hu XY, Chang A, Yan R (2004) Reticulon family members modulate BACE1 activity and amyloid-beta peptide generation. Nat Med 10:959–965

Hettema EH, Girzalsky W, van Den Berg M, Erdmann R, Distel B (2000) Saccharomyces cerevisiae pex3p and pex19p are required for proper localization and stability of peroxisomal membrane proteins. EMBO J 19:223–233

Hoepfner D, van den Berg M, Philippsen P, Tabak HF, Hettema EH (2001) A role for Vps1p, actin, and the Myo2p motor in peroxisome abundance and inheritance in Saccharomyces cerevisiae. J Cell Biol 155:979–990

Hoepfner D, Schildknegt D, Braakman I, Philippsen P, Tabak HF (2005) Contribution of the endoplasmic reticulum to peroxisome formation. Cell 122:85–95

Honsho M, Tamura S, Shimozawa N, Suzuki Y, Kondo N, Fujiki Y (1998) Mutation in PEX16 is causal in the peroxisome-deficient Zellweger syndrome of complementation group D. Am J Hum Genet 63:1622–1630

Huber A, Koch J, Kragler F, Brocard C, Hartig A (2012) A subtle interplay between three Pex11 proteins shapes de novo formation and fission of peroxisomes. Traffic 13:157–167

Iwata J, Ezaki J, Komatsu M, Yokota S, Ueno T, Tanida I, Chiba T, Tanaka K, Kominami E (2006) Excess peroxisomes are degraded by autophagic machinery in mammals. J Biol Chem 281:4035–4041

Jacquier N, Choudhary V, Mari M, Toulmay A, Reggiori F, Schneiter R (2011) Lipid droplets are functionally connected to the endoplasmic reticulum in Saccharomyces cerevisiae. J Cell Sci 124:2424–2437

Jones JM, Morrell JC, Gould SJ (2004) PEX19 is a predominantly cytosolic chaperone and import receptor for class 1 peroxisomal membrane proteins. J Cell Biol 164:57–67

Joslyn G, Carlson M, Thliveris A, Albertsen H, Gelbert L, Samowitz W, Groden J, Stevens J, Spirio L, Robertson M et al (1991) Identification of deletion mutations and three new genes at the familial polyposis locus. Cell 66:601–613

Kiel JA (2010) Autophagy in unicellular eukaryotes. Philos Trans R Soc Lond B Biol Sci 365:819–830

Kim PK, Mullen RT, Schumann U, Lippincott-Schwartz J (2006) The origin and maintenance of mammalian peroxisomes involves a de novo PEX16-dependent pathway from the ER. J Cell Biol 173:521–532

Klinger M, Diekmann H, Heinz D, Hirsch C, Hannbeck von Hanwehr S, Petrausch B, Oertle T, Schwab ME, Stuermer CA (2004) Identification of two NOGO/RTN4 genes and analysis of Nogo-A expression in Xenopus laevis. Mol Cell Neurosci 25:205–216

Knoblach B, Sun X, Coquelle N, Fagarasanu A, Poirier RL, Rachubinski RA (2013) An ER-peroxisome tether exerts peroxisome population control in yeast. EMBO J 32:2439–2453

Koch J, Brocard C (2011) Membrane elongation factors in organelle maintenance: the case of peroxisome proliferation. Biomol Concepts 2:353–364

Koch J, Brocard C (2012) PEX11 proteins attract Mff and human Fis1 to coordinate peroxisomal fission. J Cell Sci 125:3813–3826

Koch J, Pranjic K, Huber A, Ellinger A, Hartig A, Kragler F, Brocard C (2010) PEX11 family members are membrane elongation factors that coordinate peroxisome proliferation and maintenance. J Cell Sci 123:3389–3400

Kornmann B, Walter P (2010) ERMES-mediated ER-mitochondria contacts: molecular hubs for the regulation of mitochondrial biology. J Cell Sci 123:1389–1393

Kornmann B, Currie E, Collins SR, Schuldiner M, Nunnari J, Weissman JS, Walter P (2009) An ER-mitochondria tethering complex revealed by a synthetic biology screen. Science 325:477–481

Kragt A, Voorn-Brouwer T, van den Berg M, Distel B (2005) Endoplasmic reticulum-directed Pex3p routes to peroxisomes and restores peroxisome formation in a Saccharomyces cerevisiae pex3Delta strain. J Biol Chem 280:34350–34357

Kunau WH (2005) Peroxisome biogenesis: end of the debate. Curr Biol 15:R774–R776

Kuravi K, Nagotu S, Krikken AM, Sjollema K, Deckers M, Erdmann R, Veenhuis M, van der Klei IJ (2006) Dynamin-related proteins Vps1p and Dnm1p control peroxisome abundance in Saccharomyces cerevisiae. J Cell Sci 119:3994–4001

Lam SK, Yoda N, Schekman R (2010) A vesicle carrier that mediates peroxisome protein traffic from the endoplasmic reticulum. Proc Natl Acad Sci USA 107:21523–21528

Lazarow PB, Fujiki Y (1985) Biogenesis of peroxisomes. Annu Rev Cell Biol 1:489–530

Li X, Baumgart E, Dong GX, Morrell JC, Jimenez-Sanchez G, Valle D, Smith KD, Gould SJ (2002) PEX11alpha is required for peroxisome proliferation in response to 4-phenylbutyrate but is dispensable for peroxisome proliferator-activated receptor alpha-mediated peroxisome proliferation. Mol Cell Biol 22:8226–8240

Liu J, Aoki M, Illa I, Wu C, Fardeau M, Angelini C, Serrano C, Urtizberea JA, Hentati F, Hamida MB, Bohlega S, Culper EJ, Amato AA, Bossie K, Oeltjen J, Bejaoui K, McKenna-Yasek D, Hosler BA, Schurr E, Arahata K, de Jong PJ, Brown RH Jr (1998) Dysferlin, a novel skeletal muscle gene, is mutated in Miyoshi myopathy and limb girdle muscular dystrophy. Nat Genet 20:31–36

Ma C, Subramani S (2009) Peroxisome matrix and membrane protein biogenesis. IUBMB Life 61:713–722

Marshall PA, Krimkevich YI, Lark RH, Dyer JM, Veenhuis M, Goodman JM (1995) Pmp27 promotes peroxisomal proliferation. J Cell Biol 129:345–355

Matsuzono Y, Fujiki Y (2006) In vitro transport of membrane proteins to peroxisomes by shuttling receptor Pex19p. J Biol Chem 281:36–42

McGuinness MC, Wei H, Smith KD (2000) Therapeutic developments in peroxisome biogenesis disorders. Expert Opin Investig Drugs 9:1985–1992

Motley AM, Hettema EH (2007) Yeast peroxisomes multiply by growth and division. J Cell Biol 178:399–410

Motley AM, Nuttall JM, Hettema EH (2012) Atg36: the Saccharomyces cerevisiae receptor for pexophagy. Autophagy 8:1680–1681

Munck JM, Motley AM, Nuttall JM, Hettema EH (2009) A dual function for Pex3p in peroxisome formation and inheritance. J Cell Biol 187:463–471

Muntau AC, Roscher AA, Kunau WH, Dodt G (2003) Interaction of PEX3 and PEX19 visualized by fluorescence resonance energy transfer (FRET). Adv Exp Med Biol 544:221–224

Novikoff PM, Novikoff AB (1972) Peroxisomes in absorptive cells of mammalian small intestine. J Cell Biol 53:532–560

Novikoff AB, Quintana N, Mori M (1978) Studies on the secretory process in exocrine pancreas cells. II. C57 black and beige mice. J Histochem Cytochem 26:83–93

Oertle T, Schwab ME (2003) Nogo and its paRTNers. Trends Cell Biol 13:187–194

Oku M, Sakai Y (2010) Peroxisomes as dynamic organelles: autophagic degradation. FEBS J 277:3289–3294

Opalinski L, Kiel JA, Williams C, Veenhuis M, van der Klei IJ (2010) Membrane curvature during peroxisome fission requires Pex11. EMBO J 30:5–16

Opalinski L, Veenhuis M, van der Klei IJ (2011) Peroxisomes: membrane events accompanying peroxisome proliferation. Int J Biochem Cell Biol 43:847–851

Orso G, Pendin D, Liu S, Tosetto J, Moss TJ, Faust JE, Micaroni M, Egorova A, Martinuzzi A, McNew JA, Daga A (2009) Homotypic fusion of ER membranes requires the dynamin-like GTPase atlastin. Nature 460:978–983

Orth T, Reumann S, Zhang X, Fan J, Wenzel D, Quan S, Hu J (2007) The PEROXIN11 protein family controls peroxisome proliferation in Arabidopsis. Plant Cell 19:333–350

Passreiter M, Anton M, Lay D, Frank R, Harter C, Wieland FT, Gorgas K, Just WW (1998) Peroxisome biogenesis: involvement of ARF and coatomer. J Cell Biol 141:373–383

Pastori GM, Del Rio LA (1997) Natural senescence of pea leaves (an activated oxygen-mediated function for peroxisomes). Plant Physiol 113:411–418

Peretti D, Dahan N, Shimoni E, Hirschberg K, Lev S (2008) Coordinated lipid transfer between the endoplasmic reticulum and the Golgi complex requires the VAP proteins and is essential for Golgi-mediated transport. Mol Biol Cell 19:3871–3884

Perry RJ, Mast FD, Rachubinski RA (2009) Endoplasmic reticulum-associated secretory proteins Sec20p, Sec39p, and Dsl1p are involved in peroxisome biogenesis. Eukaryot Cell 8:830–843

Pinot M, Goud B, Manneville JB (2010) Physical aspects of COPI vesicle formation. Mol Membr Biol 27:428–442

Purdue PE, Lazarow PB (1995) Identification of peroxisomal membrane ghosts with an epitope-tagged integral membrane protein in yeast mutants lacking peroxisomes. Yeast 11:1045–1060

Rapp S, Saffrich R, Anton M, Jakle U, Ansorge W, Gorgas K, Just WW (1996) Microtubule-based peroxisome movement. J Cell Sci 109(Pt 4):837–849

Rottensteiner H, Kal AJ, Hamilton B, Ruis H, Tabak HF (1997) A heterodimer of the Zn2Cys6 transcription factors Pip2p and Oaf1p controls induction of genes encoding peroxisomal proteins in Saccharomyces cerevisiae. Eur J Biochem 247:776–783
Rottensteiner H, Stein K, Sonnenhol E, Erdmann R (2003) Conserved function of pex11p and the novel pex25p and pex27p in peroxisome biogenesis. Mol Biol Cell 14:4316–4328
Rottensteiner H, Kramer A, Lorenzen S, Stein K, Landgraf C, Volkmer-Engert R, Erdmann R (2004) Peroxisomal membrane proteins contain common Pex19p-binding sites that are an integral part of their targeting signals. Mol Biol Cell 15:3406–3417
Sacksteder KA, Gould SJ (2000) The genetics of peroxisome biogenesis. Annu Rev Genet 34:623–652
Sakai Y, Oku M, van der Klei IJ, Kiel JA (2006) Pexophagy: autophagic degradation of peroxisomes. Biochim Biophys Acta 1763:1767–1775
Santos MJ, Imanaka T, Shio H, Small GM, Lazarow PB (1988) Peroxisomal membrane ghosts in Zellweger syndrome–aberrant organelle assembly. Science 239:1536–1538
Saraya R, Krikken AM, Veenhuis M, van der Klei IJ (2011) Peroxisome reintroduction in Hansenula polymorpha requires Pex25 and Rho1. J Cell Biol 193:885–900
Schekman R (2005) Peroxisomes: another branch of the secretory pathway? Cell 122:1–2
Schluter A, Fourcade S, Ripp R, Mandel JL, Poch O, Pujol A (2006) The evolutionary origin of peroxisomes: an ER-peroxisome connection. Mol Biol Evol 23:838–845
Schrader M, Burkhardt JK, Baumgart E, Luers G, Spring H, Volkl A, Fahimi HD (1996) Interaction of microtubules with peroxisomes. Tubular and spherical peroxisomes in HepG2 cells and their alterations induced by microtubule-active drugs. Eur J Cell Biol 69:24–35
Schrader M, Reuber BE, Morrell JC, Jimenez-Sanchez G, Obie C, Stroh TA, Valle D, Schroer TA, Gould SJ (1998) Expression of PEX11beta mediates peroxisome proliferation in the absence of extracellular stimuli. J Biol Chem 273:29607–29614
Schulz TA, Prinz WA (2007) Sterol transport in yeast and the oxysterol binding protein homologue (OSH) family. Biochim Biophys Acta 1771:769–780
Shibata Y, Shemesh T, Prinz WA, Palazzo AF, Kozlov MM, Rapoport TA (2010) Mechanisms determining the morphology of the peripheral ER. Cell 143:774–788
Smith JJ, Marelli M, Christmas RH, Vizeacoumar FJ, Dilworth DJ, Ideker T, Galitski T, Dimitrov K, Rachubinski RA, Aitchison JD (2002) Transcriptome profiling to identify genes involved in peroxisome assembly and function. J Cell Biol 158:259–271
Snyder WB, Koller A, Choy AJ, Subramani S (2000) The peroxin Pex19p interacts with multiple, integral membrane proteins at the peroxisomal membrane. J Cell Biol 149:1171–1178
Spang A (2009) On vesicle formation and tethering in the ER-Golgi shuttle. Curr Opin Cell Biol 21:531–536
Stefan CJ, Manford AG, Baird D, Yamada-Hanff J, Mao Y, Emr SD (2011) Osh proteins regulate phosphoinositide metabolism at ER-plasma membrane contact sites. Cell 144:389–401
Tabak HF, Murk JL, Braakman I, Geuze HJ (2003) Peroxisomes start their life in the endoplasmic reticulum. Traffic 4:512–518
Tam YY, Torres-Guzman JC, Vizeacoumar FJ, Smith JJ, Marelli M, Aitchison JD, Rachubinski RA (2003) Pex11-related proteins in peroxisome dynamics: a role for the novel peroxin Pex27p in controlling peroxisome size and number in Saccharomyces cerevisiae. Mol Biol Cell 14:4089–4102
Tam YY, Fagarasanu A, Fagarasanu M, Rachubinski RA (2005) Pex3p initiates the formation of a preperoxisomal compartment from a subdomain of the endoplasmic reticulum in Saccharomyces cerevisiae. J Biol Chem 280:34933–34939
Thoms S, Gartner J (2012) First PEX11beta patient extends spectrum of peroxisomal biogenesis disorder phenotypes. J Med Genet 49:314–316
Titorenko VI, Rachubinski RA (1998) The endoplasmic reticulum plays an essential role in peroxisome biogenesis. Trends Biochem Sci 23:231–233
Toro A, Arredondo C, Cordova G, Araya C, Palacios JL, Venegas A, Morita M, Imanaka T, Santos MJ (2007) Evaluation of the role of the endoplasmic reticulum-Golgi transit in the biogenesis

of peroxisomal membrane proteins in wild type and peroxisome biogenesis mutant CHO cells. Biol Res 40:231–249

Toro AA, Araya CA, Cordova GJ, Arredondo CA, Cardenas HG, Moreno RE, Venegas A, Koenig CS, Cancino J, Gonzalez A, Santos MJ (2009) Pex3p-dependent peroxisomal biogenesis initiates in the endoplasmic reticulum of human fibroblasts. J Cell Biochem 107:1083–1096

Tower RJ, Fagarasanu A, Aitchison JD, Rachubinski RA (2011) The peroxin Pex34p functions with the Pex11 family of peroxisomal divisional proteins to regulate the peroxisome population in yeast. Mol Biol Cell 22(10):1727–1738

Underwood KW, Jacobs NL, Howley A, Liscum L (1998) Evidence for a cholesterol transport pathway from lysosomes to endoplasmic reticulum that is independent of the plasma membrane. J Biol Chem 273:4266–4274

van de Velde HJ, Roebroek AJ, Senden NH, Ramaekers FC, Van de Ven WJ (1994) NSP-encoded reticulons, neuroendocrine proteins of a novel gene family associated with membranes of the endoplasmic reticulum. J Cell Sci 107(Pt 9):2403–2416

Veenhuis M, Komori M, Salomons F, Hilbrands RE, Hut H, Baerends RJ, Kiel JA, van der Klei IJ (1996) Peroxisomal remnants in peroxisome-deficient mutants of the yeast Hansenula polymorpha [corrected]. FEBS Lett 383:114–118

Vizeacoumar FJ, Torres-Guzman JC, Tam YY, Aitchison JD, Rachubinski RA (2003) YHR150w and YDR479c encode peroxisomal integral membrane proteins involved in the regulation of peroxisome number, size, and distribution in Saccharomyces cerevisiae. J Cell Biol 161:321–332

Vizeacoumar FJ, Torres-Guzman JC, Bouard D, Aitchison JD, Rachubinski RA (2004) Pex30p, Pex31p, and Pex32p form a family of peroxisomal integral membrane proteins regulating peroxisome size and number in Saccharomyces cerevisiae. Mol Biol Cell 15:665–677

Voeltz GK, Prinz WA, Shibata Y, Rist JM, Rapoport TA (2006) A class of membrane proteins shaping the tubular endoplasmic reticulum. Cell 124:573–586

Wei H, Kemp S, McGuinness MC, Moser AB, Smith KD (2000) Pharmacological induction of peroxisomes in peroxisome biogenesis disorders. Ann Neurol 47:286–296

Wiemer EA, Wenzel T, Deerinck TJ, Ellisman MH, Subramani S (1997) Visualization of the peroxisomal compartment in living mammalian cells: dynamic behavior and association with microtubules. J Cell Biol 136:71–80

Wyles JP, McMaster CR, Ridgway ND (2002) Vesicle-associated membrane protein-associated protein-A (VAP-A) interacts with the oxysterol-binding protein to modify export from the endoplasmic reticulum. J Biol Chem 277:29908–29918

Yonekawa S, Furuno A, Baba T, Fujiki Y, Ogasawara Y, Yamamoto A, Tagaya M, Tani K (2011) Sec16B is involved in the endoplasmic reticulum export of the peroxisomal membrane biogenesis factor peroxin 16 (Pex16) in mammalian cells. Proc Natl Acad Sci USA 108:12746–12751

Zurek N, Sparks L, Voeltz G (2011) Reticulon short hairpin transmembrane domains are used to shape ER tubules. Traffic 12:28–41

Zutphen T, Veenhuis M, van der Klei IJ (2008) Pex14 is the sole component of the peroxisomal translocon that is required for pexophagy. Autophagy 4:63–66

The Functions of Pex11 Family Proteins in Peroxisome Biology

19

Chris Williams and Ida J. van der Klei

Abstract

Regulating peroxisome numbers within the cell, through synthesis and degradation events, is an essential part of cell homeostasis. New peroxisomes are made either de novo or by the division of pre-existing organelles. Elongation of the peroxisomal membrane, one of the initial steps in fission, is proposed to be followed by assembly of the membrane fission machinery at the site of tubulation, bilayer constriction and ultimately, membrane fission. A role for the peroxisomal membrane protein Pex11p in peroxisome fission has long been known. Recent reports have shed light on the mechanistic details behind Pex11p's membrane remodeling activity and the role of amphipathic helices in Pex11p. However, a number of additional tasks have been attributed to Pex11p, ranging from directing peroxisome inheritance to peroxisomal membrane protein reorganization. Furthermore, several Pex11-like proteins have been described, the functions of which differ from Pex11p. This chapter will discuss the current understanding of the role of the Pex11 protein family in peroxisome biogenesis.

Keywords

Pex11 • Pex25 • Pex27 • Peroxisome fission • Membrane elongation • Membrane remodeling • Amphipathic helix • Peroxisomal membrane proteins • Membrane proteins reorganization

C. Williams (✉) • I.J. van der Klei
Molecular Cell Biology, Groningen Biomolecular Sciences and Biotechnology Institute (GBB), University of Groningen, 9700 CC Groningen, The Netherlands
e-mail: c.p.williams@rug.nl

C. Brocard and A. Hartig (eds.), *Molecular Machines Involved in Peroxisome Biogenesis and Maintenance*, DOI 10.1007/978-3-7091-1788-0_19, © Springer-Verlag Wien 2014

19.1 Introduction

Peroxisomes house a large number of metabolic pathways, which depend on organism, tissue, and developmental stage and in fungi on the carbon and nitrogen source used for growth. Consequently, regulating peroxisome numbers, through synthesis and degradation events, is an essential part of cell homeostasis. The presence of too many or too few peroxisomes, or organelles with the improper enzyme content, can potentially disturb the metabolic balance in the cell. Our current understanding of how new peroxisomes are made suggests the existence of two different pathways: pre-existing organelles can divide, a process known as fission, or peroxisomes can form de novo (Fagarasanu et al. 2007). Peroxisome multiplication by fission can be separated in three consecutive steps (Fig. 19.1a): elongation of the peroxisome, constriction of the peroxisomal membrane and finally, the actual fission step (Thoms and Erdmann 2005). Almost 20 years ago, three publications simultaneously reported on a yeast peroxisomal membrane protein (PMP) with a role in peroxisome proliferation (Erdmann and Blobel 1995; Marshall et al. 1995; Sakai et al. 1995). Referred to as Pmp27p in *S. cerevisiae* and Pmp31p/Pmp32p in *C. boidini*, this protein was later renamed Pex11p (Distel et al. 1996). Since this time Pex11 proteins have been identified in a wide range of organisms (Table 19.1, adapted from Kiel et al. 2006). Pex11p plays a central role in peroxisome fission. However, Pex11p has since been assigned a number of novel functions, including fatty acid transport, directing the formation of PMP subdomains during fission, as well as aiding peroxisome inheritance. Furthermore, the Pex11p family now includes a number of more distantly related members, the functions of which are different from classical Pex11 proteins. This review gives an overview of the Pex11p family of proteins and their properties and discusses the role(s) they play in peroxisome biogenesis.

19.1.1 The Pex11p Family

As mentioned above, Pex11p was originally identified and characterized in the yeasts *S. cerevisiae* and *C. boidini* (Erdmann and Blobel 1995; Marshall et al. 1995; Sakai et al. 1995). Reverse genetics and homology searches have aided in the identification of Pex11p in other fungi (Kiel et al. 2005; Krikken et al. 2009), *T. brucei* (Lorenz et al. 1998), plants (Lingard and Trelease 2006; Nayidu et al. 2008) and mammals (Passreiter et al. 1998; Schrader et al. 1998b). However, many organisms contain more than one Pex11 protein. Table 19.1 (adapted from Kiel et al. 2006) presents several members of the Pex11p family of proteins from different organisms. In filamentous fungi for example, three isoforms of Pex11p have been identified, a classical Pex11p, as well as the Pex11-like proteins Pex11Bp and Pex11Cp (Table 19.1 and Kiel et al. 2006). Several yeasts, including *C. albicans, H. polymorpha,* and *Y. lipolytica* also contain, in addition to Pex11p, a Pex11Cp homologue (Kiel et al. 2006). Mammals contain three Pex11p isoforms—Pex11αp, Pex11βp and the more distantly related Pex11γp, which is

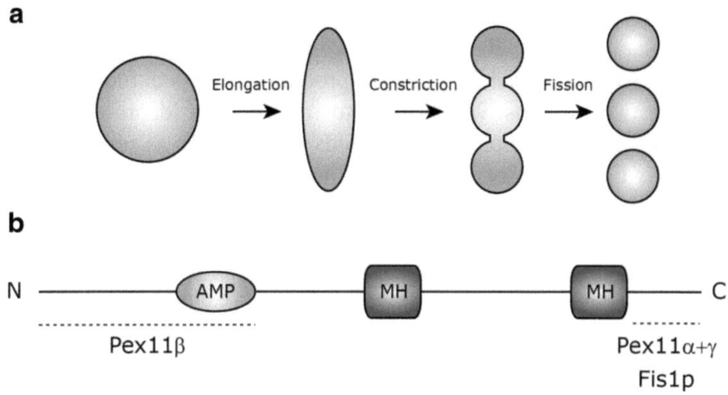

Fig. 19.1 (**a**) Model of peroxisome multiplication by fission, indicating the three steps involved in the process. See text for further details. (**b**) Schematic representation of human Pex11βp. Although a general description of Pex11p family members is not possible, Pex11βp exhibits has a number of characteristics that are found in other Pex11p and Pex11-like proteins, including an N-terminal amphipathic helix (AMP) and C-terminal hydrophobic α-helices (MH), which may represent transmembrane or membrane associated domains. Regions important for interactions with itself (Pex11βp), other Pex11p isoforms (Pex11αp + Pex11γp) and the fission protein Fis1p (Fis1p) are indicated with a *dashed line*

Table 19.1 Selected members of the Pex11p family of proteins, adapted from Kiel et al. (2006)

	Sc[a]	Pp	Hp	Pc	Hs	Tb	At
Pex11p	Q12462	C4R0W1	A4GFC5	Q6EZ50	O75192 (Pex11α) O96011 (Pex11β)	O60944 Q9N9D0 (GIM5A) Q9N9C9 (GIM5B)	Q9STY0 (Pex11b) Q9LQ73 (Pex11c) O80845 (Pex11d) Q84JW1 (Pex11e)
Pex11Bp	-	-	-	A4GFN3	-	-	-
Pex11Cp	-	-	A4GFC6	A4GFN4	Q96HA9 (Pex11γ)	-	Q9FZF1 (Pex11a)
Pex25p	Q02969	F2QX58	A4GFD0	-	-	-	-
Pex27p	Q08580	-	-	-	-	-	-

[a]Abbreviations are: Sc, *Saccharomyces cerevisiae*; Pp, *Pichia pastoris*; Hp, *Hansenula polymorpha*; Pc, *Penicillium chrysogenum*; Hs, *Homo sapiens*; Tb, *Trypanosoma brucei*; At, *Arabidopsis thaliana*. Numbers indicated the Uniprot accession numbers for each protein

similar to Pex11Cp (Schrader et al. 1998b; Li et al. 2002a)—whereas five isoforms have been identified in plants (Lingard and Trelease 2006; Nayidu et al. 2008). Although not annotated as Pex11p isoforms, GIM5A and GIM5B in *T. brucei* exhibit weak homology and a similar domain structure to other Pex11 proteins. Both are involved in peroxisome proliferation (Maier et al. 2001), which suggests they could be members of the Pex11p family. The identification of Pex25p (Smith et al. 2002; Saraya et al. 2011) and Pex27p (Rottensteiner et al. 2003b; Tam et al. 2003), two additional Pex11-like proteins, extended the Pex11p family further. These two proteins, which have so far not been identified in mammals, were not discovered using homology methods with Pex11p as search model, since the homology with Pex11p is rather weak. Instead, Pex25p was identified simultaneously in two separate reports, one using transcription profiling of *S. cerevisiae* cells grown on oleic acid and the other using mass spectrometry on isolated peroxisome fractions. Pex27p was then identified as a homologue of Pex25p using homology methods (Smith et al. 2002; Tam et al. 2003).

Members of the Pex11p family of proteins are largely α-helical and relatively small, generally between 25 and 30 kDa, although Pex25p and Pex27p are at around 45 kDa, somewhat larger. While it is not possible to present a general description that covers all members of the Pex11p family, a number of characteristics shared by many Pex11 proteins are shown in Fig. 19.1b. Nearly all proteins from the Pex11p family target to peroxisomes (Erdmann and Blobel 1995; Lorenz et al. 1998; Schrader et al. 1998b; Smith et al. 2002; Rottensteiner et al. 2003b; Lingard and Trelease 2006; Saraya et al. 2011) and are considered to be peroxisomal membrane proteins. However, Knoblauch and Rachubinski (Knoblach and Rachubinski 2010) demonstrated that Pex11p could redistribute from the ER to peroxisomes, depending on growth conditions. Furthermore, Pex11p has been observed at the ER in certain mutant yeast strains, leading to the suggestion that the protein may target via the ER to peroxisomes (Joshi et al. 2012). Pex11Bp from the filamentous fungus *P. chrysogenum* can be found exclusively at the ER (Opalinski et al. 2012). In line with membrane localization, hydrophobic α-helices can be predicted in the C-terminal region of many members of the Pex11p family, which could represent transmembrane or membrane buried domains. Topological studies indicate that the Pex11p family members are orientated with both the N- and C-termini exposed toward the cytosol and that part of the protein is inserted into the membrane (Schrader et al. 1998a; Lingard and Trelease 2006; Lorenz et al. 2006; Koch and Brocard 2011). Membrane extraction experiments suggest that in *S. cerevisiae* Pex11p (as well as Pex25p and Pex27p) is a peripheral PMP, due to its susceptibility to high salt/carbonate treatment (Marshall et al. 1995; Rottensteiner et al. 2003b; Tam et al. 2003). However, this is in contrast to most other organisms, where Pex11p is reported to be an integral membrane protein (Lorenz et al. 1998; Orth et al. 2007; Koch and Brocard 2012). This apparent difference in properties does not stop human, trypanosome and plant Pex11p complementing the *S. cerevisiae pex11* mutant phenotype (Lorenz et al. 1998; Huber et al. 2012), which may suggest that Pex11p is an integral membrane protein, but that *S. cerevisiae* Pex11p is easier to remove from the peroxisomal membrane. Indeed, Pex11βp is extracted from the membranes of paraformaldehyde-fixed human cells by treatment with the non-ionic detergent Triton X-100, a phenomenon that is not observed with other integral membrane proteins, including Pex11αp and Pex11γp (Schrader et al. 2012a), suggesting a more loose association with the peroxisomal membrane.

19.1.2 Pex11p Proteins Control Peroxisome Numbers

Peroxisome proliferation can be stimulated by a number of factors. In yeasts, peroxisome induction is usually controlled by the carbon source on which the cells are grown. Yeast cells grown on glucose contain fewer and smaller peroxisome than those grown on oleic acid or methanol, the carbon sources most commonly used to induce peroxisomes (Erdmann and Blobel 1995; Krikken et al. 2009; Joshi et al. 2012). Peroxisome numbers are directly linked to Pex11p levels. The absence of Pex11p results in a reduction in the number of peroxisomes, which are enlarged (Erdmann and Blobel 1995; Marshall et al. 1995; Tam et al. 2003; Krikken et al. 2009). Conversely, overproduction of Pex11p leads to an increase in peroxisome number and a decrease in their size (Erdmann and Blobel 1995; Marshall et al. 1995; Schrader et al. 1998b; Tam et al. 2003; Lingard and Trelease 2006; Krikken et al. 2009; Joshi et al. 2012). This is also apparently the case for the Pex11-like proteins Pex25p and Pex27p in *S. cerevisiae*, although deletion of *PEX25* in *H. polymorpha* results in a slight increase in peroxisome number (Saraya et al. 2011). Current theories define a role for Pex11p in peroxisome fission, whereas Pex25p plays a different role, namely in de novo peroxisome formation. *H. polymorpha pex11* or *pex25* cells contain peroxisomes. However, cells deleted for both *PEX11* and *PEX25* appear devoid of peroxisomes. Reintroduction of Pex25p, but not Pex11p in this strain allows peroxisomes to reform, indicating the Pex25 plays a role in de novo peroxisome formation (Saraya et al. 2011). Also, *H. polymorpha pex3 pex25* cells, which are devoid of peroxisomes, are not functionally complemented upon reintroduction of *PEX3*. Similarly, *S. cerevisiae* Pex25p was required to reintroduce peroxisomes in a *S. cerevisiae* strain deleted for *PEX11, PEX25, PEX27,* and *PEX3*, upon reintroduction of *PEX3,* confirming the role of Pex25p in de novo peroxisome formation in yeast (Huber et al. 2012).

Expression of genes encoding proteins of the Pex11p family in yeasts is usually dependent on the carbon source. For example, methanol stimulates *PEX11* and *PEX25* expression in *H. polymorpha* (van Zutphen et al. 2010) and oleic acid also induces *PEX11* expression in *P. pastoris* (Joshi et al. 2012). Transcription of *PEX11* and *PEX25* in *S. cerevisiae* is controlled by the transcription factors Adr1p and Pip2p-Oaf1p (Gurvitz et al. 2001; Rottensteiner et al. 2003a). These transcriptional regulators are also responsible for the expression of a number of genes encoding proteins involved in fatty acid oxidation (Rottensteiner et al. 1996) and are upregulated when cells grow on oleic acid (Gurvitz and Rottensteiner 2006). Interestingly *S. cerevisiae PEX27* is not induced by oleic acid, but is instead constitutively expressed (Tam et al. 2003). Transcription of *PEX11b* in Arabidopsis is regulated by the far-red light receptor phyA in combination with the bZIP transcription factor HYH (Desai and Hu 2008) whereas four of the five PEX11 isoforms in rice are upregulated under stress conditions (Nayidu et al. 2008). In mammals expression of the three different isoforms of *PEX11* is controlled separately. *PEX11β* mRNA levels are similar in all tissue types, whereas *PEX11α* and *PEX11γ* expression is tissue specific, with significant levels of mRNA detected in the kidney, liver, brain, and testis of rats (Schrader et al. 1998b; Li et al. 2002a).

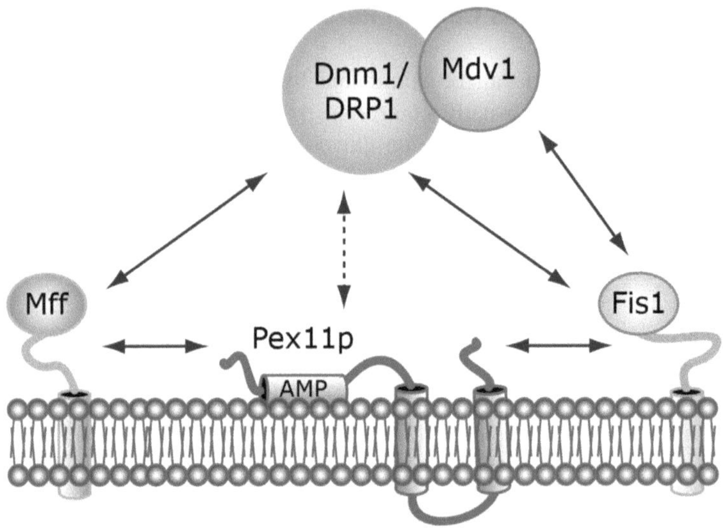

Fig. 19.2 Factors involved in peroxisome fission, adapted from Schrader et al. (2012b). The GTPase Dnml/DLP1, responsible for the scission step in peroxisome fission, associates with peroxisomes through direct interactions (*solid arrows*) with the tail-anchored membrane proteins Fis1p and, in mammals Mff. The adaptor protein Mdv1p, which also interacts directly with Fis1p, has so far only been identified in yeast. Pex11p, which causes membrane tubulation through the action of its amphipathic helix (AMP), helps in recruiting components of the fission machinery, either directly (*solid arrows*) or indirectly (*dashed arrow*)

PEX11α expression can be induced by the compounds clofibrate and 4-phenylbutyrate, whereas these compounds do not affect *PEX11β* and *PEX11γ* levels (Abe et al. 1998; Li et al. 2002a). Transcription of PEX11α is controlled by the clofibrate-inducible peroxisome proliferator activated receptor gamma (PPARγ) transcription factor (Shimizu et al. 2006). These observations suggest that Pex11αp controls peroxisome proliferation as a response to external stimuli, whereas Pex11βp is required for constitutive peroxisome proliferation.

Rat Pex11p was shown to bind coatomer (Anton et al. 2000), a large complex of proteins involved in vesicle trafficking between the endoplasmic reticulum and the Golgi (Cosson and Letourneur 1997; Wieland and Harter 1999). Coatomer binding to rat Pex11p is regulated by a di-lysine motif of the type—KxKxx, present at the C-terminus of Pex11p (Anton et al. 2000; Maier et al. 2000). Di-lysine motifs play a role in vesicle generation and it was suggested that Pex11p is involved in vesicle formation from the ER (Cosson and Letourneur 1994). However, removal of the di-lysine motif from rat Pex11p did not inhibit peroxisome proliferation. This, coupled with the observation that the di-lysine motif is not conserved argues against a general role for coatomer in Pex11p function.

Pex11p may control peroxisome proliferation events through the recruitment of additional factors involved in peroxisome fission (Fig. 19.2, adapted from Schrader et al. 2012b). Physical interactions between Pex11p and Dnm1p (DRP1 in mammals), Mdv1p, Fis1p, and Mff have been identified (Li and Gould 2003; Koch et al. 2010; Koch and Brocard 2012). Dnm1p/DRP1 is a Dynamin-related GTPase involved in the membrane scission step during mitochondria and peroxisome fission (Bleazard et al. 1999; Motley and Hettema 2007; Motley et al. 2008; Mears et al. 2011). In yeast Dnm1p requires an additional binding partner for its function, the adaptor protein Mdv1p. Cells lacking DNM1/DRP1 contain fewer peroxisomes and those present develop long extensions, indication a block in peroxisome fission (Nagotu et al. 2008). Mitochondria are also no longer able to divide in cells lacking Dnm1p/DRP1 (Otsuga et al. 1998). Overexpression of Pex11β increases the association of DRP1 with the peroxisomal membrane and Pex11β interacts with DRP1 in the mammalian two-hybrid assay, although the authors of this contribution could not identify a direct interaction between the two isolated proteins (Li and Gould 2003). This may suggest the requirement of another factor that mediates contact between DRP1 and Pex11p. Two good candidates for such a role are the tail-anchored membrane proteins Fis1p and in mammals, Mff. Both proteins are critical for mitochondrial and peroxisome fission and are thought to recruit Dnm1p/DRP1 to the membrane (Mozdy et al. 2000; Gandre-Babbe and van der Bliek 2008; Otera et al. 2010). Fis1p interacts with Pex11p in a number of species and this interaction requires the C-terminal region of Pex11p (Koch et al. 2010). Fis1p also binds to Dnm1p/DRP1 (Yoon et al. 2003; Kobayashi et al. 2007; Lingard et al. 2008; Koch et al. 2010; Joshi et al. 2012). In mammals all Pex11p isoforms interact with Fis1p and Mff, whereas plant PEX11c-e bind to Fis1p. In the yeast *P. pastoris* the Pex11p–Fis1p interaction appears to be regulated by Pex11p phosphorylation (Joshi et al. 2012). Pex11p phosphorylation was also reported to play a role in peroxisome fission in *S. cerevisiae*, suggesting a conserved role for Pex11p phosphorylation in Fis1p binding (Knoblach and Rachubinski 2010). Another phenomenon that indicates a role for Pex11p in fission is the formation of so-called juxtaposed elongated peroxisomes (JEPs) upon overexpression of Pex11p isoforms in mammalian and plant cells (Koch et al. 2010). These large structures represent a congregation of elongated peroxisomes that form due to an overload on the fission machinery. They dissociate upon overexpression of *FIS1*. Furthermore, siRNA experiments that knocked down *FIS1* also resulted in JEP formation, an effect that was greatly enhanced by concomitant overexpression of *PEX11*. Taken together, these results indicate that a fine balance between Pex11p and Fis1p needs to be held in order to allow fission to occur.

Many members of the Pex11p family can form homodimers (Fig. 19.1b). Interactions between Pex11p and itself have been reported in *S. cerevisiae*, *H. polymorpha,* and mammals (Marshall et al. 1996; Kobayashi et al. 2007; Koch et al. 2010; Saraya et al. 2011). The N-terminal region of Pex11β is required for homo-dimerization in mammals (Kobayashi et al. 2007). The Pex11p-like proteins Pex25p and Pex27p in *S. cerevisiae* are also able to form homodimers (Tam et al. 2003). The different isoforms of Pex11p in mammalian cells can all form heterodimers although unlike the Pex11β–Pex11β interaction, binding requires the C-terminal region (Koch et al. 2010). In *S. cerevisiae* Pex25p and Pex27p can interact with each other (Tam et al. 2003; Koch et al. 2010; Saraya et al. 2011), although *H. polymorpha* Pex25p and Pex11p do not (Saraya et al. 2011). How these interactions regulate protein function in vivo is not clear. In *H. polymorpha dnm1* mutant cells, Pex11p with a C-terminal GFP concentrates at the region, where the peroxisome extension, which forms in this strain due to a defect in fission, emerges from the peroxisome (Nagotu et al. 2008). Likewise Pex11β concentrates at sites of peroxisome constriction (Schrader et al. 1998b). This may suggest a role for oligomerization in marking the site where peroxisome fission occurs. In support of this, deletion of the N-terminal region of Pex11β inhibits homodimerization as well as function (Kobayashi et al. 2007). In contrast, *S. cerevisiae* Pex11p mutants that are no longer able to dimerize stimulate peroxisome proliferation (Marshall et al. 1996). Until further information on homo/heterodimerization becomes available, it remains challenging to understand its potential role.

19.1.3 Membrane Remodeling and the Role of Amphipathic α-helices

Another characteristic feature of Pex11p proteins is the presence of a conserved amphipathic α-helix in the N-terminal region (Fig. 19.1b). Amphipathic α-helices, where one face of the helix is hydrophobic and the other, polar, are involved in membrane remodeling events. One famous example of this is the bin/amphiphysin/rvs-homology (BAR) domain. Insertion of the hydrophobic surface of the amphipathic α-helix into membranes displaces lipid on one side of the membrane, resulting in curvature of the membrane (Low et al. 2008). Synthetic peptides of the amphipathic α-helix from different Pex11 proteins were able to tubulate negatively charged liposomes with a lipid content similar to the peroxisomal membrane in vitro (Opalinski et al. 2010). Furthermore, mutations that inhibit the tubulating activity of the amphipathic α-helix in vitro also disturb peroxisome fission in vivo. The conservation of this amphipathic α-helix in many members of the Pex11p family points toward a general role for the family in membrane remodeling, as was previously suggested (Fagarasanu et al. 2007). However, Koch and Brocard reported an interesting variation on this theme (Koch and Brocard 2011). Pex11γ contains, at its C-terminus, three hydrophobic α-helices. Helix 1 and 3 are suggested to insert into the membrane, whereas helix 2, which is in between the other two, has amphipathic properties. The introduction of a proline mutation into this amphipathic helix abrogated the ability of Pex11γ to remodel the peroxisomal membrane. Therefore, the contribution provided by amphipathic helices is crucial to Pex11p's ability to stimulate peroxisome fission.

19.1.4 Additional Roles of Pex11p

Pex11p has always been associated with peroxisome proliferation. However, in 2000 van Roermund and coworkers described that Pex11p is involved in fatty acid oxidation (Van Roermund et al. 2000). This was based on the observation that intact *S. cerevisiae* cells lacking *PEX11* were deficient in medium chain fatty acid β-oxidation, whereas this process was normal in whole cell lysates. The enzymes required for this process were correctly targeted to peroxisomes in the deletion strain, suggesting that Pex11p aids the transport of medium chain fatty acids into peroxisomes. The authors also showed that the peroxisomal phenotype of the *pex11* mutant strain is comparable to that of a strain lacking genes involved in medium chain fatty acid oxidation (i.e., less peroxisomes per cell) and suggested that peroxisome proliferation is dependent on fatty acid β-oxidation. It was later shown that Pex11p-dependent peroxisome proliferation could also be induced in cells grown on glucose, a condition where expression of the β-oxidation enzymes is repressed (Li and Gould 2002). Nevertheless, the reason as to why peroxisomal fatty acid oxidation is disturbed in cells lacking *PEX11* remains unknown.

Pex11p is known to recruit members of the fission machinery during the fission process. However, Pex11p also acts in redistributing other PMPs to subdomains on the peroxisomal membrane through an unknown mechanism (Cepinska et al. 2011). This was observed in wild-type *H. polymorpha* cells but became clearer in cells lacking *DNM1*, where fission is blocked at a late stage. The PMPs Pex14p, Pex8p, Pex10p, and Pex25p all target to the peroxisome extension that forms due to a lack of *DNM1*. Such redistribution was not observed in cells that lack both *DNM1* and *PEX11* but was still present in cells deleted for *EMP24* and *ERP3*, which also exhibit a fission defect, suggesting an important role for Pex11p in the process (Kurbatova et al. 2009; Cepinska et al. 2011).

Pex11p has been implicated in peroxisome inheritance. This was reported in the original publication identifying Pex11p back in 1995 (Erdmann and Blobel 1995). In *S. cerevisiae* cells lacking *PEX11* peroxisomes were not transported to the daughter during cell division but instead remained in the mother. Confusingly, the exact opposite phenotype was observed in the *H. polymorpha pex11* mutant strain—peroxisomes were transported to the bud during cell division, leaving the mother devoid of organelles (Krikken et al. 2009). How Pex11p may control inheritance is not known, although it does not involve targeting of peroxisomal inheritance protein Inp1, as this was not disturbed by a deficiency in Pex11p (Krikken et al. 2009). Additionally, why such different behavior is observed for the two organisms is not clear. It may simply be a case of which system exerts the strongest force on the single organelle in the absence of Pex11p, either peroxisome retention by the mother or peroxisome transport to the daughter (Krikken et al. 2009) and this may vary depending on the organism.

Concluding Remarks

Our understanding of the many functions carried out by members of the Pex11p family of proteins has advanced considerably over the last few years. Outlining a role for Pex25p in de novo peroxisome formation was an important milestone. Insights into the Pex11 protein linkage map have defined Pex11p as a fission machinery recruitment factor and functional characterization of the amphipathic helices show their crucial role in membrane remodeling. One of the current challenges is therefore to define the link between these two functions—is the fission machinery recruited to sites of membrane curvature that result from the insertion of the amphipathic helices, as suggested previously (Koch et al. 2004) or does the association of the fission machinery, including Fis1p and/or Mff, with the peroxisomal membrane induce amphipathic helix insertion and consequently membrane curvature? Furthermore, how does Pex11p control peroxisome inheritance? What are the mechanisms behinds its role in PMP redistribution and crucially, how these additional functions are linked to the role of Pex11p in peroxisomal fission, are all questions eagerly awaiting an answer.

In conclusion, the role of the Pex11p family in peroxisome biology is being unraveled, although the family still holds many secrets. The role of Pex11Cp for example, the only member of the Pex11p family that exclusively localizes to the ER is currently unknown (Opalinski et al. 2012). Likewise, the individual function of the five Pex11p isoforms in plants has yet to be determined. The recent identification of a 26 year old patient unable to produce the Pex11β protein (Ebberink et al. 2012) again raised many questions concerning the role of Pex11p, especially since Li and coworkers demonstrated that mice lacking Pex11β died shortly after birth (Li et al. 2002b). Why absence of this protein results in such dramatic differences in the two situations is unclear. Therefore, it is safe to state that we still do not fully comprehend the role of the Pex11p family of proteins in peroxisome biology.

Acknowledgments CW is supported by a grant from the Netherlands Organization for Scientific research (NWO), section Earth and Life Sciences.

References

Abe I, Okumoto K, Tamura S, Fujiki Y (1998) Clofibrate-inducible, 28-kDa peroxisomal integral membrane protein is encoded by PEX11. FEBS Lett 431:468–472

Anton M, Passreiter M, Lay D, Thai TP, Gorgas K, Just WW (2000) ARF- and coatomer-mediated peroxisomal vesiculation. Cell Biochem Biophys 32:27–36

Bleazard W, Mccaffery JM, King EJ, Bale S, Mozdy A, Tieu Q, Nunnari J, Shaw JM (1999) The dynamin-related GTPase Dnm1 regulates mitochondrial fission in yeast. Nat Cell Biol 1:298–304

Cepinska MN, Veenhuis M, Van Der Klei IJ, Nagotu S (2011) Peroxisome fission is associated with reorganization of specific membrane proteins. Traffic 12:925–937

Cosson P, Letourneur F (1994) Coatomer interaction with di-lysine endoplasmic reticulum retention motifs. Science 263:1629–1631

Cosson P, Letourneur F (1997) Coatomer (COPI)-coated vesicles: role in intracellular transport and protein sorting. Curr Opin Cell Biol 9:484–487

Desai M, Hu J (2008) Light induces peroxisome proliferation in Arabidopsis seedlings through the photoreceptor phytochrome A, the transcription factor HY5 HOMOLOG, and the peroxisomal protein PEROXIN11b. Plant Physiol 146:1117–1127

Distel B, Erdmann R, Gould SJ, Blobel G, Crane DI, Cregg JM, Dodt G, Fujiki Y, Goodman JM, Just WW, Kiel JA, Kunau WH, Lazarow PB, Mannaerts GP, Moser HW, Osumi T, Rachubinski RA, Roscher A, Subramani S, Tabak HF, Tsukamoto T, Valle D, Van Der

Klei I, Van Veldhoven PP, Veenhuis M (1996) A unified nomenclature for peroxisome biogenesis factors. J Cell Biol 135:1–3
Ebberink MS, Koster J, Visser G, Spronsen F, Stolte-Dijkstra I, Smit GP, Fock JM, Kemp S, Wanders RJ, Waterham HR (2012) A novel defect of peroxisome division due to a homozygous non-sense mutation in the PEX11beta gene. J Med Genet 49:307–313
Erdmann R, Blobel G (1995) Giant peroxisomes in oleic acid-induced *Saccharomyces cerevisiae* lacking the peroxisomal membrane protein Pmp27p. J Cell Biol 128:509–523
Fagarasanu A, Fagarasanu M, Rachubinski RA (2007) Maintaining peroxisome populations: a story of division and inheritance. Annu Rev Cell Dev Biol 23:321–344
Gandre-Babbe S, Van Der Bliek AM (2008) The novel tail-anchored membrane protein Mff controls mitochondrial and peroxisomal fission in mammalian cells. Mol Biol Cell 19:2402–2412
Gurvitz A, Rottensteiner H (2006) The biochemistry of oleate induction: transcriptional upregulation and peroxisome proliferation. Biochim Biophys Acta 1763:1392–1402
Gurvitz A, Hiltunen JK, Erdmann R, Hamilton B, Hartig A, Ruis H, Rottensteiner H (2001) Saccharomyces cerevisiae Adr1p governs fatty acid beta-oxidation and peroxisome proliferation by regulating POX1 and PEX11. J Biol Chem 276:31825–31830
Huber A, Koch J, Kragler F, Brocard C, Hartig A (2012) A subtle interplay between three Pex11 proteins shapes de novo formation and fission of peroxisomes. Traffic 13:157–167
Joshi S, Agrawal G, Subramani S (2012) Phosphorylation-dependent Pex11p and Fis1p interaction regulates peroxisome division. Mol Biol Cell 23:1307–1315
Kiel JA, Van Der Klei IJ, Van Den Berg MA, Bovenberg RA, Veenhuis M (2005) Overproduction of a single protein, Pc-Pex11p, results in 2-fold enhanced penicillin production by Penicillium chrysogenum. Fungal Genet Biol 42:154–164
Kiel JA, Veenhuis M, Van Der Klei IJ (2006) PEX genes in fungal genomes: common, rare or redundant. Traffic 7:1291–1303
Knoblach B, Rachubinski RA (2010) Phosphorylation-dependent activation of peroxisome proliferator protein PEX11 controls peroxisome abundance. J Biol Chem 285:6670–6680
Kobayashi S, Tanaka A, Fujiki Y (2007) Fis1, DLP1, and Pex11p coordinately regulate peroxisome morphogenesis. Exp Cell Res 313:1675–1686
Koch J, Brocard C (2011) Membrane elongation factors in organelle maintenance: the case of peroxisome proliferation. Biomol Concepts 2:353–364
Koch J, Brocard C (2012) PEX11 proteins attract Mff and human Fis1 to coordinate peroxisomal fission. J Cell Sci 125:3813–3826
Koch A, Schneider G, Luers GH, Schrader M (2004) Peroxisome elongation and constriction but not fission can occur independently of dynamin-like protein 1. J Cell Sci 117:3995–4006
Koch J, Pranjic K, Huber A, Ellinger A, Hartig A, Kragler F, Brocard C (2010) PEX11 family members are membrane elongation factors that coordinate peroxisome proliferation and maintenance. J Cell Sci 123:3389–3400
Krikken AM, Veenhuis M, Van Der Klei IJ (2009) Hansenula polymorpha pex11 cells are affected in peroxisome retention. FEBS J 276:1429–1439
Kurbatova E, Otzen M, Van Der Klei IJ (2009) p24 proteins play a role in peroxisome proliferation in yeast. FEBS Lett 583:3175–3180
Li X, Gould SJ (2002) PEX11 promotes peroxisome division independently of peroxisome metabolism. J Cell Biol 156:643–651
Li X, Gould SJ (2003) The dynamin-like GTPase DLP1 is essential for peroxisome division and is recruited to peroxisomes in part by PEX11. J Biol Chem 278:17012–17020
Li X, Baumgart E, Dong GX, Morrell JC, Jimenez-Sanchez G, Valle D, Smith KD, Gould SJ (2002a) PEX11alpha is required for peroxisome proliferation in response to 4-phenylbutyrate but is dispensable for peroxisome proliferator-activated receptor alpha-mediated peroxisome proliferation. Mol Cell Biol 22:8226–8240

Li X, Baumgart E, Morrell JC, Jimenez-Sanchez G, Valle D, Gould SJ (2002b) PEX11 beta deficiency is lethal and impairs neuronal migration but does not abrogate peroxisome function. Mol Cell Biol 22:4358–4365

Lingard MJ, Trelease RN (2006) Five Arabidopsis peroxin 11 homologs individually promote peroxisome elongation, duplication or aggregation. J Cell Sci 119:1961–1972

Lingard MJ, Gidda SK, Bingham S, Rothstein SJ, Mullen RT, Trelease RN (2008) Arabidopsis PEROXIN11c-e, FISSION1b, and DYNAMIN-RELATED PROTEIN3A cooperate in cell cycle-associated replication of peroxisomes. Plant Cell 20:1567–1585

Lorenz P, Maier AG, Baumgart E, Erdmann R, Clayton C (1998) Elongation and clustering of glycosomes in Trypanosoma brucei overexpressing the glycosomal Pex11p. EMBO J 17:3542–3555

Lorenz H, Hailey DW, Wunder C, Lippincott-Schwartz J (2006) The fluorescence protease protection (FPP) assay to determine protein localization and membrane topology. Nat Protoc 1:276–279

Low C, Weininger U, Lee H, Schweimer K, Neundorf I, Beck-Sickinger AG, Pastor RW, Balbach J (2008) Structure and dynamics of helix-0 of the N-BAR domain in lipid micelles and bilayers. Biophys J 95:4315–4323

Maier AG, Schulreich S, Bremser M, Clayton C (2000) Binding of coatomer by the PEX11 C-terminus is not required for function. FEBS Lett 484:82–86

Maier A, Lorenz P, Voncken F, Clayton C (2001) An essential dimeric membrane protein of trypanosome glycosomes. Mol Microbiol 39:1443–1451

Marshall PA, Krimkevich YI, Lark RH, Dyer JM, Veenhuis M, Goodman JM (1995) Pmp27 promotes peroxisomal proliferation. J Cell Biol 129:345–355

Marshall PA, Dyer JM, Quick ME, Goodman JM (1996) Redox-sensitive homodimerization of Pex11p: a proposed mechanism to regulate peroxisomal division. J Cell Biol 135:123–137

Mears JA, Lackner LL, Fang S, Ingerman E, Nunnari J, Hinshaw JE (2011) Conformational changes in Dnm1 support a contractile mechanism for mitochondrial fission. Nat Struct Mol Biol 18:20–26

Motley AM, Hettema EH (2007) Yeast peroxisomes multiply by growth and division. J Cell Biol 178:399–410

Motley AM, Ward GP, Hettema EH (2008) Dnm1p-dependent peroxisome fission requires Caf4p, Mdv1p and Fis1p. J Cell Sci 121:1633–1640

Mozdy AD, Mccaffery JM, Shaw JM (2000) Dnm1p GTPase-mediated mitochondrial fission is a multi-step process requiring the novel integral membrane component Fis1p. J Cell Biol 151:367–380

Nagotu S, Saraya R, Otzen M, Veenhuis M, Van Der Klei IJ (2008) Peroxisome proliferation in Hansenula polymorpha requires Dnm1p which mediates fission but not de novo formation. Biochim Biophys Acta 1783:760–769

Nayidu NK, Wang L, Xie W, Zhang C, Fan C, Lian X, Zhang Q, Xiong L (2008) Comprehensive sequence and expression profile analysis of PEX11 gene family in rice. Gene 412:59–70

Opalinski L, Kiel JA, Williams C, Veenhuis M, Van Der Klei IJ (2010) Membrane curvature during peroxisome fission requires Pex11. EMBO J 30:5–16

Opalinski L, Bartoszewska M, Fekken S, Liu H, De Boer R, Van Der Klei I, Veenhuis M, Kiel JA (2012) De novo peroxisome biogenesis in Penicillium chrysogenum is not dependent on the Pex11 family members or Pex16. PLoS One 7:e35490

Orth T, Reumann S, Zhang X, Fan J, Wenzel D, Quan S, Hu J (2007) The PEROXIN11 protein family controls peroxisome proliferation in Arabidopsis. Plant Cell 19:333–350

Otera H, Wang C, Cleland MM, Setoguchi K, Yokota S, Youle RJ, Mihara K (2010) Mff is an essential factor for mitochondrial recruitment of Drp1 during mitochondrial fission in mammalian cells. J Cell Biol 191:1141–1158

Otsuga D, Keegan BR, Brisch E, Thatcher JW, Hermann GJ, Bleazard W, Shaw JM (1998) The dynamin-related GTPase, Dnm1p, controls mitochondrial morphology in yeast. J Cell Biol 143:333–349

Passreiter M, Anton M, Lay D, Frank R, Harter C, Wieland FT, Gorgas K, Just WW (1998) Peroxisome biogenesis: involvement of ARF and coatomer. J Cell Biol 141:373–383

Rottensteiner H, Kal AJ, Filipits M, Binder M, Hamilton B, Tabak HF, Ruis H (1996) Pip2p: a transcriptional regulator of peroxisome proliferation in the yeast Saccharomyces cerevisiae. EMBO J 15:2924–2934

Rottensteiner H, Hartig A, Hamilton B, Ruis H, Erdmann R, Gurvitz A (2003a) Saccharomyces cerevisiae Pip2p-Oaf1p regulates PEX25 transcription through an adenine-less ORE. Eur J Biochem 270:2013–2022

Rottensteiner H, Stein K, Sonnenhol E, Erdmann R (2003b) Conserved function of pex11p and the novel pex25p and pex27p in peroxisome biogenesis. Mol Biol Cell 14:4316–4328

Sakai Y, Marshall PA, Saiganji A, Takabe K, Saiki H, Kato N, Goodman JM (1995) The Candida boidinii peroxisomal membrane protein Pmp30 has a role in peroxisomal proliferation and is functionally homologous to Pmp27 from Saccharomyces cerevisiae. J Bacteriol 177:6773–6781

Saraya R, Krikken AM, Veenhuis M, Van Der Klei IJ (2011) Peroxisome reintroduction in Hansenula polymorpha requires Pex25 and Rho1. J Cell Biol 193:885–900

Schrader M, Krieglstein K, Fahimi HD (1998a) Tubular peroxisomes in HepG2 cells: selective induction by growth factors and arachidonic acid. Eur J Cell Biol 75:87–96

Schrader M, Reuber BE, Morrell JC, Jimenez-Sanchez G, Obie C, Stroh TA, Valle D, Schroer TA, Gould SJ (1998b) Expression of PEX11beta mediates peroxisome proliferation in the absence of extracellular stimuli. J Biol Chem 273:29607–29614

Schrader M, Almeida M, Grille S (2012a) Postfixation detergent treatment liberates the membrane modelling protein Pex11beta from peroxisomal membranes. Histochem Cell Biol 138:541–547

Schrader M, Bonekamp NA, Islinger M (2012b) Fission and proliferation of peroxisomes. Biochim Biophys Acta 1822:1343–1357

Shimizu M, Akter MH, Emi Y, Sato R, Yamaguchi T, Hirose F, Osumi T (2006) Peroxisome proliferator-activated receptor subtypes differentially cooperate with other transcription factors in selective transactivation of the perilipin/PEX11 alpha gene pair. J Biochem 139:563–573

Smith JJ, Marelli M, Christmas RH, Vizeacoumar FJ, Dilworth DJ, Ideker T, Galitski T, Dimitrov K, Rachubinski RA, Aitchison JD (2002) Transcriptome profiling to identify genes involved in peroxisome assembly and function. J Cell Biol 158:259–271

Tam YY, Torres-Guzman JC, Vizeacoumar FJ, Smith JJ, Marelli M, Aitchison JD, Rachubinski RA (2003) Pex11-related proteins in peroxisome dynamics: a role for the novel peroxin Pex27p in controlling peroxisome size and number in Saccharomyces cerevisiae. Mol Biol Cell 14:4089–4102

Thoms S, Erdmann R (2005) Dynamin-related proteins and Pex11 proteins in peroxisome division and proliferation. FEBS J 272:5169–5181

Van Roermund CW, Tabak HF, Van Den Berg M, Wanders RJ, Hettema EH (2000) Pex11p plays a primary role in medium-chain fatty acid oxidation, a process that affects peroxisome number and size in Saccharomyces cerevisiae. J Cell Biol 150:489–498

Van Zutphen T, Baerends RJ, Susanna KA, De Jong A, Kuipers OP, Veenhuis M, Van Der Klei IJ (2010) Adaptation of Hansenula polymorpha to methanol: a transcriptome analysis. BMC Genomics 11:1

Wieland F, Harter C (1999) Mechanisms of vesicle formation: insights from the COP system. Curr Opin Cell Biol 11:440–446

Yoon Y, Krueger EW, Oswald BJ, Mcniven MA (2003) The mitochondrial protein hFis1 regulates mitochondrial fission in mammalian cells through an interaction with the dynamin-like protein DLP1. Mol Cell Biol 23:5409–5420

Dynamin-Related Proteins in Peroxisome Division

20

Kyaw Aung, Navneet Kaur, and Jianping Hu

Abstract

Peroxisomes are ubiquitous organelles in eukaryotes and essential to the survival of animals and plants. These organelles can be formed de novo in the endoplasmic reticulum and multiply through division. In the past two decades, advances have been made in elucidating the molecular mechanisms governing peroxisome division, using model organisms such as yeasts, mammalian cell lines, and plants. Studies have shown that members of the dynamin-related protein (DRP) superfamily are responsible for the fission step (and at least in some cases, membrane constriction as well) of peroxisome division, and most of these DRPs are shared with the mitochondrial fission machinery. This chapter summarizes current knowledge of the role of DRPs in peroxisomal division in various organisms, isolation of the DRP-based peroxisome division ring from the unicellular red alga *Cyanidioschyzon merolae*, and DRP's organelle receptors and adaptors identified from different model systems. Regulation of DRP's activity through posttranslational modification mechanisms will also be briefly discussed, followed by some future questions to be addressed to further dissect mechanisms underlying DRP-mediated peroxisome division.

Keywords

Dynamin-related protein • Peroxisome division • Dnm1 • Vps1 • Drp1 • DRP3 • Fis1 • Mff • Mdv1/Caf4

K. Aung • N. Kaur
Department of Energy Plant Research Laboratory, Michigan State University, East Lansing, MI 48824, USA

J. Hu (✉)
Department of Energy Plant Research Laboratory, Michigan State University, East Lansing, MI 48824, USA

Plant Biology Department, Michigan State University, East Lansing, MI 48824, USA
e-mail: huji@msu.edu

C. Brocard and A. Hartig (eds.), *Molecular Machines Involved in Peroxisome Biogenesis and Maintenance*, DOI 10.1007/978-3-7091-1788-0_20, © Springer-Verlag Wien 2014

20.1 Introduction

Peroxisomes are single-membrane-delimited organelles present in almost all eukaryotic cells and are indispensable for the viability of humans (Wanders 2013) and plants (Hu et al. 2012). Peroxisomes are often found as roughly spherical organelles, 0.1–1 μm in diameter. However, their abundance, morphology, mobility, and protein content are highly dynamic and can be modulated by internal and external cues (Schrader et al. 2012). The peroxisome is believed to be an offshoot of the endoplasmic reticulum (ER)-derived endomembrane system, whereby (at least in yeast) ER-derived preperoxisomal vesicles that contain some membrane proteins fuse, assemble the peroxisomal translocon, and import soluble proteins from the cytosol (Tabak et al. 2013). In addition, new peroxisomes also arise from pre-existing peroxisomes through division (Schrader et al. 2012).

Peroxisome division includes a few partially overlapping steps, i.e., initial elongation/tubulation, membrane constriction, and fission (Koch and Brocard 2011; Schrader et al. 2012). The evolutionarily conserved Peroxin 11 (PEX11) family of proteins governs the initial elongation step by remodeling the peroxisomal membrane and recruiting other components of the division machinery to the organelle (see Chap. 18). Following membrane constriction, which is controlled by a yet elusive mechanism, peroxisomes go through the final fission step that is executed by another conserved factor, dynamin-related protein (DRP). Unlike mitochondria in many eukaryotic systems, which undergo frequent fission and fusion (Chan 2012), peroxisome fission appears to be predominant over fusion. Although numerous proteins have been reported to affect peroxisome size and abundance, only some of them seem to play a direct role in peroxisome division and/or are functionally associated with PEX11 or DRPs.

This chapter will mainly focus on the role of DRPs in peroxisome division and their partners or adaptors at the organelle membrane and will briefly touch on the regulation of DRPs by posttranslational modification (PTM). When necessary, DRP's roles in mitochondrial and chloroplast division will also be discussed. For additional details of the action of DRPs in mitochondrial fission, which have been extensively reported, readers are referred to several recent reviews (Zhao et al. 2013; Chang and Blackstone 2010; Chan 2012; Otera et al. 2013; Lackner 2013).

20.2 Dynamin and Dynamin-Related Proteins

20.2.1 Protein Structure

Dynamins and DRPs constitute a superfamily of large GTPases present in mammals, plants, fungi, and bacteria (Bramkamp 2012; Praefcke and McMahon 2004; Ramachandran 2011). Classical dynamin proteins contain five conserved domains: the GTPase domain, middle domain (MD), GTPase-effector domain (GED), pleckstrin homology (PH) domain, and a proline- and arginine-rich domain

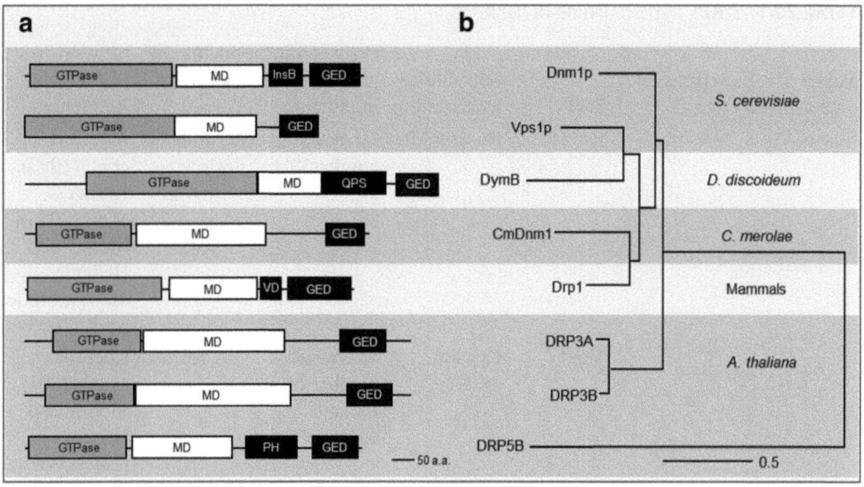

Fig. 20.1 Protein domain organization and phylogenetic analysis of DRPs involved in peroxisome division. (**a**) Linear representations of domain organization of DRPs involved in peroxisome division. GTPase, the GTPase domain; MD, middle domain; GED, GTPase-effector domain; InsB, insert B domain; QPS; glutamine, proline, and serine rich domain; VD, variable domain; PH, pleckstrin homology domain. (**b**) Phylogenetic tree of peroxisome division DRPs. Sequence alignment and the phylogenetic tree construction were performed using the "one Click" mode of Phylogeny.fr software (Dereeper et al. 2008). The scale bar represents 0.5 substitutions per nucleotide site. Proteins analyzed are as follows: Dnm1 (NM_001181821.1), Vps1 (NM_001179791.1), DymB (XM_637355.1), CmDnm1 (AY162473.1), Drp1 (NM_012063), DRP3A (At4g33650), DRP3B (At2g14120), and DRP5B (At3g19720)

(PRD). In brief, the GTPase domain binds and hydrolyzes guanosine triphosphate (GTP), and its activity is stimulated through interaction with GED. MD is composed of coiled-coil domains that together with the GED mediate the formation of dynamin-dynamin complex. The PH domain preferentially binds to negatively charged lipids, and PRD interacts with the Src-homology 3 (SH3) domain present in dynamin-binding proteins (Ferguson and De Camilli 2012).

In general, DRP is defined by the presence of the first three conserved domains, i.e., GTPase, MD, and GED (Heymann and Hinshaw 2009; Fig. 20.1a). Between MD and GED, yeast Dnm1 and mammalian Drp1 also contain an additional functional domain named insert B domain (InsB), sometimes referred to as the variable domain (VD; Fig. 20.1a) (Bui et al. 2012; Strack and Cribbs 2012). In Dnm1, a novel motif within InsB promotes Dnm1's membrane recruitment by facilitating interactions between Dnm1 and an adaptor protein Mdv1 (Bui et al. 2012). In mammalian Drp1, VD/InsB regulates the activity of the protein by mediating oligomeric assembly (Strack and Cribbs 2012; Frohlich et al. 2013).

Table 20.1 DRPs in peroxisome division

Model organism	Experimental system	DRP	Subcellular localization	Peroxisome phenotype in mutants	Function in other organelles	References
Yeasts	*S. cerevisiae*	Dnm1	Px. and Mt.	Decreased number of Px.	Mt. division	Bleazard et al. (1999), Kuravi et al. (2006), Motley and Hettema (2007), Motley et al. (2008)
		Vps1	Cytosol, TGN?	Decreased number of Px. and enlarged Px.	Vac. fusion	Hoepfner et al. (2001), Motley and Hettema (2007), Motley et al. (2008)
	S. pombe	Dnm1	n.d.	No effect	Mt. division	Jourdain et al. (2008), Guillou et al. (2005)
		Vps1	n.d.	No effect	Mt. division	Jourdain et al. (2008)
	H. polymorpha	Dnm1	Px. and Mt.	Elongated Px. and decreased number of Px.	n.d.	Nagotu et al. (2008)
Mammals	Cell lines	Drp1	Px. and Mt.	Elongated Px.	Mt. division	Koch et al. (2003), Li and Gould (2003), Smirnova et al. (2001)
Plants	*A. thaliana*	DRP3A	Px. and Mt.	Elongated and aggregated Px.	Mt. division	Arimura et al. (2004), Arimura and Tsutsumi (2002), Aung and Hu (2009), Fujimoto et al. (2009), Logan et al. (2004), Mano et al. (2004), Zhang and Hu (2009)
		DRP3B	Px. and Mt.	Slightly elongated Px.	Mt. division	Fujimoto et al. (2009), Zhang and Hu (2009)

(continued)

Table 20.1 (continued)

Model organism	Experimental system	DRP	Subcellular localization	Peroxisome phenotype in mutants	Function in other organelles	References
		DRP5B	Px. and Cpt.	Enlarged and aggregated Px.	Cpt. and Mt. division	Gao et al. (2003), Zhang and Hu (2010), Aung and Hu (2012)
Algae	*C. merolae*	CmDnm1	Px. and Mt.	Enlarged Px.	Mt. division	Imoto et al. (2013)
Protists	*D. discoideum*	DymB	Mt.	Elongated Px. and decreased number of Px.	Contractile vac. dynamics	Rai et al. (2011)

Px., Peroxisome; Mt., Mitochondrion; Cpt., Chloroplast; Vac., Vacuole; TGN, *trans*-Golgi network; n.d., not determined

20.2.2 DRPs in Organelle Division

Dynamins and DRPs participate in a wide range of cellular processes including clathrin-mediated endocytosis, vesicle scission, organelle fission and fusion, cytokinesis, and viral resistance (Heymann and Hinshaw 2009; Ferguson and De Camilli 2012). During the past two decades, the functional role of dynamins and DRPs in organelle division has been extensively studied, leading to the discovery that DRPs are conserved division factors executing the final fission of organelles, such as peroxisomes, mitochondria, and chloroplasts, across diverse organisms (Benard and Karbowski 2009; Kaur and Hu 2009; Table 20.1).

20.2.2.1 Yeast Dnm1 and Vps1

The involvement of DRPs in organelle division was first discovered in the budding yeast *Saccharomyces cerevisiae*, in which the DRP Dnm1 was found to physically associate with mitochondria at the constriction sites and form a spiral-like structure to mediate membrane fission through GTP hydrolysis (Otsuga et al. 1998; Bleazard et al. 1999; Ingerman et al. 2005). Besides its role in mitochondrial division, Dnm1 also controls the fission of peroxisomes, and the decrease in peroxisome number in *Dnm1Δ* is only apparent when cells are growing on oleate, an inducer of peroxisome proliferation in yeast (Kuravi et al. 2006; Fig. 20.2a).

Another yeast DRP, Vps1, also controls the number of peroxisomes besides mediating vacuole morphogenesis, as the *Vps1Δ* mutant contains only one or two giant peroxisomes (Fig. 20.2a; Hoepfner et al. 2001; Kuravi et al. 2006). Peroxisome numbers are further reduced in the *Vps1Δ/Dnm1Δ* double mutant (Fig. 20.2a) and reintroduction of either Vps1 or Dnm1 to the double mutant restores peroxisome fission (Hoepfner et al. 2001; Kuravi et al. 2006; Motley and Hettema 2007;

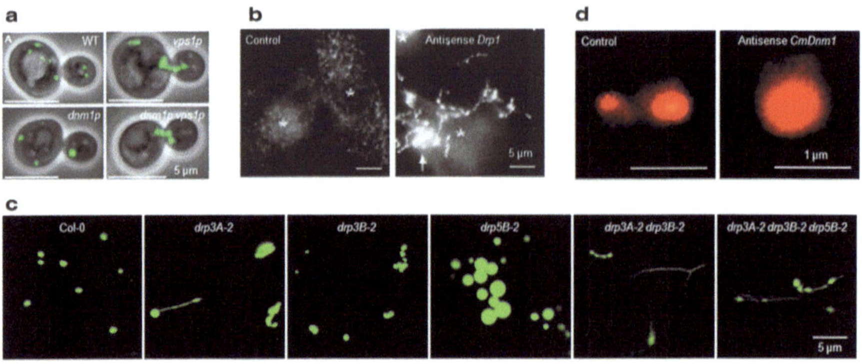

Fig. 20.2 Peroxisomal morphology in DRP deficient mutants from various organisms. (**a**) Peroxisome morphology in *S. cerevisiae Dnml* and *Vpsl* mutants grown on oleate. Peroxisomes were labeled by peroxisomal targeting signal 1 (PTS1)-tagged green fluorescent protein (GFP). Images adapted from Kuravi et al. (2006). (**b**) Immunofluorescence images showing peroxisomes in mammalian HepG2 cell lines. Control cells were treated with buffer and antisense *Drpl* cells were transfected with the *Drpl* siRNA duplex. Peroxisomes were detected by anti-catalase antibodies. Images adapted from Koch et al. (2003). (**c**) Peroxisome morphology in Arabidopsis *drp* mutants. Peroxisomes were labeled by cyan fluorescent protein (CFP)-PTS1. Images adapted from Aung and Hu (2012). (**d**) Immunofluorescence images showing peroxisomes dividing in control cells and non-dividing in *CmDnml*-down-regulated cells of *C. merolae*. Peroxisomes were detected by anti-catalase antibodies. Images were adapted from Imoto et al. (2013)

Motley et al. 2008). However, Vps1 and Dnm1 cannot compensate each other's role in vacuolar fusion or mitochondrial fission (Hoepfner et al. 2001; Kuravi et al. 2006; Motley and Hettema 2007; Motley et al. 2008). These data are consistent with the idea that Vps1 and Dnm1 play partially redundant roles in peroxisome fission in *S. cerevisiae*. Further, it is thought that Vps1 is a major player in cell division-associated peroxisome replication, while Dnm1 makes a greater contribution under peroxisome proliferating conditions (Kuravi et al. 2006; Motley et al. 2008).

In the fission yeast *Schizosaccharomyces pombe*, cells lacking both Vps1 and Dnm1 either contain no peroxisomes or have enlarged peroxisomes, whereas disruption of a single gene does not affect peroxisome abundance, leading to the conclusion that the two proteins act redundantly in peroxisome fission (Guillou et al. 2005; Jourdain et al. 2008). However, in the methylotrophic yeast, *Hansenula polymorpha*, only Dnm1 mediates peroxisome fission (Nagotu et al. 2008). In summary, at least one DRP is a key component of the peroxisome division machinery in various yeast species.

20.2.2.2 Mammalian Drp1

Mammalian cells seem to utilize a single DRP with at least six transcriptional variants (Howng et al. 2004; Chen et al. 2000) in executing the division of both peroxisomes (Koch et al. 2003; Li and Gould 2003) and mitochondria (Smirnova et al. 2001), although whether each variant plays a specific role is unclear. This mammalian protein has been named Drp1, DLP1, or DNM1L in various papers; here for clarity we refer to it as Drp1. Drp1 was found to associate with a subpopulation of peroxisomes and distribute along the length of the elongated peroxisome in cells expressing *PEX11β*, and silencing of Drp1 results in elongated peroxisomes that have already been constricted, causing the "beads-on-a-string" phenotype (Koch et al. 2003; Li and Gould 2003). Ectopic expression of *PEX11β* in cells containing antisense Drp1 results in tubulated peroxisomes without increasing the peroxisome population (Koch et al. 2003; Fig. 20.2b). These findings together suggest that Drp1 performs the final fission in peroxisomal and mitochondrial division in mammals.

Human Drp1 is vital for survival, as shown by neonatal lethality due to a heterozygous dominant negative mutation in *Drp1* (Waterham et al. 2007). The finding that brain-specific Drp1 knockout mice exhibit several developmental defects that lead to embryonic lethality further reinforced the essential role of Drp1 in mammalian growth and development (Wakabayashi et al. 2009; Ishihara et al. 2009). In addition, dysfunctional Drp1 is associated with a host of human diseases including cardiomyopathy, neurodegenerative diseases like Alzheimer's and is also implicated in resistance to apoptotic stimuli in cancer cells (Ashrafian et al. 2010; Thomas and Jacobson 2012; Manczak and Reddy 2012; DuBoff et al. 2012; Cho et al. 2009; Zhao et al. 2013).

20.2.2.3 Arabidopsis DRP3A, DRP3B, and DRP5B

The plant model system *Arabidopsis thaliana* contains over a dozen dynamins and DRPs (Hong et al. 2003), including the two subfamily-3 DRPs (DRP3A and DRP3B) that are in the same subclade as Dnm1 and Drp1 in phylogenetic analyses (Miyagishima et al. 2008; Fig. 20.1b), implying a functional conservation among DRP3A, DRP3B, and other members of this subclade. Indeed, *drp3A* and *drp3B* mutants contain elongated, enlarged, and/or clustered peroxisomes and mitochondria, suggesting their roles in the division of both organelles (Arimura et al. 2004; Mano et al. 2004; Aung and Hu 2009; Fujimoto et al. 2009; Zhang and Hu 2009; Logan et al. 2004; Arimura and Tsutsumi 2002). Consistent with the genetic data, both DRP3 proteins are dual localized, frequently observed at mitochondrial constriction sites or tips, and juxtaposed to peroxisomes (Fujimoto et al. 2009; Zhang and Hu 2009; Mano et al. 2004; Lingard et al. 2008).

Several observations support the idea that DRP3A and DRP3B are functionally equivalent in mitochondrial division but not in peroxisome division. For example, *drp3A* mutants display much stronger peroxisome division defects than *drp3B* by having drastically elongated, enlarged, and highly clustered peroxisomes (Fig. 20.2c), DRP3B fails to complement the *drp3A* peroxisomal mutant phenotype whereas DRP3A can complement *drp3B*, and overexpression of DRP3B (but not DRP3A) leads to an elongated and constricted peroxisome phenotype indicative of incomplete fission (Mano et al. 2004; Fujimoto et al. 2009; Zhang and Hu 2009). Moreover, peroxisome morphological mutant screens uncovered various *drp3A* alleles that correspond to 14 different mutations in the *DRP3A* gene, but not a single mutant of *DRP3B* was isolated, and *drp3A drp3B* double mutants only vary slightly from the *drp3A* mutant in peroxisome phenotypes (Fig. 20.2c), reinforcing the notion that DRP3A exerts a stronger effect in peroxisome division (Aung and Hu 2009; Mano et al. 2004; Zhang and Hu 2009). In line with the genetic evidence, DRP3B is not a major component of the DRP3-containing protein complex (Aung and Hu 2012). Finally, only DRP3A contributes to cell cycle-associated duplication of peroxisomes in suspension cell cultures (Lingard et al. 2008).

Taken together, studies strongly suggest that DRP3A makes unique and irreplaceable contributions to plant peroxisome division. Given that DRP3A and DRP3B interact physically by way of homo- and hetero-dimerization and their co-presence in the same protein complex (Aung and Hu 2012; Fujimoto et al. 2009), it is likely that the two proteins function in the same division machinery. One possibility is that DRP3A mediates peroxisome constriction, either alone or in conjunction with other yet-to-be identified protein(s), and acts redundantly with DRP3B in the final fission step.

DRP5B is distantly related to DRP3 in protein structure and sequence (Fig. 20.1a, b) and was originally identified as a chloroplast division protein localized to a ring structure at the chloroplast division site, with mutants showing enlarged and dumbbell-shaped chloroplasts (Gao et al. 2003). Interestingly, DRP5B was later found to be dual localized and involved in peroxisome division as well (Zhang and Hu 2010). Unlike the juxtaposed association of DRP3A and DRP3B proteins with peroxisomes, DRP5B-YFP is evenly dispersed on the surface of peroxisomes (Zhang and Hu 2010), possibly due to the presence of a PH domain that could interact directly with membrane lipids. Different from the phenotypes exhibited by *drp3A* and *drp3B*, *drp5B* contains enlarged peroxisomes in root and epidermal cells and aggregated peroxisomes in mesophyll cells (Zhang and Hu 2010; Aung and Hu 2012; Fig. 20.2c). Surprisingly, mitochondria in *drp5B* mutants are also impaired in division despite the lack of mitochondrial targeting of DRP5B-YFP, adding an unforeseen new function for DRP5B in regulating the morphology of multiple eukaryotic organelles (Aung and Hu 2012). Mitochondrial division defects are further compounded in the *drp3A drp3B drp5B* triple mutant compared with the *drp3A drp3B* double mutant, suggesting that DRP5B is partially redundant with DRP3 in mitochondrial division. In contrast, peroxisomal phenotype in the triple mutant is virtually indistinguishable from that in *drp3A drp3B* (Fig. 20.2c). Furthermore, DRP5B is not a component of the DRP3-containing protein complex (Aung and Hu 2012). These observations together argue for a DRP3-independent mode of action of DRP5B in mediating peroxisome division, which needs to be further elucidated. Thus, Arabidopsis DRP3A, DRP3B, and DRP5B make differential contributions to the division of peroxisomes, mitochondria, and chloroplasts, three organelles linked by a number of metabolic pathways in plants.

The *drp3A* null, *drp5B* null, *drp3A drp3B* double, and *drp3A drp3B drp5B* triple mutants exhibit various levels of growth defects, with *drp3A drp3B* and *drp3A drp3B drp5B* displaying dwarfness and reduced fertility (Aung and Hu 2009, 2012; Fujimoto et al. 2009; Mano et al. 2004; Zhang and Hu 2009, 2010). The viability of the triple mutant implies that Arabidopsis DRPs other than DRP3A, DRP3B, and DRP5B may also participate in the division of peroxisomes and mitochondria.

DRP1C and DRP1E were reported to be involved in mitochondrial morphogenesis (Jin et al. 2003). It remains to be determined whether these two DRPs, or any other DRPs, also take part in plant peroxisome division.

20.2.2.4 Dictyostelium DRP

The slime mold, *Dictyostelium discoideum*, is considered as a lower eukaryote that is evolutionarily equidistant from yeasts and humans and is at the junction of uni- and multicellularity (Muller-Taubenberger et al. 2013). *D. discoideum* has a five-member dynamin family composed of two dynamins and three DRPs, among which dynamin B (DymB; Fig. 20.1) modulates peroxisome dynamics and, in its absence, leads to elongated peroxisomes with fewer numbers (Rai et al. 2011).

20.2.3 Mode of Action for DRP

Dynamin assembles into spirals on liposomes, tubulates them and splits the lipid tube upon GTP hydrolysis. The two key features of dynamins, oligomerization, and GTP hydrolysis exhibit mutual interdependence, since nucleotide binding induces oligomerization and oligomerization accelerates the rate of GTP hydrolysis thus enhancing enzyme efficiency (Ferguson and De Camilli 2012). Due to the property of converting chemical energy into an actualized force, dynamins are often referred to as mechanoenzymes or molecular scissors (Morlot and Roux 2013).

Yeast Dnm1 also has GTPase activity and can self-assemble to form a spiral-like structure (Ingerman et al. 2005). Cryo-EM-derived reconstruction of lipid tube decorated with Dnm1 revealed that Dnm1 assembles into tetrameric asymmetric units that form two distinct helices, which wrap around the lipid tubes (termed two-start helices) at a distance of 3–4 nm, a gap that could potentially be occupied by adaptor proteins in vivo (Mears et al. 2011). Addition of GTP stimulates a conformational change, during which the ~120 nm Dnm1-coated lipid tube undergoes ~50 nm constriction, followed by Dnm1's disassociation from the lipid tubes (Mears et al. 2011). Similarly, negative-stain EM showed that liposomes tubulated by Drp1 are 130–150 nm in diameter and constrict to ~75 nm in the presence of GTP (Frohlich et al. 2013). In contrast, similar experiments conducted on dynamin showed that it forms a single helix made of dimeric asymmetric units, with an outer diameter of 50 nm; the oligomers are closely appressed to the lipid tubes (Chen et al. 2004; Zhang and Hinshaw 2001). Dynamin constricts the lipid tubes by 10 nm (Sweitzer and Hinshaw 1998; Danino et al. 2004) and remains bound to lipid tubes after fission (Morlot et al. 2012). The larger magnitude of constriction by Dnm1 in comparison to dynamin was attributed to a more flexible helical structure and Dnm1's loose association with the lipid bilayer (Mears et al. 2011).

High-resolution crystal structure of human dynamin1 assigned three domains to the protein: the G domain, the bundle signaling element, and the stalk. Three distinct interfaces in the stalk facilitate the criss-cross arrangement of the stalk region that invokes dimerization of dynamin and its subsequent oligomerization

(Chappie et al. 2010; Faelber et al. 2011; Ford et al. 2011). Parts of MD and GED are in the stalk region, which is consistent with the known functions of these predicted protein domains. Crystal structure analysis of Drp1 revealed that, like dynamin, Drp1 contains the same three domains and the three stalk interfaces. In addition to its described role in dimerization, the central interface-2 is also critical for membrane recruitment of Drp1. Analogous to their role on dynamin, interface-1 and -3 are important for ordered assembly of Drp1 on the membranes. A new interface (4) unique to Drp1 stalk was proposed to aid in Drp1 assembly into linear filament. Mutations in this region failed to tubulate liposomes, suggesting that this region may play a role in nucleating Drp1 molecules on organelle membranes (Frohlich et al. 2013).

Thus, despite sharing key architectural features, dynamin and DRP form polymeric structures somewhat differently.

20.3 Isolating the CmDnm1-Containing Peroxisomal Division Complex from *Cyanidioschyzon merolae*

The unicellular red alga *Cyanidioschyzon merolae* contains a basic set of eukaryotic organelles, including a single chloroplast, mitochondrion, and peroxisome. Remarkably, light/dark cycles can synchronize organelle division, and the individual organelles divide in a strictly sequential order that can be mapped to specific time-points in the mitotic cycle (Imoto et al. 2011). This unique cell biology of *C. merolae* has made this red alga an ideal model system to study organelle division at the molecular level, as exemplified by the successful isolation of plastid and mitochondrial division rings (PD and MD) and subsequent identification of components of the plastid/mitochondrial division machineries (Yoshida et al. 2010; Nishida et al. 2007). It has also enabled the recent isolation of the DRP-based peroxisomal division ring (Imoto et al. 2013).

Peroxisome division in *C. merolae* occurs after mitochondrial division, sometime during the late M (mitotic) phase, and takes about 1.5–2 h to completion (Miyagishima et al. 1999). The single peroxisome undergoes morphological changes from oval to dumb-bell shapes before splitting into two daughter peroxisomes. Using proteomic analysis of the dividing peroxisome fraction, the Kuroiwa group isolated the peroxisome-dividing (POD) ring and identified CmDnm1 as the major component of the POD (Imoto et al. 2013). Interestingly, CmDnm1 was also detected on MD, where it plays a critical role in mitochondrial division prior to its function on the peroxisome. Repressing *CmDnm1* expression resulted in a large peroxisome inhibited in division (Fig. 20.2d). CmDnm1 is the ortholog of DRP3 (Miyagishima et al. 2008; Imoto et al. 2013), suggesting that these proteins belong to an ancient subfamily of DRPs with a specific role in the fission of mitochondria and peroxisomes.

Immunofluorescence analysis detected CmDnm1 as a ring structure surrounding the peroxisome division site, with the diameter of the ring shrinking with the progression of peroxisome division (Fig. 20.3a, b), suggesting that the CmDnm1-

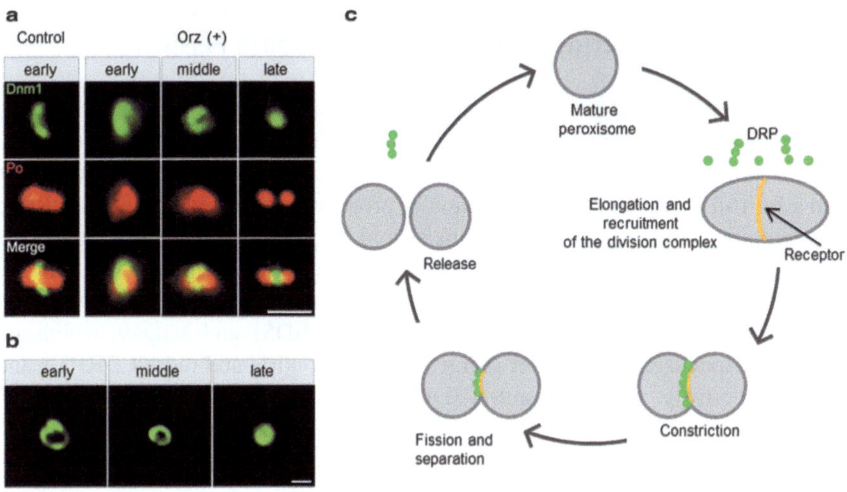

Fig. 20.3 Dynamics of the DRP-based peroxisomal division machinery. (**a**) Immunofluorescence images of the dynamin ring-like structure encircling dividing peroxisomes in control and oryzalin-treated (Orz+) cells. CmDnm1 and peroxisomes (Po) were labeled by anti-Dnm1 and anti-catalase antibodies, respectively. Scale bar = 1 μm. Images adapted from Imoto et al. (2013). (**b**) Immunofluorescence microscopy of dynamin rings isolated from peroxisomes at different phases of division. Scale bar = 0.5 μm. Images adapted from Imoto et al. (2013). (**c**) A model showing the mode of action of DRP in peroxisome division. Membrane-bound receptor (e.g., Fis1, Mff) recruits the cytosolic DRP to the division site. The DRP-based ring later constricts the division site and separates the organelles, followed by releasing of the DRPs into the cytosol upon separation of the organelles

ring is involved in constricting the division site (Imoto et al. 2013). Whole-mount EM of isolated dividing peroxisomes revealed that the POD consists of two rings: an outer dynamin ring that is composed of CmDnm1 and amorphous string and an inner filamentous ring. Immunogold particle labeling-based detection of CmDnm1 on the outer layer agrees well with the cryo-EM rendered Dnm1 3D structure, which suggested that Dnm1 is not strongly anchored to the lipid bilayer (Mears et al. 2011). The skeletal filamentous ring is rigid, ~30 nm in width, and made of a bundle of fine filaments that are 4–5 nm in diameter. In contrast to PD and MD rings, the width of the filamentous ring of POD remains unchanged during peroxisome contraction. It was suggested that the outer CmDnm1-based ring slides the filamentous ring at the division site, generating the contractile force. Following this event, the inner ring disassembles and the dynamin ring severs the membrane bridge between the daughter peroxisomes to complete fission (Imoto et al. 2013).

20.4 Anchors, Adaptors, and Other Factors Needed for the Function of DRPs in Peroxisome Fission

Most DRPs lack a lipid binding or transmembrane domain, thus they are localized in the cytosol until being recruited to the organelles by membrane-bound receptors/anchors, sometimes with the assistance of cytosolic adaptors. The last decade has been a fruitful period in the identification of proteins involved in recruiting DRP to the organelles, among them are proteins shared by peroxisomes and mitochondria as well as proteins that are organelle specific. Here we will only focus on proteins involved in peroxisome division. Proteins such as Arabidopsis ELONGATED MITOCHONDRIA1 (ELM1) and human MIEF1/MiD51 and MiD49, which are exclusively localized to the outer surface of mitochondria and recruit DRP3A and Drp1, respectively, to mitochondria (Duncan et al. 2011; Arimura et al. 2008; Palmer et al. 2011; Zhao et al. 2011), will not be discussed further.

20.4.1 FIS1 as a Receptor for DRP

FISSION1 (FIS1), a tail-anchored membrane protein dual-targeted to peroxisomes and mitochondria, is another organelle division factor conserved across eukaryotic species. FIS1 is tethered to the membranes by its C-terminal tail and contains a cytoplasmic N-terminal tetratricopeptide repeat (TPR) domain, suggesting that the DRP or DRP-containing complex is recruited to the division sites directly or indirectly through FIS1's protein–protein interaction domain (TPR) (Kobayashi et al. 2007; Koch et al. 2005; Mozdy et al. 2000). The rate-limiting role of FIS1 in peroxisomal and mitochondrial division was supported by evidence that suppression of *FIS1* by siRNA phenocopied the *Drp1* siRNA mutant, whereas ectopic expression of *FIS1* caused an increase in the number of peroxisomes and mitochondria (Kobayashi et al. 2007; Koch et al. 2005). However, FIS1 is not the only receptor for the DRPs. For example, Fis1 is required for the peroxisome targeting of Dnm1 (Kuravi et al. 2006; Motley et al. 2008; Fig. 20.3c) but not Vps1, which is actually recruited to the peroxisome by the peroxisomal membrane protein Pex19 (Hoepfner et al. 2001; Vizeacoumar et al. 2006). Studies also demonstrated that the mammalian Mff protein (see below) is an essential receptor for Drp1, whereas Fis1 is the mitochondrial receptor for another protein that controls mitochondrial morphogenesis (Onoue et al. 2013; Otera et al. 2010).

Arabidopsis contains two FIS1 orthologs, FIS1A (BIGYIN) and FIS1B, both of which are involved in peroxisomal and mitochondrial division. Loss-of-function mutants contain a reduced number of peroxisomes and mitochondria, whereas plants overexpressing each *FIS1* exhibit a significant increase in peroxisomal and mitochondrial numbers (Scott et al. 2006; Zhang and Hu 2008, 2009). Although the Arabidopsis FIS1s have been characterized as positive regulators in organelle division and shown to interact with DRP3 and DRP5B (Zhang and Hu 2010), their ability to directly or indirectly recruit DRPs has yet to be demonstrated.

20.4.2 Yeast Mdv1 and Caf4 as Adaptors for DRP

In yeast, Fis1p functions together with two WD40 repeat cytosolic adaptor proteins, Mdv1 and its paralog, Caf4 (Tieu and Nunnari 2000; Tieu et al. 2002; Hoppins et al. 2007; Motley et al. 2008). Mdv1 and Caf4 contain an N-terminal domain that interacts with Fis1, a middle coiled-coil domain involved in homodimerization, and the C-terminal WD-40 domain that heterodimerizes with Dnm1 (Koirala et al. 2010). The *mdv1* null mutant phenocopies the *dnm1* mutant by blocking the fragmentation of mitochondria, and Mdv1 and Caf4 physically interact with both Fis1p and Dnm1 (Griffin et al. 2005; Tieu and Nunnari 2000), suggesting their role as molecular adaptors in recruiting Dnm1 to mitochondria. In contrast, *Mdv1Δ/Caf4Δ* double mutant by itself has no impact on peroxisome population and morphology, but when combined with *Vps1Δ*, the triple mutants resemble *Vps1Δ/Dnm1Δ* and *Fis1Δ/Vps1Δ* mutants, i.e., containing a single undivided peroxisome in the cell, suggesting that Mdv1 and Caf4 are part of the peroxisome division machinery. Further, it was determined that Fis1 facilitates the peroxisome association of Mdv1 and Caf4, and the three proteins together are required for Dnm1-dependent peroxisomal division (Motley et al. 2008). Mammals seem to lack apparent functional or structural homologs of Mdv1 and Caf4. Plant genomes contain a few proteins structurally similar to Mdv1/Caf4, but none has proven to be involved in organelle morphogenesis (Pan and Hu 2011). Thus, Mdv1 and Caf4 are currently defined as yeast-specific factors in peroxisomal and mitochondrial division.

20.4.3 The Metazoan-Specific DRP Receptor Mff

A screen of the Drosophila RNAi library for mutants with abnormal mitochondria identified a metazoan-specific mitochondrial fission factor, Mff. Mammalian cells lacking the Mff ortholog contain interconnected mitochondria and elongated peroxisomes, demonstrating its role in the division of peroxisomes and mitochondria (Gandre-Babbe and van der Bliek 2008). Mff is a tail-anchored coiled-coil protein on the mitochondrial outer membrane and peroxisomal membrane, recruiting Drp1 to the peroxisomal and mitochondrial constriction sites (Fig. 20.3c). Mff has a positive effect in regulating the GTPase activity of Drp1, presumably through stimulating the self-assembly of Drp1 at the fission site, and interactions between Mff, PEX11 proteins and Drp1 were also proposed to coordinate peroxisome division (Otera et al. 2010; Koch and Brocard 2012; Itoyama et al. 2013).

20.4.4 Other Factors That Influence DRP's Function in Peroxisome Division

Mammalian GDAP1 (ganglioside-induced differentiation associated protein 1) is a C-terminal tail-anchored membrane protein that was initially characterized as a positive regulator of mitochondrial division, as recessive mutants show reduced mitochondrial fragmentation while gain-of-function mutation causes excessive mitochondrial fragmentation (Niemann et al. 2005, 2009; Wagner et al. 2009). A recent study established that GDAP1 is also targeted to peroxisome membrane, resulting in peroxisome elongation when its function is lost and increases in peroxisome fragmentation when the gene is overexpressed. Similar to what was found in mitochondrial division, Drp1, Mff and an intact GDAP1 hydrophobic domain are requisite for the function of GDAP1 in peroxisome fission (Huber et al. 2013). Collectively, these features make GDAP1 yet another component shared between the mitochondrial and peroxisome division machinery in mammals.

In addition to the proteins discussed above, other factors may also mediate DRP-based peroxisome division. For example, given the recent findings that endoplasmic reticulum (ER) and actin filament mark future division site by constricting mitochondria (Toulmay and Prinz 2012; Korobova et al. 2013; Kornmann et al. 2009; Friedman et al. 2011), it would also be interesting to investigate whether actin and ER are involved in peroxisomal division through similar mechanisms. Finally, there are other tail-anchored organelle membrane proteins involved in peroxisome morphogenesis but do not seem to be associated with DRPs. One such example is the Arabidopsis Peroxisome and Mitochondrial Division Factor 1 (PMD1) protein, which is dual targeted with a role in the division/morphogenesis of both peroxisomes and mitochondria (Aung and Hu 2011).

20.5 Posttranslational Regulation of DRP's Function

Multiple posttranslational modification (PTM) strategies have been shown to regulate the function of the mammalian Drp1 protein. This is not surprising, given that Drp1 is the major molecular regulator of peroxisomal and mitochondrial fission in mammals.

Drp1 is phosphorylated at multiple sites, resulting in the activation or attenuation of Drp1's activity in mitochondrial division or changes in Drp1's translocation to, or association with, mitochondria (Cereghetti et al. 2008; Chang and Blackstone 2007; Cribbs and Strack 2007; Taguchi et al. 2007; Han et al. 2008). In plants, it was recently reported that the Arabidopsis DRP3A and DRP3B proteins undergo mitotic phosphorylation, which promotes mitochondrial fission during mitosis (Wang et al. 2012). Further, DRP3A and DRP3B were experimentally identified to be phosphorylated at multiple sites (Nakagami et al. 2010; Sugiyama et al. 2008; Wang et al. 2013; Mayank et al. 2012; Durek et al. 2009; Heazlewood et al. 2008), yet how these phosphorylation events impact DRP3's function in the division of mitochondria or peroxisomes is still elusive.

The human mitochondrial membrane associated RING-CH (MARCH)-V protein (also known as MITOL) is an E3 ubiquitin ligase that physically interacts with Drp1 to promote the ubiquitination of Drp1 for protein activation, leading to elongated mitochondria when the protein function is disrupted (Yonashiro et al. 2006; Karbowski et al. 2007; Nakamura et al. 2006). In contrast, ubiquitination by another mitochondrial E3 ubiquitin ligase, Parkin, triggers Drp1's degradation through the ubiquitin proteasome system, whereby suppression of Parkin blocks the degradation of Drp1, resulting in the fragmentation of mitochondria (Wang et al. 2011).

A growing body of evidence also suggests that SUMOylation regulates Drp1's function. SUMO1 physically interacts with Drp1, and when overexpressed, can stabilize Drp1 and induce mitochondrial fragmentation (Harder et al. 2004). On the other hand, a SUMO protease, SUMO1/sentrin-specific peptidase 5 (SENP5), deSUMOylates and inactivates Drp1, resulting in elongated mitochondria due to the suppression of SUMO1-induced mitochondrial division (Zunino et al. 2007). Further, both Ubc9 (a SUMO-conjugating enzyme) and the mitochondrial-anchored protein MAPL (a SUMO E3 ligase) assist in SUMOylating Drp1 in vivo, and ectopic expression of MAPL induces mitochondrial fragmentation (Figueroa-Romero et al. 2009; Neuspiel et al. 2008). All these findings point toward a role for SUMOlyation in enhancing the activity of Drp1 in mitochondrial division.

S-Nitrosylation of Drp1 at Cys644 leads to an induction of mitochondrial fragmentation (Cho et al. 2009). Since *S*-Nitrosylation of dynamin has been shown to promote homodimerization of dynamin and increase its GTPase activity (Wang et al. 2006), it is hypothesized that *S*-Nitrosylation of Drp1 may also promote mitochondrial fission through a similar mechanism.

Drp1 also undergoes protein degradation via autophagy-dependent pathways, as inhibition of autophagy results in an accumulation of Drp1 and induction of mitochondrial division, whereas induction of autophagy promotes the degradation of Drp1 and leads to enlarged mitochondria (Purnell and Fox 2013).

Despite the large number of studies reporting the role of posttranslational modification in regulating the function of Drp1 in mitochondrial division, how these events affect Drp1's role in peroxisomal dynamics has not been reported. How protein modifications regulate the function of DRPs from other model systems also remains a critical question to be elucidated.

20.6 Questions for Future Investigations

Work from diverse model systems clearly demonstrates that DRP's action underpins the fission step of peroxisome division and at least in some cases, membrane constriction, too. However, several aspects of DRP-mediated peroxisome fission are still poorly understood, and many questions remain. For example, why do DRPs localize to the tips of peroxisomes more frequently than at constriction sites? All receptor/adaptor proteins known to be involved in recruiting DRPs to the peroxisome also work for mitochondria, are there any adaptor proteins exclusive to peroxisomes? Does DRP activity vary in spatiotemporal and tissue/development specific context? Why do DRPs show lower basal GTPase activity than dynamin? For the mammalian Drp1 protein, it will be crucial to address the specific roles for the several transcriptional variants, and more importantly, whether PTMs that influence Drp1 activity on the mitochondria exert similar effects in peroxisome fission. Many questions regarding the plant DRPs also need to be elucidated. For instance, what other proteins constitute the DRP3-containing supercomplex? How exactly does DRP5B accomplish peroxisome fission? Does DRP3A have a role in membrane constriction? Are there other plant DRPs involved in peroxisome division? Studies using the emerging model system *C. merolae* may also answer questions regarding the nature and composition of the filamentous ring of POD, why mitochondria and peroxisomes do not divide simultaneously in *C. merolae* despite using the same DRP, and the significance of the observed peroxisome-mitochondria tethering in *C. merolae* and fission yeast. Finally, since Dnm1-linked mitochondrial fission has been attributed to facilitating selective mitophagy in yeast (Mao et al. 2013), it would be interesting to address whether it functions in pexophagy in a similar fashion.

It is undisputable that DRPs are a fundamental part of the peroxisome division complex. Technological advances and innovative experimentation will allow us to answer some of these outstanding questions and uncover the identity of other molecular factors in peroxisome division and dissect the mechanism thereof.

Acknowledgment We apologize to researchers whose works we were unable to cover in this chapter due to space limitations. Peroxisome biogenesis research in the Hu laboratory was supported by the National Science Foundation (MCB-0618335 and MCB 1330441) and the Chemical Sciences, Geosciences and Biosciences Division, Office of Basic Energy Sciences, Office of Science, U.S. Department of Energy (DE-FG02-91ER20021) to JH.

References

Arimura S, Tsutsumi N (2002) A dynamin-like protein (ADL2b), rather than FtsZ, is involved in Arabidopsis mitochondrial division. Proc Natl Acad Sci USA 99:5727–5731

Arimura S, Aida GP, Fujimoto M, Nakazono M, Tsutsumi N (2004) Arabidopsis dynamin-like protein 2a (ADL2a), like ADL2b, is involved in plant mitochondrial division. Plant Cell Physiol 45:236–242

Arimura S, Fujimoto M, Doniwa Y, Kadoya N, Nakazono M, Sakamoto W, Tsutsumi N (2008) Arabidopsis ELONGATED MITOCHONDRIA1 is required for localization of DYNAMIN-RELATED PROTEIN3A to mitochondrial fission sites. Plant Cell 20:1555–1566

Ashrafian H, Docherty L, Leo V, Towlson C, Neilan M, Steeples V, Lygate CA, Hough T, Townsend S, Williams D, Wells S, Norris D, Glyn-Jones S, Land J, Barbaric I, Lalanne Z, Denny P, Szumska D, Bhattacharya S, Griffin JL, Hargreaves I, Fernandez-Fuentes N, Cheeseman M, Watkins H, Dear TN (2010) A mutation in the mitochondrial fission gene Dnm1l leads to cardiomyopathy. PLoS Genet 6:e1001000

Aung K, Hu J (2009) The Arabidopsis peroxisome division mutant pdd2 is defective in the DYNAMIN-RELATED PROTEIN3A (DRP3A) gene. Plant Signal Behav 4:542–544

Aung K, Hu J (2011) The *Arabidopsis* tail-anchored protein PEROXISOMAL AND MITOCHONDRIAL DIVISION FACTOR1 is involved in the morphogenesis and proliferation of peroxisomes and mitochondria. Plant Cell 23:4446–4461

Aung K, Hu J (2012) Differential roles of arabidopsis dynamin-related proteins DRP3A, DRP3B, and DRP5B in organelle division. J Integr Plant Biol 54:921–931

Benard G, Karbowski M (2009) Mitochondrial fusion and division: regulation and role in cell viability. Semin Cell Dev Biol 20:365–374

Bleazard W, Mccaffery JM, King EJ, Bale S, Mozdy A, Tieu Q, Nunnari J, Shaw JM (1999) The dynamin-related GTPase Dnm1 regulates mitochondrial fission in yeast. Nat Cell Biol 1:298–304

Bramkamp M (2012) Structure and function of bacterial dynamin-like proteins. Biol Chem 393:1203–1214

Bui HT, Karren MA, Bhar D, Shaw JM (2012) A novel motif in the yeast mitochondrial dynamin Dnm1 is essential for adaptor binding and membrane recruitment. J Cell Biol 199:613–622

Cereghetti GM, Stangherlin A, Martins de Brito O, Chang CR, Blackstone C, Bernardi P, Scorrano L (2008) Dephosphorylation by calcineurin regulates translocation of Drp1 to mitochondria. Proc Natl Acad Sci USA 105:15803–15808

Chan DC (2012) Fusion and fission: interlinked processes critical for mitochondrial health. Annu Rev Genet 46:265–287

Chang CR, Blackstone C (2007) Drp1 phosphorylation and mitochondrial regulation. EMBO Rep 8:1088–1089, author reply 1089–1090

Chang CR, Blackstone C (2010) Dynamic regulation of mitochondrial fission through modification of the dynamin-related protein Drp1. Ann N Y Acad Sci 1201:34–39

Chappie JS, Acharya S, Leonard M, Schmid SL, Dyda F (2010) G domain dimerization controls dynamin's assembly-stimulated GTPase activity. Nature 465:435–440

Chen CH, Howng SL, Hwang SL, Chou CK, Liao CH, Hong YR (2000) Differential expression of four human dynamin-like protein variants in brain tumors. DNA Cell Biol 19:189–194

Chen YJ, Zhang P, Egelman EH, Hinshaw JE (2004) The stalk region of dynamin drives the constriction of dynamin tubes. Nat Struct Mol Biol 11:574–575

Cho DH, Nakamura T, Fang J, Cieplak P, Godzik A, Gu Z, Lipton SA (2009) S-nitrosylation of Drp1 mediates beta-amyloid-related mitochondrial fission and neuronal injury. Science 324:102–105

Cribbs JT, Strack S (2007) Reversible phosphorylation of Drp1 by cyclic AMP-dependent protein kinase and calcineurin regulates mitochondrial fission and cell death. EMBO Rep 8:939–944

Danino D, Moon KH, Hinshaw JE (2004) Rapid constriction of lipid bilayers by the mechanochemical enzyme dynamin. J Struct Biol 147:259–267

Dereeper A, Guignon V, Blanc G, Audic S, Buffet S, Chevenet F, Dufayard JF, Guindon S, Lefort V, Lescot M, Claverie JM, Gascuel O (2008) Phylogeny.fr: robust phylogenetic analysis for the non-specialist. Nucleic Acids Res 36:W465–W469

Duboff B, Gotz J, Feany MB (2012) Tau promotes neurodegeneration via DRP1 mislocalization in vivo. Neuron 75:618–632

Duncan O, Taylor NL, Carrie C, Eubel H, Kubiszewski-Jakubiak S, Zhang B, Narsai R, Millar AH, Whelan J (2011) Multiple lines of evidence localize signaling, morphology, and lipid biosynthesis machinery to the mitochondrial outer membrane of Arabidopsis. Plant Physiol 157:1093–1113

Durek P, Schmidt R, Heazlewood JL, Jones A, Maclean D, Nagel A, Kersten B, Schulze WX (2009) PhosPhAt: the Arabidopsis thaliana phosphorylation site database. An update. Nucleic Acids Res 38:D828–D834

Faelber K, Posor Y, Gao S, Held M, Roske Y, Schulze D, Haucke V, Noe F, Daumke O (2011) Crystal structure of nucleotide-free dynamin. Nature 477:556–560

Ferguson SM, de Camilli P (2012) Dynamin, a membrane-remodelling GTPase. Nat Rev Mol Cell Biol 13:75–88

Figueroa-Romero C, Iniguez-Lluhi JA, Stadler J, Chang CR, Arnoult D, Keller PJ, Hong Y, Blackstone C, Feldman EL (2009) SUMOylation of the mitochondrial fission protein Drp1 occurs at multiple nonconsensus sites within the B domain and is linked to its activity cycle. FASEB J 23:3917–3927

Ford MG, Jenni S, Nunnari J (2011) The crystal structure of dynamin. Nature 477:561–566

Friedman JR, Lackner LL, West M, Dibenedetto JR, Nunnari J, Voeltz GK (2011) ER tubules mark sites of mitochondrial division. Science 334:358–362

Frohlich C, Grabiger S, Schwefel D, Faelber K, Rosenbaum E, Mears J, Rocks O, Daumke O (2013) Structural insights into oligomerization and mitochondrial remodelling of dynamin 1-like protein. EMBO J 32:1280–1292

Fujimoto M, Arimura S, Mano S, Kondo M, Saito C, Ueda T, Nakazono M, Nakano A, Nishimura M, Tsutsumi N (2009) Arabidopsis dynamin-related proteins DRP3A and DRP3B are functionally redundant in mitochondrial fission, but have distinct roles in peroxisomal fission. Plant J 58:388–400

Gandre-Babbe S, Van der Bliek AM (2008) The novel tail-anchored membrane protein Mff controls mitochondrial and peroxisomal fission in mammalian cells. Mol Biol Cell 19:2402–2412

Gao H, Kadirjan-Kalbach D, Froehlich JE, Osteryoung KW (2003) ARC5, a cytosolic dynamin-like protein from plants, is part of the chloroplast division machinery. Proc Natl Acad Sci USA 100:4328–4333

Griffin EE, Graumann J, Chan DC (2005) The WD40 protein Caf4p is a component of the mitochondrial fission machinery and recruits Dnm1p to mitochondria. J Cell Biol 170:237–248

Guillou E, Bousquet C, Daloyau M, Emorine LJ, Belenguer P (2005) Msp1p is an intermembrane space dynamin-related protein that mediates mitochondrial fusion in a Dnm1p-dependent manner in S. pombe. FEBS Lett 579:1109–1116

Han XJ, Lu YF, Li SA, Kaitsuka T, Sato Y, Tomizawa K, Nairn AC, Takei K, Matsui H, Matsushita M (2008) CaM kinase I alpha-induced phosphorylation of Drp1 regulates mitochondrial morphology. J Cell Biol 182:573–585

Harder Z, Zunino R, Mcbride H (2004) Sumo1 conjugates mitochondrial substrates and participates in mitochondrial fission. Curr Biol 14:340–345

Heazlewood JL, Durek P, Hummel J, Selbig J, Weckwerth W, Walther D, Schulze WX (2008) PhosPhAt: a database of phosphorylation sites in Arabidopsis thaliana and a plant-specific phosphorylation site predictor. Nucleic Acids Res 36:D1015–D1021

Heymann JA, Hinshaw JE (2009) Dynamins at a glance. J Cell Sci 122:3427–3431

Hoepfner D, Van den Berg M, Philippsen P, Tabak HF, Hettema EH (2001) A role for Vps1p, actin, and the Myo2p motor in peroxisome abundance and inheritance in *Saccharomyces cerevisiae*. J Cell Biol 155:979–990

Hong Z, Bednarek SY, Blumwald E, Hwang I, Jurgens G, Menzel D, Osteryoung KW, Raikhel NV, Shinozaki K, Tsutsumi N, Verma DP (2003) A unified nomenclature for Arabidopsis dynamin-related large GTPases based on homology and possible functions. Plant Mol Biol 53:261–265

Hoppins S, Lackner L, Nunnari J (2007) The machines that divide and fuse mitochondria. Annu Rev Biochem 76:751–780

Howng SL, Sy WD, Cheng TS, Lieu AS, Wang C, Tzou WS, Cho CL, Hong YR (2004) Genomic organization, alternative splicing, and promoter analysis of human dynamin-like protein gene. Biochem Biophys Res Commun 314:766–772

Hu J, Baker A, Bartel B, Linka N, Mullen RT, Reumann S, Zolman BK (2012) Plant peroxisomes: biogenesis and function. Plant Cell 24:2279–2303

Huber N, Guimaraes S, Schrader M, Suter U, Niemann A (2013) Charcot-Marie-Tooth disease-associated mutants of GDAP1 dissociate its roles in peroxisomal and mitochondrial fission. EMBO Rep 14:545–552

Imoto Y, Yoshida Y, Yagisawa F, Kuroiwa H, Kuroiwa T (2011) The cell cycle, including the mitotic cycle and organelle division cycles, as revealed by cytological observations. J Electron Microsc (Tokyo) 60(Suppl 1):S117–S136

Imoto Y, Kuroiwa H, Yoshida Y, Ohnuma M, Fujiwara T, Yoshida M, Nishida K, Yagisawa F, Hirooka S, Miyagishima SY, Misumi O, Kawano S, Kuroiwa T (2013) Single-membrane-bounded peroxisome division revealed by isolation of dynamin-based machinery. Proc Natl Acad Sci USA 110:9583–9588

Ingerman E, Perkins EM, Marino M, Mears JA, Mccattery JM, Hinshaw JE, Nunnari J (2005) Dnm1 forms spirals that are structurally tailored to fit mitochondria. J Cell Biol 170:1021–1027

Ishihara N, Nomura M, Jofuku A, Kato H, Suzuki SO, Masuda K, Otera H, Nakanishi Y, Nonaka I, Goto Y, Taguchi N, Morinaga H, Maeda M, Takayanagi R, Yokota S, Mihara K (2009) Mitochondrial fission factor Drp1 is essential for embryonic development and synapse formation in mice. Nat Cell Biol 11:958–966

Itoyama A, Michiyuki S, Honsho M, Yamamoto T, Moser A, Yoshida Y, Fujiki Y (2013) Mff functions with Pex11pβ and DLP1 in peroxisomal fission. Biol Open. doi:10.1242/bio.20135298

Jin JB, Bae H, Kim SJ, Jin YH, Goh CH, Kim DH, Lee YJ, Tse YC, Jiang L, Hwang I (2003) The Arabidopsis dynamin-like proteins ADL1C and ADL1E play a critical role in mitochondrial morphogenesis. Plant Cell 15:2357–2369

Jourdain I, Sontam D, Johnson C, Dillies C, Hyams JS (2008) Dynamin-dependent biogenesis, cell cycle regulation and mitochondrial association of peroxisomes in fission yeast. Traffic 9:353–365

Karbowski M, Neutzner A, Youle RJ (2007) The mitochondrial E3 ubiquitin ligase MARCH5 is required for Drp1 dependent mitochondrial division. J Cell Biol 178:71–84

Kaur N, Hu J (2009) Dynamics of peroxisome abundance: a tale of division and proliferation. Curr Opin Plant Biol 12:781–788

Kobayashi S, Tanaka A, Fujiki Y (2007) Fis1, DLP1, and Pex11p coordinately regulate peroxisome morphogenesis. Exp Cell Res 313:1675–1686

Koch J, Brocard C (2011) Membrane elongation factors in organelle maintenance: the case of peroxisome proliferation. Biomol Concepts 2:353–364

Koch J, Brocard C (2012) PEX11 proteins attract Mff and human Fis1 to coordinate peroxisomal fission. J Cell Sci 125:3813–3826

Koch A, Thiemann M, Grabenbauer M, Yoon Y, Mcniven MA, Schrader M (2003) Dynamin-like protein 1 is involved in peroxisomal fission. J Biol Chem 278:8597–8605

Koch A, Yoon Y, Bonekamp NA, Mcniven MA, Schrader M (2005) A role for Fis1 in both mitochondrial and peroxisomal fission in mammalian cells. Mol Biol Cell 16:5077–5086

Koirala S, Bui HT, Schubert HL, Eckert DM, Hill CP, Kay MS, Shaw JM (2010) Molecular architecture of a dynamin adaptor: implications for assembly of mitochondrial fission complexes. J Cell Biol 191:1127–1139

Kornmann B, Currie E, Collins SR, Schuldiner M, Nunnari J, Weissman JS, Walter P (2009) An ER-mitochondria tethering complex revealed by a synthetic biology screen. Science 325:477–481

Korobova F, Ramabhadran V, Higgs HN (2013) An actin-dependent step in mitochondrial fission mediated by the ER-associated formin INF2. Science 339:464–467

Kuravi K, Nagotu S, Krikken AM, Sjollema K, Deckers M, Erdmann R, Veenhuis M, Van der Klei IJ (2006) Dynamin-related proteins Vps1p and Dnm1p control peroxisome abundance in *Saccharomyces cerevisiae*. J Cell Sci 119:3994–4001

Lackner LL (2013) Determining the shape and cellular distribution of mitochondria: the integration of multiple activities. Curr Opin Cell Biol 25:471–476

Li X, Gould SJ (2003) The dynamin-like GTPase DLP1 is essential for peroxisome division and is recruited to peroxisomes in part by PEX11. J Biol Chem 278:17012–17020

Lingard MJ, Gidda SK, Bingham S, Rothstein SJ, Mullen RT, Trelease RN (2008) Arabidopsis PEROXIN11c-e, FISSION1b, and DYNAMIN-RELATED PROTEIN3A cooperate in cell cycle-associated replication of peroxisomes. Plant Cell 20:1567–1585

Logan DC, Scott I, Tobin AK (2004) ADL2a, like ADL2b, is involved in the control of higher plant mitochondrial morphology. J Exp Bot 55:783–785

Manczak M, Reddy PH (2012) Abnormal interaction between the mitochondrial fission protein Drp1 and hyperphosphorylated tau in Alzheimer's disease neurons: implications for mitochondrial dysfunction and neuronal damage. Hum Mol Genet 21:2538–2547

Mano S, Nakamori C, Kondo M, Hayashi M, Nishimura M (2004) An Arabidopsis dynamin-related protein, DRP3A, controls both peroxisomal and mitochondrial division. Plant J 38:487–498

Mao K, Wang K, Liu X, Klionsky DJ (2013) The scaffold protein Atg11 recruits fission machinery to drive selective mitochondria degradation by autophagy. Dev Cell 26:9–18

Mayank P, Grossman J, Wuest S, Boisson-Dernier A, Roschitzki B, Nanni P, Nuhse T, Grossniklaus U (2012) Characterization of the phosphoproteome of mature Arabidopsis pollen. Plant J 72:89–101

Mears JA, Lackner LL, Fang S, Ingerman E, Nunnari J, Hinshaw JE (2011) Conformational changes in Dnm1 support a contractile mechanism for mitochondrial fission. Nat Struct Mol Biol 18:20–26

Miyagishima S-Y, Itoh R, Toda K, Kuroiwa H, Nishimura M, Kuroiwa T (1999) Microbody proliferation and segregation cycle in the single microbody-alga Cyanidioschyzon merolae. Planta 208:326–336

Miyagishima SY, Kuwayama H, Urushihara H, Nakanishi H (2008) Evolutionary linkage between eukaryotic cytokinesis and chloroplast division by dynamin proteins. Proc Natl Acad Sci USA 105:15202–15207

Morlot S, Roux A (2013) Mechanics of dynamin-mediated membrane fission. Annu Rev Biophys 42:629–649

Morlot S, Galli V, Klein M, Chiaruttini N, Manzi J, Humbert F, Dinis L, Lenz M, Cappello G, Roux A (2012) Membrane shape at the edge of the dynamin helix sets location and duration of the fission reaction. Cell 151:619–629

Motley AM, Hettema EH (2007) Yeast peroxisomes multiply by growth and division. J Cell Biol 178:399–410

Motley AM, Ward GP, Hettema EH (2008) Dnm1p-dependent peroxisome fission requires Caf4p, Mdv1p and Fis1p. J Cell Sci 121:1633–1640

Mozdy AD, Mccaffery JM, Shaw JM (2000) Dnm1p GTPase-mediated mitochondrial fission is a multi-step process requiring the novel integral membrane component Fis1p. J Cell Biol 151:367–380

Muller-Taubenberger A, Kortholt A, Eichinger L (2013) Simple system–substantial share: the use of Dictyostelium in cell biology and molecular medicine. Eur J Cell Biol 92:45–53

Nagotu S, Saraya R, Otzen M, Veenhuis M, Van der Klei IJ (2008) Peroxisome proliferation in Hansenula polymorpha requires Dnm1p which mediates fission but not de novo formation. Biochim Biophys Acta 1783:760–769

Nakagami H, Sugiyama N, Mochida K, Daudi A, Yoshida Y, Toyoda T, Tomita M, Ishihama Y, Shirasu K (2010) Large-scale comparative phosphoproteomics identifies conserved phosphorylation sites in plants. Plant Physiol 153:1161–1174

Nakamura N, Kimura Y, Tokuda M, Honda S, Hirose S (2006) MARCH-V is a novel mitofusin 2- and Drp1-binding protein able to change mitochondrial morphology. EMBO Rep 7:1019–1022

Neuspiel M, Schauss AC, Braschi E, Zunino R, Rippstein P, Rachubinski RA, Andrade-Navarro MA, Mcbride HM (2008) Cargo-selected transport from the mitochondria to peroxisomes is mediated by vesicular carriers. Curr Biol 18:102–108

Niemann A, Ruegg M, La Padula V, Schenone A, Suter U (2005) Ganglioside-induced differentiation associated protein 1 is a regulator of the mitochondrial network: new implications for Charcot-Marie-Tooth disease. J Cell Biol 170:1067–1078

Niemann A, Wagner KM, Ruegg M, Suter U (2009) GDAP1 mutations differ in their effects on mitochondrial dynamics and apoptosis depending on the mode of inheritance. Neurobiol Dis 36:509–520

Nishida K, Yagisawa F, Kuroiwa H, Yoshida Y, Kuroiwa T (2007) WD40 protein Mda1 is purified with Dnm1 and forms a dividing ring for mitochondria before Dnm1 in Cyanidioschyzon merolae. Proc Natl Acad Sci USA 104:4736–4741

Onoue K, Jofuku A, Ban-Ishihara R, Ishihara T, Maeda M, Koshiba T, Itoh T, Fukuda M, Otera H, Oka T, Takano H, Mizushima N, Mihara K, Ishihara N (2013) Fis1 acts as a mitochondrial recruitment factor for TBC1D15 that is involved in regulation of mitochondrial morphology. J Cell Sci 126:176–185

Otera H, Wang C, Cleland MM, Setoguchi K, Yokota S, Youle RJ, Mihara K (2010) Mff is an essential factor for mitochondrial recruitment of Drp1 during mitochondrial fission in mammalian cells. J Cell Biol 191:1141–1158

Otera H, Ishihara N, Mihara K (2013) New insights into the function and regulation of mitochondrial fission. Biochim Biophys Acta 1833:1256–1268

Otsuga D, Keegan BR, Brisch E, Thatcher JW, Hermann GJ, Bleazard W, Shaw JM (1998) The dynamin-related GTPase, Dnm1p, controls mitochondrial morphology in yeast. J Cell Biol 143:333–349

Palmer CS, Osellame LD, Laine D, Koutsopoulos OS, Frazier AE, Ryan MT (2011) MiD49 and MiD51, new components of the mitochondrial fission machinery. EMBO Rep 12:565–573

Pan R, Hu J (2011) The conserved fission complex on peroxisomes and mitochondria. Plant Signal Behav 6:870–872

Praefcke GJ, Mcmahon HT (2004) The dynamin superfamily: universal membrane tubulation and fission molecules? Nat Rev Mol Cell Biol 5:133–147

Purnell PR, Fox HS (2013) Autophagy-mediated turnover of dynamin-related protein 1. BMC Neurosci 14:86

Rai A, Nothe H, Tzvetkov N, Korenbaum E, Manstein DJ (2011) Dictyostelium dynamin B modulates cytoskeletal structures and membranous organelles. Cell Mol Life Sci 68:2751–2767

Ramachandran R (2011) Vesicle scission: dynamin. Semin Cell Dev Biol 22:10–17

Schrader M, Bonekamp NA, Islinger M (2012) Fission and proliferation of peroxisomes. Biochim Biophys Acta. doi:10.1016/j.bbadis.2011.12.014

Scott I, Tobin AK, Logan DC (2006) BIGYIN, an orthologue of human and yeast FIS1 genes functions in the control of mitochondrial size and number in *Arabidopsis thaliana*. J Exp Bot 57:1275–1280

Smirnova E, Griparic L, Shurland DL, Van der Bliek AM (2001) Dynamin-related protein Drp1 is required for mitochondrial division in mammalian cells. Mol Biol Cell 12:2245–2256

Strack S, Cribbs JT (2012) Allosteric modulation of Drp1 mechanoenzyme assembly and mitochondrial fission by the variable domain. J Biol Chem 287:10990–11001

Sugiyama N, Nakagami H, Mochida K, Daudi A, Tomita M, Shirasu K, Ishihama Y (2008) Large-scale phosphorylation mapping reveals the extent of tyrosine phosphorylation in Arabidopsis. Mol Syst Biol 4:193

Sweitzer SM, Hinshaw JE (1998) Dynamin undergoes a GTP-dependent conformational change causing vesiculation. Cell 93:1021–1029

Tabak HF, Braakman I, Van der Zand A (2013) Peroxisome formation and maintenance are dependent on the endoplasmic reticulum. Annu Rev Biochem 82:723–744

Taguchi N, Ishihara N, Jofuku A, Oka T, Mihara K (2007) Mitotic phosphorylation of dynamin-related GTPase Drp1 participates in mitochondrial fission. J Biol Chem 282:11521–11529

Thomas KJ, Jacobson MR (2012) Defects in mitochondrial fission protein dynamin-related protein 1 are linked to apoptotic resistance and autophagy in a lung cancer model. PLoS One 7:e45319

Tieu Q, Nunnari J (2000) Mdv1p is a WD repeat protein that interacts with the dynamin-related GTPase, Dnm1p, to trigger mitochondrial division. J Cell Biol 151:353–366

Tieu Q, Okreglak V, Naylor K, Nunnari J (2002) The WD repeat protein, Mdv1p, functions as a molecular adaptor by interacting with Dnm1p and Fis1p during mitochondrial fission. J Cell Biol 158:445–452

Toulmay A, Prinz WA (2012) A conserved membrane-binding domain targets proteins to organelle contact sites. J Cell Sci 125:49–58

Vizeacoumar FJ, Vreden WN, Fagarasanu M, Eitzen GA, Aitchison JD, Rachubinski RA (2006) The dynamin-like protein Vps1p of the yeast Saccharomyces cerevisiae associates with peroxisomes in a Pex19p-dependent manner. J Biol Chem 281:12817–12823

Wagner KM, Ruegg M, Niemann A, Suter U (2009) Targeting and function of the mitochondrial fission factor GDAP1 are dependent on its tail-anchor. PLoS One 4:e5160

Wakabayashi J, Zhang Z, Wakabayashi N, Tamura Y, Fukaya M, Kensler TW, Iijima M, Sesaki H (2009) The dynamin-related GTPase Drp1 is required for embryonic and brain development in mice. J Cell Biol 186:805–816

Wanders RJ (2013) Peroxisomes in human health and disease: metabolic pathways, metabolite transport, interplay with other organelles and signal transduction. Subcell Biochem 69:23–44

Wang G, Moniri NH, Ozawa K, Stamler JS, Daaka Y (2006) Nitric oxide regulates endocytosis by S-nitrosylation of dynamin. Proc Natl Acad Sci USA 103:1295–1300

Wang H, Song P, Du L, Tian W, Yue W, Liu M, Li D, Wang B, Zhu Y, Cao C, Zhou J, Chen Q (2011) Parkin ubiquitinates Drp1 for proteasome-dependent degradation: implication of dysregulated mitochondrial dynamics in Parkinson disease. J Biol Chem 286:11649–11658

Wang F, Liu P, Zhang Q, Zhu J, Chen T, Arimura S, Tsutsumi N, Lin J (2012) Phosphorylation and ubiquitination of dynamin-related proteins (AtDRP3A/3B) synergically regulate mitochondrial proliferation during mitosis. Plant J 72:43–56

Wang X, Bian Y, Cheng K, Gu LF, Ye M, Zou H, Sun SS, He JX (2013) A large-scale protein phosphorylation analysis reveals novel phosphorylation motifs and phosphoregulatory networks in Arabidopsis. J Proteomics 78:486–498

Waterham HR, Koster J, Van Roermund CW, Mooyer PA, Wanders RJ, Leonard JV (2007) A lethal defect of mitochondrial and peroxisomal fission. N Engl J Med 356:1736–1741

Yonashiro R, Ishido S, Kyo S, Fukuda T, Goto E, Matsuki Y, Ohmura-Hoshino M, Sada K, Hotta H, Yamamura H, Inatome R, Yanagi S (2006) A novel mitochondrial ubiquitin ligase plays a critical role in mitochondrial dynamics. EMBO J 25:3618–3626

Yoshida Y, Kuroiwa H, Misumi O, Yoshida M, Ohnuma M, Fujiwara T, Yagisawa F, Hirooka S, Imoto Y, Matsushita K, Kawano S, Kuroiwa T (2010) Chloroplasts divide by contraction of a bundle of nanofilaments consisting of polyglucan. Science 329:949–953

Zhang P, Hinshaw JE (2001) Three-dimensional reconstruction of dynamin in the constricted state. Nat Cell Biol 3:922–926

Zhang XC, Hu JP (2008) FISSION1A and FISSION1B proteins mediate the fission of peroxisomes and mitochondria in Arabidopsis. Mol Plant 1:1036–1047

Zhang X, Hu J (2009) Two small protein families, DYNAMIN-RELATED PROTEIN3 and FISSION1, are required for peroxisome fission in Arabidopsis. Plant J 57:146–159

Zhang X, Hu J (2010) The Arabidopsis chloroplast division protein DYNAMIN-RELATED PROTEIN5B also mediates peroxisome division. Plant Cell 22:431–442

Zhao J, Liu T, Jin S, Wang X, Qu M, Uhlen P, Tomilin N, Shupliakov O, Lendahl U, Nister M (2011) Human MIEF1 recruits Drp1 to mitochondrial outer membranes and promotes mitochondrial fusion rather than fission. EMBO J 30:2762–2778

Zhao J, Lendahl U, Nister M (2013) Regulation of mitochondrial dynamics: convergences and divergences between yeast and vertebrates. Cell Mol Life Sci 70:951–976

Zunino R, Schauss A, Rippstein P, Andrade-Navarro M, Mcbride HM (2007) The SUMO protease SENP5 is required to maintain mitochondrial morphology and function. J Cell Sci 120:1178–1188

Plant Peroxisome Dynamics: Movement, Positioning and Connections

21

Imogen Sparkes and Hongbo Gao

Abstract

Compartmentalization of metabolic functions in membrane bounded organelles is a defining characteristic of eukaryotes. Movement, positioning and morphology of such organelles are key determinants for function, maintenance and inheritance. For example, mutations in the molecular motors (myosin) that drive organelle movement in plants result in dwarf plants with reduced seed set. Therefore, movement and positioning of organelles are key factors for plant development and growth.

Peroxisomes are both functionally and morphologically pleomorphic. Several metabolic processes span more than one organelle, highlighting the importance of coordinated movement and positioning of these organelles. Here, we deal with peroxisome dynamics in terms of movement, positioning and how these dynamics may relate to their functional role. In order to understand the potential role of such movement, a brief discussion of the functional role of plant peroxisomes is provided.

Keywords

Peroxisome • Peroxule • Movement • Myosin • Plant

I. Sparkes (✉) • H. Gao
Biosciences, College of Life and Environmental Sciences, University of Exeter, Geoffrey Pope building, Stocker Road, Exeter, Devon EX4 4QD, UK
e-mail: I.Sparkes@exeter.ac.uk

C. Brocard and A. Hartig (eds.), *Molecular Machines Involved in Peroxisome Biogenesis and Maintenance*, DOI 10.1007/978-3-7091-1788-0_21, © Springer-Verlag Wien 2014

21.1 Functional Role of Plant Peroxisomes

21.1.1 Tissue-Specific Peroxisomes

Since the initial biochemical characterisation of mammalian peroxisomes by De Duve and Baudin in 1966, a wealth of data relating to form and function has been revealed. In plants, peroxisomes have been observed in many tissues (Mano et al. 2002; Frederick and Newcomb 1969), and based on their biochemical role and location, were even divided up into different classifications; leaf type, root nodule type, glyoxysomes and unspecialised peroxisomes. Glyoxysomes, so called based on their predominant functional role in the glyoxylate cycle, were assigned to seedlings and senescent material where scavenging and break down of fatty acids is key. More recently, the break down into classifications has been argued to be a redundant feature with peroxisomes in many tissues being able to 'adapt' and change their functional complement of enzymes to cater for the requirement of certain peroxisomal metabolic function(s). For example, so-called leaf type peroxisomes can import enzymatic components required to generate a functional glyoxylate cycle, which would result in the 'leaf type' peroxisome becoming a 'glyoxysome', and vice versa. This metabolic plasticity has been shown to occur during the postgerminative transition to 'leaf type' peroxisomes and back to 'glyoxysomes' during senescence (Titus and Becker 1985; Nishimura et al. 1986, 1993; Sautter 1986). It is now more widely accepted that peroxisomes in all tissues are termed peroxisomes which therefore caters for plasticity in metabolic function and helps to avoid confusion within the scientific community (Pracharoenwattana and Smith 2008; Hu et al. 2012).

21.1.2 Role of Peroxisomes in Plants

Several Arabidopsis *pex* mutants, mutants defective in genes required for peroxisome biogenesis, have been isolated and display various developmental defects. A few of these mutants are embryo lethal highlighting the critical functional nature of the organelle in plant development (Sparkes et al. 2003; Schumann et al. 2003; Hu et al. 2002; Nito et al. 2007; Fan et al. 2005).

Peroxisomal metabolism has been unpicked through proteomic, transcriptomic, biochemical and genetic analysis (see reviews by Hu et al. 2012; Theodoulou and Eastmond 2012 and references therein). The biochemical roles commonly associated with plant peroxisomes are the glyoxylate cycle, β-oxidation and photorespiration. Furthermore, roles in producing phytohormones jasmonic acid (JA) and indole-3-acetic acid (IAA), generation and detoxification of signalling molecules such as reactive oxygen species (ROS) and nitric oxide (NO) (Del Rio 2011), urate metabolism and degradation of branched chain amino acids have been highlighted. In addition, peroxisomes are also important in the biosynthesis of phyloquinone, isoprenoids, biotin and S-allantoin.

Photorespiration results in the production of 3-phosphoglycerate from 2-phosphoglycolate. The latter is produced due to the oxygenase activity of Ribulose-1,5-bisphosphate (Rubisco). The pathway spans peroxisomes, chloroplasts and mitochondria requiring efficient shuttling of metabolites and co-factors between compartments and the cytosol.

β-oxidation produces acetyl CoA through the cleavage of free fatty acids formed from the lipolysis of lipids such as triacylglycerol (TAG) from oil bodies. Through the glyoxylate cycle, acetyl CoA is then converted into succinate and oxaloacetate required for the TCA cycle and gluconeogenesis respectively. In oil seeds these metabolic processes provide energy and carbon skeletons required for germination and growth prior to the development of photosynthetic capacity. Oil mobilisation is also implicated in several other important physiological roles in plants; pollen development, pollen tube growth, fertilisation, senescence and embryogenesis (Theodoulou and Eastmond 2012). β-oxidation also results in the synthesis of IAA and JA. IAA, the main form of auxin, is synthesised from IBA. IAA is important in many aspects of development from lateral root formation, cotyledon expansion to stamen, hypocotyl and root hair elongation (Strader and Bartel 2011). JA is implicated in several processes including defence and development (Acosta and Farmer 2010; Li et al. 2005; Koo and Howe 2007). Similarly, ROS levels are implicated in various physiological processes where levels determine a capacity for signalling versus oxidative damage (Sandalio et al. 2013; Mittler et al. 2011). It is worth mentioning though that peroxisomes are not the sole site of ROS synthesis and degradation in plants.

Plant peroxisomes therefore house several key metabolic processes and produce signalling molecules required for plants to grow and adapt to environmental conditions (Nyathi and Baker 2006). Some of these functions span more than just one organelle. For example photorespiration spans peroxisomes, chloroplasts and mitochondria, whereas unlike in mammals, in plants, peroxisomes are the sole site for β-oxidation. Based on the metabolic co-ordination between organelles, one would assume positioning of organelles to also be regulated (see next section).

21.2 Plant Peroxisome Dynamics: Movement and Positioning

21.2.1 Plant Organelle Dynamics

With the advent of fluorescent protein technology, subject to imaging constraints, the range of motions specific organelles undergo in any one cell type can now be assessed. In plants, organelle movement is an extremely dynamic process with spheroid organelles reaching speeds of up to 8 μm/s in leaf epidermal cells, and up to 100 μm/s in cytoplasmic streams in Characean internodal cells. Cytoplasmic streams represent areas of fast flow suggested to be due to the movement of myosin bound cargo along actin required for cytoplasmic mixing in highly vacuolated cells. The actual biophysical principles that govern these streams is unclear, with suggestions ranging from shear forces generated from the movement of the cortical

ER through the cytoplasm to movement of myosin-tethered spheroid organelles (Nothangel and Webb 1982; Kachar and Reese 1988; Wolff et al. 2012). During cell division organelles migrate and position themselves along the developing cell plate (Collings et al. 2003; Nebenfuhr 2007; Sheahan et al. 2007), whereas movement in cells which are not undergoing division and undergo diffuse growth, movement appears almost erratic and chaotic. In contrast, in pollen tubes in angiosperms and root hairs undergoing polarized tip growth, movement occurs through a process called reverse fountain streaming (Hepler et al. 2001). Here, movement towards the tip occurs along the periclinal walls and flows back towards the base of the cell along the central axis of the cell. Organelle movement in non-vascular plants such as the moss *Physcomitrella pattens*, tip growth is evident in protonemata and rhizoids tissues. Here, movement does not occur through cytoplasmic streaming and in general appears to be at least three orders of magnitude slower with organelles moving on the nm/s scale rather than μm/s (Furt et al. 2012; Vidali and Bezanilla 2012).

The molecular machinery which controls organelle movement in plants is actin-myosin based (see reviews and references therein by Sparkes 2010, 2011; Vick and Nebenfuhr 2012). By using cytoskeletal drugs that perturb the polymerisation of either actin (latrunculin b and cytochalasin) or microtubules (nocodazole/oryzalin), it has been shown that filamentous actin plays a major role in organelle movement in plants. A role for myosins, the molecular motors that traverse actin, has been highlighted through studies of Arabidopsis T-DNA insertional mutants, RNAi and overexpression of truncated myosin fragments. BDM has also been used to inhibit myosin activity; however, the validity of BDM as an inhibitor has been brought into question (McCurdy 1999).

Organelle movement is an extremely important aspect of plant biology as mutations in key myosins results in developmental defects in Arabidopsis; short stature, delayed flowering, reduced fecundity and cell size (Prokhnevsky et al. 2008; Peremyslov et al. 2010; Ojangu et al. 2007, 2012). Whilst movement appears to be an essential process, we are currently unable to pinpoint the specific role that it plays in plant development. Organelle movement has been proposed to be required for mixing the cytoplasmic content in cells with large vacuoles. However, the erratic nature of organelle movement where any one peroxisome, mitochondria or Golgi body can display multiple types of movement characteristics over a short time frame brings into question a mere requirement for cytoplasmic mixing. Differences in patterns of organelle movement may indicate a requirement for delivery of new material for growth or maintenance of cell shape and function. For example, tip growth allows channeled delivery to the tip, and during cell division delivery occurs at the cell plate. In certain myosin mutants root hair length is affected indicating a defect in polarized tip growth (Peremyslov et al. 2008, 2010; Prokhnevsky et al. 2008). Interestingly, during wounding and pathogen invasion there appears to be a correlation with organelle positioning around these sites perhaps indicating a requirement for delivery of metabolites/material to ameliorate such biotic and abiotic stresses (Takemoto et al. 2003; Hardham et al. 2008; Lipka et al. 2005). In terms of determining the role of organelle movement, the best studied system is that of chloroplast movement. Here, chloroplast movement appears to be an adaptive response to reduce photodamage under high light (see review by Morita and Nakamura 2012 and references therein). Future studies will undoubtedly start to uncover and pinpoint the functional requirement for organelle movement. Below is a summary of peroxisome morphology and dynamics in plants and the potential roles of such.

21.2.2 Plant Peroxisome Morphology

Plant peroxisomes are 0.2–1.5 μm in diameter and are delimited by a single membrane. Whilst they tend to be spherical in nature, a 'dumb bell' form and even membranous protrusions termed peroxules, reported to extend upto 7 μm from the peroxisome body, have been observed (See Fig. 21.1a–c and Cutler et al. 2000; Mano et al. 2002; Scott et al. 2007; Sinclair et al. 2009). Peroxule extensions appear to be prevalent in peroxisomes in hypocotyl cells and in dark grown tissue. They have been proposed to be a requirement for increased surface area to cope with increased metabolic demand and could even reflect an intermediary in the division process during peroxisome biogenesis from pre-existing peroxisomes (see Chaps. 10 and 20). Mutants defective in the molecular machinery required for peroxisome division can produce extremely long membranous protrusions (Fig. 21.1d and Mano et al. 2004). More recently, peroxules have been linked to responses in stress and were even suggested to co-align with the cortical ER (Sinclair et al. 2009). However, given the fact that both ER and peroxisomes share the same common underlying cytoskeletal 'tracks' for movement (actin), it is unclear if the co-alignment reflects bona fide connections between the two organelles or mere occupation of the same underlying actin filaments. In addition, if peroxules do in fact represent division intermediaries then connection with the ER could reflect a role in this process (peroxisome division is not reviewed here and you are referred to Chap. 20). It is important to note that membranous extensions also emanate from other organelles (stromule from chloroplasts and matrixules form from mitochondria) and may not reflect a peroxisome specific event (Mathur et al. 2012).

21.2.3 Plant Peroxisome Movement, Positioning and Organelle Associations

The pleomorphic nature of peroxisomes, in terms of morphology and movement characteristics, has been unveiled through the development and use of fluorescent probes and live cell imaging techniques. The movement characteristics of peroxisomes in several Arabidopsis tissues, tobacco leaf epidermis and moss have been reported (See Table 21.1). Quantitative data have revealed that an individual peroxisome can display several movement characteristics and that there also appears to be tissue specific differences. For instance, Table 21.1 highlights

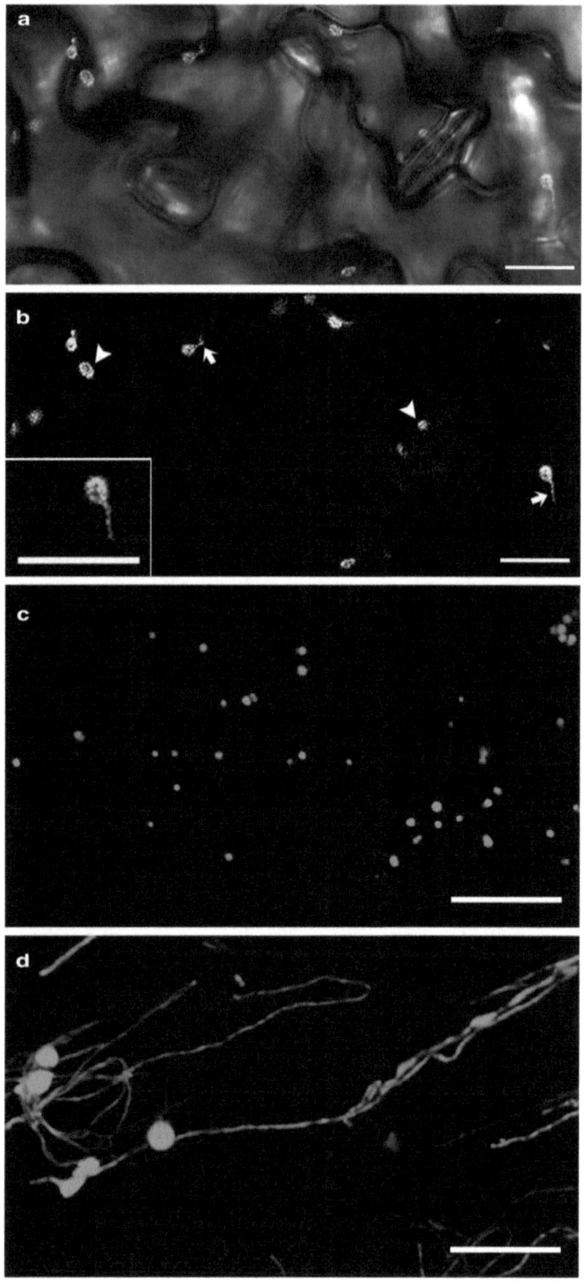

Fig. 21.1 Peroxisome morphology in Arabidopsis. Peroxisomes in leaf (**a**, **b**) and root epidermal cells (**c**) are ovoid (*arrowhead*) and range in size with some displaying peroxules (*arrow* and see *inset*). *apml–6*, a mutant defective in DRP3a, exhibits long peroxules (**d**) (Mano et al. 2004). Images were taken with a confocal microscope. Scale bar 10 μm. We thank Prof. M. Nishimura for kindly providing *apml–6* seed

reported peroxisome speeds in different tissues where movement appears to be faster on average in roots and the phloem compared to leaves. Figure 21.2 consists of representative images of peroxisome movement in Arabidopsis leaf and root

Table 21.1 Peroxisome movement rates in various plant tissues

Organism	Plant organ	Average speed (μm/s)	Reference
Arabidopsis thaliana	Young leaf	0.18	Rodriguez-Serrano et al. (2009)
	Mature rosette leaf	0.9	
	Root	Vibrational (value not given) and fast movement (2.0–3.25)	Mano et al. (2002)
	Leaf	Mostly vibrational movement (value not given)	
	Root hair of 5d seedling	~1	Peremyslov et al. (2008)
	Leaf of 5d seedling	~1.3	
	Leaf epidermal cells	70 % Brownian; 15 % average speed 0.7 ± 0.2; 10 % bidirectional movement up to 4	Mathur et al. (2002)
	Hypocotyl cortical cells of 3d dark grown seedling	0–6	Jedd and Chua (2002)
	Root phloem of 3d dark grown seedling	Peak 9	
Nicotiana benthamiana	Leaf epidermal cells	~0.9	Avisar et al. (2008)
Nicotiana tabacum	Leaf epidermal cells	1.1	Sparkes et al. (2008)
Physcomitrella patens	Protonemal cells	0.058	Furt et al. (2012)

Peroxisomes display a range of movements, and average speeds therefore vary. References for average peroxisome speeds are given in the table

epidermal cells. Individual images from time lapse movies are shown and the movement of peroxisomes displaying different types of behavior marked: saltatory (Fig. 21.2, arrowhead), fast unidirectional (Fig. 21.2, arrow) and a meandering type motion whereby the organelle trajectory changes several times (Fig. 21.2, asterisk). The movement was tracked using automated tracking software and representative tracks over the entire length of movie are depicted (Fig. 21.2e, j for movies from leaf and root epidermal cells respectively). Jedd and Chua even reported that two peroxisomes could move in opposite directions along the same trajectory suggestive of parallel filamentous actin with opposing polarities (Jedd and Chua 2002). It was also noted by Sinclair et al. that peroxules do not appear to always form at the trailing edge of movement and can actually form at the leading edge (Sinclair et al. 2009). Based on these erratic types of movement, which all spheroid organelles undergo, average rates of speed of an entire population are insufficient in describing the nature of their movement, and better analytical tools need to be developed and employed to understand the true nature of peroxisome dynamics and their relationship with surrounding organelles. Steps are being taken towards this end with movement characteristics now being displayed as cumulative distribution frequency plots which display the average movement of an individual organelle

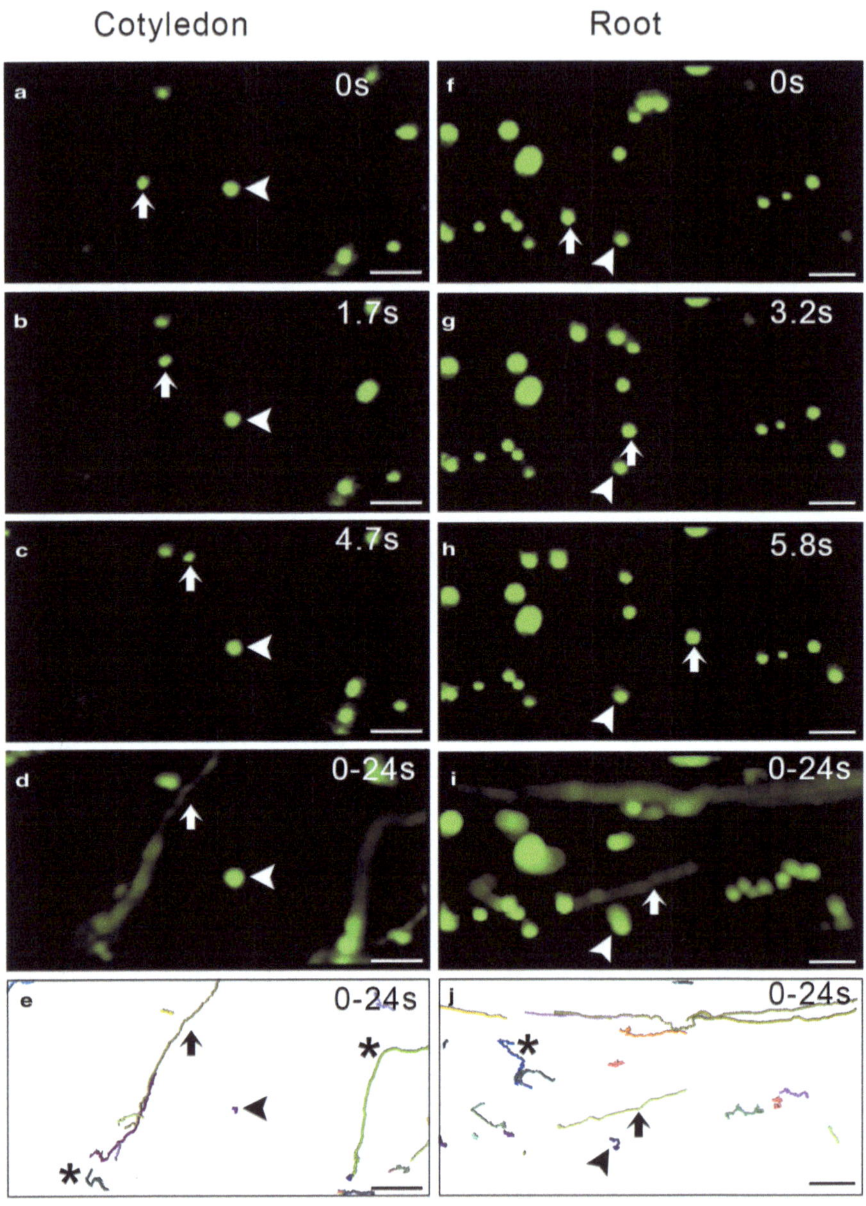

Fig. 21.2 Peroxisome movement in Arabidopsis leaf and root epidermal cells. Sequential images were taken of peroxisomes in Arabidopsis cotyledon (**a**–**e**) and root (**f**–**j**) epidermal cells using a confocal microscope. Organelles were tracked over 24 s using Imaris tracking software (Bitplane) and tracks displayed (**e**, **j**). Comparison between the tracks (**e**, **j**) and the projection of all the images taken (**d**, **i**) highlights that the latter cannot discriminate between individual organelles in regions of fast movement where several organelles can occupy the same space over a given time frame. Several types of motion are evident; saltatory (*arrowhead*), fast directional (*arrow*) and meandering (*asterisk*). Scale bar 5 μm

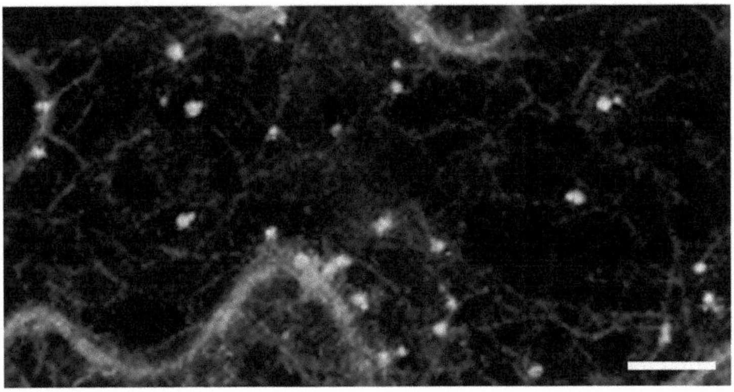

Fig. 21.3 Peroxisomes align along actin microfilaments. Peroxisomes (puncta) associate with actin microfilaments (filamentous structure) in tobacco epidermal cells. Image was taken with a confocal microscope. Scale bar 10 μm

rather than the whole population; however, this does not take in to account the erratic behavior of any one organelle.

Plant peroxisomes align with filamentous actin (Fig. 21.3). Cytoskeletal drug inhibition studies published by a few groups in quick succession highlighted the requirement for actin in peroxisome movement (Jedd and Chua 2002; Mano et al. 2002; Mathur et al. 2002). Several studies have alluded to potential myosin specificity, but key experiments showing myosin localisation and affects specifically on peroxisome dynamics have yet to be documented. Plants encode for two myosin families, class XI and class VIII. Based on domain architecture class XI are similar to class V which is involved in organelle movement in mammals and yeast. Myosins are composed of an actin binding ATP hydrolysis domain at the amino terminus, a regulatory neck region and the carboxy terminus containing tail domain proposed to bind cargo. Immunofluorescence data has indicated that MYA2, a class XI myosin, may bind to peroxisomes in Arabidopsis leaf epidermal cells and guard cells (Hashimoto et al. 2005). A yeast-2-hybrid screen isolated RabC2a as a potential binding partner with the tail domain of MYA2 (Hashimoto et al. 2008). The authors suggested that RabC2a collocates with a peroxisomal marker in Arabidopsis leaf protoplasts. In independent studies, transient expression of regions of the myosin tail domain has highlighted that MYA1, MYA2, XI-I and XI-K may also bind to the peroxisome surface; however, some of these fragments were also seen to collocate to Golgi bodies (Li and Nebenfuhr 2007; Reisen and Hanson 2007) . Therefore, whilst there is sparse data indicating that perhaps MYA2 may collocate to peroxisomes, expression of dominant negative forms and T-DNA insertional analysis has not shown a specific role in solely perturbing peroxisome movement. In all of the studies reported thus far, one of six class XI myosins (XIC, XIE, XII, XIK, MYA1 and MYA2 also known as XI-1 and XI-2 respectively) seems to perturb the movement of several classes of organelle, not just one (Avisar et al. 2008, 2009; Peremyslov et al. 2008, 2010; Prokhnevsky et al. 2008). This is not that unsurprising since, in other model systems, myosins can control the movement of several cargoes regulated through specific multiprotein recruitment complexes, lipid composition and changes in myosin conformation through proteolysis, phosphorylation and calcium signalling (see review and references therein by Akhmanova and Hammer 2010). At present, the identity of factors required for myosin recruitment and regulation in plants is scarce and requires further study.

A role for microtubules in peroxisome movement has also been investigated through drug inhibition studies with no apparent effect being reported (Mano et al. 2002; Jedd and Chua 2002; Mathur et al. 2002). Whilst it is commonly thought that the majority of plant organelle movement is driven by actin–myosin processes, a requirement for microtubules in potentially 'slowing down' organelles and affecting their positioning has started to emerge. Microtubules appear to affect ER dynamics in characean internodal cells undergoing expansion (Foissner et al. 2009) and the movement and positioning of mitochondria in Arabidopsis root hairs (Zheng et al. 2009). More recently, high tempo-spatial imaging using a spinning disc microscope was able to visualize dynamic interactions between F-actin and microtubules revealing that F-actin organization is in part controlled by microtubules in Arabidopsis hypocotyls (Sampathkumar et al. 2011). In light of recent reviews on the role of microtubules and kinesins in plant dynamics (Brandizzi and Wasteneys 2013; Cai and Cresti 2013), in combination with advances in automated organelle tracking and increased tempo-spatial resolution imaging techniques, a role for microtubule based processes in peroxisome movement should now be revisited. Potential roles for microtubules in peroxisome protein trafficking have been suggested through the identification of interactions between peroxisomal Multifunctional Protein (MFP) from Rice and microtubules (see reviews by Muench and Mullen 2003; Chuong et al. 2002, 2005).

Static snapshots provided through ultrastructural studies of fixed plant tissues have revealed peroxisomes closely associated with chloroplasts and mitochondria in aerial tissues (Frederick and Newcomb 1969). The relationship between these organelles was proposed to represent a requirement for efficient shuttling of metabolites through the photorespiratory pathway. The molecular factors controlling peroxisome division have been identified (PEX11, DRPs and Fis proteins; see Chap. 10). Interestingly, it is important to note that whilst peroxisomes are functionally linked to mitochondria and chloroplasts through the photorespiratory pathway, they also share some common components in their division machineries (see Chap. 10).

Chloroplast unusual positioning 1 (CHUP1), a chloroplast membrane protein, controls chloroplast positioning through controlling the dynamics of the short actin filament associated with chloroplasts (Kadota et al. 2009; Oikawa et al. 2008). In mesophyll cells, chloroplasts move from the periclinal to the anticlinal walls under high light irradiance to prevent photodamage. This process does not occur in *chup1* mutants where chloroplasts tend to reside on the basal side of the mesophyll cells. In *chup1* mutants peroxisomes appeared to move with chloroplasts in mesophyll cells, whereas mitochondria movement appeared to be unaffected. The authors concluded that the abnormal peroxisome distribution was 'likely a secondary effect of chloroplast mislocalization' (Oikawa et al. 2003). CHUP1 studies therefore provide indirect evidence that peroxisomes and chloroplasts may be physically tethered. Additional evidence provided through density centrifugation highlighted a chloroplast–peroxisome-enriched fraction (Schnarrenberger and Burkhard 1977). An Arabidopsis *pex10* mutant with a defective RING finger domain displays defects in photorespiration, and contains multilobed peroxisomes which appear to no longer be closely appressed to chloroplasts (Schumann et al. 2007). Here, the authors proposed that PEX10 is involved in anchoring peroxisomes and chloroplasts to one another, and that by affecting the connective bridge metabolite shuttling for photorespiration was impaired. However, further studies highlighted additional PEX10 mutants (TLGEEY and P126S), which do not appear to possess a defect in peroxisome–chloroplast contacts, yet exhibit defects in photorespiratory metabolite levels (Prestele et al. 2010). Therefore, static snapshots of peroxisome–chloroplast juxtapositioning implicating a direct role in metabolite shuttling between the two compartments are inconclusive. Whilst it is tempting to speculate that PEX10, a peroxisomal membrane protein, may act as bridge or a tether between the two organelles, there is no direct evidence supporting this role. Independent studies of mutations in PEX10 indicated that it affects cuticular wax deposition and alters the ER network (Kamigaki et al. 2009). Studies relating to PEX10 trafficking have resulted in conflicting reports on whether it traffics through the ER (Sparkes et al. 2005; Flynn et al. 2005). Therefore, PEX10 could perhaps have multiple functions in the import of peroxisomal matrix proteins (see Chap. 14) and affecting tethering either directly to chloroplasts or through interaction with the ER. The interaction of peroxisomes with the ER and a role in peroxisome biogenesis is a contentious topic and readers are directed to reviews on the topic (Hu et al. 2012; Mullen and Trelease 2006; Kaur et al. 2009) and chapters in this edition.

The positioning of peroxisomes relative to the ER, chloroplasts and mitochondria in tobacco leaf epidermal cells is highlighted in Fig. 21.4. It can be seen that peroxisomes, to a certain extent, appear to be juxtaposed to the ER, chloroplasts and mitochondria. It should be noted that close association between organelles in highly vacuolated plant cells can be a result of the physical constraints placed upon the cytoplasmic contents by the large central vacuole. Therefore, ascertaining true physical connections between organelles cannot be merely drawn from static snapshots of juxtaposed organelles. In addition, biochemical extraction does not allow for the investigation of in vivo interactions in real time. A bona fide assessment of true physical connectivity between two organelles in vivo can be determined though biophysical studies using optical tweezers. Optical tweezers allow objects with a different refractive index to the surrounding media to be physically trapped and constrained within the optical trapping beam, subsequent movement of which concomitantly moves the trapped organelle/structure. Such studies have been carried out to highlight a physical association between Golgi bodies and the ER in Arabidopsis leaf epidermal cells (Sparkes et al. 2009).

Peroxisome abundance in plant cells is controlled by metabolic and environmental stresses such as clofibrate, herbicide, ozone, salt and high light (Castillo and León 2008; Castillo et al. 2008; Lopez-Huertas et al. 2000; Morre et al. 1990; Palma

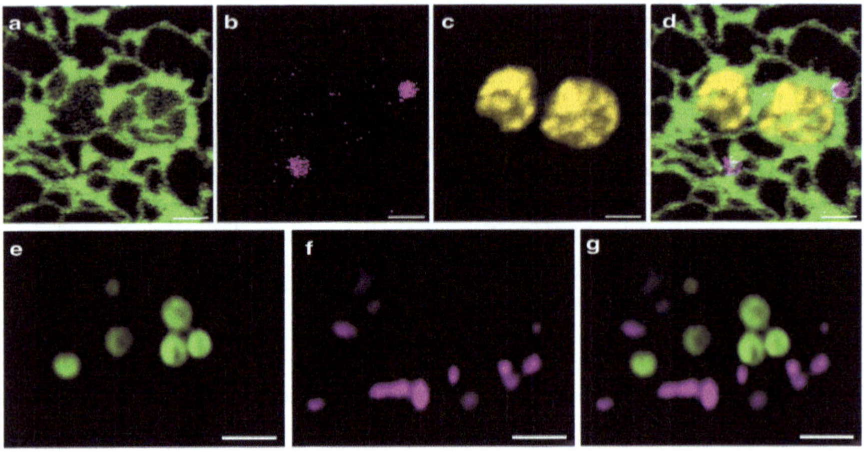

Fig. 21.4 Peroxisome positioning within tobacco leaf epidermal cells. The cortical ER (**a** and **d**, *green*) is a polygonal web of tubules and cisternae that pervades throughout the cytoplasm. It encapsulates chloroplasts (**c** and **d**, *yellow*), and peroxisomes appear closely associated with both ER and chloroplasts (**b** and **d**, *magenta*). Peroxisomes (**e**, *green*) can also appear closely associated with mitochondria (**f**, *magenta*). Images were taken with a confocal microscope. Scale bar 2 μm

et al. 1991; Oksanen et al. 2003; Ferreira et al. 1989; de Felipe et al. 1988; Nila et al. 2006; Mitsuya et al. 2010).

Recently, it has been shown that exposure to cadmium increases peroxisome movement in Arabidopsis leaf epidermal cell (Rodriguez-Serrano et al. 2009). It is unclear what triggers this effect, but the hypothesis proposed was that increased movement may be required for increased ROS detoxification; cadmium exposure induced ROS production, and the effects on peroxisome dynamics were ameliorated through addition of ascorbate, an anti-oxidant. Intriguingly, Sinclair et al. observed that exogenously added ROS reduced peroxisomal movement with a concomitant alteration in shape resulting in the initial extension of peroxules followed by peroxisome elongation (Sinclair et al. 2009). It is unclear why these reports are conflicting regarding the effects of ROS on peroxisome dynamics, and could be due to differences in experimental systems whereby ROS production is induced endogenously rather than added from an exogenous source.

During pathogen invasion, and in response to wounding, actin dynamics in the plant cell alters and organelles aggregate around the site (Hardham et al. 2008; Koh and Somerville 2006; Lipka et al. 2008). The functional role, and regulation of such, is not fully understood. Studies of peroxisome movement, however, have highlighted a functional role in the potential delivery of glycosyl hydrolase products to sites of *Blumeria graminis* f. sp. *hordei* (a powdery mildew) infection in Arabidopsis. PEN2, a glycosyl hydrolase, associates with the periphery of peroxisomes which cluster around the sites of infection (Lipka et al. 2005). Similar peroxisome clustering has also been observed in *Erysiphe cichoracearum* (a powdery mildew) infection in Arabidopsis (Koh et al. 2005).

21.3 Perspectives

Peroxisomes perform many essential functions required throughout plant development, and their enzymatic content can readily change. Based on the range of motions peroxisomes display within any one cell it is difficult to ascribe a function with a particular motion. Generation of probes highlighting metabolite generation and exchange in coordination with movement will help unlock and directly relate function with movement responses. Here, we have highlighted connections between peroxisome positioning in relation to other organelles and function. A pattern is emerging that organelle movement, morphology and inter-organellar connections appear to feed into a cells capacity to cope with changes in cellular environment (Grimm 2012; Rowland and Voeltz 2012). Organelles should no longer be seen as discrete, static membrane bounded structures housing discrete metabolic pathways. The cell is 'enigmatic' and dynamic, and development of tools, techniques and analytical methods will undoubtedly highlight a network and plethora of interactions between organelles required for cellular maintenance and growth.

References

Acosta IF, Farmer EE (2010) Jasmonates. Arabidopsis Book 8:e0129. doi:10.1199/tab.0129

Akhmanova A, Hammer JA (2010) Linking molecular motors to membrane cargo. Curr Opin Cell Biol 22(4):479–487

Avisar D, Prokhnevsky AI, Makarova KS, Koonin EV, Dolja VV (2008) Myosin XI-K is required for rapid trafficking of Golgi stacks, peroxisomes and mitochondria in leaf cells of Nicotiana benthamiana. Plant Physiol 146(3):1098–1108

Avisar D, Abu-Abied M, Belausov E, Sadot E, Hawes C, Sparkes IA (2009) A comparative study of the involvement of 17 Arabidopsis myosin family members on the motility of Golgi and other organelles. Plant Physiol 150(2):700–709

Brandizzi F, Wasteneys GO (2013) Cytoskeleton-dependent endomembrane organization in plant cells: an emerging role for microtubule. Plant J. doi:10.1111/tpj.12227

Cai G, Cresti M (2013) Are kinesins required for organelle trafficking in plant cells? Front Plant Cell Biol 3:170. doi:10.3389/fpls.2012.00170

Castillo MC, León J (2008) Expression of the β-oxidation gene 3-ketoacyl-CoA thiolase 2 (KAT2) is required for the timely onset of natural and dark-induced leaf senescence in Arabidopsis. J Exp Biol 59(8):2171–2179

Castillo MC, Sandalio LM, Del Rio LA, Leon J (2008) Peroxisome proliferation, wound-activated responses and expression of peroxisome-associated genes are cross-regulated but uncoupled in Arabidopsis thaliana. Plant Cell Environ 31(4):492–505

Chuong SDX, Mullen RT, Muench DG (2002) Identification of a rice RNA- and microtubule-binding protein as the multifunctional protein, a peroxisomal enzyme involved in the β-oxidation of fatty acids. J Biol Chem 277(4):2419–2429

Chuong SDX, Park NL, Freeman MC, Mullen RT, Muench DG (2005) The peroxisomal multifunctional protein interacts with cortical microtubules in plant cells. BMC Cell Biol 6:40. doi:10.1186/1471-2121-1186-1140

Collings DA, Harper JDI, Vaughn KC (2003) The association of peroxisomes with the developing cell plate in dividing onion root cells depends on actin microfilaments and myosin. Planta 218 (2):204–216

Cutler SR, Ehrhardt DW, Griffitts JS, Somerville CR (2000) Random GFP::cDNA fusions enable visualization of subcellular structures in cells of Arabidopsis at a high frequency. Proc Natl Acad Sci USA 97(7):3718–3723

De Duve C, Baudin P (1966) Peroxisomes (microbodies and related particles). Physiol Rev 46 (2):323–357

de Felipe M, Lucas MM, Pozuelo JM (1988) Cytochemical study of catalase and peroxidase in the mesophyll of Lolium rigidum plants treated with isoproturon. J Plant Physiol 132:67–73

Del Rio LA (2011) Peroxisomes as a cellular source of reactive nitrogen species signal molecules. Arch Biochem Biophys 506(1):1–11

Fan J, Quan S, Orth T, Awai C, Chory J, Hu J (2005) The Arabidopsis PEX12 gene is required for peroxisome biogenesis and is essential for development. Plant Physiol 139(1):231–239

Ferreira M, Bird B, Davies DD (1989) The effect of light on the structure and organization of Lemna peroxisomes. J Exp Biol 40:1029–1035

Flynn CR, Heinze M, Schumann U, Gietl C, Trelease RN (2005) Compartmentalization of the plant peroxin, AtPex10p, within subdomain(s) of ER. Plant Sci 168(3):635–652

Foissner I, Menzel D, Wasteneys GO (2009) Microtubule-dependent motility and orientation of the cortical endoplasmic reticulum in elongating characean internodal cells. Cell Motil Cytoskeleton 66(3):142–155

Frederick SE, Newcomb EH (1969) Microbody-like organelles in leaf cells. Science 163 (3873):1353–1355

Furt F, Lemoi K, Tuzel E, Vidali L (2012) Quantitative analysis of organelle distribution and dynamics in Physcomitrella patens protonemal cells. BMC Plant Biol 12:70. doi:10.1186/1471-2229-1112-1170

Grimm S (2012) The ER-mitochondria interface: the social network of cell death. Biochim Biophys Acta 1823(2):327–334

Hardham AR, Takemoto D, White RG (2008) Rapid and dynamic subcellular reorganisation following mechanical stimulation of Arabidopsis epidermal cells mimics responses to fungal and oomycete attack. BMC Plant Biol 8:63. doi:10.1186/1471-2229-1188-1163

Hashimoto K, Igarashi H, Mano S, Nishimura M, Shimmen T, Yokota E (2005) Peroxisomal localization of a myosin XI isoform in Arabidopsis thaliana. Plant Cell Physiol 46(5):782–789

Hashimoto K, Igarashi H, Mano S, Takenaka C, Shiina T, Yamaguchi M, Demura T, Nishimura M, Shimmen T, Yokota E (2008) An isoform of Arabidopsis myosin XI interacts with small GTPases in its C-terminal tail region. J Exp Bot 59(13):3523–3531

Hepler PK, Vidali L, Cheung AY (2001) Polarized cell growth in higher plants. Annu Rev Cell Dev Biol 17:159–187

Hu JP, Aguirre M, Peto C, Alonso J, Ecker J, Chory J (2002) A role for peroxisomes in photomorphogenesis and development of Arabidopsis. Science 297(5580):405–409

Hu J, Baker A, Bartel B, Linka N, Mullen RT, Reumann S, Zolman BK (2012) Plant peroxisomes: biogenesis and function. Plant Cell 24(6):2279–2303

Jedd G, Chua NH (2002) Visualization of peroxisomes in living plant cells reveals acto-myosin-dependent cytoplasmic streaming and peroxisome budding. Plant Cell Physiol 43(4):384–392

Kachar B, Reese TS (1988) The mechanism of cytoplasmic streaming in characean algal cells: sliding of endoplasmic reticulum along actin filaments. J Cell Biol 106(5):1545–1552

Kadota A, Yamada N, Suetsugu N, Hirose M, Saito C, Shoda K, Ichikawa S, Kagawa T, Nakano A, Wada M (2009) Short actin-based mechanism for light-directed chloroplast movementin Arabidopsis. Proc Natl Acad Sci USA 106(31):13106–13111

Kamigaki A, Kondo M, Mano S, Hayashi M, Nishimura M (2009) Suppression of peroxisome biogenesis factor 10 reduces cuticular wax accumulation by disrupting the ER network in Arabidopsis thaliana. Plant Cell Physiol 50(12):2034–2046

Kaur N, Reumann S, Hu J (2009) Peroxisome biogenesis and function. Arabidopsis Book 7:e0123. doi:10.1199/tab.0123

Koh S, Somerville S (2006) Show and tell: cell biology of pathogen invasion. Curr Opin Plant Biol 9(4):406–413

Koh S, Andre AL, Edwards H, Ehrhardt D, Somerville S (2005) Arabidopsis thaliana subcellular responses to compatible Erysiphe cichoracearum infections. Plant J 44(3):516–529
Koo AJK, Howe GA (2007) Role of peroxisomal β-oxidation in the production of plant signaling compounds. Plant Signal Behav 2(1):20–22
Li J, Nebenfuhr A (2007) Organelle targeting of myosin XI is mediated by two globular tail domains with separate cargo binding sites. J Biol Chem 282(28):20593–20602
Li C, Schilmiller AL, Liu G, Lee GI, Javantry S, Sageman C, Vrebalov J, Giovannoni JJ, Yagi K, Kobayashi Y, Howe GA (2005) Role of β-oxidation in jasmonate biosynthesis and systemic wound signaling in tomato. Plant Cell 17(3):971–986
Lipka V, Dittgen J, Bednarek P, Bhat R, Wiermer M, Stein M, Landtag J, Brandt W, Rosahl S, Scheel D, Llorente F, Molina A, Parker J, Somerville S, Schulze-Lefert P (2005) Pre- and postinvasion defences both contribute to nonhost resistance in Arabidopsis. Science 310 (5751):1180–1183
Lipka U, Fuchs R, Lipka V (2008) Arabidopsis non-host resistance to powdery mildews. Curr Opin Plant Biol 11(4):404–411
Lopez-Huertas E, Charlton WL, Johnson B, Graham IA, Baker A (2000) Stress induces peroxisome biogenesis genes. EMBO J 19(24):6770–6777
Mano S, Nakamori C, Hayashi M, Kato A, Kondo M, Nishimura M (2002) Distribution and characterization of peroxisomes in arabidopsis by visualization with GFP: dynamic morphology and actin- dependent movement. Plant Cell Physiol 43(3):331–341
Mano S, Nakamori C, Kondo M, Hayashi M, Nishimura M (2004) An Arabidopsis dynamin-related protein, DRP3A, controls both peroxisomal and mitochondrial division. Plant J 38 (3):487–498
Mathur J, Mathur N, Hulskamp M (2002) Simultaneous visualisation of peroxisomes and cytoskeletal elements reveals actin and not microtubule-based peroxisome motility in plants. Plant Physiol 128(3):1031–1045
Mathur J, Mammone A, Barton KA (2012) Organelle extensions in plant cells. J Integr Plant Biol 54(11):851–867
McCurdy DW (1999) Is 2,3-butanedione monoxime an effective inhibitor of myosin-based activities in plant cells? Protoplasma 209(1–2):120–125
Mitsuya S, El-shami M, Sparkes IA, Charlton W, Lousa CM, Johnson B, Baker A (2010) Salt stress causes peroxisome proliferation, but inducing peroxisome proliferation does not improve NaCl tolerance in Arabidopsis thaliana. PLoS One 5:e9408. doi:10.1371/journal.pone.0009408
Mittler R, Vanderauwera S, Suzuki N, Miller GAD, Tognetti VB, Vandepoele K, Gollery M, Shulaev V, Van Breusegem F (2011) ROS signaling: the new wave? Trends Plant Sci 16 (6):300–309
Morita MT, Nakamura M (2012) Dynamic behavior of plastids related to environmental response. Curr Opin Plant Biol 15(6):722–728
Morre DJ, Selldén G, Ojanperä K, Sandelius AS, Egger A, Morré DM, Chalko CM, Chalki RA (1990) Peroxisome proliferation in Norway spruce induced by ozone. Protoplasma 155(1–3):58–65
Muench DG, Mullen RT (2003) Peroxisome dynamics in plant cells: a role for the cytoskeleton. Plant Sci 164(3):307–315
Mullen RT, Trelease RN (2006) The ER-peroxisome connection in plants: development of the "ER semi-autonomous peroxisome maturation and replication" model for plant peroxisome biogenesis. Biochim Biophys Acta 1763(12):1655–1668
Nebenfuhr A (2007) Organelle dynamics during cell division. Plant Cell Monogr. doi:10.1007/7089_2007_1129
Nila AG, Sandalio LM, Lopez MG, Gomez M, del Rio LA, Gomez-Lim MA (2006) Expression of a peroxisome proliferator-activated receptor gene (xPPARα) from Xenopus laevis in tobacco (Nicotiana tabacum) plants. Planta 224(3):569–581
Nishimura M, Yamaguchi S, Mori T, Akazawa T, Yokota S (1986) Immunocytochemical analysis shows that glyoxysomes are directly transformed to leaf peroxisomes during greening of pumpkin cotyledons. Plant Physiol 81(1):313–316

Nishimura M, Takeuchi Y, Debellis L, Haranishimura I (1993) Leaf peroxisomes are directly transformed to glyoxysomes during senescence of pumpkin cotyledons. Protoplasma 175(3–4):131–137

Nito K, Kamigaki A, Kondo M, Hayashi M, Nishimura M (2007) Functional classification of Arabidopsis peroxisome biogenesis factors proposed from analyses of knockdown mutants. Plant Cell Physiol 48(6):763–774

Nothangel EA, Webb WW (1982) Hydrodynamic models of viscous coupling between motile myosin and endoplasm in Characean algae. J Cell Biol 94(2):444–454

Nyathi Y, Baker A (2006) Plant peroxisomes as a source of signalling molecules. Biochim Biophys Acta 1763(12):1478–1495

Oikawa K, Kasahara M, Kiyosue T, Kagawa T, Suetsugu N, Takahashi F, Kanegae T, Niwa Y, Kadota A, Wada M (2003) CHLOROPLAST UNUSUAL POSITIONING1 is essential for proper chloroplast positioning. Plant Cell 15(12):2805–2815

Oikawa K, Yamasato A, Kong SG, Kasahara M, Nakai M, Takahashi F, Ogura Y, Kagawa T, Wada M (2008) Chloroplast outer envelope protein CHUP1 Is essential for chloroplast anchorage to the plasma membrane and chloroplast movement. Plant Physiol 148(2):829–842

Ojangu EL, Jarve K, Paves H, Truve E (2007) Arabidopsis thaliana myosin XIK is involved in root hair as well as trichome morphogenesis on stems and leaves. Protoplasma 230(3–4):193–202

Ojangu EL, Tanner K, Pata P, Jarve K, Holweg C, Truve E, Paves H (2012) Myosins XI-K, XI-1, and XI-2 are required for development of pavement cells, trichomes, and stigmatic papillae in Arabidopsis. BMC Plant Biol 12:81. doi:10.1186/1471-2229-1112-1181

Oksanen E, Haikio E, Sober J, Karnosky DF (2003) Ozone-induced H2O2 accumulation in field-grown aspen and birch is linked to foliar ultrastructure and peroxisomal activity. New Phytol 161(3):791–799

Palma JM, Garrido M, Rodriguez Garcia MI, del Rio LA (1991) Peroxisome proliferation and oxidative stress mediated by activated oxygen species in plant peroxisomes. Arch Biochem Biophys 287(1):68–74

Peremyslov VV, Prokhnevsky AI, Avisar D, Dolja VV (2008) Two class XI myosins function in organelle trafficking and root hair development in Arabidopsis thaliana. Plant Physiol 146 (3):1109–1116

Peremyslov VV, Prokhnevsky AI, Dolja VV (2010) Class XI myosins are required for development, cell expansion, and F-actin organization in Arabidopsis. Plant Cell 22(6):1883–1897

Pracharoenwattana I, Smith SM (2008) When is a peroxisome not a peroxisome? Trends Plant Sci 13(10):522–525

Prestele J, Hierl G, Scherling C, Hetkamp S, Schwechheimer C, Isono E, Weckwerth W, Wanner G, Gietl C (2010) Different functions of the C3HC4 zinc RING finger peroxisome PEX10, PEX2 and PEX12 in peroxisome formation and matrix protein import. Proc Natl Acad Sci USA 107(33):14915–14920

Prokhnevsky AI, Peremyslov VV, Dolja VV (2008) Overlapping functions of the four class XI myosins in Arabidopsis growth, root hair elongation and organelle motility. Proc Natl Acad Sci USA 105(50):19744–19749

Reisen D, Hanson MR (2007) Association of six YFP-myosin XI-tail fusions with mobile plant cell organelles. BMC Plant Biol 7:6. doi:10.1186/1471-2229-1187-1186

Rodriguez-Serrano M, Romero-Puertas MC, Sparkes IA, Hawes C, del Rio LA, Sandalio LM (2009) Peroxisome dynamics in Arabidopsis plants under oxidative stress induced by cadmium. Free Radic Biol Med 47(11):1632–1639

Rowland AA, Voeltz G (2012) Endoplasmic reticulum-mitochondria contacts: function of the junction. Nat Rev Mol Cell Biol 13(10):607–615

Sampathkumar A, Lindeboom JJ, Debolt S, Gutierrez R, Ehrhardt DW, Ketelaar T, Persson S (2011) Live cell imaging reveals structural associations between the actin and microtubule cytoskeleton in Arabidopsis. Plant Cell 23(6):2302–2313

Sandalio LM, Rodríguez-Serrano M, Romero-Puertas MC, del Río LA (2013) Role of peroxisomes as a source of reactive oxygen species (ROS) signaling molecules. In: Del Rio LA (ed) Peroxisomes and their key role in cellular signaling and metabolism. Springer, Alemania

Sautter C (1986) Microbody transition in greening watermelon cotyledons. Double immunocytochemical labeling of isocitrate lyase and hydroxypyruvate reductase. Planta 167(4):491–503

Schnarrenberger C, Burkhard C (1977) In-vitro interaction between chloroplasts and peroxisomes as controlled by inorganic phosphate. Planta 134(2):109–114

Schumann U, Wanner G, Veenhuis M, Schmid M, Gietl C (2003) AthPEX10, a nuclear gene essential for peroxisome and storage organelle formation during Arabidopsis embryogenesis. Proc Natl Acad Sci USA 100(16):9626–9631

Schumann U, Prestele J, O'Geen H, Brueggeman R, Wanner G, Gietl C (2007) Requirement of the C3HC4 zinc RING finger of the Arabidopsis PEX10 for photorespiration and leaf peroxisome contact with chloroplasts. Proc Natl Acad Sci USA 104(3):1069–1074

Scott I, Sparkes IA, Logan DC (2007) The missing link: inter-organellar connections in mitochondria and peroxisomes? Trends Plant Sci 12(9):380–381

Sheahan MB, Rose RJ, McCurdy DW (2007) Mechanisms of organelle inheritance in dividing plant cells. J Integr Plant Biol 49(8):1208–1218

Sinclair AM, Trobacher CP, Mathur N, Greenwood JS, Mathur J (2009) Peroxule extension over ER-defined paths constitutes a rapid subcellular response to hydroxyl stress. Plant J 59(2):231–242

Sparkes I (2010) Motoring around the plant cell: insights from plant myosins. Biochem Soc Trans 38(3):833–838

Sparkes I (2011) Recent advances in understanding plant myosin function: life in the fast lane. Mol Plant 4(5):805–812

Sparkes IA, Brandizzi F, Slocombe SP, El-Shami M, Hawes C, Baker A (2003) An Arabidopsis pex10 null mutant is embryo lethal, implicating peroxisomes in an essential role during plant embryogenesis. Plant Physiol 133(4):1809–1819

Sparkes IA, Hawes C, Baker A (2005) AtPEX2 and AtPEX10 are targeted to peroxisomes independently of known endoplasmic reticulum trafficking routes. Plant Physiol 139(2):690–700

Sparkes IA, Teanby NA, Hawes C (2008) Truncated myosin XI tail fusions inhibit peroxisome, Golgi, and mitochondrial movement in tobacco leaf epidermal cells: a genetic tool for the next generation. J Exp Biol 59(9):2499–2512

Sparkes IA, Ketelaar T, Ruijter NC, Hawes C (2009) Grab a Golgi: laser trapping of Golgi bodies reveals in vivo interactions with the endoplasmic reticulum. Traffic 10(5):567–571

Strader LC, Bartel B (2011) Transport and metabolism of the endogenous auxin precursor indole-3-butryic acid. Mol Plant 4(3):477–486

Takemoto D, Jones DA, Hardham AR (2003) GFP-tagging of cell components reveals the dynamics of subcellular re-organisation in response to infection of Arabidopsis by oomycete pathogens. Plant J 33(4):775–792

Theodoulou F, Eastmond P (2012) Seed storage oil catabolism: a story of give and take. Curr Opin Plant Biol 15(3):322–328

Titus DE, Becker WM (1985) Investigation of the glyoxysome-peroxisome transition in germinating cucumber cotyledons using double-label immunoelectron microscopy. J Cell Biol 101(4):1288–1299

Vick JK, Nebenfuhr A (2012) Putting the breaks on: regulating organelle movements in plant cells. J Integr Plant Biol 54(11):868–874

Vidali L, Bezanilla M (2012) Physcomitrella patens: a model for tip cell growth and differentiation. Curr Opin Plant Biol 15(6):625–631

Wolff K, Marenduzzo D, Cates ME (2012) Cytoplasmic streaming in plant cells: the role of wall slip. J R Soc Interface 9(71):1398–1408

Zheng M, Beck M, Muller J, Chen T, Wang X, Wang F, Wang Q, Wang Y, Baluska F, Logan DC, Samaj J, Lin J (2009) Actin turnover is required for myosin-dependent mitochondrial movements in Arabidopsis root hairs. PLoS One 4(6):e5961

Part VI

Specific Degradation Through Pexophagy

Molecular Machines Involved in Pexophagy 22

Taras Y. Nazarko and Jean-Claude Farré

Abstract

Pexophagy, the selective autophagy of peroxisomes, is a membrane trafficking pathway that delivers damaged or superfluous peroxisomes from the cytosol to the vacuole/lysosome for degradation and recycling. Pexophagosomes, the peroxisome-containing autophagosomes, are the double-membrane carriers in the pexophagy pathway. Like autophagosomes, they are built by the core autophagic machinery, a set of autophagy-related (Atg) proteins that get together at the phagophore assembly site (PAS). The pexophagy-specific receptor protein complex (RPC) plays a central role in pexophagy by bridging the peroxisomal membrane with the core autophagic machinery and, later, with the growing isolation membrane or phagophore to ensure the high selectivity of peroxisome sequestration. This process is tightly regulated by phosphorylation and involves the cytoskeleton. Additional mechanisms secure the pexophagy of large peroxisomes. Finally, the fusion machineries support the membrane flow at different stages of the pexophagy pathway.

Keywords

Selective autophagy • Peroxisome degradation • Pexophagy • receptor • Atg30 • Atg36 • NBR1

22.1 Introduction

Macroautophagy (hereafter autophagy) is a major pathway utilized by the eukaryotic cell to degrade the cytosol, intracellular organelles, and pathogens. It is achieved by delivering them to the lytic compartment, lysosome in mammalian

T.Y. Nazarko (✉) • J.-C. Farré (✉)
Section of Molecular Biology, Division of Biological Sciences, UC San Diego, La Jolla, CA 92093-0322, USA
e-mail: tnazarko@ucsd.edu; jfarre@ucsd.edu

C. Brocard and A. Hartig (eds.), *Molecular Machines Involved in Peroxisome Biogenesis and Maintenance*, DOI 10.1007/978-3-7091-1788-0_22, © Springer-Verlag Wien 2014

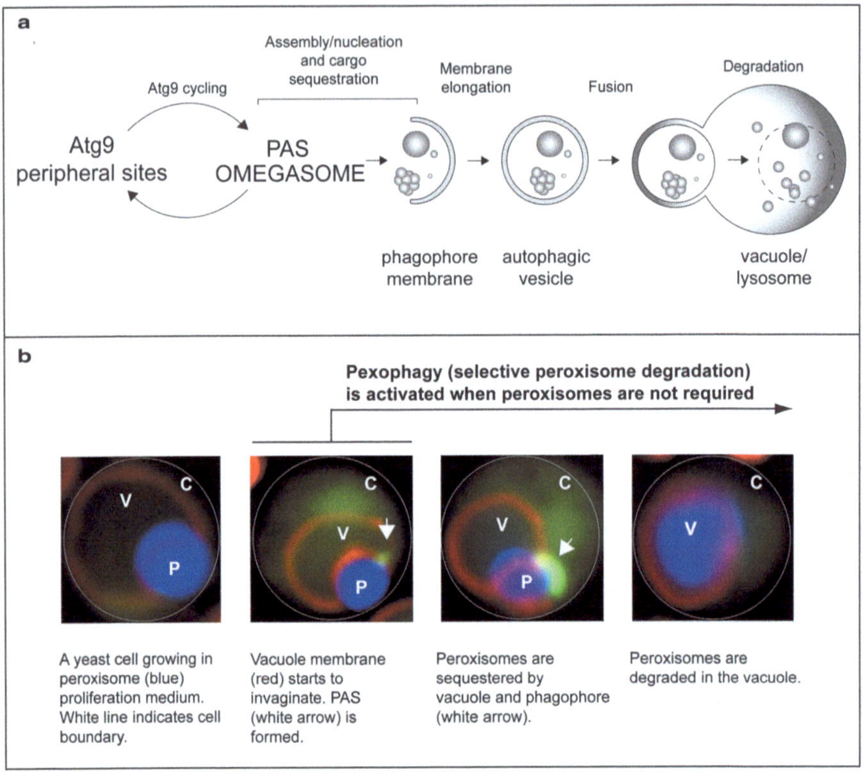

Fig. 22.1 General overview of autophagy and morphology of pexophagy. (**a**) Autophagy begins with the nucleation of the phagophore assembly site (PAS) in yeast and omegasome in mammals (assembly/nucleation and cargo sequestration step). The concerted action of the autophagy core machinery proteins at the PAS/omegasome leads to the expansion of the phagophore into the autophagosome (membrane elongation step). The autophagosome can engulf bulk cytoplasm non-specifically, including entire organelles, or target cargos specifically. The outer membrane of the autophagosome fuses with the vacuole/lysosome (fusion step). Finally, the sequestered material is degraded inside the vacuole/lysosome (degradation step) and recycled. (**b**) Morphological intermediates of the pexophagy pathway in *P. pastoris*. *V* vacuole, *P* peroxisome, *C* cytosol

cells or vacuole in yeasts. Autophagy culminates by formation of the double-membrane vesicles called autophagosomes. These vesicles are the membrane carriers of the autophagy pathway, which trap the cytosolic cargos and deliver them to the lysosome/vacuole for degradation and recycling (Fig. 22.1a). Formation of autophagosomes is mediated by a set of core Atg proteins. These proteins organize the PAS, which expands into the "isolation membrane" or the phagophore, an intermediate in autophagosome formation (Nakatogawa et al. 2009; Yang and Klionsky 2009). Pexophagy is the selective autophagy pathway, which sequesters the peroxisomes from the cytosol and delivers them to the lumen of the lysosome/vacuole (Fig. 22.1b) (Kraft et al. 2009; Manjithaya et al. 2010b). The peroxisome was the first organelle proven to be degraded by the selective autophagy and the first

Table 22.1 Core autophagic machinery

Yeast protein	Function	Mammalian homolog	Required for pexophagy	References for pexophagy
Atg1 kinase complex				
Atg1	Serine/threonine kinase, interacts with Atg13	ULK1-2	+	Stromhaug et al. (2001)
Atg11	Scaffold for selective autophagy	RB1CC1/FIP200	+	Kim et al. (2001)
Atg13	Regulatory subunit of Atg1 kinase	ATG13	NT	
Atg17	Scaffold for induced autophagy	RB1CC1/FIP200	+	Nazarko et al. (2009)
Atg29	Part of the Atg17 scaffold complex	–	+[a]	Stasyk et al. (2006)
Atg31	Part of the Atg17 scaffold complex	–	+[a]	Stasyk et al. (2006)
–		ATG101	NT	
PI3K complexes				
Atg6/Vps30	Subunit of the PI3K complexes I and II	BECN1	+	Farre et al. (2010)
Atg14	Localizes the PI3K complex to the PAS	ATG14/BARKOR	NT	
Vps15	Regulatory subunit of the PI3K complexes	PIK3R4	+	Stasyk et al. (1999)
Vps34	PI3K catalytic subunit	VPS34	+	Kiel et al. (1999)
Vps38	Functions in the vacuolar protein sorting pathway, not required for autophagy in yeast	UVRAG	–	Farre et al. (2010)
Atg9 cycling system				
Atg2p	Required for the Atg9 recycling from the PAS	ATG2	+	Stromhaug et al. (2001)
Atg9	Integral membrane protein, putative lipid carrier	ATG9	+	Chang et al. (2005)
Atg18	Binds the PI3P and is required for the Atg9 recycling from the PAS	WIPI1-2	+	Guan et al. (2001)
UBLP conjugation systems				
Atg3	E2-like enzyme for Atg8	ATG3	+	Dunn et al. (2005)
Atg4	Cysteine protease that processes Atg8p and Atg8-PE	ATG4	+	Mukaiyama et al. (2002)
Atg5	Conjugated with Atg12, the E3-like enzyme for Atg8	ATG5	+	Personal communication
Atg7	E1-like enzyme for both Atg8 and Atg12	ATG7	+	Yuan et al. (1999)
Atg8	Ubiquitin-like protein, conjugated to PE, regulates autophagosome size	LC3	+	Mukaiyama et al. (2002)
Atg10	E2-like enzyme for Atg12	ATG10	NT	

(continued)

Table 22.1 (continued)

Yeast protein	Function	Mammalian homolog	Required for pexophagy	References for pexophagy
Atg12	Ubiquitin-like protein, conjugated to Atg5	ATG12	NT	
Atg16	Component of the Atg12-Atg5-Atg16 complex	ATG16	+	Mukaiyama et al. (2002)

NT not tested

[a]Atg29 and Atg31 have not been tested for pexophagy in *S. cerevisiae*, but the putative *P. pastoris* homolog, Atg28, is required for pexophagy

organelle for which the organelle-specific autophagy receptor, Atg30, was identified (Farre et al. 2008; Bellu and Kiel 2003).

22.2 Pexophagy Relies on the Core Autophagic Machinery

Although autophagy was discovered in mammalian cells more than 60 years ago (De Duve and Wattiaux 1966), the first Atg proteins were identified in yeast only 20 years ago (Tsukada and Ohsumi 1993). Afterwards, several conserved Atg proteins were also found in higher eukaryotes. To date, over 30 Atg proteins have been identified in different species. The Atg proteins required for efficient autophagosome formation are referred to as the core autophagic machinery. Most of the core machinery is conserved from yeast to mammals (Table 22.1) and can be organized in at least four functional groups: (1) the Atg1/ULK1-2 kinase complex (Atg1, Atg13 and the two partially redundant scaffolds, Atg11 and Atg17–Atg31–Atg29), (2) the class III phosphatidylinositol-3 kinase (PI3K) complex I (Atg6/Vps30/BECN1, Atg14, Vps15/PIK3R4 and Vps34), (3) the Atg9 cycling system (Atg2, Atg9 and Atg18/WIPI1-2), and (4) the ubiquitin-like protein (UBLP) conjugation system (Atg3, Atg4, Atg5, Atg7, Atg8/LC3, Atg10, Atg12 and Atg16). Out of these four functional groups, we will highlight the role of the Atg1 kinase complex and UBLP conjugation system, since some of their components link the core machinery with the selective machinery required for the selective organelle degradation, such as pexophagy.

The Atg1 kinase complex is the first group of the core autophagic machinery to be required for autophagosome formation and is under the regulation of four nutrient-sensing kinases, target of rapamycin (TOR) kinase, protein kinase A (PKA), protein kinase B (Akt/PKB) homolog Sch9, and sucrose non-fermenting 1 or AMP-regulated kinase (Snf1/AMPK; Wong et al. 2013). Atg13 is the regulatory subunit of Atg1 and is a direct target of the TOR complex 1 (TORC1). Under nutrient-rich conditions, Atg13 is highly phosphorylated by TORC1 preventing autophagy, whereas in starvation conditions TORC1 is inactive, Atg13 is hypophosphorylated and autophagy is induced (Hara et al. 2008; Kraft et al. 2012). In yeast, the phagophore membrane is generated from the PAS to ultimately become a double-membrane vesicle. The PAS is organized by the scaffold components of the Atg1 kinase complex, Atg11 and Atg17–Atg31–Atg29 (Kawamata et al. 2008; Shintani and Klionsky 2004; Suzuki et al. 2007). These two alternative scaffolds are localized to the PAS in both fed and starved cells (Geng et al. 2008). However, Atg17–Atg31–Atg29 complex is activated by starvation and, therefore, essential exclusively for induced autophagy. In starvation conditions, Atg17 binds to the hypophosphorylated Atg13 and organizes the PAS (Kabeya et al. 2005). On the other hand, Atg11 is required in nutrient-rich and starvation conditions for all selective autophagy pathways, but dispensable for general autophagy. Atg11 does not interact with Atg13, but rather with Atg1, and the Atg11–Atg1 binding is probably independent of Atg13 phosphorylation status and, therefore, independent of TORC1 activity (Kim et al. 2001; Kamada et al. 2000). Interestingly the selective degradation of peroxisomes, mitochondria and nucleus requires both scaffolds (Farre et al. 2009; Nazarko et al. 2009). It is conceivable that the two scaffold proteins cooperate to organize the PAS during organelle degradation, which is more complex than bulk degradation by its selective nature and by the complex morphology of the cargo. Following the same line, recent evidence suggested that Atg11 could recruit Atg17 to the PAS in the absence of Atg1 and Atg13 through interaction with Atg29–Atg31 (Mao et al. 2013a). However, the significance of this interaction needs to be determined, since Atg17 does not need Atg11 to form the PAS in the presence of Atg1 and Atg13 (Kawamata et al. 2008). Atg11 is not considered as a part of the core autophagic machinery, because it is not needed for general autophagy in the presence of Atg17. However, we decided to include it here, since (1) new data suggest that Atg11 may direct Atg17 to the PAS (Mao et al. 2013a), and (2) in the absence of Atg17, Atg11 is essential for autophagy (Suzuki et al. 2007). As a final point on Atg1 kinase complex, it is interesting to note that the Atg1 kinase (and its regulator, Atg13) is among the first proteins recruited to the PAS, but its kinase activity is not required to recruit other Atg proteins. Instead, the Atg1 kinase activity seems to be required at the later steps in the pathway (Cheong et al. 2008). Probably, the first Atg1 function is related to its ability to bind highly curved vesicles. It has been suggested that two Atg1 complexes tether two vesicles together and help prime them for subsequent SNARE-dependent fusion (Ragusa et al. 2012). Such tethering function might be needed at several steps during the pathway, like the PAS organization, phagophore membrane elongation and sealing. The Atg1 kinase activity seems to play a role in disassembly of the PAS, since the mutants of *ATG1* defective in the kinase activity accumulate higher levels of the PAS marker proteins (Cheong et al. 2008).

In higher eukaryotes, the PAS has not been clearly defined, but a structure named the omegasome seems to be at the origin of the phagophore formation. However, it is not known, if this structure is the PAS equivalent or an intermediate during phagophore elongation (Axe et al. 2008). Curiously, the omegasome localizes at the endoplasmic reticulum (ER), whereas the PAS is in close proximity to the vacuole. As we mentioned before, most of the components of the core autophagic machinery are conserved in mammals, but the principal autophagy organizers, the scaffold proteins, Atg11 and Atg17, were missing. Recently, several lines of evidence suggested that RB1CC1/FIP200 is the human counterpart (Hara and Mizushima 2009; Hara et al. 2008). Since RB1CC1 is required for general autophagy, it might function as Atg17 but, surprisingly, it shares sequence similarity with Atg11. Interestingly, in metazoan, the RB1CC1 homolog, EPG-7, seems to be dispensable for the starvation-induced autophagy, similar to Atg11 in *S. cerevisiae* (Lin et al. 2013). Remarkably, EPG-7 binds to the cargo receptor SQSTM1/p62 homolog, SQST-1, resembling receptor-scaffold interaction described in yeast (see Sect. 22.3). Finally, some of the components of the Atg1 kinase complex have been described only in mammals (despite present in some yeasts), such as ATG101 (Hosokawa et al. 2009; Mercer et al. 2009) or only in *S. cerevisiae*, such as Atg29 and Atg31 [Atg28 in *P. pastoris* (Nazarko et al. 2009)].

Another early event in autophagy is the generation of phosphatidylinositol 3-phosphate (PI3P) by the autophagy-specific PI3K complex. Several epistasis studies in yeast and mammals have positioned the Atg1 kinase complex just upstream of the PI3K complex (Itakura and Mizushima 2010; Suzuki et al. 2007). Atg14 is the PI3K-specific component required exclusively for the autophagy-related pathways. It is responsible for the localization of the complex at the PAS (Kametaka et al. 1998; Obara et al. 2006). In agreement with the epistasis studies, Atg14 localization depends on the HORMA domain of the Atg1 regulator, Atg13 (Jao et al. 2013). PI3K complex participates in various membrane trafficking events by phosphorylating PI to PI3P. The latter phospholipid localizes at the specific membrane compartments and plays a role in recruiting proteins and in signal transduction. Vps34 is the only PI3K in yeast and is present in at least two distinct complexes, complex I and II (Kihara et al. 2001). These complexes contain three common components, Vps34, Vps15 and Atg6, and one specific subunit, Atg14 or Vps38/UVRAG, respectively. In yeast, only complex I with Atg14 is required for autophagic pathways and acts early in phagophore formation, whereas in mammals, the complex II with UVRAG is also required for autophagic pathways, but at later stage during autophagosome maturation (Liang et al. 2006; Matsunaga et al. 2009; Takahashi et al. 2007). The autophagic pool of PI3P can be bind by at least five autophagy-related proteins: Atg18/WIPI1-2, Atg20, Atg21, Atg24/SNX4 and Hsv2/WIPI3-4 (Noda et al. 2010b). Atg18, Atg21 and Hsv2 are highly homologous, but specific for different autophagic pathways (Barth et al. 2001; Guan et al. 2001; Stromhaug et al. 2004; Polson et al. 2010). Atg18 is the only PI3P-binding protein essential for all autophagic pathways and, therefore, part of the core machinery. Atg18 belongs to the β-propellers that bind polyphosphoinositides (PROPPINs) and, consequently, its localization at the PAS depends on the PI3K complex I, but also on Atg1 kinase complex, Atg2 and the transmembrane protein, Atg9 (Guan et al. 2001; Obara et al. 2008; Krick et al. 2012). Intriguingly, Atg2 localization depends on Atg18 and it has been proposed that these two proteins constitutively form a complex, which is directed to the PAS by Atg18 (Obara et al. 2008).

The two scaffold proteins recruit additional Atg proteins, other than Atg1 and Atg13, such as Atg9 (He et al. 2006; Sekito et al. 2009), probably via the transport protein particle III (TRAPIII) and the Rab GTPase, Ypt1 (Lipatova et al. 2012;

Wang et al. 2013). Atg9 is the only membrane protein in the core machinery. It is essential for all autophagic pathways and may carry lipids from the peripheral sites to the PAS for phagophore elongation (Reggiori et al. 2004). The nature of the peripheral sites is still under debate in yeast and mammals. Several cellular compartments, including the ER, mitochondria, Golgi, endosomes, and the plasma membrane, have been implicated as the possible source of the autophagosomal membrane (Tooze and Yoshimori 2010). Furthermore, a recent study in mammalian cells has added the ER–mitochondria interface to this already long list (Hamasaki et al. 2013). Atg9 is absent on the completed double-membrane vesicles, suggesting that it is retrieved before the vesicle sealing/completion step. Atg9 interacts with the Atg2–Atg18 complex. In the cells lacking Atg2 or Atg18, Atg9 is confined to the PAS, indicating that Atg2–Atg18 is implicated in the retrograde transport of this protein (Reggiori et al. 2004; Chang et al. 2005). Hence, protein complexes that are implicated in the localization of Atg2–Atg18, such as the Atg1–Atg13 and PI3K complexes, are also required for the cycling of Atg9. In mammalians cells, ATG9 anterograde transport depends on ULK1 and PI3K complex (Young et al. 2006), and its retrieval depends on WIPI2, the mammalian Atg18 homolog, but is independent of ULK1 (Orsi et al. 2012).

The last group of the core autophagic machinery is the UBLP conjugation system, which contains Atg8, the one of the two UBLPs required for autophagosome formation (Mizushima et al. 1998; Ichimura et al. 2000). Atg8 localizes on the phagophore and autophagosome and is a reliable marker to study autophagy progression. Atg8 is first cleaved by a specific cysteine protease, Atg4, to expose its C-terminal glycine (Kirisako et al. 2000). The cleaved Atg8 protein serves as a substrate in the UBLP conjugation reaction catalyzed by Atg7 and Atg3, which correspond to the E1 and E2 enzymes of the ubiquitination system, respectively (Ichimura et al. 2000). Remarkably, unlike other UBLP conjugation systems, Atg8 is conjugated to the lipid, phosphatidylethanolamine (PE), which anchors Atg8 to the membrane (Ichimura et al. 2000; Kirisako et al. 2000). The second UBLP conjugate, Atg12–Atg5, acts as the E3 ligase to conjugate Atg8 to PE. Atg12–Atg5 forms a complex with Atg16 that recruits Atg8 to the PAS (Hanada et al. 2007; Ichimura et al. 2000). The precise function of Atg8–PE is not clear yet, but it is required for phagophore expansion, regulates the autophagosome size and interacts with the selective autophagy receptors (Mizushima et al. 2001; Xie et al. 2008). Several controversial evidences suggest that Atg8–PE can act as a fusogen by promoting membrane tethering and fusion (Nair et al. 2011; Nakatogawa et al. 2007; Weidberg et al. 2011). The conjugation of Atg8 to PE is reversible and Atg4 also functions as a deconjugation enzyme resulting in the release of soluble Atg8 from the outer autophagosome membrane (Kirisako et al. 2000). The localization of the Atg12–Atg5–Atg16 complex to the PAS relies on the PI3K complex I, but specific factors responsible for its membrane association are unknown. In mammals, WDFY3/ALFY, might be one of them at least during the selective autophagy of protein aggregates, aggrephagy (Clausen et al. 2010; Hocking et al. 2010), since it can bind PI3P via its FYVE domain and the UBLP conjugation system through ATG5 (see Sect. 22.3).

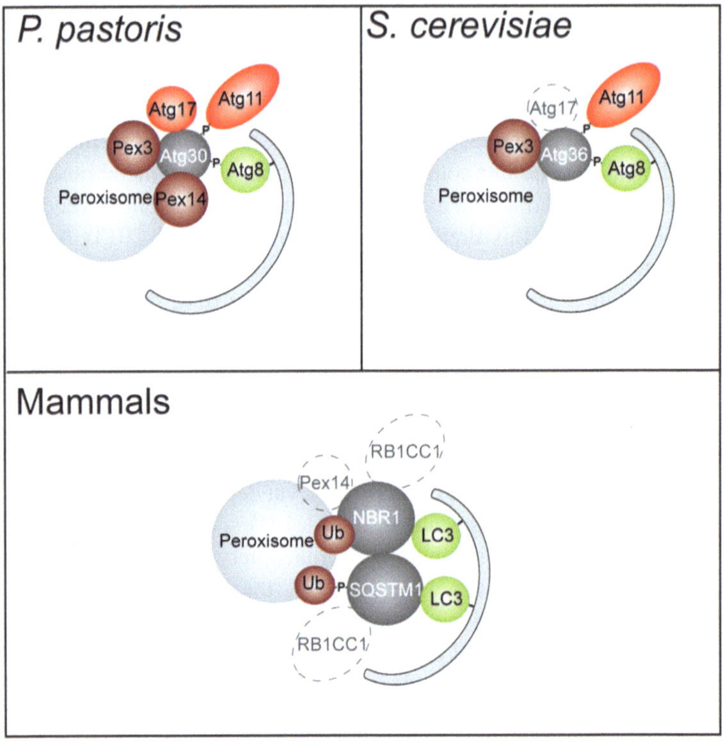

Fig. 22.2 Pexophagic receptor protein complexes in yeasts and mammals. Please note that the illustrated protein–protein interactions do not necessarily happen at the same time

22.3 The Core Autophagic Machinery Is Engaged by the Pexophagic RPCs

To engage autophagic machinery, the peroxisome is first recognized and tagged for degradation by the pexophagy receptor (Mijaljica et al. 2012). Three pexophagy receptors have been identified so far: *P. pastoris* Atg30, *S. cerevisiae* Atg36 and mammalian NBR1 with its co-receptor, SQSTM1 (Fig. 22.2 and Table 22.2) (Farre et al. 2008; Motley et al. 2012; Deosaran et al. 2013). Atg30 binds the peroxisomal membrane protein (PMP), Pex3, and is also associated with Pex14. The levels of Atg30 are diminished and the protein is localized to the nucleus in *pex3* mutant. Peroxisome remnants in either *pex3* or *pex14* mutant are not the subjects of Atg30 overexpression-induced pexophagy underlying the importance of both ligands in the recruitment of the pexophagy receptor in *P. pastoris* (Farre et al. 2008). Although Pex3 is also required to recruit Atg36 to peroxisomes, peroxisome remnants in the *pex14* mutant of *S. cerevisiae* are degraded normally suggesting that Pex14 might not be the pexophagy ligand in baker's yeast (Motley et al. 2012). The studies in *H. polymorpha* and Chinese hamster ovary (CHO) cells indicate that

Table 22.2 Components of the pexophagic receptor protein complexes

Organism	Ligand	Receptor	Scaffold	Atg8 family protein	References
P. pastoris	Pex3, Pex14	Atg30	Atg11, Atg17	Atg8	Farre et al. (2008, 2013)
S. cerevisiae	Pex3	Atg36	Atg11, Atg17[a]	Atg8	Motley et al. (2012)
Mammals	Ub, PEX14[a]	NBR1	RB1CC1[a]	LC3	Deosaran et al. (2013), Kim et al. (2008)
	Ub	SQSTM1 (co-receptor)	RB1CC1[a]	LC3	
	PEX14	–	–	LC3	Hara-Kuge and Fujiki (2008)

[a]Hypothetical

the N-terminus of PEX14 might be involved in pexophagy (Bellu et al. 2001; Hara-Kuge and Fujiki 2008) through the association with the lipidated form of LC3 (Hara-Kuge and Fujiki 2008). However, the pexophagy receptor, which could bridge this interaction, was not described in these model systems yet.

The human pexophagy receptor, NBR1, recognizes peroxisomes via the combination of membrane- and ubiquitin binding by its J and UBA domains, respectively (Deosaran et al. 2013). It is consistent with the finding that attaching ubiquitin to the PMPs induces pexophagy (Kim et al. 2008). However, UBA domain by itself is not sufficient to localize the protein to peroxisomes. Therefore, SQSTM1, which does not have its own J domain, must rely on the interaction with NBR1 (via PB1 domains of two proteins) for recruitment to peroxisomes (Deosaran et al. 2013). Interestingly, knockdown of *PEX14* affects binding of NBR1 to peroxisomes and pexophagy, but the role of *PEX14* is not clear yet (Deosaran et al. 2013). Since J domain can bind various membranes and UBA domain can bind different proteins tagged with ubiquitin, it is possible that NBR1 might play a more general role in the selective autophagy of membrane organelles. Similarly, NBR1, together with SQSTM1, plays a general role in the selective autophagy of ubiquitinated protein aggregates by acting as aggrephagy receptors (Kirkin et al. 2009). However, NBR1 also serves as the midbophagy receptor via direct binding to the midbody protein, CEP55 (Kuo et al. 2011). Therefore, it would not be surprising, if NBR1 directly bound PEX3 and/or PEX14 to provide a higher level of selectivity for mammalian pexophagy. In summary, the peroxisome is recognized by the pexophagy receptor through a direct binding to at least one ligand (PMP or ubiquitin) in the peroxisomal membrane.

At the next step, the pexophagy receptor binds a scaffold protein that brings the core autophagic machinery to the peroxisomal membrane. Atg30 binds two autophagic scaffolds, Atg11 and Atg17, while Atg36 binds Atg11 and the interaction with Atg17 remains to be tested (Farre et al. 2008, 2013; Motley et al. 2012). Since both Atg11 and Atg17 are required for pexophagy in yeasts (Kim et al. 2001; Cheong et al. 2005; Nazarko et al. 2009), Atg36 might also rely on two scaffolds.

The inability of Atg30 to bind Atg11 leads to as strong pexophagy defect, as the lack of Atg11 (Nazarko et al. 2009). However, in contrast to *atg11* mutant, the Atg11-binding site mutant of Atg30 has normal phagophore formation (Farre et al. 2013), suggesting that direct Atg30–Atg17 (Farre et al. 2008) and indirect Atg17–Atg11 (Mao et al. 2013a) interactions might be enough to support phagophore elongation, but not enough to accomplish the delivery of the peroxisome to the vacuole. Therefore, the Atg30–Atg11 interaction might have an additional role in pexophagy beyond phagophore formation. The Atg30–Atg17 binding is mediated by two coiled-coil (CC) motifs in Atg30 and CC2 motif in Atg17 (Farre et al. 2008), but the exact role of this interaction in pexophagy is unknown.

It is not clear how NBR1 and/or SQSTM1 recruit the core autophagic machinery to mammalian peroxisomes. However, several lines of evidence suggest that at least the Atg11-dependent mechanism is conserved from yeast to metazoan: (1) the homolog of SQSTM1 in worms, SQST-1, binds the scaffold protein, EPG-7, which is similar to yeast Atg11 and human RB1CC1; EPG-7 brings multiple ATG proteins to the SQST-1 aggregates (Lin et al. 2013), (2) knockout of mouse RB1CC1 prevents co-localization of SQSTM1 with the early core autophagic protein, ULK1 (Itakura and Mizushima 2011). Therefore, like Atg11 in yeasts, EPG-7/RB1CC1 could act as a scaffold during pexophagy in metazoa. Another protein that could serve this role is WDFY3. During aggrephagy, SQSTM1 brings WDFY3 to the ubiquitinated protein aggregates (Clausen et al. 2010; Hocking et al. 2010), where WDFY3 recruits the PI3P-containing membranes and ATG12–ATG5–ATG16L complex (Filimonenko et al. 2010; Simonsen et al. 2004). However, it remains to be tested, if RB1CC1 and WDFY3 are involved in pexophagy and how are they recruited to the pexophagic RPCs in mammals? Self-oligomerizations of NBR1 (via its CC1 domain) and/or SQSTM1 (via its PB1 domain) might play a role in the recruitment of scaffold protein(s), since without self-oligomerization (1) SQSTM1 does not co-localize with ULK1 in mouse cells (Itakura and Mizushima 2011) and (2) NBR1 is not able to induce pexophagy in human cells (Deosaran et al. 2013). Taken together, the pexophagy receptor interacts with at least one scaffold protein, which brings the core autophagic machinery to the receptor-tagged peroxisomes.

Besides interaction with the scaffold, both yeast and mammalian pexophagy receptors bind the UBLP, Atg8/LC3, through the Atg8-family interacting motif (AIM) or LC3-interacting region (LIR) (Farre et al. 2013; Motley et al. 2012; Waters et al. 2009; Kirkin et al. 2009; Pankiv et al. 2007; Komatsu et al. 2007; Ichimura et al. 2008; Noda et al. 2008). In Atg30 and Atg36, the AIM precedes the Atg11-binding site and, like the latter, is regulated by phosphorylation (see Sect. 22.4). The same molecule of the pexophagy receptor must interact with both Atg8 and Atg11, but only with one protein at a time due to the close proximity of Atg8- and Atg11-binding sites and competition of Atg8 and Atg11 for binding to the receptor (Farre et al. 2013). In contrast to Atg8 protein, which is essential for pexophagy (Hutchins et al. 1999; Mukaiyama et al. 2002), the Atg30–Atg8 or Atg36–Atg8 interaction does not play an essential role. It is rather partially required to facilitate phagophore elongation around large mature peroxisomes (Farre et al. 2013). Alternatively, Atg8 could have another binding partner in the peroxisomal membrane that would partially substitute for the missing interaction of Atg8 with the pexophagy receptor.

NBR1 also directly binds the mammalian Atg8-family proteins (Waters et al. 2009; Kirkin et al. 2009). It has two LIRs, LIR1 and LIR2, which have major and minor roles in the binding of mammalian Atg8 homologs, respectively (Kirkin et al. 2009). Similar to yeasts, the NBR1–LC3 interaction in human cells is only partially required for pexophagy (Deosaran et al. 2013), possibly due to redundancy of the NBR1–LC3 interaction with the SQSTM1–LC3 binding (Pankiv et al. 2007; Komatsu et al. 2007). SQSTM1 has a single LIR responsible for the SQSTM1–LC3 interaction (Pankiv et al. 2007; Ichimura et al. 2008; Noda et al. 2008). Interestingly, NBR1 and SQSTM1 independently co-localize with ULK1 even in the presence of the PI3K inhibitor, wortmannin, which blocks recruitment of proteins downstream of PI3K complex, including LC3. Moreover, co-localization of SQSTM1 and ULK1 does not depend on the ability of SQSTM1 to interact with LC3 (Itakura and Mizushima 2011). These results suggest that the interactions of NBR1 and SQSTM1 with scaffold protein(s) might precede their interactions with the Atg8-family protein(s).

In summary, the pexophagy receptor forms the pexophagic RPC by recognizing the ligand in the peroxisomal membrane and recruiting the autophagic scaffold protein. Scaffold protein engages the core autophagic machinery, including Atg8/LC3, which localizes to the phagophore membrane and binds the pexophagy receptor to ensure the selective sequestration of the peroxisome from the cytosol.

22.4 Assembly of the Pexophagic RPCs Is Regulated by Phosphorylation

Peroxisome number and size can change rapidly in response to environmental and/or physiological clues. Generally, the conditions that require peroxisome metabolism will stimulate proliferation of these organelles and a switch to the conditions that do not require peroxisome metabolism will induce their degradation by pexophagy. In yeasts, pexophagy was studied mostly following the degradation of superfluous peroxisomes. In mammalian cells, most of the research was focused on basal pexophagy. Mammalian peroxisomes have a short half-life (1.5–2 days) (Nordgren et al. 2013; Poole et al. 1969) and their turnover depends on basal pexophagy (Deosaran et al. 2013; Nordgren et al. 2013). However, both kinds of pexophagy, basal and induced, may exist in both yeast and mammals (Iwata et al. 2006), though the signaling for these pexophagy modes might be different.

In yeast, the pexophagy receptors are present in a non-active hypophosphorylated form in the peroxisome proliferation conditions, and the change to a carbon source (with or without nitrogen), which utilization does not require peroxisome metabolism, will induce their activation via hyperphosphorylation (Farre et al. 2008, 2013; Motley et al. 2012). The signals triggering the phosphorylation are unknown, as well as the kinase(s). However, some evidences from the *S. cerevisiae* mitophagy receptor, Atg32, which shares motif organization and phosphoregulation with the pexophagy receptors, Atg30 and Atg36, suggest the mitogen-activated protein kinase (MAPK) pathways might be involved (Aoki et al. 2011). Mitophagy and pexophagy, but not autophagy, are regulated by the MAPK pathways (Manjithaya et al. 2010a; Mao et al. 2011). The high osmolarity glycerol (HOG) pathway regulates mitophagy and the cell wall integrity (CWI) pathway is essential for both mitophagy and pexophagy. Mitophagy and pexophagy require the complete CWI signal transduction pathway, including the MAPK kinase kinase kinase (PKC1), MAPK kinase kinase (Bck1), MAPK kinase (Mkk1/Mkk2), MAPK (Slt2), and even the cell surface sensors, Wsc1 and Mid2. It has been suggested that the changes in carbon source, which induce pexophagy, might influence the cell wall composition that is directly or indirectly sensed by the cell wall sensor to activate Slt2 (Manjithaya et al. 2010a). The Slt2 activity is needed for the formation of the PAS (Mao et al. 2011) and the phenotype of *slt2* cells resembles the phenotype of the Atg11-binding site phosphomutant of Atg32 (Aoki et al. 2011). Interestingly, Atg32 phosphorylation defect was observed in the mutants of the HOG pathway. However, Atg32 is not a direct target of the Hog1p kinase (Aoki et al. 2011). Therefore, the cross talk between the CWI and HOG pathways was suggested (Bermejo et al. 2008). It is possible that the CWI pathway is involved in the phosphorylation of both mitophagy and pexophagy receptors.

Atg30 and Atg36 are highly phosphorylated proteins and at least two phosphorylation events are involved in pexophagy, one in the Atg8- and another in Atg11-binding sites (Fig. 22.3). Simultaneous mutation of both phosphosites has the phenotype of *atg30* or *atg36* deletion strain (Farre et al. 2013). Most of the AIMs/LIRs have a conserved short motif (W/F/Y)xx(L/I/V) surrounded by at least one (often more than one) proximal acidic residue (Johansen and Lamark 2011). Structural studies show that acidic residues form electrostatic interactions with the N-terminal arm of the Atg8 homologs (Noda et al. 2010a). The pexophagy receptors, Atg30 and Atg36, are phosphorylated two residues upstream of the AIM (Farre et al. 2013). The phosphorylated residue mimics an acidic residue and is essential for the binding of the pexophagy receptor to Atg8. In many cases, AIM/LIR does not need a phosphoregulation and is able to bind constitutively to Atg8/LC3, but the AIM in Atg30 does not follow the strict AIM consensus and contains phenylalanine (F) in the fourth position. Interestingly, Atg36 and OPTN, the autophagic receptor for clearance of cytosolic Salmonella, also contain the F, but in the first position of the AIM/LIR, and both are also regulated by phosphorylation (Farre et al. 2013; Wild et al. 2011). The presence of F in the AIM/LIR is probably the reason for a low or no affinity of the non-phosphorylated pexophagy receptors to Atg8, as suggested for OPTN (Rogov et al. 2013).

The interaction of the pexophagy receptors with Atg11 is also mediated by phosphorylation in the conserved motif, (I/V)LSx(S/T) (Farre et al. 2008, 2013). In addition, Atg8- and Atg11-binding sites are juxtaposed (0–33 aa apart) excluding simultaneous binding with the receptor. Notably, such binding sites organization

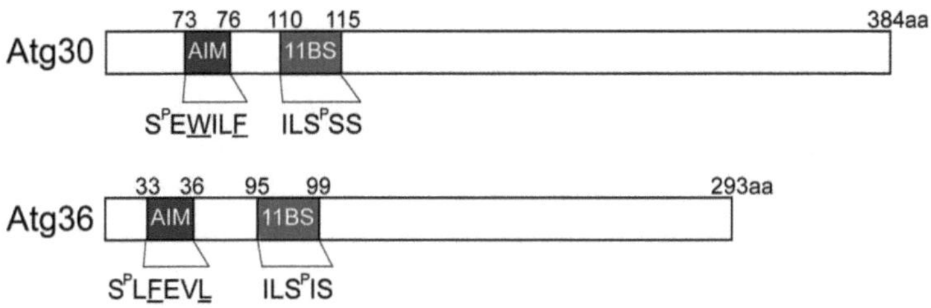

Fig. 22.3 Yeast pexophagy receptors are regulated by phosphorylation of the Atg8-family interacting motif (AIM) and the Atg11-binding site (11BS). The conserved residues of the WxxL-like sequences are *underlined*

and phosphoregulation is also true for the mitophagy receptor, Atg32 (Aoki et al. 2011; Farre et al. 2013). The regulation by phosphorylation is an essential event in activation of the pexophagy and mitophagy receptors. It might also dictate the order of the receptor interactions with the scaffold and Atg8/LC3 proteins.

The mammalian pexophagy co-receptor, SQSTM1, is also phosphorylated in the proximity of the LIR (6 residues upstream), but this phosphorylation is not needed to bind LC3 (Pankiv et al. 2007). Although NBR1 has two serines upstream of the LIR1, the phosphorylation of NBR1 in the proximity of the LIR has not been described yet. Interestingly, SQSTM1 is phosphorylated in the UBA domain by the casein kinase 2 (CK2) and this modification increases its affinity to polyubiquitin chains (Matsumoto et al. 2011). This is different from the yeast pexophagy receptors, which do not require any known modifications to interact with their ligands in the peroxisomal membrane.

22.5 The Role of Cytoskeleton in Pexophagy

The cytoskeleton is a network of microtubules, microfilaments, and intermediate filaments, the protein polymers that play multiple roles in cell shape and movement, compartmentalization and trafficking pathways, including autophagy. Microtubules and microfilaments form dynamic protein fibers that associate with molecular motors, which can transport the components of autophagic machinery, as well as the completed autophagosomes (Monastyrska et al. 2009). Several stages of pexophagy might require an engagement or disengagement of microtubules and microfilaments, such as the peroxisome clustering and the pexophagic PAS organization (see below). Cytoskeleton might also play an important role in the pexophagosome motility, maturation and fusion with the lysosome, as it does in the autophagosome motility, maturation, and fusion (Mackeh et al. 2013; Tumbarello et al. 2013), but these aspects of pexophagy await future studies.

Overexpression of NBR1, but not SQSTM1, in mammalian cells induces clustering of peroxisomes, which precedes their sequestration and delivery to the lysosomes. Although recognition of peroxisomes by and self-oligomerization of NBR1 are required for peroxisome clustering (Deosaran et al. 2013), nothing is known about the involvement of cytoskeletal elements in this process. We suggest that motor proteins (kinesin, dynein, and the dynein activator complex, dynactin), which move peroxisomes along the microtubule tracks and support the uniform distribution of peroxisomes in mammalian and Drosophila cells (Kural et al. 2005; Schrader et al. 2000), might need to be disengaged during pexophagy, since depolymerization of microtubules with nocodazole has a similar peroxisome clustering effect, as overexpression of NBR1 (Wiemer et al. 1997; Schrader et al. 2000). Interestingly, a potential ligand of NBR1 in the peroxisomal membrane, PEX14 (N-terminus of which was implicated in pexophagy), directly binds β-tubulin by its N-terminal domain in human cells (Bharti et al. 2011). Therefore, by binding to PEX14, NBR1 could displace peroxisomes from microtubules to prevent peroxisome motility and promote peroxisome clustering necessary for pexophagy. Despite actin microfilaments and not microtubules are used for peroxisome movements in yeast cells, a similar disengagement of peroxisomes from the cytoskeleton might be required for pexophagy. In this case, Atg30 and Atg36 would need to displace the peroxisome retention factor, Inp1, or the class V myosin motor, Myo2, from the pexophagy ligand, Pex3 (Munck et al. 2009; Chang et al. 2009). Otherwise, the peroxisome would be anchored to the cell cortex via the Pex3p–Inp1p binding or propelled to the bud via the Pex3–Myo2 interaction and would not be available for sequestration from the cytosol by the phagophore membrane.

Actin and actin-related proteins, Act1 and Arp2, also play an important positive role in yeast pexophagy most probably at a stage of the pexophagic PAS organization (Reggiori et al. 2005; Monastyrska et al. 2008). In the Cvt pathway, Act1 and Arp2 are both required for the Atg11-dependent delivery of Atg9 from the peripheral compartment to the Cvt-specific PAS (Reggiori et al. 2005; He et al. 2006; Monastyrska et al. 2008). Interestingly, the Atg11–Atg9 binding, which is essential for the anterograde trafficking of Atg9 to the PAS during the Cvt pathway in growth conditions, is not essential for both pexophagy and autophagy pathways in starvation conditions (He et al. 2006; He and Klionsky 2007). It might be explained by the fact that Atg17 recruits most of Atg9 to the PAS in starvation conditions (Sekito et al. 2009). However, the latter process is independent of actin cytoskeleton (Reggiori et al. 2005). Since actin is essential for pexophagy, but not autophagy (Reggiori et al. 2005), actin might play a more important role in the pexophagic PAS organization than recruitment of Atg9. It is possible that Act1 and Arp2 deliver Atg11, an important pexophagic, but not autophagic scaffold (Nazarko et al. 2009), to the pexophagy-specific PAS. Such scenario is in agreement with: (1) the role of Act1 in the delivery of Atg11 to the Cvt-specific PAS (He et al. 2006) and (2) the plausible role of Atg11 in bridging the Arp2–Atg9 interaction (Monastyrska et al. 2008). Interestingly, actin polymerization and activation of the actin-associated motor protein, myosin II, are required for the starvation-induced autophagosome formation in Drosophila, CHO, and human cells (Tang et al. 2011; Aguilera et al. 2012). In Drosophila, myosin II is activated via phosphorylation of the myosin light chain kinase-like protein, Sqa, by Atg1. Moreover, human myosin II and Sqa homolog, DAPK3/ZIPK, regulate ATG9 trafficking (Tang et al. 2011). In addition, actin fibers in CHO cells co-localize with: (1) the components of class III PI3K complex I, ATG14 and BECN1, and (2) the PI3P-containing membranes and the PI3P-binding protein, ZFYVE1/DFCP1 (Aguilera et al. 2012). Therefore, the actomyosin network might play a conserved role in the pexophagic PAS organization from yeast to human.

Intriguingly, microtubules and their molecular motors also play an important role in autophagosome formation in the mammalian, but not in yeast, cells (Mackeh et al. 2013; Kirisako et al. 1999). However, their role in pexophagosome formation was not established yet.

22.6 Pexophagy of Large Peroxisomes

The size of peroxisomes in yeasts depends on the nature of peroxisome inducer and induction time. For example, the *P. pastoris* peroxisomes induced by methanol for 15 h have an average area of 0.4 μm^2 and are 31 times larger than those induced by methanol for only 0.5 h or 5–7 times larger than peroxisomes induced for 15 h by oleate or primary amines (Nazarko et al. 2009). The larger size of a mature methanol-induced peroxisome requires synthesis of the larger phagophore to sequester it from the cytosol. At least three mechanisms ensure that the pexophagy-specific PAS meets the demand. The first mechanism was partially discussed above and is related to the ability of the pexophagy receptor, Atg30, to engage two pexophagic scaffold proteins, Atg11 and Atg17. Although Atg11 and Atg17 are redundant in the pexophagy of medium peroxisomes, Atg11 becomes essential for the pexophagy of large peroxisomes. Interestingly, the essential role of Atg11 can be narrowed down to the Atg30–Atg11 interaction (Nazarko et al. 2009). Since phagophore formation is unaffected in the Atg11-binding site mutant of Atg30 (Farre et al. 2013), it remains a mystery what essential activity is contributed by Atg11 for the pexophagy of large peroxisomes? However, this unknown activity of Atg11 might regulate pexophagy in a dose-dependent manner, since the levels of Atg11 increase concomitant with the size of peroxisomes on methanol, similar to the levels of most peroxins (Nazarko et al. 2009).

The second mechanism is mediated by the UDP-glucose:sterol glucosyl-transferase, Atg26/Ugt51, that converts sterol to sterol glucoside (Warnecke et al. 1999). Atg26 is not required for pexophagy in *S. cerevisiae* or *Yarrowia lipolytica*, which can form only small-to-medium peroxisomes (Stasyk et al. 2003; Cao and Klionsky 2007; Nazarko et al. 2007). However, Atg26 is required for pexophagy in the species, which are able to form large peroxisomes, like the methylotrophic yeast *P. pastoris* (Oku et al. 2003; Stasyk et al. 2003; Nazarko et al. 2007), or the species that depend on efficient pexophagy for their pathogenicity, like the cucumber anthracnose fungus *Colletotrichum orbiculare* (Asakura et al. 2009). But even in *P. pastoris*, Atg26 is critical only for the pexophagy of large peroxisomes, which were induced either by methanol in wild-type cells or by oleate in *pex11* mutant cells deficient in peroxisome division (Nazarko et al. 2009).

Both the GRAM and truncated GRAM-PH domains of *P. pastoris* Atg26 bind specifically to PI4P, which is generated mainly by the PI4K, Pik1, and delivered to the pexophagy-specific PAS presumably from the Golgi (Yamashita et al. 2006). Although the Golgi-localized Pik1 is required for Atg9 trafficking from the Golgi to the PAS in *S. cerevisiae* (Wang et al. 2012), the pexophagy-specific PAS formation is not affected in the *pik1* mutant of *P. pastoris*, since multiple post-Atg9 proteins, like Atg5, Atg8 and Atg16 (just not Atg26) are able to reach the PAS (Yamashita et al. 2006). Once at the PAS, the sterol glucosyltransferase activity of Atg26 drives the elongation of this dot-like structure into the cup-shaped phagophore, but the mechanism of this process is unknown (Yamashita et al. 2006).

The third mechanism to degrade large peroxisomes is to divide them into the smaller organelles for efficient sequestration by the phagophore. Indeed, the proteins required for peroxisome fission, Dnm1 and Pex11, are also required for both glucose-induced and basal pexophagy in *H. polymorpha* (Manivannan et al. 2013). Moreover, consistent with the redundant role of the *S. cerevisiae* dynamin-related proteins, Dnm1 and Vps1, in peroxisome fission (Kuravi et al. 2006; Motley et al. 2008), they also play a redundant role in pexophagy (Manivannan et al. 2013). Additionally, coupling of peroxisome fission and pexophagy is also utilized by *H. polymorpha* cells to remove large intra-peroxisomal protein aggregates. First, the protein aggregate is separated from the mother peroxisome by the Dnm1- and Pex11-dependent asymmetric fission. Then, the smaller aggregate-containing peroxisome is degraded by the Atg1- and Atg11-dependent pexophagy (Manivannan et al. 2013). Interestingly, peroxisomes share their fission machinery with mitochondria (Kuravi et al. 2006; Motley et al. 2008). Similar to peroxisome fission and pexophagy, mitochondria fission is important for the progression of mitophagy. Dnm1 is recruited to the degrading mitochondria via its interaction with Atg11 and this interaction is crucial for mitophagy (Mao et al. 2013b). If similar mechanism operated during pexophagy, it would explain an essential role of Atg11 in the pexophagy of large peroxisomes (see above). The role of Dnm1- and Fis1-mediated mitochondria fission in mitophagy is conserved from yeast to mammals (Twig et al. 2008; Tanaka et al. 2010) suggesting that peroxisome fission in mammals might potentially contribute to efficient pexophagy, like it does in yeasts.

22.7 Fusion Machineries Involved in Pexophagy

Membrane fusion is required at multiple stages of the autophagy-related pathways, but not all the mechanisms are clearly defined. Autophagy (and most likely other autophagic pathways) requires vesicles fusion proteins for the PAS and autophagosome biogenesis. The work in yeast suggests that the exocytic SNAREs, Sso1/Sso2 and Sec9, and endosomal SNARE, Tlg2, are required for correct organization of the Atg9-positive tubulovesicular clusters (a likely intermediate in membrane delivery to the PAS; Nair et al. 2011). Similarly, in mammalian cells, SNAREs are also required for the organization of the PAS-like structure (Moreau et al. 2011). Once completed, the double-membrane vesicle is transported to the vacuole and the outer membrane of the vesicle fuses with the vacuolar membrane. This fusion step is thought to proceed in an essentially identical fashion, as the endocytic fusion. This process was studied for autophagy and several other autophagy-related pathways, including pexophagy. It depends on (1) the SNARE proteins, Vam3, Vti1, and Vam7; (2) the NSF, Sec18; (3) the α-SNAP, Sec17; (4) the Rab GTPase, Ypt7, and its GEF complex, Ccz1-Mon1; and (5) the class C Vps/HOPS tethering complex (Darsow et al. 1997; Fischer von Mollard and Stevens 1999; Ishihara et al. 2001; Kirisako et al. 1999; Polupanov et al. 2011; Rieder and Emr 1997; Sato et al. 1998, 2000; Scott et al. 1997; Seals et al. 2000; Stevens et al. 2005; Wang et al. 2002). Likewise, the SNARE proteins, Rab7, and the HOPS complex have been implicated in mammalian autophagy (Fader et al. 2009; Furuta et al. 2010; Gutierrez et al. 2004; Itakura et al. 2012; Jager et al. 2004; Liang et al. 2008; Pankiv et al. 2010).

The fusion of autophagosome with the vacuole/lysosome is modulated by PI3P. In mammalians cells; this fusion is positively regulated by the UVRAG–VSP34–BECN1 PI3K complex and negatively regulated by the Rubicon–UVRAG–VSP34–BECN1 PI3K complex (Itakura et al. 2008; Liang et al. 2006; Matsunaga et al. 2009; Takahashi et al. 2007; Zhong et al. 2009). In yeast, the PI3P phosphatase, Ymr1, promotes the hydrolysis of PI3P to PI and is required for the dissociation of Atg proteins from the mature autophagosome prior to the fusion with the vacuole (Cebollero et al. 2012). At the same time, PI3P is required to direct the soluble SNARE, Vam7, to the vacuolar membrane (Boeddinghaus et al. 2002). Moreover, the pexophagy and Cvt pathways need additional PI3P-binding proteins, Atg20/Snx41 and Atg24/Snx4, for fusion of the selective autophagy vesicles with the vacuole (Ano et al. 2005; Deng et al. 2012; Nice et al. 2002). Atg24 localizes to the vertex ring of the contact area between the pexophagosome and the vacuole membrane. It is not required for phagophore elongation and *atg24* mutant cells accumulate mature pexophagosomes in the cytosol. Paradoxically, both the lack of the PI3P-binding protein (*atg24* mutant) and the excess of the PI3P (*ymr1* mutant) create the same fusion defect in the pexophagy and autophagy/Cvt pathways, respectively, suggesting that a fine-tuned regulation of the PI3P levels is essential at the late stage of all autophagy-related pathways. The role of Atg20 and Atg24 in fusion is not clear, but these proteins also function in a distinct pathway that mediates the sorting of proteins from the endosomes (Bonifacino and Rojas 2006). Atg24 binds to Atg20 and Snx42, and forms the complex, which is directly involved in the sorting of the SNARE, Snc1, from the endosomes to the Golgi (Hettema et al. 2003). Similarly, the mammalian Atg24 homolog, SNX4, also functions in the recycling of the cell surface proteins (Traer et al. 2007). Interestingly, Atg20 and Atg24 interact with the scaffold proteins, Atg11 and Atg17, respectively (Yorimitsu and Klionsky 2005; Nice et al. 2002), suggesting that these PI3P-binding proteins might be involved in the recycling of Atg proteins before the fusion of the pexophagosome with the vacuole membrane.

Acknowledgments T.Y.N. and J.C.F were supported by the NIH grants DK094843 and GM069373, respectively.

References

Aguilera MO, Beron W, Colombo MI (2012) The actin cytoskeleton participates in the early events of autophagosome formation upon starvation induced autophagy. Autophagy 8 (11):1590–1603

Ano Y, Hattori T, Oku M, Mukaiyama H, Baba M, Ohsumi Y, Kato N, Sakai Y (2005) A sorting nexin PpAtg24 regulates vacuolar membrane dynamics during pexophagy via binding to phosphatidylinositol-3-phosphate. Mol Biol Cell 16(2):446–457

Aoki Y, Kanki T, Hirota Y, Kurihara Y, Saigusa T, Uchiumi T, Kang D (2011) Phosphorylation of Serine 114 on Atg32 mediates mitophagy. Mol Biol Cell 22(17):3206–3217

Asakura M, Ninomiya S, Sugimoto M, Oku M, Yamashita S, Okuno T, Sakai Y, Takano Y (2009) Atg26-mediated pexophagy is required for host invasion by the plant pathogenic fungus Colletotrichum orbiculare. Plant Cell 21(4):1291–1304

Axe EL, Walker SA, Manifava M, Chandra P, Roderick HL, Habermann A, Griffiths G, Ktistakis NT (2008) Autophagosome formation from membrane compartments enriched in phosphatidylinositol 3-phosphate and dynamically connected to the endoplasmic reticulum. J Cell Biol 182(4):685–701

Barth H, Meiling-Wesse K, Epple UD, Thumm M (2001) Autophagy and the cytoplasm to vacuole targeting pathway both require Aut10p. FEBS Lett 508(1):23–28

Bellu AR, Kiel JA (2003) Selective degradation of peroxisomes in yeasts. Microsc Res Tech 61 (2):161–170

Bellu AR, Komori M, van der Klei IJ, Kiel JA, Veenhuis M (2001) Peroxisome biogenesis and selective degradation converge at Pex14p. J Biol Chem 276(48):44570–44574

Bermejo C, Rodriguez E, Garcia R, Rodriguez-Pena JM, Rodriguez de la Concepcion ML, Rivas C, Arias P, Nombela C, Posas F, Arroyo J (2008) The sequential activation of the yeast HOG and SLT2 pathways is required for cell survival to cell wall stress. Mol Biol Cell 19 (3):1113–1124

Bharti P, Schliebs W, Schievelbusch T, Neuhaus A, David C, Kock K, Herrmann C, Meyer HE, Wiese S, Warscheid B, Theiss C, Erdmann R (2011) PEX14 is required for microtubule-based peroxisome motility in human cells. J Cell Sci 124(Pt 10):1759–1768

Boeddinghaus C, Merz AJ, Laage R, Ungermann C (2002) A cycle of Vam7p release from and PtdIns 3-P-dependent rebinding to the yeast vacuole is required for homotypic vacuole fusion. J Cell Biol 157(1):79–89

Bonifacino JS, Rojas R (2006) Retrograde transport from endosomes to the trans-Golgi network. Nat Rev Mol Cell Biol 7(8):568–579

Cao Y, Klionsky DJ (2007) Atg26 is not involved in autophagy-related pathways in *S. cerevisiae*. Autophagy 3(1):17–20

Cebollero E, van der Vaart A, Zhao M, Rieter E, Klionsky DJ, Helms JB, Reggiori F (2012) Phosphatidylinositol-3-phosphate clearance plays a key role in autophagosome completion. Curr Biol 22(17):1545–1553

Chang T, Schroder LA, Thomson JM, Klocman AS, Tomasini AJ, Stromhaug PE, Dunn WA Jr (2005) PpATG9 encodes a novel membrane protein that traffics to vacuolar membranes, which sequester peroxisomes during pexophagy in *P. pastoris*. Mol Biol Cell 16(10):4941–4953

Chang J, Mast FD, Fagarasanu A, Rachubinski DA, Eitzen GA, Dacks JB, Rachubinski RA (2009) Pex3 peroxisome biogenesis proteins function in peroxisome inheritance as class V myosin receptors. J Cell Biol 187(2):233–246

Cheong H, Yorimitsu T, Reggiori F, Legakis JE, Wang CW, Klionsky DJ (2005) Atg17 regulates the magnitude of the autophagic response. Mol Biol Cell 16(7):3438–3453

Cheong H, Nair U, Geng J, Klionsky DJ (2008) The Atg1 kinase complex is involved in the regulation of protein recruitment to initiate sequestering vesicle formation for nonspecific autophagy in *S. cerevisiae*. Mol Biol Cell 19(2):668–681

Clausen TH, Lamark T, Isakson P, Finley K, Larsen KB, Brech A, Overvatn A, Stenmark H, Bjorkoy G, Simonsen A, Johansen T (2010) p62/SQSTM1 and ALFY interact to facilitate the formation of p62 bodies/ALIS and their degradation by autophagy. Autophagy 6(3):330–344

Darsow T, Rieder SE, Emr SD (1997) A multispecificity syntaxin homologue, Vam3p, essential for autophagic and biosynthetic protein transport to the vacuole. J Cell Biol 138(3):517–529

De Duve C, Wattiaux R (1966) Functions of lysosomes. Annu Rev Physiol 28:435–492

Deng YZ, Qu Z, He Y, Naqvi NI (2012) Sorting nexin Snx41 is essential for conidiation and mediates glutathione-based antioxidant defense during invasive growth in Magnaporthe oryzae. Autophagy 8(7):1058–1070

Deosaran E, Larsen KB, Hua R, Sargent G, Wang Y, Kim S, Lamark T, Jauregui M, Law K, Lippincott-Schwartz J, Brech A, Johansen T, Kim PK (2013) NBR1 acts as an autophagy receptor for peroxisomes. J Cell Sci 126(Pt 4):939–952

Dunn WA Jr, Cregg JM, Kiel JA, van der Klei IJ, Oku M, Sakai Y, Sibirny AA, Stasyk OV, Veenhuis M (2005) Pexophagy: the selective autophagy of peroxisomes. Autophagy 1(2):75–83

Fader CM, Sanchez DG, Mestre MB, Colombo MI (2009) TI-VAMP/VAMP7 and VAMP3/cellubrevin: two v-SNARE proteins involved in specific steps of the autophagy/multivesicular body pathways. Biochim Biophys Acta 1793(12):1901–1916

Farre JC, Manjithaya R, Mathewson RD, Subramani S (2008) PpAtg30 tags peroxisomes for turnover by selective autophagy. Dev Cell 14(3):365–376

Farre JC, Krick R, Subramani S, Thumm M (2009) Turnover of organelles by autophagy in yeast. Curr Opin Cell Biol 21(4):522–530

Farre JC, Mathewson RD, Manjithaya R, Subramani S (2010) Roles of *P. pastoris* Uvrag in vacuolar protein sorting and the phosphatidylinositol 3-kinase complex in phagophore elongation in autophagy pathways. Autophagy 6(1):86–99

Farre JC, Burkenroad A, Burnett SF, Subramani S (2013) Phosphorylation of mitophagy and pexophagy receptors coordinates their interaction with Atg8 and Atg11. EMBO Rep 14 (5):441–449

Filimonenko M, Isakson P, Finley KD, Anderson M, Jeong H, Melia TJ, Bartlett BJ, Myers KM, Birkeland HC, Lamark T, Krainc D, Brech A, Stenmark H, Simonsen A, Yamamoto A (2010) The selective macroautophagic degradation of aggregated proteins requires the PI3P-binding protein Alfy. Mol Cell 38(2):265–279

Fischer von Mollard G, Stevens TH (1999) The *S. cerevisiae* v-SNARE Vti1p is required for multiple membrane transport pathways to the vacuole. Mol Biol Cell 10(6):1719–1732

Furuta N, Fujita N, Noda T, Yoshimori T, Amano A (2010) Combinational soluble N-ethylmaleimide-sensitive factor attachment protein receptor proteins VAMP8 and Vti1b mediate fusion of antimicrobial and canonical autophagosomes with lysosomes. Mol Biol Cell 21 (6):1001–1010

Geng J, Baba M, Nair U, Klionsky DJ (2008) Quantitative analysis of autophagy-related protein stoichiometry by fluorescence microscopy. J Cell Biol 182(1):129–140

Guan J, Stromhaug PE, George MD, Habibzadegah-Tari P, Bevan A, Dunn WA Jr, Klionsky DJ (2001) Cvt18/Gsa12 is required for cytoplasm-to-vacuole transport, pexophagy, and autophagy in *S. cerevisiae* and *P. pastoris*. Mol Biol Cell 12(12):3821–3838

Gutierrez MG, Munafo DB, Beron W, Colombo MI (2004) Rab7 is required for the normal progression of the autophagic pathway in mammalian cells. J Cell Sci 117(Pt 13):2687–2697

Hamasaki M, Furuta N, Matsuda A, Nezu A, Yamamoto A, Fujita N, Oomori H, Noda T, Haraguchi T, Hiraoka Y, Amano A, Yoshimori T (2013) Autophagosomes form at ER-mitochondria contact sites. Nature 495(7441):389–393

Hanada T, Noda NN, Satomi Y, Ichimura Y, Fujioka Y, Takao T, Inagaki F, Ohsumi Y (2007) The Atg12-Atg5 conjugate has a novel E3-like activity for protein lipidation in autophagy. J Biol Chem 282(52):37298–37302

Hara T, Mizushima N (2009) Role of ULK-FIP200 complex in mammalian autophagy: FIP200, a counterpart of yeast Atg17? Autophagy 5(1):85–87

Hara T, Takamura A, Kishi C, Iemura S, Natsume T, Guan JL, Mizushima N (2008) FIP200, a ULK-interacting protein, is required for autophagosome formation in mammalian cells. J Cell Biol 181(3):497–510

Hara-Kuge S, Fujiki Y (2008) The peroxin Pex14p is involved in LC3-dependent degradation of mammalian peroxisomes. Exp Cell Res 314(19):3531–3541

He C, Klionsky DJ (2007) Atg9 trafficking in autophagy-related pathways. Autophagy 3(3):271–274

He C, Song H, Yorimitsu T, Monastyrska I, Yen WL, Legakis JE, Klionsky DJ (2006) Recruitment of Atg9 to the preautophagosomal structure by Atg11 is essential for selective autophagy in budding yeast. J Cell Biol 175(6):925–935

Hettema EH, Lewis MJ, Black MW, Pelham HR (2003) Retromer and the sorting nexins Snx4/41/42 mediate distinct retrieval pathways from yeast endosomes. EMBO J 22(3):548–557

Hocking LJ, Mellis DJ, McCabe PS, Helfrich MH, Rogers MJ (2010) Functional interaction between sequestosome-1/p62 and autophagy-linked FYVE-containing protein WDFY3 in human osteoclasts. Biochem Biophys Res Commun 402(3):543–548

Hosokawa N, Sasaki T, Iemura S, Natsume T, Hara T, Mizushima N (2009) Atg101, a novel mammalian autophagy protein interacting with Atg13. Autophagy 5(7):973–979

Hutchins MU, Veenhuis M, Klionsky DJ (1999) Peroxisome degradation in *S. cerevisiae* is dependent on machinery of macroautophagy and the Cvt pathway. J Cell Sci 112 (Pt 22):4079–4087

Ichimura Y, Kirisako T, Takao T, Satomi Y, Shimonishi Y, Ishihara N, Mizushima N, Tanida I, Kominami E, Ohsumi M, Noda T, Ohsumi Y (2000) A ubiquitin-like system mediates protein lipidation. Nature 408(6811):488–492

Ichimura Y, Kumanomidou T, Sou YS, Mizushima T, Ezaki J, Ueno T, Kominami E, Yamane T, Tanaka K, Komatsu M (2008) Structural basis for sorting mechanism of p62 in selective autophagy. J Biol Chem 283(33):22847–22857

Ishihara N, Hamasaki M, Yokota S, Suzuki K, Kamada Y, Kihara A, Yoshimori T, Noda T, Ohsumi Y (2001) Autophagosome requires specific early Sec proteins for its formation and NSF/SNARE for vacuolar fusion. Mol Biol Cell 12(11):3690–3702

Itakura E, Mizushima N (2010) Characterization of autophagosome formation site by a hierarchical analysis of mammalian Atg proteins. Autophagy 6(6):764–776

Itakura E, Mizushima N (2011) p62 Targeting to the autophagosome formation site requires self-oligomerization but not LC3 binding. J Cell Biol 192(1):17–27

Itakura E, Kishi C, Inoue K, Mizushima N (2008) Beclin 1 forms two distinct phosphatidylinositol 3-kinase complexes with mammalian Atg14 and UVRAG. Mol Biol Cell 19(12):5360–5372

Itakura E, Kishi-Itakura C, Mizushima N (2012) The hairpin-type tail-anchored SNARE syntaxin 17 targets to autophagosomes for fusion with endosomes/lysosomes. Cell 151(6):1256–1269

Iwata J, Ezaki J, Komatsu M, Yokota S, Ueno T, Tanida I, Chiba T, Tanaka K, Kominami E (2006) Excess peroxisomes are degraded by autophagic machinery in mammals. J Biol Chem 281 (7):4035–4041

Jager S, Bucci C, Tanida I, Ueno T, Kominami E, Saftig P, Eskelinen EL (2004) Role for Rab7 in maturation of late autophagic vacuoles. J Cell Sci 117(Pt 20):4837–4848

Jao CC, Ragusa MJ, Stanley RE, Hurley JH (2013) A HORMA domain in Atg13 mediates PI 3-kinase recruitment in autophagy. Proc Natl Acad Sci USA 110(14):5486–5491

Johansen T, Lamark T (2011) Selective autophagy mediated by autophagic adapter proteins. Autophagy 7(3):279–296

Kabeya Y, Kamada Y, Baba M, Takikawa H, Sasaki M, Ohsumi Y (2005) Atg17 functions in cooperation with Atg1 and Atg13 in yeast autophagy. Mol Biol Cell 16(5):2544–2553

Kamada Y, Funakoshi T, Shintani T, Nagano K, Ohsumi M, Ohsumi Y (2000) Tor-mediated induction of autophagy via an Apg1 protein kinase complex. J Cell Biol 150(6):1507–1513

Kametaka S, Okano T, Ohsumi M, Ohsumi Y (1998) Apg14p and Apg6/Vps30p form a protein complex essential for autophagy in the yeast, *S. cerevisiae*. J Biol Chem 273(35):22284–22291

Kawamata T, Kamada Y, Kabeya Y, Sekito T, Ohsumi Y (2008) Organization of the pre-autophagosomal structure responsible for autophagosome formation. Mol Biol Cell 19 (5):2039–2050

Kiel JA, Rechinger KB, van der Klei IJ, Salomons FA, Titorenko VI, Veenhuis M (1999) The *H. polymorpha* PDD1 gene product, essential for the selective degradation of peroxisomes, is a homologue of *S. cerevisiae* Vps34p. Yeast 15(9):741–754

Kihara A, Noda T, Ishihara N, Ohsumi Y (2001) Two distinct Vps34 phosphatidylinositol 3-kinase complexes function in autophagy and carboxypeptidase Y sorting in *S. cerevisiae.* J Cell Biol 152(3):519–530

Kim J, Kamada Y, Stromhaug PE, Guan J, Hefner-Gravink A, Baba M, Scott SV, Ohsumi Y, Dunn WA Jr, Klionsky DJ (2001) Cvt9/Gsa9 functions in sequestering selective cytosolic cargo destined for the vacuole. J Cell Biol 153(2):381–396

Kim PK, Hailey DW, Mullen RT, Lippincott-Schwartz J (2008) Ubiquitin signals autophagic degradation of cytosolic proteins and peroxisomes. Proc Natl Acad Sci USA 105(52):20567–20574

Kirisako T, Baba M, Ishihara N, Miyazawa K, Ohsumi M, Yoshimori T, Noda T, Ohsumi Y (1999) Formation process of autophagosome is traced with Apg8/Aut7p in yeast. J Cell Biol 147 (2):435–446

Kirisako T, Ichimura Y, Okada H, Kabeya Y, Mizushima N, Yoshimori T, Ohsumi M, Takao T, Noda T, Ohsumi Y (2000) The reversible modification regulates the membrane-binding state of Apg8/Aut7 essential for autophagy and the cytoplasm to vacuole targeting pathway. J Cell Biol 151(2):263–276

Kirkin V, Lamark T, Sou YS, Bjorkoy G, Nunn JL, Bruun JA, Shvets E, McEwan DG, Clausen TH, Wild P, Bilusic I, Theurillat JP, Overvatn A, Ishii T, Elazar Z, Komatsu M, Dikic I, Johansen T (2009) A role for NBR1 in autophagosomal degradation of ubiquitinated substrates. Mol Cell 33(4):505–516

Komatsu M, Waguri S, Koike M, Sou YS, Ueno T, Hara T, Mizushima N, Iwata J, Ezaki J, Murata S, Hamazaki J, Nishito Y, Iemura S, Natsume T, Yanagawa T, Uwayama J, Warabi E, Yoshida H, Ishii T, Kobayashi A, Yamamoto M, Yue Z, Uchiyama Y, Kominami E, Tanaka K (2007) Homeostatic levels of p62 control cytoplasmic inclusion body formation in autophagy-deficient mice. Cell 131(6):1149–1163

Kraft C, Reggiori F, Peter M (2009) Selective types of autophagy in yeast. Biochim Biophys Acta 1793(9):1404–1412

Kraft C, Kijanska M, Kalie E, Siergiejuk E, Lee SS, Semplicio G, Stoffel I, Brezovich A, Verma M, Hansmann I, Ammerer G, Hofmann K, Tooze S, Peter M (2012) Binding of the Atg1/ULK1 kinase to the ubiquitin-like protein Atg8 regulates autophagy. EMBO J 31 (18):3691–3703

Krick R, Busse RA, Scacioc A, Stephan M, Janshoff A, Thumm M, Kuhnel K (2012) Structural and functional characterization of the two phosphoinositide binding sites of PROPPINs, a beta-propeller protein family. Proc Natl Acad Sci USA 109(30):E2042–E2049

Kuo TC, Chen CT, Baron D, Onder TT, Loewer S, Almeida S, Weismann CM, Xu P, Houghton JM, Gao FB, Daley GQ, Doxsey S (2011) Midbody accumulation through evasion of autophagy contributes to cellular reprogramming and tumorigenicity. Nat Cell Biol 13 (10):1214–1223

Kural C, Kim H, Syed S, Goshima G, Gelfand VI, Selvin PR (2005) Kinesin and dynein move a peroxisome in vivo: a tug-of-war or coordinated movement? Science 308(5727):1469–1472

Kuravi K, Nagotu S, Krikken AM, Sjollema K, Deckers M, Erdmann R, Veenhuis M, van der Klei IJ (2006) Dynamin-related proteins Vps1p and Dnm1p control peroxisome abundance in *S. cerevisiae*. J Cell Sci 119(Pt 19):3994–4001

Liang C, Feng P, Ku B, Dotan I, Canaani D, Oh BH, Jung JU (2006) Autophagic and tumour suppressor activity of a novel Beclin1-binding protein UVRAG. Nat Cell Biol 8(7):688–699

Liang C, Lee JS, Inn KS, Gack MU, Li Q, Roberts EA, Vergne I, Deretic V, Feng P, Akazawa C, Jung JU (2008) Beclin1-binding UVRAG targets the class C Vps complex to coordinate autophagosome maturation and endocytic trafficking. Nat Cell Biol 10(7):776–787

Lin L, Yang P, Huang X, Zhang H, Lu Q (2013) The scaffold protein EPG-7 links cargo-receptor complexes with the autophagic assembly machinery. J Cell Biol 201(1):113–129

Lipatova Z, Belogortseva N, Zhang XQ, Kim J, Taussig D, Segev N (2012) Regulation of selective autophagy onset by a Ypt/Rab GTPase module. Proc Natl Acad Sci USA 109(18):6981–6986

Mackeh R, Perdiz D, Lorin S, Codogno P, Pous C (2013) Autophagy and microtubules—new story, old players. J Cell Sci 126(Pt 5):1071–1080

Manivannan S, de Boer R, Veenhuis M, van der Klei IJ (2013) Lumenal peroxisomal protein aggregates are removed by concerted fission and autophagy events. Autophagy 9(7):1044–1056

Manjithaya R, Jain S, Farre JC, Subramani S (2010a) A yeast MAPK cascade regulates pexophagy but not other autophagy pathways. J Cell Biol 189(2):303–310

Manjithaya R, Nazarko TY, Farre JC, Subramani S (2010b) Molecular mechanism and physiological role of pexophagy. FEBS Lett 584(7):1367–1373

Mao K, Wang K, Zhao M, Xu T, Klionsky DJ (2011) Two MAPK-signaling pathways are required for mitophagy in *S. cerevisiae*. J Cell Biol 193(4):755–767

Mao K, Chew LH, Inoue-Aono Y, Cheong H, Nair U, Popelka H, Yip CK, Klionsky DJ (2013a) Atg29 phosphorylation regulates coordination of the Atg17-Atg31-Atg29 complex with the Atg11 scaffold during autophagy initiation. Proc Natl Acad Sci USA 110(31):E2875–E2884

Mao K, Wang K, Liu X, Klionsky DJ (2013b) The scaffold protein Atg11 recruits fission machinery to drive selective mitochondria degradation by autophagy. Dev Cell 26(1):9–18

Matsumoto G, Wada K, Okuno M, Kurosawa M, Nukina N (2011) Serine 403 phosphorylation of p62/SQSTM1 regulates selective autophagic clearance of ubiquitinated proteins. Mol Cell 44 (2):279–289

Matsunaga K, Saitoh T, Tabata K, Omori H, Satoh T, Kurotori N, Maejima I, Shirahama-Noda K, Ichimura T, Isobe T, Akira S, Noda T, Yoshimori T (2009) Two Beclin 1-binding proteins, Atg14L and Rubicon, reciprocally regulate autophagy at different stages. Nat Cell Biol 11 (4):385–396

Mercer CA, Kaliappan A, Dennis PB (2009) A novel, human Atg13 binding protein, Atg101, interacts with ULK1 and is essential for macroautophagy. Autophagy 5(5):649–662

Mijaljica D, Nazarko TY, Brumell JH, Huang WP, Komatsu M, Prescott M, Simonsen A, Yamamoto A, Zhang H, Klionsky DJ, Devenish RJ (2012) Receptor protein complexes are in control of autophagy. Autophagy 8(11):1701–1705

Mizushima N, Noda T, Yoshimori T, Tanaka Y, Ishii T, George MD, Klionsky DJ, Ohsumi M, Ohsumi Y (1998) A protein conjugation system essential for autophagy. Nature 395 (6700):395–398

Mizushima N, Yamamoto A, Hatano M, Kobayashi Y, Kabeya Y, Suzuki K, Tokuhisa T, Ohsumi Y, Yoshimori T (2001) Dissection of autophagosome formation using Apg5-deficient mouse embryonic stem cells. J Cell Biol 152(4):657–668

Monastyrska I, He C, Geng J, Hoppe AD, Li Z, Klionsky DJ (2008) Arp2 links autophagic machinery with the actin cytoskeleton. Mol Biol Cell 19(5):1962–1975

Monastyrska I, Rieter E, Klionsky DJ, Reggiori F (2009) Multiple roles of the cytoskeleton in autophagy. Biol Rev Camb Philos Soc 84(3):431–448

Moreau K, Ravikumar B, Renna M, Puri C, Rubinsztein DC (2011) Autophagosome precursor maturation requires homotypic fusion. Cell 146(2):303–317

Motley AM, Ward GP, Hettema EH (2008) Dnm1p-dependent peroxisome fission requires Caf4p, Mdv1p and Fis1p. J Cell Sci 121(Pt 10):1633–1640

Motley AM, Nuttall JM, Hettema EH (2012) Pex3-anchored Atg36 tags peroxisomes for degradation in *S. cerevisiae*. EMBO J 31(13):2852–2868

Mukaiyama H, Oku M, Baba M, Samizo T, Hammond AT, Glick BS, Kato N, Sakai Y (2002) Paz2 and 13 other PAZ gene products regulate vacuolar engulfment of peroxisomes during micropexophagy. Genes Cells 7(1):75–90

Munck JM, Motley AM, Nuttall JM, Hettema EH (2009) A dual function for Pex3p in peroxisome formation and inheritance. J Cell Biol 187(4):463–471

Nair U, Jotwani A, Geng J, Gammoh N, Richerson D, Yen WL, Griffith J, Nag S, Wang K, Moss T, Baba M, McNew JA, Jiang X, Reggiori F, Melia TJ, Klionsky DJ (2011) SNARE proteins are required for macroautophagy. Cell 146(2):290–302

Nakatogawa H, Ichimura Y, Ohsumi Y (2007) Atg8, a ubiquitin-like protein required for autophagosome formation, mediates membrane tethering and hemifusion. Cell 130(1):165–178

Nakatogawa H, Suzuki K, Kamada Y, Ohsumi Y (2009) Dynamics and diversity in autophagy mechanisms: lessons from yeast. Nat Rev Mol Cell Biol 10(7):458–467

Nazarko TY, Polupanov AS, Manjithaya RR, Subramani S, Sibirny AA (2007) The requirement of sterol glucoside for pexophagy in yeast is dependent on the species and nature of peroxisome inducers. Mol Biol Cell 18(1):106–118

Nazarko TY, Farre JC, Subramani S (2009) Peroxisome size provides insights into the function of autophagy-related proteins. Mol Biol Cell 20(17):3828–3839

Nice DC, Sato TK, Stromhaug PE, Emr SD, Klionsky DJ (2002) Cooperative binding of the cytoplasm to vacuole targeting pathway proteins, Cvt13 and Cvt20, to phosphatidylinositol 3-phosphate at the pre-autophagosomal structure is required for selective autophagy. J Biol Chem 277(33):30198–30207

Noda NN, Kumeta H, Nakatogawa H, Satoo K, Adachi W, Ishii J, Fujioka Y, Ohsumi Y, Inagaki F (2008) Structural basis of target recognition by Atg8/LC3 during selective autophagy. Genes Cells 13(12):1211–1218

Noda NN, Ohsumi Y, Inagaki F (2010a) Atg8-family interacting motif crucial for selective autophagy. FEBS Lett 584(7):1379–1385

Noda T, Matsunaga K, Taguchi-Atarashi N, Yoshimori T (2010b) Regulation of membrane biogenesis in autophagy via PI3P dynamics. Semin Cell Dev Biol 21(7):671–676

Nordgren M, Wang B, Apanasets O, Fransen M (2013) Peroxisome degradation in mammals: mechanisms of action, recent advances, and perspectives. Front Physiol 4:145

Obara K, Sekito T, Ohsumi Y (2006) Assortment of phosphatidylinositol 3-kinase complexes–Atg14p directs association of complex I to the pre-autophagosomal structure in *S. cerevisiae*. Mol Biol Cell 17(4):1527–1539

Obara K, Sekito T, Niimi K, Ohsumi Y (2008) The Atg18-Atg2 complex is recruited to autophagic membranes via phosphatidylinositol 3-phosphate and exerts an essential function. J Biol Chem 283(35):23972–23980

Oku M, Warnecke D, Noda T, Muller F, Heinz E, Mukaiyama H, Kato N, Sakai Y (2003) Peroxisome degradation requires catalytically active sterol glucosyltransferase with a GRAM domain. EMBO J 22(13):3231–3241

Orsi A, Razi M, Dooley HC, Robinson D, Weston AE, Collinson LM, Tooze SA (2012) Dynamic and transient interactions of Atg9 with autophagosomes, but not membrane integration, are required for autophagy. Mol Biol Cell 23(10):1860–1873

Pankiv S, Clausen TH, Lamark T, Brech A, Bruun JA, Outzen H, Overvatn A, Bjorkoy G, Johansen T (2007) p62/SQSTM1 binds directly to Atg8/LC3 to facilitate degradation of ubiquitinated protein aggregates by autophagy. J Biol Chem 282(33):24131–24145

Pankiv S, Alemu EA, Brech A, Bruun JA, Lamark T, Overvatn A, Bjorkoy G, Johansen T (2010) FYCO1 is a Rab7 effector that binds to LC3 and PI3P to mediate microtubule plus end-directed vesicle transport. J Cell Biol 188(2):253–269

Polson HE, de Lartigue J, Rigden DJ, Reedijk M, Urbe S, Clague MJ, Tooze SA (2010) Mammalian Atg18 (WIPI2) localizes to omegasome-anchored phagophores and positively regulates LC3 lipidation. Autophagy 6(4):506–522

Polupanov AS, Nazarko VY, Sibirny AA (2011) CCZ1, MON1 and YPT7 genes are involved in pexophagy, the Cvt pathway and non-specific macroautophagy in the methylotrophic yeast *P. pastoris*. Cell Biol Int 35(4):311–319

Poole B, Leighton F, De Duve C (1969) The synthesis and turnover of rat liver peroxisomes. II. Turnover of peroxisome proteins. J Cell Biol 41(2):536–546

Ragusa MJ, Stanley RE, Hurley JH (2012) Architecture of the Atg17 complex as a scaffold for autophagosome biogenesis. Cell 151(7):1501–1512

Reggiori F, Tucker KA, Stromhaug PE, Klionsky DJ (2004) The Atg1-Atg13 complex regulates Atg9 and Atg23 retrieval transport from the pre-autophagosomal structure. Dev Cell 6(1):79–90

Reggiori F, Monastyrska I, Shintani T, Klionsky DJ (2005) The actin cytoskeleton is required for selective types of autophagy, but not nonspecific autophagy, in the yeast *S. cerevisiae*. Mol Biol Cell 16(12):5843–5856

Rieder SE, Emr SD (1997) A novel RING finger protein complex essential for a late step in protein transport to the yeast vacuole. Mol Biol Cell 8(11):2307–2327

Rogov VV, Suzuki H, Fiskin E, Wild P, Kniss A, Rozenknop A, Kato R, Kawasaki M, McEwan DG, Lohr F, Guntert P, Dikic I, Wakatsuki S, Dotsch V (2013) Structural basis for phosphorylation-triggered autophagic clearance of Salmonella. Biochem J 454(3):459–466

Sato TK, Darsow T, Emr SD (1998) Vam7p, a SNAP-25-like molecule, and Vam3p, a syntaxin homolog, function together in yeast vacuolar protein trafficking. Mol Cell Biol 18(9):5308–5319

Sato TK, Rehling P, Peterson MR, Emr SD (2000) Class C Vps protein complex regulates vacuolar SNARE pairing and is required for vesicle docking/fusion. Mol Cell 6(3):661–671

Schrader M, King SJ, Stroh TA, Schroer TA (2000) Real time imaging reveals a peroxisomal reticulum in living cells. J Cell Sci 113(Pt 20):3663–3671

Scott SV, Baba M, Ohsumi Y, Klionsky DJ (1997) Aminopeptidase I is targeted to the vacuole by a nonclassical vesicular mechanism. J Cell Biol 138(1):37–44

Seals DF, Eitzen G, Margolis N, Wickner WT, Price A (2000) A Ypt/Rab effector complex containing the Sec1 homolog Vps33p is required for homotypic vacuole fusion. Proc Natl Acad Sci USA 97(17):9402–9407

Sekito T, Kawamata T, Ichikawa R, Suzuki K, Ohsumi Y (2009) Atg17 recruits Atg9 to organize the pre-autophagosomal structure. Genes Cells 14(5):525–538

Shintani T, Klionsky DJ (2004) Cargo proteins facilitate the formation of transport vesicles in the cytoplasm to vacuole targeting pathway. J Biol Chem 279(29):29889–29894

Simonsen A, Birkeland HC, Gillooly DJ, Mizushima N, Kuma A, Yoshimori T, Slagsvold T, Brech A, Stenmark H (2004) Alfy, a novel FYVE-domain-containing protein associated with protein granules and autophagic membranes. J Cell Sci 117(Pt 18):4239–4251

Stasyk OV, van der Klei IJ, Bellu AR, Shen S, Kiel JA, Cregg JM, Veenhuis M (1999) A *P. pastoris* VPS15 homologue is required in selective peroxisome autophagy. Curr Genet 36(5):262–269

Stasyk OV, Nazarko TY, Stasyk OG, Krasovska OS, Warnecke D, Nicaud JM, Cregg JM, Sibirny AA (2003) Sterol glucosyltransferases have different functional roles in *P. pastoris* and Yarrowia lipolytica. Cell Biol Int 27(11):947–952

Stasyk OV, Stasyk OG, Mathewson RD, Farre JC, Nazarko VY, Krasovska OS, Subramani S, Cregg JM, Sibirny AA (2006) Atg28, a novel coiled-coil protein involved in autophagic degradation of peroxisomes in the methylotrophic yeast *P. pastoris*. Autophagy 2(1):30–38

Stevens P, Monastyrska I, Leao-Helder AN, van der Klei IJ, Veenhuis M, Kiel JA (2005) *H. polymorpha* Vam7p is required for macropexophagy. FEMS Yeast Res 5(11):985–997

Stromhaug PE, Bevan A, Dunn WA Jr (2001) GSA11 encodes a unique 208-kDa protein required for pexophagy and autophagy in *P. pastoris*. J Biol Chem 276(45):42422–42435

Stromhaug PE, Reggiori F, Guan J, Wang CW, Klionsky DJ (2004) Atg21 is a phosphoinositide binding protein required for efficient lipidation and localization of Atg8 during uptake of aminopeptidase I by selective autophagy. Mol Biol Cell 15(8):3553–3566

Suzuki K, Kubota Y, Sekito T, Ohsumi Y (2007) Hierarchy of Atg proteins in pre-autophagosomal structure organization. Genes Cells 12(2):209–218

Takahashi Y, Coppola D, Matsushita N, Cualing HD, Sun M, Sato Y, Liang C, Jung JU, Cheng JQ, Mule JJ, Pledger WJ, Wang HG (2007) Bif-1 interacts with Beclin 1 through UVRAG and regulates autophagy and tumorigenesis. Nat Cell Biol 9(10):1142–1151

Tanaka A, Cleland MM, Xu S, Narendra DP, Suen DF, Karbowski M, Youle RJ (2010) Proteasome and p97 mediate mitophagy and degradation of mitofusins induced by Parkin. J Cell Biol 191(7):1367–1380

Tang HW, Wang YB, Wang SL, Wu MH, Lin SY, Chen GC (2011) Atg1-mediated myosin II activation regulates autophagosome formation during starvation-induced autophagy. EMBO J 30(4):636–651

Tooze SA, Yoshimori T (2010) The origin of the autophagosomal membrane. Nat Cell Biol 12 (9):831–835

Traer CJ, Rutherford AC, Palmer KJ, Wassmer T, Oakley J, Attar N, Carlton JG, Kremerskothen J, Stephens DJ, Cullen PJ (2007) SNX4 coordinates endosomal sorting of TfnR with dynein-mediated transport into the endocytic recycling compartment. Nat Cell Biol 9(12):1370–1380

Tsukada M, Ohsumi Y (1993) Isolation and characterization of autophagy-defective mutants of *S. cerevisiae*. FEBS Lett 333(1–2):169–174

Tumbarello DA, Kendrick-Jones J, Buss F (2013) Myosin VI and its cargo adaptors—linking endocytosis and autophagy. J Cell Sci 126(Pt 12):2561–2570

Twig G, Elorza A, Molina AJ, Mohamed H, Wikstrom JD, Walzer G, Stiles L, Haigh SE, Katz S, Las G, Alroy J, Wu M, Py BF, Yuan J, Deeney JT, Corkey BE, Shirihai OS (2008) Fission and selective fusion govern mitochondrial segregation and elimination by autophagy. EMBO J 27 (2):433–446

Wang CW, Stromhaug PE, Shima J, Klionsky DJ (2002) The Ccz1-Mon1 protein complex is required for the late step of multiple vacuole delivery pathways. J Biol Chem 277(49):47917–47927

Wang K, Yang Z, Liu X, Mao K, Nair U, Klionsky DJ (2012) Phosphatidylinositol 4-kinases are required for autophagic membrane trafficking. J Biol Chem 287(45):37964–37972

Wang J, Menon S, Yamasaki A, Chou HT, Walz T, Jiang Y, Ferro-Novick S (2013) Ypt1 recruits the Atg1 kinase to the preautophagosomal structure. Proc Natl Acad Sci USA 110(24):9800–9805

Warnecke D, Erdmann R, Fahl A, Hube B, Muller F, Zank T, Zahringer U, Heinz E (1999) Cloning and functional expression of UGT genes encoding sterol glucosyltransferases from *S. cerevisiae*, Candida albicans, Pichia pastoris, and Dictyostelium discoideum. J Biol Chem 274(19):13048–13059

Waters S, Marchbank K, Solomon E, Whitehouse C, Gautel M (2009) Interactions with LC3 and polyubiquitin chains link nbr1 to autophagic protein turnover. FEBS Lett 583(12):1846–1852

Weidberg H, Shpilka T, Shvets E, Abada A, Shimron F, Elazar Z (2011) LC3 and GATE-16 N termini mediate membrane fusion processes required for autophagosome biogenesis. Dev Cell 20(4):444–454

Wiemer EA, Wenzel T, Deerinck TJ, Ellisman MH, Subramani S (1997) Visualization of the peroxisomal compartment in living mammalian cells: dynamic behavior and association with microtubules. J Cell Biol 136(1):71–80

Wild P, Farhan H, McEwan DG, Wagner S, Rogov VV, Brady NR, Richter B, Korac J, Waidmann O, Choudhary C, Dotsch V, Bumann D, Dikic I (2011) Phosphorylation of the autophagy receptor optineurin restricts Salmonella growth. Science 333(6039):228–233

Wong PM, Puente C, Ganley IG, Jiang X (2013) The ULK1 complex: sensing nutrient signals for autophagy activation. Autophagy 9(2):124–137

Xie Z, Nair U, Klionsky DJ (2008) Atg8 controls phagophore expansion during autophagosome formation. Mol Biol Cell 19(8):3290–3298

Yamashita S, Oku M, Wasada Y, Ano Y, Sakai Y (2006) PI4P-signaling pathway for the synthesis of a nascent membrane structure in selective autophagy. J Cell Biol 173(5):709–717

Yang Z, Klionsky DJ (2009) An overview of the molecular mechanism of autophagy. Curr Top Microbiol Immunol 335:1–32

Yorimitsu T, Klionsky DJ (2005) Atg11 links cargo to the vesicle-forming machinery in the cytoplasm to vacuole targeting pathway. Mol Biol Cell 16(4):1593–1605

Young AR, Chan EY, Hu XW, Kochl R, Crawshaw SG, High S, Hailey DW, Lippincott-Schwartz-J, Tooze SA (2006) Starvation and ULK1-dependent cycling of mammalian Atg9 between the TGN and endosomes. J Cell Sci 119(Pt 18):3888–3900

Yuan W, Stromhaug PE, Dunn WA Jr (1999) Glucose-induced autophagy of peroxisomes in *P. pastoris* requires a unique E1-like protein. Mol Biol Cell 10(5):1353–1366

Zhong Y, Wang QJ, Li X, Yan Y, Backer JM, Chait BT, Heintz N, Yue Z (2009) Distinct regulation of autophagic activity by Atg14L and Rubicon associated with Beclin 1-phosphatidylinositol-3-kinase complex. Nat Cell Biol 11(4):468–476

Pexophagy Sensing and Signaling in the Methylotrophic Yeasts

23

Andriy A. Sibirny

Abstract

Methylotrophic yeasts are unique eukaryotic organisms capable of utilizing the one-carbon toxic substrate, methanol. During methylotrophic growth, peroxisomes occupy 30–80 % of the cellular volume. A shift of methanol-grown cells to media with the alternative carbon sources, glucose or ethanol, induces massive peroxisome degradation. In *Pichia pastoris*, two morphologically distinct events have been observed namely, macro- and microautophagy. In other species, macroautophagy was mostly noted under massive peroxisome degradation. It was found that genes involved in non-specific autophagy (most of them known as *ATG* genes) also participate in carbon source-induced pexophagy. Many *ATG* genes have been discovered using methylotrophic yeasts models, mainly in *P. pastoris*, due to convenient and easy methods to monitor pexophagy. However, the mechanisms of glucose and ethanol sensing and signaling, which initiate the subsequent events of micro- and macroautophagy are poorly understood. Similarly, the nature of the low-molecular-weight effectors, derivatives of glucose and ethanol, which induce pexophagy, has not been identified.

P. pastoris possesses a single glucose sensor, Gss1p, ortholog of the *S. cerevisiae* high- and low-affinity glucose sensors Snf3p and Rgt2p, respectively. The Gss1 protein participates in glucose sensing and is involved in pexophagy and glucose catabolite repression. In the yeast *P. pastoris*, the orthologs of GPCR signaling proteins, Gpr1p and Gpa2p, do not participate in pexophagy upon glucose signaling. In this yeast species, the α-subunit of phosphofructokinase Pfk1p and the mitogen-activated protein kinase (MAPK) Slt2p are involved in glucose signaling of pexophagy. Ethanol signaling studied

A.A. Sibirny (✉)
Institute of Cell Biology, NAS of Ukraine, Lviv 79005, Ukraine

University of Rzeszow, Zelwerowicza 4, Rzeszow 35-601, Poland
e-mail: sibirny@cellbiol.lviv.ua

C. Brocard and A. Hartig (eds.), *Molecular Machines Involved in Peroxisome Biogenesis and Maintenance*, DOI 10.1007/978-3-7091-1788-0_23,

in *P. methanolica* mutants defective in ethanol catabolism suggested that glyoxylic acid is the most probable substance, which triggers pexophagy.

Keywords
Methylotrophic yeasts • Pexophagy • *ATG* genes • Glucose sensing • Gss1 glucose sensor

23.1 Peroxisomes and Their Functions

Peroxisomes are ubiquitous organelles present in virtually all eukaryotic cells, with the exception of *Archaezoa* (Michels et al. 2005; Brown and Baker 2008). Peroxisomes also known as microbodies (specific types of these organelles are also named glyoxysomes and glycosomes) are organelles surrounded by a single membrane. Their size is in average 0.5–1.5 μm. They do not contain DNA, RNA, or ribosomes. Cells can contain from 1 to 2 peroxisomes (e.g., yeast growing on glucose) to several hundred peroxisomes such as mammalian cells (Till et al. 2012). According to their name, peroxisomes harbor H_2O_2-producing oxidases and catalase decomposing the latter compound. However, peroxisomes are extremely versatile organelles sometimes specializing in different functions. Interestingly a peculiarity of the peroxisomal catabolizing enzymes is their inability to produce ATP, which differ they from the catabolic enzymes located in mitochondria (Mast et al. 2010). Liver peroxisomes contain enzymes enable to metabolize very long-chain fatty acids and enzymes of the β-oxidation of fatty acids and bile acid precursors. With enzymes involved in the oxidation of ingested ethanol to acetaldehyde they account for as much as 50 % of the total metabolism of ethanol. In yeasts, peroxisomes are responsible for the initial steps of methanol and fatty acid catabolism (Veenhuis et al. 1983; van der Klei et al. 2006). In addition to catabolic functions, peroxisomes fulfill biosynthetic roles. In mammals, peroxisomes harbor enzymes participating in synthesis of bile acids, cholesterol, and plasmalogens (Wanders et al. 2010). In fungi, peroxisomes are involved in lysine biosynthesis in yeasts and penicillin biosynthesis in mycelial fungi (Schrader and Fahimi 2008; Aksam et al. 2009; Meijer et al. 2010). In parasitic protozoa of the genera *Trypanosoma* and *Leishmania*, glycolytic enzymes occur in a specialized peroxisome known as glycosome (Michels et al. 2006). The compartmentalization of glycolytic enzymes is essential for the survival of these organsims. Woronin bodies, which serve to plug the septal pores in mycelial fungi, are also specialized peroxisomes. Plant peroxisomes are classified into three groups: glyoxisomes, leaf peroxisomes, and unspecialized peroxisomes. There are approximately 50 proteins in animal and fungal peroxisomes and approximately 100 proteins in plant peroxisomes. Proteomic and genetic studies continuously reveal new functions for peroxisomes (Michels et al. 2005; Lanyon-Hogg et al. 2010).

Defects in peroxisome structure and functions underlie many human diseases. The so-called Zellweger syndrome is the best known peroxisomal inheritable disease. Patients with Zellweger syndrome fall into four groups with different defects in protein transport to peroxisomes (see Chap. 2 and Subramani 1997). Peroxisome damage has serious consequences and is often fatal, causing death within the first year of life (Steinberg et al. 2006). It is of interest that similar genetic defects were observed in yeasts with defective peroxisome biogenesis (*pex* mutants; Subramani 1998). In summary, peroxisomes are surprisingly dynamic organelles, whose dimensions, number in the cell, and protein content change in response to environmental changes. Peroxisome biogenesis is accompanied by other processes, including signal transduction (Saleem et al. 2008), chromatin modification (Wan et al. 2009), reorganization of transcriptional networks (Smith et al. 2002), and changes in the peroxisomal proteome (Marelli et al. 2004; Saleem et al. 2006).

Yeasts provide a convenient model to study the mechanisms of peroxisome biogenesis because a cell transfer from glucose-containing medium to a medium containing oleate, or methanol in the case of methylotrophic yeasts, induces the synthesis of peroxisomal enzymes and growth and division of peroxisomes. Peroxisomes may occupy up to 80 % of the cellular volume in cells growing in the presence of methanol (Veenhuis et al. 1983; Sibirny 2012). When cells growing in the presence of methanol or oleate are transferred to glucose-containing medium or from methanol- to ethanol-containing medium, the transfer is rapidly followed by autophagic degradation of the majority of peroxisomes (pexophagy), while one peroxisome somehow avoids this degradation in a way that is still unclear (Dunn et al. 2005). Methylotrophic yeasts appear to be one of the most convenient model for studying peroxisome biogenesis and degradation due to the ability to induce massive propagation of peroxisomes with methanol. As a result, the one or two small peroxisomes present in cells during growth in glucose are substituted by numerous large peroxisomes, which occupy nearly 30 % of cellular volume during batch cultivation and up to 80 % of cellular volume under continuous cultivation with a low dilution rate in methanol as sole carbon and energy source (Veenhuis et al. 1983). An inverse shift of methanol-grown cells to glucose (or ethanol) causes major reorganization of intracellular structures leading to degradation of the majority of peroxisomes due to autophagic process. Consequently, 30–80 % of cell volume is degraded. Methods of classical and molecular genetics are well developed for several species of methylotrophic yeasts (Cregg et al. 2008; Faber et al. 1995; Lahtchev et al. 2002; Tolstorukov and Cregg 2007) and genome sequence of several type strains are publicly available (http://www.genome.jp/kegg-bin/show_organism?org=ppa or http://www.pichiagenome.org/ for *P. pastoris* and http://genomeportal.jgi-psf.org/Hanpo2/Hanpo2.info.html for *H. polymorpha*). Thus, the use of these available tools permits the mechanistic description of events, which occur during autophagic degradation of peroxisomes (pexophagy) in methylotrophic yeasts.

23.2 General Characteristics of Pexophagy

Pexophagy is the special type of autophagy, namely, autophagic vacuolar (lysosomal) degradation of peroxisomes. Autophagy could be involved in degradation of cytosolic components and some of cellular organelles (e.g., mitochondria, nucleus, endoplasmic reticulum) and macromolecular complexes, e.g., ribosomes. These specific types of autophagy have their own names, such as mitophagy, piecemeal microautophagy of the nucleus (PMN), ER-phagy, and ribophagy (Kiel 2010; Sibirny 2011).

Autophagy of cytosolic cell components, which mostly occurs due to nonspecific process though specific autophagy is proved to be responsible for degradation of fructose-1,6-bisphosphatase and malate dehydrogenase in the baker's yeast *S. cerevisiae*. A shift of the methylotrophic yeasts from methanol to glucose medium leads, in addition to autophagy degradation of peroxisome (pexophagy), also to inactivation of cytosolic enzymes of methanol metabolism (formaldehyde dehydrogenase, formate dehydrogenase, fructose-1,6-bisphosphatase) and FAD synthesis (riboflavin kinase, FAD synthetase; Brooke et al. 1986). Inactivation of fructose-1,6-bisphosphatase in *P. pastoris* apparently occurs due to the degradation process (O. Dmytruk, A. Sibirny, unpublished). However, it is not known to date whether the mentioned enzyme inactivation is a result of the autophagic process.

Pexophagy can occur as part of the nonspecific general autophagy mechanism. Apparently it takes place during yeast propagation in each medium as component of cell constituent maintenance, housekeeping, or turnover mechanism (Aksam et al. 2007). However, during shift from some cultivation conditions to the others, massive pexophagy occurs. The last type of pexophagy is the specific one. Peroxisome degradation in *H. polymorpha*, similarly to mammal cells, could also occur in a process which is unrelated to autophagy, but involves permeabilization of the peroxisomal membrane mediated by 15-lipoxygenase (Baerends et al. 1996; Yokota 2003). Upon lysis, the content of the peroxisome becomes digested by cytosolic proteases. In *H. polymorpha*, such kind of peroxisome disintegration was observed in an engineered yeast strain in which the levels of the peroxin Pex3p had been strongly reduced. This suggests that the loss of certain peroxisomal membrane proteins may destabilize the peroxisomal membrane, resulting in its lysis. The genes involved in pexophagy in methylotrophic yeasts are homologous to those found in *S. cerevisiae* (van Zutphen et al. 2008; Polupanov et al. 2011; Till et al. 2012; Suzuki 2013).

Methods for isolation of the mutants defective in pexophagy have been developed in methylotrophic yeasts. All of them belong to negative selection methods: when a few mutant colonies grow on plates among a huge number of wild-type colonies, mutants are identified directly in colonies using peroxisome enzyme analysis (Stasyk et al. 2008a). Apart from mutagenesis under standard mutagen treatment, insertion mutagenesis using DNA fragments was proposed, which substantially facilitates further cloning of mutant genes (Mukaiyama et al. 2002).

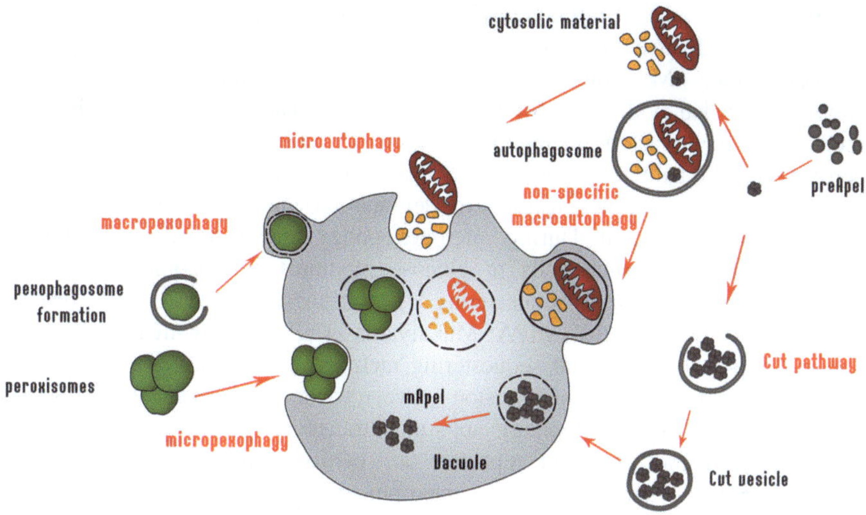

Fig. 23.1 Scheme of main autophagic processes

Most steps and genes involved in specific pexophagy also participate in general (nonspecific) autophagy. The steps of autophagy and participating genes code for the following proteins (Manjithaya et al. 2010; Fig. 23.1):

1. Signaling proteins required for autophagy induction: protein kinase Tor1p, protein kinase A, Sch9p, Tap42p, and phosphatase type 2A.
2. Packaging proteins or organelle transported for degradation (Atg19p, Atg11p, and Atg8p).
3. Formation of preautophagosomal structure (Atg1p, Atg11p, Atg13p, Atg17p, Atg29p, and Atg31p).
4. Vesicle nucleation (Atg6p, Atg9p, and phosphatidylinositol 3-kinase).
5. Vesicle expansion and completion (Atg3p-5p, Atg6p, Atg7p, Atg8p, Atg10p, Atg12p, Atg14p, and Atg16p).
6. Protein retrieval (Atg1p, Atg2p, Atg18p, Atg23p, and Atg27p).
7. Homotypical fusion of isolation membrane (Tlg2p).
8. Transport and heterotypical fusion of autophagosome and vacuoles (v- and t-SNAREs, Ccz1p, Mon1p, and HOPS complex).
9. Intravacuolar vesicle degradation (Atg15p, proteinase A, and proteinase B).

From 36 Atg proteins currently known, only 17 are necessary for all types of autophagy (selective and nonselective), whereas the other 19 are specific: either used in special pathways of selective autophagy or representing species-specific modifications. Specific pexophagy pathways utilize several specific proteins, which do not participate in non-specific autophagy. During pexophagy, the specific PAS is formed, distinct from PAS structures which are produced during other types of selective autophagy. The pexophagy-specific PAS is organized by Atg11p, Atg17p, and Atg30p (Farré et al. 2008; Nazarko et al. 2009).

Studying pexophagy in the methylotrophic yeast *P. pastoris* revealed two morphologically diverse pexophagy processes, called macropexophagy and micropexophagy (Manjithaya et al. 2010; Farre and Subramani 2004; Tuttle and Dunn 1995; Sakai et al. 1998; Sibirny 2011). During macropexophagy initiated by transferring cells from methanol medium to ethanol medium, individual peroxisomes are gathered in double membrane structures called pexophagosomes that merge with vacuoles leading to degradation and repeated usage of the pexophagosomal content. During micropexophagy (occurring after transferring methylotrophically grown cells to glucose medium) peroxisome clusters are engulfed by vacuolar sequestering membranes (VSM) and specific micropexophagy apparatus (MIPA; Mukaiyama et al. 2004), which forms a cap above a cup-shaped vacuolar sequestering membrane surrounding a peroxisome (Farré et al. 2009). Heterotypical fusion between vacuolar sequestering membranes and the specific micropexophagy apparatus transports peroxisomes inside the vacuole for degradation and repeated use of its components. The specific micropexophagy apparatus and pexophagosomes originate from the pre-autophagosomal structure PAS. Glucose and ethanol were shown to be specific inducers of micro- and macropexophagy, correspondingly.

There are several specific proteins involved only in pexophagy and not in the other types of autophagy: Atg24p, Atg26p, Atg28p, and Atg30p. In *P. pastoris* Atg24p localizes to the pexophagosome–vacuole fusion complex during macropexophagy. This protein contains a PtdIns3P-binding module (Ano et al. 2005a). A defect in *Pp*Atg24p blocked pexophagy after pexophagosome formation and before fusion to the vacuole. Apparently *Pp*Atg24p is involved in pexophagosome fusion with the vacuole. Micropexophagy is also impaired in *atg24* mutant cells. *ATG26* encodes an enzyme, sterol glucoside transferase (Oku et al. 2003; Stasyk et al. 2003), which is involved in pexophagy in *P. pastoris* but not in alkane-utilizing yeast *Yarrowia lipolytica*. It was found that in *P. pastoris ATG26* was necessary for pexophagy of large peroxisomes, which accumulated in methanol medium. Pexophagy of small peroxisomes in this species, induced by oleate or methylamine, only partially depends on *ATG26* and its product, sterol glucoside (Nazarko et al. 2007a, b). It was also shown that the *P. pastoris* Atg26p was required for elongation of the pre-autophagosomal structure (PAS) into MIPA during micropexophagy (Yamashita et al. 2006). It was hypothesized that in *P. pastoris*, sterol glucosides acquired a new function during evolution related to facilitation of the elongation of the double membranes from the PAS. The enhancer function of sterol glucosides becomes essential when cells are challenged with elongation of the extremely large double membranes, i.e., during biogenesis of the MIPA or pexophagosome, around methanol-induced peroxisomes (Nazarko et al. 2007b).

P. pastoris ATG28 also encodes a pexophagy-specific protein as its deficiency impairs both pexophagic mechanisms (macro- and micropexophagy) and only partially affects the general (nonspecific bulk turnover) autophagy induced by nitrogen starvation (Stasyk et al. 2006; Nazarko et al. 2009). Atg28p contains a coiled-coil domain that overlaps with a putative leucine-zipper motif. This coiled-coil region in Atg28p may be involved in oligomerization and protein-protein interactions. It is functionally important, as modified Atg28p lacking the coiled-coil region is not functional. Atg28p is involved in the formation of one or more protein complexes specific for pexophagy and its interaction with the micropexophagy-specific protein Atg35p was experimentally demonstrated (Nazarko et al. 2011, see below). Atg28p exhibits a complex intracellular localization pattern. In most methanol-induced cells, this protein was localized to the cytosol. However, in some cells, the fusion protein was also localized to punctate structures of unknown nature associated with vacuoles and to the vacuolar membrane. In rare cases, Atg28p could be seen localized to the vacuolar matrix.

Another pexophagy-specific protein is Atg30p. Two other proteins specifically involved in pexophagy and not in general autophagy or other types of specific autophagy are Pex3p and Pex14p, known as peroxins also involved in peroxisome biogenesis. In *P. pastoris*, Atg30p interacts with Pex3p and Pex14p both localized on the peroxisomal membrane (Farré et al. 2008). Effective peroxisome homeostasis probably requires their biogenesis and degradation to be coordinated. It was shown that interacting partners of Atg30p are proteins participating in peroxisome biogenesis. Thus, Pex3p is important for peroxisome biogenesis, and Pex14p—for protein import to peroxisomal matrix (Ma and Subramani 2009). In *H. polymorpha* Pex14p, more exactly the 64 N-terminal amino acid residues, are necessary for pexophagy (Bellu et al. 2001a; van Zutphen et al. 2008). Also, it was shown that during macropexophagy in *H. polymorpha* Pex3p is removed from peroxisomes and does not undergo degradation (Bellu et al. 2002). The way Pex3p is removed from peroxisomes is unknown. Pex3p is known to be required for stabilization of a complex of proteins with a RING finger domain (Really Interesting New Gene, structural domain similar to protein zinc finger) of peroxisome importer (Hazra et al. 2002). Therefore, at this stage, besides inhibiting peroxisome biogenesis, also destabilization of some complexes occurs in the peroxisomal membrane.

In *P. pastoris*, a gene designated *PDG1* (Peroxisome DeGradation) was identified whose mutations led to disturbances in peroxisome degradation (Dunn et al. 2005; O Stasyk and A Sibirny, unpublished data). Moreover, such mutations disturbed localization of peroxisomal proteins that, apart from peroxisomes, were also localized in the cytosol, indicating disturbance in peroxisome biogenesis in *pdg1* mutants. The corresponding protein Pdg1p is a membrane protein, which confirms its role in peroxisome biogenesis.

In *H. polymorpha*, the transcriptional repressor Tup1p was shown to be essential for macropexophagy (Leão-Helder et al. 2004). Defects in orthologs of presumable corepressors involved in glucose catabolite repression, *MIG1* and *MIG2*, also showed impairment in pexophagy (Stasyk et al. 2007). As mutants defective in *MIG1* and *MIG2* were not affected in glucose catabolite repression, one may assume that the functions of their products are different between baker's and methylotrophic yeasts.

23.3 Micro- and Macropexophagy in *Pichia pastoris* and *Hansenula polymorpha*

As was pointed out above, macropexophagy could be observed in *P. pastoris* after a shift of methanol-grown cells to a medium with ethanol whereas micropexophagy is observed when methanol-grown cells are transferred into medium with glucose (Tuttle and Dunn 1995). Another methylotrophic yeast, *H. polymorpha*, is characterized by carbon source independent macropexophagy, which induces pexophagy (van Zutphen et al. 2008). During macropexophagy, multiple membrane layers sequester a single peroxisome resulting in the formation of a pexophagosome from which the outer membrane layer fuses with the vacuole where the peroxisome becomes hydrolyzed. Micropexophagy involves the uptake of a cluster of peroxisomes through the direct engulfment by the vacuolar membrane (Fig. 23.1). Three main steps could be outlined for macropexophagy: recognition of the organelle destined for degradation, formation of the pexophagosome, and fusion with the vacuole (Fig. 23.1). For micropexophagy, the following steps could be distinguished: vacuolar engulfment of peroxisomes, formation of the MIPA at the peroxisomal surface, and vacuolar membrane fusion (Sakai et al. 2006).

Micropexophagy turned out to be more sensitive to a decrease in intracellular ATP compared to macropexophagy; In other words, intracellular ATP pool plays a more important role in defining the pexophagy pathway than the nature of the carbon substrate (Ano et al. 2005b). However, it is not known whether the ATP concentration is the reason of the observed type of pexophagy or pexophagy is the consequence of some other triggering mechanisms. In other methylotrophic yeasts, e.g., *H. polymorpha*, a shift of methanol-grown cells either to glucose or ethanol leads to morphological changes described as macropexophagy. In *H. polymorpha*, nitrogen limitation leads to peroxisome degradation by a mechanism similar to micropexophagy. However, this process occurs due to a non-specific autophagic mechanism, by which cytosolic components are taken up by vacuoles concomitantly with peroxisomes, and was therefore named microautophagy of peroxisomes (Bellu et al. 2001b; van Zutphen et al. 2008).

During the last years, genes have been identified that are specifically involved in macro- and micropexophagy. The *ATG25* gene in *H. polymorpha* is specifically involved in macropexophagy. It codes for a coiled-coil protein that acts as selectivity factor during macropexophagy (Monastyrska et al. 2005). This protein is located in the pexophagosomes and moved there via the PAS. Atg25p is involved in the completion of sequestration of peroxisomes or in the fusion of pexophagosomes with the vacuolar membrane (Sakai et al. 2006). For the latter process, the SNARE Vam7p and the GTPase Ypt7p are also essential in *H. polymorpha* (Stevens et al. 2005).

The presence of a specific morphological structure in the micropexophagy process, the micropexophagy apparatus MIPA in *P. pastoris*, suggests the existence of specific genes and proteins participating in this process. The gene *PFK1* codes for phosphofructokinase 1 α-subunit, which is required for peroxisome engulfment by vacuoles after transferring *P. pastoris* cells from methanol to glucose medium

(Yuan et al. 1997). The participation of phosphofructokinase 1 α-subunit in micropexophagy does not depend on its ability to phosphorylate fructose-6-phosphate since a catalytically inactive form of this enzyme allows for normal pexophagy. Moreover, the *VAC8* gene (VACuole related) was identified whose product is a 60–64 kDa protein with so-called armadillo repeat that specifically participates in micro- but not macropexophagy (Fry et al. 2006; Nazarko et al. 2007a). In mutant cells, vacuolar sequestering membrane during micropexophagy is not formed. A hybrid protein Vac8p–GFP localized on the vacuolar membrane independently of Atg1p, Atg9p, or Atg11p. Deletion of the armadillo repeats did not alter Vac8p localization, but the protein lost its functionality. Vac8p probably participates in early (formation of sequestering membrane) and late (membrane fusion after formation of micropexophagy apparatus) micropexophagy stages. Micropexophagy and vacuole inheritance were shown to be dependent on different Vac8p domains (Oku et al. 2006). Mutations in the genes *PpGCN1*, *PpGCN2*, *PpGCN3*, or *PpGCN4* involved in general amino acid control regulation, specifically inhibits micropexophagy after incorporation of the peroxisomes into the vacuole (Mukaiyama et al. 2002; Sakai et al. 2006), but the detailed functions of the Gcn proteins are not clear. It is known that *GCN1* regulates translation elongation; *GCN2* codes for protein kinase and regulates translation initiation (eIF2 kinase); *GCN3* codes for a translation initiation factor (eIF2B), whereas *GCN4* codes for basic leucine-zipper (bZIP) transcriptional activator of amino acid biosynthetic genes in response to amino acid starvation. The exact function of the mentioned genes in micropexophagy remains unknown.

The new micropexophagy-specific protein Atg35p, the first autophagy protein with nuclear localization, was identified during the analysis of partners interacting with the protein Atg28p from *P. pastoris* (Stasyk et al. 2006). To search for such Atg proteins, a yeast two hybrid (YTH) screening system was used for the first time. YTH screening of the genome database of *P. pastoris* DNA was carried out in *S. cerevisiae* cells using *Pp*Atg28p as "bait" (Nazarko et al. 2011). Identified in this way Atg35p consists of 463 amino acids and incorporates two putative domains: a RING finger and a PHD (Plant Homeo Domain). Testing *P. pastoris atg35* mutants showed that macropexophagy is normal, whereas micropexophagy is impaired. It was found that Atg35p is necessary only for micropexophagy at the stage of micropexophagy apparatus formation (Nazarko et al. 2011). It is interesting that overexpression of *ATG35* as well as deletion of this gene both inhibit micropexophagy but do not influence macropexophagy. Atg35p contains a putative nuclear localization signal. Atg35p localization on single dot-like structures of the nuclear membrane in glucose medium was found to be dependent on Atg17p and is significant for the micropexophagy process. Atg28p is known to interact with Atg17p (Nazarko et al. 2007a) and Atg35p (Nazarko et al. 2011). It was shown that interaction between Atg17p and Atg35p occurs due to Atg28 protein (Fig. 23.2). Thus, Atg35p is the first revealed nuclear Atg protein participating in autophagy in yeasts. Deletion and overexpression of this gene lead to specific disturbance of micropexophagy alone. The Atg35 protein functions through

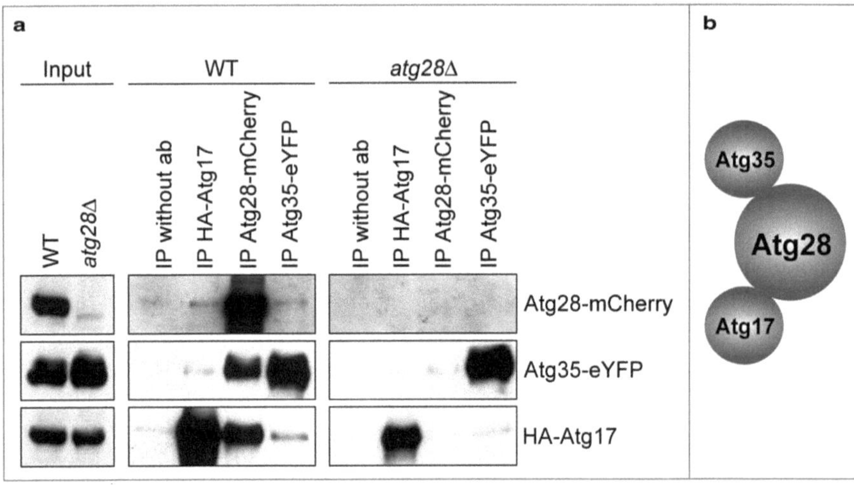

Fig. 23.2 Interaction between Atg35p, Atg28p, and Atg17p. (**a**) co-immunoprecipitation of Atg17p, Atg28p, and Atg35p. (**b**) Schematic view of Atg17p, Atg28p and Atg35p interactions (from Nazarko et al. 2011)

interaction with Atg17p and Atg28p, the latter protein playing a central role in this interaction (Nazarko et al. 2011).

23.4 Glucose Sensing and Signaling Mechanisms and Pexophagy in *H. polymorpha* and *P. pastoris*

When cells are transferred from methanol to glucose medium, micropexophagy (*P. pastoris*) or macropexophagy (*H. polymorpha*) occur. Cells in some way recognize (sense) glucose and triggers glucose signal to activate all specific to micropexophagy, other pexophagy-specific, and many general autophagy Atg proteins ended by peroxisome degradation. Mechanisms of glucose sensing and signaling during (micro)pexophagy is poorly understood, especially in methylotrophic yeasts.

Below we consider available data on glucose sensing and signaling connected to pexophagy.

23.4.1 Sensing

Mechanisms of glucose sensing have been studied in detail in *S. cerevisiae* as glucose induces complex regulatory responses, which include induction of glucose transporters, catabolite repression of hundreds of genes, catabolite inactivation of several enzymes including proteasomal and autophagic degradation of some of them. Besides, glucose is sensed for subsequent trehalose mobilization and other responses to stress factors. Still many aspects of glucose sensing in *S. cerevisiae* remain to be elucidated. The reader is referred to the corresponding reviews (Ozcan and Johnston 1999; Santangelo 2006; Gancedo 2008; Rubio-Texeira et al. 2010). Briefly, there are two types of glucose sensors in *S. cerevisiae*. One is involved in glucose-dependent stress response and the other is responsible for glucose induction and glucose catabolite repression phenomena.

The plasma membrane contains many proteins capable of glucose binding and part of them act as glucose sensors. There are 20 glucose transporters (Wieczorke et al. 1999); however, they are apparently not all involved in glucose sensing (Gancedo 2008). Specific glucose sensors can be divided in three groups. The first class of sensors comprises the classical receptor proteins or G protein-coupled receptors (GPCRs), which, in yeast, detect the presence of glucose and sucrose. It is responsible for glucose and sucrose control of the protein kinase A (PKA) pathway (Thevelein and de Winde 1999), which plays a central role in the nutritional control of metabolism, stress resistance, cell cycle, growth, and transcription. All these properties are tightly controlled by the availability of nutrients in the medium, especially by the presence of a rapidly fermentable sugar such as glucose. In addition to rapidly fermentable sugars, for derepression yeast cells trigger an immediate increase in the cAMP levels, which in turn causes the rapid activation of PKA, resulting in drastic changes in its multiple targets. The sugar-sensing GPCR system consists of the receptor Gpr1p and the Gα protein Gpa2p (Colombo et al. 1998). The second class of glucose sensors in *S. cerevisiae* is represented by two non-transporting transceptors Snf3p and Rgt2p, which are sugar transporter homologs. The high-affinity sensor Snf3p and low-affinity glucose sensor Rgt2p generate the intracellular signal required for the induction of the hexose transporter genes in response to glucose (Gancedo 2008). They, however, are not involved in glucose sensing for catabolite repression. The third class of glucose sensor is represented by an intracellular protein the enzyme hexokinase II Hxk2p. Glucose sensing by Hxk2p depends in some way on its ability to phosphorylate glucose (Rose et al. 1991). Hxk2p sensor is involved in glucose catabolite repression; it was shown that a small proportion of Hxk2p is located in the nucleus (Herrero et al. 1998) and that under conditions for which Hxk2p does not enter the nucleus, glucose repression does not take place. These results indicate a non-metabolic role for Hxk2p.

The role of the GPCR sensor proteins Gpr1p and Gpa2p and the non-transporting transceptors Snf3p and Rgt2p in glucose-induced pexophagy has been studied in *S. cerevisiae*. For this, the fate of the peroxisomal protein thiolase, involved in fatty acid β-oxidation, was analyzed. This enzyme is induced in the medium with the peroxisome proliferator oleic acid. Pexophagy is started after shift of the induced cells in the medium with glucose. It was shown that knockout of either *GPR1* or *GPA2* leads to strong defects in glucose-activated autophagic degradation of peroxisomal thiolase (Nazarko et al. 2008a). Knockout of *SNF3* or *RGT2* individually led to only insignificant defects in pexophagy, whereas double knockout of both *SNF3* and *RGT2* resulted in practically total defect in thiolase autophagic degradation (Nazarko et al. 2008b). It was concluded that both glucose signaling

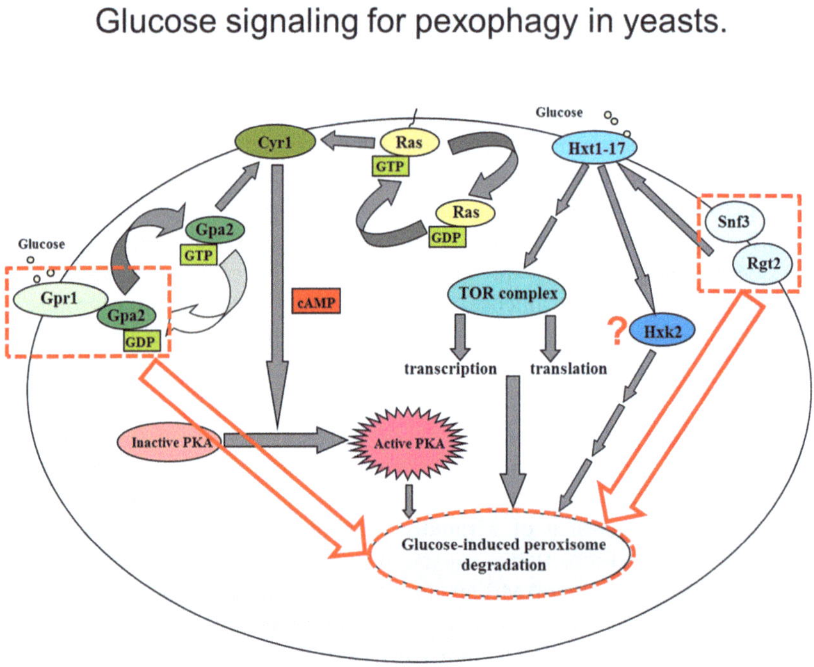

Fig. 23.3 Scheme of glucose signaling for pexophagy in yeasts (from Nazarko et al. 2008a, b)

components are involved in glucose sensing for pexophagy (Fig. 23.3). Defects in Hxk2p, however, have no effect on pexophagy in *S. cerevisiae* (V. Nazarko, A. Sibirny, unpublished).

In methylotrophic yeasts, we know much less on glucose sensing and signaling. In *H. polymorpha*, two glucose sensors, Gcr1p and Hxs1p, have been identified, along with the glucose (hexose) transporter Hxt1p (Stasyk et al. 2004, 2008b). *P. pastoris*, on the other hand, apparently possesses only one glucose sensor, designated Gss1p (Polupanov et al. 2011). Point mutations or deletion in the *GCR1* gene of *H. polymorpha* affected glucose catabolite repression and led to the constitutive presence of peroxisomes in glucose medium (Stasyk et al. 2004). However, the *GCR1* gene is apparently not directly involved in pexophagy. A decrease in specific activity and protein levels of peroxisomal enzyme alcohol oxidase was observed in *gcr1* mutant cells upon glucose adaptation. But residual alcohol oxidase levels were higher in the *gcr1* mutants relative to the wild type. However, these data do not demonstrate the direct involvement of the Gcr1 protein in pexophagy since in *gcr1* mutant strains, *de novo* peroxisome synthesis occurred due to the defect in glucose repression. A time course examination of cell morphology revealed clear signs that pexophagy proceeds in *gcr1* mutants. Some peroxisomes were observed sequestered by additional membrane layers typical for the initial stages of macroautophagic peroxisome degradation in *H. polymorpha* (Veenhuis et al. 2000). Also, in *gcr1* mutant cells with fluorescently labeled peroxisomes, the pexophagic process was evident upon glucose adaptation. Shortly after the shift, the GFP fluorescence was observed in vacuoles, while in methanol-growing cells it is confined to peroxisomes. These data led to the conclusion that Gcr1p is not directly involved in pexophagy. Both point or missense mutations and *gcr1*-deletion continued to exhibit normal wildtype peroxisome degradation in response to ethanol.

In contrast to that, knockout mutation in another hexose sensor gene *HXS1* did not lead to defect in glucose catabolite repression and led to defect in glucose transport capacity (Stasyk et al. 2008b). To study, whether *HXS1* was involved in pexophagy, it was observed that in methanol-preincubated *hxs1*-mutant cells, alcohol oxidase activity and protein level decreased upon glucose adaptation. This occurred at a rate similar to that of the wild type strain. The *H. polymorpha tup1* mutant, deficient in pexophagy, has been utilized as a positive control (Leão-Helder et al. 2004; Stasyk et al. 2007). When methanol-preinduced *hxs1*-mutant cells were shifted to fructose or ethanol, they also did not differ from the wild type strain in their rate of alcohol oxidase degradation. Therefore, similarly to Gcr1p Hxs1p is not essential for glucose signaling in pexophagy in *H. polymorpha*. Thus, both identified glucose sensors in this organism are involved in several regulatory processes exerted by glucose but not in glucose recognition for pexophagy. Thus, the specific glucose sensor involved in glucose-induced macropexophagy in *H. polymorpha* still needs to be found.

The situation looks different in *P. pastoris*. In this organism, orthologs of the GPCR sensor proteins Gpr1p and Gpa2p and glucose transceptor sensor proteins Snf3p/Rgt2p have been identified. One potential ortholog of the *GPR1* gene and one of the *GPA2* gene that exhibit 60 % and 65 % similarity to their *S. cerevisiae* counterparts, respectively. It was found that knockout of *P. pastoris GPR1* and *GPA2* orthologs had no apparent effect on degradation of peroxisomal thiolase, inactivation of peroxisomal alcohol oxidase and general autophagy (Nazarko et al. 2008b). Thus, in contrast to *S. cerevisiae*, *Pp*Gpr1p and *Pp*Gpa2p are not involved in glucose signaling for pexophagy in *P. pastoris*. It is known that in contrast to *S. cerevisiae*, *Candida albicans GPR1* and *GPA2* are not involved in a transient cAMP burst after glucose addition (Maidan et al. 2005). Similarly, *Pp*Gpr1p and *Pp*Gpa2p could be not involved in regulation of cAMP production and it looks as if in *P. pastoris* glucose for pexophagy is sensed by other components of the PKA-cAMP signaling pathway or by distinct sensors, which are not involved in this pathway.

Two hexose transporters were recently identified in the yeast *P. pastoris*, Hxt1p and Hxt2p, which are transcriptionally regulated by glucose. Deletion of *PpHXT1* but not *PpHXT2*, led to the expression of alcohol oxidase in glucose medium due to impairment in glucose catabolite repression. However, mutants lacking *Pphxt1* were normal in glucose utilization and peroxisome degradation (Zhang et al. 2010). *P. pastoris* contains only one ortholog of the *S. cerevisiae* genes *SNF3* and *RGT2*, designated as *GSS1* (from GlucoSe Sensor) with 57 % identity and 71 % of similarity to ScSnf3p, and 46 % identity and 63 % similarity to ScRgt2p (Polupanov et al. 2012). *Pp*Gss1p also reveals high level of homology to

H. polymorpha proteins Hxs1p (62 % identity and 77 % similarity) and Gcr1p (42 % identity and 60 % similarity). Like the *S. cerevisiae* Snf3p and Rgt2p sensors, *Pp*Gss1p possesses 12 transmembrane domains, a long C-terminal extension, a distinguishing characteristic for glucose sensors (Ozcan et al. 1998), but lacks of N-terminal peptide (52 amino acids) present in the *S. cerevisiae* proteins. The strain with knockout of the gene *GSS1* had impaired growth on glucose. Despite the high homology between HpHxs1p and *Pp*Gss1p, the Hxs1p deficiency had a moderate effect on glucose growth and utilization in *H. polymorpha* (Stasyk et al. 2008b). Deletion of the *GSS1* gene affects glucose catabolite repression in the methylotrophic yeast *P. pastoris* as was found in *H. polymorpha gcr1* mutants to be resistant to 2-deoxy-D-glucose (Stasyk et al. 2004).

Alcohol oxidase (AOX) replica plate overlay assay was used as preliminary examination of micropexophagy in *gss1*-mutant. *P. pastoris* wild-type, *gss1* and *pep4, prb1*-mutants strains with defect in vacuolar proteinases were grown on methanol minimal medium for 2 days and then replica plated to glucose minimal medium to induce micropexophagy. Residual alcohol oxidase activity led to the red-colored cells on the plates with glucose indicating impairment of peroxisome degradation (Sibirny and Titorenko 1986; Stasyk et al. 2008a). The cells of *gss1*-mutant strain, similarly to that of *pep4, prb1*-mutant, showed residual alcohol oxidase activity suggesting a block in pexophagy, unlike the wild-type strain with normal inactivation of the enzyme (Fig. 23.4). These results support the hypothesis that Gss1p is important for micropexophagy in the methylotrophic yeast *P. pastoris* (Polupanov et al. 2012). Pexophagy also was monitored by Western blot analysis for the alcohol oxidase protein in strains cultivated in methanol or oleate medium and transferred to glucose. Unlike wild types, *gss1*-mutant cells maintained stable alcohol oxidase. To validate these results, *P. pastoris* cells were analyzed by fluorescent microscopy after shift from methanol to glucose medium. For this, *gss1*-and *atg1*-mutant and wild-type cells were used with GFP-SKL labeled peroxisomes. During glucose adaptation, wild-type cells showed peroxisome degradation in contrast to the *gss1*- and *atg1*-mutants, which possessed peroxisome clusters surrounded by the arm-like structures of vacuolar membrane typical for micropexophagy. Thus, the gene *GSS1* seemed to be important for micropexophagy (Polupanov et al. 2012). During incubation of methanol-grown cells in ethanol medium, *gss1*-mutant cells showed a drop in their amounts of alcohol oxidase protein. However, the process was slower than in wild-type cells. Thus, the product of the *GSS1* gene is only partially involved in macropexophagy. Fluorescent observations supported this conclusion. It was also shown that *GSS1* is not involved in Cvt pathway and general (non-specific) autophagy (Polupanov et al. 2012).

It was found that that deletion of 150 residues of Gss1p leads to phenotypic alterations although still maintaining the signaling function of Gss1p. At the same time, substitution of one conserved amino acid R180K of the Gss1 protein has no visible phenotype, in contrast to corresponding changes in glucose sensors from other yeast species. It has been suggested that the C-terminal cytoplasmic extension of *Pp*Gss1p plays a role different to that of its homologs in *Saccharomyces cerevisiae* and *Hansenula polymorpha* (Polupanov and Sibirny 2014). Thus, the

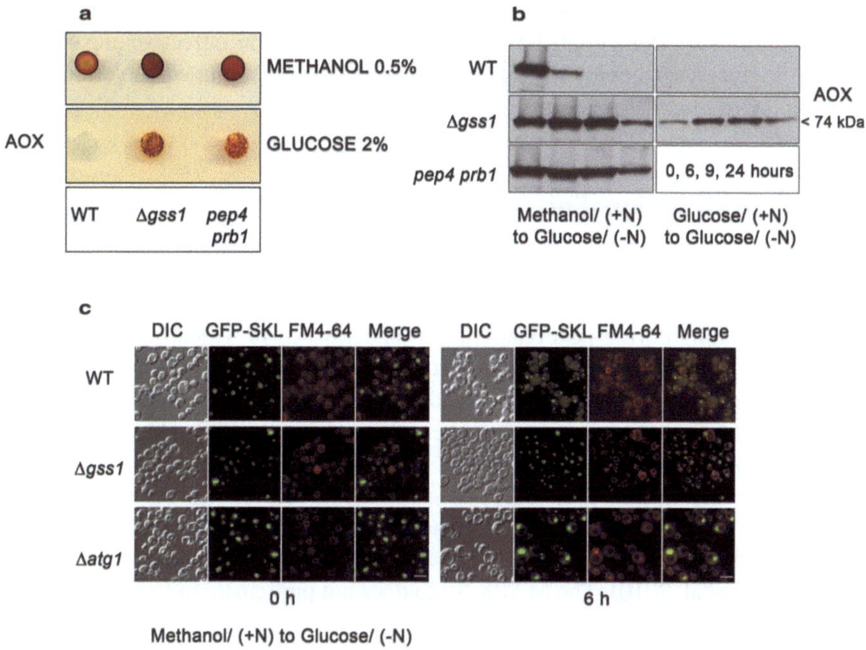

Fig. 23.4 The deletion of *P. pastoris GSS1* gene impairs micropexophagy and glucose catabolite repression. (**a**) Alcohol oxidase plate overlay assay. *P. pastoris* WT, *gss1*, and *pep4 prb1* mutant (Δ) strains were transferred from methanol medium to glucose medium to induce micropexophagy. Residual activity of alcohol oxidase reveals pexophagy deficiency. (**b**) *P. pastoris* WT, *gss1*, and *pep4 prb1* mutant cells were induced in methanol medium and transferred to glucose medium to induce micropexophagy. (**c**) Fluorescence microscopy studying of micropexophagy. *P. pastoris* WT (STN017), *gss1* (SAP01), and *atg1* (SAP02) mutant cells with GFP-labeled peroxisomes were induced in methanol/(+N) medium and transferred to glucose/(−N) medium. After 6 h of glucose adaptation pexophagy was monitored by fluorescence microscopy. Peroxisomes were labeled with GFP-SKL and vacuolar membranes—with FM4-64 (from Polupanov et al. 2012)

mechanism and amino acid residues responsible for glucose sensing by the Gss1 protein remain to be elucidated in future studies. The specific homolog of glucose transporters, transceptor sensor Gss1p was found to be involved in glucose sensing for micropexophagy. It is also involved in glucose catabolite repression.

23.4.2 Low Molecular Weight Effector Which Triggers Glucose Signal for Pexophagy

It is not known at the moment, which metabolite is the immediate signaling molecule initiating pexophagy in glucose medium. It could be glucose or its metabolite. The observation that enzymatically inactive phosphofructokinase restored micropexophagy in glucose medium without restoration of cell growth suggests that such metabolite has to be upstream of fructose-1,6-bisphosphate (Yuan et al. 1997; Dunn et al. 2005). Studying other mutants defective in particular steps of glycolysis could help the identification of the immediate effector of pexophagy in glucose medium.

23.4.3 Glucose Signaling for Pexophagy

Mechanisms of glucose signaling in *S. cerevisiae* have been studied in details (Santangelo 2006; Gancedo 2008; Rubio-Texeira et al. 2010). The scheme of glucose signaling during pexophagy in this species was provided before (Nazarko et al. 2008b). Our knowledge is quite restrictive regarding glucose signaling during pexophagy in methylotrophic yeasts. Studying thiolase and the bifunctional enzyme Fox3p degradation as peroxisomal markers showed that the Slt2p (Mpk1) mitogen-activated protein kinase (MAPK) is necessary for pexophagy but not for pexophagosome formation or other nonselective and selective forms of autophagy. It was also showed that several upstream components of its signal transduction pathway (Pkc1p, Bkc1p, Mkk1p, and Mkk2p) are involved in glucose signaling (Manjithaya et al. 2010). The MAPK Slt2p does not participate in Cvt pathway and general (nonspecific) autophagy.

It was proposed that pexophagy requires the simultaneous activation of this MAPK pathway and an hexose-sensing mechanism acting through protein kinase A and cyclic adenosine monophosphate. Data showing that the orthologs of the *S. cerevisiae* Mig1p and Mig2p are apparently not involved in glucose catabolite repression (Stasyk et al. 2007) suggest possible strong differences in the mechanism of glucose signaling between baker's and methylotrophic yeasts.

23.5 Ethanol Sensing for Pexophagy in Methylotrophic Yeasts

Ethanol signaling for pexophagy apparently exists only in methylotrophic yeasts. In other yeast species used for pexophagy studies (*S. cerevisiae*, *Y. lipolytica*) ethanol does not induce pexophagy of oleate-induced peroxisomes. However, practically nothing is known on ethanol sensing in yeasts, including in *S. cerevisiae*. Nevertheless, there have to be several quite specific mechanisms of ethanol sensing and signaling. It is known that ethanol specifically and strongly induces several proteins in *S. cerevisiae*, glucokinase being induced near 25-fold (Herrero et al. 1999). In *S. cerevisiae*, ethanol represses *PDC1* coding for pyruvate decarboxylase through ERA regulatory sequence (Liesen et al. 1996) and in *Kluyveromyces lactis*, ethanol specifically represses the expression of *ADH3* coding for mitochondrial alcohol dehydrogenase (Saliola et al. 2007). In methylotrophic yeasts, ethanol specifically activates the repression of the synthesis of enzymes involved in methanol metabolism in addition to pexophagy (Tolstorukov et al. 1989). We do not know if there are specific ethanol sensors in cytoplasmic membrane, until now no such protein was reported. Possibly ethanol is sensed by some intracellular specific sensors and/or ethanol metabolizing enzymes. The *ecr1* and *adh1* mutants of the methylotrophic yeast *Pichia methanolica* (*Pichia pinus* MH4) are known in which ethanol is unable to repress the synthesis of peroxisomal enzymes involved in methanol catabolism (Sibirny et al. 1987; Sibirny et al. 1991). In *adh1* mutant cells, ethanol and methanol are utilized simultaneously and hybrid peroxisomes are produced, which apparently maintain enzymes for both methanol and ethanol metabolism whereas in *ecr1* mutant cells, methanol is utilized first from the mixture of both alcohols. Though the genes were not isolated, *adh1* mutation apparently tagged one of alcohol dehydrogenases, whereas the *ECR1* gene possibly codes for a protein involved in ethanol sensing. In *P. methanolica*, attempts were made to identify derivative of ethanol initiating pexophagy in ethanol medium. Mutants defective in distinct steps of ethanol utilization have been isolated (Tolstorukov et al. 1989; Sibirny 1990). It was found that pexophagy was affected in *icl1*-mutants defective in isocitrate lyase suggesting that isocitrate is an immediate ethanol metabolite initiating pexophagy.

Thus, the mechanisms of sensing and signaling in glucose- and ethanol-induced pexophagy in yeast in general and in methylotrophic yeasts in particular are far from being understood. Presently, we do not know the exact glucose sensors and signaling components to the pexophagy machinery. In the case of ethanol-induced pexophagy, our knowledge is at the initial stage. It could be envisaged that studies in this field will be more active in nearest future and we will have soon a mechanistic picture for pexophagy sensing and signaling by glucose and ethanol in methylotrophic yeasts.

Acknowledgment The work in the field of pexophagy in the author's laboratory was supported by grant № F46.1/012 from the State Foundation for Basic Research "Key State Laboratory of Molecular and Cellular Biology", Ukraine.

References

Aksam EB, Koek A, Kiel JA et al (2007) A peroxisomal lon protease and peroxisome degradation by autophagy play key roles in vitality of Hansenula polymorpha cells. Autophagy 3:96–105

Aksam EB, de Vries B, van der Klei IJ et al (2009) Preserving organelle vitality: peroxisomal quality control mechanisms in yeast. FEMS Yeast Res 9:808–820

Ano Y, Hattori T, Oku M et al (2005a) A sorting nexin PpAtg24 regulates vacuolar membrane dynamics during pexophagy via binding to phosphatidylinositol-3-phosphate. Mol Biol Cell 16:446–457

Ano Y, Hattori T, Kato N et al (2005b) Intracellular ATP correlates with mode of pexophagy in Pichia pastoris. Biosci Biotechnol Biochem 69:1527–1533

Baerends RJ, Rasmussen SW, Hilbrands RE et al (1996) The Hansenula polymorpha PER9 gene encodes a peroxisomal membrane protein essential for peroxisome assembly and integrity. J Biol Chem 271:8887–8894

Bellu AR, Komori M, van der Klei IJ et al (2001a) Peroxisome biogenesis and selective degradation converge at Pex14p. J Biol Chem 276:44570–44574

Bellu AR, Kram AM, Kiel JA et al (2001b) Glucose-induced and nitrogen-starvation-induced peroxisome degradation are distinct processes in Hansenula polymorpha that involve both common and unique genes. FEMS Yeast Res 1:23–31

Bellu AR, Salomons FA, Kiel JA, Veenhuis M, Van Der Klei IJ (2002) Removal of Pex3p is an important initial stage in selective peroxisome degradation in Hansenula polymorpha. J Biol Chem 277:42875–42880

Brooke AG, Dijkhuizen L, Harder W (1986) Regulation of flavin biosynthesis in the methylotrophie yeast Hansenula polymorpha. Arch Microbiol 145:62–70

Brown LA, Baker A (2008) Shuttles and cycles: transport of proteins into the peroxisome matrix (review). Mol Membr Biol 25:363–375

Colombo S, Ma P, Cauwenberg L et al (1998) Involvement of distinct G-proteins, Gpa2 and Ras, in glucose- and intracellular acidification-induced cAMP signalling in the yeast *Saccharomyces cerevisiae*. EMBO J 17:3326–3341

Cregg JM, Tolstorukov I, Kusari A et al (2008) Expression in the yeast Pichia pastoris. Methods Enzymol 463:169–189

Dunn WA Jr, Cregg JM, Kiel JA et al (2005) Pexophagy: the selective autophagy of peroxisomes. Autophagy 1:75–83

Faber KN, Harder W, Ab G et al (1995) Review: methylotrophic yeasts as factories for the production of foreign proteins. Yeast 11:1331–1344

Farre JC, Subramani S (2004) Peroxisome turnover by micropexophagy: an autophagy-related process. Trends Cell Biol 14:515–523

Farré JC, Manjithaya R, Mathewson RD, Subramani S (2008) PpAtg30 tags peroxisomes for turnover by selective autophagy. Dev Cell 14:365–376

Farré JC, Krick R, Subramani S et al (2009) Turnover of organelles by autophagy in yeast. Curr Opin Cell Biol 21:522–530

Fry MR, Thomson JM, Tomasini AJ et al (2006) Early and late molecular events of glucose-induced pexophagy in Pichia pastoris require Vac8. Autophagy 2:280–288

Gancedo JM (2008) The early steps of glucose signalling in yeast. FEMS Microbiol Rev 32:673–704

Hazra PP, Suriapranata I, Snyder WB et al (2002) Peroxisome remnants in pex3-delta cells and the requirement of Pex3p for interactions between the peroxisomal docking and translocation subcomplexes. Traffic 3:560–574

Herrero P, Martínez-Campa C, Moreno F (1998) The hexokinase 2 protein participates in regulatory DNA-protein complexes necessary for glucose repression of the SUC2 gene in Saccharomyces cerevisiae. FEBS Lett 434:71–76

Herrero P, Flores L, de la Cera T et al (1999) Functional characterization of transcriptional regulatory elements in the upstream region of the yeast GLK1 gene. Biochem J 343:319–325

Kiel JA (2010) Autophagy in unicellular eukaryotes. Philos Trans R Soc Lond B Biol Sci 365:819–830

Lahtchev KL, Semenova VD, Tolstorukov II et al (2002) Isolation and properties of genetically defined strains of the methylotrophic yeast Hansenula polymorpha CBS4732. Arch Microbiol 177:150–158

Lanyon-Hogg T, Warriner SL, Baker A (2010) Getting a camel through the eye of a needle: the import of folded proteins by peroxisomes. Biol Cell 102:245–263

Leão-Helder AN, Krikken AM, Lunenborg MG et al (2004) Hansenula polymorpha Tup1p is important for peroxisome degradation. FEMS Yeast Res 4:789–794

Liesen T, Hollenberg CP, Heinisch JJ (1996) ERA, a novel cis-acting element required for autoregulation and ethanol repression of PDC1 transcription in Saccharomyces cerevisiae. Mol Microbiol 21:621–632

Ma C, Subramani S (2009) Peroxisome matrix and membrane protein biogenesis. IUBMB Life 61:713–722

Maidan MM, De Rop L, Serneels J et al (2005) The G protein-coupled receptor Gpr1 and the Gα protein Gpa2 act through the cAMP-protein kinase A pathway to induce morphogenesis in Candida albicans. Mol Biol Cell 16:1971–1986

Manjithaya R, Nazarko TY, Farré JC et al (2010) Molecular mechanism and physiological role of pexophagy. FEBS Lett 584:1367–1373

Marelli M, Smith JJ, Jung S et al (2004) Quantitative mass spectrometry reveals a role for the GTPase Rho1p in actin organization on the peroxisomemembrane. J Cell Biol 167:1099–1112

Mast FD, Fagarasanu A, Knoblach B et al (2010) Peroxisome biogenesis: something old, something new, something borrowed. Physiology (Bethesda) 25:347–356

Meijer WH, Gidijala L, Fekken S et al (2010) Peroxisomes are required for efficient penicillin biosynthesis in Penicillium chrysogenum. Appl Environ Microbiol 76:5702–5709

Michels PA, Moyersoen J, Krazy H et al (2005) Peroxisomes, glyoxysomes and glycosomes (review). Mol Membr Biol 22:133–145

Michels PAM, Bringaud F, Herman M et al (2006) Metabolic functions of glycosomes in trypanosomatids. Biochim Biophys Acta 1763:1463–1477

Monastyrska I, Kiel JA, Krikken AM et al (2005) The Hansenula polymorpha ATG25 gene encodes a novel coiled-coil protein that is required for macropexophagy. Autophagy 1:92–100

Mukaiyama H, Oku M, Baba M et al (2002) Paz2 and 13 other PAZ gene products regulate vacuolar engulfment of peroxisomes during micropexophagy. Genes Cells 7:75–90

Mukaiyama H, Baba M, Osumi M et al (2004) Modification of a ubiquitin-like protein Paz2 conducted micropexophagy through formation of a novel membrane structure. Mol Biol Cell 15:58–70

Nazarko TY, Polupanov AS, Manjithaya RR et al (2007a) The requirement of sterol glucoside for pexophagy in yeast is dependent on the species and nature of peroxisome inducers. Mol Biol Cell 18:106–118

Nazarko TY, Farré JC, Polupanov AS et al (2007b) Autophagy-related pathways and specific role of sterol glucoside in yeasts. Autophagy 3:263–265

Nazarko VY, Thevelein JM, Sibirny AA (2008a) G-protein-coupled receptor Gpr1 and G-protein Gpa2 of cAMP-dependent signaling pathway are involved in glucose-induced pexophagy in the yeast Saccharomyces cerevisiae. Cell Biol Int 32:502–504

Nazarko VY, Futej KO, Thevelein JM et al (2008b) Differences in glucose sensing and signaling for pexophagy between the baker's yeast Saccharomyces cerevisiae and the methylotrophic yeast Pichia pastoris. Autophagy 4:381–384

Nazarko TY, Farré JC, Subramani S (2009) Peroxisome size provides insights into the function of autophagy-related proteins. Mol Biol Cell 20:3828–3839

Nazarko VY, Nazarko TY, Farré JC et al (2011) Atg35, a micropexophagy-specific protein that regulates micropexophagic apparatus formation in Pichia pastoris. Autophagy 7:375–385

Oku M, Warnecke D, Noda T et al (2003) Peroxisome degradation requires catalytically active sterol glucosyltransferase with a GRAM domain. EMBO J 22:3231–3241

Oku M, Nishimura T, Hattori T et al (2006) Role of Vac8 in formation of the vacuolar sequestering membrane during micropexophagy. Autophagy 2:272–279

Ozcan S, Johnston M (1999) Function and regulation of yeast hexose transporters. Microbiol Mol Biol Rev 63:554–569

Ozcan S, Dover J, Johnston M (1998) Glucose sensing and signaling by two glucose receptors in the yeast Saccharomyces cerevisiae. EMBO J 17:2566–2573

Polupanov AS, Sibirny AA (2014) Cytoplasmic extension peptide of Pichia pastoris glucose sensor Gss1 is not compulsory for glucose signalling. Cell Biol Int 38(2):172–178. doi:10.1002/cbin.10189

Polupanov AS, Nazarko VY, Sibirny AA (2011) CCZ1, MON1 and YPT7 genes are involved in pexophagy, the Cvt pathway and non-specific macroautophagy in the methylotrophic yeast Pichia pastoris. Cell Biol Int 35:311–319

Polupanov AS, Nazarko VY, Sibirny AA (2012) Gss1 protein of the methylotrophic yeast Pichia pastoris is involved in glucose sensing, pexophagy and catabolite repression. Int J Biochem Cell Biol 44:1906–1918
Rose M, Albig W, Entian KD (1991) Glucose repression in Saccharomyces cerevisiae is directly associated with hexose phosphorylation by hexokinases PI and PII. Eur J Biochem 199:511–518
Rubio-Texeira M, Van Zeebroeck G, Voordeckers K et al (2010) Saccharomyces cerevisiae plasma membrane nutrient sensors and their role in PKA signaling. FEMS Yeast Res 10:134–149
Sakai Y, Koller A, Rangell LK et al (1998) Peroxisome degradation by microautophagy in Pichia pastoris: identification of specific steps and morphological intermediates. J Cell Biol 141:625–636
Sakai Y, Oku M, van der Klei IJ et al (2006) Pexophagy: autophagic degradation of peroxisomes. Biochim Biophys Acta 1763:1767–1775
Saleem RA, Smith JJ, Aitchison JD (2006) Proteomics of the peroxisome. Biochim Biophys Acta 1763:1541–1551
Saleem RA, Knoblach B, Mast FD et al (2008) Genome-wide analysis of signaling networks regulating fatty acid-induced gene expression and organelle biogenesis. J Cell Biol 181:281–292
Saliola M, Getuli C, Mazzoni C (2007) A new regulatory element mediates ethanol repression of KlADH3, a Kluyveromyces lactis gene coding for a mitochondrial alcohol dehydrogenase. FEMS Yeast Res 7:693–701
Santangelo GM (2006) Glucose signaling in Saccharomyces cerevisiae. Microbiol Mol Biol Rev 70:253–282
Schrader M, Fahimi HD (2008) The peroxisome: still a mysterious organelle. Histochem Cell Biol 129:421–440
Sibirny AA (1990) Genetic control of methanol and ethanol metabolism in the yeast Pichia pinus. In: Heslot H (ed) Proceedings of the 6th international symposium on genetics of industrial microorganisms. Starsbourg, France. Soc. Franc. Microbiol., v.1, pp 545–554
Sibirny AA (2011) Mechanisms of autophagy and pexophagy in yeasts. Biochem Mosc 76:1279–1290
Sibirny AA (2012) Molecular mechanisms of peroxisome biogenesis in yeasts. Mol Biol (Mosc) 46:11–26
Sibirny AA, Titorenko VI (1986) The method for qualitative assay of alcohol oxidase and catalase in yeast colonies. Ukr Biokhim Zh 58:65–68 (in Russian)
Sibirny AA, Titorenko VI, Efremov BD et al (1987) Multiplicity of mechanisms of carbon catabolite repression involved in the synthesis of alcohol oxidase in the methylotrophic yeast Pichia pinus. Yeast 3:233–241
Sibirny AA, Titorenko VI, Teslyar GE et al (1991) Methanol and ethanol utilization in methylotrophic yeast Pichia pinus wild-type and mutant strains. Arch Microbiol 156:455–462
Smith JJ, Marelli M, Christmas RH et al (2002) Transcriptome profiling to identify genes involved in peroxisome assembly and function. J Cell Biol 158:259–271
Stasyk OV, Nazarko TY, Stasyk OG et al (2003) Sterol glucosyltransferases have different roles in Pichia pastoris and Yarrowia lipolytica. Cell Biol Int 27:947–952
Stasyk OV, Stasyk OG, Komduur J et al (2004) A hexose transporter homologue controls glucose repression in the methylotrophic yeast Hansenula polymorpha. J Biol Chem 279:8116–8125
Stasyk OV, Stasyk OG, Mathewson RD et al (2006) Atg28, a novel coiled-coil protein involved in autophagic degradation of peroxisomes in the methylotrophic yeast Pichia pastoris. Autophagy 2:30–38
Stasyk OG, van Zutphen T, Kang HA et al (2007) The role of Hansenula polymorpha MIG1 homologues in catabolite repression and pexophagy. FEMS Yeast Res 7:1103–1113
Stasyk OV, Nazarko TY, Sibirny AA (2008a) Methods of plate pexophagy monitoring and positive selection for ATG gene cloning in yeasts. Methods Enzymol 451:229–239

Stasyk OG, Maidan MM, Stasyk OV et al (2008b) Identification of hexose transporter-like sensor HXS1 and functional hexose transporter HXT1 in the methylotrophic yeast Hansenula polymorpha. Eukaryot Cell 7:735–746

Steinberg SJ, Dodt G, Raymond GV (2006) Peroxisome biogenesis disorders. Biochim Biophys Acta 1763:1733–1748

Stevens P, Monastyrska I, Leão-Helder AN et al (2005) Hansenula polymorpha Vam7p is required for macropexophagy. FEMS Yeast Res 5:985–997

Subramani S (1997) PEX genes on the rise. Nat Genet 15:331–333

Subramani S (1998) Components involved in peroxisome import, biogenesis, proliferation, turnover, and movement. Physiol Rev 78:171–188

Suzuki K (2013) Selective autophagy in budding yeast. Cell Death Differ 20:43–48

Thevelein JM, de Winde JH (1999) Novel sensing mechanisms and targets for the cAMP-protein kinase A pathway in the yeast Saccharomyces cerevisiae. Mol Microbiol 33:904–918

Till A, Lakhani R, Burnett SF, Subramani S (2012) Pexophagy: the selective degradation of peroxisomes. Int J Cell Biol 2012:512721

Tolstorukov I, Cregg JM (2007) Classical genetics. Methods Mol Biol 389:189–202

Tolstorukov II, Efremov BD, Benevolensky SV et al (1989) Mutants of methylotrophic yeast Pichia pinus defective in C2 metabolism. Yeast 5:179–186

Tuttle DL, Dunn WA Jr (1995) Divergent modes of autophagy in the methylotrophic yeast Pichia pastoris. J Cell Sci 108:25–35

van der Klei IJ, Yurimoto H, Sakai Y et al (2006) The significance of peroxisomes in methanol metabolism in methylotrophic yeast. Biochim Biophys Acta 1763:1453–1462

van Zutphen T, van der Klei IJ, Kiel JA (2008) Pexophagy in Hansenula polymorpha. Methods Enzymol 451:197–215

Veenhuis M, Van Dijken JP, Harder W (1983) The significance of peroxisomes in the metabolism of one-carbon compounds in yeasts. Adv Microb Physiol 24:1–82

Veenhuis M, Salomons FA, van der Klei IJ (2000) Peroxisome biogenesis and degradation in yeast: a structure/function analysis. Microsc Res Tech 51:584–600

Wan Y, Saleem RA, Ratushny AV et al (2009) Role of the histone variant H2A.Z/Htz1p in TBP recruitment, chromatin dynamics, and regulated expression of oleate-responsive genes. Mol Cell Biol 29:2346–2358

Wanders RJ, Duran M, Loupatty FJ (2010) Enzymology of the branched-chain amino acid oxidation disorders: the valine pathway. J Inherit Metab Dis 35:5–12

Wieczorke R, Krampe S, Weierstall T (1999) Concurrent knock-out of at least 20 transporter genes is required to block uptake of hexoses in Saccharomyces cerevisiae. FEBS Lett 464:123–128

Yamashita S, Oku M, Wasada Y et al (2006) PI4P-signaling pathway for the synthesis of a nascent membrane structure in selective autophagy. J Cell Biol 173:709–717

Yokota S (2003) Degradation of normal and proliferated peroxisomes in rat hepatocytes: regulation of peroxisomes quantity in cells. Microsc Res Tech 61:151–160

Yuan W, Tuttle DL, Shi YJ (1997) Glucose-induced microautophagy in Pichia pastoris requires the alpha-subunit of phosphofructokinase. J Cell Sci 110:1935–1945

Zhang P, Zhang W, Zhou X et al (2010) Catabolite repression of Aox in Pichia pastoris is dependent on hexose transporter PpHxt1 and pexophagy. Appl Environ Microbiol 76:6108–6118

Index

C. Brocard and A. Hartig (eds.), *Molecular Machines Involved in Peroxisome Biogenesis and Maintenance*, DOI 10.1007/978-3-7091-1788-0, © Springer-Verlag Wien 2014

MIX
Papier aus verantwortungsvollen Quellen
Paper from responsible sources
FSC® C105338

If you have any concerns about our products,
you can contact us on
ProductSafety@springernature.com

In case Publisher is established outside the EU,
the EU authorized representative is:
Springer Nature Customer Service Center GmbH
Europaplatz 3, 69115 Heidelberg, Germany

Printed by Libri Plureos GmbH
in Hamburg, Germany